10. $\dfrac{d}{dx} e^x = e^x$, $\qquad \dfrac{d}{dx} e^{f(x)} = f'(x)e^{f(x)}$

11. $\dfrac{d}{dx} \ln x = \dfrac{1}{x}$, $\quad x > 0$;

$\dfrac{d}{dx} \ln f(x) = f'(x) \cdot \dfrac{1}{f(x)}$, $\quad f(x) > 0$

12. $\dfrac{d}{dx} \ln |x| = \dfrac{1}{x}$, $\quad x < 0$;

$\dfrac{d}{dx} \ln |f(x)| = f'(x) \cdot \dfrac{1}{f(x)}$, $\quad f(x) < 0$

13. If $\dfrac{dP}{dt} = kP$, then $P(t) = P_0 e^{kt}$.

14. If $\dfrac{dP}{dt} = -kP$, then $P(t) = P_0 e^{-kt}$.

15. $\dfrac{d}{dx} a^x = (\ln a)a^x$

16. $\dfrac{d}{dx} \log_a x = \dfrac{1}{\ln a} \cdot \dfrac{1}{x}$, $\quad x > 0$;

$\dfrac{d}{dx} \log_a |x| = \dfrac{1}{\ln a} \cdot \dfrac{1}{x}$, $\quad x < 0$

17. $a^x = e^{x(\ln a)}$

$5^x = e^{x(\ln 5)}$

$\ln 5 \cdot e^x$

APPLIED CALCULUS

SECOND EDITION

MARVIN L. BITTINGER • BERNARD B. MORREL

Indiana University—Purdue University at Indianapolis

APPLIED CALCULUS

SECOND EDITION

Addison-Wesley Publishing Company

Reading, Massachusetts · Menlo Park, California
New York · Don Mills, Ontario · Wokingham, England
Amsterdam · Sydney · Singapore · Tokyo · Madrid
Bogotá · Santiago · San Juan

CALCULATOR NOTE TO THE STUDENT

Having a scientific calculator will enhance your study of this text. Be sure that it has a power key $\boxed{y^x}$, a natural logarithm key $\boxed{\ln}$, and possibly trigonometric keys. An $\boxed{e^x}$ key would also be helpful.

Sponsoring Editor	Elizabeth Burr
Production Supervisor	Herbert Nolan
Design, Editorial, and Production Services	Quadrata, Inc.
Illustrator	Hardlines
Art Consultant	Loretta Bailey
Manufacturing Supervisor	Roy Logan
Cover Designer	Marshall Henrichs

Library of Congress Cataloging-in-Publication Data

Bittinger, Marvin L.
    Applied calculus.

    Includes index.
        1. Calculus. I. Morrel, Bernard B.
II. Title.
QA303.B6647   1988   515   87-19582
ISBN 0-201-12211-1

Photo credits appear on page 728.

ABCDEFGHIJ–DO–8987

PREFACE

Appropriate for a two-term calculus course, this text is an introduction to calculus as applied to business, economics, the behavioral sciences, the social sciences, biology, and medicine. For those who need a one-term text, *Calculus*, by Marvin L. Bittinger, covers Chapters 1–8 of this text, and includes a section on differential equations in Chapter 6. A basic course in algebra is a prerequisite for the text, although Chapter 1 provides sufficient review to unify the diverse backgrounds of most students.

Content Features

Intuitive approach. Although the word "intuitive" has many meanings and interpretations, its use here means "experience based." Throughout the text, when a particular concept is discussed, its presentation is designed so that the students' learning process is based either on their earlier mathematical experience or on a new experience presented by the authors before the concept is formalized.

- When maximum problems involving volume are introduced (see p. 195), a function is derived that is to be maximized. Instead of forging ahead with the standard calculus solution, the student is first asked to stop, compute the function values, graph the equation, and then estimate the maximum value. This experience provides students with more insight into the problem—recognition not only that different dimensions yield different volumes, but also that the dimensions yielding the maximum volume may be conjectured or estimated as a result of the calculations.

- The definition of the derivative in Chapter 2 is presented in the context of a discussion of average rates of change (see p. 97). This presentation is more accessible and realistic than the strictly geometric idea of slope.

- The concept behind the definition of the number e in Chapter 4 is explained both graphically and through a discussion of continuously compounded interest (see p. 239).

Applications. Relevant and factual applications drawn from a broad range of fields are included throughout the text as applied problems and exercises and in application sections. These have been updated and expanded in this edition.

- An early discussion of compound interest, cost and profit, and supply-and-demand functions in Chapter 1 sets the stage for later applications. The notions of total revenue and cost and profit, together with their derivatives (marginal functions), are threads that run throughout the text (see pp. 43, 132, and 193).

- All techniques of differentiation, including the Extended Power Rule, the Chain Rule, and material on higher derivatives are covered in Chapter 2 before applications of differentiation are introduced in Chapter 3. This allows for many interesting applications of differentiation to be presented in Chapter 3, which includes separate applications sections for business and biology, as well as a section on implicit differentiation and related rates (see pp. 204–231).

- When the exponential model is studied in Chapter 4, other applications, such as continuously compounded interest and the demand for natural resources, are also considered (see pp. 264–286). Growth and decay are covered in separate sections in Chapter 4 to allow room for the many worthwhile applications that relate to these concepts.

- Applications of integration are covered in a separate chapter and include a thorough coverage of probability (see pp. 367–418).

Coverage. Many aspects of the treatment and coverage in this text have been commended by reviewers. The following are a few examples.

- Trouble areas in algebra are integrated into coverage of traditional precalculus topics in Chapter 1.

- Relative maxima and minima and absolute maxima and minima are covered in separate sections in Chapter 3 so that students always understand which they are studying (see pp. 157–189).

- Chapter 5 on techniques of integration includes a section on integration by tables, which will prepare students for using calculus in real-life situations (see pp. 351–354).
- Coverage of differential equations, Taylor series and polynomials, and numerical methods is extensive and includes material on l'Hôpital's Rule, Newton's method, and Euler's method.

Pedagogical Features

Interactive approach.   As each new section begins, its objectives are stated in the margin. These can be spotted easily by the student, and when the typical question, "What material am I responsible for?", arises, the objectives provide an answer. They may also help take the fear out of the word "calculus." In the margins on each page, sample, developmental, and exploratory exercises are placed near the related text material. As students work through the material, they are encouraged to do the margin exercises. This involves them actively in the development of each topic and gives them a deeper understanding of the material they are studying. All margin exercises have answers in the text. It is recommended that students work out all these problems, stopping to do them when the text so indicates (see pp. 146 and 442).

Variety of Exercises

- *Calculator exercises.* Exercises and examples geared to the use of a calculator are included throughout the text, although students who do not have a calculator can still achieve their goals. Calculator exercises are highlighted by the symbol ▦ (see pp. 206 and 240).
- *Extension exercises.* Extension exercises are included in most exercise sets and on all chapter tests. They require students to go beyond the immediate objectives of the section or chapter, and are designed both to challenge students and to make them think about what they are learning (see pp. 149 and 346).
- *Exploratory exercises.* These are extended exercises that are included in the exercise sets or in the margin exercises. They emphasize analysis of data produced from sets of equations, home experiments, or theoretical situations, and are designed to build students' mathematical experience and conceptual understanding (see pp. 280 and 314).
- *Applications.* A section of applied problems is included in most exercise sets. These give the students extensive practice in applying the material they are learning to real-life situations (see pp. 200 and 273).

- *Computer software exercises.* Many exercises throughout *Applied Calculus* can be done on the computer using the software utilities package that accompanies the text, *Cactusplot*, Student Edition. Step-by-step procedures for working out exercises designated with the symbol CSS are contained in the Cactusplot Student Supplement.

Tests and Reviews

- *Summary and Review.* New to this edition, the Summary and Review section is found at the end of each chapter and is designed to provide students with all the material they need for successful review. The objectives of the chapter are summarized in boldface type and followed by a set of review exercises. Answers are at the back of the book, together with section references so that students can easily find the correct material to restudy if they miss an exercise (see pp. 153 and A-11).

- *Tests.* Each chapter ends with a chapter test, which includes challenge questions. There is a cumulative review at the end of the text, which can also serve as a final examination; the answers to the chapter tests and the cumulative review are at the back of the book. Four additional forms of each of the tests appear, ready for classroom use, in the Instructor's Manual, and six additional chapter tests are available in the Printed Test Bank. The Instructor's Manual also includes four different forms of a very comprehensive final examination.

What's New in the Second Edition?

The style, format, and approach of the first edition have been retained in this new edition, but the text has been polished in many places on the basis of extensive reviewer feedback. New applications have been added, and many of the former applications have been updated.

- A Summary and Review section has been added to the end of each chapter (see the description above).

- In Chapter 2, the discussion of limits has been split into two sections, and is now presented before the material on continuity.

- In Chapter 3, the material on determining absolute maxima and minima has been extensively revised, and is now presented in a section separate from the discussion on relative maxima and minima. Relative maxima and minima are now covered in Section 3.3, where such concepts are directly useful to graphing functions.

- Chapter 4 has been revised under the assumption that all students have calculators and no longer need tables to find values of the exponential and logarithmic functions.

- Chapter 5 now contains a complete definition of differential notation, together with exercises to reinforce the concept. Integration by tables is now covered in a separate section.

- In Section 6.2 the concept of continuous money flow replaces the finite process of determining the amount of an annuity.

- In Chapter 7, finding function values for functions of several variables has been separated from finding partial derivatives to ease the length of the first section.

- The trigonometry material in Chapter 8 has been related more directly to right triangles and degrees as well as to radians.

- In the later chapters, a number of interesting new problems have been added, and the presentation has been improved in many places. These alterations, while minor in nature, should make the text more teachable for the instructor and more readable for the student.

Supplements for the Instructor

Instructor's Manual. The Instructor's Manual contains four alternative test forms with answers for each chapter test and four comprehensive final examinations. The tests have been revised and are now completely different from those for the first edition. The Instructor's Manual also has answers for the even-numbered exercises in the exercise sets, which can be copied and handed out to students. The answers to the odd-numbered exercises are at the back of the text.

Printed Test Bank. This contains over 1200 test items that have been generated from Addison-Wesley's computerized test generator, "AWTest." The test bank items are arranged in a convenient chapter-test format with six different tests for each chapter. Instructors can pick and choose test items to create their own tests or tear out ready-to-use chapter tests complete with answer sheets and answer keys.

Videotapes. Videotape reviews that cover important topics in the book have been prepared. John Jobe of Oklahoma State University speaks to students, works out examples, and provides lucid explanations. Although the videotapes do not provide an entire course on television, they have many uses, among which are to supplement lectures, to provide partial lectures, and to offer self-study opportunities for students.

Software

- *Computerized testing.* "AWTest," Addison-Wesley's random-number test-generating system, is available with this text for the IBM PC. Using AWTest, instructors can generate up to 99 variations of any particular test with a few keystrokes. They can also choose test items by number from a bank of over 200 test items or request tests to be printed out in chapter-test format. AWTest will support almost any completely compatible IBM PC printer.

- *Computer Software Supplement.* The Student Version of Cactusplot Software by John Losse of Scottsdale Community College is available exclusively with this text in both Apple II series and IBM PC versions. Cactusplot is an easy-to-use software utilities package that can perform many mathematical operations, such as graphing functions, processing tables, solving equations, and finding the area under curves. A site license to the Student Version of Cactusplot is available from Addison-Wesley and is free to qualifying adopters.

    Cactusplot is accompanied by a student software supplement containing directions for solving several key exercises in each chapter using a computer. These exercises are designated in the text by the symbol CSS . The manual also gives the students additional activities for each exercise, such as changing the parameters of a graph, which are designed to strengthen their intuitive understanding of the concepts involved. Cactusplot will also print out graphs during the course of an exercise so that students can show their work to their instructor. The Software Supplement is for sale to students at a nominal cost.

    The Student Version of Cactusplot is available only from Addison-Wesley. A professional version of Cactusplot with many enhanced capabilities is available directly from John Losse, The Cactusplot Company, 1442 N. McAllister, Tempe, AZ 85281 (602) 945-1667.

Related Addison-Wesley Software Titles   *The Calculus Toolkit* is a calculus utilities package available for the Apple II series and IBM PC. It is designed to accompany Thomas/Finney's *Calculus and Analytic Geometry*, Seventh Edition (Addison-Wesley, 1988), but can be used with any calculus text. *MathCAD, Student Edition* is an electronic scratch-pad for the IBM PC that allows you to write equations on the PC exactly as you would on paper. MathCAD automatically calculates and displays results as numbers or graphs. *Lotus 1-2-3, Student Version* is a fully functional version of Lotus 1-2-3 with a 64-column-by-256-row spreadsheet that includes all the database, calculating, and graphics capabilities of Lotus 1-2-3 with teaching materials and enhancements particularly suited for educational use. Contact your Addison-Wesley sales representative for more information about these titles.

Supplements for the Student

Student's Solutions Manual by Judith A. Beecher and Judith A. Penna   Complete worked-out solutions with extra hints and suggestions are provided in this booklet for all odd-numbered problems in the exercise sets. This supplement is available to instructors and is for sale to students.

Acknowledgments

The authors wish to express their appreciation to the many people who helped with the development of this book: to their students for providing suggestions and criticisms so willingly during the preceding editions; to Judy Penna for her helpful suggestions, proofreading, and preparation of the Instructor's Manual and the Student's Solutions Manual; to Judy Beecher for her helpful suggestions and for writing the Student's Solutions Manual; to John Jobe of Oklahoma State University for preparing the videotapes; to Mike Penna of IUPUI for his help with the computer graphics; and to Barbara Miller, Michael Dagg, Judy Penna, Barbara Johnson, and Karen Anderson for their precise proofreading of the manuscript.

In addition, we wish to thank the following professors for their thorough reviewing:

Stephen Bernfeld, *University of Texas at Arlington*
John Bishir, *North Carolina State University*
Michael Bleicher, *University of Wisconsin*
John Busovicki, *Indiana University of Pennsylvania*
Ming San Chen, *George Mason University*
Samuel Councilman, *California State University at Long Beach*
Maureen H. Fenrick, *Wichita State University*
Linda Holden, *Indiana University at Bloomington*
Jody Lockhart, *Indiana University at South Bend*
J. Glenn Maxwell, *Kent State University*
Mel A. Mitchell, *Clarion University*
William Rettig, *Indiana University of Pennsylvania*
David Smith, *Central Connecticut State University*
W. Arlene Starwalt-Jeskey, *Rose State College*
Bert K. Waits, *Ohio State University*
Mary Winter, *Michigan State University*

M.L.B.
B.A.M.

INDEX OF APPLICATIONS

CONTENTS

3

APPLICATIONS OF DIFFERENTIATION   157

4

EXPONENTIAL AND LOGARITHMIC FUNCTIONS   233

5

INTEGRATION   301

6

APPLICATIONS OF INTEGRATION   367

7

FUNCTIONS OF SEVERAL VARIABLES   419

8

TRIGONOMETRIC FUNCTIONS   505

9

DIFFERENTIAL EQUATIONS   539

10

TAYLOR SERIES AND L'HÔPITAL'S RULE   609

11

NUMERICAL TECHNIQUES   665

1

ALGEBRA REVIEW, FUNCTIONS, AND MODELING

This chapter is mainly a review of the basic concepts of algebra used in calculus. It does introduce many topics that we will consider several times throughout the text: supply and demand, total cost, revenue, and profit, and the concept of a mathematical model.

AN APPLICATION

The number of tangerines in a pile of the type shown here is approximated by the function

$$f(x) = \tfrac{1}{6}x^3 + \tfrac{1}{2}x^2 + \tfrac{1}{3}x,$$

where $f(x)$ is the number of tangerines and x is the number of layers. Find the number of tangerines when the number of layers is 7, 10, or 12.

THE MATHEMATICS

To solve the problem we substitute 7 for x, 10 for x, and 12 for x in the equation

$$f(x) = \tfrac{1}{6}x^3 + \tfrac{1}{2}x^2 + \tfrac{1}{3}x.$$

This is a formula for a *function*.

1.1

EXPONENTS, MULTIPLYING, AND FACTORING

OBJECTIVES

a) Rename an exponential expression without exponents.

b) Multiply exponential expressions by adding exponents.

c) Divide exponential expressions by subtracting exponents.

d) Raise a power to a power by multiplying exponents.

e) Multiply algebraic expressions.

f) Factor algebraic expressions.

g) Solve applied problems involving the comparison of a power like $(3.1)^2$ with 3^2.

h) Solve applied problems involving compound interest.

Rename without exponents.

1. 3^4

2. $(-3)^2$

3. $(1.02)^3$

4. $\left(\dfrac{1}{4}\right)^2$

Exponential Notation

Let us review the meaning of an expression

$$a^n,$$

where a is any real number and n is an integer; that is, n is a number in the set $\ldots, -3, -2, -1, 0, 1, 2, 3, \ldots$. The number a is called the *base* and n is called the *exponent*. When n is larger than 1, then

$$a^n = \underbrace{a \cdot a \cdot a \cdots a}_{n \text{ factors}}.$$

In other words, a^n is the product of n factors, each of which is a.

In later sections of the text we will consider a^n when n is any real number.

Example 1   Rename without exponents.

a) $4^3 = 4 \cdot 4 \cdot 4$, or 64

b) $(-2)^5 = (-2)(-2)(-2)(-2)(-2)$, or -32

c) $(1.08)^2 = 1.08 \times 1.08$, or 1.1664

d) $\left(\dfrac{1}{2}\right)^3 = \dfrac{1}{2} \cdot \dfrac{1}{2} \cdot \dfrac{1}{2}$, or $\dfrac{1}{8}$

DO EXERCISES 1–4. (EXERCISES ARE IN THE MARGIN.)

We define an exponent of 1 as follows:

$$a^1 = a, \quad \text{for any real number } a.$$

That is, any real number to the first power is that number itself.

We define an exponent of 0 as follows:

$$a^0 = 1, \quad \text{for any nonzero real number } a.$$

That is, any nonzero real number a to the 0 power is 1.

Rename without exponents.

5. $(5t)^0$

6. $(5t)^1$

7. k^0

8. m^1

9. $\left(\dfrac{1}{4}\right)^1$

10. $\left(\dfrac{1}{4}\right)^0$

Rename without negative exponents.

11. 2^{-4}

12. 10^{-2}

13. $\left(\dfrac{1}{4}\right)^{-3}$

14. t^{-7}

15. e^{-t}

16. M^{-1}

17. $(x + 1)^{-2}$

Example 2   Rename without exponents.

a) $(-2x)^0 = 1$

b) $(-2x)^1 = -2x$

c) $\left(\dfrac{1}{2}\right)^0 = 1$

d) $e^0 = 1$

e) $e^1 = e$

f) $\left(\dfrac{1}{2}\right)^1 = \dfrac{1}{2}$

DO EXERCISES 5–10.

The meaning of a negative integer as an exponent is as follows:

$$a^{-n} = \frac{1}{a^n}, \text{ for any nonzero real number } a.$$

That is, any nonzero real number a to the $-n$ power is the reciprocal of a^n.

Example 3   Rename without negative exponents.

a) $2^{-5} = \dfrac{1}{2^5} = \dfrac{1}{2 \cdot 2 \cdot 2 \cdot 2 \cdot 2} = \dfrac{1}{32}$

b) $10^{-3} = \dfrac{1}{10^3} = \dfrac{1}{10 \cdot 10 \cdot 10} = \dfrac{1}{1000}$, or 0.001

c) $\left(\dfrac{1}{4}\right)^{-2} = \dfrac{1}{\left(\dfrac{1}{4}\right)^2} = \dfrac{1}{\dfrac{1}{4} \cdot \dfrac{1}{4}} = \dfrac{1}{\dfrac{1}{16}} = 1 \cdot \dfrac{16}{1} = 16$

d) $x^{-5} = \dfrac{1}{x^5}$

e) $e^{-k} = \dfrac{1}{e^k}$

f) $t^{-1} = \dfrac{1}{t^1} = \dfrac{1}{t}$

DO EXERCISES 11–17.

Properties of Exponents

Note the following:

$$b^5 \cdot b^{-3} = (b \cdot b \cdot b \cdot b \cdot b) \cdot \frac{1}{b \cdot b \cdot b} = \frac{b \cdot b \cdot b}{b \cdot b \cdot b} \cdot b \cdot b = 1 \cdot b \cdot b = b^2.$$

Multiply.

18. $t^4 \cdot t^5$

19. $t^{-4} \cdot t$

20. $10e^{-4} \cdot 5e^{-9}$

21. $t^{-3} \cdot t^{-4} \cdot t$

22. $4b^5 \cdot 6b^{-2}$

The result could have been obtained by adding the exponents. This is true in general.

THEOREM 1

For any real number a and any integers n and m,

$$a^n \cdot a^m = a^{n+m}.$$

(To multiply when the bases are the same, add the exponents.)

Example 4 Multiply.

a) $x^5 \cdot x^6 = x^{5+6} = x^{11}$

b) $x^{-5} \cdot x^6 = x^{-5+6} = x$

c) $2x^{-3} \cdot 5x^{-4} = 10x^{-3+(-4)} = 10x^{-7}$

d) $r^2 \cdot r = r^{2+1} = r^3$

DO EXERCISES 18–22.

Note the following:

$$b^5 \div b^2 = \frac{b^5}{b^2} = \frac{b \cdot b \cdot b \cdot b \cdot b}{b \cdot b} = \frac{b \cdot b}{b \cdot b} \cdot b \cdot b \cdot b = 1 \cdot b \cdot b \cdot b = b^3.$$

The result could have been obtained by subtracting the exponents. This is true in general.

THEOREM 2

For any nonzero real number a and any integers n and m,

$$\frac{a^n}{a^m} = a^{n-m}.$$

(To divide when the bases are the same, subtract the exponent in the denominator from the exponent in the numerator.)

Example 5 Divide.

a) $\dfrac{a^3}{a^2} = a^{3-2} = a^1 = a$

b) $\dfrac{x^7}{x^7} = x^{7-7} = x^0 = 1$

Divide.

23. $\dfrac{x^6}{x^2}$

24. $\dfrac{x^2}{x^6}$

25. $\dfrac{e^t}{e^t}$

26. $\dfrac{e^2}{e^k}$

27. $\dfrac{e^5}{e^{-7}}$

28. $\dfrac{e^{-5}}{e^{-7}}$

Simplify.

29. $(x^{-4})^3$

30. $(e^2)^2$

31. $(e^x)^3$

32. $(5x^3y^5)^2$

33. $(4x^{-5}y^{-6}z^2)^{-4}$

c) $\dfrac{e^3}{e^{-4}} = e^{3-(-4)} = e^{3+4} = e^7$

d) $\dfrac{e^{-4}}{e^{-1}} = e^{-4-(-1)} = e^{-4+1} = e^{-3}$, or $\dfrac{1}{e^3}$

DO EXERCISES 23–28.

Note the following:
$$(b^2)^3 = b^2 \cdot b^2 \cdot b^2$$
$$= b^{2+2+2} = b^6.$$

The result could have been obtained by multiplying the exponents. This is true in general.

THEOREM 3

For any real number a and any integers n and m,
$$(a^n)^m = a^{nm}.$$

(To raise a power to a power, multiply the exponents.)

Example 6 Simplify.

a) $(x^{-2})^3 = x^{-2 \cdot 3} = x^{-6}$, or $\dfrac{1}{x^6}$

b) $(e^x)^2 = e^{2x}$

c) $(2x^4y^{-5}z^3)^{-3} = 2^{-3}(x^4)^{-3}(y^{-5})^{-3}(z^3)^{-3}$
$$= \dfrac{1}{2^3}x^{-12}y^{15}z^{-9}, \text{ or } \dfrac{y^{15}}{8x^{12}z^9}$$

DO EXERCISES 29–33.

Multiplication

The distributive laws are important in multiplying. The laws are as follows:

For any numbers A, B, and C,
$$A(B + C) = AB + AC \quad \text{and} \quad A(B - C) = AB - AC.$$

Multiply.

34. $2(x + 7)$

35. $P(1 - i)$

36. $(x - 4)(x + 7)$

37. $(a - b)(a - b)$

38. $(a - b)(a + b)$

Multiply.

39. $(x - h)^2$

40. $(3x + t)^2$

41. $(5t - m)(5t + m)$

Example 7　Multiply.

a)　$3(x - 5) = 3 \cdot x - 3 \cdot 5 = 3x - 15$

b)　$P(1 + i) = P \cdot 1 + P \cdot i = P + Pi$

c)　$(x - 5)(x + 3) = (x - 5)x + (x - 5)3$
$$= x \cdot x - 5x + 3x - 5 \cdot 3$$
$$= x^2 - 2x - 15$$

d)　$(a + b)(a + b) = (a + b)a + (a + b)b$
$$= a \cdot a + ba + ab + b \cdot b$$
$$= a^2 + 2ab + b^2$$

DO EXERCISES 34–38.

　　The following formulas, which are obtained using the distributive laws, are useful in multiplying.

$$(A + B)^2 = A^2 + 2AB + B^2 \qquad (1)$$
$$(A - B)^2 = A^2 - 2AB + B^2 \qquad (2)$$
$$(A - B)(A + B) = A^2 - B^2 \qquad (3)$$

Example 8　Multiply.

a)　$(x + h)^2 = x^2 + 2xh + h^2$

b)　$(2x - t)^2 = (2x)^2 - 2(2x)t + t^2 = 4x^2 - 4xt + t^2$

c)　$(3c + d)(3c - d) = (3c)^2 - d^2 = 9c^2 - d^2$

DO EXERCISES 39–41.

Factoring

Factoring is the reverse of multiplication. That is, to factor an expression, we find an equivalent expression that is a product. Always remember to look first for a common factor.

Example 9　Factor.

a)　$P + Pi = P \cdot 1 + P \cdot i = P(1 + i)$　　　We used a distributive law.

b)　$2xh + h^2 = h(2x + h)$

c)　$x^2 - 6xy + 9y^2 = (x - 3y)^2$

d)　$x^2 - 5x - 14 = (x - 7)(x + 2)$　　　Here we looked for factors of -14 whose sum is -5.

Factor.

42. $P - Pi$

43. $x^2 + 10xy + 25y^2$

44. $4x^2 + 28x + 40$

45. $25c^2 - d^2$

46. $12y^2 + 13y - 14$

47. How close is $(5.1)^2$ to 5^2?

e) $6x^2 + 7x - 5 = (2x - 1)(3x + 5)$ Here we first considered ways of factoring the first coefficients. For example, $(2x \quad)(3x \quad)$. Then we looked for factors of -5 such that when we multiply we obtain the given expression.

f) $x^2 - 9t^2 = (x - 3t)(x + 3t)$ We used $(A - B)(A + B) = A^2 - B^2$.

DO EXERCISES 42–46.

In later work we will consider expressions like

$$(x + h)^2 - x^2.$$

To simplify this, first note that

$$(x + h)^2 = x^2 + 2xh + h^2.$$

Subtracting x^2 on both sides of this equation gives us

$$(x + h)^2 - x^2 = 2xh + h^2.$$

Factoring out an h on the right side, we get

$$(x + h)^2 - x^2 = h(2x + h). \tag{4}$$

Let us now use this result to compare two squares.

Example 10   How close is $(3.1)^2$ to 3^2?

Solution   Substituting $x = 3$ and $h = 0.1$ in Eq. (4), we get

$$(3.1)^2 - 3^2 = 0.1(2 \cdot 3 + 0.1) = 0.1(6.1) = 0.61.$$

So $(3.1)^2$ differs from 3^2 by 0.61.

DO EXERCISE 47.

Compound Interest

Suppose we invest P dollars at interest rate i, compounded annually. The amount A_1 in the account at the end of one year is given by

$$A_1 = P + Pi = P(1 + i) = Pr,$$

where, for convenience,

$$r = 1 + i.$$

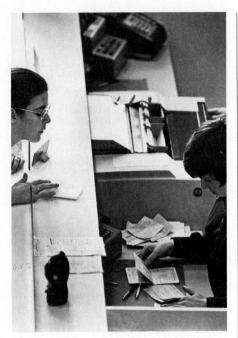

There is a formula for finding the amount in a savings account after a certain period of time.

48. Suppose $1000 is invested at 9% compounded annually. How much is in the account at the end of 2 years?

Going into the second year we have Pr dollars, so by the end of the second year we would have the amount A_2 given by

$$A_2 = A_1 \cdot r = (Pr)r = Pr^2 = P(1 + i)^2.$$

Going into the third year we have Pr^2 dollars, so by the end of the third year we would have the amount A_3 given by

$$A_3 = A_2 \cdot r = (Pr^2)r = Pr^3 = P(1 + i)^3.$$

In general, we have the following.

THEOREM 4

> **If an amount P is invested at interest rate i, compounded annually, in t years it will grow to the amount A given by**
>
> $$A = P(1 + i)^t.$$

Example 11 Suppose $1000 is invested at 8% compounded annually. How much is in the account at the end of 2 years?

Solution We substitute into the equation $A = P(1 + i)^t$ and get

$$A = 1000(1 + 0.08)^2 = 1000(1.08)^2 = 1000(1.1664) = \$1166.40.$$

DO EXERCISE 48.

For interest that is compounded quarterly, we can find a formula like the one above as follows:

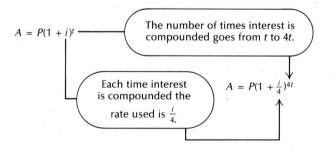

In general, the following theorem applies.

49. Suppose $1000 is invested at 11% compounded semiannually $(n = 2)$. How much is in the account at the end of 3 years?

THEOREM 5

If a principal P is invested at interest rate i, compounded n times a year, in t years it will grow to the amount A given by

$$A = P\left(1 + \frac{i}{n}\right)^{nt}.$$

Example 12 Suppose $1000 is invested at 8% compounded quarterly. How much is in the account at the end of 2 years?

Solution We use the equation $A = P(1 + i/n)^{nt}$, substituting 1000 for P, 0.08 for i, 4 for n (compounding quarterly), and 2 for t. Then we get

$$A = 1000\left(1 + \frac{0.08}{4}\right)^{4 \times 2} = 1000(1 + 0.02)^8 = 1000(1.02)^8$$

$$\approx 1000(1.171659381) \qquad \text{Using a calculator to approximate } (1.02)^8$$

$$= 1171.659381$$

$$\approx \$1171.66.^*$$

DO EXERCISE 49.

———————

* *A calculator note:* A calculator with a $\boxed{y^x}$ key and a ten-digit readout was used to find $(1.02)^8$ in Example 12. The number of places on a calculator may affect the accuracy of the answer. Thus you may occasionally find that your answers do not agree with those at the back of the book, which were found on a calculator with a ten-digit readout. In general, when using a calculator, do all your computations and round only at the end, as in Example 12. Usually, your answer should agree to at least four digits. It might be wise to consult with your instructor on the accuracy required.

EXERCISE SET 1.1

Rename without exponents.

1. 5^3 2. 7^2 3. $(-7)^2$ 4. $(-5)^3$ 5. $(1.01)^2$

6. $(1.01)^3$ 7. $\left(\frac{1}{2}\right)^4$ 8. $\left(\frac{1}{4}\right)^3$ 9. $(6x)^0$ 10. $(6x)^1$

11. t^1 **12.** t^0 **13.** $\left(\dfrac{1}{3}\right)^0$ **14.** $\left(\dfrac{1}{3}\right)^1$

Rename without negative exponents.

15. 3^{-2} **16.** 4^{-2} **17.** $\left(\dfrac{1}{2}\right)^{-3}$ **18.** $\left(\dfrac{1}{2}\right)^{-2}$ **19.** 10^{-1}

20. 10^{-4} **21.** e^{-b} **22.** t^{-k} **23.** b^{-1} **24.** h^{-1}

Multiply.

25. $x^2 \cdot x^3$ **26.** $t^3 \cdot t^4$ **27.** $x^{-7} \cdot x$ **28.** $x^5 \cdot x$ **29.** $5x^2 \cdot 7x^3$

30. $4t^3 \cdot 2t^4$ **31.** $x^{-4} \cdot x^7 \cdot x$ **32.** $x^{-3} \cdot x \cdot x^3$ **33.** $e^{-t} \cdot e^t$ **34.** $e^k \cdot e^{-k}$

Divide.

35. $\dfrac{x^5}{x^2}$ **36.** $\dfrac{x^7}{x^3}$ **37.** $\dfrac{x^2}{x^5}$ **38.** $\dfrac{x^3}{x^7}$ **39.** $\dfrac{e^k}{e^k}$

40. $\dfrac{t^k}{t^k}$ **41.** $\dfrac{e^t}{e^4}$ **42.** $\dfrac{e^k}{e^3}$ **43.** $\dfrac{t^6}{t^{-8}}$ **44.** $\dfrac{t^5}{t^{-7}}$

45. $\dfrac{t^{-9}}{t^{-11}}$ **46.** $\dfrac{t^{-11}}{t^{-7}}$

Simplify.

47. $(t^{-2})^3$ **48.** $(t^{-3})^4$ **49.** $(e^x)^4$ **50.** $(e^x)^5$

51. $(2x^2y^4)^3$ **52.** $(2x^2y^4)^5$ **53.** $(3x^{-2}y^{-5}z^4)^{-4}$ **54.** $(5x^3y^{-7}z^{-5})^{-3}$

55. $(-3x^{-8}y^7z^2)^2$ **56.** $(-5x^4y^{-5}z^{-3})^4$

Multiply.

57. $5(x-7)$ **58.** $4(x-3)$ **59.** $x(1-t)$

60. $x(1+t)$ **61.** $(x-5)(x-2)$ **62.** $(x-4)(x-3)$

63. $(a-b)(a^2+ab+b^2)$ **64.** $(x^2-xy+y^2)(x+y)$ **65.** $(2x+5)(x-1)$

66. $(3x+4)(x-1)$ **67.** $(a-2)(a+2)$ **68.** $(3x-1)(3x+1)$

69. $(5x+2)(5x-2)$ **70.** $(t-1)(t+1)$ **71.** $(a-h)^2$

72. $(a+h)^2$ **73.** $(5x+t)^2$ **74.** $(7a-c)^2$

75. $5x(x^2+3)^2$ **76.** $-3x^2(x^2-4)(x^2+4)$

Use the following equation (Eq. 1) for Exercises 77–80.

$$\begin{aligned}
(x+h)^3 &= (x+h)(x+h)^2 = (x+h)(x^2+2xh+h^2) \\
&= (x+h)x^2 + (x+h)2xh + (x+h)h^2 \\
&= x^3 + x^2h + 2x^2h + 2xh^2 + xh^2 + h^3 \\
&= x^3 + 3x^2h + 3xh^2 + h^3 \tag{1}
\end{aligned}$$

77. $(a+b)^3$ **78.** $(a-b)^3$ **79.** $(x-5)^3$ **80.** $(2x+3)^3$

Factor.

81. $x - xt$

82. $x + xh$

83. $x^2 + 6xy + 9y^2$

84. $x^2 - 10xy + 25y^2$

85. $x^2 - 2x - 15$

86. $x^2 + 8x + 15$

87. $x^2 - x - 20$

88. $x^2 - 9x - 10$

89. $49x^2 - t^2$

90. $9x^2 - b^2$

91. $36t^2 - 16m^2$

92. $25y^2 - 9z^2$

93. $a^3b - 16ab^3$

94. $2x^4 - 32$

95. $a^8 - b^8$

96. $36y^2 + 12y - 35$

97. $10a^2x - 40b^2x$

98. $x^3y - 25xy^3$

99. $2 - 32x^4$

100. $2xy^2 - 50x$

101. $9x^2 + 17x - 2$

102. $6x^2 - 23x + 20$

103. $x^3 + 8$

104. $a^3 - 27b^3$

[*Hint:* See Exercise 64.]

[*Hint:* See Exercise 63.]

105. $y^3 - 64t^3$

106. $m^3 + 1000p^3$

107. Use the following: $(x + h)^2 - x^2 = h(2x + h)$.

   a)   How close is $(4.1)^2$ to 4^2?

   b)   How close is $(4.01)^2$ to 4^2?

   c)   How close is $(4.001)^2$ to 4^2?

108. From Eq. (1) it follows that

$$(x + h)^3 - x^3 = h(3x^2 + 3xh + h^2).$$

Use the equation for the following.

   a)   How close is $(4.1)^3$ to 4^3?

   b)   How close is $(4.01)^3$ to 4^3?

   c)   How close is $(4.001)^3$ to 4^3?

The symbol ▦ indicates an exercise designed to be done using a calculator.

109. *Business: Compound interest.* Suppose $1000 is invested at 16%. How much is in the account at the end of 1 year, if interest is compounded

   a)   annually?

   b)   semiannually?

   c)   quarterly?

   d)   daily? (▦ with $\boxed{y^x}$ key)

   e)   hourly?

110. *Business: Compound interest.* Suppose $1000 is invested at 10%. How much is in the account at the end of 1 year, if interest is compounded

   a)   annually?

   b)   semiannually?

   c)   quarterly?

   d)   daily? (▦ with $\boxed{y^x}$ key)

   e)   hourly?

Business: Determining monthly payments on a loan. P dollars are borrowed on a home mortgage. The monthly payment M is given by

$$M = P\left[\frac{\dfrac{i}{12}\left(1 + \dfrac{i}{12}\right)^n}{\left(1 + \dfrac{i}{12}\right)^n - 1}\right],$$

where i is the interest rate and n is the total number of monthly payments.

111. ▦ The mortgage on a house is $43,000. The interest rate is $8\frac{3}{4}$%. The loan period is 25 years. What is the monthly payment?

112. ▦ The mortgage on a house is $100,000. The interest rate is $9\frac{1}{2}$%. The loan period is 30 years. What is the monthly payment?

1.2

EQUATIONS, INEQUALITIES, AND INTERVAL NOTATION

a) Solve equations like

$$-5x + 7 = 8x + 4$$

and

$$2t^2 = 9 + t.$$

b) Solve inequalities like

$$-5x + 7 < 8x + 4.$$

c) Solve applied problems.

d) Write interval notation for a given graph or inequality.

Equations

Basic to the solution of many equations are two simple principles. We can add the same number on both sides of a true equation and still obtain a true equation. We can also multiply on both sides of a true equation by any number and obtain a true equation.

THE ADDITION PRINCIPLE

If an equation $a = b$ is true, then the equation $a + c = b + c$ is true for any number c.

THE MULTIPLICATION PRINCIPLE

If an equation $a = b$ is true, then the equation $ac = bc$ is true for any number c.

When solving an equation we use these equation-solving principles and other properties of real numbers to get the variable alone on one side. Then it is easy to determine the solution.

Example 1   Solve $-\frac{5}{6}x + 10 = \frac{1}{2}x + 2$.

Solution   We first multiply on both sides by 6 to clear of fractions.

$$6\left(-\tfrac{5}{6}x + 10\right) = 6\left(\tfrac{1}{2}x + 2\right) \qquad \text{Using the Multiplication Principle}$$

$$6\left(-\tfrac{5}{6}x\right) + 6 \cdot 10 = 6\left(\tfrac{1}{2}x\right) + 6 \cdot 2 \qquad \text{Using the Distributive Law}$$

$$-5x + 60 = 3x + 12 \qquad \text{Simplifying}$$

$$60 = 8x + 12 \qquad \begin{array}{l}\text{Using the Addition Principle:}\\ \text{We add } 5x \text{ on both sides.}\end{array}$$

$$48 = 8x \qquad \text{Adding } -12 \text{ on both sides}$$

$$\tfrac{1}{8} \cdot 48 = \tfrac{1}{8} \cdot 8x \qquad \text{Multiplying by } \tfrac{1}{8} \text{ on both sides}$$

$$6 = x$$

1. Solve:

$$-\tfrac{7}{8}x + 5 = \tfrac{1}{4}x - 2.$$

2. An investment is made at 14%, compounded annually. It grows to $826.50 at the end of 1 year. How much was invested originally?

The variable is now alone, and we see that 6 is the solution. We can check by substituting 6 into the original equation.

DO EXERCISE 1.

To solve applied problems, we first translate to mathematical language, usually an equation. Then we solve the equation and check to see whether the solution to the equation is a solution to the problem.

Example 2   After a 5% gain in weight, an animal weighs 693 lb. What was its original weight?

Solution   We first translate to an equation:

$$\underbrace{\text{(Original weight)}}_{w} + \underbrace{5\%}_{+\ 5\%}\underbrace{\text{(Original weight)}}_{w} = 693$$
$$= 693.$$

Now we solve the equation:

$$w + 5\%w = 693$$
$$1 \cdot w + 0.05w = 693$$
$$(1 + 0.05)w = 693$$
$$1.05w = 693$$
$$w = \frac{693}{1.05} = 660.$$

Check: $660 + 5\% \times 660 = 660 + 0.05 \times 660 = 660 + 33 = 693.$

DO EXERCISE 2.

The third principle for solving equations is the *Principle of Zero Products.*

THE PRINCIPLE OF ZERO PRODUCTS

For any numbers a and b, if $ab = 0$, then $a = 0$ or $b = 0$; and if $a = 0$ or $b = 0$, then $ab = 0$.

An equation being solved using this principle must have a 0 on one side and a product on the other. The solutions are then obtained by setting each factor equal to 0 and solving the resulting equations.

3. Solve:

$$5x(x + 2)(2x - 3) = 0.$$

Solve.

4. $x^2 + x = 12$

5. $x^3 = x$

Example 3　Solve $3x(x - 2)(5x + 4) = 0$.

Solution

$$3x(x - 2)(5x + 4) = 0$$

$3x = 0$　　or　$x - 2 = 0$　or　$5x + 4 = 0$　　**Using the Principle of Zero Products**

$\frac{1}{3} \cdot 3x = \frac{1}{3} \cdot 0$　or　　　　$x = 2$　or　　　　$5x = -4$　　**Solving each separately**

$x = 0$　　or　　　　$x = 2$　or　　　　$x = -\frac{4}{5}$

The solutions are 0, 2, and $-\frac{4}{5}$.

　　Note that the Principle of Zero Products can be applied only when a product is 0. For example, because we know that $ab = 8$, we *do not know* that $a = 8$ or $b = 8$.

DO EXERCISE 3.

Example 4　Solve $4x^3 = x$.

Solution

$$4x^3 = x$$
$$4x^3 - x = 0 \qquad \text{Adding } -x$$
$$x(4x^2 - 1) = 0$$
$$x(2x - 1)(2x + 1) = 0 \qquad \text{Factoring}$$

$x = 0$　or　$2x - 1 = 0$　or　$2x + 1 = 0$　　**Using the Principle of Zero Products**

$x = 0$　or　　　$2x = 1$　or　　　$2x = -1$

$x = 0$　or　　　　$x = \frac{1}{2}$　or　　　　$x = -\frac{1}{2}$

The solutions are 0, $\frac{1}{2}$, and $-\frac{1}{2}$.

DO EXERCISES 4 AND 5.

Inequalities

The principles for solving inequalities are similar to those for solving equations. We can add the same number on both sides of an inequality. We can also multiply on both sides by the same nonzero number, but if that number is negative, we must reverse the inequality sign. The following is a reformulation of the inequality-solving principles.

Solve.

6. $3x < 11 - 2x$

7. $16 - 7x \leqslant 10x - 4$

8. In Example 6, determine the number of suits the firm must sell so that its total revenue will be more than $40,000.

INEQUALITY-SOLVING PRINCIPLES

If the inequality $a < b$ is true, then:

  i) $a + c < b + c$ is true, for any c;

  ii) $a \cdot c < b \cdot c$, for any positive c;

  iii) $a \cdot c > b \cdot c$, for any negative c.

Similar principles hold when $<$ is replaced by $\leqslant$ and $>$ is replaced by $\geqslant$.

Example 5 Solve $17 - 8x \geqslant 5x - 4$.

Solution

$$17 - 8x \geqslant 5x - 4$$
$$-8x \geqslant 5x - 21 \qquad \text{Adding } -17$$
$$-13x \geqslant -21 \qquad \text{Adding } -5x$$
$$-\tfrac{1}{13}(-13x) \leqslant -\tfrac{1}{13}(-21) \qquad \textbf{Multiplying by } -\tfrac{1}{13} \textbf{ and } \textit{reversing} \textbf{ the inequality sign}$$
$$x \leqslant \tfrac{21}{13}$$

Any number less than or equal to $\tfrac{21}{13}$ is a solution.

DO EXERCISES 6 AND 7.

Example 6 Raggs, Ltd., a clothing firm, determines that its total revenue, in dollars, from the sale of x suits is given by

$$2x + 50.$$

Determine the number of suits the firm must sell so that its total revenue will be more than $70,000.

Solution We translate to an inequality and solve:

$$2x + 50 > 70{,}000$$
$$2x > 69{,}950 \qquad \text{Adding } -50$$
$$x > 34{,}975. \qquad \text{Multiplying by } \tfrac{1}{2}$$

Thus the company's total revenue will exceed $70,000 when it sells more than 34,975 suits.

DO EXERCISE 8.

Intervals between houses are analogous to intervals on a number line.

Interval Notation

The set of real numbers corresponds to the set of points on a line.

For real numbers a and b such that $a < b$ (a is to the left of b on a number line), we define the *open interval* (a, b) to be the set of numbers between, but not including, a and b. That is,

(a, b) = the set of all numbers x such that $a < x < b$.

The graph of (a, b) is shown above in color. The open circles and the parentheses indicate that a and b are not included. The numbers a and b are called *endpoints*.

DO EXERCISES 9 AND 10.

9. Write interval notation for each graph.

a)
b)

The *closed interval* $[a, b]$ is the set of numbers between and including a and b. That is,

$[a, b]$ = the set of all numbers x such that $a \leqslant x \leqslant b$.

The graph of $[a, b]$ is shown above in color. The solid circles and the brackets indicate that a and b are included.

There are two kinds of *half-open intervals* defined as follows:

$(a, b]$ = the set of all numbers x such that $a < x \leqslant b$.

10. Write interval notation for each of the following.

a) The set of all numbers x such that $-1 < x < 4$

b) The set of all numbers x such that $-\frac{1}{4} < x < \frac{1}{4}$

11. Write interval notation for each graph.

a)

b)

c)

d)

12. Write interval notation for each of the following.

   a) The set of all numbers x such that $-\sqrt{2} < x < \sqrt{2}$

   b) The set of all numbers x such that $0 \leqslant x < 1$

   c) The set of all numbers x such that $-6.7 < x \leqslant -4.2$

   d) The set of all numbers x such that $3 \leqslant x \leqslant 7\frac{1}{2}$

13. Write interval notation for each graph.

   a)

   b)

   c)

   d)

14. Write interval notation for each of the following.

   a) The set of all numbers x such that $x \geqslant 8$

   b) The set of all numbers x such that $x < -7$

   c) The set of all numbers x such that $x > 10$

   d) The set of all numbers x such that $x \leqslant -0.78$

The open circle and the parenthesis indicate that a is not included. The solid circle and the bracket indicate that b is included. Also,

$[a, b)$ = the set of all numbers x such that $a \leqslant x < b$.

The solid circle and the bracket indicate that a is included. The open circle and the parenthesis indicate that b is not included.

DO EXERCISES 11 AND 12.

Some intervals are of unlimited extent in one or both directions. In such cases we use the infinity symbol ∞. For example,

$[a, \infty)$ = the set of all numbers x such that $x \geqslant a$.

Note that ∞ is not a number.

(a, ∞) = the set of all numbers x such that $x > a$.

$(-\infty, b]$ = the set of all numbers x such that $x \leqslant b$.

$(-\infty, b)$ = the set of all numbers x such that $x < b$.

We can name the entire set of real numbers using $(-\infty, \infty)$.

DO EXERCISES 13 AND 14.

Any point in an interval that is not an endpoint is an *interior point.*

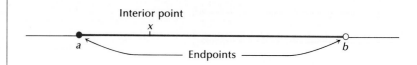

Note that all the points in an open interval are interior points.

EXERCISE SET 1.2

Solve.

1. $-7x + 10 = 5x - 11$

2. $-8x + 9 = 4x - 70$

3. $5x - 17 - 2x = 6x - 1 - x$

4. $5x - 2 + 3x = 2x + 6 - 4x$

5. $x + 0.8x = 216$

6. $x + 0.5x = 210$

7. $x + 0.08x = 216$

8. $x + 0.05x = 210$

APPLICATIONS

9. *Biology.* After a 6% gain in weight, an animal weighs 508.8 lb. What was its original weight?

10. *Biology.* After a 7% gain in weight, an animal weighs 363.8 lb. What was its original weight?

11. *Business.* An investment is made at 11%, compounded annually. It grows to $721.50 at the end of 1 year. How much was invested originally?

12. *Business.* An investment is made at 13%, compounded annually. It grows to $904 at the end of 1 year. How much was invested originally?

13. *Sociology.* After a 2% increase, the population of a city is 826,200. What was the former population?

14. *Sociology.* After a 3% increase, the population of a city is 741,600. What was the former population?

Solve.

15. $2x(x + 3)(5x - 4) = 0$

16. $7x(x - 2)(2x + 3) = 0$

17. $x^2 + 1 = 2x + 1$

18. $2t^2 = 9 + t^2$

19. $t^2 - 2t = t$

20. $6x - x^2 = x$

21. $6x - x^2 = -x$

22. $2x - x^2 = -x$

23. $9x^3 = x$

24. $16x^3 = x$

25. $(x - 3)^2 = x^2 + 2x + 1$

26. $(x - 5)^2 = x^2 + x + 3$

27. $3 - x \leqslant 4x + 7$

28. $x + 6 \leqslant 5x - 6$

29. $5x - 5 + x > 2 - 6x - 8$

30. $3x - 3 + 3x > 1 - 7x - 9$

31. $-7x < 4$

32. $-5x \geqslant 6$

33. $5x + 2x \leqslant -21$

34. $9x + 3x \geqslant -24$

35. $2x - 7 < 5x - 9$

36. $10x - 3 \geqslant 13x - 8$

37. $8x - 9 < 3x - 11$

38. $11x - 2 \geqslant 15x - 7$

39. $8 < 3x + 2 < 14$

40. $2 < 5x - 8 \leqslant 12$

41. $3 \leqslant 4x - 3 \leqslant 19$

42. $9 \leqslant 5x + 3 < 19$

43. $-7 \leqslant 5x - 2 \leqslant 12$

44. $-11 \leqslant 2x - 1 < -5$

APPLICATIONS

45. *Business.* A firm determines that the total revenue, in dollars, from the sale of x units of a product is

$$3x + 1000.$$

Determine the number of units that must be sold so that its total revenue will be more than \$22,000.

46. *Business.* A firm determines that the total revenue, in dollars, from the sale of x units of a product is

$$5x + 1000.$$

Determine the number of units that must be sold so that its total revenue will be more than \$22,000.

47. To get a B in a course a student's average must be greater than or equal to 80% (at least 80%) and less than 90%. On the first three tests the student scores 78%, 90%, and 92%. Determine the scores on the fourth test that will guarantee a B.

48. To get a C in a course a student's average must be greater than or equal to 70% and less than 80%. On the first three tests the student scores 65%, 83%, and 82%. Determine the scores on the fourth test that will guarantee a C.

Write interval notation for each graph in Exercises 49–56.

49.

50.

51.

52.

53.

54.

55.

56.

Write interval notation for Exercises 57–62.

57. The set of all numbers x such that $-3 \leqslant x \leqslant 3$

58. The set of all numbers x such that $-4 < x < 4$

59. The set of all numbers x such that $-14 \leqslant x < -11$

60. The set of all numbers x such that $6 < x \leqslant 20$

61. The set of all numbers x such that $x \leqslant -4$

62. The set of all numbers x such that $x > -5$

1.3

OBJECTIVES

a) Given a function and several inputs, find the outputs.

b) Graph a given equation or function.

c) Decide whether a graph is that of a function.

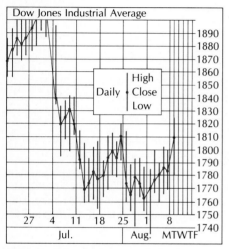

A graph of the Dow Jones Industrial Average.

GRAPHS AND FUNCTIONS

Graphs

Each point in the plane corresponds to an ordered pair of numbers. Note that the pair $(2, 5)$ is different from the pair $(5, 2)$. This is why we call $(2, 5)$ an *ordered pair*. The first member, 2, is called the *first coordinate* of the point, and the second member, 5, is called the *second coordinate*. Together these are called the *coordinates of the point*. The vertical line is often called the *y-axis*, and the horizontal line is often called the *x-axis*.

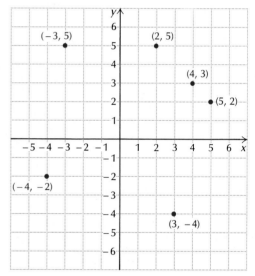

DO EXERCISE 1 ON THE FOLLOWING PAGE.

Graphs of Equations

A *solution* of an equation in two variables is an ordered pair of numbers that, when substituted alphabetically for the variables, gives a true sentence. For example, $(-1, 2)$ is a solution to the equation $3x^2 + y = 5$, because when we substitute -1 for x and 2 for y we get a true sentence:

$$3x^2 + y = 5$$

$$\begin{array}{c|c} 3(-1)^2 + 2 & 5 \\ 3 + 2 & \\ 5 & \end{array}$$

1. Plot the ordered pairs $(2, 0)$, $(0, 2)$, $(-1, 3)$, $(4, 3)$, and $(-2, -3)$.

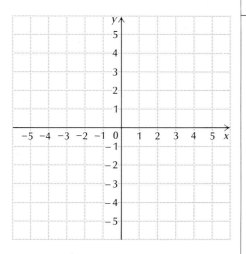

2. Decide whether each pair is a solution of

$$x^2 - 2y = 6.$$

a) $(-2, -1)$

b) $(3, 0)$

DO EXERCISE 2.

DEFINITION

The *graph* of an equation is a geometric representation of all its solutions.

The graph of an equation is obtained by plotting enough ordered pairs (which are solutions) to see a pattern. The graph could be a line, curve (or curves), or some other configuration.

Example 1   Graph $y = 2x + 1$.

Solution   We find some ordered pairs that are solutions and arrange them in a table. To find an ordered pair, we can choose *any* number for x and then determine y. For example, if we choose -2 for x, then $y = 2(-2) + 1 = -4 + 1 = -3$. We substituted -2 for x in the equation $y = 2x + 1$. For balance, we make some negative choices for x, as well as some positive choices. If a number takes us off the graph paper, we generally do not use it.

x	y	(x, y)
-2	-3	$(-2, -3)$
-1	-1	$(-1, -1)$
0	1	$(0, 1)$
1	3	$(1, 3)$
2	5	$(2, 5)$

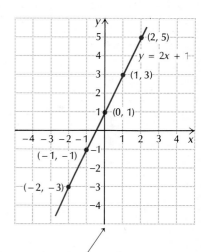

(1)   Choose any x.
(2)   Compute y.
(3)   Form the pair (x, y).
(4)   Plot the points.

After we plot the points we look for a pattern in the graph. If we had enough of the points they would make a solid line. We can draw the line with a ruler and label it $y = 2x + 1$.

3. Graph $y = -2x + 1$.

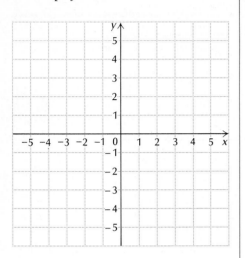

4. Graph $y = x^2 - 3$.

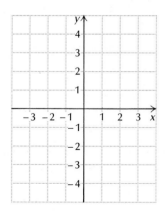

DO EXERCISE 3.

Example 2   Graph $y = x^2 - 1$.

x	y	(x, y)
-2	3	$(-2, 3)$
-1	0	$(-1, 0)$
0	-1	$(0, -1)$
1	0	$(1, 0)$
2	3	$(2, 3)$

(1)   Choose any x. ──┐
(2)   Compute y. ────────
(3)   Form the pair (x, y). ──────
(4)   Plot the points. ────────

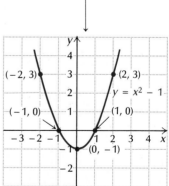

This time the pattern of the points is a curve called a *parabola*. We fill in the pattern and obtain the graph. Note that enough points must be plotted to see a pattern.

DO EXERCISE 4.

Example 3   Graph $x = y^2$.

5. Graph $x = y^2 + 1$.

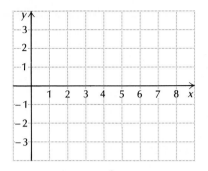

Solution This time x is expressed in terms of the variable y. Thus we choose numbers for y and compute x.

x	y	(x, y)
4	-2	$(4, -2)$
1	-1	$(1, -1)$
0	0	$(0, 0)$
1	1	$(1, 1)$
4	2	$(4, 2)$

(1) Choose any y.
(2) Compute x.
(3) Form the pair (x, y).
(4) Plot the points.

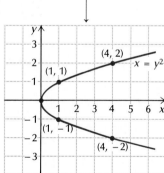

We plot these points, keeping in mind that x is still the first coordinate and y the second. We look for a pattern and complete the graph.

DO EXERCISE 5.

Functions

A *relation* is any set of ordered pairs. Thus the solutions of an equation in two variables form a relation.

A *function* is a special kind of relation. Such relations are of fundamental importance in calculus.

6. The operation of "taking the reciprocal" is a function. That is, the operation of going from x to $1/x$ is a function defined for all numbers except 0. Thus the domain is the set of all nonzero real numbers.

Complete this table.

Input	Output
5	
$-\frac{2}{3}$	
$\frac{1}{4}$	
$\dfrac{1}{a}$	
k	
$1 + t$	

A Function as an Input–Output Relation

DEFINITION

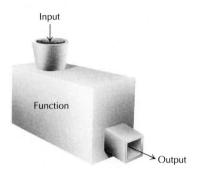

A *function* is a relation that assigns to each "input," or first coordinate, a unique "output," or second coordinate. The set of all input numbers is called the *domain*. The set of all output numbers is called the *range*.

Example 4   Squaring numbers is a function. We can take any number x as an input. We square that number to find the output, x^2.

Input	Output
-3	9
1.73	2.9929
k	k^2
$\sqrt{a}$	a
$1 + t$	$(1 + t)^2$, or $1 + 2t + t^2$

Think of a function as a machine. Think of putting a member of the domain (an *input*) into the machine. The machine squares the input and gives you the *output*, a member of the range.

The domain of this function is the set of all real numbers, because any real number can be squared.

DO EXERCISE 6.

7. The reciprocal function is given by

$$f(x) = \frac{1}{x}.$$

Find $f(5), f(-2), f(\frac{1}{4}), f(1/a), f(k),$ $f(1 + t),$ and $f(x + h)$.

It is customary to use letters such as f and g to represent functions. Suppose f is a function and x is a number in its domain. For the input x, we can name the output as

$$f(x), \text{ read "}f\text{ of }x\text{," or "the value of }f\text{ at }x\text{."}$$

If f is the squaring function, then $f(3)$ is the output for the input 3. Thus $f(3) = 3^2 = 9$.

Example 5 The squaring function is given by

$$f(x) = x^2.$$

Find $f(-3), f(1), f(k), f(\sqrt{k}), f(1 + t),$ and $f(x + h)$.

Solution

$$
\begin{aligned}
f(-3) &= (-3)^2 = 9, \\
f(1) &= 1^2 = 1, \\
f(k) &= k^2, \\
f(\sqrt{k}) &= (\sqrt{k})^2 = k, \\
f(1 + t) &= (1 + t)^2 = 1 + 2t + t^2, \\
f(x + h) &= (x + h)^2 = x^2 + 2xh + h^2
\end{aligned}
$$

To find $f(x + h)$, remember what the function does: It squares the input. Thus $f(x + h) = (x + h)^2 = x^2 + 2xh + h^2$. This amounts to replacing x on both sides of $f(x) = x^2$ by $x + h$.

DO EXERCISE 7.

8. A function t is given by

$$t(x) = x + x^2.$$

Find $t(5), t(-5),$ and $t(x + h)$.

Example 6 A function f subtracts the square of an input from the input. A description of f is given by

$$f(x) = x - x^2.$$

Find $f(4)$ and $f(x + h)$.

Solution We replace the x's on both sides by the inputs. Thus

$$
\begin{aligned}
f(4) &= 4 - 4^2 = 4 - 16 = -12; \\
f(x + h) &= (x + h) - (x + h)^2 = x + h - (x^2 + 2xh + h^2) \\
&= x + h - x^2 - 2xh - h^2.
\end{aligned}
$$

DO EXERCISE 8.

The number of people at the beach at a given time is a function of the temperature, although a formula may not be readily available.

9. Subtracting 3 from a number and then taking the reciprocal is a function f given by

$$f(x) = \frac{1}{x - 3}.$$

a) What is the domain of this function? Explain.

b) Find $f(5)$, $f(4)$, $f(2.5)$, and $f(x + h)$.

Taking square roots is *not* a function, because an input can have more than one output. For example, the input 4 has two outputs, 2 and -2.

When a function is given by a formula, and nothing is said about its domain, its domain is understood to be the set of all numbers that can be substituted into the formula. For example, consider the reciprocal function

$$f(x) = \frac{1}{x}.$$

The only number that cannot be substituted into the formula is 0. We say that f *is not defined at* 0, or $f(0)$ *does not exist.* The domain consists of all nonzero real numbers.

Example 7   Taking principal square roots (nonnegative roots) is a function. Let g be this function. Then g can be described as

$$g(x) = \sqrt{x}.$$

Recall from algebra that the symbol $\sqrt{a}$ represents the nonnegative square root of a. There is only one such real-number root.

a)   Find the domain of this function.

b)   Find $g(0)$, $g(2)$, $g(a)$, $g(16)$, and $g(t + h)$.

Solution

a)   The domain consists of numbers that can be substituted into the formula. We can take the principal square root of any nonnegative number, but the principal square root of a negative number is not a real number. (Taking square roots of negative numbers would require us to consider complex numbers, which we will not do in this text.) Thus the domain consists of all nonnegative numbers.

b)
$$g(0) = \sqrt{0} = 0,$$
$$g(2) = \sqrt{2},$$
$$g(a) = \sqrt{a},$$
$$g(16) = \sqrt{16} = 4,$$
$$g(t + h) = \sqrt{t + h}$$

DO EXERCISE 9.

A Function as a Mapping

We can also think of a function as a "mapping" of one set to another.

DEFINITION

A *function* is a mapping that associates with each number x in one set (called the domain) a unique number y in another set.

The set to which numbers are mapped can be either totally different from the domain or the same set.

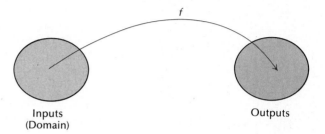

Inputs
(Domain)

Outputs

For example, the squaring function maps members of the set of real numbers to members of the set of nonnegative numbers.

The statement

$$y = f(x)$$

means that the number x is mapped to the number y by the function f. Functions are often implicit in certain equations. For example, consider

$$xy = 2.$$

For any nonzero x there is a unique number y satisfying the equation. This yields a function that is given explicitly by

$$y = f(x) = \frac{2}{x}.$$

On the other hand, consider the equation

$$x = y^2.$$

A number x would be related to two values of y, namely $\sqrt{x}$ and $-\sqrt{x}$. Thus this equation is not an implicit description of a function that maps inputs x to outputs y.

Graphs of Functions

Consider again the squaring function. The input 3 is associated with the output 9. The input–output pair $(3, 9)$ is one point on the *graph* of this function.

DEFINITION

The *graph* of a function f is a geometric representation of all its input–output pairs $(x, f(x))$. In cases where the function is given by an equation, the graph of a function is the graph of the equation $y = f(x)$.

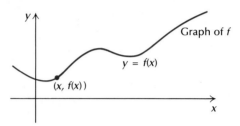

It is customary to locate input values (the domain) on the horizontal axis and output values on the vertical axis.

Example 8   Graph $f(x) = x^2 - 1$.

Solution

x	$f(x)$	$(x, f(x))$
-2	3	$(-2, 3)$
-1	0	$(-1, 0)$
0	-1	$(0, -1)$
1	0	$(1, 0)$
2	3	$(2, 3)$

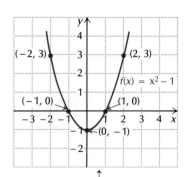

(1) Choose any x.
(2) Compute $f(x)$.
(3) Form the pair (x, y).
(4) Plot the points.

10. Graph $f(x) = -2x + 1$.

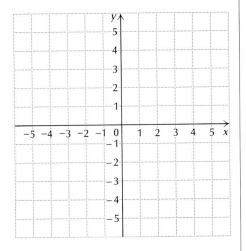

We plot the input–output pairs from the table and, in this case, draw a curve to complete the graph.

DO EXERCISES 10 AND 11.

The following figure illustrates how the idea of a mapping is connected with the graph of a function.

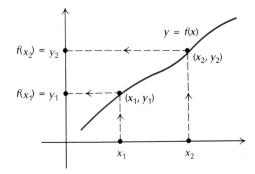

11. Graph $g(x) = x^2 - 3$.

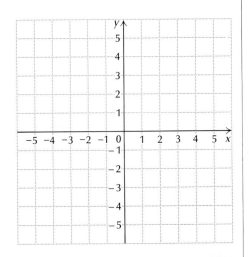

Let us now determine how we can look at a graph and decide whether it is a graph of a function. We already know that

$$x = y^2$$

does not yield a function that maps a number x to a unique number y. Look at its graph.

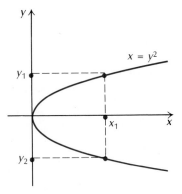

Note that there is a point x_1 that has two outputs. Equivalently, we have a vertical line that meets the graph in more than one place.

12. Which of the following are graphs of functions?

a)

b)

c)

d)

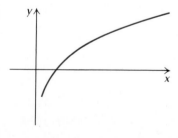

THE VERTICAL-LINE TEST

A graph is that of a function provided no vertical line meets the graph more than once.

Examples Which of the following are graphs of functions?

a)

b)

c)

d)

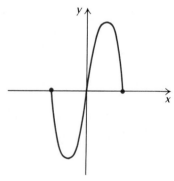

Solution

a) A function. No vertical line meets the graph more than once.

b) Not a function. A vertical line (in fact, many) meets the graph more than once.

c) Not a function.

d) A function.

DO EXERCISE 12.

13. Graph the function defined as follows:

$$f(x) = \begin{cases} x + 3 & \text{for } x \leqslant -2, \\ 1 & \text{for } -2 < x \leqslant 3, \\ x^2 - 10 & \text{for } 3 < x. \end{cases}$$

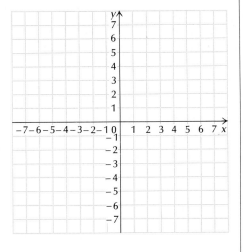

Functions Defined Piecewise

Sometimes functions are defined piecewise. That is, we have different output formulas for different parts of the domain.

Example 9 Graph the function defined as follows.

$$f(x) = \begin{cases} 4, & \text{for } x \leqslant 0 \\ & \text{(This means that for any input } x \text{ less than or equal to 0, the output is 4.)} \\ 4 - x^2, & \text{for } 0 < x \leqslant 2 \\ & \text{(This means that for any input } x \text{ greater than 0 and less than or equal to 2, the output is } 4 - x^2.) \\ 2x - 6, & \text{for } x > 2 \\ & \text{(This means that for any input } x \text{ greater than 2, the output is } 2x - 6.) \end{cases}$$

Solution See the following graph.

a) We graph $f(x) = 4$ for inputs less than or equal to 0 (that is, $x \leqslant 0$).

b) We graph $f(x) = 4 - x^2$ for inputs greater than 0 and less than or equal to 2 (that is, $0 < x \leqslant 2$).

c) We graph $f(x) = 2x - 6$ for inputs greater than 2 (that is, $x > 2$).

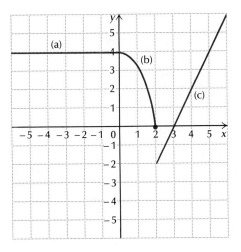

DO EXERCISE 13.

Some Final Remarks

Almost all the functions in this text can be described by equations. Some functions, however, cannot. For example, there will be a function that assigns grades to students in this course, but that function will most likely not have a formula.

We sometimes use the terminology *y is a function of x*. This means that x is an input and y is an output. We often refer to x as the *independent variable* when it represents inputs and y as the *dependent variable* when it represents outputs. We may refer to "the function $y = x^2$," without naming it with a letter f. We may simply refer to x^2 (alone) as a function.

In calculus we will be studying how the outputs of a function change when the inputs change.

EXERCISE SET 1.3

1. A function f is given by

$$f(x) = 2x + 3.$$

This function takes a number x, multiplies it by 2, and adds 3.

a) Complete this table.

Input	Output
4.1	
4.01	
4.001	
4	

CSS

b) Find $f(5)$, $f(-1)$, $f(k)$, $f(1 + t)$, and $f(x + h)$.

3. A function g is given by

$$g(x) = x^2 - 3.$$

This function takes a number x, squares it, and subtracts 3. Find $g(-1)$, $g(0)$, $g(1)$, $g(5)$, $g(u)$, $g(a + h)$, and $g(1 - h)$.

CSS

2. A function f is given by

$$f(x) = 3x - 1.$$

This function takes a number x, multiplies it by 3, and subtracts 1.

a) Complete this table.

Input	Output
5.1	
5.01	
5.001	
5	

b) Find $f(4)$, $f(-2)$, $f(k)$, $f(1 + t)$, and $f(x + h)$.

4. A function g is given by

$$g(x) = x^2 + 4.$$

This function takes a number x, squares it, and adds 4. Find $g(-3)$, $g(0)$, $g(-1)$, $g(7)$, $g(v)$, $g(a + h)$, and $g(1 - t)$.

5. A function f is given by

$$f(x) = (x - 3)^2.$$

This function takes a number x, subtracts 3 from it, and squares the result.

   a) Find $f(4), f(-2), f(0), f(a), f(t + 1), f(t + 3)$, and $f(x + h)$.

   b) Note that f could also be given by

$$f(x) = x^2 - 6x + 9.$$

    Explain what this does to an input number x.

6. A function f is given by

$$f(x) = (x + 4)^2.$$

This function takes a number x, adds 4 to it, and squares the result.

   a) Find $f(3), f(-6), f(0), f(k), f(t - 1), f(t - 4)$, and $f(x + h)$.

   b) Note that f could also be given by

$$f(x) = x^2 + 8x + 16.$$

    Explain what this does to an input number x.

Graph each of the following functions.

  7. $f(x) = 2x + 3$ **8.** $f(x) = 3x - 1$ **9.** $g(x) = -4x$ **10.** $g(x) = -2x$

11. $f(x) = x^2 - 1$ **12.** $f(x) = x^2 + 4$ **13.** $g(x) = x^3$ **14.** $g(x) = \dfrac{1}{2}x^3$

Which of the following are graphs of functions?

15.

16.

17.

18.

19.

20.

21.

22.

23. **a)** Graph $x = y^2 - 1$.

     b) Is this a function?

25. For $f(x) = x^2 - 3x$, find $f(x + h)$.

24. **a)** Graph $x = y^2 - 3$.

     b) Is this a function?

26. For $f(x) = x^2 + 4x$, find $f(x + h)$.

APPLICATIONS

27. *Business: Revenue.* Raggs, Ltd., a clothing firm, determines that its total revenue from the sale of x suits is given by the function

$$R(x) = 2x + 50,$$

where $R(x)$ is the revenue, in dollars, from the sale of x units. Find $R(10)$ and $R(100)$.

28. *Business: Compound interest.* The amount of money in a savings account at 14%, compounded annually, depends on the initial investment x and is given by the function

$$A(x) = x + 14\%x,$$

where $A(x) = $ the amount in the account at the end of one year. Find $A(100)$ and $A(1000)$.

Graph.

29. $f(x) = \begin{cases} 1 & \text{for } x < 0, \\ -1 & \text{for } x \geqslant 0 \end{cases}$

CSS

30. $f(x) = \begin{cases} 2 & \text{for } x \text{ an integer}, \\ -2 & \text{for } x \text{ not an integer} \end{cases}$

31. $f(x) = \begin{cases} -3 & \text{for } x = -2, \\ x^2 & \text{for } x \neq -2 \end{cases}$

32. $f(x) = \begin{cases} -2x - 6 & \text{for } x \leqslant -2, \\ 2 - x^2 & \text{for } -2 < x < 2, \\ 2x - 6 & \text{for } x \geqslant 2 \end{cases}$

CSS

EXTENSION EXERCISES

Solve each of the following for y in terms of x. Decide whether each of the resulting equations represents a function.

33. $2x + y - 16 = 4 - 3y + 2x$ 34. $2y^2 + 3x = 4x + 5$ 35. $(4y^{2/3})^3 = 64x$ 36. $(3y^{3/2})^2 = 72x$

1.4

OBJECTIVES

a) Graph equations of the type $y = b$ and $x = a$.

b) Graph linear functions.

c) Find an equation of a line given its slope and one point on the line.

d) Find the slope of the line containing a given pair of points.

e) Find an equation of the line containing a given pair of points.

f) Given an equation of a line, find the slope and the y-intercept.

g) Solve problems involving linear functions.

STRAIGHT LINES AND LINEAR FUNCTIONS

Horizontal and Vertical Lines

Let us consider graphs of equations $y = b$ and $x = a$.

Example 1

a) Graph $y = 4$.

b) Decide whether the relation is a function.

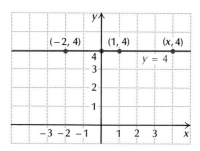

Solution

a) The graph consists of all ordered pairs whose second coordinate is 4. To see how a pair such as $(-2, 4)$ could be a solution of $y = 4$, we can consider the equation above in the form

$$0x + y = 4.$$

1. **a)** Graph $y = 3$.
   b) Decide whether it is a function.

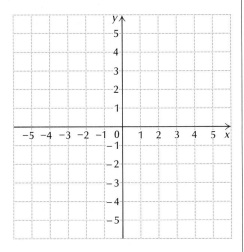

2. **a)** Graph $x = 1$.
   b) Decide whether it is a function.

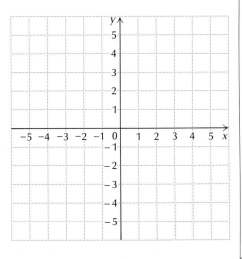

Then $(-2, 4)$ is a solution because

$$0(-2) + 4 = 4$$

is true.

b) The vertical-line test holds. Thus this is a function.

DO EXERCISE 1.

Example 2

a) Graph $x = -3$.
b) Decide whether it is a function.

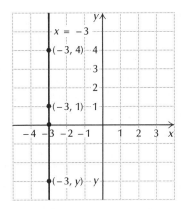

Solution

a) The graph consists of all ordered pairs whose first coordinate is -3.

b) This is *not* a function. It fails the vertical-line test. The line itself meets the graph more than once—in fact, infinitely many times.

DO EXERCISE 2.

In general, we have the following.

THEOREM 6

The graph of $y = b$, a horizontal line, is the graph of a function.
The graph of $x = a$, a vertical line, is not the graph of a function.

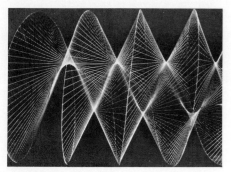

Lines of various slopes.

The Equation $y = mx$

Consider the following table of numbers and look for a pattern.

x	1	-1	$-\frac{1}{2}$	2	-2	3	-7	5
y	3	-3	$-\frac{3}{2}$	6	-6	9	-21	15

Note that the ratio of the bottom number to the top number is 3. That is,

$$\frac{y}{x} = 3, \quad \text{or } y = 3x.$$

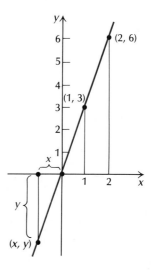

Ordered pairs from the table can be used to graph the equation $y = 3x$. Note that this is a function.

THEOREM 7

The graph of the function given by

$$y = mx \quad \text{or} \quad f(x) = mx$$

is the straight line through the origin $(0, 0)$ and the point $(1, m)$. The constant m is called the *slope* of the line.

---

3. a) Graph $y = -2x$.

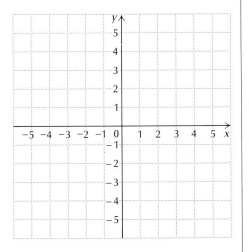

b) Is this a function?

c) What is the slope?

4. Consider these lines.

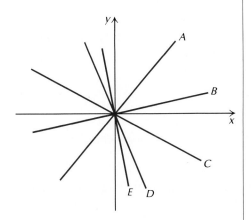

a) Which lines have positive slope?

b) Which lines have negative slope?

c) Which line has the largest slope?

d) Which line has the smallest slope?

DO EXERCISE 3.

Various graphs of $y = mx$ for positive m are shown below (part a). Note that such graphs slant up from left to right. A line with large positive slope rises faster than a line with smaller positive slope.

a)   **b)**

c)

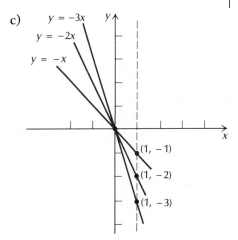

When $m = 0$, $y = 0x$, or $y = 0$. Graph (b) is a graph of $y = 0$. Note that this is the x-axis and is a horizontal line. Graphs of $y = mx$ for negative m are shown in part (c). Note that such graphs slant down from left to right.

DO EXERCISE 4.

Hair will grow 6 inches in 12 months.

5. *Ecology: Newspaper recycling.* The number T of trees saved by recycling is directly proportional to the height h of a stack of recycled newspaper.

a) It is known that a stack of newspaper 36 in. high will save 1 tree. Find an equation of variation expressing T as a function of h.

b) How many trees are saved by a stack of paper 162 in. (13.5 ft) high?

Direct Variation

There are many applications involving equations like $y = mx$, where m is some positive number. In such situations we say that we have *direct variation,* and m (the slope) is called the *variation constant,* or *constant of proportionality.* Generally, only positive values of x and y are considered.

DEFINITION

The variable y *varies directly* as x if there is some positive constant m such that $y = mx$. We also say that y is *directly proportional* to x.

Example 3 *Biomedical: Hair growth.* The number N of inches that human hair will grow is directly proportional to the time t in months. Hair will grow 6 inches in 12 months.

a) Find an equation of variation.

b) How many months does it take for hair to grow 10 inches?

Solution

a) Since $N = mt$, then $6 = m(12)$ and $\frac{1}{2} = m$. Thus $N = \frac{1}{2}t$.

b) To find how many months it takes for hair to grow 10 inches we solve

$$10 = \tfrac{1}{2}t$$

and get

$$20 = t.$$

Thus it takes 20 months for hair to grow 10 inches.

DO EXERCISE 5.

The Equation $y = mx + b$

Compare the graphs of the equations

$$y = 3x \quad \text{and} \quad y = 3x - 2.$$

Note that the graph of $y = 3x - 2$ is a shift downward 2 units of the graph of $y = 3x$, and that $y = 3x - 2$ has y-intercept $(0, -2)$. Note also that the graph of $y = 3x - 2$ is a graph of a function.

6. a) Using the same axes, graph

$$y = 3x$$

and

$$y = 3x + 1.$$

b) How can the graph of $y = 3x + 1$ be obtained from the graph of $y = 3x$?

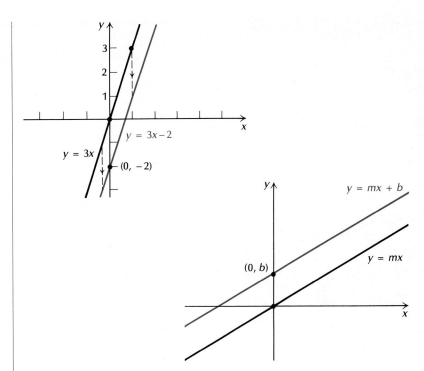

DO EXERCISE 6.

A *linear function* is given by

$$y = mx + b \quad \text{or} \quad f(x) = mx + b$$

and has a graph that is the straight line parallel to $y = mx$ with y-intercept $(0, b)$. The constant m is called the *slope*.

When $m = 0$, $y = 0x + b = b$, and we have what is known as a *constant function*. The graph of such a function is a horizontal line.

The Slope–Intercept Equation

Any nonvertical line l is uniquely determined by its slope m and its y-intercept $(0, b)$. In other words, the slope describes the "slant" of the line, and the y-intercept is the point at which it crosses the y-axis. Thus we have the following definition.

7. Find the slope and the y-intercept of $2x + 3y - 6 = 0$.

DEFINITION

$y = mx + b$ is called the *slope–intercept equation* of a line.

Example 4 Find the slope and the y-intercept of $2x - 4y - 7 = 0$.

Solution We solve for y:

$$-4y = -2x + 7$$
$$y = \tfrac{1}{2}x - \tfrac{7}{4}$$

Slope: $\tfrac{1}{2}$ y-intercept: $(0, -\tfrac{7}{4})$

DO EXERCISE 7.

The Point–Slope Equation

Suppose we know the slope of a line and some point of the line other than the y-intercept. We can still find an equation of the line.

Example 5 Find an equation of the line with slope 3 containing the point $(-1, -5)$.

8. Find an equation of the line with slope -4 containing the point $(2, -7)$.

Solution From the slope–intercept equation we have

$$y = 3x + b,$$

so we must determine b. Since $(-1, -5)$ is on the line, it follows that

$$-5 = 3(-1) + b,$$

so

$$-2 = b \quad \text{and} \quad y = 3x - 2.$$

DO EXERCISE 8.

If a point (x_1, y_1) is on the line

$$y = mx + b, \tag{1}$$

it must follow that

$$y_1 = mx_1 + b. \tag{2}$$

Lines of the same slope.

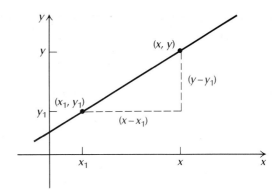

Subtracting Eq. (2) from Eq. (1) eliminates the b's, and we have

$$y - y_1 = (mx + b) - (mx_1 + b)$$
$$= mx + b - mx_1 - b$$
$$= mx - mx_1 = m(x - x_1).$$

DEFINITION

$y - y_1 = m(x - x_1)$ is called the *point–slope equation* of a line.

This definition allows us to write an equation of a line given its slope and the coordinates of *any* point on it.

Example 6 Find an equation of the line with slope 3 containing the point $(-1, -5)$.

Solution Substituting in

$$y - y_1 = m(x - x_1),$$

we get

$$y - (-5) = 3[x - (-1)].$$

Simplifying and solving for y, we get the slope–intercept equation as found in Example 5:

$$y + 5 = 3(x + 1)$$
$$y = 3x + 3 - 5$$
$$y = 3x - 2.$$

DO EXERCISE 9.

9. Find an equation of the line with slope -4 containing the point $(2, -7)$.

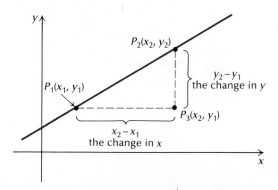

We now determine a method of computing the slope of a line when we know the coordinates of two of its points. Suppose (x_1, y_1) and (x_2, y_2) are the coordinates of two different points, P_1 and P_2, respectively, on a line that is not vertical. Consider a right triangle as shown, with legs parallel to the axes. The point P_3 with coordinates (x_2, y_1) is the third vertex of the triangle. As we move from P_1 to P_2, y changes from y_1 to y_2. The change in y is $y_2 - y_1$. Similarly, the change in x is $x_2 - x_1$. The ratio of these changes is the slope. To see this, consider the point–slope equation

$$y - y_1 = m(x - x_1).$$

Since (x_2, y_2) is on the line, it must follow that

$$y_2 - y_1 = m(x_2 - x_1).$$

Since the line is not vertical, the two x-coordinates must be different, so $x_2 - x_1$ is nonzero and we can divide by it to get the following theorem.

THEOREM 8

$$m = \frac{y_2 - y_1}{x_2 - x_1} = \frac{\text{change in } y}{\text{change in } x} = \begin{array}{l}\text{slope of line containing points} \\ (x_1, y_1) \text{ and } (x_2, y_2).\end{array}$$

Example 7   Find the slope of the line containing the points $(-2, 6)$ and $(-4, 9)$.

Solution   We have

$$m = \frac{y_2 - y_1}{x_2 - x_1} = \frac{6 - 9}{-2 - (-4)} = \frac{-3}{2} = -\frac{3}{2}.$$

Note that it does not matter which point is taken first, so long as we

Find the slope of the line containing each pair of points.

10. $(1, 3)$, $(2, 5)$

11. $(-6, 4)$, $(2, 5)$

12. $(4, 7)$, $(6, -10)$

13. $(3, 5)$, $(-1, 5)$

Find the slope, if it exists, of the line containing each pair of points.

14. $(4, -7)$, $(-2, -7)$

15. $(4, -7)$, $(4, -9)$

subtract the coordinates in the same order. In this example we can also find m as follows:

$$m = \frac{9 - 6}{-4 - (-2)} = \frac{3}{-2} = -\frac{3}{2}.$$

DO EXERCISES 10–13.

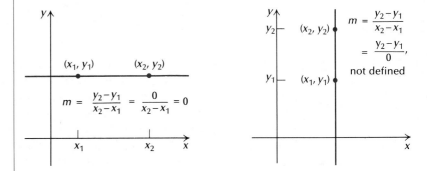

If a line is horizontal, the change in y for any two points is 0. Thus a horizontal line has slope 0. If a line is vertical, the change in x for any two points is 0. Thus the slope is not defined because we cannot divide by 0. A vertical line has no slope. Thus "0 slope" and "no slope" are two very distinct concepts.

DO EXERCISES 14 AND 15.

Applications of Linear Functions

Many applications are modeled by linear functions.

Example 8 *Business: Total cost.* Raggs, Ltd., a clothing firm, has *fixed costs* of $10,000 per year. These costs, such as rent, maintenance, and so on, must be paid no matter how much the company produces. To produce x units of a certain kind of suit it costs $20 per unit in addition to the fixed costs. That is, the *variable costs* for producing x of these units is $20x$ dollars. These are costs that are directly related to production, such as material, wages, fuel, and so on. Then the *total cost, $C(x)$,* of producing x suits in a year is given by a function C:

$$C(x) = (\text{Variable costs}) + (\text{Fixed costs}) = 20x + 10,000.$$

a) Graph the variable-cost, the fixed-cost, and the total-cost functions.

b) What is the total cost of producing 100 suits? 400 suits?

c) How much more does it cost to produce 400 suits than 100 suits?

Solution

a) The variable-cost and the fixed-cost functions appear in the first graph below. The total-cost function is shown in the second graph. From a practical standpoint, the domains of these functions are nonnegative integers 0, 1, 2, 3, and so on, since it does not make sense to make a negative number of suits or a fractional number of suits. Nevertheless, it is common practice to draw the graphs as though the domains were the entire set of nonnegative real numbers.

b) The total cost of producing 100 suits is

$$C(100) = 20 \cdot 100 + 10{,}000 = \$12{,}000.$$

The total cost of producing 400 suits is

$$C(400) = 20 \cdot 400 + 10{,}000 = \$18{,}000.$$

c) The extra cost of producing 400 suits rather than 100 suits is given by

$$C(400) - C(100) = \$18{,}000 - \$12{,}000 = \$6000.$$

16. Rework Example 8, given that variable costs $= 30x$, fixed costs $= \$15,000$, and total costs $= C(x) = 30x + 15,000$.

DO EXERCISE 16.

Example 9 *Business: Profit-and-loss analysis.* In reference to Example 8, Raggs, Ltd., determines that its total revenue from the sale of x suits is $80 per suit. That is, the total revenue $R(x)$ is given by the function

$$R(x) = 80x.$$

a) Graph $R(x)$ and $C(x)$ using the same axes.

b) The total profit $P(x)$ is given by a function P:

$$P(x) = (\text{Total revenue}) - (\text{Total costs}) = R(x) - C(x).$$

Determine $P(x)$ and draw its graph using the same axes.

c) The company will *break even* at that value of x for which $P(x) = 0$ (that is, no profit and no loss). This is where $R(x) = C(x)$. Find the break-even value of x.

Solution

a) The graphs of $R(x) = 80x$ and $C(x) = 20x + 10,000$ are shown here. When $C(x)$ is above $R(x)$ a loss will occur. This is shown by the color-shaded region. When $R(x)$ is above $C(x)$ a gain will occur. This is shown by the gray-shaded region.

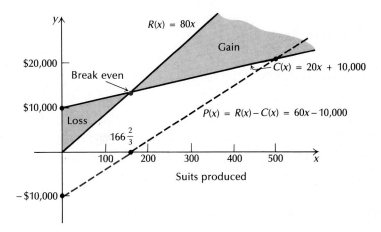

b) We see that

$$P(x) = R(x) - C(x) = 80x - (20x + 10,000) = 60x - 10,000.$$

The graph of $P(x)$ is shown by the dashed line. The color dashed line shows a "negative" profit, or loss. The black dashed line shows a "positive" profit, or gain.

17. Rework Example 9, given that

$$C(x) = 30x + 15{,}000$$

and

$$R(x) = 90x.$$

c) To find the break-even value we solve $R(x) = C(x)$:

$$R(x) = C(x)$$
$$80x = 20x + 10{,}000$$
$$60x = 10{,}000$$
$$x = 166\tfrac{2}{3}.$$

How do we interpret the fractional answer, since it is not possible to produce $\tfrac{2}{3}$ of a suit? We simply round to 167. Estimates of break-even points are usually sufficient since companies want to operate well away from break-even points in order to maximize profit.

DO EXERCISE 17.

EXERCISE SET 1.4

Graph.

1. $y = -4$ **2.** $y = -3.5$ **3.** $x = 4.5$ **4.** $x = 10$

Graph. Find the slope and the y-intercept.

5. $y = -3x$ **6.** $y = -0.5x$ **7.** $y = 0.5x$ **8.** $y = 3x$

9. $y = -2x + 3$ **10.** $y = -x + 4$ **11.** $y = -x - 2$ **12.** $y = -3x + 2$

Find the slope and the y-intercept.

13. $2x + y - 2 = 0$ **14.** $2x - y + 3 = 0$ **15.** $2x + 2y + 5 = 0$ **16.** $3x - 3y + 6 = 0$

Find an equation of the line:

17. with $m = -5$, containing $(1, -5)$.

18. with $m = 7$, containing $(1, 7)$.

19. with $m = -2$, containing $(2, 3)$.

20. with $m = -3$, containing $(5, -2)$.

21. with y-intercept $(0, -6)$ and slope $\tfrac{1}{2}$.

22. with y-intercept $(0, 7)$ and slope $\tfrac{4}{3}$.

23. with slope 0, containing $(2, 3)$.

24. with slope 0, containing $(4, 8)$.

Find the slope of the line containing each pair of points.

25. $(-4, -2), (-2, 1)$

26. $(-2, 1), (6, 3)$

27. $(2, -4), (4, -3)$

28. $(-5, 8), (5, -3)$

29. $(3, -7), (3, -9)$

30. $(-4, 2), (-4, 10)$

31. $(2, 3), (-1, 3)$

32. $(-6, \tfrac{1}{2}), (-7, \tfrac{1}{2})$

33. $(x, 3x), (x + h, 3(x + h))$

34. $(x, 4x), (x + h, 4(x + h))$

35. $(x, 2x + 3), (x + h, 2(x + h) + 3)$

36. $(x, 3x - 1), (x + h, 3(x + h) - 1)$

DEFINITION

The two-point equation of the nonvertical line containing the points (x_1, y_1) and (x_2, y_2) is given by

$$y - y_1 = \frac{y_2 - y_1}{x_2 - x_1}(x - x_1).$$ Two-point equation

This can be proved by replacing m in the point–slope equation $y - y_1 = m(x - x_1)$ by $(y_2 - y_1)/(x_2 - x_1)$ in Theorem 8.

37.–48. Find an equation of the line containing each pair of points in Exercises 25–36.

APPLICATIONS

49. *Ecology: Energy conservation.* The R-factor of home insulation is directly proportional to its thickness T.

   a) Find an equation of variation if $R = 12.51$ when $T = 3$ in.

   b) What is the R-factor for insulation that is 6 in. thick?

51. *Biomedical: Brain weight.* The weight B of a human's brain is directly proportional to his or her body weight W.

   a) It is known that a person who weighs 200 lb has a brain that weighs 5 lb. Find an equation of variation expressing B as a function of W.

   b) Express the variation constant as a percent and interpret the resulting equation.

   c) What is the weight of the brain of a person who weighs 120 lb?

53. *Business: Investment.* A person makes an investment of P dollars at 14%. After 1 year it grows to an amount A.

   a) Show that A is directly proportional to P.

   b) Find A when $P = \$100$.

   c) Find P when $A = \$273.60$.

54. *Sociology: Urban population.* The population of a town is P. After a growth of 2%, its new population is N.

   a) Assuming that N is directly proportional to P, find an equation of variation.

   b) Find N when $P = 200,000$.

   c) Find P when $N = 367,200$.

50. *Biomedical: Nerve impulse speed.* Impulses in nerve fibers travel at a speed of 293 ft/sec. The distance D traveled in t sec is given by $D = 293t$. How long would it take an impulse to travel from the brain to the toes of a person who is 6 ft tall?

52. *Biomedical: Muscle weight.* The weight M of the muscles in a human is directly proportional to his or her body weight W.

   a) It is known that a person who weighs 200 lb has 80 lb of muscles. Find an equation of variation expressing M as a function of W.

   b) Express the variation constant as a percent and interpret the resulting equation.

   c) What is the muscle weight of a person who weighs 120 lb?

The muscle weight is directly proportional to body weight.

55. *Stopping distance on glare ice.* The stopping distance (at some fixed speed) of regular tires is given by a linear function of the air temperature F,

$$D(F) = 2F + 115,$$

where $D(F)$ = the stopping distance, in feet, when the air temperature is F, in degrees Fahrenheit.

a) Find $D(0°)$, $D(-20°)$, $D(10°)$, and $D(32°)$.

b) Graph $D(F)$.

c) Explain why the domain should be restricted to the interval $[-57.5°, 32°]$.

56. *Sociology: Median age of women at first marriage.* Our society is marrying at a later age. The median age of women at first marriage can be approximated by the linear function

$$A(t) = 0.08t + 19.7,$$

where $A(t)$ = the median age of women at first marriage the tth year after 1950. Thus $A(0)$ is the median age of women at first marriage in the year 1950, $A(30)$ is the median age in 1980, and so on.

a) Find $A(0)$, $A(1)$, $A(10)$, $A(30)$, and $A(40)$.

b) What will be the median age of women at first marriage in 1996?

c) Graph $A(t)$.

57. *Biology: Spread of an organism.* A certain kind of organism is released over an area of 2 mi². It grows and spreads over more area. The area covered by the organism after time t is given by a linear function

$$A(t) = 1.1t + 2,$$

where $A(t)$ = the area covered, in square miles, after time t, in years.

a) Find $A(0)$, $A(1)$, $A(4)$, and $A(10)$.

b) Graph $A(t)$.

c) Why should the domain be restricted to the interval $[0, \infty)$?

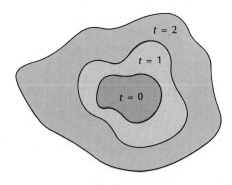

58. *Business: Profit-and-loss analysis.* A ski manufacturer is planning a new line of skis. For the first year, the fixed costs for setting up the new production line are $22,500. The variable costs for producing each pair of skis are estimated at $40. The sales department projects that 3000 pairs can be sold during the first year at a price of $85 per pair.

a) Formulate a function $C(x)$ for the total cost of producing x pairs of skis.

b) Formulate a function $R(x)$ for the total revenue from the sale of x pairs of skis.

c) Formulate a function $P(x)$ for the total profit from the production and sale of x pairs of skis.

d) What profit or loss will the company realize if the expected sales of 3000 pairs occurs?

e) How many pairs must the company sell in order to break even?

59. *Business: Profit-and-loss analysis.* Boxowitz, Inc., a computer firm, is planning to sell a new minicalculator. For the first year, the fixed costs for setting up the new production line are $100,000. The variable costs for producing each calculator are estimated at $20. The sales department projects that 150,000 calculators can be sold during the first year at a price of $45 each.

a) Formulate a function $C(x)$ for the total cost of producing x calculators.

b) Formulate a function $R(x)$ for the total revenue from the sale of x calculators.

c) Formulate a function $P(x)$ for the total profit from the production and sale of x calculators.

d) What profit or loss will the firm realize if the expected sales of 150,000 calculators occurs?

e) How many calculators must the firm sell in order to break even?

61. *Anthropology: Estimating heights.* An anthropologist can use certain linear functions to estimate the height of a male or female, given the length of certain bones. A *humerus* is the bone from the elbow to the shoulder. Let $x =$ the length of the humerus in centimeters. Then the height, in centimeters, of a male with a humerus of length x is given by

$$M(x) = 2.89x + 70.64.$$

The height, in centimeters, of a female with a humerus of length x is given by $F(x) = 2.75x + 71.48$. A 45-cm humerus was uncovered in a ruins.

a) If we assume it was from a male, how tall was he?

b) If we assume it was from a female, how tall was she?

60. *Business: Straight-line depreciation.* A company buys an office machine for $5200 on January 1 of a given year. The machine is expected to last for 8 years, at the end of which time its *trade-in*, or *salvage value* will be $1100. If the company figures the decline in value to be the same each year, then the *book value*, or *salvage value*, after t years, $0 \leqslant t \leqslant 8$, is given by the linear function

$$V(t) = C - t\left(\frac{C - S}{N}\right),$$

where $C =$ the original cost of the item ($5200), $N =$ the years of expected life (8), and $S =$ the salvage value ($1100).

a) Find the linear function for the straight-line depreciation of the office machine.

b) Find the salvage value after 0 years, 1 year, 2 years, 3 years, 4 years, 7 years, 8 years.

62. *Business: Sales commissions.* A person applying for a sales position is offered alternative salary plans:

Plan A: A base salary of $600 per month plus a commission of 4% of the gross sales for the month.

Plan B: A base salary of $700 per month plus a commission of 6% of the gross sales for the month in excess of $10,000.

a) For each plan, formulate a function that expresses monthly earnings as a function of gross sales x.

b) For what gross-sales values is plan B preferable?

Humerus —

1.5

OTHER TYPES OF FUNCTIONS

OBJECTIVES

a) Graph a given function.

b) Convert from radical notation to fractional exponents and from fractional exponents to radical notation.

c) Simplify expressions with fractional exponents.

d) Determine the domain of a function.

e) Find the equilibrium point, given a supply function and a demand function.

Quadratic Functions

DEFINITION

A *quadratic function f* is given by

$$f(x) = ax^2 + bx + c, \quad \text{where } a \neq 0.$$

We have already considered some such functions—for example, $f(x) = x^2$ and $g(x) = x^2 - 1$. Graphs of quadratic functions are always cup-shaped, like those in Example 1. Each has a dashed line of symmetry (not part of the graph).

Example 1   Graph $y = x^2 - 2x - 3$ and $y = -2x^2 + 4x + 1$.

Solutions $y = x^2 - 2x - 3$ $y = -2x^2 + 4x + 1$

x	y
0	−3
1	−4
2	−3
3	0
4	5
−1	0
−2	5

x	y
0	1
1	3
2	1
3	−5
−1	−5

1. Using the same axes, graph $y = x^2$ and $y = -x^2$. (*Note:* $-x^2$ means $-1 \cdot x^2$.) CSS

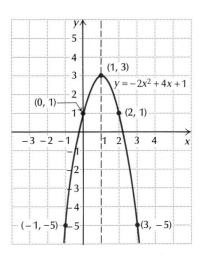

If the coefficient a is positive, the graph opens upward. If a is negative, the graph opens downward.

DO EXERCISES 1 AND 2.

First coordinates of points at which a quadratic function intersects the x-axis (x-intercepts), if they exist, can be found by solving the quadratic equation $ax^2 + bx + c = 0$. If real-number solutions exist, they can be found using the quadratic formula.

THEOREM 9

The Quadratic Formula

The solutions of any quadratic equation $ax^2 + bx + c = 0$, $a \neq 0$, are given by

$$x = \frac{-b \pm \sqrt{b^2 - 4ac}}{2a}.$$

---

2. a) Using the same axes, graph $y = x^2$ and $y = (x - 3)^2$.

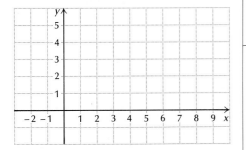

b) How could the graph of $y = (x - 3)^2$ be obtained from the graph of $y = x^2$? CSS

When solving a quadratic equation, first try to factor and then use the Principle of Zero Products, as we did in Section 1.2. When factoring is not possible or seems difficult, try the quadratic formula. It will always give the solutions. When $b^2 - 4ac < 0$, there are no real-number solutions, but there are solutions in an expanded number system called the *complex numbers*. In this text we will be considering only real-number solutions.

3. Solve $3x^2 = 7 - 2x$.

Example 2   Solve $3x^2 - 4x = 2$.

Solution   We first find standard form $ax^2 + bx + c = 0$ and then determine a, b, and c.

$$3x^2 - 4x - 2 = 0,$$
$$a = 3, \quad b = -4, \; c = -2$$

We then use the quadratic formula:

$$x = \frac{-b \pm \sqrt{b^2 - 4ac}}{2a} = \frac{-(-4) \pm \sqrt{(-4)^2 - 4(3)(-2)}}{2 \cdot 3}$$
$$= \frac{4 \pm \sqrt{16 + 24}}{6} = \frac{4 \pm \sqrt{40}}{6}$$
$$= \frac{4 \pm \sqrt{4 \cdot 10}}{6} = \frac{4 \pm 2\sqrt{10}}{6}$$
$$= \frac{2(2 \pm \sqrt{10})}{2 \cdot 3} = \frac{2 \pm \sqrt{10}}{3}.$$

The solutions are $(2 + \sqrt{10})/3$ and $(2 - \sqrt{10})/3$.

DO EXERCISE 3.

Polynomial Functions

Linear and quadratic functions are part of a general class of polynomial functions.

DEFINITION

A *polynomial function* f is given by

$$f(x) = a_n x^n + a_{n-1}x^{n-1} + \cdots + a_2 x^2 + a_1 x^1 + a_0,$$

where n is a nonnegative integer and a_n, a_{n-1}, ..., a_1, a_0 are real numbers, called the *coefficients* of the polynomial.

The following are examples of polynomial functions:

$f(x) = -5,$	(A constant function)
$f(x) = 4x + 3,$	(A linear function)
$f(x) = -x^2 + 2x + 3,$	(A quadratic function)
$f(x) = 2x^3 - 4x^2 + x + 1.$	(A cubic function)

4. Using the same set of axes, graph $y = x^2$ and $y = x^4$.

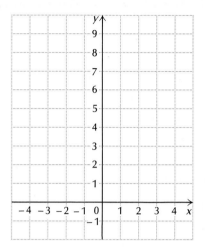

In general, graphing polynomial functions other than linear and quadratic functions is difficult. We use calculus to sketch such graphs in Section 3.3. Some *power functions,* such as

$$y = ax^n,$$

are relatively easy to graph.

Example 3   Using the same set of axes, graph $y = x^2$ and $y = x^3$.

Solution

x	-2	-1	$-\frac{1}{2}$	0	$\frac{1}{2}$	1	2
x^2	4	1	$\frac{1}{4}$	0	$\frac{1}{4}$	1	4
x^3	-8	-1	$-\frac{1}{8}$	0	$\frac{1}{8}$	1	8

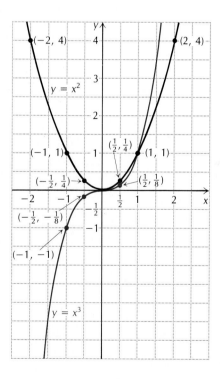

DO EXERCISE 4.

5. Determine the domain of each function.

a) $g(x) = \dfrac{x^2 - 16}{x + 4}$

b) $t(x) = \dfrac{x + 7}{x^2 + 4x - 5}$

c) $k(x) = \dfrac{1}{x - 5}$

Rational Functions

DEFINITION

Functions given by the quotient, or ratio, of two polynomials are called *rational functions*.

The following are examples of rational functions:

$$f(x) = \frac{x^2 - 9}{x - 3},$$

$$g(x) = \frac{x^2 - 16}{x + 4},$$

$$h(x) = \frac{x - 3}{x^2 - x - 2}.$$

The domain of a rational function is restricted to those input values that do not result in division by 0. Thus for f, the domain consists of all real numbers except 3. To determine the domain of h, we set the denominator equal to 0 and solve:

$$x^2 - x - 2 = 0$$
$$(x + 1)(x - 2) = 0$$
$$x = -1 \quad \text{or} \quad x = 2.$$

Therefore, -1 and 2 are not in the domain. The domain consists of all real numbers except -1 and 2.

DO EXERCISE 5.

One important class of rational functions is given by $y = k/x$.

Example 4   Graph $y = 1/x$.

Solution

x	-3	-2	-1	$-\frac{1}{2}$	$-\frac{1}{4}$	$\frac{1}{4}$	$\frac{1}{2}$	1	2	3
y	$-\frac{1}{3}$	$-\frac{1}{2}$	-1	-2	-4	4	2	1	$\frac{1}{2}$	$\frac{1}{3}$

6. Graph $y = \dfrac{-1}{x}$.

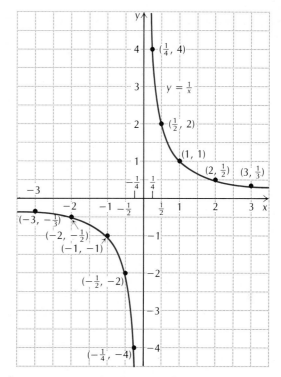

Note that 0 is not in the domain of this function since it would yield a denominator of 0. The function is decreasing over the intervals $(-\infty, 0)$ and $(0, \infty)$. It is an example of inverse variation.

DEFINITION

y varies inversely as *x* if there is some positive number *k* such that $y = k/x$. We also say that *y* is *inversely proportional* to *x*.

DO EXERCISE 6.

Example 5 *Stocks and gold.* Certain economists theorize that stock prices are inversely proportional to the price of gold. That is, when the price of gold goes up, the prices of stock go down; and when the price of gold goes down, the prices of stock go up. Let us assume that the Dow-Jones Industrial Average, *D*, an index of the overall price of stock, is inversely proportional to the price of gold, *G*, in dollars per ounce. One day the Dow-Jones was 818 and the price of gold was $520 per ounce. What will the Dow-Jones be if the price of gold drops to $490?

7. *Demand.* The price p of a certain kind of radio is found to be inversely proportional to the number sold, x. It was found that 240,000 radios will be sold when the price per radio is $12.50. How many will be sold if the price is $18.75?

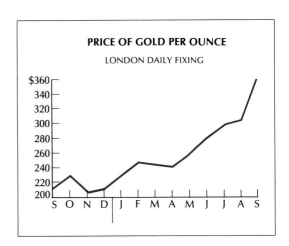

PRICE OF GOLD PER OUNCE

LONDON DAILY FIXING

Solution

a) We know that $D = k/G$, so $818 = k/520$ and $k = 425{,}360$. Thus $D = 425{,}360/G$.

b) We substitute 490 for G and compute D:

$$D = \frac{425{,}360}{490} \approx 868.1.$$

Warning! Do not put too much "stock" in the equation of Example 5. It is meant to give us an idea of economic relationships. An equation to predict the stock market accurately has not been found!

DO EXERCISE 7.

Absolute-Value Functions

The following is an example of an absolute-value function and its graph. The absolute value of a number is its distance from 0. We denote the absolute value of a number x as $|x|$.

Example 6   Graph $f(x) = |x|$.

Solution

x	-3	-2	-1	0	1	2	3
$f(x)$	3	2	1	0	1	2	3

8. Graph $f(x) = |x - 1|$. To find an output, take an input, subtract 1 from it, and then take the absolute value.

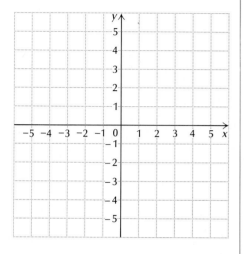

9. Graph $f(x) = \sqrt{x}$.

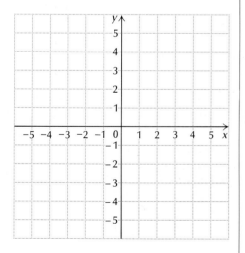

10. Find the domain of $y = \sqrt{2x + 3}$.
    (*Hint:* Solve $2x + 3 \geqslant 0$.)

We can think of this function as being defined piecewise by considering the definition of absolute value:

$$f(x) = |x| = \begin{cases} x, & \text{if } x \geqslant 0, \\ -x, & \text{if } x < 0. \end{cases}$$

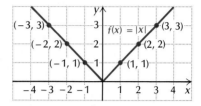

DO EXERCISE 8.

Square-Root Functions

The following is an example of a square-root function and its graph.

Example 7 Graph $f(x) = -\sqrt{x}$.

Solution The domain of this function is just the nonnegative numbers—the interval $[0, \infty)$. Table 1 at the back of the book contains approximate values of square roots of certain numbers. You can, of course, find them on your calculator.

x	0	1	2	3	4	5	10
$f(x)$, or $-\sqrt{x}$	0	-1	-1.4	-1.7	-2	-2.2	-3.2

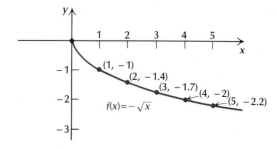

DO EXERCISES 9 AND 10.

Convert to fractional exponents.

11. $\sqrt[4]{t^3}$

12. $\sqrt[5]{y}$

13. $\dfrac{1}{\sqrt[5]{x^2}}$

14. $\dfrac{1}{\sqrt[3]{t}}$

15. $\sqrt{x^6}$

16. $\sqrt{x^7}$

Convert to radical notation.

17. $y^{1/7}$

18. $x^{3/2}$

19. $t^{-3/2}$

20. $b^{-1/2}$

Simplify.

21. $4^{5/2}$

22. $27^{2/3}$

Power Functions with Fractional Exponents

We are motivated to define fractional exponents so that the laws we discussed in Section 1.1 still hold. For example, if the laws of exponents are to hold, we would have

$$a^{1/2} \cdot a^{1/2} = a^{1/2+1/2} = a^1 = a.$$

Thus we are led to define $a^{1/2}$ as $\sqrt{a}$. Similarly, we are led to define $a^{1/3}$ as the cube root of a, $\sqrt[3]{a}$. In general,

$$a^{1/n} = \sqrt[n]{a}.$$

Again, if the laws of exponents are to hold, we would have

$$\sqrt[n]{a^m} = (a^m)^{1/n} = (a^{1/n})^m = a^{m/n}.$$

An expression $a^{-m/n}$ is defined by

$$a^{-m/n} = \frac{1}{a^{m/n}} = \frac{1}{\sqrt[n]{a^m}}.$$

Example 8   Convert to fractional exponents.

a) $\sqrt[3]{x^2} = x^{2/3}$ b) $\sqrt[4]{y} = y^{1/4}$

c) $\dfrac{1}{\sqrt[3]{b^5}} = \dfrac{1}{b^{5/3}} = b^{-5/3}$ d) $\dfrac{1}{\sqrt{x}} = \dfrac{1}{x^{1/2}} = x^{-1/2}$

e) $\sqrt{x^8} = x^{8/2}$, or x^4

DO EXERCISES 11–16.

Example 9   Convert to radical notation.

a) $x^{1/3} = \sqrt[3]{x}$ b) $t^{6/7} = \sqrt[7]{t^6}$

c) $x^{-2/3} = \dfrac{1}{x^{2/3}} = \dfrac{1}{\sqrt[3]{x^2}}$ d) $e^{-1/4} = \dfrac{1}{e^{1/4}} = \dfrac{1}{\sqrt[4]{e}}$

DO EXERCISES 17–20.

Example 10   Simplify.

a) $8^{5/3} = (8^{1/3})^5 = (\sqrt[3]{8})^5 = 2^5 = 32$

b) $81^{3/4} = (81^{1/4})^3 = (\sqrt[4]{81})^3 = 3^3 = 27$

DO EXERCISES 21 AND 22.

Earlier when we graphed $f(x) = \sqrt{x}$, we were also graphing $f(x) = x^{1/2}$, or $f(x) = x^{0.5}$. The power functions

$$f(x) = ax^k, \qquad k \text{ fractional},$$

do arise in application. For example, the *home range* of an animal is defined as the region to which the animal confines its movements. It has been hypothesized in statistical studies[*] that the area H of that region can be approximated using the body weight W of an animal by the function

$$H = W^{1.41}.$$

We can approximate function values using a power key $\boxed{y^x}$ on a calculator.

W	0	10	20	30	40	50
H	0	26	68	121	182	249

Note that

$$H = W^{1.41} = W^{141/100} = \sqrt[100]{W^{141}}.$$

The graph is shown here. Note that this is an increasing function. As body weight increases, the area over which the animal moves increases.

Caribou in their territorial area.

[*] See J. M. Emlen, *Ecology: An Evolutionary Approach*, p. 200 (Reading, MA: Addison-Wesley, 1973).

Supply and Demand Functions

Supply and demand in economics are modeled by increasing and decreasing functions. Although specific scientific formulas for these concepts are not generally known, the notions of increasing and decreasing help us understand the ideas.

   Demand Functions.   Look at the following table.

This table shows the relationship between the price p per bag of sugar and the quantity x of 5-lb bags that the consumer will buy at that price. Note that as the price per bag increases, the quantity demanded by the consumer decreases; and as the price per bag decreases, the quantity demanded by the consumer increases. Thus it is natural to think of x as a function of p. In our later work it will be more convenient to think of p as a function of x. Thus, for a *demand function D*, $D(x)$ is the price per unit of an item when x units are demanded by the consumer. The following figure is the graph of a demand function for sugar using the data from the preceding table.

DEMAND SCHEDULE	
Quantity (x) of 5-lb bags in millions	**Price (p) per bag**
4	$5
5	4
7	3
10	2
15	1

Supply Functions.   Look at the following table.

SUPPLY SCHEDULE	
Quantity (x) of 5-lb bags in millions	Price (p) per bag
0	$1
10	2
15	3
20	4
24	5

This table shows the relationship between the price p per bag of sugar and the quantity x of 5-lb bags that the seller is willing to supply at that price. Note that as the price per bag increases, the more the seller is willing to supply; and as the price per bag decreases, the less the seller is willing to supply. Again, it is natural to think of x as a function of p, but for our later work it is more convenient to think of p as a function of x. Thus, for a *supply function S, S(x)* is the price per unit of an item at which the seller is willing to supply x units of a product to the consumer. The following figure is the graph of a supply function for sugar using the data from the preceding table.

Let us now look at these curves together. Note that as supply increases, demand decreases; and as supply decreases, demand increases. The point of intersection of the two curves (x_E, p_E) is called the *equilibrium point*. The equilibrium price p_E (in this case, $2 per bag) is the point at which the amount x_E (in this case, 10 million bags) that the seller willingly supplies is the same as the amount that the consumer willingly demands. The situation is analogous to a buyer and seller haggling over the sale of an item. The equilibrium point, or selling price, is what they finally agree on.

Example 11　Find the equilibrium point for the demand and supply functions

$$D(x) = (x - 6)^2 \quad \text{and} \quad S(x) = x^2 + x + 10.$$

Solution　To find the equilibrium point, we set $D(x) = S(x)$ and solve:

$$(x - 6)^2 = x^2 + x + 10$$
$$x^2 - 12x + 36 = x^2 + x + 10$$
$$-12x + 36 = x + 10$$
$$-13x = -26$$
$$x = \frac{-26}{-13}$$
$$x = 2.$$

23. Given $D(x) = (x - 5)^2$ and $S(x) = x^2 + x + 3$, find the equilibrium point.

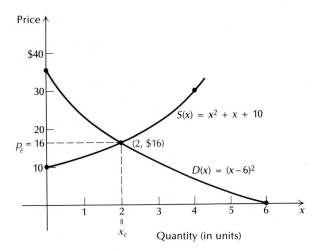

Thus $x_E = 2$ (units). To find p_E we substitute x_E into either $D(x)$ or $S(x)$. We use $D(x)$. Then

$$p_E = D(x_E) = D(2) = (2 - 6)^2 = (-4)^2 = \$16.$$

Thus the equilibrium price is $16 per unit and the equilibrium point is $(2, \$16)$.

DO EXERCISE 23.

EXERCISE SET 1.5

CSS

Using the same set of axes, graph each pair of equations.

1. $y = \frac{1}{2}x^2, y = -\frac{1}{2}x^2$

2. $y = \frac{1}{4}x^2, y = -\frac{1}{4}x^2$

3. $y = x^2, y = (x - 1)^2$

4. $y = x^2, y = (x - 3)^2$

5. $y = x^2, y = (x + 1)^2$

6. $y = x^2, y = (x + 3)^2$

7. $y = |x|, y = |x + 3|$

8. $y = |x|, y = |x + 1|$

9. $y = x^3, y = x^3 + 1$

10. $y = x^3, y = x^3 - 1$

11. $y = \sqrt{x}, y = \sqrt{x + 1}$

12. $y = \sqrt{x}, y = \sqrt{x - 2}$

Graph.

13. $y = x^2 - 4x + 3$

14. $y = x^2 - 6x + 5$

15. $y = -x^2 + 2x - 1$

16. $y = -x^2 - x + 6$

17. $y = \dfrac{2}{x}$

18. $y = \dfrac{3}{x}$

19. $y = \dfrac{-2}{x}$

20. $y = \dfrac{-3}{x}$

21. $y = \dfrac{1}{x^2}$

22. $y = \dfrac{1}{x - 1}$

23. $y = \sqrt[3]{x}$ (*Hint:* ▦ or use Table 1 at the end of the book.)

24. $y = \dfrac{1}{|x|}$

Solve.

25. $x^2 - 2x = 2$ **26.** $x^2 - 2x + 1 = 5$ **27.** $x^2 + 6x = 1$ **28.** $x^2 + 4x = 3$

29. $4x^2 = 4x + 1$ **30.** $-4x^2 = 4x - 1$ **31.** $3y^2 + 8y + 2 = 0$ **32.** $2p^2 - 5p = 1$

Convert to fractional exponents.

33. $\sqrt{x^3}$ **34.** $\sqrt{x^5}$ **35.** $\sqrt[5]{a^3}$ **36.** $\sqrt[4]{b^2}$ **37.** $\sqrt[7]{t}$ **38.** $\sqrt[8]{c}$

39. $\dfrac{1}{\sqrt[3]{t^4}}$ **40.** $\dfrac{1}{\sqrt[5]{b^6}}$ **41.** $\dfrac{1}{\sqrt{t}}$ **42.** $\dfrac{1}{\sqrt{m}}$ **43.** $\dfrac{1}{\sqrt{x^2 + 7}}$ **44.** $\sqrt{x^3 + 4}$

Convert to radical notation.

45. $x^{1/5}$ **46.** $t^{1/7}$ **47.** $y^{2/3}$ **48.** $t^{2/5}$

49. $t^{-2/5}$ **50.** $y^{-2/3}$ **51.** $b^{-1/3}$ **52.** $b^{-1/5}$

53. $e^{-17/6}$ **54.** $m^{-19/6}$ **55.** $(x^2 - 3)^{-1/2}$ **56.** $(y^2 + 7)^{-1/4}$

Simplify.

57. $9^{3/2}$ **58.** $16^{5/2}$ **59.** $64^{2/3}$ **60.** $8^{2/3}$

61. $16^{3/4}$ **62.** $25^{5/2}$

Determine the domain of each function.

63. $f(x) = \dfrac{x^2 - 25}{x - 5}$ **64.** $f(x) = \dfrac{x^2 - 4}{x + 2}$ **65.** $f(x) = \dfrac{x^3}{x^2 - 5x + 6}$

66. $f(x) = \dfrac{x^4 + 7}{x^2 + 6x + 5}$ **67.** $f(x) = \sqrt{5x + 4}$ `CSS`   **68.** $f(x) = \sqrt{2x - 6}$ `CSS`

Find the equilibrium point for each of the following demand and supply functions.

69. $D(x) = -2x + 8, S(x) = x + 2$

70. $D(x) = -\frac{5}{6}x + 10, S(x) = \frac{1}{2}x + 2$

71. $D(x) = (x - 3)^2, S(x) = x^2 + 2x + 1$

72. $D(x) = (x - 4)^2, S(x) = x^2 + 2x + 6$

73. $D(x) = (x - 4)^2, S(x) = x^2$

74. $D(x) = (x - 6)^2, S(x) = x^2$

APPLICATIONS

75. ▤ It is theorized that the dividends paid on utilities stocks are inversely proportional to the prime (interest) rate. Recently, the dividends D on the stock of Indianapolis Power and Light were $2.09 per share and the prime rate was 19%. The prime rate R dropped to 17.5%. What dividends would be paid if the assumption of inverse proportionality is correct?

76. (🖩 with $\boxed{y^x}$ key) *Biology.* The *territorial area* of an animal is defined to be its defended region, or exclusive region. For example, a lion has a certain region over which it is ruler. It has been hypothesized in statistical studies (see the footnote on p. 59) that the area T of that region can be approximated using body weight W by the power function

$$T = W^{1.31}.$$

Complete the table of approximate function values and graph the function. CSS

W	0	10	20	30	40	50	100	150
T	0	20						

77. The number of tangerines in a pile of the type shown in the photo is approximated by the function

$$f(x) = \tfrac{1}{6}x^3 + \tfrac{1}{2}x^2 + \tfrac{1}{3}x,$$

where $f(x)$ is the number of tangerines and x is the number of layers. Find the number of tangerines when the number of layers is 7, 10, and 12.

1.6

MATHEMATICAL MODELING

a) Use curve fitting to find a model for a set of data; then use the model to make predictions.

What Is a Mathematical Model?

When the essential parts of a problem are described in mathematical language, we say that we have a *mathematical model*. For example, the arithmetic of the natural numbers constitutes a mathematical model for situations in which counting is the essential ingredient. Situations in which calculus can be brought to bear often require the use of equations and functions, and typically there is concern with the way a change in one variable affects a change in another.

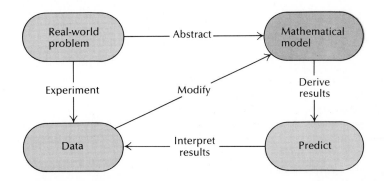

Mathematical models are abstracted from real-world situations (see the diagram). Procedures within the mathematical model then give results that allow one to predict what will happen in that real-world situation. To the extent that these predictions are inaccurate or the results of experimentation do not conform to the model, the model is in need of modification.

The diagram above indicates that mathematical modeling is an ongoing, possibly everchanging, process. This is often the case. For example, finding a mathematical model that will enable accurate prediction of population growth is not a simple problem. Surely any population model one might devise will need to be altered as further relevant information is acquired.

Although models can reveal worthwhile information, one must always be cautious using them. An interesting case in point is a study* showing that world records in *any* running race can be modeled by a linear function. In particular, for the mile run,

$$R = -0.00582x + 15.3476,$$

where R is the world record in minutes and x is the year. Roger Bannister shocked the world in 1954 by breaking the 4-min mile. Had people been aware of this model they would not have been shocked, for when we substitute 1954 for x we get

$$R = -0.00582(1954) + 15.3476 = 3.97532 \approx 3:58.5.$$

The actual record was 3:59.4. Although this model will continue for 40 to 50 years to be worthwhile in predicting the world record in the mile run, we see that we can't get meaningful answers to some questions. For example, we could use the model to find when the 1-min mile will be broken. We set $R = 1$ and solve for x:

$$1 = -0.00582x + 15.3476$$
$$2465 = x.$$

Most track people would assure us that the 1-min mile is beyond human capability. In fact, at the time of this writing, experienced runners think it will never reach 3:40.0, the current world record being 3:48.8. Going to an even further extreme, we see that the model predicts that the 0-min mile will be run in 2637. In conclusion, one must be careful in the use of any model. (You will develop this model in Exercise Set 7.5.)

---

* H. W. Ryder, H. J. Carr, and P. Herget, "Future Performance in Footracing," *Scientific American,* **234** (June 1976): 109–119.

Use the model for the world record in the mile run.

1. What will the world record be in 1988?

2. When will the world record be 3:45.0? (*Hint:* 3:45.0 = 3.75.)

DO EXERCISES 1 AND 2.

In Section 1.5 we saw an example of a mathematical model using supply and demand functions. In general, the idea is to find a function that fits observations and theoretical reasoning (including common sense) as well as possible. In Section 7.4, we will see how calculus can be used to develop and analyze models. For now we will consider one type of modeling procedure using a somewhat oversimplified procedure that we call *curve fitting*.

Curve Fitting

These four types of function fit many situations.

Constant function   Linear function

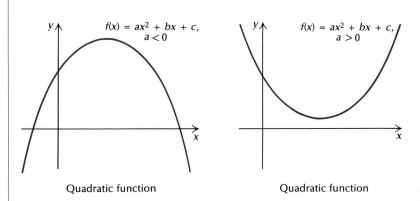

Quadratic function   Quadratic function

The following is a procedure that sometimes works for finding mathematical models.

Curve Fitting

Given a set of data:

1. Graph the data.
2. Look at the data and determine whether a known function seems to fit.
3. Find a function that fits the data by using data points to derive the constants.

The following problem is based on factual data.

Example 1 *Business: Taxes from each dollar earned.* From the given set of data, (a) find a model and (b) use the model to find that part of each dollar earned in 1989 that will go for taxes.

Year (x)	Part of each dollar earned that goes for taxes (T)
1929	12¢
1950	29¢
1960	34¢
1977	42¢

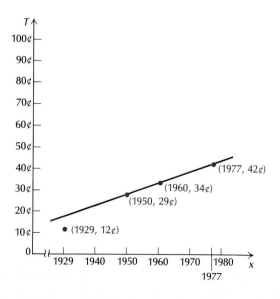

3. Rework Example 1 using the data points (1950, 29¢) and (1960, 34¢). How does the answer to part (b) here compare to the answer to part (b) in Example 1?

Solution

a) The graph is shown on the preceding page. It looks as though a linear function fits these data fairly well:

$$T = mx + b. \tag{1}$$

To derive the constants (or parameters) m and b, we choose two data points. Although this choice is subjective, you should not pick a point that deviates greatly from the general pattern. We choose the points (1950, 29¢) and (1977, 42¢). Since the points are to be solutions to Eq. (1) it follows that

$$42 = m \cdot 1977 + b \quad \text{or} \quad 42 = 1977m + b, \tag{2}$$
$$29 = m \cdot 1950 + b \quad \text{or} \quad 29 = 1950m + b. \tag{3}$$

This is a system of equations. We subtract Eq. (3) from Eq. (2) to eliminate b:

$$13 = 27m.$$

Then we have

$$\tfrac{13}{27} = m.$$

Substituting $\tfrac{13}{27}$ for m in Eq. (3), we get

$$29 = 1950 \cdot \tfrac{13}{27} + b$$
$$29 = 938.9 + b \qquad \text{We approximate } 1950 \cdot \tfrac{13}{27}.$$
$$-909.9 = b.$$

Substituting these values of m and b into Eq. (1), we get the function (model) given by

$$T = \tfrac{13}{27}x - 909.9$$

Since we are interested only in estimates, we use an approximation for $\tfrac{13}{27}$ and get

$$T = 0.481x - 909.9. \tag{4}$$

b) That part of the earned dollar that will go for taxes in 1989 is found by letting $x = 1989$ in Eq. (4):

$$T = 0.481(1989) - 909.9 \approx 47¢.$$

DO EXERCISE 3.

To repeat, the technique of curve fitting illustrated here is an oversimplified method of finding models. Other techniques such as the least-squares method (Section 7.4) and computer simulations are used for more thorough research.

Example 2   For the given set of factual data, (a) find a model and (b) use the model to find the death rate for those who sleep 2 hr; 8 hr; 10 hr.

Average number of hours of sleep (x)	Death rate per 100,000 males (y)
5	1121
7	626
9	967

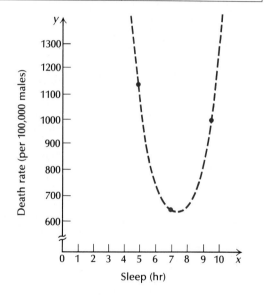

Solution

a) The graph is shown above. It looks as though a quadratic function might fit:

$$y = ax^2 + bx + c. \tag{1}$$

To derive the constants (or parameters) a, b, and c, we use the three data points $(5, 1121)$, $(7, 626)$, and $(9, 967)$. Since these points are to be solutions to Eq. (1), it follows that

$$1121 = a \cdot 5^2 + b \cdot 5 + c \quad \text{or} \quad 1121 = 25a + 5b + c,$$
$$626 = a \cdot 7^2 + b \cdot 7 + c \quad \text{or} \quad 626 = 49a + 7b + c,$$
$$967 = a \cdot 9^2 + b \cdot 9 + c \quad \text{or} \quad 967 = 81a + 9b + c.$$

4.

Age (x) of driver in years	Number of daytime accidents (A) committed by driver of age x
20	420
40	150
60	210
70	400

a) For the given set of data, graph the data and find a quadratic function that fits the data. Use the data points $(20, 420)$, $(40, 150)$, and $(70, 400)$. First give exact fractional values for a, b, and c. Then give decimal values rounded to the nearest thousandth.

b) How many daytime accidents are committed by a driver of age 16?

We solve this system of three equations in three variables using procedures of algebra and get

$$a = 104.5, \quad b = -1501.5, \quad \text{and} \quad c = 6016.$$

Substituting these values of a, b, and c into Eq. (1), we get the function (model) given by

$$y = 104.5x^2 - 1501.5x + 6016.$$

b) The death rate for 2 hr is given by

$$y = 104.5(2)^2 - 1501.5(2) + 6016 = 3431.$$

The death rate for 8 hr is given by

$$y = 104.5(8)^2 - 1501.5(8) + 6016 = 692.$$

The death rate for 10 hr is given by

$$y = 104.5(10)^2 - 1501.5(10) + 6016 = 1451.$$

DO EXERCISE 4.

EXERCISE SET 1.6

For each exercise, parts (a) and (b) are as follows:

a) Graph the data given.

b) Look at the data and determine whether one of the four models discussed in this section seems to fit.

1. *Business.* Raggs, Ltd., keeps track of its total costs of producing x items of a certain suit. These data are shown below.

Number of suits (x)	Total cost (C) of producing x suits
0	$10,000
1	10,030
2	10,059
3	10,094

c) Use the data points $(1, \$10,030)$ and $(3, \$10,094)$ to find a linear function that fits the data.

d) Predict the total cost of producing 4 suits; 10 suits.

e) Use the data points $(0, \$10,000)$ and $(2, \$10,059)$ to find a linear function that fits the data.

f) Use the model of part (e) to predict the total cost of producing 4 suits; 10 suits.

2. *Business.* Pizza, Unltd., keeps track of its total costs of producing x pizzas. These data are shown below.

Number of pizzas (x)	Total cost (C) of producing x pizzas
0	$1000.00
1	1001.00
2	1001.80
3	1002.50

c) Use the data points $(1, \$1001.00)$ and $(3, \$1002.50)$ to find a linear function that fits the data.

d) Predict the total cost of producing 4 pizzas; 100 pizzas.

e) Use the data points $(0, \$1000.00)$ and $(2, \$1001.80)$ to find a linear function that fits the data.

f) Use the model of part (e) to predict the total cost of producing 4 pizzas; 100 pizzas.

The problems in Exercises 3 and 4 are based on factual data.

3. The instruction booklet for a video cassette recorder (VCR) includes a table relating the counter readings and the time the tape has run.

Time (t) tape has run (in hours)	Counter readings (N)
0	000
1	300
2	500
3	675
4	800

c) Use the data points $(0, 0)$, $(1, 300)$, and $(2, 500)$ to find a quadratic equation that fits the data.

d) The counter readings do not give times for programs on the half hour. Use the function to find the counter reading after the tape has run for $\frac{1}{2}$ hr; $1\frac{1}{2}$ hr; $2\frac{1}{2}$ hr.

4.

Travel speed (x) in mph	Number (N) of vehicles involved in an accident at nighttime (for every 100 million miles of travel)
20	10,000
30	2,000
40	400
50	250
60	250
70	350
80	1,500

c) Use the data points $(20, 10{,}000)$, $(50, 250)$, and $(80, 1500)$ to find a quadratic function that fits the data.

d) Use the model to find the number of vehicles involved in an accident at 30 mph. Check this with the data.

5. *Business.*

Year (t)	Total sales (S) in dollars
1	$100,310
2	100,290
3	100,305
4	100,280

c) Use the data points $(1, \$100{,}310)$ and $(2, \$100{,}290)$ to find a linear function that fits the data.

d) This data set approximates a constant function. What procedure, apart from that of part (c), could you use to find the constant?

6. *Business.*

Year (t)	Sales (S) in dollars
1	$10,000
2	21,000
3	27,000
4	37,000

c) Use the data points $(1, \$10,000)$ and $(4, \$37,000)$ to find a linear function that fits the data.

d) Predict the sales of the company in the 5th year.

SUMMARY AND REVIEW: CHAPTER 1

The following contains a summary of what you should be able to do after completing this chapter. The review exercises are for practice. Answers are at the back of the book. If you miss an exercise, restudy the section indicated alongside the answers.

You should be able to:

Rename an exponential expression without exponents. Multiply exponential expressions by adding exponents. Divide exponential expressions by subtracting exponents. Raise a power to a power by multiplying exponents.

1. Rename without an exponent: $(2.01)^2$.

2. Divide: $\dfrac{y^{-2}}{y^{-5}}$.

3. Multiply: $y^{-2} \cdot y^{-5}$.

4. Simplify: $(3t^{-2}m^4)^5$.

Multiply and factor algebraic expressions. Solve applied problems involving compound interest.

Multiply.

5. $(3x - t)^2$

6. $(3x - t)(3x + t)$

7. $(3x - t)(4x + 5t)$

Factor.

8. $x^2 - 2x - 48$

9. $25x^2 - 16t^2$

10. $a^2 - 8ab + 16b^2$

11. $21x^2 - 19x + 4$

12. Suppose $1100 is invested at 10%. How much is in the account at the end of one year if interest is compounded semiannually?

13. Suppose $4000 is invested at 12% compounded annually. How much is in the account at the end of two years?

Solve equations and inequalities like $-5x + 7 = 8x + 4$, $2t^2 = 9 + t$, **and** $-5x + 7 < 8x + 4$. **Solve applied problems involving such equations and inequalities. Write interval notation for a given graph or inequality.**

Solve.

14. $3 - 6x \leqslant 5x - 8$

15. $3 - 6x = 5x - 8$

16. $16x^2 - 25 = 0$

17. $x(x - 3)(2x + 5) = 0$

18. Write interval notation for the set of all numbers x such that $-6 < x \leqslant 1$.

19. Write interval notation for the set of all positive numbers.

Given a function and several inputs, find the outputs. Graph a given equation or function. Decide whether a graph is that of a function.

A function f is given by $f(x) = 2x^2 - x + 3$. Find each of the following.

20. $f(-2)$ **21.** $f(1 + h)$ **22.** $f(0)$

A function f is given by $f(x) = (1 - x)^2$. Find each of the following.

23. $f(-5)$ **24.** $f(2 - h)$ **25.** $f(4)$

Graph.

26. $y = |x + 1|$ **27.** $f(x) = (x - 2)^2$

Which of the following are graphs of functions?

28.

29.

30.

Graph equations of the type $y = b$ and $x = a$. Graph linear functions. Find the slope and an equation of a line containing a given pair of points. Given a linear equation, find the slope and the y-intercept.

Graph.

31. $x = -2$ **32.** $y = 4 - 2x$

33. Write an equation of the line containing the points $(4, -2)$ and $(-7, 5)$.

34. Write an equation of the line with slope 8, containing the point $(\frac{1}{2}, 11)$.

35. For the linear equation $y = 3 - \frac{1}{6}x$, find the slope and the y-intercept.

36. The weight M of muscles in a human is directly proportional to its body weight W. It is known that a person who weighs 150 lb has 60 lb of muscles. What is the muscle weight of a person who weighs 180 lb?

37. A furniture manufacturer has fixed costs of \$80,000 for producing a certain type of classroom chair for universities. The variable costs for producing x chairs is $16x$ dollars. The chairs will be sold at a price of \$28 each.

   a) Formulate a function $R(x)$ for the total revenue from the sale of x chairs.

   b) Formulate a function $C(x)$ for the total cost of producing x chairs.

   c) Formulate a function $P(x)$ for the total profit from the production and sale of x chairs.

   d) What profit or loss will the company realize from expected sales of 8000 chairs?

   e) How many chairs must the company sell in order to break even?

Convert from radical notation to fractional exponents, and vice versa. Determine the domain of a function. Find the equilibrium point, given a demand function and a supply function.

38. Convert to radical notation: $y^{1/6}$.

39. Convert to fractional exponents: $\sqrt[20]{x^3}$.

40. Simplify: $25^{3/2}$.

Find the domain of each function.

41. $f(x) = \sqrt{18 - 6x}$

42. $g(x) = \dfrac{1}{x^2 - 1}$

43. Find the equilibrium point if $D(x) = x^2 - 11x + 51$ and $S(x) = x^2 + 18$.

Use curve fitting to find a model for a set of data. Then use the model to make predictions.

44. Derive a linear function that fits the data points $(6, -1)$ and $(0, 10)$.

45. Derive a quadratic function that fits the data points $(1, 4)$, $(3, 8)$, and $(-1, 8)$.

Graph a function that is defined piecewise.

46. Graph:

$$f(x) = \begin{cases} x - 5, & \text{for } x > 2, \\ x + 3, & \text{for } x \leqslant 2. \end{cases}$$

EXTENSION EXERCISE

47. Simplify: $(64^{5/3})^{-1/2}$.

TEST: CHAPTER 1

1. Rename without a negative exponent: e^{-k}.

2. Divide: $\dfrac{e^{-5}}{e^8}$.

3. Multiply: $(x + h)^2$.

4. Factor: $25x^2 - t^2$.

5. A person makes an investment at 13% compounded annually. It grows to $1039.60 at the end of one year. How much was originally invested?

6. Solve: $-3x < 12$.

7. A function is given by $f(x) = x^2 - 4$. Find (a) $f(-3)$; (b) $f(x + h)$.

8. What are the slope and the y-intercept of $y = -3x + 2$?

9. Find an equation of the line with slope $\frac{1}{4}$, containing the point $(8, -5)$.

10. Find the slope of the line containing the points $(-2, 3)$ and $(-4, -9)$.

11. The weight F of fluids in a human is directly proportional to body weight W. It is known that a person who weighs 180 lb has 120 lb of fluids. Find an equation of variation expressing F as a function of W.

12. A record company has fixed costs of \$10,000 for producing a record master. Thereafter, the variable costs are \$0.50 per record for duplicating from the record master. Revenue from each record is expected to be \$1.30.
    a) Formulate a function $C(x)$ for the total cost of producing x records.
    b) Formulate a function $R(x)$ for the total revenue from the sale of x records.
    c) Formulate a function $P(x)$ for the total profit from the production and sale of x records.
    d) How many records must the company sell in order to break even?

13. Find the equilibrium point for the demand and supply functions

$$D(x) = (x - 7)^2 \quad \text{and} \quad S(x) = x^2 + x + 4.$$

14. Graph: $y = 4/x$.

15. Convert to fractional exponents: $1/\sqrt{t}$.

16. Convert to radical notation: $t^{-3/5}$.

Determine the domain of each function.

17. $f(x) = \dfrac{x^2 + 20}{(x - 2)(x + 7)}$

18. $f(x) = \sqrt{5x + 10}$

19. Find a linear function that fits the data points $(1, 3)$ and $(2, 7)$.

20. Find a quadratic function that fits the data points $(1, 5)$, $(2, 9)$, and $(3, 4)$.

21. Write interval notation for this graph.

$$c \qquad\qquad\qquad\qquad d$$

22. Graph:

$$f(x) = \begin{cases} x^2 + 2, & \text{for } x \geq 0, \\ x^2 - 2, & \text{for } x < 0. \end{cases}$$

EXTENSION EXERCISE

23. The demand function for a product is given by

$$p = D(x) = \sqrt[3]{800 - x}.$$

    a) Find the price per unit when 9 units are sold.
    b) Find the number of units sold when the price per unit is \$6.50.

2

DIFFERENTIATION

This chapter begins our study of calculus. The first concepts we consider are those of limits and continuity. Then we apply those concepts to establishing the first of the two building blocks of calculus: *differentiation*. Differentiation is a process that takes a formula for a function and derives a formula for another function, called a *derivative*, which allows one to find the slope of the tangent line to a curve at a point. We also find that a derivative can represent an instantaneous rate of change. Throughout the chapter we will learn various techniques for finding derivatives.

AN APPLICATION

A firm estimates that it will sell N units of a product after spending a dollars on advertising, where

$$N(a) = -a^2 + 300a + 6$$

and a is measured in thousands of dollars. What is the rate of change of the number of units sold with respect to the amount spent on advertising?

THE MATHEMATICS

The rate of change $N'(a)$ with respect to the amount spent on advertising is given by

$$N'(a) = -2a + 300.$$

└──────This is a *derivative*.

OBJECTIVES

a) Find

$$\lim_{x \to a} f(x),$$

if it exists.

b) Determine whether a graph is that of a continuous function.

c) Determine whether a function is continuous at a given point a.

2.1

LIMITS AND CONTINUITY

In this section we give an intuitive (meaning "based on prior and present experience") treatment of two important concepts: limits and continuity.

Limits

One important aspect of the study of calculus is the analysis of how function values, or outputs, change when inputs change. Basic to this study is the notion of a *limit*. Suppose the inputs get closer and closer to some number. If the corresponding outputs get closer and closer to a number, that number is called a *limit*.

Consider the function f given by

$$f(x) = 2x + 3.$$

Suppose we select input numbers x closer and closer to the number 4, and look at the output numbers $2x + 3$. Study the following input–output table and the graph on the following page.

| | These inputs approach 4 from the left. | 4 | These inputs approach 4 from the right. | |

x	2	3.6	3.9	3.99	3.999	4.001	4.01	4.1	4.8	5
$f(x)$	7	10.2	10.8	10.98	10.998	11.002	11.02	11.2	12.6	13

| These outputs approach 11. | 11 | These outputs approach 11. |

In the table and the graph, as input numbers approach 4 from the left, output numbers approach 11. As input numbers approach 4 from the right, output numbers approach 11. Thus we say:

As x *approaches* 4, $2x + 3$ *approaches* 11.

An arrow, $\longrightarrow$, is often used for the word "approaches." Thus the statement above can be written

As $x \longrightarrow 4$, $2x + 3 \longrightarrow 11$.

1. Consider $f(x) = 3x - 1$.

   a) Complete this table. (📱 helpful, though not necessary)

x	$f(x)$
5	
5.8	
5.9	
5.99	
5.999	
6 ←←	→→ ? 7
6.001	
6.01	
6.1	
6.4	
7	

   b) Find $\lim\limits_{x \to 6} f(x)$.

   c) Graph $f(x) = 3x - 1$:

   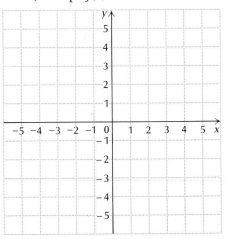

   Use the graph to find each of the following limits.

   d) $\lim\limits_{x \to -1} f(x)$

   e) $\lim\limits_{x \to 2} f(x)$

   f) $\lim\limits_{x \to 0} f(x)$

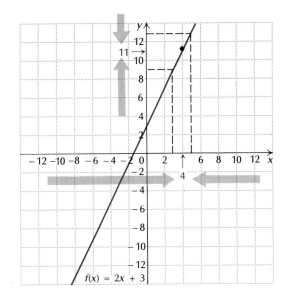

$f(x) = 2x + 3$

The number 11 is said to be the *limit* of $2x + 3$ as x approaches 4. We can abbreviate this statement as follows:

$$\lim_{x \to 4} (2x + 3) = 11.$$

This is read, "The limit, as x approaches 4, of $2x + 3$ is 11."

DEFINITION

A function f has the *limit L* as x approaches a, written

$$\lim_{x \to a} f(x) = L,$$

if we can get $f(x)$ as close to L as we wish by restricting x to a sufficiently small interval about a but excluding a.

DO EXERCISE 1.

Example 1 Consider the function

$$F(x) = \frac{1}{x}.$$

Find each of the following limits, if they exist.

a) $\lim\limits_{x \to 3} F(x)$ b) $\lim\limits_{x \to 0} F(x)$

2. Consider the function $g(x) = \dfrac{1}{x-3}$.

a) Complete this table. (▤ helpful)

Inputs, x	Outputs, $g(x)$
1	
2	
2.9	
2.99	
2.999	
3 ←——————→ ?	
3.001	
3.01	
3.1	
3.8	
4	

b) Find $\lim\limits_{x \to 3} g(x)$.

c) Graph $g(x) = \dfrac{1}{x-3}$.

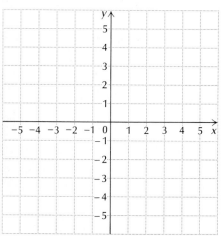

Use the graph to find each of these limits, if they exist.

d) $\lim\limits_{x \to 3} g(x)$ **e)** $\lim\limits_{x \to 1} g(x)$

f) $\lim\limits_{x \to 4} g(x)$

Solution

a) From the following graph, we see that as inputs x approach 3 from either the left or the right, outputs $1/x$ approach $\frac{1}{3}$. We can also check this using an input–output table. Thus we have

$$\lim_{x \to 3} F(x) = \tfrac{1}{3}.$$

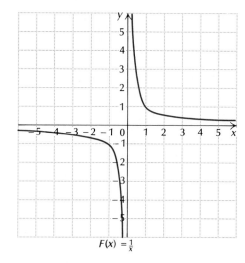

$F(x) = \frac{1}{x}$

b) From either the graph or an input–output table, we see that as inputs x approach 0 from the left, outputs become more and more negative without bound. Similarly, as inputs x approach 0 from the right, outputs get larger and larger without bound. We say that

$$\lim_{x \to 0} \frac{1}{x} \quad \text{does not exist.}$$

DO EXERCISE 2.

The following is very important to keep in mind when determining whether a limit exists.

> In order for a limit to exist, both of the limits from the left and the right must exist and be the same.

3. Consider the following graph.

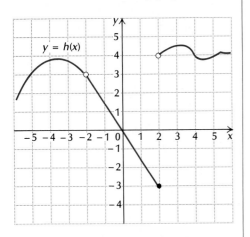

Find each of the following limits, if they exist.

a) $\lim\limits_{x \to 2^+} h(x)$

b) $\lim\limits_{x \to 2^-} h(x)$

c) $\lim\limits_{x \to 2} h(x)$

d) $\lim\limits_{x \to 0^+} h(x)$

e) $\lim\limits_{x \to 0^-} h(x)$

f) $\lim\limits_{x \to 0} h(x)$

g) $\lim\limits_{x \to -2^+} h(x)$

h) $\lim\limits_{x \to -2^-} h(x)$

i) $\lim\limits_{x \to -2} h(x)$

We use the notation

$$\lim_{x \to a^+} f(x) \quad \text{to indicate the limit from the right}$$

and

$$\lim_{x \to a^-} f(x) \quad \text{to indicate the limit from the left.}$$

Then in order for a limit to exist, both of the above limits must exist and be the same.

Example 2 Consider the function H defined as follows:

$$H(x) = \begin{cases} 2x + 2, & \text{for } x < 1, \\ 2x - 2, & \text{for } x \geqslant 1. \end{cases}$$

Graph the function and find each of the following limits, if they exist.

a) $\lim\limits_{x \to 1} H(x)$

b) $\lim\limits_{x \to -3} H(x)$

Solution The graph is shown here.

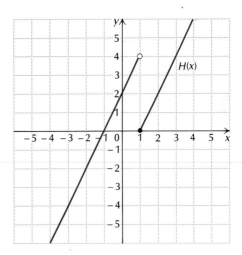

a) As inputs x approach 1 from the left, outputs $H(x)$ approach 4. Thus the limit from the left is 4. That is,

$$\lim_{x \to 1^-} H(x) = 4.$$

4. Consider the following function:

$$f(x) = \begin{cases} 2 - x^2, & \text{for } x \geq 0, \\ x^2 - 2, & \text{for } x < 0. \end{cases}$$

a) Graph $y = f(x)$.

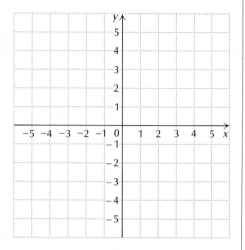

Find each of the following limits, if they exist.

b) $\lim\limits_{x \to 0} f(x)$

c) $\lim\limits_{x \to -2} f(x)$

But as inputs x approach 1 from the right, outputs $H(x)$ approach 0. Thus the limit from the right is 0. That is,

$$\lim_{x \to 1^+} H(x) = 0.$$

Since the limit from the left, 4, is not the same as the limit from the right, 0, we say that

$$\lim_{x \to 1} H(x) \quad \text{does not exist.}$$

b) As inputs x approach -3 from the left, outputs $H(x)$ approach -4, so the limit from the left is -4. That is,

$$\lim_{x \to -3^-} H(x) = -4.$$

As inputs x approach -3 from the right, outputs $H(x)$ approach -4, so the limit from the right is -4. That is,

$$\lim_{x \to -3^+} H(x) = -4.$$

Since the limits from the left and from the right exist and are the same, we have

$$\lim_{x \to -3} H(x) = -4.$$

DO EXERCISES 3 AND 4.

The following is also important in finding limits.

> The limit at a point *a does not depend* on the function value at *a* even if that function value, $f(a)$, exists. That is, whether or not a limit exists at *a* has nothing to do with the function value $f(a)$.

Example 3 Consider the function G defined as follows:

$$G(x) = \begin{cases} 5, & \text{for } x = 1, \\ x + 1, & \text{for } x \neq 1. \end{cases}$$

Graph the function and find each of the following limits, if they exist.

a) $\lim\limits_{x \to 1} G(x)$

b) $\lim\limits_{x \to 3} G(x)$

5. Consider the following function:

$$f(x) = \begin{cases} -3, & \text{for } x = -2, \\ x^2, & \text{for } x \neq -2. \end{cases}$$

a) Graph $y = f(x)$.

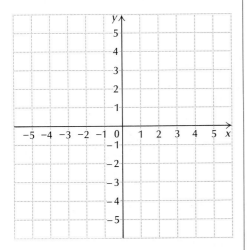

Find each of the following limits, if they exist.

b) $\lim\limits_{x \to -2^+} f(x)$

c) $\lim\limits_{x \to -2^-} f(x)$

d) $\lim\limits_{x \to -2} f(x)$

e) $\lim\limits_{x \to 2^+} f(x)$

f) $\lim\limits_{x \to 2^-} f(x)$

g) Does $\lim\limits_{x \to -2} f(x) = f(-2)$?

h) Does $\lim\limits_{x \to 2} f(x) = f(2)$?

Solution The graph is shown here.

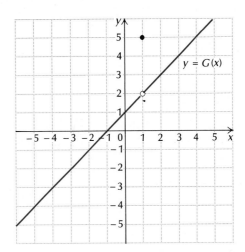

a) As inputs x approach 1 from the left, outputs $G(x)$ approach 2, so the limit from the left is 2. As inputs x approach 1 from the right, outputs $G(x)$ approach 2, so the limit from the right is 2. Since the limit from the left, 2, is the same as the limit from the right, 2, we have

$$\lim\limits_{x \to 1} G(x) = 2.$$

Note that the limit, 2, is not the same as the function value at 1, $G(1)$, which is 5.

b) We have $\lim\limits_{x \to 3} G(x) = 4 = G(3)$. In this case, the function value and the limit are the same.

DO EXERCISES 5 AND 6. (EXERCISE 6 IS ON THE FOLLOWING PAGE.)

Continuity

When the limit of a function is the same as its function value, it satisfies a condition called "continuity at a point." We now consider the concept of continuity.

The following are graphs of functions that are *continuous* over the whole real line $(-\infty, \infty)$.

6. Consider the following graph.

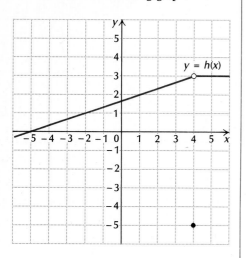

Find each of the following limits, if they exist.

a) $\lim_{h \to 1} h(x)$

b) $\lim_{h \to 4} h(x)$

c) Does $\lim_{h \to 1} h(x) = h(1)$?

d) Does $\lim_{h \to 4} h(x) = h(4)$?

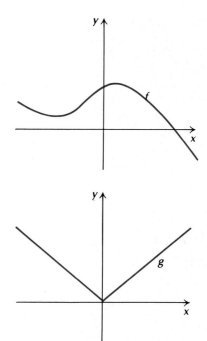

Note that there are no "jumps" or holes in the graphs. For now we will use a somewhat intuitive definition of continuity, which we will refine. We say that a function is *continuous* over, or on, some interval of the real line if its graph can be traced without lifting a pencil from the paper. The following are graphs of functions that are *not* continuous over the whole real line.

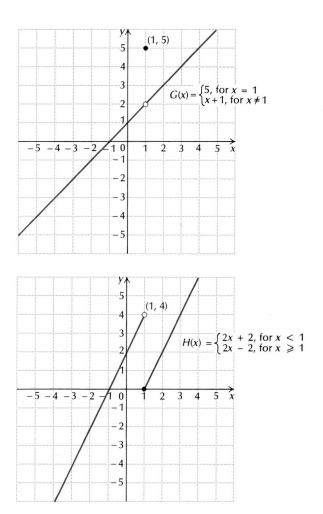

$$G(x) = \begin{cases} 5, \text{ for } x = 1 \\ x+1, \text{ for } x \neq 1 \end{cases}$$

$$H(x) = \begin{cases} 2x + 2, \text{ for } x < 1 \\ 2x - 2, \text{ for } x \geqslant 1 \end{cases}$$

A continuous curve.

For functions G and H, the open circle indicates that the circled point is not part of the graph.

In each case, the graph *cannot* be traced without lifting the pencil from the paper. However, each case represents a different situation. Let us discuss why each case fails to be continuous over the whole real line.

The function F fails to be continuous over the *whole* real line $(-\infty, \infty)$. Since F is not defined at $x = 0$, the point $x = 0$ is not part of the domain, so $f(0)$ does not exist and there is no point $(0, f(0))$ on the graph. Thus there is no point to trace at $x = 0$. However, F is continuous on the intervals $(-\infty, 0)$ and $(0, \infty)$.

The function G is not continuous over the whole real line since it is not continuous at $x = 1$. Let us trace the graph of G to the left of

7. Which of the following functions
are continuous?

a)

b)

c)

d)

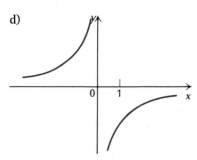

8. a) Decide whether the function in
Margin Exercise 1(c) is continuous at −2; at 1.

b) Decide whether the function in
Margin Exercise 1(d) is continuous at 1; at 0.

$x = 1$. As x approaches 1, $G(x)$ seems to approach 2. However, at $x = 1$, $G(x)$ *jumps* up to 5, whereas to the right of $x = 1$, $G(x)$ *jumps* back to some value close to 2. Thus G is discontinuous at $x = 1$.

The function H is not continuous over the whole real line since it is not continuous at $x = 1$. Let us trace the graph of H starting to the left of $x = 1$. As x approaches 1, $H(x)$ seems to approach 4. However, at $x = 1$, $H(x)$ *jumps* down to 0, whereas just to the right of $x = 1$, $H(x)$ is close to 0. Thus $H(x)$ is discontinuous at $x = 1$.

DO EXERCISES 7 AND 8.

Limits and Continuity

After working through Example 3, we might ask, "When can we sub-stitute to find a limit?" The answer lies in the following definition, which also provides a more formal definition of continuity.

DEFINITION

A function f is *continuous* at $x = a$ if:

i) $f(a)$ exists,

ii) $\lim_{x \to a} f(x)$ exists, and

iii) $\lim_{x \to a} f(x) = f(a)$.

A function is *continuous over an interval I* if it is continuous at each point in I.

Note that we are also using the notation "$\lim_{x \to a} f(x)$" for a limit.

Example 4   Determine whether the function given by

$$f(x) = 2x + 3$$

is continuous at $x = 4$.

Solution   This function is continuous at 4 because:

i) $f(4)$ exists ($f(4) = 11$),

ii) $\lim_{x \to 4} f(x)$ exists [$\lim_{x \to 4} f(x) = 11$ (as shown earlier)],

iii) $\lim_{x \to 4} f(x) = 11 = f(4)$.

In fact, $f(x) = 2x + 3$ is continuous at any point on the real line.

9. Consider

$$f(x) = 3x - 1$$

(see Margin Exercise 1).

a) Does $f(6)$ exist? If so, what is it?

b) Does $\lim_{x \to 6} f(x)$ exist? If so, what is it?

c) Does $\lim_{x \to 6} f(x) = f(6)$?

d) Is f continuous at 6?

10. Determine whether the function in Margin Exercise 2 is continuous at each of the following.

a) $x = 3$

b) $x = -2$

11. Determine whether the function in Margin Exercise 4 is continuous at $x = 0$.

12. Determine whether the function in Margin Exercise 5 is continuous at $x = -2$.

DO EXERCISE 9.

Example 5 Determine whether the function in Example 1 is continuous at $x = 0$.

Solution The function is *not* continuous at $x = 0$ because $f(0)$ does not exist.

Example 6 Determine whether the function in Example 2 is continuous at $x = 1$.

Solution The function is not continuous at $x = 1$ because $\lim_{x \to 1} H(x)$ does not exist.

Example 7 Determine whether the function in Example 3 is continuous at $x = 1$.

Solution The function is not continuous at $x = 1$ because $G(1) = 5$, but $\lim_{x \to 1} G(x) = 2$.

DO EXERCISES 10–12.

Continuity Principles

The following continuity principles, which we will not prove, allow us to build up continuous functions.

i) Any constant function is continuous (such a function never varies).

ii) For any positive integer n, x^n and $\sqrt[n]{x}$ are continuous. When n is even, the inputs of $\sqrt[n]{x}$ are restricted to $[0, \infty)$.

iii) If $f(x)$ and $g(x)$ are continuous, then so are $f(x) + g(x)$, $f(x) - g(x)$, and $f(x) \cdot g(x)$.

iv) If $f(x)$ is continuous, so is $1/f(x)$, so long as the inputs x do not yield outputs $f(x) = 0$.

Example 8 Provide an argument to show that

$$f(x) = x^2 - 3x + 2$$

is continuous.

13. Provide an argument to show that the function given by

$$f(x) = \frac{\sqrt[3]{x} - 7x^2}{x - 2}$$

is continuous so long as $x \neq 2$.

Solution   We know that x^2 is continuous by (ii). The constant function 3 is continuous by (i), and the function x is continuous by (ii), so the product $3x$ is continuous by (iii). Thus $x^2 - 3x$ is continuous by (iii), and since the constant 2 is continuous, we can apply (iii) again to show that $x^2 - 3x + 2$ is continuous.

In similar fashion, we can show that any polynomial, such as

$$f(x) = x^4 - 5x^3 + x^2 - 7,$$

is continuous. A rational function is a quotient of two polynomials

$$r(x) = \frac{f(x)}{q(x)}.$$

Thus by (iv), a rational function is continuous so long as the inputs x are not such that $q(x) = 0$.

DO EXERCISE 13.

EXERCISE SET 2.1

Which of the following are continuous?

1.
   $y = f(x)$

2.
   $y = g(x)$

3.

4.

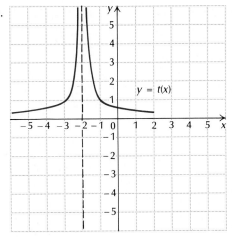

Use the graphs and functions in Exercises 1–4 to answer each of the following.

5. **a)** Find $\lim_{x \to 1^+} f(x)$, $\lim_{x \to 1^-} f(x)$, and $\lim_{x \to 1} f(x)$.

   b) Find $f(1)$.

   c) Is f continuous at $x = 1$?

   d) Find $\lim_{x \to -2} f(x)$.

   e) Find $f(-2)$.

   f) Is f continuous at $x = -2$?

7. **a)** Find $\lim_{x \to 1} h(x)$.

   b) Find $h(1)$.

   c) Is h continuous at $x = 1$?

   d) Find $\lim_{x \to -2} h(x)$.

   e) Find $h(-2)$.

   f) Is h continuous at $x = -2$?

6. **a)** Find $\lim_{x \to 1^+} g(x)$, $\lim_{x \to 1^-} g(x)$, and $\lim_{x \to 1} g(x)$.

   b) Find $g(1)$.

   c) Is g continuous at $x = 1$?

   d) Find $\lim_{x \to -2} g(x)$.

   e) Find $g(-2)$.

   f) Is g continuous at $x = -2$?

8. **a)** Find $\lim_{x \to 1} t(x)$.

   b) Find $t(1)$.

   c) Is t continuous at $x = 1$?

   d) Find $\lim_{x \to -2} t(x)$.

   e) Find $t(-2)$.

   f) Is t continuous at $x = -2$?

The postage function. Postal rates are 22¢ for the first ounce and 17¢ for each additional ounce or fraction thereof. Formally speaking, if x is the weight of a letter in ounces, then $p(x)$ is the cost of mailing the letter, where

$$p(x) = 22\text{¢}, \text{ if } 0 < x \leqslant 1,$$
$$p(x) = 39\text{¢}, \text{ if } 1 < x \leqslant 2,$$
$$p(x) = 56\text{¢}, \text{ if } 2 < x \leqslant 3,$$

and so on, up to 12 ounces (at which point postal cost also depends on distance). The graph of p is shown here.

9. Is p continuous at 1? at $1\frac{1}{2}$? at 2? at 2.53?

Using the graph, find each limit, if it exists.

11. $\lim\limits_{x \to 1^-} p(x),\ \lim\limits_{x \to 1^+} p(x),\ \lim\limits_{x \to 1} p(x)$

13. $\lim\limits_{x \to 2.3} p(x)$

10. Is p continuous at 3? at $3\frac{1}{4}$? at 4? at 3.98?

12. $\lim\limits_{x \to 2^-} p(x),\ \lim\limits_{x \to 2^+} p(x),\ \lim\limits_{x \to 2} p(x)$

14. $\lim\limits_{x \to 1/2} p(x)$

OBJECTIVES

a) Find

$$\lim_{x \to a} f(x),$$

if it exists, using input–output tables, algebra, or graphs.

b) Find limits at infinity.

c) Find a limit like

$$\lim_{h \to 0} (3x^2 + 3xh + h^2).$$

Find the limit, if it exists.

1. $\lim\limits_{x \to -2} (x^4 - 5x^3 + x^2 - 7)$

2. $\lim\limits_{x \to 1} \sqrt{x^2 + 3x + 4}$

2.2

MORE ON LIMITS

Using Limit Principles

If a function is continuous at a, we can substitute to find the limit.

Example 1 Find $\lim_{x \to 2} (x^4 - 5x^3 + x^2 - 7)$.

Solution It follows from the Continuity Principles that $x^4 - 5x^3 + x^2 - 7$ is continuous. Thus the limit can be found by substitution:

$$\lim_{x \to 2} (x^4 - 5x^3 + x^2 - 7) = 2^4 - 5 \cdot 2^3 + 2^2 - 7$$
$$= 16 - 40 + 4 - 7$$
$$= -27.$$

Example 2 Find $\lim_{x \to 0} \sqrt{x^2 - 3x + 2}$.

Solution Using the Continuity Principles, we have shown that $x^2 - 3x + 2$ is continuous, and so long as x is restricted to values for which $x^2 - 3x + 2$ is nonnegative, it follows from Principle (ii) that $\sqrt{x^2 - 3x + 2}$ is continuous. Thus we can substitute to find the limit:

$$\lim_{x \to 0} \sqrt{x^2 - 3x + 2} = \sqrt{0^2 - 3 \cdot 0 + 2}$$
$$= \sqrt{2}.$$

DO EXERCISES 1 AND 2.

There are Limit Principles that correspond to the Continuity Principles. We can use them to find limits when we are uncertain of the continuity of a function at a given point.

Limit Principles

If $\lim_{x \to a} f(x) = L$ and $\lim_{x \to a} g(x) = M$, then we have the following.

L1. $\lim_{x \to a} c = c$.

    (The limit of a constant is the constant.)

L2. $\lim_{x \to a} x^n = a^n$, $\lim_{x \to a} \sqrt[n]{x} = \sqrt[n]{a}$, for any positive integer n.

    (When n is even, the inputs of $\sqrt[n]{x}$ must be restricted to $[0, \infty)$.)

L3. $\lim_{x \to a} [f(x) \pm g(x)] = \lim_{x \to a} f(x) \pm \lim_{x \to a} g(x) = L \pm M$.

    (The limit of a sum or difference is the sum or difference of the limits.)

    $\lim_{x \to a} [f(x) \cdot g(x)] = \left[\lim_{x \to a} f(x)\right] \cdot \left[\lim_{x \to a} g(x)\right] = L \cdot M$.

    (The limit of a product is the product of the limits.)

L4. $\lim_{x \to a} \dfrac{1}{f(x)} = \dfrac{1}{\lim_{x \to a} f(x)} = \dfrac{1}{L}$, provided $L \neq 0$.

    (The limit of a reciprocal is the reciprocal of the limit.)

L5. $\lim_{x \to a} cf(x) = c \cdot \lim_{x \to a} f(x) = c \cdot L$

    (The limit of a constant times a function is the constant times the limit.)

Principle L5 can actually be proved using Principles L1 and L3, but we state it for emphasis.

Example 3 Find

$$\lim_{x \to -3} \frac{x^2 - 9}{x + 3}.$$

Solution The function $(x^2 - 9)/(x + 3)$ is not continuous at $x = -3$. We use some algebraic simplification and then some limit principles:

$$\lim_{x \to -3} \frac{x^2 - 9}{x + 3} = \lim_{x \to -3} \frac{(x + 3)(x - 3)}{x + 3}$$

$$= \lim_{x \to -3} (x - 3) \qquad \text{Simplifying, assuming } x \neq -3$$

$$= \lim_{x \to -3} x - \lim_{x \to -3} 3 \qquad \text{By L3}$$

$$= -3 - 3 = -6$$

Find the limit, if it exists. **CSS**

3. $\lim\limits_{x \to -4} \dfrac{x^2 - 16}{x + 4}$

4. $\lim\limits_{x \to 3} \dfrac{x - 3}{x^2 - 9}$

5. **a)** Complete the table.

h	$2x + h$
1	
0.7	
0.4	
0.1	
0.01	
0.001	

b) Find

$$\lim\limits_{h \to 0} (2x + h).$$

DO EXERCISES 3 AND 4.

In Section 2.3 we will encounter expressions with two variables, x and h. Our interest is in limits where x is fixed as a constant and h approaches zero.

Example 4 Find $\lim\limits_{h \to 0} (3x^2 + 3xh + h^2)$.

Solution We treat x as a constant since we are interested only in the way in which the expression varies when $h \longrightarrow 0$. We use the Limit Principles to find that

$$\lim\limits_{h \to 0} (3x^2 + 3xh + h^2) = 3x^2 + 3x(0) + 0^2 = 3x^2.$$

The reader can check any limit about which there is uncertainty by using an input–output table. Below is a table for this limit.

h	$3x^2 + 3xh + h^2$
1	$3x^2 + 3x \cdot 1 + 1^2$, or $3x^2 + 3x + 1$
0.8	$3x^2 + 3x(0.8) + (0.8)^2$, or $3x^2 + 2.4x + 0.64$
0.5	$3x^2 + 3x(0.5) + (0.5)^2$, or $3x^2 + 1.5x + 0.25$
0.1	$3x^2 + 3x(0.1) + (0.1)^2$, or $3x^2 + 0.3x + 0.01$
0.01	$3x^2 + 3x(0.01) + (0.01)^2$, or $3x^2 + 0.03x + 0.0001$
0.001	$3x^2 + 3x(0.001) + (0.001)^2$, or $3x^2 + 0.003x + 0.000001$

From the pattern in the table, it appears that

$$\lim\limits_{h \to 0} (3x^2 + 3xh + h^2) = 3x^2.$$

DO EXERCISE 5.

Limits and Infinity

In Example 1 of Section 2.1 we discussed the fact that the limit

$$\lim\limits_{x \to 0} \frac{1}{x} \quad \text{does not exist.}$$

6. *Earned-run average.* A pitcher's earned-run average (the average number of runs given up every 9 innings, or 1 game) is given by

$$E = 9 \cdot \frac{n}{i},$$

where n = the number of earned runs allowed and i = the number of innings pitched. Suppose we fix the number of earned runs allowed at 4 and let i vary. We get a function given by

$$E(i) = 9 \cdot \frac{4}{i}.$$

a) Complete the following table, rounding to two decimal places.

Innings pitched (i)	Earned-run average (E)
9	
8	
7	
6	
5	
4	
3	
2	
1	
$\frac{2}{3}$ (2 outs)	
$\frac{1}{3}$ (1 out)	

b) Find

$$\lim_{i \to 0} E(i).$$

c) On the basis of parts (a) and (b), determine what a pitcher's earned run average might be if 4 runs were allowed and there were 0 outs?

If you look back at the graph, you will note that as x approaches 0 from the right, the outputs increase without bound. These numbers do not approach any real number, though it might be said that the limit from the right is ∞ (infinity). As x approaches 0 from the left, the outputs become more and more negative without bound. These numbers do not approach any real number, though it might be said that the limit from the left is $-\infty$ (negative infinity). Keep in mind that ∞ and $-\infty$ are *not* real numbers. We associate ∞ with numbers increasing without bound in a positive direction and $-\infty$ with numbers decreasing without bound in a negative direction.

DO EXERCISE 6.

Limits at Infinity

Occasionally we need to determine limits when the inputs get larger and larger without bound, that is, when they approach infinity. In such cases we are finding *limits at infinity*. Such a limit is expressed as

$$\lim_{x \to \infty} f(x).$$

Example 5   Find

$$\lim_{x \to \infty} \left(\frac{3x - 1}{x} \right).$$

Solution   One way to find such a limit is to use an input–output table, as follows.

Inputs, x	1	10	50	100	2000
Outputs, $\dfrac{3x - 1}{x}$	2.0	2.9	2.98	2.99	2.9995

As the inputs get larger and larger without bound, the outputs get closer and closer to 3. Thus,

$$\lim_{x \to \infty} \left(\frac{3x - 1}{x} \right) = 3.$$

Another way to do this is to use some algebra and the fact that as $x \to \infty$, $b/ax^n \to 0$, for any positive integer n.

94 DIFFERENTIATION

7. Consider [CSS]

$$f(x) = \frac{2x + 5}{x}.$$

a) Complete this table. (helpful)

Inputs, x	Outputs, $\dfrac{2x + 5}{x}$
4	
20	
80	
200	
1,000	
10,000	

b) Find

$$\lim_{x \to \infty} \frac{2x + 5}{x}.$$

8. Find

$$\lim_{x \to \infty} \frac{2x^2 + x - 7}{3x^2 - 4x + 1}.$$

We multiply by 1, using $(1/x) \div (1/x)$. This amounts to dividing both the numerator and the denominator by x:

$$\lim_{x \to \infty} \frac{3x - 1}{x} = \lim_{x \to \infty} \frac{3x - 1}{x} \cdot \frac{(1/x)}{(1/x)}$$

$$= \lim_{x \to \infty} \frac{(3x - 1)\dfrac{1}{x}}{x \cdot \dfrac{1}{x}}$$

$$= \lim_{x \to \infty} \frac{3x \cdot \dfrac{1}{x} - 1 \cdot \dfrac{1}{x}}{1}$$

$$= \lim_{x \to \infty} \left(3 - \frac{1}{x}\right) = 3 - 0 = 3.$$

DO EXERCISE 7.

Example 6 Find

$$\lim_{x \to \infty} \frac{3x^2 - 7x + 2}{7x^2 + 5x + 1}.$$

Solution The highest power of x in the denominator is x^2. We divide both the numerator and the denominator by x^2:

$$\lim_{x \to \infty} \frac{3x^2 - 7x + 2}{7x^2 + 5x + 1} = \lim_{x \to \infty} \frac{3 - \dfrac{7}{x} + \dfrac{2}{x^2}}{7 + \dfrac{5}{x} + \dfrac{1}{x^2}}$$

$$= \frac{3 - 0 + 0}{7 + 0 + 0} = \frac{3}{7}.$$

DO EXERCISE 8.

EXERCISE SET 2.2

Find each of the following limits. Use any method: algebra, graphs, or input–output tables. [CSS]

1. $\lim_{x \to 1} (x^2 - 3)$ **2.** $\lim_{x \to 1} (x^2 + 4)$ **3.** $\lim_{x \to 0} \frac{3}{x}$ **4.** $\lim_{x \to 0} \frac{-4}{x}$

5. $\lim_{x \to 3} (2x + 5)$

6. $\lim_{x \to 4} (5 - 3x)$

7. $\lim_{x \to -5} \dfrac{x^2 - 25}{x + 5}$

8. $\lim_{x \to -4} \dfrac{x^2 - 16}{x + 4}$

9. $\lim_{x \to -2} \dfrac{5}{x}$

10. $\lim_{x \to -5} \dfrac{-2}{x}$

11. $\lim_{x \to 2} \dfrac{x^2 + x - 6}{x - 2}$

12. $\lim_{x \to -4} \dfrac{x^2 - x - 20}{x + 4}$

13. $\lim_{x \to 5} \sqrt[3]{x^2 - 17}$

14. $\lim_{x \to 2} \sqrt{x^2 + 5}$

15. $\lim_{x \to 1} (x^4 - x^3 + x^2 + x + 1)$

16. $\lim_{x \to 2} (2x^5 - 3x^4 + x^3 - 2x^2 + x + 1)$

17. $\lim_{x \to 2} \dfrac{1}{x - 2}$

18. $\lim_{x \to 1} \dfrac{1}{(x - 1)^2}$

19. $\lim_{x \to 2} \dfrac{3x^2 - 4x + 2}{7x^2 - 5x + 3}$

20. $\lim_{x \to -1} \dfrac{4x^2 + 5x - 7}{3x^2 - 2x + 1}$

21. $\lim_{x \to 2} \dfrac{x^2 + x - 6}{x^2 - 4}$

22. $\lim_{x \to 4} \dfrac{x^2 - 16}{x^2 - x - 12}$

23. $\blacksquare$ $\lim_{x \to 1} \dfrac{1 - \sqrt{x}}{1 - x}$

24. $\blacksquare$ $\lim_{x \to 4} \dfrac{\sqrt{x} - 2}{x - 4}$

25. $\lim_{h \to 0} (6x^2 + 6xh + 2h^2)$

26. $\lim_{h \to 0} (10x + 5h)$

27. $\lim_{h \to 0} \dfrac{-2x - h}{x^2(x + h)^2}$

28. $\lim_{h \to 0} \dfrac{-5}{x(x + h)}$

29. $\lim_{x \to \infty} \dfrac{2x - 4}{5x}$

30. $\lim_{x \to \infty} \dfrac{3x + 1}{4x}$

31. $\lim_{x \to \infty} \left(5 - \dfrac{2}{x} \right)$

32. $\lim_{x \to \infty} \left(7 + \dfrac{3}{x} \right)$

33. $\lim_{x \to \infty} \dfrac{2x - 5}{4x + 3}$

34. $\lim_{x \to \infty} \dfrac{6x + 1}{5x - 2}$

35. $\lim_{x \to \infty} \dfrac{2x^2 - 5}{3x^2 - x + 7}$

36. $\lim_{x \to \infty} \dfrac{4 - 3x - 12x^2}{1 + 5x + 3x^2}$

37. Consider

$$f(x) = \begin{cases} 1, & \text{for } x \neq 2, \\ -1, & \text{for } x = 2. \end{cases}$$

Find each of the following.

a) $\lim_{x \to 0} f(x)$

b) $\lim_{x \to 2^-} f(x)$

c) $\lim_{x \to 2^+} f(x)$

d) $\lim_{x \to 2} f(x)$

e) Is f continuous at 0? at 2?

38. Consider

$$g(x) = \begin{cases} -4, & \text{for } x = 3, \\ 2x + 5, & \text{for } x \neq 3. \end{cases}$$

Find each of the following.

a) $\lim_{x \to 3^-} g(x)$

b) $\lim_{x \to 3^+} g(x)$

c) $\lim_{x \to 3} g(x)$

d) $\lim_{x \to 2} g(x)$

e) Is g continuous at 3? at 2?

APPLICATIONS

39. *Business: Depreciation.* A new conveyor system costs $10,000. In any year it depreciates 8% of its value at the beginning of that year.

a) What is the annual depreciation in each of the first five years?

b) What is the total depreciation at the end of ten years?

c) What is the limit of the sum of the annual depreciation costs?

40. *Business: Depreciation.* A new car costs $6000. In any year it depreciates 30% of its value at the beginning of that year.

a) What is the annual depreciation in each of the first five years?

b) What is the total depreciation at the end of ten years?

c) What is the limit of the sum of the annual depreciation costs?

41. Inside its own 5-yd line, a defensive football team is penalized half the distance to the goal. Suppose a defensive team keeps getting such a penalty. What is the limit of the distance of the offensive team from the goal? Can the offensive team ever score a touchdown in this manner?

EXTENSION EXERCISES

Find the limit, if it exists.

42. $\lim\limits_{x \to 0} \dfrac{|x|}{x}$

43. ▦ $\lim\limits_{x \to 1} \dfrac{2 - \sqrt{x + 3}}{x - 1}$

44. $\lim\limits_{x \to 1} \dfrac{x^3 - 1}{x^2 - 1}$

45. $\lim\limits_{x \to \infty} \dfrac{4 - 3x}{5 - 2x^2}$

46. $\lim\limits_{x \to \infty} \dfrac{6x^5 - x^4}{4x^2 - 3x^3}$

2.3

OBJECTIVES

a) Compute an average rate of change of one variable with respect to another.

b) Find a simplified difference quotient.

AVERAGE RATES OF CHANGE

The following graph shows the total production of suits by Raggs, Ltd., during one morning of work. Industrial psychologists have found curves like this typical of the production of factory workers.

(8 A.M.)   (9 A.M.)   (10 A.M.) (11 A.M.) (12 P.M.)

1. Refer to the graph of the number
   of suits produced by Raggs, Ltd.

   a) Find the number of suits pro-
      duced per hour from

      8 A.M. to 9 A.M.;
      9 A.M. to 10 A.M.;
      10 A.M. to 11 A.M.;
      11 A.M. to 12 P.M.

   b) Which interval in part (a) had
      the highest number?

   c) Why do you think this
      happened?

   d) Which interval in part (a) had
      the lowest number?

   e) Why do you think this
      happened?

   f) What was the average number
      of suits produced per hour from
      8 A.M. to 12 P.M.?

Example 1   What is the number of suits produced from 9 A.M. to
10 A.M.?

Solution   At 10 A.M., 55 suits have been produced. At 9 A.M., 20 suits
have been produced. In the hour from 9 A.M. to 10 A.M., the number of
suits produced was

$$55 \text{ suits} - 20 \text{ suits}, \quad \text{or } 35 \text{ suits}.$$

Note that this is the slope of the line from P to Q.

Example 2   What was the average number of suits produced per hour
from 9 A.M. to 11 A.M.?

Solution   We have

$$\frac{64 \text{ suits} - 20 \text{ suits}}{11 \text{ A.M.} - 9 \text{ A.M.}} = \frac{44 \text{ suits}}{2 \text{ hr}} = 22 \frac{\text{suits}}{\text{hr}} \text{ (suits per hour)}.$$

This is the slope of the line from P to R. It is not shown in the graph.

DO EXERCISE 1.

Let us consider a function $y = f(x)$ and two inputs x_1 and x_2. The
change in input, or the *change in x,* is

$$x_2 - x_1.$$

The *change in output,* or the *change in y,* is

$$y_2 - y_1.$$

DEFINITION

The *average rate of change of y with respect to x,* as x changes
from x_1 to x_2, is the ratio of the change in output to the change
in input:

$$\frac{y_2 - y_1}{x_2 - x_1}, \quad \text{where } x_2 \neq x_1.$$

If we look at a graph of the function, we see that

$$\frac{y_2 - y_1}{x_2 - x_1} = \frac{f(x_2) - f(x_1)}{x_2 - x_1}$$

and that this is the slope of the line from $P(x_1, y_1)$ to $Q(x_2, y_2)$. The line PQ is called a *secant line*.

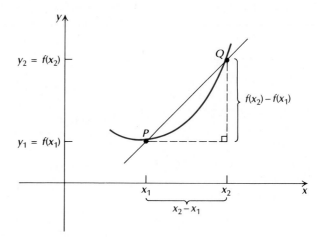

Example 3 For $y = f(x) = x^2$, find the average rates of change as:

a) x changes from 1 to 3;

b) x changes from 1 to 2;

c) x changes from 2 to 3.

Solution The following graph is not necessary to the computations, but gives us a look at two of the secant lines whose slopes are being computed.

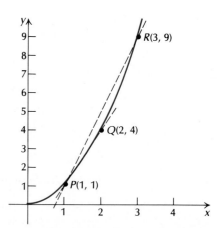

2. For

$$f(x) = x^3,$$

find the average rates of change and sketch the secant lines as:

a) x changes from 1 to 4;

b) x changes from 1 to 2;

c) x changes from 2 to 4;

d) x changes from -1 to -4.

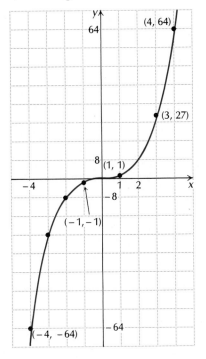

a) When $x_1 = 1$,

$$y_1 = f(x_1) = f(1) = 1^2 = 1;$$

and when $x_2 = 3$,

$$y_2 = f(x_2) = f(3) = 3^2 = 9.$$

The average rate of change is

$$\frac{y_2 - y_1}{x_2 - x_1} = \frac{f(x_2) - f(x_1)}{x_2 - x_1} = \frac{9 - 1}{3 - 1} = \frac{8}{2} = 4.$$

b) When $x_1 = 1$,

$$y_1 = f(x_1) = f(1) = 1^2 = 1;$$

and when $x_2 = 2$,

$$y_2 = f(x_2) = f(2) = 2^2 = 4.$$

The average rate of change is

$$\frac{4 - 1}{2 - 1} = \frac{3}{1} = 3.$$

c) When $x_1 = 2$,

$$y_1 = f(x_1) = f(2) = 2^2 = 4;$$

and when $x_2 = 3$,

$$y_2 = f(x_2) = f(3) = 3^2 = 9.$$

The average rate of change is

$$\frac{9 - 4}{3 - 2} = \frac{5}{1} = 5.$$

DO EXERCISES 2 AND 3. (EXERCISE 3 IS ON THE FOLLOWING PAGE.)

For a linear function the average rates of change are the same for any choice of x_1 and x_2; that is, they are equal to the slope m of the line. As we saw in Example 3 and in Margin Exercise 2, a function that is not linear has average rates of change that vary with the choice of x_1 and x_2.

Difference Quotients

Let us now simplify our notation a bit by eliminating the subscripts. Instead of x_1, we will simply write x.

3. For

$$f(x) = \tfrac{1}{2}x + 1,$$

find the average rates of change and sketch the secant lines as:

a) x changes from 2 to 4;

b) x changes from 2 to 3;

c) x changes from -1 to 4.

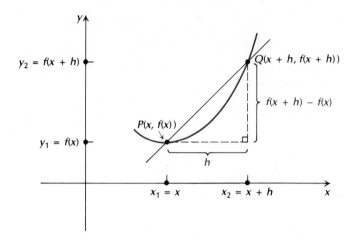

To get from x_1, or x, to x_2 we move a distance h. Thus $x_2 = x + h$. Then the average rate of change, also called a *difference quotient*, is given by

$$\frac{y_2 - y_1}{x_2 - x_1} = \frac{f(x_2) - f(x_1)}{x_2 - x_1}$$

$$= \frac{f(x + h) - f(x)}{(x + h) - x}$$

$$= \frac{f(x + h) - f(x)}{h}.$$

DEFINITION

The average rate of change of f with respect to x is also called the *difference quotient*. It is given by

$$\frac{f(x + h) - f(x)}{h}, \quad \text{where } h \neq 0.$$

The difference quotient is equal to the slope of the line from $P(x, f(x))$ to $Q(x + h, f(x + h))$.

Example 4 For $f(x) = x^2$, find the difference quotient when:

a) $x = 5$ and $h = 3$;

b) $x = 5$ and $h = 0.1$.

CSS

4. For

$$f(x) = 4x^2,$$

complete the following table to find the difference quotients.

x	h	$x+h$	$f(x)$	$f(x+h)$	$f(x+h)-f(x)$	$\dfrac{f(x+h)-f(x)}{h}$
3	2					
3	1					
3	0.1					
3	0.01					
3	0.001					

Solution

a) We substitute $x = 5$ and $h = 3$ into the formula:

$$\frac{f(x+h) - f(x)}{h} = \frac{f(5+3) - f(5)}{3} = \frac{f(8) - f(5)}{3}.$$

Now $f(8) = 8^2 = 64$ and $f(5) = 5^2 = 25$, and we have

$$\frac{f(8) - f(5)}{3} = \frac{64 - 25}{3} = \frac{39}{3} = 13.$$

b) We substitute $x = 5$ and $h = 0.1$ into the formula:

$$\frac{f(x+h) - f(x)}{h} = \frac{f(5+0.1) - f(5)}{0.1} = \frac{f(5.1) - f(5)}{0.1}.$$

Now $f(5.1) = (5.1)^2 = 26.01$ and $f(5) = 25$, and we have

$$\frac{f(5.1) - f(5)}{0.1} = \frac{26.01 - 25}{0.1} = \frac{1.01}{0.1} = 10.1.$$

DO EXERCISE 4.

For the function in Example 4, let us find a general form of the difference quotient. This will allow more efficient computations.

Example 5 For $f(x) = x^2$, find a simplified form of the difference quotient. Then find the value of the difference quotient when $x = 5$ and $h = 0.1$.

Solution We have

$$f(x) = x^2,$$

so

$$f(x+h) = (x+h)^2 = x^2 + 2xh + h^2.$$

Then

$$f(x+h) - f(x) = (x^2 + 2xh + h^2) - x^2 = 2xh + h^2.$$

So

$$\frac{f(x+h) - f(x)}{h} = \frac{2xh + h^2}{h} = \frac{h(2x + h)}{h} = 2x + h, \quad h \neq 0.$$

5. For

$$f(x) = 4x^2,$$

find a simplified form of the difference quotient by completing steps (a) through (c). Then complete the table in (d) using the simplified form.

a) Find $f(x + h)$.

b) Find $f(x + h) - f(x)$.

c) Find $[f(x + h) - f(x)]/h$ and simplify.

d) Complete the following table.

CSS

x	h	$\dfrac{f(x + h) - f(x)}{h}$
6	-3	
6	-2	
6	-1	
6	-0.1	
6	-0.01	
6	-0.001	

6. a) For

$$f(x) = 4x^3,$$

find a simplified difference quotient.

b) Complete the following table.

x	h	$\dfrac{f(x + h) - f(x)}{h}$
-2	1	
-2	0.1	
-2	0.01	
-2	0.001	

It is important to note that a difference quotient is defined only when $h \neq 0$. The simplification above is valid only for nonzero values of h. When $x = 5$ and $h = 0.1$,

$$\frac{f(x + h) - f(x)}{h} = 2x + h = 2 \cdot 5 + 0.1 = 10 + 0.1 = 10.1.$$

DO EXERCISE 5.

Example 6 For $f(x) = x^3$, find a simplified form of the difference quotient.

Solution Now $f(x) = x^3$, so

$$f(x + h) = (x + h)^3 = x^3 + 3x^2h + 3xh^2 + h^3.$$

(This is shown in Exercise Set 1.1.) Then

$$f(x + h) - f(x) = (x^3 + 3x^2h + 3xh^2 + h^3) - x^3$$
$$= 3x^2h + 3xh^2 + h^3.$$

So

$$\frac{f(x + h) - f(x)}{h} = \frac{3x^2h + 3xh^2 + h^3}{h}$$

$$= \frac{h(3x^2 + 3xh + h^2)}{h} = 3x^2 + 3xh + h^2, \quad h \neq 0.$$

Again, this is true *only* for $h \neq 0$.

DO EXERCISE 6.

Example 7 For $f(x) = 3/x$, find a simplified form of the difference quotient.

Solution Now

$$f(x) = \frac{3}{x},$$

so

$$f(x + h) = \frac{3}{x + h}.$$

7. a) For

$$f(x) = \frac{1}{x},$$

find a simplified difference quotient.

b) Complete the following table.

x	h	$\dfrac{f(x + h) - f(x)}{h}$
2	3	
2	1	
2	0.1	
2	0.01	
2	0.001	

Then

$$f(x + h) - f(x) = \frac{3}{x + h} - \frac{3}{x}$$

$$= \frac{3}{x + h} \cdot \frac{x}{x} - \frac{3}{x} \cdot \frac{x + h}{x + h}$$

Here we are multiplying by 1 to get a common denominator.

$$= \frac{3x - 3(x + h)}{x(x + h)}$$

$$= \frac{3x - 3x - 3h}{x(x + h)} = \frac{-3h}{x(x + h)}.$$

Thus

$$\frac{f(x + h) - f(x)}{h} = \frac{\dfrac{-3h}{x(x + h)}}{h} = \frac{-3h}{x(x + h)} \cdot \frac{1}{h} = \frac{-3}{x(x + h)}, \quad h \neq 0.$$

This is true only for $h \neq 0$.

DO EXERCISE 7.

EXERCISE SET 2.3

APPLICATIONS

1. *Economics: Utility.* Utility is a type of function that occurs in economics. When a consumer receives x units of a certain product, a certain amount of pleasure, or utility, U, is derived from them. At right is a typical graph of a utility function.

a) Find the average rate of change of U as x changes from 0 to 1; from 1 to 2; from 2 to 3; from 3 to 4.

b) Why do you think the average rates of change are decreasing?

2. *Business: Advertising results.* The following graph shows a typical response to advertising. After an amount a is spent on advertising, the company sells $N(a)$ units of a product.

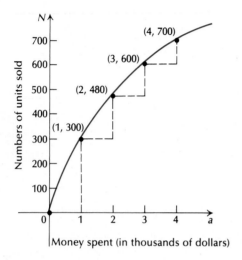

Money spent (in thousands of dollars)

a) Find the average rate of change of N as a changes from 0 to 1; from 1 to 2; from 2 to 3; from 3 to 4.

b) Why do you think the average rates of change are decreasing?

3. *Psychology: Memory.* The total number of words $M(t)$ that a person can memorize in time t, in minutes, is shown in the following graph.

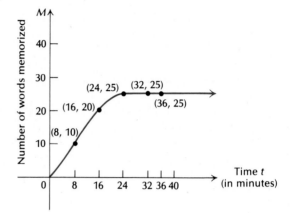

a) Find the average rate of change of M as t changes from 0 to 8; from 8 to 16; from 16 to 24; from 24 to 32; from 32 to 36.

b) Why do the average rates of change become 0 after 24 min?

4. *Biomedical: Temperature during an illness.* The °F temperature T of a patient during an illness is given by the graph at right, where t = time in days.

a) Find the average rate of change of T as t changes from 1 to 10. Using this rate of change, would you know that the person was sick?

b) Find the average rate of change of T with respect to t as t changes from 1 to 2; from 2 to 3; from 3 to 4; from 4 to 5; from 5 to 6; from 6 to 7; from 7 to 8; from 8 to 9; from 9 to 10; from 10 to 11.

c) When do you think the temperature began to rise?

d) When do you think the temperature reached its peak?

e) When do you think the temperature began to subside?

f) When was the temperature back to normal?

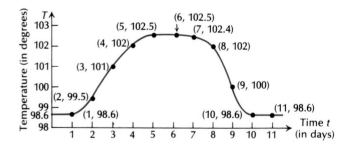

5. *Sociology: Population growth.* The two curves shown here describe the number of people in each of two countries at time *t*, in years.

   a) Find the average rate of change of each population (the number of people in the population) with respect to time *t* as *t* changes from 0 to 4. This is often called an *average growth rate.*

   b) If the calculation in part (a) were the only one made, would we detect the fact that the populations were growing differently?

   c) Find the average rates of change of each population as *t* changes from 0 to 1; from 1 to 2; from 2 to 3; from 3 to 4.

   d) For which population does the statement "the population grew 125 million each year" convey the least information about what really took place?

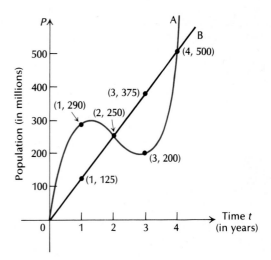

6. *Business: Cost.* A firm determines that the total cost *C* of producing *x* units of a certain product is given by

$$C(x) = -0.05x^2 + 50x,$$

where *C(x)* is in dollars.

   a) Find $C(301)$.

   b) Find $C(300)$.

   c) Find $C(301) - C(300)$.

   d) Find $\dfrac{C(301) - C(300)}{301 - 300}$.

7. *Business: Revenue.* A firm determines that the total revenue from the sale of *x* units of a certain product is given by

$$R(x) = -0.01x^2 + 1000x,$$

where *R(x)* is in dollars.

   a) Find $R(301)$.

   b) Find $R(300)$.

   c) Find $R(301) - R(300)$.

   d) Find $\dfrac{R(301) - R(300)}{301 - 300}$.

8. *Average velocity.* A car is at a distance *s* (in miles) from its starting point in *t* hours, given by

$$s(t) = 10t^2.$$

   a) Find $s(2)$ and $s(5)$.

   b) Find $s(5) - s(2)$. What does this represent?

   c) Find the average rate of change of distance with respect to time as *t* changes from $t_1 = 2$ to $t_2 = 5$. This is known as *average velocity,* or *speed.*

9. *Average velocity.* An object is dropped from a certain height. It is known that it will fall a distance *s* (in feet) in *t* seconds, given by

$$s(t) = 16t^2.$$

   a) How far will the object fall in 3 sec?

   b) How far will the object fall in 5 sec?

   c) What is the average rate of change of distance with respect to time during the time from 3 to 5 sec? This is also *average velocity,* or *speed.*

10. *Sociology: Divorce rate.* It is known that in 1960 there were 393,000 divorces. In 1982 there were 1,180,000 divorces. Find the average rate of change in the number of divorces with respect to time. This is called an *average divorce rate.*

12. *Gasoline mileage.* At the beginning of a trip, the odometer on a car reads 30,680 and the car has a full tank of gas. At the end of the trip the odometer reads 30,970. It takes 20 gal of gas to fill the tank again.

    a) What is the average rate of consumption (the rate of change of the number of miles with respect to the number of gallons)?

    b) What is the average rate of change of the number of gallons with respect to the number of miles?

13. *Business: Total revenue.* In 1979 the total revenue of Chi-Chi's, Inc., a national Mexican food franchise, was $1,047,000. In 1982 it was $40,500,000 and in 1986 it was $258,766,000.

    a) Find the average rate of change of total revenue from 1979 to 1982.

    b) Find the average rate of change of total revenue from 1982 to 1986.

11. *Sociology: Marriage rate.* It is known that in 1960 there were 1,523,000 marriages. In 1982 there were 2,495,000 marriages. Find the average rate of change of the number of marriages with respect to time. This is called an *average marriage rate.*

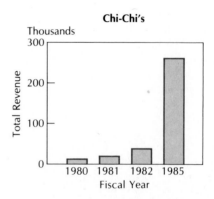

For the functions in each of Exercises 14–25, (a) find a simplified difference quotient; (b) complete the table shown here.

14. $f(x) = 5x^2$

15. $f(x) = 7x^2$

16. $f(x) = -5x^2$

17. $f(x) = -7x^2$

18. $f(x) = 5x^3$

19. $f(x) = 7x^3$

20. $f(x) = \dfrac{4}{x}$

21. $f(x) = \dfrac{5}{x}$

22. $f(x) = 2x + 3$

23. $f(x) = -2x + 5$

24. $f(x) = x^2 + x$

25. $f(x) = x^2 - x$

x	h	$\dfrac{f(x + h) - f(x)}{h}$
4	2	
4	1	
4	0.1	
4	0.01	

CSS

EXTENSION EXERCISES

Find the simplified difference quotient.

26. $f(x) = mx + b$

27. $f(x) = ax^2 + bx + c$

28. $f(x) = ax^3 + bx^2$

29. $f(x) = \sqrt{x}$

30. $f(x) = x^4$

31. $f(x) = \dfrac{1}{x^2}$

32. $f(x) = \dfrac{1}{1 - x}$

33. $f(x) = \dfrac{x}{1 + x}$

2.4

DIFFERENTIATION USING LIMITS

OBJECTIVE

a) Given a formula of a function, find a formula for its derivative, and then find various values of the derivative.

Tangent Lines

A line tangent to a circle is a line that touches the circle exactly once.

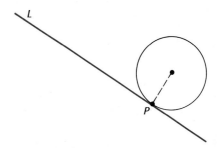

This definition becomes unworkable with other curves. For example, consider the following curve. Line *L* touches the curve at point *P* but meets the curve at other places. It is considered a tangent line, but "touching at one point" cannot be its definition.

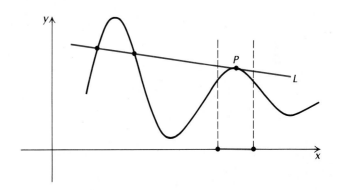

Note in the preceding figure that over a small interval containing *P*, line *L* does touch the curve exactly once. This is still not a suitable definition of a tangent line because it allows a line like *M* in the following figure to be a tangent line, which we will not accept.

The path of the loose ski follows a tangent line to the ski chute at the point where it breaks loose.

1. a) Which of the following appear to be tangent lines?

b) Below is a graph of $y = x^2$. Tangent lines are drawn at various points on the graph. Let $m(x) =$ the slope at the point $(x, f(x))$.

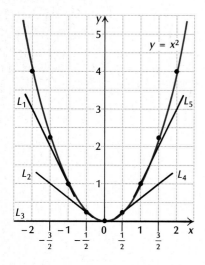

(Exercise 1 is continued at the top of the following page.)

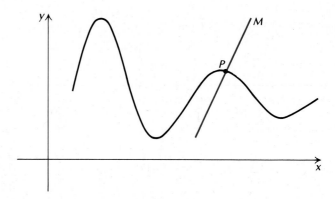

Later we will give a definition of a tangent line, but for now we will rely on intuition. In the following figure, lines L_1 and L_2 are not tangents. All the others are tangent lines.

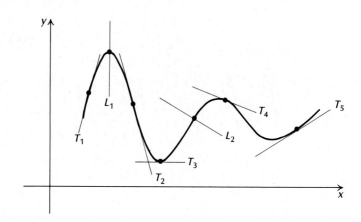

DO EXERCISE 1.

Why Do We Study Tangent Lines?

The reason for our study of tangent lines will become evident in Chapter 3. For now, look at the following graph of a total-profit function.

Note that the largest (or maximum) value of the function occurs where the graph has a horizontal tangent; that is, where the tangent line has slope 0.

Estimate the slope of each line and complete this table.

Lines	x	$m(x)$
L_1	-1	
L_2	$-\frac{1}{2}$	
L_3	0	
L_4	$\frac{1}{2}$	
L_5	1	

c) Derive a formula for $m(x)$.

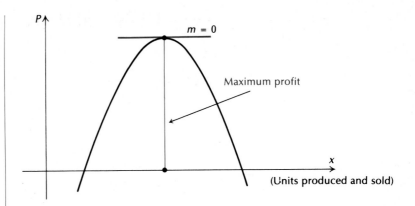

Differentiation Using Limits

We will define *tangent line* in such a way that it makes sense for *any* curve. To do this we use the notion of a limit.

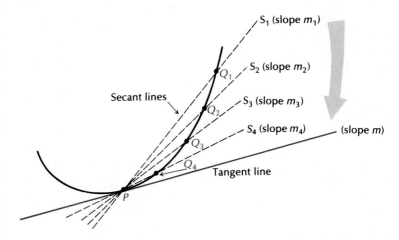

In the figure above, we obtain the line tangent to the curve at point P by considering secant lines through P and neighboring points Q_1, Q_2, and so on. As the points Q approach P, the secant lines approach the tangent line. Each secant has a slope. The slopes of the secant lines approach the slope of the tangent line. In fact, we *define* the *tangent line* as the line that contains the point P and has slope m, where m is the limit of the slopes of the secant lines as the points Q approach P.

How might we calculate the limit m?

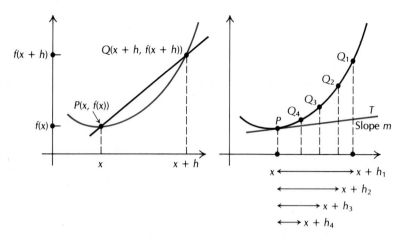

Suppose P has coordinates $(x, f(x))$. Then the first coordinate of Q is x plus some number h, or $x + h$. The coordinates of Q are $(x + h, f(x + h))$. From Section 2.3, we know that the slope of the secant line $\overrightarrow{PQ}$ is given by the difference quotient

$$\frac{f(x + h) - f(x)}{h}.$$

Now, as we see in the figure on the right above, as the points Q approach P, $x + h$ approaches x. That is, h approaches 0. Thus we have the following.

The slope of the tangent line $= m = \lim_{h \to 0} \frac{f(x + h) - f(x)}{h}.$

The formal definition of the *derivative of a function f* can now be given. We will designate the derivative at x as $f'(x)$, rather than $m(x)$.

DEFINITION

For a function $y = f(x)$, its *derivative* at x is defined as

$$f'(x) = \lim_{h \to 0} \frac{f(x + h) - f(x)}{h},$$

provided the limit exists. If $f'(x)$ exists, we say that f is *differentiable* at x.

This is the basic definition of *differential calculus*.

Let us now calculate some formulas for derivatives. That is, given a formula for a function f, we will be trying to find a formula for f'.

"Nothing in this world is so powerful as an idea whose time has come."

Victor Hugo

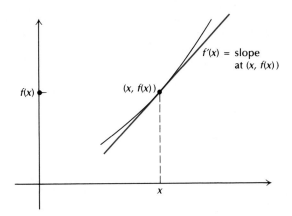

There are three steps in calculating a derivative.

1. Write down the difference quotient $[f(x + h) - f(x)]/h$.
2. Simplify the difference quotient.
3. Find the limit as h approaches 0.

A formula for the derivative of a linear function

$$f(x) = mx + b$$

is

$$f'(x) = m.$$

Let us verify this using the definition.

Example 1   For $f(x) = mx + b$, find $f'(x)$.

Solution   We have

1. $$\frac{f(x + h) - f(x)}{h} = \frac{[m(x + h) + b] - (mx + b)}{h}$$

2. $$= \frac{mx + mh + b - mx - b}{h}$$

$$= \frac{mh}{h} = m, \quad h \neq 0$$

3. $$\lim_{h \to 0} \frac{f(x + h) - f(x)}{h} = \lim_{h \to 0} m = m,$$

since m represents a constant. Thus,

$$f'(x) = m.$$

2. For $f(x) = x^2$, find $f'(5)$ using the definition of a derivative.

In Margin Exercise 1 you may have conjectured that the function

$$f(x) = x^2$$

has derivative

$$f'(x) = 2x.$$

This would mean that the tangent line at $x = 4$ has slope $f'(4) = 8$. Let us verify this particular case and then verify the general formula.

Example 2   For $f(x) = x^2$, find $f'(4)$.   CSS

Solution   We have

1. $\dfrac{f(4 + h) - f(4)}{h} = \dfrac{(4 + h)^2 - 4^2}{h}$

2. $\dfrac{f(4 + h) - f(4)}{h} = \dfrac{16 + 8h + h^2 - 16}{h} = \dfrac{8h + h^2}{h}$

    $\qquad\qquad = \dfrac{h(8 + h)}{h} = 8 + h, \quad h \neq 0$

3. $\displaystyle\lim_{h \to 0} \dfrac{f(4 + h) - f(4)}{h} = \lim_{h \to 0} (8 + h) = 8.$

Thus $f'(4) = 8$.

DO EXERCISE 2.

Example 3   For $f(x) = x^2$, find (the general formula) $f'(x)$.

Solution

1. We have

    $$\frac{f(x + h) - f(x)}{h} = \frac{(x + h)^2 - x^2}{h}.$$

2. In Example 5 of Section 2.3, we showed how this difference quotient can be simplified to

    $$\frac{f(x + h) - f(x)}{h} = 2x + h.$$

3. We want to find

    $$\lim_{h \to 0} \frac{f(x + h) - f(x)}{h} = \lim_{h \to 0} (2x + h).$$

3. For $f(x) = 4x^2$, find $f'(x)$. Then find $f'(5)$ and interpret the meaning.

As $h \longrightarrow 0$, we see that $2x + h \longrightarrow 2x$. Thus,

$$\lim_{h \to 0} (2x + h) = 2x,$$

and we have

$$f'(x) = 2x,$$

which tells us, for example, that at $x = -3$, the curve has a tangent line whose slope is

$$f'(-3) = 2(-3), \quad \text{or} \ -6.$$

We may say, simply, "The curve has slope -6."

DO EXERCISE 3.

Example 4   For $f(x) = x^3$, find $f'(x)$. Then find $f'(-1)$ and $f'(10)$.

Solution

1.   We have

$$\frac{f(x + h) - f(x)}{h} = \frac{(x + h)^3 - x^3}{h}.$$

4. For $f(x) = 4x^3$, find $f'(x)$. Then find $f'(-5)$ and $f'(0)$.

2.   In Example 6 of Section 2.3, we showed how this difference quotient can be simplified to

$$\frac{f(x + h) - f(x)}{h} = 3x^2 + 3xh + h^2.$$

3.   We then have

$$\lim_{h \to 0} \frac{f(x + h) - f(x)}{h} = \lim_{h \to 0} (3x^2 + 3xh + h^2) = 3x^2.$$

An input–output table for this is shown in Example 4 of Section 2.2. Thus, for $f(x) = x^3$, we have $f'(x) = 3x^2$. Then

$$f'(-1) = 3(-1)^2 = 3 \quad \text{and} \quad f'(10) = 3(10)^2 = 300.$$

DO EXERCISE 4.

Example 5   For $f(x) = 3/x$, find $f'(x)$. Then find $f'(1)$ and $f'(2)$.

Solution

1.   We have

$$\frac{f(x + h) - f(x)}{h} = \frac{[3/(x + h)] - (3/x)}{h}.$$

5. For

$$f(x) = \frac{1}{x},$$

find $f'(x)$. Then find $f'(-10)$ and $f'(-2)$.

2. In Example 7 of Section 2.3, we showed that this difference quotient can be simplified to

$$\frac{f(x + h) - f(x)}{h} = \frac{-3}{x(x + h)}.$$

3. We want to find

$$\lim_{h \to 0} \frac{f(x + h) - f(x)}{h} = \lim_{h \to 0} \frac{-3}{x(x - h)}.$$

As $h \longrightarrow 0$, $x + h \longrightarrow x$, so we have

$$f'(x) = \lim_{h \to 0} \frac{-3}{x(x + h)} = \frac{-3}{x^2}.$$

Then

$$f'(1) = \frac{-3}{1^2} = -3 \quad \text{and} \quad f'(2) = \frac{-3}{2^2} = -\frac{3}{4}.$$

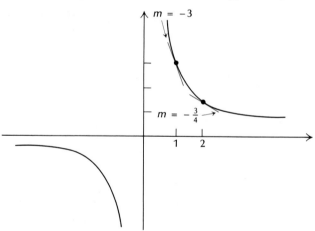

Note that $f'(0)$ does not exist because $f(0)$ does not exist. We say, "f is not differentiable at 0." When a function is not defined at a point, it is not differentiable at that point. In fact, if a function is discontinuous at a point, it is not differentiable at that point.

DO EXERCISE 5.

It can happen that a function f is defined and continuous at a point but that its derivative f' is not. The function f given by

$$f(x) = |x|$$

is an example. Note that

$$f(0) = |0| = 0,$$

so the function is defined at 0.

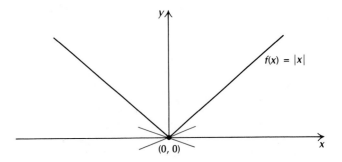

Suppose we tried to draw a tangent line at $(0, 0)$. A function like this with a corner (not smooth) would seem to have many tangent lines at $(0, 0)$, and thus many slopes. The derivative at such a point would not be unique. Let us try to calculate the derivative at 0.

Since

$$f'(x) = \lim_{h \to 0} \frac{|x + h| - |x|}{h},$$

at $x = 0$, we have

$$f'(0) = \lim_{h \to 0} \frac{|0 + h| - |0|}{h} = \lim_{h \to 0} \frac{|h|}{h}.$$

h	$\dfrac{|h|}{h}$	h	$\dfrac{|h|}{h}$
2	$\dfrac{|2|}{2}$, or $\dfrac{2}{2}$, or 1	-2	$\dfrac{|-2|}{-2}$, or $\dfrac{2}{-2}$, or -1
1	1	-1	-1
0.1	1	-0.1	-1
0.01	1	-0.01	-1
0.001	1	-0.001	-1

6. List the points at which the following function is not differentiable.

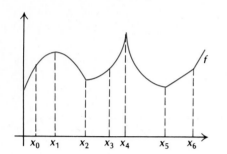

Look at the input–output tables. Note that as h approaches 0 from the right, $|h|/h$ approaches 1, but as h approaches 0 from the left, $|h|/h$ approaches -1. Thus,

$$\lim_{h \to 0} \frac{|h|}{h} \quad \text{does not exist,}$$

so

$$f'(0) \text{ does not exist.}$$

If a function has a "sharp point" or "corner," it will not have a derivative at that point.

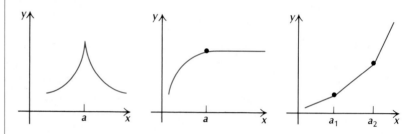

DO EXERCISE 6.

A function may also fail to be differentiable at a point by having a vertical tangent at that point. The following function has a vertical tangent at point a.

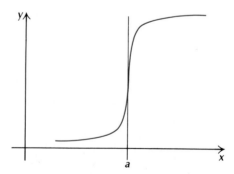

Recall that slopes of vertical lines are undefined; thus there is no derivative at such a point.

The function $f(x) = |x|$ illustrates that a function may be continuous at each point in an interval I but may not be differentiable at each point in I. That is, continuity does not imply differentiability. On the other hand, if we know that a function is differentiable at each point in an interval I, then it is continuous over I. That is, if $f'(a)$ exists, then f is continuous at a. The function $f(x) = x^2$ is an example. Also, if a function is discontinuous at some point a, then it is not differentiable at a. Thus when we know a function is differentiable over an interval, it is *smooth* in the sense that there are no "sharp points," "corners," or "breaks" in the graph.

EXERCISE SET 2.4

For each function, find $f'(x)$. Then find $f'(-2), f'(-1), f'(0), f'(1),$ and $f'(2)$, if they exist. CSS

1. $f(x) = 5x^2$
2. $f(x) = 7x^2$
3. $f(x) = -5x^2$
4. $f(x) = -7x^2$

5. $f(x) = 5x^3$
6. $f(x) = 7x^3$
7. $f(x) = 2x + 3$
8. $f(x) = -2x + 5$

9. $f(x) = -4x$
10. $f(x) = \frac{1}{2}x$
11. $f(x) = x^2 + x$
12. $f(x) = x^2 - x$

13. $f(x) = \dfrac{4}{x}$
14. $f(x) = \dfrac{5}{x}$
15. $f(x) = mx$
16. $f(x) = ax^2 + bx + c$

17. List the points in the graph shown here at which the function is not differentiable.

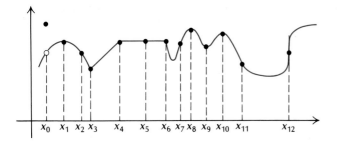

18. *The postage function.* Consider the postage function defined in Exercise Set 2.1. At what values is the function not differentiable?

19. Consider the function f given by
$$f(x) = \frac{x^2 - 9}{x + 3}.$$
For what values is this function not differentiable?

EXTENSION EXERCISES

Find $f'(x)$.

20. $f(x) = x^4$

21. $f(x) = \dfrac{1}{x^2}$

22. $f(x) = \dfrac{1}{1 - x}$

23. $f(x) = \dfrac{x}{1 + x}$

24. $f(x) = \sqrt{x}\left(\text{Multiply by 1, using } \dfrac{\sqrt{x + h} + \sqrt{x}}{\sqrt{x + h} + \sqrt{x}}.\right)$

2.5

a) Differentiate using the Power Rule, the Sum–Difference Rule, or the rule for differentiating a constant or a constant times a function.

b) Find the points on the graph of a function at which the tangent line has a given slope.

DIFFERENTIATION TECHNIQUES: THE POWER AND SUM–DIFFERENCE RULES

Leibniz's Notation

When y is a function of x, we will also designate the derivative, $f'(x)$, as*

$$\frac{dy}{dx},$$

which is read "the derivative of y with respect to x." This notation was invented by the German mathematician Leibniz. It does *not* mean dy divided by dx! (That is, we cannot interpret dy/dx as a quotient until meanings are given to dy and dx, which we will not do here.) For example, if $y = x^2$, then

$$\frac{dy}{dx} = 2x.$$

We may also write

$$\frac{d}{dx} f(x)$$

to denote the derivative of f with respect to x. For example,

$$\frac{d}{dx} x^2 = 2x.$$

The value of dy/dx when $x = 5$ can be denoted by

$$\left.\frac{dy}{dx}\right|_{x=5}$$

Thus for $dy/dx = 2x$,

$$\left.\frac{dy}{dx}\right|_{x=5} = 2 \cdot 5, \quad \text{or } 10.$$

In general, for $y = f(x)$,

$$\left.\frac{dy}{dx}\right|_{x=a} = f'(a).$$

Historical Note: The German mathematician and philosopher Gottfried Wilhelm von Leibniz (1646–1716) and the English mathematician, philosopher, and physicist Sir Isaac Newton (1642–1727) are both credited with the invention of the calculus, though each made the invention independent of the other. Newton used the dot notation $\dot{y}$ for dy/dt, where y is a function of time, and this notation is still used, though it is not as prevalent as Leibniz's notation.

---

* The notation $D_x y$ is also used.

1. For

$$y = x^3,$$

use the results of previous work to find each of the following.

a) $\dfrac{dy}{dx}$

b) $\dfrac{d}{dx} x^3$

c) $\dfrac{dy}{dx}\Big|_{x=4}$

DO EXERCISE 1.

The Power Rule

In the remainder of this section we will develop rules and techniques for efficient differentiation.

Look for a pattern in this table, which contains functions and derivatives that we have found in previous work.

Function	Derivative
x^2	$2x^1$
x^3	$3x^2$
x^{-1}, or $\dfrac{1}{x}$	$-1 \cdot x^{-2}$, or $-\dfrac{1}{x^2}$

Perhaps you have discovered the following.

THEOREM 1

The Power Rule

For any real number a,

$$\frac{d}{dx} x^a = a \cdot x^{a-1}.$$

Note that this rule holds no matter what the exponent. That is, to differentiate x^a, we write the exponent a as the coefficient, followed by x with an exponent 1 less than a.

①   ① Write the exponent as the coefficient.

$$\begin{array}{l} \rule{0pt}{12pt}\llcorner\!\!\!-\!\!-x^a \\ \rule{0pt}{12pt} \quad\quad ② \\ \rule{0pt}{12pt}\downarrow \\ a \cdot x^{a-1} \end{array}$$

② Subtract 1 from the exponent.

Example 1 $\dfrac{d}{dx} x^5 = 5x^4$

Find $\dfrac{dy}{dx}$ (differentiate).

2. $y = x^6$

Example 2 $\dfrac{d}{dx} x = 1 \cdot x^{1-1}$

$$= 1 \cdot x^0 = 1$$

Example 3 $\dfrac{d}{dx} x^{-4} = -4 \cdot x^{-4-1}$

$$= -4x^{-5}, \quad \text{or} \quad -4 \cdot \dfrac{1}{x^5}, \quad \text{or} \quad -\dfrac{4}{x^5}$$

The Power Rule allows us to differentiate $\sqrt{x}$.

3. $y = x^{-7}$

Example 4 $\dfrac{d}{dx} \sqrt{x} = \dfrac{d}{dx} x^{1/2} = \dfrac{1}{2} \cdot x^{(1/2)-1}$

$$= \dfrac{1}{2} x^{-1/2}, \quad \text{or} \quad \dfrac{1}{2} \cdot \dfrac{1}{x^{1/2}}, \quad \text{or} \dfrac{1}{2} \cdot \dfrac{1}{\sqrt{x}}, \quad \text{or} \dfrac{1}{2\sqrt{x}}$$

Example 5 $\dfrac{d}{dx} x^{-2/3} = -\dfrac{2}{3} x^{(-2/3)-1}$

$$= -\dfrac{2}{3} x^{-5/3}, \quad \text{or} \quad -\dfrac{2}{3} \dfrac{1}{x^{5/3}}, \quad \text{or} \quad -\dfrac{2}{3\sqrt[3]{x^5}}$$

4. $y = \sqrt[3]{x}$

DO EXERCISES 2–5.

The Derivative of a Constant Function

Look at the graph of the constant function $F(x) = c$. What is the slope at each point on the graph?

5. $y = x^{-1/4}$

We now have the following.

6. Find $g'(x)$ if $g(x) = -14$.

THEOREM 2   CSS

The derivative of a constant function is 0.

Proof.   Let F be the function given by $F(x) = c$. Then

$$\frac{F(x + h) - F(x)}{h} = \frac{c - c}{h} = \frac{0}{h} = 0.$$

The difference quotient is always 0. Thus, as h approaches 0, the limit of the difference quotient approaches 0, so $F'(x) = 0$.

DO EXERCISE 6.

The Derivative of a Constant Times a Function

Now let us consider differentiating functions such as

$$f(x) = 5x^2 \quad \text{and} \quad g(x) = -7x^4.$$

Note that we already know how to differentiate x^2 and x^4. Let us again look for a pattern in the results of Exercise Set 2.3.

Function	Derivative
$5x^2$	$10x$
$-4x$	-4
$-7x^2$	$-14x$
$5x^3$	$15x^2$

Perhaps you have discovered the following.

THEOREM 3

The derivative of a constant times a function is the constant times the derivative of the function. Using derivative notation, we can write this as

$$\frac{d}{dx}[c \cdot f(x)] = c \cdot f'(x).$$

Find $\dfrac{dy}{dx}$.

7. $y = 5x^{20}$

8. $y = -\dfrac{3}{x}$

9. $y = -8\sqrt{x}$

10. $y = 0.16x^{6.25}$

Proof. Let F be the function given by $F(x) = cf(x)$. Then

$$\frac{F(x + h) - F(x)}{h} = \frac{cf(x + h) - cf(x)}{h} = c\left[\frac{f(x + h) - f(x)}{h}\right].$$

As h approaches 0, the limit on the right is the same as c times $f'(x)$. Thus $F'(x) = cf'(x)$.

Combining this rule with the Power Rule allows us to find many derivatives.

Example 6 $\dfrac{d}{dx} 5x^4 = 5\dfrac{d}{dx} x^4$

$$= 5 \cdot 4 \cdot x^{4-1} = 20x^3$$

Example 7 $\dfrac{d}{dx}(-9x) = -9\dfrac{d}{dx} x$

$$= -9 \cdot 1 = -9$$

With practice you will be able to differentiate many such functions in one step.

Example 8 $\dfrac{d}{dx}\dfrac{-4}{x^2} = \dfrac{d}{dx}(-4x^{-2}) = -4 \cdot \dfrac{d}{dx} x^{-2}$

$$= -4(-2)x^{-2-1}$$

$$= 8x^{-3}, \quad \text{or } \frac{8}{x^3}$$

Example 9 $\dfrac{d}{dx}(-x^{0.7}) = -1 \cdot \dfrac{d}{dx} x^{0.7}$

$$= -1 \cdot 0.7 \cdot x^{0.7-1}$$

$$= -0.7x^{-0.3}$$

DO EXERCISES 7–10.

The Derivative of a Sum or Difference

In Exercise 11 of Exercise Set 2.4, you found that the derivative of

$$f(x) = x^2 + x$$

is

$$f'(x) = 2x + 1.$$

Note that the derivative of x^2 is $2x$ and the derivative of x is 1; and the sum of these derivatives is $f'(x)$. This illustrates the following.

THEOREM 4

The Sum–Difference Rule

i)   The derivative of a sum is the sum of the derivatives:

$$\text{If } F(x) = f(x) + g(x), \quad \text{then } F'(x) = f'(x) + g'(x).$$

ii)   The derivative of a difference is the difference of the derivatives:

$$\text{If } F(x) = f(x) - g(x), \quad \text{then } F'(x) = f'(x) - g'(x).$$

Proof.   For (i), we have

$$\frac{F(x+h) - F(x)}{h} = \frac{[f(x+h) + g(x+h)] - [f(x) + g(x)]}{h}$$

$$= \frac{f(x+h) - f(x)}{h} + \frac{g(x+h) - g(x)}{h}.$$

As h approaches 0, the two terms on the right approach $f'(x)$ and $g'(x)$, respectively, so their sum approaches $f'(x) + g'(x)$. Thus $F'(x) = f'(x) + g'(x)$.

The proof of (ii) is similar.

Any function that is a sum or difference of several terms can be differentiated term by term.

Example 10 $\dfrac{d}{dx}(3x + 7) = \dfrac{d}{dx}(3x) + \dfrac{d}{dx}(7)$

$$= 3\,\frac{d}{dx}x + 0 = 3 \cdot 1 = 3$$

Example 11 $\dfrac{d}{dx}(5x^3 - 3x^2) = \dfrac{d}{dx}(5x^3) - \dfrac{d}{dx}(3x^2)$

$$= 5\,\frac{d}{dx}x^3 - 3\,\frac{d}{dx}x^2$$

$$= 5 \cdot 3x^2 - 3 \cdot 2x$$

$$= 15x^2 - 6x$$

Find $\dfrac{dy}{dx}$ (differentiate).

11. $y = -\frac{1}{4}x - 9$

12. $y = 7x^4 + 6x^2$

13. $y = 15x^2 + \dfrac{4}{x} + \sqrt{x}$

Example 12

$$\frac{d}{dx}\left(24x - \sqrt{x} + \frac{2}{x}\right) = \frac{d}{dx}(24x) - \frac{d}{dx}(\sqrt{x}) + \frac{d}{dx}\left(\frac{2}{x}\right)$$

$$= 24 \cdot \frac{d}{dx}x - \frac{d}{dx}x^{1/2} + 2 \cdot \frac{d}{dx}x^{-1}$$

$$= 24 \cdot 1 - \frac{1}{2}x^{(1/2)-1} + 2(-1)x^{-1-1}$$

$$= 24 - \frac{1}{2}x^{-1/2} - 2x^{-2}$$

$$= 24 - \frac{1}{2\sqrt{x}} - \frac{2}{x^2}$$

DO EXERCISES 11–13.

A word of caution! The derivative of

$$f(x) + c,$$

a function plus a constant, is just the derivative of the function

$$f'(x).$$

The derivative of

$$c \cdot f(x),$$

a function times a constant, is the constant times the derivative

$$c \cdot f'(x).$$

That is, for a product the constant is retained, but for a sum it is not.

It is important to be able to determine points at which the tangent line to a curve has a certain slope—that is, points at which the derivative attains a certain value.

Example 13 Find the points on the graph of $y = -x^3 + 6x^2$ at which the tangent line is horizontal.

Solution A horizontal tangent line has slope 0. Thus we seek the values of x for which $dy/dx = 0$. That is, we want to find x such that

$$-3x^2 + 12x = 0.$$

We factor and solve:

$$x(-3x + 12) = 0$$

$$x = 0 \quad \text{or} \quad -3x + 12 = 0$$

$$x = 0 \quad \text{or} \quad -3x = -12$$

$$x = 0 \quad \text{or} \quad x = 4.$$

14. Find the points on the graph of

$$y = \tfrac{1}{3}x^3 - 2x^2 + 4x$$

at which the tangent line is horizontal.

CSS

We are to find the points *on the graph,* so we must determine the second coordinates from the original equation $y = -x^3 + 6x^2$.

For $x = 0$, $y = -0^3 + 6 \cdot 0^2 = 0$.

For $x = 4$, $y = -(4)^3 + 6 \cdot 4^2 = -64 + 96 = 32$.

Thus the points we are seeking are $(0, 0)$ and $(4, 32)$.

DO EXERCISE 14.

Example 14 Find the points on the graph of $y = -x^3 + 6x^2$ at which the tangent line has slope 6.

Solution We want to find values of x for which $dy/dx = 6$. That is, we want to find x such that

$$-3x^2 + 12x = 6.$$

To solve, we add -6 and get

$$-3x^2 + 12x - 6 = 0.$$

We can simplify this equation by multiplying by $-\tfrac{1}{3}$ since each term has a common factor of -3. We get

$$x^2 - 4x + 2 = 0.$$

This is a quadratic equation, not readily factorable, so we use the Quadratic Formula, where $a = 1$, $b = -4$, and $c = 2$:

$$x = \frac{-b \pm \sqrt{b^2 - 4ac}}{2a} = \frac{-(-4) \pm \sqrt{(-4)^2 - 4 \cdot 1 \cdot 2}}{2 \cdot 1} = \frac{4 \pm \sqrt{8}}{2}$$

$$= \frac{2 \cdot 2 \pm 2\sqrt{2}}{2 \cdot 1}$$

$$= \frac{2}{2} \cdot \frac{2 \pm \sqrt{2}}{1} = 2 \pm \sqrt{2}.$$

The solutions are $2 + \sqrt{2}$ and $2 - \sqrt{2}$. We determine the second coordinates from the original equation. For $x = 2 + \sqrt{2}$,

$$y = -(2 + \sqrt{2})^3 + 6(2 + \sqrt{2})^2$$

$$= -[(2 + \sqrt{2})^2(2 + \sqrt{2})] + 6(4 + 4\sqrt{2} + 2)$$

$$= -[(6 + 4\sqrt{2})(2 + \sqrt{2})] + 6(6 + 4\sqrt{2})$$

$$= -[12 + 6\sqrt{2} + 8\sqrt{2} + 8] + 36 + 24\sqrt{2}$$

$$= -[20 + 14\sqrt{2}] + 36 + 24\sqrt{2}$$

$$= -20 - 14\sqrt{2} + 36 + 24\sqrt{2} = 16 + 10\sqrt{2}.$$

15. Find the points on the graph of

$$y = \tfrac{1}{3}x^3 - 2x^2 + 4x$$

at which the tangent line has slope 3.

Similarly, for $x = 2 - \sqrt{2}$,

$$y = 16 - 10\sqrt{2}.$$

Thus the points we are seeking are $(2 + \sqrt{2}, \; 16 + 10\sqrt{2})$ and $(2 - \sqrt{2}, \; 16 - 10\sqrt{2})$.

DO EXERCISE 15.

We illustrate the results of Examples 13 and 14 in the following graph. You will not be asked to sketch such graphs at this time.

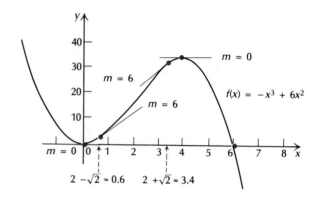

EXERCISE SET 2.5

Find $\dfrac{dy}{dx}$.

1. $y = x^7$	**2.** $y = x^8$	**3.** $y = 15$	**4.** $y = 78$
5. $y = 4x^{150}$	**6.** $y = 7x^{200}$	**7.** $y = x^3 + 3x^2$	**8.** $y = x^4 - 7x$
9. $y = 8\sqrt{x}$	**10.** $y = 4\sqrt{x}$	**11.** $y = x^{0.07}$	**12.** $y = x^{0.78}$
13. $y = \tfrac{1}{2}x^{4/5}$	**14.** $y = -4.8x^{1/3}$	**15.** $y = x^{-3}$	**16.** $y = x^{-4}$
17. $y = 3x^2 - 8x + 7$	**18.** $y = 4x^2 - 7x + 5$	**19.** $y = \sqrt[4]{x} - \dfrac{1}{x}$	**20.** $y = \sqrt[5]{x} - \dfrac{2}{x}$

Find $f'(x)$.

21. $f(x) = 0.64x^{2.5}$ **22.** $f(x) = 0.32x^{12.5}$

23. $f(x) = \dfrac{5}{x} - x$

24. $f(x) = \dfrac{4}{x} - x$

25. $f(x) = 4x - 7$

26. $f(x) = 7x + 11$

27. $f(x) = 4x + 9$

28. $f(x) = 7x - 14$

29. $f(x) = \dfrac{x^4}{4}$

30. $f(x) = \dfrac{x^3}{3}$

31. $f(x) = -0.01x^2 - 0.5x + 70$

32. $f(x) = -0.01x^2 + 0.4x + 50$

33. $f(x) = 3x^{-2/3} + x^{3/4} + x^{6/5} + \dfrac{8}{x^3}$

34. $f(x) = x^{-3/4} - 3x^{2/3} + x^{5/4} + \dfrac{2}{x^4}$

For each function, find the points on the graph at which the tangent line is horizontal.   CSS

35. $y = x^2$

36. $y = -x^2$

37. $y = -x^3$

38. $y = x^3$

39. $y = 3x^2 - 5x + 4$

40. $y = 5x^2 - 3x + 8$

41. $y = -0.01x^2 - 0.5x + 70$

42. $y = -0.01x^2 + 0.4x + 50$

43. $y = 2x + 4$

44. $y = -2x + 5$

45. $y = 4$

46. $y = -3$

47. $y = -x^3 + x^2 + 5x - 1$

48. $y = -\frac{1}{3}x^3 + 6x^2 - 11x - 50$

49. $y = \frac{1}{3}x^3 - 3x + 2$

50. $y = x^3 - 6x + 1$

For each function, find the points on the graph at which the tangent line has slope 1.

51. $y = 20x - x^2$

52. $y = 6x - x^2$

53. $y = -0.025x^2 + 4x$

54. $y = -0.01x^2 + 2x$

55. $y = \frac{1}{3}x^3 + 2x^2 + 2x$

56. $y = \frac{1}{3}x^3 - x^2 - 4x + 1$

EXTENSION EXERCISES

57. Find the points on the graph of

$$y = x^4 - \tfrac{4}{3}x^2 - 4$$

at which the tangent line is horizontal.

58. Find the points on the graph of

$$y = 2x^6 - x^4 - 2$$

at which the tangent line is horizontal

Find dy/dx. Each of the following can be differentiated using the rules developed in this section, but some algebra may be required beforehand.

59. $y = x(x - 1)$

60. $y = (x - 1)(x + 1)$

61. $y = (x - 2)(x + 3)$

62. $y = \dfrac{5x^2 - 8x + 3}{8}$

63. $y = \dfrac{x^5 + x}{x^2}$

64. $y = (5x)^2$

65. $y = (-4x)^3$

66. $y = \sqrt{7x}$

67. $y = \sqrt[3]{8x}$

68. $y = (x - 3)^2$

69. $y = (x + 1)^3$

70. $y = (x - 2)^3(x + 1)$

71. Prove Theorem 4(ii).

OBJECTIVES

a) Given a distance function $s(t)$, find a formula for the velocity $v(t)$ and the acceleration $a(t)$, and evaluate $s(t)$, $v(t)$, and $a(t)$ for given values of t.

b) Given y as a function of x, find the rate of change of y with respect to x, and evaluate this rate of change for values of x.

An instantaneous velocity.

2.6

APPLICATIONS AND RATES OF CHANGE

Instantaneous Rate of Change

A car travels 110 miles in 2 hours. Its *average speed* (or velocity) is 110 mi/2 hr, or 55 mi/hr. This is the *average rate of change* of distance with respect to time. At various times during the trip the speedometer did not read 55, however. Thus we say that 55 is the *average*. A snapshot of the speedometer taken at any instant would indicate *instantaneous speed,* or rate of change.

Average rates of change are given by difference quotients. If distance s is a function of time t, then average velocity is given by

$$\text{Average velocity} = \frac{s(t+h) - s(t)}{h}.$$

Instantaneous rates of change are found by letting h approach 0. Thus,

$$\text{Instantaneous velocity} = \lim_{h \to 0} \frac{s(t+h) - s(t)}{h} = s'(t).$$

Example 1 An object travels in such a way that distance s (in miles) from the starting point is a function of time t (in hours) as follows:

$$s(t) = 10t^2.$$

a) Find the average velocity between times $t = 2$ and $t = 5$.

b) Find the (instantaneous) velocity when $t = 4$.

1. For the function in Example 1: CSS
   a) find the average velocity between the times $t = 1$ and $t = 6$;
   b) find the (instantaneous) velocity when $t = 5$.

2. An object is dropped from a certain height. It will fall downward a distance of s feet in t seconds as given by
$$s(t) = 16t^2.$$
   a) Find the velocity $v(t)$.
   b) Find the object's velocity 2 sec after it has been dropped.
   c) Find its velocity 10 sec after it has been dropped.

3. Referring to Margin Exercise 2, find $a(t)$. In what units should it be expressed?

Some of the newer automobiles describe fuel economy by giving *average* miles per gallon and *instantaneous* miles per gallon.

Solution

a) From $t = 2$ to $t = 5$, $h = 3$, so
$$\frac{s(t + h) - s(t)}{h} = \frac{s(2 + 3) - s(2)}{3} = \frac{s(5) - s(2)}{3}$$
$$= \frac{10 \cdot 5^2 - 10 \cdot 2^2}{3} = \frac{250 - 40}{3}$$
$$= \frac{210}{3} = 70 \, \frac{\text{mi}}{\text{hr}}.$$

b) The instantaneous velocity $= s'(t) = 20t$. Thus,
$$s'(4) = 20 \cdot 4 = 80 \, \frac{\text{mi}}{\text{hr}}.$$

DO EXERCISE 1.

We generally use the letter v for velocity. Thus we have
$$v(t) = \lim_{h \to 0} \frac{s(t + h) - s(t)}{h} = s'(t).$$

The rate of change of velocity is called *acceleration*. We generally use the letter a for acceleration. Thus,
$$\text{Acceleration} = a(t) = v'(t).$$

Example 2 For $s(t) = 10t^2$, find $v(t)$ and $a(t)$.

Solution We have
$$v(t) = s'(t) = 20t,$$
$$a(t) = v'(t) = 20.$$

For an automobile we may give the velocity in *miles per hour,* and the acceleration, which is the change in velocity per unit time, in (*miles per hour*) *per hour.* We abbreviate this as mi/hr^2. Thus in Example 2, the acceleration is a constant, 20 mi/hr^2.

DO EXERCISES 2 AND 3.

In general, derivatives give instantaneous rates of change.

4. The volume V of a cubical carton with a side of length s, in feet, is given by

$$V = s^3.$$

a) Find the rate of change of the volume V with respect to the length s of a side.

b) Find the rate of change of the volume when $s = 10$ ft.

DEFINITION

If y is a function of x, then the (instantaneous) *rate of change of y with respect to x* is given by the derivative

$$\frac{dy}{dx}, \quad \text{or } f'(x).$$

Example 3   The spherical volume V of a tumor is given by

$$V = \tfrac{4}{3}\pi r^3,$$

where r is the radius of the tumor, in centimeters.

a) Find the rate of change of the volume with respect to the radius.

b) Find the rate of change of the volume at $r = 1.2$ cm.

Solution

a) $\dfrac{dV}{dr} = V'(r) = \dfrac{4}{3} \cdot 3 \cdot \pi r^2 = 4\pi r^2$

   (This turns out to be the surface area.)

b) $V'(1.2) = 4\pi(1.2)^2 = 5.76\pi \approx 18 \dfrac{\text{cm}^3}{\text{cm}} = 18 \text{ cm}^2$

DO EXERCISE 4.

Example 4   The initial population in a bacteria colony is 10,000. After t hours the colony grows to a number $P(t)$ given by

$$P(t) = 10{,}000(1 + 0.86t + t^2).$$

a) Find the rate of change of the population P with respect to time t. This is also known as the *growth rate*.

b) Find the number of bacteria present after 5 hours. Also find the growth rate when $t = 5$.

Solution

a) Note that $P(t) = 10{,}000 + 8600t + 10{,}000t^2$. Then

$$P'(t) = 8600 + 20{,}000t.$$

b) The number of bacteria present when $t = 5$ is given by

$$P(5) = 10{,}000 + 8600 \cdot 5 + 10{,}000 \cdot 5^2 = 303{,}000.$$

5. The initial population of a bacteria colony is 10,000. After t hours the colony grows to a number $P(t)$ given by

$$P(t) = 10,000(1 + 0.97t + t^2).$$

a) Find the growth rate of the population.

b) Find the number of bacteria present (the population) when $t = 5$ hr. Find the growth rate when $t = 5$.

c) Find the number of bacteria present when $t = 6$ hr. Find the growth rate when $t = 6$.

The growth rate when $t = 5$ is given by

$$P'(5) = 8600 + 20,000 \cdot 5 = 108,600 \; \frac{\text{bacteria}}{\text{hr}}.$$

Thus at $t = 5$, there are 303,000 bacteria present, and the colony is growing at the rate of 108,600 bacteria per hour.

DO EXERCISE 5.

Rates of Change in Economics

In the study of economics we are frequently interested in how such quantities as cost, revenue, and profit change with an increase in product quantity. In particular, we are interested in what is called *marginal*** cost* or profit (or whatever). This term is used to signify *rate of change with respect to quantity*. Thus, if

$C(x) =$ the *total cost* of producing x units of a product (usually considered in some time period),

then

$C'(x) =$ the *marginal cost*

$\quad\quad =$ the rate of change of the total cost with respect to the number of units, x, produced.

Let us think about these interpretations. The total cost of producing 5 units of a product is $C(5)$. The rate of change $C'(5)$ is the cost per unit at that stage in the production process. That this cost per unit does not include fixed costs is seen in this example:

$$C(x) = \underbrace{(x^2 + 4x)}_{\text{Variable costs}} + \underbrace{\$10,000.}_{\text{Fixed costs (constant)}}$$

Because the derivative of a constant is 0,

$$C'(x) = 2x + 4.$$

This verifies an economic principle stating that the fixed costs of a company have no effect on marginal cost.

---

* The term "marginal" comes from the Marginalist School of Economic Thought, which originated in Austria for the purpose of applying mathematics and statistics to the study of economics.

Following are some other marginal functions. Recall that

$$R(x) = \text{the } total \ revenue \text{ from the sale of } x \text{ units.}$$

Then

$R'(x) = \text{the } marginal \ revenue$

= the rate of change of the total revenue with respect to the number of units, x, sold.

Also,

$P(x) = \text{the } total \ profit \text{ from the production and sale of } x \text{ units of a product}$

$= R(x) - C(x).$

Then

$P'(x) = \text{the } marginal \ profit$

= the rate of change of the total profit with respect to the number of units, x, produced and sold

$= R'(x) - C'(x).$

Example 5   Given

$$R(x) = 50x,$$
$$C(x) = 2x^3 - 12x^2 + 40x + 10,$$

find each of the following.

a) $P(x)$

b) $R(2), C(2), P(2)$

c) $R'(x), C'(x), P'(x)$

d) $R'(2), C'(2), P'(2)$

Solution

a) $P(x) = R(x) - C(x) = 50x - (2x^3 - 12x^2 + 40x + 10)$
$$= -2x^3 + 12x^2 + 10x - 10$$

b) $R(2) = 50 \cdot 2 = \$100$ (the total revenue from the sale of the first 2 units)

$C(2) = 2 \cdot 2^3 - 12 \cdot 2^2 + 40 \cdot 2 + 10 = \$58$ (the total cost of producing the first 2 units)

$P(2) = R(2) - C(2) = \$100 - \$58 = \$42$ (the total profit from the production and sale of the first 2 units)

6. Given

$$R(x) = 50x - 0.5x^2,$$
$$C(x) = 10x + 3,$$

find each of the following.

a) $P(x)$

b) $R(40), C(40), P(40)$

c) $R'(x), C'(x), P'(x)$

d) $R'(40), C'(40), P'(40)$

e) Is the marginal revenue constant?

c) $R'(x) = 50,$

$C'(x) = 6x^2 - 24x + 40,$

$P'(x) = R'(x) - C'(x) = 50 - (6x^2 - 24x + 40)$

$\qquad = -6x^2 + 24x + 10$

d) $R'(2) = \$50$ per unit,

$C'(2) = 6 \cdot 2^2 - 24 \cdot 2 + 40 = \$16$ per unit,

$P'(2) = \$50 - \$16 = \$34$ per unit

Note that marginal revenue is constant. No matter how much is produced and sold, the revenue per unit stays the same. This may not always be the case. Also note that $C'(2)$, or $16 per unit, is not the average cost per unit, which is given by

$$\frac{\text{Total cost of producing 2 units}}{2 \text{ units}} = \frac{\$58}{2} = \$29 \text{ per unit.}$$

In general,

$$A(x) = \text{the } \textit{average cost} \text{ of producing } x \text{ units} = \frac{C(x)}{x}.$$

DO EXERCISE 6.

Let us look at a typical marginal-cost function C' and its associated total-cost function C.

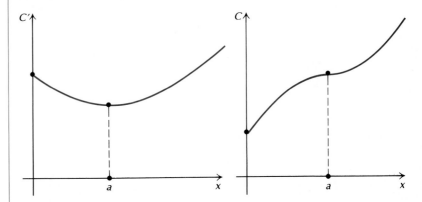

Marginal cost normally decreases as more units are produced until it reaches some minimum value at a; then it increases. (This is probably due to something like having to pay overtime or buying more machinery. Since $C'(x)$ represents the slope of $C(x)$ and is positive and decreasing up to a, the graph of $C(x)$ turns downward as x goes from 0 to a. Then past a, it turns upward.

EXERCISE SET 2.6

1. Given

$$s(t) = t^3 + t,$$

where s is measured in feet and t in seconds, find each of the following.

a) $v(t)$

b) $a(t)$

c) The velocity and acceleration when $t = 4$ sec

2. Given

$$s(t) = 3t + 10,$$

where s is measured in miles and t in hours, find each of the following.

a) $v(t)$

b) $a(t)$

c) The velocity and acceleration when $t = 2$ hr. When the distance function is given by a linear function, we have what is called *uniform motion*.

APPLICATIONS

3. *View to the horizon.* The view V, or distance, in miles that one can see to the horizon from a height h, in feet, is given by

$$V = 1.22\sqrt{h}.$$

a) Find the rate of change of V with respect to h.

b) How far can one see to the horizon from an airplane window from a height of 40,000 ft?

c) Find the rate of change at $h = 40,000$.

5. *Sociology: Median age of women at first marriage.* The median age of women at first marriage can be approximated by the linear function

$$A(t) = 0.08t + 19.7,$$

where $A(t) =$ the median age of women at first marriage the tth year after 1950. Find the rate of change of the median age A with respect to time t.

7. *Biomedical: Healing wound.* The circumference C, in centimeters, of a healing wound is given by

$$C = 2\pi r,$$

where r is the radius, in centimeters. Find the rate of change of the circumference with respect to the radius.

4. *Stopping distance on glare ice.* The stopping distance (at some fixed speed) of regular tires is given by a linear function of the air temperature F,

$$D(F) = 2F + 115,$$

where $D(F) =$ the stopping distance, in feet, when the air temperature is F, in degrees Fahrenheit. Find the rate of change of the stopping distance D with respect to the air temperature F.

6. *Biomedical: Healing wound.* The circular area A, in square centimeters, of a healing wound is given by

$$A = \pi r^2,$$

where r is the radius, in centimeters. Find the rate of change of the area with respect to the radius.

8. *Sociology: Population growth rate.* The population of a city grows from an initial size of 100,000 to an amount P given by

$$P = 100,000 + 2000t^2,$$

where t is measured in years.

a) Find the growth rate.

b) Find the number of people in the city after 10 years (at $t = 10$ yr).

c) Find the growth rate at $t = 10$.

9. *Biomedical: Fever.* The temperature T of a person during an illness is given by

$$T(t) = -0.1t^2 + 1.2t + 98.6,$$

where T is the temperature (°F) at time t, measured in days.

a) Find the rate of change of the temperature with respect to time.

b) Find the temperature at $t = 1.5$ days.

c) Find the rate of change at $t = 1.5$ days.

11. *Biomedical: Blood pressure.* For a certain dosage of x cubic centimeters (cc) of a drug, the resulting blood pressure B is approximated by

$$B(x) = 0.05x^2 - 0.3x^3.$$

Find the rate of change of the blood pressure with respect to the dosage.

13. *Ecology: Territorial area.* The territorial area T of an animal is defined as its defended, or exclusive, region. The area T of that region can be approximated using its body weight W by

$$T = W^{1.31}$$

(see Section 1.5). Find dT/dW.

10. *Business: Advertising.* A firm estimates that it will sell N units of a product after spending a dollars on advertising, where

$$N(a) = -a^2 + 300a + 6$$

and a is measured in thousands of dollars.

a) What is the rate of change of the number of units sold with respect to the amount spent on advertising?

b) How many units will be sold after spending $10,000 on advertising?

c) What is the rate of change at $a = 10$?

12. *Ecology: Home range.* The home range H of an animal is defined as the region to which the animal confines its movements. The area of that region can be approximated using its body weight W by

$$H = W^{1.41}$$

(see Section 1.5). Find dH/dW.

These lions may be determining territorial area.

14. Given

$$R(x) = 50x - 0.5x^2,$$
$$C(x) = 4x + 10,$$

find each of the following.

a) $P(x)$

b) $R(20), C(20), P(20)$

c) $R'(x), C'(x), P'(x)$

d) $R'(20), C'(20), P'(20)$

15. Given

$$R(x) = 5x,$$
$$C(x) = 0.001x^2 + 1.2x + 60,$$

find each of the following.

a) $P(x)$

b) $R(100), C(100), P(100)$

c) $R'(x), C'(x), P'(x)$

d) $R'(100), C'(100), P'(100)$

2.7

DIFFERENTIATION TECHNIQUES: THE PRODUCT AND QUOTIENT RULES

a) Differentiate using the Product and the Quotient Rules.

The derivative of a sum is the sum of the derivatives, but the derivative of a product is not the product of the derivatives. To see this, consider x^2 and x^5. The product is x^7, and the derivative of this product is $7x^6$. The individual derivatives are $2x$ and $5x^4$, and the product of these derivatives is $10x^5$, which is not $7x^6$.

The following is the rule for finding the derivative of a product.

THEOREM 5

The Product Rule

Suppose $F(x) = f(x) \cdot s(x)$, where $f(x)$ is the "first" function and $s(x)$ is the "second" function. Then

$$F'(x) = f(x) \cdot s'(x) + f'(x) \cdot s(x).$$

The derivative of a product is the first function times the derivative of the second function, plus the derivative of the first function times the second function.

Let us check this for $x^2 \cdot x^5$. There are five steps.

1. Write down the first factor.
2. Multiply it by the derivative of the second factor.
3. Write the derivative of the first factor.
4. Multiply it by the second factor.
5. Add the result of steps (1) and (2) to the result of steps (3) and (4).

$$x^2 \quad \bullet \quad x^5$$
$$\text{①} \quad \text{②} \qquad \text{③} \quad \text{④}$$
$$x^2 \cdot 5x^4 + 2x \cdot x^5$$
$$= 5x^6 + 2x^6 \qquad \text{⑤}$$
$$= 7x^6$$

Example 1

$$\frac{d}{dx}(x^4 - 2x^3 - 7)(3x^2 - 5x) = (x^4 - 2x^3 - 7)(6x - 5)$$
$$+ (4x^3 - 6x^2)(3x^2 - 5x)$$

Note that we could have multiplied the polynomials and then differentiated, avoiding the use of the Product Rule, but this would have been more work.

Use the Product Rule to find $f'(x)$.

1. $f(x) = 3x^8 \cdot x^{10}$

DO EXERCISES 1 AND 2.

The derivative of a quotient is *not* the quotient of the derivatives. To see why, consider x^5 and x^2. The quotient x^5/x^2 is x^3, and the derivative of this quotient is $3x^2$. The individual derivatives are $5x^4$ and $2x$, and the quotient of these derivatives $5x^4/2x$ is $(\frac{5}{2})x^3$, which is not $3x^2$. The rule for differentiating quotients is as follows.

THEOREM 6

The Quotient Rule

If

$$q(x) = \frac{n(x)}{d(x)},$$

then

$$q'(x) = \frac{d(x) \cdot n'(x) - d'(x) \cdot n(x)}{[d(x)]^2}.$$

The derivative of a quotient is the denominator times the derivative of the numerator, minus the derivative of the denominator times the numerator, all divided by the square of the denominator.

---

2. $f(x) = (9x^3 + 4x^2 + 10)(-7x^2 + x^4)$

Another way to remember this is shown below. It starts with squaring the denominator. The denominator is also used as the first factor of the first term above.

$$\frac{g(x) \cdot f'(x) - g'(x) \cdot f(x)}{[g(x)]^2}$$

1. Write down the denominator.
2. Multiply the denominator by the derivative of the numerator.
3. Write a minus sign.
4. Find the derivative of the denominator.
5. Multiply it by the numerator.
6. Divide by the square of the denominator.

Example 2 For $q(x) = x^5/x^3$, find $q'(x)$.

3. For

$$f(x) = \frac{x^9}{x^5},$$

find $f'(x)$ using the Quotient Rule.

Differentiate.

4. $f(x) = \dfrac{1 - x^2}{x^5}$

5. $f(x) = \dfrac{x^2 - 1}{x^3 + 1}$

Solution

$$q'(x) = \frac{x^3 \cdot 5x^4 - 3x^2 \cdot x^5}{[x^3]^2}$$

$$= \frac{5x^7 - 3x^7}{x^6}$$

$$= \frac{2x^7}{x^6} = 2x$$

DO EXERCISE 3.

Example 3 Differentiate

$$f(x) = \frac{1 + x^2}{x^3}.$$

Solution

$$f'(x) = \frac{x^3 \cdot 2x - 3x^2(1 + x^2)}{(x^3)^2} = \frac{2x^4 - 3x^2 - 3x^4}{x^6} = \frac{-x^4 - 3x^2}{x^6}$$

$$= \frac{-x^2 \cdot x^2 - 3x^2}{x^6} = \frac{x^2(-x^2 - 3)}{x^6} = \frac{-x^2 - 3}{x^4}$$

Example 4 Differentiate

$$f(x) = \frac{x^2 - 3x}{x - 1}.$$

Solution

$$f'(x) = \frac{(x - 1)(2x - 3) - 1(x^2 - 3x)}{(x - 1)^2}$$

$$= \frac{2x^2 - 5x + 3 - x^2 + 3x}{(x - 1)^2}$$

$$= \frac{x^2 - 2x + 3}{(x - 1)^2}$$

It is not necessary to multiply out $(x - 1)^2$.

DO EXERCISES 4 AND 5.

An Application

We discussed earlier that it is typical for a total-revenue function to vary depending on the number of units x sold. Let us see what can determine this. Recall the consumer's demand function, or price func-

6. A company determines that the demand function for a certain product is given by

$$p = D(x) = 200 - x.$$

a) Find an expression for the total revenue $R(x)$.

b) Find the marginal revenue $R'(x)$.

tion, $p = D(x)$, discussed in Section 1.5. It is the price p a seller must charge in order to sell exactly x units of a product. This is typically a decreasing function.

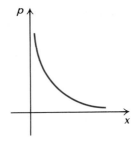

The total revenue from the sale of x units is then given by

$$R(x) = \text{(Number of units sold)} \cdot \text{(Price charged to sell the units)},$$

or

$$R(x) = x \cdot p = x D(x).$$

A typical graph of a revenue function is shown below.

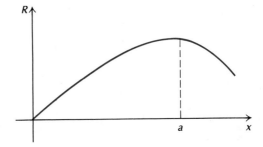

To sell more units, $D(x)$ decreases. Because we have a product $x \cdot D(x)$, the revenue typically rises for a while as x increases, but tapers off as $D(x)$ gets smaller and smaller.

Using the Product Rule, we can obtain an expression for the marginal revenue $R'(x)$ in terms of x and $D'(x)$ as

$$R(x) = x D(x),$$

so

$$R'(x) = 1 \cdot D(x) + x \cdot D'(x) = D(x) + x D'(x).$$

You need not memorize this. You can merely repeat the Product Rule where necessary.

DO EXERCISE 6.

EXERCISE SET 2.7

Differentiate.

1. $y = x^3 \cdot x^8$; two ways

2. $y = x^4 \cdot x^9$; two ways

3. $y = \dfrac{-1}{x}$; two ways

4. $y = \dfrac{1}{x}$; two ways

5. $y = \dfrac{x^8}{x^5}$; two ways

6. $y = \dfrac{x^9}{x^5}$; two ways

7. $y = (8x^5 - 3x^2 + 20)(8x^4 - 3\sqrt{x})$

8. $f(x) = (7x^6 + 4x^3 - 50)(9x^{10} - 7\sqrt{x})$

9. $f(x) = x(300 - x)$; two ways

10. $f(x) = x(400 - x)$; two ways

11. $f(x) = \dfrac{x}{300 - x}$

12. $f(x) = \dfrac{x}{400 - x}$

13. $f(x) = \dfrac{3x - 1}{2x + 5}$

14. $f(x) = \dfrac{2x + 3}{x - 5}$

15. $y = \dfrac{x^2 + 1}{x^3 - 1}$

16. $y = \dfrac{x^3 - 1}{x^2 + 1}$

17. $y = \dfrac{x}{1 - x}$

18. $y = \dfrac{x}{3 - x}$

19. $y = \dfrac{x - 1}{x + 1}$

20. $y = \dfrac{x + 2}{x - 2}$

21. $f(x) = \dfrac{1}{x - 3}$

22. $f(x) = \dfrac{1}{x + 2}$

23. $f(x) = \dfrac{3x^2 + 2x}{x^2 + 1}$

24. $f(x) = \dfrac{3x^2 - 5x}{x^2 - 1}$

25. $f(x) = \dfrac{3x^2 - 5x}{x^8}$; two ways

26. $f(x) = \dfrac{3x^2 + 2x}{x^5}$; two ways

In each of Exercises 27–30, a demand function $p = D(x)$ is given. Find (a) the total revenue $R(x)$ and (b) the marginal revenue $R'(x)$.

27. $D(x) = 400 - x$

28. $D(x) = 500 - x$

29. $D(x) = \dfrac{4000}{x} + 3$

30. $D(x) = \dfrac{3000}{x} + 5$

31. In Section 2.6, we defined the average cost of producing x units of a product in terms of the total cost $C(x)$ by

$$A(x) = \frac{C(x)}{x}.$$

Use the Quotient Rule to find a general expression for *marginal average cost* $A'(x)$.

32. In this section we determined that

$$R(x) = x D(x).$$

Then

$$D(x) = \frac{R(x)}{x} = \text{the average revenue from the sale of } x \text{ units.}$$

Use the Quotient Rule to find a general expression for *marginal average revenue* $D'(x)$.

EXTENSION EXERCISES

Differentiate each function.

33. $f(x) = \dfrac{x^3}{\sqrt{x - 5}}$

34. $g(t) = \dfrac{1 + \sqrt{t}}{t^5 + 3}$

35. $f(v) = \dfrac{3}{1 + v + v^2}$

36. $g(z) = \dfrac{1 + z + z^2}{1 - z + z^2}$

37. $p(t) = \dfrac{t}{1 - t + t^2 - t^3}$

38. $f(x) = \dfrac{\dfrac{2}{3x} - 1}{\dfrac{3}{x^2} + 5}$

39. $h(x) = \dfrac{x^3 + 5x^2 - 2}{\sqrt{x}}$

40. $y(t) = 5t(t - 1)(2t + 3)$

41. $f(x) = x(3x^3 + 6x - 2)(3x^4 + 7)$

42. $g(x) = (x^3 - 8) \cdot \dfrac{x^2 + 1}{x^2 - 1}$

43. $f(t) = (t^5 + 3) \cdot \dfrac{t^3 - 1}{t^3 + 1}$

44. $f(x) = \dfrac{(x^2 + 3x)(x^5 - 7x^2 - 3)}{x^4 - 3x^3 - 5}$

45. $f(x) = \dfrac{(2x^2 + 3)(4x^3 - 7x + 2)}{x^7 - 2x^6 + 9}$

46. $s(t) = \dfrac{5t^8 - 2t^3}{(t^5 - 3)(t^4 + 7)}$

2.8

THE CHAIN RULE

OBJECTIVES

a) **Differentiate using the Extended Power Rule and the Chain Rule.**

b) **Find the composition of functions.**

The Extended Power Rule

How do we differentiate more complicated functions such as

$$y = (1 + x^2)^3, \qquad y = (1 + x^2)^{89}, \quad \text{or} \quad y = (1 + x^2)^{1/3}?$$

For $(1 + x^2)^3$ we can expand and then differentiate. Although this could be done for $(1 + x^2)^{89}$, it would certainly be time-consuming, and such an expansion of the Power Rule would not work for $(1 + x^2)^{1/3}$. Not knowing this, we might conjecture that the derivative of the function $y = (1 + x^2)^3$ is

$$3(1 + x^2)^2. \tag{1}$$

To check this, we expand $(1 + x^2)^3$ and then differentiate. From Section 1.1, we know that $(a + h)^3 = a^3 + 3a^2h + 3ah^2 + h^3$, so

$$(1 + x^2)^3 = 1^3 + 3 \cdot 1^2 \cdot (x^2)^1 + 3 \cdot 1 \cdot (x^2)^2 + (x^2)^3$$
$$= 1 + 3x^2 + 3x^4 + x^6.$$

(We could also have done this by finding $(1 + x^2)^2$ and then multiplying again by $1 + x^2$.) It follows that

$$\frac{dy}{dx} = 6x + 12x^3 + 6x^5 = (1 + 2x^2 + x^4)6x$$

$$= 3(1 + x^2)^2 \cdot 2x. \tag{2}$$

Comparing this with Eq. (1), we see that the Power Rule is not sufficient for such a differentiation. Note that the factor $2x$ in the actual

Differentiate.

1. $f(x) = (1 + x^2)^{10}$

2. $y = (1 - x^2)^{1/2}$

derivative, Eq. (2), is the derivative of the "inside" function, $1 + x^2$. This is consistent with the following new rule.

THEOREM 7

The Extended Power Rule

Suppose $g(x)$ is a function of x. Then for any real number a,

$$\frac{d}{dx}[g(x)]^a = a[g(x)]^{a-1} \cdot \frac{d}{dx}g(x).$$

Let us differentiate $(1 + x^3)^5$. There are three steps to carry out.

$(\boxed{1 + x^3})^5$ 1. Mentally block out the "inside" function, $1 + x^3$.

$5(\boxed{1 + x^3})^4$ 2. Differentiate the "outside" function, $(\boxed{1 + x^3})^5$.

$5(1 + x^3)^4 \cdot 3x^2$
$= 15x^2(1 + x^3)^4$ 3. Multiply by the derivative of the "inside" function.

Step (3) is most commonly overlooked. Try not to forget it!

Example 1

$$\frac{d}{dx}(1 + x^3)^{1/2} = \frac{1}{2}(\boxed{1 + x^3})^{1/2 - 1} \cdot 3x^2$$

$$= \frac{1}{2}(1 + x^3)^{-1/2} \cdot 3x^2$$

$$= \frac{3x^2}{2\sqrt{1 + x^3}}$$

DO EXERCISES 1 AND 2.

Example 2   Differentiate $y = (1 - x^2)^3 - (1 - x^2)^2$.

Solution   Here we combine the Difference Rule and the Extended Power Rule:

$$\frac{dy}{dx} = 3(1 - x^2)^2(-2x) - 2(1 - x^2)(-2x)$$ We differentiate each term using the Extended Power Rule.

3. Differentiate:

$$f(x) = (1 + x^2)^2 - (1 + x^2)^3.$$

Thus

$$\frac{dy}{dx} = -6x(1 - x^2)^2 + 4x(1 - x^2)$$
$$= x(1 - x^2)[-6(1 - x^2) + 4] \qquad \textbf{Here we factor out } x(1 - x^2).$$
$$= x(1 - x^2)[-6 + 6x^2 + 4]$$
$$= x(1 - x^2)(6x^2 - 2) = 2x(1 - x^2)(3x^2 - 1).$$

DO EXERCISE 3.

Example 3   Differentiate $f(x) = (x - 5)^4(7 - x)^{10}$.

Solution   Here we combine the Product Rule and the Extended Power Rule:

$$f'(x) = (x - 5)^4 \cdot 10(7 - x)^9(-1) + 4(x - 5)^3(7 - x)^{10}$$
$$= -10(x - 5)^4(7 - x)^9 + 4(x - 5)^3(7 - x)^{10}$$
$$= (x - 5)^3(7 - x)^9[-10(x - 5) + 4(7 - x)] \qquad \textbf{We factored out}$$
$$\qquad\qquad\qquad\qquad\qquad\qquad\qquad\qquad\qquad\qquad \textbf{(x - 5)^3(7 - x)^9}.$$
$$= (x - 5)^3(7 - x)^9[-10x + 50 + 28 - 4x]$$
$$= (x - 5)^3(7 - x)^9(78 - 14x)$$
$$= 2(x - 5)^3(7 - x)^9(39 - 7x).$$

4. Differentiate:

$$y = (x - 4)^5(6 - x)^3.$$

DO EXERCISE 4.

Example 4   Differentiate

$$f(x) = \sqrt[4]{\frac{x + 3}{x - 1}}.$$

Solution   We must use the Quotient Rule to differentiate the inside function, $(x + 3)/(x - 1)$:

$$\frac{d}{dx}\sqrt[4]{\frac{x + 3}{x - 1}} = \frac{d}{dx}\left(\frac{x + 3}{x - 1}\right)^{1/4} = \frac{1}{4}\left(\frac{x + 3}{x - 1}\right)^{1/4 - 1}\left[\frac{(x - 1)1 - 1(x + 3)}{(x - 1)^2}\right]$$
$$= \frac{1}{4}\left(\frac{x + 3}{x - 1}\right)^{-3/4}\left[\frac{x - 1 - x - 3}{(x - 1)^2}\right]$$
$$= \frac{1}{4}\left(\frac{x - 3}{x - 1}\right)^{-3/4} \cdot \frac{-4}{(x - 1)^2}$$
$$= \left(\frac{x + 3}{x - 1}\right)^{-3/4} \cdot \frac{-1}{(x - 1)^2}.$$

5. Differentiate:

$$y = \sqrt[3]{\frac{x+5}{x-4}}.$$

DO EXERCISE 5.

Composition of Functions and the Chain Rule

The Extended Power Rule is a special case of a general rule called the *Chain Rule*. Before discussing it, we define the *composition* of functions.

There is a function g that gives a correspondence between women's shoe sizes in the United States and those in Italy. The function is given by $g(x) = 2(x + 12)$, where x is a shoe size in the United States and $g(x)$ is a shoe size in Italy. For example, a shoe size of 4 in the United States corresponds to a shoe size of $g(4) = 2(4 + 12)$, or 32 in Italy.

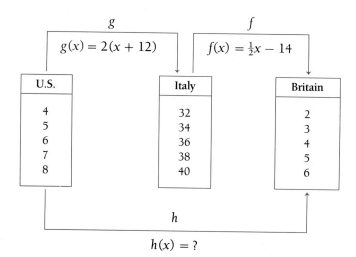

$$h(x) = ?$$

There is also a function f that gives a correspondence between women's shoe sizes in Italy and those in Britain. The function is given by $f(x) = \frac{1}{2}x - 14$, where x is a shoe size in Italy and $f(x)$ is the corresponding shoe size in Britain. For example, a shoe size of 32 in Italy corresponds to a shoe size of $f(32) = \frac{1}{2}(32) - 14$, or 2 in Britain.

It seems reasonable to assume that a shoe size of 4 in the United States corresponds to a shoe size of 2 in Britain and that there is a function h that describes this correspondence. Can we find a formula for h? Looking at the tables, we might guess that such a formula is $h(x) = x - 2$, and that is indeed correct. But, for more complicated formulas we would need to do some algebra.

A shoe size x in the United States corresponds to a shoe size $g(x)$ in Italy, where

$$g(x) = 2(x + 12).$$

Now $2(x + 12)$ is a shoe size in Italy. If we replace x in $f(x)$ by $2(x + 12)$, we can find the corresponding shoe size in Britain:

$$f(g(x)) = \tfrac{1}{2}[2(x + 12)] - 14$$
$$= \tfrac{1}{2}[2x + 24] - 14$$
$$= x + 12 - 14$$
$$= x - 2$$

This gives us a formula for h: $h(x) = x - 2$. Thus a shoe size of 4 in the United States corresponds to a shoe size of $h(4) = 4 - 2$, or 2 in Britain. The function h is called the *composition* of f and g and is denoted $f \circ g$.

DEFINITION

The *composed* function $f \circ g$, the *composition* of f and g, is defined as follows:

$$f \circ g(x) = f(g(x)).$$

We can visualize the composition of functions as follows.

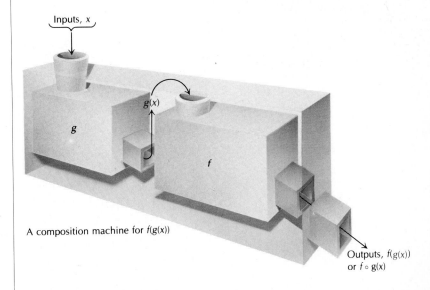

A composition machine for $f(g(x))$

To find $f \circ g(x)$, we substitute $g(x)$ for x in $f(x)$.

Example 5 Given $f(x) = x^3$ and $g(x) = 1 + x^2$, find $f \circ g(x)$ and $g \circ f(x)$.

6. Given $f(x) = 3x$ and $g(x) = x^2 - 1$, find $f \circ g(x)$ and $g \circ f(x)$.

Solution   Consider each function separately:

$$f(x) = x^3 \qquad \text{This function cubes each input.}$$

and

$$g(x) = 1 + x^2. \qquad \text{This function adds 1 to the square of each input.}$$

a) $f \circ g$ first does what g does (adds 1 to the square) and then does what f does (cubes). We find $f \circ g(x)$ by substituting $g(x)$ for x:

$$f \circ g(x) = f(g(x)) = f(1 + x^2) \qquad \text{Substituting } 1 + x^2 \text{ for } x$$
$$= (1 + x^2)^3$$
$$= 1 + 3x^2 + 3x^4 + x^6.$$

b) $g \circ f$ first does what f does (cubes) and then does what g does (adds 1 to the square). We find $g(f(x))$ by substituting $f(x)$ for x:

$$g \circ f(x) = g(f(x)) = g(x^3) \qquad \text{Substituting } x^3 \text{ for } x$$
$$= 1 + (x^3)^2$$
$$= 1 + x^6.$$

DO EXERCISE 6.

Example 6   Given $f(x) = \sqrt{x}$ and $g(x) = x - 1$, find $f \circ g(x)$ and $g \circ f(x)$.

Solution

$$f \circ g(x) = f(g(x)) = f(x - 1) = \sqrt{x - 1},$$
$$g \circ f(x) = g(f(x)) = g(\sqrt{x}) = \sqrt{x} - 1$$

DO EXERCISE 7.

7. Given $f(x) = 4x + 5$ and $g(x) = \sqrt[3]{x}$, find $f \circ g(x)$ and $g \circ f(x)$.

How do we differentiate the composition of functions? The following theorem tells us.

THEOREM 8

The Chain Rule

The derivative of the composition $f \circ g$ is given by

$$\frac{d}{dx}[f \circ g(x)] = \frac{d}{dx}[f(g(x))] = f'(g(x)) \cdot \frac{d}{dx} g(x).$$

8. Given $y = u^2$ and $u = x^3 + 2$, find $\dfrac{dy}{du}, \dfrac{du}{dx}$, and $\dfrac{dy}{dx}$.

Note that the Extended Power Rule is a special case. Consider the function $f(x) = x^a$. Then for any other function $g(x)$, $f \circ g(x) = [g(x)]^a$ and the derivative of the composition is

$$\frac{d}{dx}\,[g(x)]^a = a[g(x)]^{a-1} \cdot \frac{d}{dx}\,g(x).$$

The Chain Rule often appears in another form. Suppose $y = f(u)$ and $u = g(x)$. Then

$$\frac{dy}{dx} = \frac{dy}{du} \cdot \frac{du}{dx}.$$

Example 7   Given $y = 2 + \sqrt{u}$ and $u = x^3 + 1$, find dy/du, du/dx, and dy/dx.

Solution   First we find dy/du and du/dx:

$$\frac{dy}{du} = \frac{1}{2}\,u^{-1/2} \quad \text{and} \quad \frac{du}{dx} = 3x^2.$$

Then

$$\frac{dy}{dx} = \frac{dy}{du} \cdot \frac{du}{dx} = \frac{1}{2\sqrt{u}} \cdot 3x^2$$

$$= \frac{3x^2}{2\sqrt{x^3 + 1}}. \qquad \text{Substituting } x^3 + 1 \text{ for } u$$

DO EXERCISE 8.

9. Find $f(x)$ and $g(x)$ such that $h(x) = f \circ g(x)$. Answers may vary.
    a) $h(x) = \sqrt[3]{x^2 + 1}$

    b) $h(x) = \dfrac{1}{(x + 5)^4}$

It is important to be able to recognize how a function can be expressed as a composition.

Example 8   Find $f(x)$ and $g(x)$ such that $h(x) = f \circ g(x)$:

$$h(x) = (4x^3 - 7)^5.$$

Solution   This is $4x^3 - 7$ to the 5th power. Two functions that can be used for the composition are $f(x) = x^5$ and $g(x) = 4x^3 - 7$. We can check by forming the composition:

$$h(x) = f \circ g(x) = f(g(x)) = f(4x^3 - 7) = (4x^3 - 7)^5.$$

This is the most "obvious" answer to the question. There can be other less obvious answers—for example, $f(x) = x^3$ and $g(x) = (4x^3 - 7)^2$.

DO EXERCISE 9.

EXERCISE SET 2.8

Differentiate.

1. $y = (1 - x)^{55}$ **2.** $y = (1 - x)^{100}$ **3.** $y = \sqrt{1 + 8x}$ **4.** $y = \sqrt{1 - x}$

5. $y = \sqrt{3x^2 - 4}$ **6.** $y = \sqrt{4x^2 + 1}$ **7.** $y = (3x^2 - 6)^{-40}$ **8.** $y = (4x^2 + 1)^{-50}$

9. $y = x\sqrt{2x + 3}$ **10.** $y = x\sqrt{4x - 7}$ **11.** $y = x^2\sqrt{x - 1}$ **12.** $y = x^3\sqrt{x + 1}$

13. $y = \dfrac{1}{(3x + 8)^2}$ **14.** $y = \dfrac{1}{(4x + 5)^2}$ **15.** $f(x) = (1 + x^3)^3 - (1 + x^3)^4$

16. $f(x) = (1 + x^3)^5 - (1 + x^3)^4$ **17.** $f(x) = x^2 + (200 - x)^2$ **18.** $f(x) = x^2 + (100 - x)^2$

19. $f(x) = (x + 6)^{10}(x - 5)^4$ **20.** $f(x) = (x - 4)^8(x + 3)^9$ **21.** $f(x) = (x - 4)^8(3 - x)^4$

22. $f(x) = (x + 6)^{10}(5 - x)^9$ **23.** $f(x) = -4x(2x - 3)^3$ **24.** $f(x) = -5x(3x + 5)^6$

25. $f(x) = \sqrt{\dfrac{1 - x}{1 + x}}$ **26.** $f(x) = \sqrt{\dfrac{3 + x}{2 - x}}$

In Exercises 27 and 28, find $\dfrac{dy}{du}, \dfrac{du}{dx}$, and $\dfrac{dy}{dx}$.

27. $y = \sqrt{u}$ and $u = x^2 - 1$.

28. $y = \dfrac{15}{u^3}$ and $u = 2x + 1$.

29. Consider

$$f(x) = \frac{x^2}{(1 + x)^5}.$$

a) Find $f'(x)$ using the Quotient Rule and the Extended Power Rule.

b) Note that $f(x) = x^2(1 + x)^{-5}$. Find $f'(x)$ using the Product Rule and the Extended Power Rule.

c) Compare answers to parts (a) and (b).

30. Consider

$$g(x) = (x^3 + 5x)^2.$$

a) Find $g'(x)$ using the Extended Power Rule.

b) Note that $g(x) = x^6 + 10x^4 + 25x^2$. Find $g'(x)$.

c) Compare answers to parts (a) and (b).

31. A total-cost function is given by

$$C(x) = 1000\sqrt{x^3 + 2}.$$

Find the marginal cost $C'(x)$.

32. A total-revenue function is given by

$$R(x) = 2000\sqrt{x^2 + 3}.$$

Find the marginal revenue $R'(x)$.

APPLICATIONS

33. *Business: Compound interest.* If $1000 is invested at interest rate i, compounded annually, in 3 years it will grow to an amount A, given by

$$A = \$1000(1 + i)^3$$

(see Section 1.1). Find the rate of change dA/di.

34. *Business: Compound interest.* If $1000 is invested at interest rate i, compounded quarterly, in 5 years it will grow to an amount A, given by

$$A = \$1000\left(1 + \frac{i}{4}\right)^{20}.$$

Find the rate of change dA/di.

In each of Exercises 35–40, find $f \circ g(x)$ and $g \circ f(x)$.

35. $f(x) = 3x^2 + 2, g(x) = 2x - 1$

36. $f(x) = 4x + 3, g(x) = 2x^2 - 5$

37. $f(x) = 4x^2 - 1, g(x) = \dfrac{2}{x}$

38. $f(x) = \dfrac{3}{x}, g(x) = 2x^2 + 3$

39. $f(x) = x^2 + 1, g(x) = x^2 - 1$

40. $f(x) = \dfrac{1}{x^2}, g(x) = x + 2$

In each of Exercises 41–44, find $f(x)$ and $g(x)$ such that $h(x) = f \circ g(x)$. Answers may vary.

41. $h(x) = (3x^2 - 7)^5$

42. $h(x) = \dfrac{1}{\sqrt{7x + 2}}$

43. $h(x) = \dfrac{x^3 + 1}{x^3 - 1}$

44. $h(x) = (\sqrt{x} + 5)^4$

EXTENSION EXERCISES

Differentiate each of the following functions.

45. $y = \sqrt[3]{x^3 - 6x + 1}$

46. $s = \sqrt[4]{t^4 + 3t^2 + 8}$

47. $y = \dfrac{x}{\sqrt{x - 1}}$

48. $y = \dfrac{(x + 1)^2}{(x^2 + 1)^3}$

49. $u = \dfrac{(1 + 2v)^4}{v^4}$

50. $y = x\sqrt{1 + x^2}$

51. $y = \dfrac{\sqrt{1 - x^2}}{1 - x}$

52. $w = \dfrac{u}{\sqrt{1 + u^2}}$

53. $y = \left(\dfrac{x^2 - x - 1}{x^2 + 1}\right)^3$

54. $y = \sqrt{1 + \sqrt{x}}$

55. $s = \dfrac{\sqrt{t - 1}}{\sqrt{t + 1}}$

56. $y = x^{2/3} \cdot \sqrt[3]{1 + x^2}$

2.9

OBJECTIVE

a)   Find a higher-order derivative.

HIGHER-ORDER DERIVATIVES

Consider the function given by

$$y = f(x)$$
$$= x^5 - 3x^4 + x.$$

Its derivative f' is given by

$$y' = f'(x)$$
$$= 5x^4 - 12x^3 + 1.$$

The function f' can be differentiated. We use the notation f'' for the derivative $(f')'$. We call f'' the *second derivative* of f. It is given by

$$y'' = f''(x) = 20x^3 - 36x^2.$$

1. Find the first six derivatives of

$$f(x) = 2x^6 - x^5 + 10.$$

2. For

$$y = x^7 - x^3,$$

find each of the following.

a) $\dfrac{dy}{dx}$

b) $\dfrac{d^2y}{dx^2}$

c) $\dfrac{d^3y}{dx^3}$

d) $\dfrac{d^4y}{dx^4}$

Continuing in this manner, we have

$$f'''(x) = 60x^2 - 72x, \qquad \text{The third derivative of } f$$
$$f''''(x) = 120x - 72, \qquad \text{The fourth derivative of } f$$
$$f'''''(x) = 120. \qquad \text{The fifth derivative of } f$$

When notation like $f'''(x)$ gets lengthy we abbreviate it using a numeral in parentheses. Thus $f^{(n)}(x)$ is the nth derivative. For the above function,

$$f^{(4)}(x) = 120x - 72,$$
$$f^{(5)}(x) = 120,$$
$$f^{(6)}(x) = 0, \quad \text{and}$$
$$f^{(n)}(x) = 0, \quad \text{for any } n \geqslant 6.$$

DO EXERCISE 1.

Leibniz's notation for the second derivative of a function given by $y = f(x)$ is

$$\frac{d^2y}{dx^2}, \quad \text{or} \quad \frac{d}{dx}\left(\frac{dy}{dx}\right),$$

read "the second derivative of y with respect to x." The 2's in this notation *are not* exponents. If $y = x^5 - 3x^4 + x$, then

$$\frac{d^2y}{dx^2} = 20x^3 - 36x^2.$$

Leibniz's notation for the third derivative is d^3y/dx^3; for the fourth derivative, d^4y/dx^4; and so on:

$$\frac{d^3y}{dx^3} = 60x^2 - 72x, \qquad \frac{d^4y}{dx^4} = 120x - 72, \qquad \frac{d^5y}{dx^5} = 120.$$

DO EXERCISE 2.

Example 1   For $y = 1/x$, find d^2y/dx^2.

Solution   We have $y = x^{-1}$, so

$$\frac{dy}{dx} = -1 \cdot x^{-1-1} = -x^{-2}, \quad \text{or} \quad -\frac{1}{x^2}.$$

3. For $y = \dfrac{2}{x}$, find $\dfrac{d^2y}{dx^2}$.

Then

$$\frac{d^2y}{dx^2} = (-2)(-1)x^{-2-1} = 2x^{-3}, \quad \text{or} \quad \frac{2}{x^3}.$$

DO EXERCISE 3.

Example 2 For $y = (x^2 + 10x)^{20}$, find y' and y''.

Solution To find y', we use the Extended Power Rule:

$$y' = 20(x^2 + 10x)^{19}(2x + 10) = 20(x^2 + 10x)^{19} \cdot 2(x + 5)$$
$$= 40(x^2 + 10x)^{19}(x + 5).$$

To find y'', we use the Product Rule and the Extended Power Rule:

4. For $y = (x^2 - 12x)^{30}$, find y' and y''.

$$y'' = 19 \cdot 40(x^2 + 10x)^{18}(2x + 10)(x + 5) + 40(x^2 + 10x)^{19}(1)$$
$$= 760(x^2 + 10x)^{18} \cdot 2(x + 5)(x + 5) + 40(x^2 + 10x)^{19}$$
$$= 1520(x^2 + 10x)^{18}(x + 5)^2 + 40(x^2 + 10x)^{19}$$
$$= 40(x^2 + 10x)^{18}[38(x + 5)^2 + (x^2 + 10x)]$$
$$= 40(x^2 + 10x)^{18}[38(x^2 + 10x + 25) + x^2 + 10x]$$
$$= 40(x^2 + 10x)^{18}[39x^2 + 390x + 950].$$

DO EXERCISE 4.

5. For $s(t) = 3t + t^4$, find the acceleration $a(t)$.

Acceleration can be regarded as a second derivative. As an object moves, its distance from a fixed point after time t is some function of the time, say, $s(t)$. Then

$$v(t) = s'(t) = \text{the velocity at time } t$$

and

$$a(t) = v'(t) = s''(t) = \text{the acceleration at time } t.$$

Whenever a quantity is a function of time, the first derivative gives the rate of change with respect to time and the second derivative gives the acceleration. For example, if $y = P(t)$ gives the number of people in a population at time t, then $P'(t)$ represents how fast the size of the population is changing and $P''(t)$ gives the acceleration in the size of the population.

DO EXERCISE 5.

EXERCISE SET 2.9

In each of Exercises 1–18, find d^2y/dx^2.

1. $y = 3x + 5$

2. $y = -4x + 7$

3. $y = -\dfrac{1}{x}$

4. $y = -\dfrac{3}{x}$

5. $y = x^{1/4}$

6. $y = \sqrt{x}$

7. $y = x^4 + \dfrac{4}{x}$

8. $y = x^3 - \dfrac{3}{x}$

9. $y = x^{-3}$

10. $y = x^{-4}$

11. $y = x^n$

12. $y = x^{-n}$

13. $y = x^4 - x^2$

14. $y = x^4 + x^3$

15. $y = \sqrt{x - 1}$

16. $y = \sqrt{x + 1}$

17. $y = ax^2 + bx + c$

18. $y = (x^3 + 15x)^{20}$

19. For $y = x^4$, find d^4y/dx^4.

20. For $y = x^5$, find d^4y/dx^4.

21. For $y = x^6 - x^3 + 2x$, find d^5y/dx^5.

22. For $y = x^7 - 8x^2 + 2$, find d^6y/dx^6.

23. For $y = (x^2 - 5)^{10}$, find d^2y/dx^2.

24. For $y = x^k$, find d^5y/dx^5.

25. If s is a distance given by $s(t) = t^3 + t^2 + 2t$, find the acceleration.

26. If s is a distance given by $s(t) = t^4 + t^2 + 3t$, find the acceleration.

27. A population grows from an initial size of 100,000 to an amount $P(t)$, given by

$$P(t) = 100{,}000(1 + 0.6t + t^2).$$

What is the acceleration in the size of the population?

28. A population grows from an initial size of 100,000 to an amount $P(t)$, given by

$$P(t) = 100{,}000(1 + 0.4t + t^2).$$

What is the acceleration in the size of the population?

EXTENSION EXERCISES

Find y', y'', and y'''.

29. $y = x^{-1} + x^{-2}$

30. $y = \dfrac{1}{1 - x}$

31. $y = x\sqrt{1 + x^2}$

32. $y = 3x^5 + 8\sqrt{x}$

33. $y = \dfrac{3x - 1}{2x + 3}$

34. $y = \dfrac{1}{\sqrt{x - 1}}$

35. $y = \dfrac{x}{\sqrt{x - 1}}$

36. $y = \dfrac{\sqrt{x} - 1}{\sqrt{x} + 1}$

Find $f''(x)$.

37. $f(x) = \dfrac{x}{x - 1}$

38. $f(x) = \dfrac{1}{1 + x^2}$

SUMMARY AND REVIEW: CHAPTER 2

The following contains a summary of what you should be able to do after completing this chapter. The review exercises are for practice. Answers are at the back of the book. If you miss an exercise, restudy the section indicated alongside the answers. A summary of the important formulas for this chapter is given on the inside front cover.

You should be able to:

Find a limit if it exists. Determine whether a graph is that of a continuous function. Determine whether a function is continuous at a given point.

Find the limit, if it exists.

1. $\lim\limits_{x \to -2} \dfrac{8}{x}$

2. $\lim\limits_{x \to 1} (4x^3 - x^2 + 7x)$

3. $\lim\limits_{x \to -7} \dfrac{x^2 + 4x - 21}{x + 7}$

4. $\lim\limits_{x \to \infty} \dfrac{5x + 6}{x}$

Determine whether the graph is continuous.

5.

6.

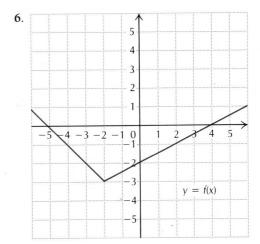

For the function in Exercise 5, answer the following.

7. Find $\lim\limits_{x \to 1} g(x)$.

8. Find $g(1)$.

9. Is g continuous at 1?

10. Find $\lim\limits_{x \to -2} g(x)$.

11. Find $g(-2)$.

12. Is g continuous at -2?

Compute an average rate of change of one variable with respect to another. Find a simplified difference quotient.

13. For $f(x) = x^3 - 2x$, find the average rate of change as x changes from -2 to 1.

14. Find a simplified difference quotient for $g(x) = -3x + 5$.

15. Find a simplified difference quotient for $f(x) = 2x^2 - 3$.

Find the points on the graph of a function at which the tangent line has a given slope.

16. Find the points on the graph of $y = -x^2 + 8x - 11$ at which the tangent line is horizontal.

17. Find the points on the graph of $y = 5x^2 - 49x + 12$ at which the tangent line has slope 1.

Differentiate using the Power Rule, the Sum–Difference Rule, the rule for differentiating a constant or a constant times a function, the Product Rule, the Quotient Rule, the Extended Power Rule, or the Chain Rule.

Find dy/dx.

18. $y = 4x^5$

19. $y = 3\sqrt[3]{x}$

20. $y = \dfrac{-8}{x^8}$

21. $y = 15x^{2/5}$

22. $y = 0.1x^7 - 3x^4 - x^3 + 6$

Differentiate.

23. $f(x) = \dfrac{1}{6}x^6 + 8x^4 - 5x$

24. $y = \dfrac{x^3 + x}{x}$

25. $y = \dfrac{x^2 + 8}{8 - x}$

26. $g(x) = (5 - x)^2(2x - 1)^5$

27. $f(x) = (x^5 - 2)^7$

28. $f(x) = x^2(4x + 3)^{3/4}$

Find higher-order derivatives.

29. For $y = x^3 - \dfrac{2}{x}$, find $\dfrac{d^5y}{dx^5}$.

30. For $y = x^7 + 3x^2$, find $\dfrac{d^4y}{dx^4}$.

Given a distance function $s(t)$, find a formula for the velocity $v(t)$ and the acceleration $a(t)$, and evaluate $s(t)$, $v(t)$, and $a(t)$ for given values of t. Given y as a function of x, find the rate of change of y with respect to x, and evaluate this rate of change for values of x.

31. Given $s(t) = t + t^4$, find each of the following.
    a) $v(t)$
    b) $a(t)$
    c) The velocity and acceleration when $t = 2$ sec

32. Given $R(x) = 40x$ and $C(x) = 8x^2 - 7x - 10$, find each of the following.
    a) $P(x)$
    b) $R(20)$, $C(20)$, $P(20)$
    c) $R'(x)$, $C'(x)$, $P'(x)$
    d) $R'(20)$, $C'(20)$, $P'(20)$

33. The population of a city grows from an initial size of 10,000 to an amount P, given by $P = 10,000 + 50t^2$, where t is measured in years.
    a) Find the growth rate.
    b) Find the number of people in the city after 20 years (at $t = 20$ yr).
    c) Find the growth rate at $t = 20$.

Find the composition of functions.

34. Find $f \circ g(x)$ and $g \circ f(x)$, given that $f(x) = x^2 + 5$, $g(x) = 1 - 2x$.

EXTENSION EXERCISES

35. Find $\displaystyle\lim_{x\to\infty} \frac{2 - 5x^5}{4 + 3x^6}$.

36. Differentiate $y = \dfrac{x\sqrt{1 + 3x}}{1 + x^3}$.

TEST: CHAPTER 2

Determine whether the function is continuous.

1.

2.

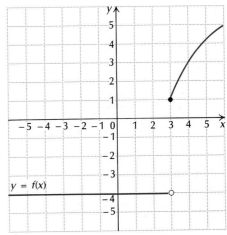

For the function in Question 2, answer each of the following.

3. Find $\displaystyle\lim_{x\to 3} f(x)$.

4. Find $f(3)$.

5. Is f continuous at 3?

6. Find $\displaystyle\lim_{x\to 4} f(x)$.

7. Find $f(4)$.

8. Is f continuous at 4?

Find the limit, if it exists.

9. $\displaystyle\lim_{x\to 1} 3x^4 - 2x^2 + 5$

10. $\displaystyle\lim_{x\to 1} \frac{x - 1}{x^2 - 1}$

11. $\displaystyle\lim_{x\to 0} \frac{7}{x}$

12. $\displaystyle\lim_{x\to\infty} \frac{4x - 3}{x}$

13. Find a simplified difference quotient for

$$f(x) = 3x^2 + 1.$$

14. Find the points on the graph of

$$y = x^3 - 3x^2$$

at which the tangent line is horizontal.

Find dy/dx.

15. $y = x^{84}$

16. $y = 10\sqrt{x}$

17. $y = \dfrac{-10}{x}$

18. $y = x^{5/4}$

19. $y = -0.5x^2 + 0.61x + 90$

Differentiate.

20. $y = \dfrac{1}{3}x^3 - x^2 + 2x + 4$

21. $y = \dfrac{2x - 5}{x^4}$

22. $f(x) = \dfrac{x}{5 - x}$

23. $f(x) = (x + 3)^4(7 - x)^5$

24. $y = (x^5 - 4x^3 + x)^{-5}$

25. $f(x) = x\sqrt{x^2 + 5}$

26. For $y = x^4 - 3x^2$, find d^3y/dx^3.

27. Given $R(x) = 50x$ and $C(x) = 0.001x^2 + 1.2x + 60$, find each of the following.
  a) $P(x)$
  b) $R(10), C(10), P(10)$
  c) $R'(x), C'(x), P'(x)$
  d) $R'(10), C'(10), P'(10)$

28. In a certain memory experiment a person is able to memorize M words after t minutes, where
$$M = -0.001t^3 + 0.1t^2.$$
  a) Find the rate of change of the number of words memorized with respect to time.
  b) How many words are memorized the first 10 min (at $t = 10$)?
  c) What is the memory rate at $t = 10$ min?

29. Find $f \circ g(x)$ and $g \circ f(x)$, given that $f(x) = x + x^2$ and $g(x) = x^3$.

EXTENSION EXERCISES

30. Find $\lim\limits_{x \to 2} \dfrac{x^3 - 8}{x - 2}$.

31. Differentiate $y = (1 - 3x)^{2/3}(1 + 3x)^{1/3}$.

3

APPLICATIONS OF DIFFERENTIATION

In this chapter we learn many applications of differentiation. We learn to find maximum and minimum values of functions, and that skill allows us to solve many kinds of problems where we need to find the largest and/or smallest value of a function. We also apply our differentiation skills to graphing functions and finding the maximum sustainable harvest of a population of animals.

AN APPLICATION

A container firm is designing an open-top rectangular box with a square base that will hold 3000 cubic inches. What dimensions yield the minimum surface area? What is the minimum surface area?

THE MATHEMATICS

Suppose x represents the length of a side of the base. Then the surface area of the box can be expressed as

$$S = x^2 + \frac{12,000}{x}.$$

If we find the first derivative, set it equal to 0, and solve, that value of x will yield the dimensions that will minimize surface area:

$$\frac{dS}{dx} = 2x - \frac{12,000}{x^2} = 0.$$

3.1

OBJECTIVE

a) Find absolute maximum and minimum values of functions using Maximum–Minimum Principle 1.

USING FIRST DERIVATIVES TO FIND ABSOLUTE MAXIMUM AND MINIMUM VALUES

Finding the largest and smallest values of a function, that is, the maximum and minimum values, has extensive application. The first and second derivatives of a function are tools of calculus that give us information about the shape of a graph that may be relevant in finding maximum and minimum values of functions. Throughout this section we will assume that the functions are continuous.

Increasing and Decreasing Functions

If the graph of a function rises from left to right on an interval I, it is said to be *increasing* on I. If the graph drops from left to right, it is said to be *decreasing* on I. We can describe this mathematically as follows.

DEFINITION

A function is *increasing* on I, if for every a and b in I,

if $a < b$, then $f(a) < f(b)$.

A function is *decreasing* on I, if for every a and b in I,

if $a < b$, then $f(a) > f(b)$.

Note that the directions of the inequalities stay the same for an increasing function, but they differ for a decreasing function.

Examples

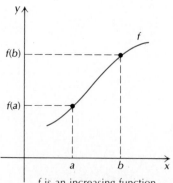

f is an increasing function.
If $a < b$, then $f(a) < f(b)$.
$f'(x) > 0$ for all x.

1. Consider this graph.

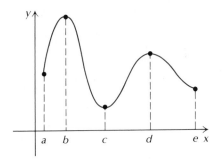

a) Over what intervals is the function increasing?

b) Over what intervals is the derivative positive? To determine this, locate a straightedge at points on the graph and decide whether the slopes of the tangent lines are positive.

c) Over what intervals is the function decreasing?

d) Over what intervals is the derivative negative?

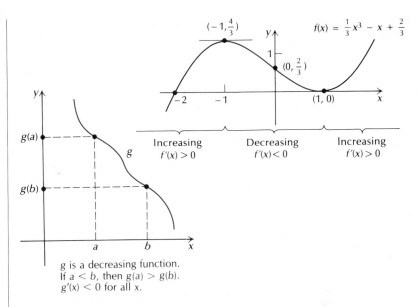

g is a decreasing function.
If $a < b$, then $g(a) > g(b)$.
$g'(x) < 0$ for all x.

In Chapter 1 we saw how the slope of a linear function determines whether that function is increasing or decreasing (or neither). For a general function, the derivative yields similar information. Let us investigate in the margin exercise how this happens.

DO EXERCISE 1.

The following theorem shows how we can use derivatives to determine whether a function is increasing or decreasing.

THEOREM 1

If $f'(x) > 0$, for all x in an interval I, then f is increasing on I.

If $f'(x) < 0$, for all x in an interval I, then f is decreasing on I.

Critical Points

DEFINITION

A *critical point* of a function is an interior point c of its domain at which the function has a horizontal tangent or at which the derivative does not exist. That is, c is a critical point if

$$f'(c) = 0 \quad \text{or} \quad f'(c) \text{ does not exist.}$$

---

2. Consider this graph.

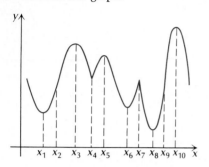

a) At which points are there horizontal tangent lines?

b) At which points does the derivative not exist?

c) Which are critical points?

3. Try to draw a graph of a continuous function from P to Q that increases on part or parts of $[a, b]$ and decreases on part or parts of $[a, b]$.

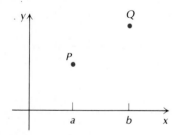

Does the function you drew have any critical points between a and b?

Consider the following graph.

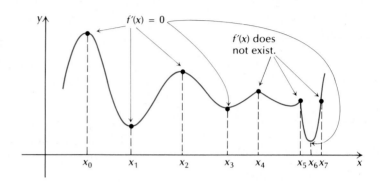

The points x_0, x_1, x_2, x_3, and x_6 are all critical points because the derivative is 0 at each of these points. The points x_4, x_5, and x_7 are all critical points because the derivative does not exist at these points.

DO EXERCISE 2.

The Shape of a Graph Between Critical Points and Endpoints

Suppose we have a continuous function defined over an interval $[a, b]$.

DO EXERCISES 3 AND 4. (EXERCISE 4 IS ON THE FOLLOWING PAGE.)

Consider the following graph.

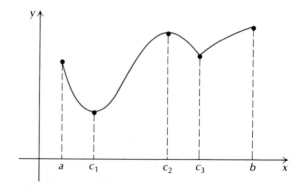

4. Now try to draw a graph of a continuous function from P to Q that increases on part or parts of $[a, b]$ and decreases on part or parts of $[a, b]$, but in such a way that no critical points occur between a and b.

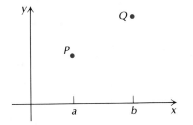

We have three critical points: c_1, c_2, and c_3. Consider these, together with the endpoints

$$a, \quad c_1, \quad c_2, \quad c_3, \quad b.$$

Note, in the preceding graph, that between any two of these points the function is either increasing or decreasing.

This graph and our experience with Margin Exercises 2 and 3 lead us to the following principle.

THE SHAPE PRINCIPLE

Suppose f is a continuous function over an interval $[a, b]$ that has a finite number of critical points $c_1, c_2, \ldots, c_n$. Then, consecutively, between the endpoints and the critical points

$$a, c_1, c_2, c_3, \ldots, c_n, b,$$

the function is either increasing or decreasing.

Finding Absolute Maximum and Minimum Values

Consider the function f whose graph over the interval $[a, b]$ is shown here. The function value of $f(c_1)$ is called an *absolute minimum value* of the function on the interval $[a, b]$, and $f(b)$ is called an *absolute maximum value* of the function on $[a, b]$.

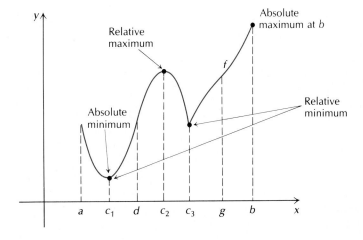

5. In each graph, find the points at which absolute maximum and minimum values occur on $[a, b]$.

a)

b)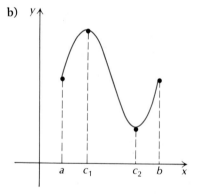

DEFINITION

A function f on an interval I has an *absolute maximum* $f(x_0)$ at x_0 if

$$f(x) \leqslant f(x_0), \quad \text{for all } x \text{ in } I.$$

A function f on an interval I has an *absolute minimum* $f(x_0)$ at x_0 if

$$f(x_0) \leqslant f(x), \quad \text{for all } x \text{ in } I.$$

In other words, f has an absolute maximum value on I if there is a number M greater than or equal to all outputs of the function and there is an input x_0 such that $f(x_0) = M$. That is, the function achieves its absolute maximum value at some point in the interval. Also, f has an absolute minimum value in I if there is a number m less than or equal to all outputs of the function and there is an input x_0 such that $f(x_0) = m$. That is, the function achieves its absolute minimum value at some point in the interval.

DO EXERCISE 5.

You may have discovered two theorems, whose proofs we will not consider. Each of the functions in Margin Exercise 5 did indeed have an absolute maximum and an absolute minimum value. This leads us to one of the theorems.

THEOREM 2

The Extreme-Value Theorem

A continuous function f defined on a closed interval $[a, b]$ must have an absolute maximum and minimum value at points in $[a, b]$.

Look carefully at each graph in Margin Exercise 5 and consider the critical points and the endpoints. In part (a) the graph starts at $f(a)$ and falls to $f(c_1)$. Then it rises from $f(c_1)$ to $f(c_2)$. From there it falls to $f(b)$. In part (b) the graph starts at $f(a)$ and rises to $f(c_1)$. Then it falls from $f(c_1)$ to $f(c_2)$. From there it rises to $f(b)$. It seems reasonable that whatever the maximum and minimum values are, they occur among the function values $f(a)$, $f(c_1)$, $f(c_2)$, and $f(b)$. This leads us to the second theorem, which is a modification of the Shape Principle.

THEOREM 3

Maximum–Minimum Principle 1

Suppose f is a continuous function over an interval $[a, b]$. To find the absolute maximum and minimum values of the function on $[a, b]$:

a) First find $f'(x)$.

b) Then find the critical points. That is, find all points c for which

$$f'(c) = 0 \quad \text{or} \quad f'(c) \text{ does not exist.}$$

c) Determine the critical points and the endpoints of the interval,

$$a, c_1, c_2, \ldots, c_n, b.$$

d) Find the function values at the points in part (c):

$$f(a), f(c_1), f(c_2), \ldots, f(c_n), f(b).$$

The largest of these is the *absolute maximum* of f on the interval $[a, b]$. The smallest of these is the *absolute minimum* of f on the interval $[a, b]$.

Example 1 Find the absolute maximum and minimum values of

$$f(x) = x^3 - 3x + 2$$

on the interval $[-2, \frac{3}{2}]$.

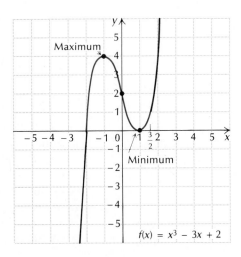

$f(x) = x^3 - 3x + 2$

6. Find the absolute maximum and minimum values of

$$f(x) = x^3 - x^2 - x + 2$$

on the interval $[-1, 2]$.

Solution

a) Find $f'(x)$:

$$f'(x) = 3x^2 - 3.$$

b) Find the critical points. The derivative exists for all real numbers. Thus we merely solve $f'(x) = 0$:

$$3x^2 - 3 = 0$$
$$3x^2 = 3$$
$$x^2 = 1$$
$$x = \pm 1.$$

c) Determine the critical points and the endpoints. These points are $-2, -1, 1, \frac{3}{2}$.

d) Find the function values at the points in part (c):

$$f(-2) = (-2)^3 - 3(-2) + 2 = -8 + 6 + 2 = 0; \quad \text{Minimum}$$
$$f(-1) = (-1)^3 - 3(-1) + 2 = -1 + 3 + 2 = 4; \quad \text{Maximum}$$
$$f(1) = (1)^3 - 3(1) + 2 = 1 - 3 + 2 = 0; \quad \text{Minimum}$$
$$f\left(\frac{3}{2}\right) = \left(\frac{3}{2}\right)^3 - 3\left(\frac{3}{2}\right) + 2$$
$$= \frac{27}{8} - \frac{9}{2} + 2$$
$$= \frac{27}{8} - \frac{36}{8} + \frac{16}{8}$$
$$= \frac{7}{8}.$$

The largest of these values, 4, is the maximum. It occurs at $x = -1$. The smallest of these values is 0. It occurs twice at $x = -2$ and $x = 1$. Thus on the interval $[-2, \frac{3}{2}]$ the

$$\text{absolute maximum} = 4 \text{ at } x = -1$$

and the

$$\text{absolute minimum} = 0 \text{ at } x = -2 \text{ and } x = 1.$$

Note that an absolute maximum or minimum value can occur at more than one point.

DO EXERCISE 6.

7. Find the absolute maximum and minimum values of

$$f(x) = x^3 - x^2 - x + 2$$

on the interval $[5, 6]$.

Example 2   Find the absolute maximum and minimum values of

$$f(x) = x^3 - 3x + 2$$

on the interval $[-3, -\frac{3}{2}]$.

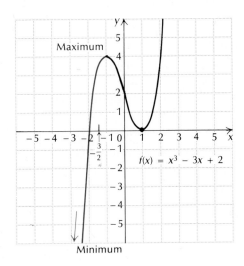

Solution   As in Example 1, the derivative is 0 at -1 and 1. But neither -1 nor 1 is in the interval $[-3, -\frac{3}{2}]$, so there are no critical points in this interval. Thus the maximum and minimum values occur at the endpoints:

$$f(-3) = (-3)^3 - 3(-3) + 2$$
$$= -27 + 9 + 2 = -16; \quad \text{Minimum}$$

$$f\left(-\frac{3}{2}\right) = \left(-\frac{3}{2}\right)^3 - 3\left(-\frac{3}{2}\right) + 2$$

$$= -\frac{27}{8} + \frac{9}{2} + 2 = \frac{25}{8} = 3\frac{1}{8}. \quad \text{Maximum}$$

Thus, on the interval $[-3, -\frac{3}{2}]$, the

$$\text{absolute maximum} = 3\tfrac{1}{8} \text{ at } x = -\tfrac{3}{2}$$

and the

$$\text{absolute minimum} = -16 \text{ at } x = -3.$$

DO EXERCISE 7.

EXERCISE SET 3.1

The curves on the graph at the right show the gasoline mileage obtained when traveling at a constant speed for an average-size car and for a compact car.

1. Consider the graph for the average-size car over the interval [20, 80].

   a) Estimate the speed at which the absolute maximum gasoline mileage is obtained.

   b) Estimate the speed at which the absolute minimum gasoline mileage is obtained.

   c) What is the mileage obtained at 70 mph?

   d) What is the mileage obtained at 55 mph?

   e) What percent increase in mileage is there by traveling at 55 mph rather than at 70 mph?

2. Answer the questions in Exercise 1 for the compact car.

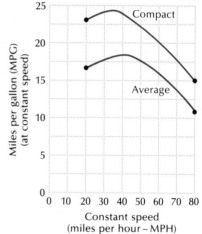

For each of the following functions, find the absolute maximum and minimum values, if they exist, over the indicated interval.

3. $f(x) = 5 + x - x^2$; [0, 2]

4. $f(x) = 4 + x - x^2$; [0, 2]

5. $f(x) = x^3 - x^2 - x + 2$; [0, 2]

6. $f(x) = x^3 + \frac{1}{2}x^2 - 2x + 5$; [0, 1]

7. $f(x) = x^3 - x^2 - x + 2$; [-1, 0]

8. $f(x) = x^3 + \frac{1}{2}x^2 - 2x + 5$; [-2, 0]

9. $f(x) = 3x - 2$; [-1, 1]

10. $f(x) = 2x + 4$; [-1, 1]

11. $f(x) = 7 - 4x$; [-2, 5]

12. $f(x) = -2 + 8x$; [-10, 10]

13. $f(x) = -5$; [-1, 1]

14. $g(x) = 24$; [4, 13]

3.2

OBJECTIVE

a) Find absolute maximum and minimum values using Maximum–Minimum Principle 2.

USING SECOND DERIVATIVES TO FIND ABSOLUTE MAXIMUM AND MINIMUM VALUES

Concavity: Increasing and Decreasing Derivatives

Shown on the following page are the graphs of two functions. The graph on the left is turning upward and the other is turning downward. Let's see if we can relate this to their derivatives.

Consider the graph of f. Take a ruler, or straightedge, and draw tangent lines as you move along the curve from left to right. What

1. Consider this graph.

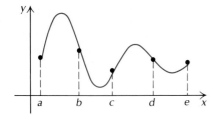

a) Over what intervals is the derivative increasing?

b) Over what intervals is the second derivative positive?

c) Over what intervals is the derivative decreasing?

d) Over what intervals is the second derivative negative?

happens to the slopes of the tangent lines? Do the same for the graph of g. Look for a pattern.

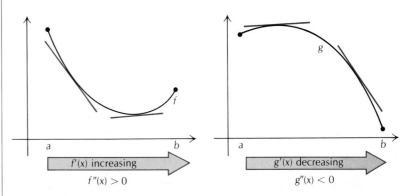

In the graph on the left the slopes are increasing. That is, f'' is positive so f' is increasing on the interval. In the graph on the right the slopes are decreasing. That is, g'' is negative, so g' is decreasing on the interval.

DO EXERCISE 1.

We now have the following theorem.

THEOREM 4

1. If $f''(x) > 0$ on an interval I, then f is turning upward on I (since f' is increasing on I). Such a graph is said to be *concave up* over I.

2. If $f''(x) < 0$ on an interval I, then f is turning downward on I (since f' is decreasing on I). Such a graph is said to be *concave down* over I.

The following figure is a helpful memory device.

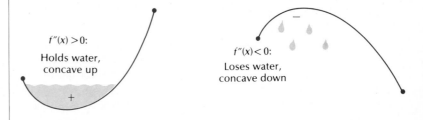

$f''(x) > 0$:

Holds water, concave up

$f''(x) < 0$:

Loses water, concave down

2. Consider this graph.

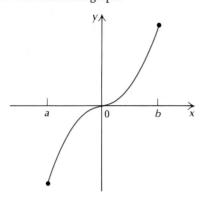

a) Over what intervals is the graph concave up?

b) Over what intervals is the graph concave down?

3. What are the points of inflection?

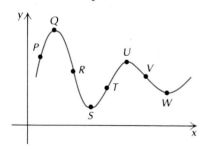

DO EXERCISE 2.

A *point of inflection,* or an *inflection point,* is a point across which the direction of concavity changes. For example, in the figure below, point P is an inflection point of the first graph. Points P, Q, R, and S are inflection points of the second graph. In Margin Exercise 2, the point $(0, 0)$ is an inflection point.

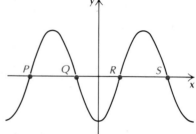

DEFINITION

The point c is a *point of inflection* if for some open interval containing c, either

a) f' is decreasing to the left of c and increasing to the right of c (that is, $f''(x) < 0$ for $x < c$ and $f''(x) > 0$ for $x > c$); or

b) f' is increasing to the left of c and decreasing to the right of c (that is, $f''(x) > 0$ for $x < c$ and $f''(x) < 0$ for $x > c$).

DO EXERCISE 3.

Just knowing the values of f' and f'' at some specific point x_0 can yield much information about the shape of the graph over some (possibly small) interval containing x_0 as an interior point (assuming f'' exists and is continuous over the interval).

	Derivatives	Effect on the graph at $x = x_0$	Shape of the graph at $x = x_0$
1.	$f'(x_0) > 0$, $f''(x_0) > 0$	f is increasing at x_0. The graph of f is concave up (over some interval containing x_0).	
2.	$f'(x_0) > 0$, $f''(x_0) < 0$	f is increasing at $x = x_0$. The graph of f is concave down.	
3.	$f'(x_0) < 0$, $f''(x_0) > 0$	f is decreasing at $x = x_0$. The graph of f is concave up.	
4.	$f'(x_0) < 0$, $f''(x_0) < 0$	f is decreasing at $x = x_0$. The graph of f is concave down.	
5.	$f'(x_0) = 0$, $f''(x_0) > 0$	f' is negative to the left of x_0 and positive to the right of x_0. The graph of f is concave up.	
6.	$f'(x_0) = 0$, $f''(x_0) < 0$	f' is positive to the left of x_0 and negative to the right of x_0. The graph of f is concave down.	

When there is only one critical point c_0 in I, we may not need to check endpoint values to determine whether the function has an absolute maximum or minimum value at that point. Consider the following cases.

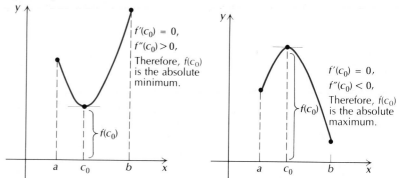

When $f'(c_0) = 0$ and $f''(c_0) > 0$, $f'(x)$ changes from negative to positive as x goes from the left of c_0 to the right. That is, the function f is decreasing to the left of c_0 and increasing to the right of c_0. It follows that $f(c_0)$ is the absolute minimum value of f on I. Similarly, if $f'(c_0) = 0$ and $f''(c_0) < 0$, $f'(x)$ changes from positive to negative as x goes from the left of c_0 to the right. That is, the function f is increasing to the left of c_0 and decreasing to the right of c_0. It follows that $f(c_0)$ is the maximum value of f on I. The above turns out to hold no matter what the interval I is, whether open, closed, or extending to infinity. This leads us to another theorem for finding absolute maximum and minimum values.

THEOREM 5

Maximum–Minimum Principle 2

Suppose f is a function such that $f'(x)$ exists for every x in an interval I, and that there is *exactly one* (critical) point c, interior to I, for which $f'(c) = 0$. Then:

$\qquad f(c)$ is the absolute maximum value on I if $f''(c) < 0$

or

$\qquad f(c)$ is the absolute minimum value on I if $f''(c) > 0$.

If $f''(c) = 0$, we would have to use Maximum–Minimum Principle 1, or we would have to know more about the behavior of the function on the given interval.

4. Find the absolute maximum and minimum values of

$$f(x) = x^2 - 4x.$$

Example 1   Find the absolute maximum and minimum values of

$$f(x) = 4x - x^2.$$

Solution   When no interval is specified, we consider the entire domain of the function. In this case, the domain is the set of all real numbers.

a)   Find $f'(x)$:

$$f'(x) = 4 - 2x.$$

b)   Find the critical points. The derivative exists for all real numbers. Thus we merely solve $f'(x) = 0$:

$$4 - 2x = 0$$
$$-2x = -4$$
$$x = 2.$$

c)   Since there is only one critical point, we can apply Maximum–Minimum Principle 2 using the second derivative:

$$f''(x) = -2.$$

Since the second derivative is constant, $f''(2) = -2$, and since this is negative, we have the

absolute maximum $= f(2) = 4 \cdot 2 - 2^2 = 8 - 4 = 4$ at $x = 2$.

The function has no minimum, as the graph shown here indicates.

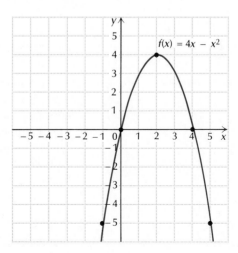

DO EXERCISE 4.

5. Find the absolute maximum and minimum values of

$$f(x) = x^2 - 4x$$

on the interval [0, 4].

Example 2 Find the absolute maximum and minimum values of $f(x) = 4x - x^2$ on the interval [0, 4].

Solution By the reasoning in Example 1 we know that the absolute maximum value of f on $(-\infty, \infty)$ is $f(2)$, or 4. Since 2 is in the interval [0, 4], the absolute maximum of f on [0, 4] will occur at 2. We need not check the endpoints. We must check, however, for the absolute minimum:

$$f(0) = 4 \cdot 0 - 0^2 = \mathbf{0} \quad \text{and} \quad f(4) = 4 \cdot 4 - 4^2 = 0.$$

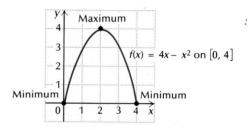

We see that the minimum is 0. It occurs twice at $x = 0$ and $x = 4$. Thus the

$$\text{absolute maximum} = 4 \text{ at } x = 2$$

and the

$$\text{absolute minimum} = 0 \text{ at } x = 0 \text{ and } x = 4.$$

DO EXERCISE 5.

We have thus far restricted the use of Maximum–Minimum Principle 2 to intervals with one critical point. Suppose a closed interval contains two critical points. Then we could break the interval up into two subintervals, consider maximums and minimums on those subintervals, and compare. But values at the endpoints would have to be considered and since we would, in effect, be using Maximum–Minimum Principle 1, we may as well use it at the outset.

A Strategy for Finding Maximum and Minimum Values

The following general strategy can be used when finding maximum and minimum values of functions.

> ### A Strategy for Finding Absolute Maximum and Minimum Values
>
> To find absolute maximum and minimum values of a continuous function on an interval:
>
> a) Find $f'(x)$.
>
> b) Find the critical points.
>
> c) If the interval is closed and there is more than one critical point, then use Maximum–Minimum Principle 1.
>
> d) If the interval is closed and there is exactly one critical point, use either Maximum–Minimum Principle 1 or Maximum–Minimum Principle 2. If the function is easy to differentiate, use Maximum–Minimum Principle 2.
>
> e) If the interval is not closed, does not have endpoints, or does not contain its endpoints, such as $(-\infty, \infty), (0, \infty)$, or (a, b), and the function has only one critical point, use Maximum–Minimum Principle 2. In such a case, if the function has a maximum, it will have no minimum; and if it has a minimum, it will have no maximum.

The case of finding maximum and minimum values when more than one critical point occurs in an interval described in part (e) above must be dealt with by a detailed graph or by techniques beyond the scope of this book.

Example 3 Find the absolute maximum and minimum values of

$$f(x) = (x - 1)^3 + 2.$$

Solution

a) Find $f'(x)$:

$$f'(x) = 3(x - 1)^2.$$

b) Find the critical points. The derivative exists for all real numbers. Thus we solve $f'(x) = 0$:

$$3(x - 1)^2 = 0$$
$$(x - 1)^2 = 0$$
$$x - 1 = 0$$
$$x = 1.$$

6. Find the absolute maximum and minimum values of

$$f(x) = x^3.$$

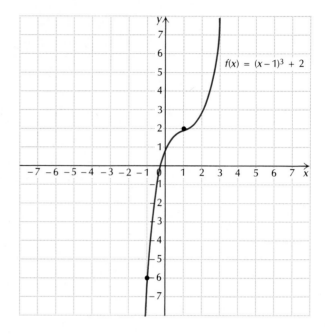

$$f(x) = (x-1)^3 + 2$$

7. Find the absolute maximum and minimum values of

$$f(x) = x^3$$

on the interval $[-2, 2]$. (*Hint:* This function must have maximum and minimum values because it is restricted to a closed interval.) What are the only numbers at which these can occur?

e) Since there is only one critical point and there are no endpoints, we can try to apply Maximum–Minimum Principle 2 using the second derivative:

$$f''(x) = 6(x - 1).$$

Now

$$f''(1) = 6(1 - 1)$$
$$= 0,$$

so Maximum–Minimum Principle 2 fails. We cannot use Maximum–Minimum Principle 1 because there are no endpoints. But note that $f'(x) = 3(x - 1)^2$ is never negative. Thus $f(x)$ is increasing everywhere except at $x = 1$, so there is no maximum or minimum. At $x = 1$, the function has a *point of inflection.*

DO EXERCISES 6 AND 7.

Example 4   Find the absolute maximum and minimum values of $f(x) = 5x + (35/x)$ on the interval $(0, \infty)$.

8. Find the absolute maximum and minimum values of

$$f(x) = 10x + \frac{1}{x}$$

on $(0, \infty)$.

Solution

a) Find $f'(x)$. We first express $f(x)$ as

$$f(x) = 5x + 35x^{-1}.$$

Then

$$f'(x) = 5 - 35x^{-2} = 5 - \frac{35}{x^2}.$$

b) Find the critical points. Now $f'(x)$ exists for all values of x in $(0, \infty)$. Thus the only critical points are those for which $f'(x) = 0$:

$$5 - \frac{35}{x^2} = 0$$

$$5 = \frac{35}{x^2}$$

$$5x^2 = 35 \qquad \text{Multiplying by } x^2, \text{ since } x \neq 0$$

$$x^2 = 7$$

$$x = \pm\sqrt{7}.$$

e) The interval is not closed and is $(0, \infty)$. The only critical point is $\sqrt{7}$. Thus we can apply Maximum–Minimum Principle 2 using the second derivative,

$$f''(x) = 70x^{-3}$$

$$= \frac{70}{x^3},$$

to determine whether we have a maximum or a minimum. Now $f''(x)$ is positive for all values of x in $(0, \infty)$, so $f''(\sqrt{7}) > 0$, and the

$$\text{absolute minimum} = f(\sqrt{7}) = 5 \cdot \sqrt{7} + \frac{35}{\sqrt{7}}$$

$$= 5\sqrt{7} + \frac{35}{\sqrt{7}} \cdot \frac{\sqrt{7}}{\sqrt{7}} = 5\sqrt{7} + \frac{35\sqrt{7}}{7}$$

$$= 5\sqrt{7} + 5\sqrt{7} = 10\sqrt{7}$$

at $x = \sqrt{7}$.

The function has no maximum value.

DO EXERCISE 8.

EXERCISE SET 3.2

For each of the following functions, find the absolute maximum and minimum values, if they exist, over the indicated interval. When no interval is specified, use the real line $(-\infty, \infty)$.

1. $f(x) = x(70 - x)$

2. $f(x) = x(50 - x)$

3. $f(x) = 2x^2 - 40x + 400$

4. $f(x) = 2x^2 - 20x + 100$

5. $f(x) = x - \frac{4}{3}x^3;\ (0, \infty)$

6. $f(x) = 16x - \frac{4}{3}x^3;\ (0, \infty)$

7. $f(x) = 17x - x^2$

8. $f(x) = 27x - x^2$

9. $f(x) = \frac{1}{3}x^3 - 3x;\ [-2, 2]$

10. $f(x) = \frac{1}{3}x^3 - 5x;\ [-3, 3]$

11. $f(x) = -0.001x^2 + 4.8x - 60$

12. $f(x) = -0.01x^2 + 1.4x - 30$

13. $f(x) = -\frac{1}{3}x^3 + 6x^2 - 11x - 50;\ (0, 3)$

14. $f(x) = -x^3 + x^2 + 5x - 1;\ (0, \infty)$

15. $f(x) = 15x^2 - \frac{1}{2}x^3;\ [0, 30]$

16. $f(x) = 4x^2 - \frac{1}{2}x^3;\ [0, 8]$

17. $f(x) = 2x + \dfrac{72}{x};\ (0, \infty)$

18. $f(x) = x + \dfrac{3600}{x};\ (0, \infty)$

19. $f(x) = x^2 + \dfrac{432}{x};\ (0, \infty)$

20. $f(x) = x^2 + \dfrac{250}{x};\ (0, \infty)$

21. $f(x) = 2x^4 - x;\ [-1, 1]$

22. $f(x) = 2x^4 + x;\ [-1, 1]$

23. $f(x) = \sqrt[3]{x};\ [0, 8]$

24. $f(x) = \sqrt{x};\ [0, 4]$

25. $f(x) = (x + 1)^3$

26. $f(x) = (x - 1)^3$

27. $f(x) = 2x - 3;\ [-1, 1]$

28. $f(x) = 9 - 5x;\ [-10, 10]$

29. $f(x) = 2x - 3$

30. $f(x) = 9 - 5x$

31. See Exercise 10 in Exercise Set 2.6. What is the maximum number of units sold? What must be spent on advertising in order to sell that number of units?

32. See Exercise 9 in Exercise Set 2.6. What is the maximum temperature during the illness and on what day does it occur?

33. See Exercise 55 in Exercise Set 1.4.

    a) What is the maximum distance it takes to stop on glare ice? At what air temperature does this occur?

    b) What is the minimum distance it takes to stop on glare ice? At what air temperature does this occur?

34. See Exercise 5 in Exercise Set 2.6. Consider the function over the interval [0, 40], that is, the years 1950 to 1990.

    a) What is the maximum median age of women at first marriage and in what year does it occur?

    b) What is the minimum median age of women at first marriage and in what year does it occur?

35. In Exercise 4 of Exercise Set 1.6, we determined that at travel speed (constant velocity) x there are y accidents at nighttime for every 100 million miles of travel, where y is given by

$$y = 6.1x^2 - 752x + 22{,}620.$$

At what travel speed does the fewest number of accidents occur?

36. At travel speed (constant velocity) x, the cost y, in cents per mile, of operating a car is approximated by

$$y = 0.02x^2 - 1.3x + 30.$$

At what travel speed is the cost of operating a car a minimum?

EXTENSION EXERCISES

Find the absolute maximum and minimum values, if they exist, over the indicated interval. When no interval is specified, use the real line $(-\infty, \infty)$.

37. $g(x) = x\sqrt{x + 3}; [-3, 3]$

38. $h(x) = x\sqrt{1 - x}; [0, 1]$

39. $f(x) = x^{2/3}; [-1, 1]$

40. $g(x) = x^{2/3}$

41. $f(x) = \frac{1}{3}x^3 - x + \frac{2}{3}$

42. $f(x) = \frac{1}{3}x^3 - \frac{1}{2}x^2 - 2x + 1$

43. $f(x) = \frac{1}{3}x^3 - 2x^2 + x; [0, 4]$

44. $g(x) = \frac{1}{3}x^3 + 2x^2 + x; [-4, 0]$

45. $t(x) = x^4 - 2x^2$

46. $f(x) = 2x^4 - 4x^2 + 2$

47. *Business.* Several costs in a business environment can be separated into two components: those that increase with volume and those that decrease with volume. Quality of customer service, although more expensive as it is increased, has part of its increased cost offset by customer goodwill. A firm has determined that its cost of service is given by the following function of "quality units,"

$$C(x) = (2x + 4) + \left(\frac{2}{x - 6}\right), \quad x > 6.$$

Find the number of "quality units" the firm should use in order to minimize its total cost of service.

48. Let

$$y = (x - a)^2 + (x - b)^2.$$

For what value of x is y a minimum?

3.3

OBJECTIVES

a) Find the relative maxima and minima of a function.

b) Sketch the graph of a function.

RELATIVE MAXIMA AND MINIMA AND GRAPH SKETCHING*

Our goal in this section is to use derivatives to sketch graphs. Before doing this we need to consider the concepts of *relative maximum* and *relative minimum*.

Relative Maxima and Minima

Apart from the (absolute) maxima (plural of "maximum") and minima (plural of "minimum") of a function, there can be smaller intervals over which there are "peaks" and "valleys." Over those subintervals a function can have a maximum or a minimum. We call each of these a *relative maximum* or a *relative minimum*.

Consider the following graph.

---

* This section can be omitted without loss of continuity.

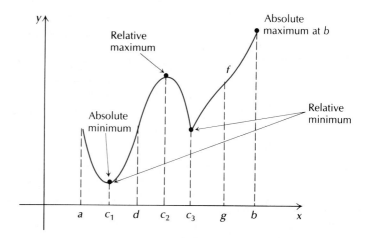

At c_2 the function has a relative maximum. Note that the absolute maximum occurs at the endpoint b. At c_1 and c_3 the function has a relative minimum. The function's absolute minimum occurs at c_1.

DEFINITION

A function f has a *relative maximum* at x_0 in an interval I if there is an open interval in I containing x_0 on which f has a maximum at x_0.

A function f has a *relative minimum* at x_0 in an interval I if there is an open interval in I containing x_0 on which f has a minimum at x_0.

There are two tests that we can use to determine whether a function has relative maxima or minima. They are quite similar to Maximum–Minimum Principles 1 and 2.

THEOREM 6

The First Derivative Test

For any continuous function f that has a critical point c in interval I:

i) If $f'(x) > 0$ on an open interval (u, c) to the left of c and $f'(x) < 0$ on an open interval (c, v) to the right of c, then f has a relative maximum at c. That is, if f is increasing to the left of c and decreasing to the right of c, then f has a relative maximum at c.

ii) If $f'(x) < 0$ on an open interval (u, c) to the left of c and $f'(x) > 0$ on an open interval (c, v) to the right of c, then f has a relative minimum at c. That is, if f is decreasing to the left of c and increasing to the right of c, then f has a relative minimum at c.

iii)   If $f'(x)$ has the same sign on an open interval (u, c) to the left of c and on an open interval (c, v) to the right of c, then f has neither a relative maximum nor a relative minimum at c.

THEOREM 7

The Second Derivative Test

Suppose f is a function such that $f'(x)$ exists for every x in an open interval (a, b) contained in an interval I, and that there is a critical point c in (a, b) for which $f'(c) = 0$. Then

$$f(c) \text{ is a relative maximum on } I \text{ if } f''(c) < 0$$

or

$$f(c) \text{ is a relative minimum on } I \text{ if } f''(c) > 0.$$

The Second Derivative Test for relative maxima or minima differs from Maximum–Minimum Principle 2 in that there is no restriction that there must be *exactly* one critical point.

Let us see how we can use these tests to find relative maxima and minima and to sketch graphs. We will use the following procedure.

Procedure for Finding Relative Maxima and Minima and for Sketching Graphs

a)   Find $f'(x)$ and $f''(x)$.

b)   Find the critical points of f by finding where $f'(x)$ does not exist and by solving $f'(x) = 0$. Find the function values at these points.

c)   Use the critical points. Find the intervals on which f is increasing by solving $f'(x) > 0$. Find the intervals on which f is decreasing by solving $f'(x) < 0$. Use this information to determine the relative maxima and minima.

d)   Find the critical points of f' by finding where $f''(x)$ does not exist and by solving $f''(x) = 0$. Find the function values at these points.

e)   Find the intervals on which f is concave up by solving $f''(x) > 0$. Find the intervals on which f is concave down by solving $f''(x) < 0$.

f)   Sketch the graph using the information from steps (a) through (e), plotting extra points (computing them with your calculator) if the need arises.

Example 1 For the following function, find the relative maxima and minima and sketch the graph.

$$f(x) = x^3 - 3x + 2$$

Solution

a) Find $f'(x)$ and $f''(x)$:

$$f'(x) = 3x^2 - 3,$$
$$f''(x) = 6x.$$

b) Find the critical points of f by finding where $f'(x)$ does not exist and by solving $f'(x) = 0$. Since $f'(x) = 3x^2 - 3$ exists for all values of x, the only critical points are where

$$3x^2 - 3 = 0$$
$$3x^2 = 3$$
$$x^2 = 1$$
$$x = \pm 1.$$

Now $f(-1) = 4$ and $f(1) = 0$. These give the points $(-1, 4)$ and $(1, 0)$ on the graph.

c) Use the critical points. Find the intervals on which f is increasing by solving $f'(x) > 0$. Find the intervals on which f is decreasing by solving $f'(x) < 0$. Whether $f'(x) = 3x^2 - 3 = 3(x + 1)(x - 1)$ is positive or negative depends on the positiveness or negativeness (signs) of the factors $x + 1$ and $x - 1$. We can determine this efficiently using the following diagram.

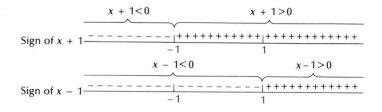

To set up the diagram, we solve $x + 1 > 0$ to get $x > -1$. Thus $x + 1$ is positive for all numbers to the right of -1. We indicate that with the $+$ signs. Accordingly, $x + 1 < 0$ for all numbers to the left of -1. We indicate that with $-$ signs.

Similarly, we solve $x - 1 > 0$ to get $x > 1$. Thus $x - 1$ is positive for all numbers to the right of 1. Accordingly, $x - 1$ is negative for all numbers to the left of 1. We indicate these facts with the $+$ and $-$ signs.

Finally, we determine the signs of the product by using the rules for multiplication. In order for the product $3(x + 1)(x - 1)$ to be negative, one factor must be positive and the other negative. The only situation in the diagram for which this happens is in the interval $(-1, 1)$, when $-1 < x < 1$. In order for the product to be positive, both factors must be positive or both must be negative. Both factors are negative on the interval $(-\infty, -1)$, when $x < -1$. Both factors are positive on the interval $(1, \infty)$, when $x > 1$.

Since $f'(x)$ is positive in an interval to the left of -1 (f increasing) and negative in an interval to the right of -1 (f decreasing), the function has a relative maximum at $(-1, 4)$. Since $f'(x)$ is negative in an interval to the left of 1 (f decreasing) and positive in an interval to the right of 1 (f increasing), the function has a relative minimum at $(1, 0)$.

d) Find the critical points of f' by finding where $f''(x)$ does not exist and by solving $f''(x) = 0$. Since $f''(x) = 6x$ exists for all values of x, the only critical points of f' are where

$$6x = 0$$
$$x = 0.$$

Now $f(0) = 2$. This gives us another point, $(0, 2)$, that lies on the graph.

e) Find the intervals at which f is concave up by solving $f''(x) > 0$. Find the intervals at which f is concave down by solving $f''(x) < 0$. We see that $6x < 0$ when $x < 0$ and that $6x > 0$ when $x > 0$, so f is concave down on the interval $(-\infty, 0)$ and concave up on the interval $(0, \infty)$. Since $(0, 2)$ is a point across which the concavity changes, it is a point of inflection.

f) Sketch the graph using the preceding information.

1. For the following function, find the relative maxima and minima and sketch the graph.

$$f(x) = \tfrac{1}{3}x^3 - \tfrac{1}{2}x^2 - 2x + 1$$

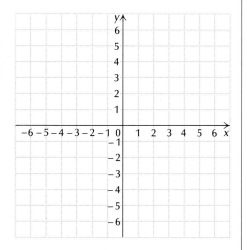

DO EXERCISE 1.

Example 2   For the following function, find the relative maxima and minima and sketch the graph.

$$f(x) = x^4 - 2x^2$$

Solution

a) Find $f'(x)$ and $f''(x)$:

$$f'(x) = 4x^3 - 4x, \qquad f''(x) = 12x^2 - 4.$$

b) Since $f'(x) = 4x^3 - 4x$ exists for all values of x, the only critical points are where

$$4x^3 - 4x = 0$$
$$4x(x^2 - 1) = 0$$
$$4x = 0 \quad \text{or} \quad x^2 - 1 = 0$$
$$x = 0 \quad \text{or} \quad x^2 = 1$$
$$x = \pm 1.$$

Now $f(0) = 0$, $f(-1) = -1$, and $f(1) = -1$. These give the points $(0, 0)$, $(-1, -1)$, and $(1, -1)$ on the graph.

c) Whether $f'(x) = 4x^3 - 4x = 4x(x + 1)(x - 1)$ is positive or negative depends on the signs of the factors $4x$, $x + 1$, and $x - 1$. We can determine this efficiently using the following diagram.

There is a relative maximum at $(0, 0)$ and two relative minima at $(-1, -1)$ and $(1, -1)$.

d) Since $f''(x) = 12x^2 - 4$ exists for all values of x, the only critical points of f' are where

$$12x^2 - 4 = 0$$
$$4(3x^2 - 1) = 0$$
$$3x^2 - 1 = 0$$
$$3x^2 = 1$$
$$x^2 = \frac{1}{3}$$
$$x = \pm\sqrt{\frac{1}{3}} = \pm\frac{1}{\sqrt{3}}.$$

Now

$$f\left(\frac{1}{\sqrt{3}}\right) = \left(\frac{1}{\sqrt{3}}\right)^4 - 2\left(\frac{1}{\sqrt{3}}\right)^2 = \frac{1}{9} - \frac{2}{3} = -\frac{5}{9}$$

and

$$f\left(-\frac{1}{\sqrt{3}}\right) = -\frac{5}{9}.$$

These give the points

$$\left(-\frac{1}{\sqrt{3}}, -\frac{5}{9}\right) \quad \text{and} \quad \left(\frac{1}{\sqrt{3}}, -\frac{5}{9}\right)$$

on the graph. These are $(-0.6, -0.6)$ and $(0.6, -0.6)$, approximately.

e) Whether $f''(x) = 12x^2 - 4 = 4(3x^2 - 1) = 4(\sqrt{3}x + 1)(\sqrt{3}x - 1)$ is positive or negative depends on the signs of the factors $\sqrt{3}x + 1$ and $\sqrt{3}x - 1$. We can determine this using the following diagram.

Sign of $\sqrt{3}x + 1$ $\underline{- - - - - - - -|+ + + + + + + + + + +|+ + + + + + + + +}$
$\phantom{Sign of \sqrt{3}x + 1 xxxxx}-1/\sqrt{3} \approx -0.6 1/\sqrt{3} \approx 0.6$

Sign of $\sqrt{3}x - 1$ $\underline{- - - - - - - -|- - - - - - - - - - -|+ + + + + + + + +}$
$\phantom{Sign of \sqrt{3}x - 1 xxxxx}-1/\sqrt{3} 1/\sqrt{3}$

Sign of
$f''(x) = 4(\sqrt{3}x + 1)(\sqrt{3}x - 1)$ $\underline{+ + + + + + + + + +|- - - - - - - - - - -|+ + + + + + + + +}$
$\phantom{f''(x) = 4(\sqrt{3}x + 1)(\sqrt{3}x - 1) xx}-1/\sqrt{3} 1/\sqrt{3}$

We see that f is concave up on the intervals

$$\left(-\infty, -\frac{1}{\sqrt{3}}\right) \quad \text{and} \quad \left(\frac{1}{\sqrt{3}}, \infty\right)$$

and concave down on the interval

$$\left(-\frac{1}{\sqrt{3}}, \frac{1}{\sqrt{3}}\right).$$

Since each of the points

$$\left(-\frac{1}{\sqrt{3}}, -\frac{5}{9}\right) \quad \text{and} \quad \left(\frac{1}{\sqrt{3}}, -\frac{5}{9}\right)$$

is a point on the graph across which the concavity changes, each is a point of inflection.

f) Sketch the graph using the preceding information. By solving $x^4 - 2x^2 = 0$ we can find the x-intercepts easily. They are $(-\sqrt{2}, 0)$, $(0, 0)$, and $(\sqrt{2}, 0)$. This also aids the graphing.

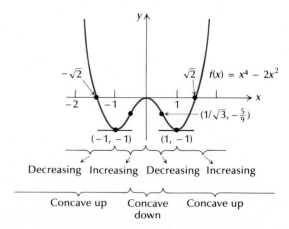

2. For the following function, find the relative maxima and minima and sketch the graph.

$$f(x) = 2x^4 - 4x^2 + 2$$

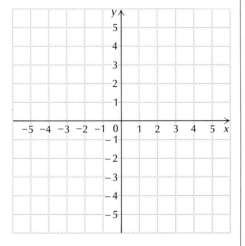

DO EXERCISE 2.

Example 3 For the following function, find the relative maxima and minima and sketch the graph.

$$f(x) = x^{2/3}$$

Solution First note that $f(x) = x^{2/3} = \sqrt[3]{x^2} = (\sqrt[3]{x})^2$ and that $f(x)$ exists for all x in $(-\infty, \infty)$.

a) Find $f'(x)$ and $f''(x)$:

$$f'(x) = \frac{2}{3}x^{-1/3}$$

$$= \frac{2}{3\sqrt[3]{x}},$$

$$f''(x) = -\frac{2}{9}x^{-4/3}$$

$$= -\frac{2}{9\sqrt[3]{x^4}},$$

b) Since $f'(0)$ does not exist, 0 is a critical point. The equation $f'(x) = 0$ has no solution, so the only critical point is 0. Now $f(0) = 0^{2/3} = 0$. This gives the point $(0, 0)$ on the graph.

c) Now when $x < 0$, $\sqrt[3]{x} < 0$, so $f'(x) = 2/3\sqrt[3]{x} < 0$. When $x > 0$, $\sqrt[3]{x} > 0$, so

$$f'(x) = \frac{2}{3\sqrt[3]{x}} > 0.$$

Thus f is decreasing on the interval $(-\infty, 0)$ and increasing on the interval $(0, \infty)$. There is a relative minimum at $(0, 0)$.

d) Since $f''(0)$ does not exist, 0 is a critical point of f'. The equation $f''(x) = 0$ has no solution, so the only critical point of f' is 0. We have already found $f(0)$ in step (b).

e) For any $x \neq 0$, $x^4 > 0$, so $\sqrt[3]{x^4} > 0$. Thus,

$$f''(x) = -\frac{2}{9\sqrt[3]{x^4}} < 0$$

for any $x \neq 0$. Therefore, f is concave down on each of the intervals $(-\infty, 0)$ and $(0, \infty)$, and there is no point of inflection.

f) Sketch the graph using the preceding information. It helps to use the values $f(1) = 1$ and $f(-1) = 1$ and the resulting points $(1, 1)$ and $(-1, 1)$ to draw the graph.

3. For the following function, find the relative maxima and minima and sketch the graph.

$$f(x) = 3x^{2/3}$$

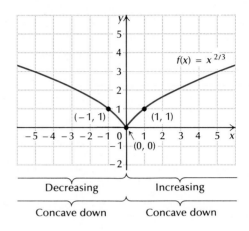

Decreasing   Increasing

Concave down   Concave down

DO EXERCISE 3.

Example 4 For the following function, find the relative maxima and minima and sketch the graph.

$$f(x) = x + 4/x$$

Solution First note that 0 is not in the domain of the function.

a) Find $f'(x)$ and $f''(x)$:

$$f'(x) = 1 - 4x^{-2} = 1 - \frac{4}{x^2},$$

$$f''(x) = 8x^{-3} = \frac{8}{x^3}.$$

b) Since 0 is not in the domain of f, it is not a critical point. It is also not in the domain of f' or f''. Thus the only critical points are where

$$1 - \frac{4}{x^2} = 0$$

$$1 = \frac{4}{x^2}$$

$$x^2 = 4$$

$$x = \pm 2.$$

Now $f(-2) = -4$ and $f(2) = 4$. These give the points $(-2, -4)$ and $(2, 4)$ on the graph.

c) Note that

$$f'(x) = 1 - \frac{4}{x^2} = \frac{x^2 - 4}{x^2}.$$

For any $x \neq 0$, $x^2 > 0$, so the positiveness or negativeness of $f'(x)$ depends on $x^2 - 4$ and its factors $x + 2$ and $x - 2$.

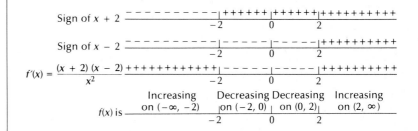

There is a relative minimum at $(2, 4)$ and a relative maximum at $(-2, -4)$.

d) Since 0 is not in the domain of f' and f'', the only critical points of f' are where

$$\frac{8}{x^3} = 0.$$

But this equation has no solution. Thus f' has no critical points.

e) For any $x < 0$, $x^3 < 0$, so

$$f''(x) = \frac{8}{x^3} < 0.$$

For any $x > 0$, $x^3 > 0$, so

$$f''(x) = \frac{8}{x^3} > 0.$$

Thus f is concave down on the interval $(-\infty, 0)$ and concave up on the interval $(0, \infty)$. There is no point of inflection since 0 is not in the domain of f.

f) Sketch the graph using the preceding information. Note that as $x \longrightarrow \infty$, $4/x \longrightarrow 0$, and $x + (4/x)$ gets closer and closer to the line $y = x$. A similar thing happens as $x \longrightarrow -\infty$. Note too that as $x \longrightarrow 0$ from the left, $f(x) \longrightarrow -\infty$, and that as $x \longrightarrow 0$ from the right, $f(x) \longrightarrow \infty$. Because of these limiting characteristics, lines such as $y = x$ and the y-axis in this example are called *asymptotes*.

4. For the following function, find the relative maxima and minima and sketch the graph.

$$f(x) = x + \frac{1}{x}$$

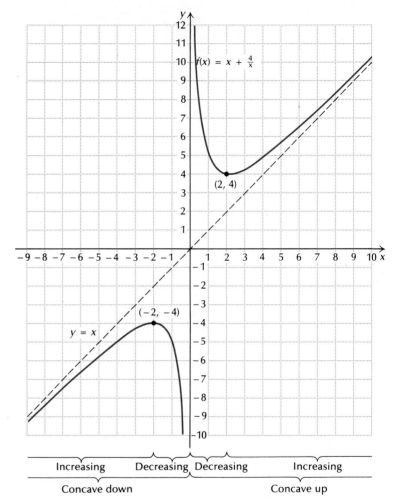

Increasing   Decreasing Decreasing   Increasing

Concave down   Concave up

DO EXERCISE 4.

   It is very helpful in curve sketching to look for asymptotes. A *vertical asymptote* of a function f is a line $x = a$ for which $f(x) \longrightarrow \infty$ or $f(x) \longrightarrow -\infty$ as $x \longrightarrow a$. That is, the function gets larger and larger or smaller and smaller without bound as x gets closer and closer to a. The y-axis is a vertical asymptote in Example 4. A *horizontal asymptote* is a line $y = L$ for which $f(x) \longrightarrow L$ as $x \longrightarrow \infty$ or $x \longrightarrow -\infty$. That is, as x gets larger and larger or smaller and smaller without bound, the function values approach L. The line $y = 1$ is a horizontal asymptote of $f(x) = x/(x - 3)$. An *oblique asymptote* is any other line $y = mx + b$ that $f(x)$ approaches as $x \longrightarrow \infty$ or $x \longrightarrow -\infty$. In Example 4, the line $y = x$ is an oblique asymptote.

EXERCISE SET 3.3

For each function, find the relative maxima and minima and sketch the graph.

1. $f(x) = 2 - x^2$ **2.** $f(x) = 3 - x^2$ **3.** $f(x) = x^2 + x - 1$ **4.** $f(x) = x^2 - x$

5. $f(x) = \frac{8}{3}x^3 - 2x + \frac{1}{3}$ **6.** $f(x) = x^3 - 3x + 2$ **7.** $f(x) = (x - 1)^3$ **8.** $f(x) = (x + 2)^3$

9. $f(x) = (x + 1)^{2/3}$ **10.** $f(x) = (x - 1)^{2/3}$ **11.** $f(x) = x^4 - 6x^2$ **12.** $f(x) = 2x^2 - x^4$

13. $f(x) = x + \dfrac{9}{x}$ **14.** $f(x) = x + \dfrac{2}{x}$ **15.** $f(x) = x^3 - 2x^2 - 4x + 3$

16. $f(x) = x^3 - 6x^2 + 9x + 1$ **17.** $f(x) = 3x^4 + 4x^3$ **18.** $f(x) = x^4 - 2x^3$

19. $f(x) = x^3 - 3x^2 + 4$ **20.** $f(x) = x^3 - 3x^2 - 144x - 140$ **21.** $f(x) = x^3 - 6x^2 - 135x$

22. $g(x) = \dfrac{1}{x^2}$ **23.** $f(x) = \dfrac{x}{x - 3}$ **24.** $g(x) = \dfrac{x}{x + 2}$

EXTENSION EXERCISES

Using the same set of axes, sketch the graphs of the total-revenue, total-cost, and total-profit functions.

25. $R(x) = 50x - 0.5x^2, C(x) = 4x + 10$ **26.** $R(x) = 50x - 0.5x^2, C(x) = 10x + 3$

Sketch the graph of each function.

27. $f(x) = x - \sqrt{x}$ **28.** $f(x) = x^2 + \dfrac{1}{x^2}$ **29.** $f(x) = \dfrac{1}{x^2 - 1}$

30. $f(x) = \dfrac{1}{x^2 + 1}$ **31.** $y = \dfrac{2x^2}{x^2 - 16}$ **32.** $y = (x - 1)^{2/3} - (x + 1)^{2/3}$

3.4

OBJECTIVE	MAXIMUM–MINIMUM PROBLEMS

OBJECTIVE

a) Solve maximum–minimum problems.

MAXIMUM–MINIMUM PROBLEMS

One very important application of the differential calculus is the solving of maximum–minimum problems, that is, finding the maximum or minimum value of some varying quantity Q and the point at which that maximum or minimum occurs.

Example 1 A hobby store has 20 ft of fencing to fence off a rectangular area for an electric train in one corner of its display room. The two sides up against the wall require no fence. What dimensions of the rectangle will maximize the area? What is the maximum area?

1. *Exploratory exercises*

   a)　Complete this table.

x	y $20 - x$	A $x(20 - x)$
0		
4		
6.5		
8		
10		
12		
13.2		
20		

   b)　Make a graph of x versus A, that is, of points (x, A), from the table; and connect them with a smooth curve.

   c)　Does it matter what dimensions we use?

   d)　Make a conjecture about what the maximum might be and where it would occur.

How can the area for the electric train be maximized?

Exploratory Solution　Intuitively, one might think that it does not matter what dimensions one uses: They will all yield the same area. To show that this is not true, as well as to conjecture a possible solution, consider the exploratory exercises in Margin Exercise 1. Before doing so, however, let us express the area in terms of one variable. If we let $x =$ the length of one side and $y =$ the length of the other, then since the sum of the lengths must be 20 ft, we have

$$x + y = 20 \quad \text{and} \quad y = 20 - x.$$

Then the area is given by

$$A = xy$$
$$A = x(20 - x) = 20x - x^2.$$

DO EXERCISE 1.

Calculus Solution　We are trying to find the maximum value of

$$A = 20x - x^2 \quad \text{on the interval} \quad (0, 20).$$

We consider the interval $(0, 20)$ because x is the length of one side and cannot be negative. Since there is only 20 ft of fencing, x cannot

2. A rancher has 50 ft of fencing to fence off a rectangular animal pen in the corner of a barn. What dimensions of the rectangle will yield the maximum area? What is the maximum area?

be greater than 20. Also, x cannot be 20 because then the length of y would be 0.

a) We first find $A'(x)$, where $A(x) = 20x - x^2$:

$$A'(x) = 20 - 2x.$$

b) This derivative exists for all values of x in $(0, 20)$. Thus the only critical points are where

$$A'(x) = 20 - 2x = 0$$
$$-2x = -20$$
$$x = 10.$$

Since there is only one critical point in the interval, we can use the second derivative to determine whether we have a maximum. Note that

$$A''(x) = -2,$$

which is a constant. Thus $A''(10)$ is negative, so $A(10)$ is a maximum. Now

$$A(10) = 10(20 - 10) = 10 \cdot 10 = 100.$$

Thus the maximum area of 100 ft² is obtained using 10 ft for the length of one side and $20 - 10$, or 10 ft, for the other. Note that although you may have conjectured this in Margin Exercise 1, the tools of calculus allowed us to prove it.

DO EXERCISE 2.

Example 2 A stereo manufacturer determines that in order to sell x units of a new stereo its price per unit must be

$$p = D(x) = 1000 - x.$$

It also determines that the total cost of producing x units is given by

$$C(x) = 3000 + 20x.$$

a) Find the total revenue $R(x)$.

b) Find the total profit $P(x)$.

c) How many units must the company produce and sell in order to maximize profit?

d) What is the maximum profit?

e) What price per unit must be charged in order to make this maximum profit?

3. A company determines that in order to sell x units of a certain product its price per unit must be

$$p = D(x) = 200 - x.$$

It also determines that its total cost of producing x units is given by

$$C(x) = 5000 + 8x.$$

a) Find the total revenue $R(x)$.

b) Find the total profit $P(x)$.

c) How many units must the company produce and sell in order to maximize profit?

d) What is the maximum profit?

e) What price per unit must be charged in order to make this maximum profit?

Solution

a) $R(x) =$ Total revenue $=$ (Number of units) $\cdot$ (Price per unit)

$$= \qquad x \qquad \cdot \qquad p$$

$$= x(1000 - x) = 1000x - x^2$$

b) $P(x) = R(x) - C(x) = (1000x - x^2) - (3000 + 20x)$

$$= -x^2 + 980x - 3000$$

c) To find the maximum value of $P(x)$ we first find $P'(x)$:

$$P'(x) = -2x + 980.$$

This is defined for all real numbers (actually we are interested in numbers x in $[0, \infty)$ only, since we cannot produce a negative number of stereos). Thus we solve

$$P'(x) = -2x + 980 = 0$$

$$-2x = -980$$

$$x = 490.$$

Since there is only one critical point, we can try to use the second derivative to determine whether we have a maximum. Note that

$$P''(x) = -2, \quad \text{a constant.}$$

Thus $P''(490)$ is negative, so $P(490)$ is a maximum.

d) The maximum profit is given by

$$P(490) = -(490)^2 + 980 \cdot 490 - 3000 = \$237{,}100.$$

Thus the stereo manufacturer makes a maximum profit of \$237,100 by producing and selling 490 stereos.

e) The price per unit needed to make the maximum profit is

$$p = 1000 - 490 = \$510.$$

DO EXERCISE 3.

Marginal Analysis

Let us take a general look at the total-profit function and its related functions.

In the first graph below we have the total-cost and the total-revenue functions. We can estimate what the maximum profit might be by looking for the widest gap between $R(x)$ and $C(x)$. Points B_0 and B_2 are "break-even" points.

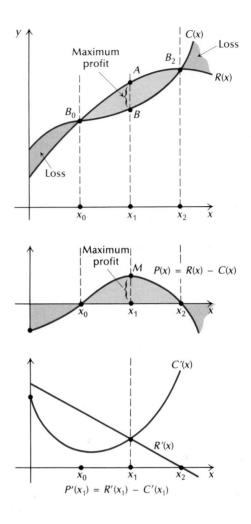

The second graph shows the total-profit function. Note that when production is too low ($< x_0$) there is a loss because of high fixed or initial costs and low revenue. When production is too high ($> x_2$), there is also a loss due to high marginal costs and low marginal revenues (as shown in the third graph).

The business operates at a profit everywhere between x_0 and x_2. Note that maximum profit occurs at a critical point x_1 of $P(x)$. If we assume that $P'(x)$ exists for all x in some interval, usually $[0, \infty)$, this critical point occurs at some number x such that

$$P'(x) = 0 \quad \text{and} \quad P''(x) < 0.$$

Since $P(x) = R(x) - C(x)$, it follows that

$$P'(x) = R'(x) - C'(x) \quad \text{and} \quad P''(x) = R''(x) - C''(x).$$

Thus the maximum profit occurs at some number x such that

$$R'(x) - C'(x) = 0 \quad \text{and} \quad R''(x) - C''(x) < 0$$

or

$$R'(x) = C'(x) \quad \text{and} \quad R''(x) < C''(x).$$

In summary, we have the following theorem.

THEOREM 8

Maximum profit is achieved when marginal revenue equals marginal cost and the rate of change of marginal revenue is less than the rate of change of marginal cost:

$$R'(x) = C'(x) \quad \text{and} \quad R''(x) < C''(x).$$

Here is a general strategy for solving maximum–minimum problems. While it may not guarantee success, it should certainly improve one's chances.

A Strategy for Solving Maximum–Minimum Problems

1. Read the problem carefully. If relevant, draw a picture.
2. Label the picture with appropriate variables and constants, noting what varies and what stays fixed.
3. Translate the problem to an equation involving a quantity Q to be maximized or minimized. Try to represent Q in terms of the variables of step (2).
4. Try to express Q as a function of *one* variable. Use the procedures developed in Sections 3.1 and 3.2 to determine the maximum or minimum values and the points at which they occur.

Example 3   From a thin piece of cardboard 8 in. by 8 in., square corners are cut out so that the sides can be folded up to make a box. What dimensions will yield a box of maximum volume? What is the maximum volume?

4. *Exploratory exercises*

a) Complete this table.

x	h $\frac{1}{2}(8-x)$	V $x \cdot x \cdot \frac{1}{2}(8-x)$
0		
1		
2		
3		
4		
4.6		
5		
6		
6.8		
7		
8		

b) Make a graph of x versus V.

c) Make a conjecture about what the maximum might be and where it would occur.

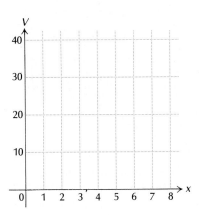

Exploratory Solution   One might again think that it does not matter what the dimensions are, but our experience with Example 1 should lead us to think otherwise. We make a drawing, as shown below.

When squares of length h on a side are cut out of the corners, we are left with a square base of length x. The volume of the resulting box is

$$V = lwh = x \cdot x \cdot h.$$

We want to express V in terms of one variable. Note that the overall length of a side of the cardboard is 8 in. We see from the figure that

$$h + x + h = 8,$$

or

$$x + 2h = 8.$$

Solving for h we get

$$2h = 8 - x$$
$$h = \tfrac{1}{2}(8-x) = \tfrac{1}{2} \cdot 8 - \tfrac{1}{2}x = 4 - \tfrac{1}{2}x.$$

Thus,

$$V = x \cdot x \cdot (4 - \tfrac{1}{2}x) = x^2(4 - \tfrac{1}{2}x) = 4x^2 - \tfrac{1}{2}x^3.$$

In Margin Exercise 4 you will compute some values of V.

DO EXERCISE 4.

Calculus Solution   You probably noted in Margin Exercise 4 that it was a bit more difficult to conjecture where the maximum occurs than it was in Example 1. At the least it seems reasonable that the maximum occurs for some x between 5 and 6. Let us find out for certain, using

5. Repeat Example 3, but for a piece of cardboard that is 10 in. by 10 in.

calculus. We are trying to find the maximum value of

$$V(x) = 4x^2 - \tfrac{1}{2}x^3 \quad \text{on the interval } (0, 8).$$

We first find $V'(x)$:

$$V'(x) = 8x - \tfrac{3}{2}x^2.$$

Now $V'(x)$ exists for all x in the interval $(0, 8)$, so we set it equal to 0 to find the critical values:

$$V'(x) = 8x - \tfrac{3}{2}x^2 = 0$$
$$x(8 - \tfrac{3}{2}x) = 0$$
$$x = 0 \quad \text{or} \quad 8 - \tfrac{3}{2}x = 0$$
$$x = 0 \quad \text{or} \quad -\tfrac{3}{2}x = -8$$
$$x = 0 \quad \text{or} \quad x = -\tfrac{2}{3}(-8) = \tfrac{16}{3}.$$

The only critical point in $(0, 8)$ is $\tfrac{16}{3}$. Thus we can use the second derivative,

$$V''(x) = 8 - 3x,$$

to determine whether we have a maximum. Since

$$V''(\tfrac{16}{3}) = 8 - 3 \cdot \tfrac{16}{3}$$
$$= -8,$$

$V''(\tfrac{16}{3})$ is negative, so $V(\tfrac{16}{3})$ is a maximum, and

$$V(\tfrac{16}{3}) = 4 \cdot (\tfrac{16}{3})^2 - \tfrac{1}{2}(\tfrac{16}{3})^3 = \tfrac{1024}{27} = 37\tfrac{25}{27}.$$

The maximum volume is $37\tfrac{25}{27}$ in^3. The dimensions that yield this maximum volume are

$$x = \tfrac{16}{3} = 5\tfrac{1}{3} \text{ in.,} \quad \text{by } x = 5\tfrac{1}{3} \text{ in.,} \quad \text{by } h = 4 - \tfrac{1}{2}(\tfrac{16}{3}) = 1\tfrac{1}{3} \text{ in.}$$

It would surely have been difficult to guess this from Margin Exercise 4.

DO EXERCISE 5.

In the following problem, an open-top container of fixed volume is to be constructed. We want to determine the dimensions that will allow it to be built with the least amount of material. Such a problem could be important from an ecological standpoint.

Example 4 A container firm is designing an open-top rectangular box, with a square base, that will hold 108 cubic centimeters (cc). What dimensions yield the minimum surface area? What is the minimum surface area?

Solution   The surface area of the box is

$$S = x^2 + 4xy.$$

The volume must be 108 cc, and is given by

$$V = x^2y = 108.$$

To express S in terms of one variable, we solve $x^2y = 108$ for y:

$$y = \frac{108}{x^2}.$$

Then

$$S = x^2 + 4x\left(\frac{108}{x^2}\right) = x^2 + \frac{432}{x}.$$

Now S is defined only for positive numbers, and the problem dictates that the length x be positive, so we are minimizing S on the interval $(0, \infty)$. We first find dS/dx:

$$\frac{dS}{dx} = 2x - \frac{432}{x^2}.$$

Since dS/dx exists for all x in $(0, \infty)$, the only critical points occur where $dS/dx = 0$. Thus we solve the following equation:

$$2x - \frac{432}{x^2} = 0$$

$$x^2\left(2x - \frac{432}{x^2}\right) = x^2 \cdot 0 \qquad \text{We multiply by } x^2 \text{ to clear of fractions.}$$

$$2x^3 - 432 = 0$$

$$2x^3 = 432$$

$$x^3 = 216$$

$$x = 6.$$

6. Repeat Example 4, but for a fixed volume of 500 cc.

This is the only critical point, so we can use the second derivative to determine whether we have a minimum:

$$\frac{d^2S}{dx^2} = 2 + \frac{864}{x^3}.$$

Note that this is positive for all positive values of x. Thus we have a minimum at $x = 6$. When $x = 6$, it follows that $\hat{y} = 3$:

$$y = \frac{108}{6^2} = \frac{108}{36} = 3.$$

Thus the surface area is minimized when $x = 6$ cm (centimeters) and $y = 3$ cm. The minimum surface area is

$$S = 6^2 + 4 \cdot 6 \cdot 3 = 108 \text{ cm}^2.$$

By coincidence, this is the same number as the fixed volume.

DO EXERCISE 6.

Example 5 *Determining a ticket price.* Fight promoters ride a thin line between profit and loss, especially in determining the price to charge for admission to closed-circuit television showings in local theaters. By keeping records, a theater determines that if the admission price is $20, it averages 1000 people in attendance. But for every increase of $1, it loses 100 customers from the average number. Every customer spends an average of $0.80 on concessions. What admission price should the theater charge in order to maximize total revenue?

Solution Let $x =$ the amount by which the price of $20 should be increased (if x is negative, the price would be decreased). We first express the total revenue R as a function of x. Note that

$$R(x) = (\text{Revenue from tickets}) + (\text{Revenue from concessions})$$
$$= (\text{Number of people}) \cdot (\text{Ticket price})$$
$$+ \$0.80(\text{Number of people})$$
$$= (1000 - 100x)(20 + x) + 0.80(1000 - 100x)$$
$$= 20{,}000 - 2000x + 1000x - 100x^2 + 800 - 80x$$
$$R(x) = -100x^2 - 1080x + 20{,}800.$$

We are trying to find the maximum value of R over the set of all real numbers. To find x such that $R(x)$ is a maximum, we first find $R'(x)$:

$$R'(x) = -200x - 1080.$$

This derivative exists for all real numbers x. Thus the only critical

7. Transit companies also ride a thin line between profit and loss. A company determines that at a fare of 30¢, it will average 10,000 fares per day. For every increase of 10¢, it loses 2000 customers. What fare should be charged in order to maximize revenue? [*Hint:* Let x = the number of 10¢ fare increases (if x is negative, the fare would be decreased). Then the new fare would be $30 + 10x$.]

points are where $R'(x) = 0$, so we solve that equation:

$$-200x - 1080 = 0$$
$$-200x = 1080$$
$$x = -5.4 = -\$5.40.$$

Since this is the only critical point, we can use the second derivative,

$$R''(x) = -200,$$

to determine whether we have a maximum. Since $R''(-5.4)$ is negative, $R(-5.4)$ is a maximum. Therefore, in order to maximize revenue the theater should charge

$$\$20 + (-\$5.40), \quad \text{or} \quad \$14.60 \text{ per ticket.}$$

That is, this reduced ticket price will get more people into the theater,

$$1000 - 100(-5.4), \quad \text{or} \quad 1540,$$

and will result in maximum revenue.

DO EXERCISE 7.

EXERCISE SET 3.4

1. Of all the numbers whose sum is 50, find the two that have the maximum product. That is, maximize $Q = xy$, where $x + y = 50$.

2. Of all the numbers whose sum is 70, find the two that have the maximum product. That is, maximize $Q = xy$, where $x + y = 70$.

3. In Exercise 1, can there be a minimum product? Explain.

4. In Exercise 2, can there be a minimum product? Explain.

5. Of all numbers whose difference is 4, find the two that have the minimum product.

6. Of all numbers whose difference is 6, find the two that have the minimum product.

7. Maximize $Q = xy^2$, where x and y are positive numbers, such that $x + y^2 = 1$.

8. Maximize $Q = xy^2$, where x and y are positive numbers, such that $x + y^2 = 4$.

9. Minimize $Q = x^2 + y^2$, where $x + y = 20$.

10. Minimize $Q = x^2 + y^2$, where $x + y = 10$.

11. Maximize $Q = xy$, where x and y are positive numbers, such that $\frac{4}{3}x^2 + y = 16$.

12. Maximize $Q = xy$, where x and y are positive numbers, such that $x + \frac{4}{3}y^2 = 1$.

APPLICATIONS

13. A rancher wants to build a rectangular fence next to a river, using 120 yd of fencing. What dimensions of the rectangle will maximize the area? What is the maximum area? Note that the rancher need not fence in the side next to the river.

14. A rancher wants to enclose two rectangular areas near a river, one for sheep and one for cattle. There is 240 yd of fencing available. What is the largest total area that can be enclosed?

15. A carpenter is building a rectangular room with a fixed perimeter of 54 ft. What are the dimensions of the largest room that can be built? What is its area?

16. Of all rectangles that have a perimeter of 34 ft, find the dimensions of the one with the largest area. What is its area?

Business: Maximizing profit. Find the maximum profit and the number of units that must be produced and sold in order to yield the maximum profit.

17. $R(x) = 50x - 0.5x^2$, $C(x) = 4x + 10$

18. $R(x) = 50x - 0.5x^2$, $C(x) = 10x + 3$

19. $R(x) = 2x$, $C(x) = 0.01x^2 + 0.6x + 30$

20. $R(x) = 5x$, $C(x) = 0.001x^2 + 1.2x + 60$

21. $R(x) = 9x - 2x^2$, $C(x) = x^3 - 3x^2 + 4x + 1$; $R(x)$ and $C(x)$ are in thousands of dollars, and x is in thousands of units.

22. $R(x) = 100x - x^2$, $C(x) = \frac{1}{3}x^3 - 6x^2 + 89x + 100$; $R(x)$ and $C(x)$ are in thousands of dollars, and x is in thousands of units.

23. Raggs, Ltd., a clothing firm, determines that in order to sell x suits, its price per suit must be

$$p = D(x) = 150 - 0.5x.$$

It also determines that its total cost of producing x suits is given by

$$C(x) = 4000 + 0.25x^2.$$

a) Find the total revenue $R(x)$.

b) Find the total profit $P(x)$.

c) How many suits must the company produce and sell in order to maximize profit?

d) What is the maximum profit?

e) What price per suit must be charged in order to make this maximum profit?

24. An appliance firm is marketing a new refrigerator. It determines that in order to sell x refrigerators, its price per refrigerator must be

$$p = D(x) = 280 - 0.4x.$$

It also determines that its total cost of producing x refrigerators is given by

$$C(x) = 5000 + 0.6x^2.$$

a) Find the total revenue $R(x)$.

b) Find the total profit $P(x)$.

c) How many refrigerators must the company produce and sell in order to maximize profit?

d) What is the maximum profit?

e) What price per refrigerator must be charged in order to make this maximum profit?

25. From a thin piece of cardboard 30 in. by 30 in., square corners are cut out so the sides can be folded up to make a box. What dimensions will yield a box of maximum volume? What is the maximum volume?

26. From a thin piece of cardboard 20 in. by 20 in., square corners are cut out so the sides can be folded up to make a box. What dimensions will yield a box of maximum volume? What is the maximum volume?

27. A container company is designing an open-top, square-based, rectangular box that will have a volume of 62.5 in³. What dimensions yield the minimum surface area? What is the minimum surface area?

28. A soup company is constructing an open-top, rectangular metal tank with a square base that will have a volume of 32 ft³. What dimensions yield the minimum surface area? What is the minimum surface area?

29. A university is trying to determine what price to charge for football tickets. At a price of $6 per ticket it averages 70,000 people per game. For every increase of $1 it loses 10,000 people from the average number. Every person at the game spends an average of $1.50 on concessions. What price per ticket should be charged in order to maximize revenue? How many people will attend at that price?

30. Suppose you are the owner of a 30-unit motel. All units are occupied when you charge $20 a day per unit. For every increase of x dollars in the daily rate, there are x units vacant. Each occupied room costs $2 per day to service and maintain. What should you charge per unit in order to maximize profit?

31. An apple farm yields an average of 30 bushels of apples per tree when 20 trees are planted on an acre of ground. Each time 1 more tree is planted per acre, the yield decreases 1 bu per tree due to the extra congestion. How many trees should be planted in order to get the highest yield?

32. When a theater owner charges $3 for admission, there is an average attendance of 100 people. For every $0.10 increase in admission, there is a loss of 1 customer from the average. What admission should be charged in order to maximize revenue?

33. The postal service places a limit of 84 in. on the combined length and girth (distance around) of a package to be sent parcel post. What dimensions of a rectangular box with square cross section will contain the largest volume that can be mailed? [*Hint:* There are two different girths.]

34. A rectangular play area is to be fenced off in a person's back lot and is to contain 48 yd^2. The neighbor agrees to pay half the cost of the frence on the side of the play area that lines the lot. What dimensions will minimize the cost of the fence?

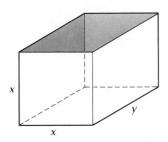

35. For what positive number is the sum of its reciprocal and five times its square a minimum?

36. For what positive number is the sum of its reciprocal and four times its square a minimum?

37. A rectangular box with a volume of 320 ft^3 is to be constructed with a square base and top (see the figure at right). The cost per square foot for the bottom is 15¢, for the top is 10¢, and for the sides is 2.5¢. What dimensions will minimize the cost?

38. A merchant who was purchasing a display sign from a salesclerk said, "I want a sign 10 ft by 10 ft." The salesclerk responded, "That's just what we'll give you; only to make it more aesthetically pleasing, why don't we change it to 7 ft by 13 ft?" Comment.

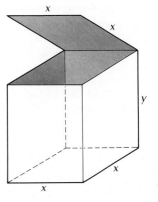

39. A Norman window is a rectangle with a semicircle on top. Suppose the perimeter of a particular Norman window is to be 24 ft. What should its dimensions be in order to allow the maximum amount of light to enter through the window?

A Norman window.

40. Solve Exercise 39, but this time assume the semicircle is to be stained glass, which transmits only half as much light as the semicircle in Exercise 39.

41. The amount of money deposited in a financial institution in savings accounts is directly proportional to the interest rate the financial institution pays on the money. Suppose a financial institution can loan *all* the money it takes in on its savings accounts at an interest rate of 18%. What interest rate should it pay on its savings accounts in order to maximize profit?

42. 📱 A page in this book is 73.125 in². On the average there is a 0.75-in. margin at the top and at the bottom of each page and a 0.5-in. margin on each of the sides. What should the outside dimensions of each page be so that the printed area is a maximum? Measure the outside dimensions to see whether the actual dimensions maximize the printed area.

EXTENSION EXERCISES

43. A 24-in. piece of string is cut in two pieces. One piece is used to form a circle and the other to form a square. How should the string be cut so that the sum of the areas is a minimum? a maximum?

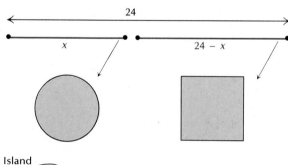

44. A power line is to be constructed from a power station at point *A* (see the figure) to an island at point *C*, which is directly 1 mile out in the water from a point *B* on the shore. Point *B* is 4 miles downshore from the power station at *A*. It costs $5000 per mile to lay the power line under water and $3000 per mile to lay the line under ground. At what point *S* downshore from *A* should the line come to the shore in order to minimize cost? Note that *S* could very well be *B* or *A*. [*Hint:* The length of *CS* is $\sqrt{1 + x^2}$.]

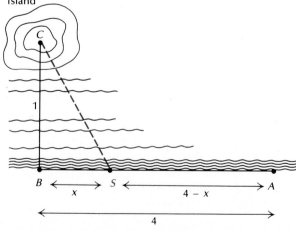

45. ▦ *Biology: Flights of homing pigeons.* It is known that homing pigeons tend to avoid flying over water in the daytime, perhaps because the downdrafts of air over water make flying difficult. Suppose a homing pigeon is released on an island at point C, which is 3 miles directly out in the water from a point B on shore. Point B is 8 miles downshore from the pigeon's home loft at point A. Assume that a pigeon requires 1.28 times the rate of energy over land to fly over water. Toward what point S downshore from A should the pigeon fly in order to minimize the total energy required to get to home loft A? Assume that

(Total energy)

  = (Energy rate over water) · (Distance over water) +
   (Energy rate over land) · (Distance over land).

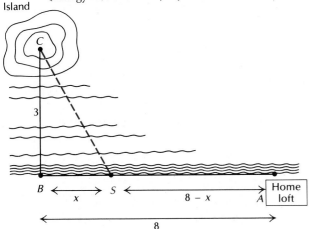

46. A road is to be built between two cities C_1 and C_2, which are on opposite sides of a river of uniform width r. Because of the river, a bridge must be built. C_1 is a units from the river, and C_2 is b units from the river; $a \leqslant b$. Where should the bridge be located in order to minimize the total distance between the cities? Give a general solution using the constants a, b, p, and r in the figure shown here.

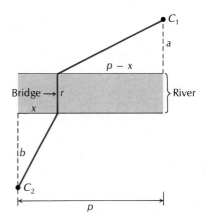

47. The total-cost function for producing x units of a certain product is given by

$$C(x) = 8x + 20 + \frac{x^3}{100}.$$

a) Find the marginal cost $C'(x)$.

b) Find the average cost $A(x) = C(x)/x$.

c) Find the *marginal average cost* $A'(x)$.

d) Find the minimum of $A(x)$ and the value x_0 at which it occurs. Find the marginal cost at x_0.

e) Compare $A(x_0)$ and $C'(x_0)$.

49. Minimize $Q = x^3 + 2y^3$, where x and y are positive numbers, such that $x + y = 1$.

48. Consider $A(x) = C(x)/x$.

a) Find $A'(x)$ in terms of $C'(x)$ and $C(x)$.

b) Show that $A(x)$ has a minimum at that value of x_0 such that

$$C'(x_0) = A(x_0) = \frac{C(x_0)}{x_0}.$$

This shows that when marginal cost and average cost are the same, a product is being produced at the least average cost.

50. Minimize $Q = 3x + y^3$, where $x^2 + y^2 = 2$.

3.5

OBJECTIVE

a) Given certain inventory costs, find how many times a year a store should reorder a product, and in what lot size, in order to minimize total inventory costs.

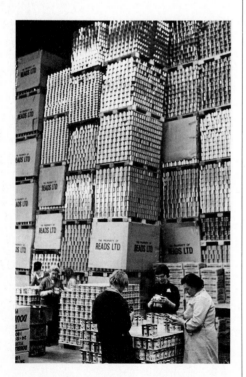

How can inventory costs be minimized?

BUSINESS APPLICATIONS: MINIMIZING INVENTORY COSTS

A retail outlet of a business is concerned about inventory costs. Suppose, for example, an appliance store sells 2500 television sets per year. It *could* operate by ordering all the sets at once. But then the owners would face the carrying costs (insurance, building space, and so on) of storing them all. Thus they might make several smaller orders, say 5, so that the largest number they would ever have to store is 500. On the other hand, each time they reorder there are costs for paperwork, delivery charges, manpower, and so on. It would seem, therefore, that there must be some balance between carrying costs and reorder costs. We will see how calculus can help to determine what that balance might be. We will be trying to minimize the following function:

$$\text{Total inventory costs} = \left(\begin{array}{c}\text{Yearly carrying}\\\text{costs}\end{array}\right) + \left(\begin{array}{c}\text{Yearly reorder}\\\text{costs}\end{array}\right).$$

The *lot size x* refers to the largest amount ordered each reordering period. Note the following graphs. If x is ordered each period, then during that time there is somewhere between 0 and x units in stock. To have a representative expression for the amount in stock at any one time in the period we can use the average, $x/2$. This represents the average amount held in stock over the course of the year. We get a feeling for this using the following graphs. If the lot size is 2500, then during the period between orders there is somewhere between 0 and 2500 units in stock. On the average there is $\frac{2500}{2}$, or 1250 units in stock. If the lot size is 1250, then during the period between orders there is somewhere between 0 and 1250 units in stock. On the average there is $\frac{1250}{2}$, or 625 units in stock.

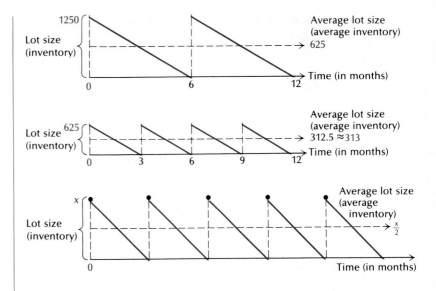

Example 1 A retail appliance store sells 2500 television sets per year. It costs $10 to store one set for a year. To reorder, there is a fixed cost of $20 plus $9 for each set. How many times per year should the store reorder, and in what lot size, in order to minimize inventory costs?

Solution Let x = the lot size. Inventory costs are given by

$$C(x) = \text{(Yearly carrying costs)} + \text{(Yearly reorder costs)}.$$

We consider each separately.

a) *Yearly carrying costs.* The average amount held in stock is $x/2$, and it costs $10 per set for storage. Thus,

$$\text{Yearly carrying costs} = \begin{pmatrix} \text{Yearly cost} \\ \text{per item} \end{pmatrix} \cdot \begin{pmatrix} \text{Average number} \\ \text{of items} \end{pmatrix}$$

$$= 10 \cdot \frac{x}{2}.$$

b) *Yearly reorder costs.* Now x = the lot size, and suppose there are N reorders each year. Then $Nx = 2500$, and $N = 2500/x$. Thus,

$$\text{Yearly reorder costs} = \begin{pmatrix} \text{Cost of each} \\ \text{order} \end{pmatrix} \cdot \begin{pmatrix} \text{Number of} \\ \text{reorders} \end{pmatrix}$$

$$= (20 + 9x)\frac{2500}{x}.$$

1. ▦ Without a knowledge of calculus one might make a rough estimate of the lot size that will minimize total inventory costs using a table like the following. Complete the table and make such an estimate.

Lot size, x	Number of reorders, $\dfrac{2500}{x}$	Average inventory, $\dfrac{x}{2}$	Carrying costs, $10\cdot\dfrac{x}{2}$	Cost of each order, $20+9x$	Reorder costs, $(20+9x)\dfrac{2500}{x}$	Total inventory costs $C(x)$, $10\cdot\dfrac{x}{2}+(20+9x)\dfrac{2500}{x}$
2500	1	1250	$12,500	$22,520	$22,520	$35,020
1250	2	625	$6,250	$11,270	$22,540	
500	5	250	$2,500	$4,520		
250	10	125				
167	15	84				
125	20					
100	25					
90	28					
50	50					

c) Thus,

$$C(x) = 10\cdot\frac{x}{2} + (20+9x)\,\frac{2500}{x}$$

$$C(x) = 5x + \frac{50{,}000}{x} + 22{,}500.$$

DO EXERCISE 1.

d) We want to find a minimum value of C on the interval $[1, 2500]$. We first find $C'(x)$:

$$C'(x) = 5 - \frac{50{,}000}{x^2}.$$

e) Now $C'(x)$ exists for all x in $[1, 2500]$, so the only critical points are those x such that $C'(x) = 0$. We solve $C'(x) = 0$:

$$5 - \frac{50{,}000}{x^2} = 0$$

$$5 = \frac{50{,}000}{x^2}$$

$$5x^2 = 50{,}000$$

$$x^2 = 10{,}000$$

$$x = \pm 100$$

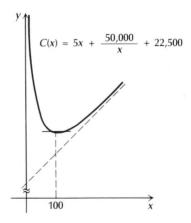

$$C(x) = 5x + \frac{50{,}000}{x} + 22{,}500$$

Now there is only one critical point in the interval $[1, 2500]$, $x = 100$, so we can use the second derivative to see whether we

EXTENSION EXERCISES

7. *Minimizing inventory costs: A general solution.* A store sells Q units of a product per year. It costs a dollars to store one unit for one year. To reorder units, there is a fixed cost of b dollars, plus c dollars for each unit. In what lot size should the store reorder in order to minimize inventory costs?

8. Use the general solution found in Exercise 7 to find how many times per year a store should reorder, and in what lot size, when $Q = 2500$, $a = \$10$, $b = \$20$, and $c = \$9$.

3.6

OBJECTIVES

a) Given a reproduction curve and an initial population P_0, locate population values for subsequent years.

b) Given a reproduction cycle described by a formula, find the population at which the maximum sustainable harvest occurs and the maximum sustainable harvest.

A BIOLOGICAL APPLICATION: MAXIMUM SUSTAINABLE HARVEST

Assuming a population of salmon is growing, what is the most number of salmon that can be harvested without depleting its initial population? Calculus can be used to find an answer.

Reproduction Curves

In certain situations biologists are able to determine what is called a *reproduction curve.* This is a function

$$y = f(P)$$

such that if P is the population at a certain time t, then the population one year later, at time $t + 1$, is $f(P)$. Such a curve is shown in Fig. 1. The line $y = P$ is significant for two reasons. First, if $y = P$ is a descrip-

A salmon population.

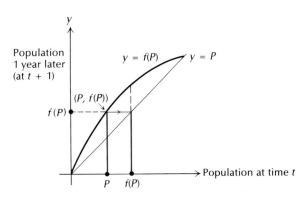

FIGURE 1

1. Below is a reproduction curve.

   a) Using P_0 as an initial population,
       locate P_1, P_2, P_3, P_4, and P_5 on
       both the vertical and horizontal
       axes.

   b) Use a ruler to transfer these
       values to the vertical axis of the
       second graph. Plot the points
       (t, P_t) and connect them with a
       smooth curve. Estimate the equi-
       librium level and draw a line to
       represent it.

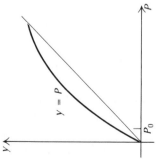

tion of f, then we know that the population stays the same from year
to year. But the graph of f in Fig. 1 lies mostly above the line. Thus
the population is increasing. Now if we start with a population P, we
know that $f(P)$ is the population one year later. Given $f(P)$ we move
horizontally from $(P, f(P))$ until we hit the line $y = P$; then we move
down to the horizontal axis. This locates $f(P)$ on the horizontal axis,
so we can find the population two years later. In this way we can
generate a sequence of population values over a period of years.

To see how this happens, look at Fig. 2. Suppose P_0 is some initial
population. The population one year later is given by

$$P_1 = f(P_0).$$

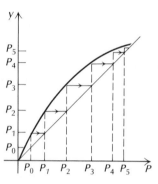

FIGURE 2

The population two years later is given by

$$P_2 = f(P_1),$$

and so on. We thus obtain a sequence of population values

$$P_0, P_1, P_2, P_3, \ldots.$$

If we transfer values from the horizontal axis of Fig. 2 to the vertical
axis of Fig. 3 and plot points (t, P_t), where t represents time, we obtain
a graph of population versus time.

Note in Fig. 3 that the population tends toward a *limiting value*
L, or *equilibrium level*. This curve is called an *S-shaped curve*, or *logistic
curve*, and is encountered in situations where the growth is inhibited
by certain environmental factors, such as resources or the size of an
ecosystem. For example, a colony of bacteria in a Petri dish will grow
to a certain size and then stop growing due to waste contamination.
We will consider this type of growth in more detail in Chapter 7.

DO EXERCISE 1.

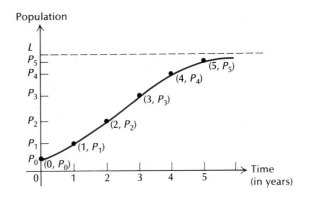

FIGURE 3

An inhibited population growth.

In Fig. 4 the reproduction curve falls below the line $y = P$. Note that as time passes, the population curve oscillates about the equilibrium level.

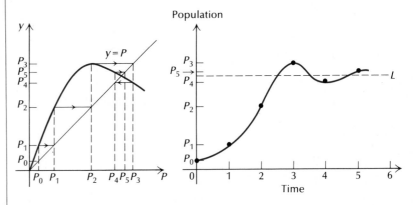

FIGURE 4

An oscillating population growth.

In Fig. 5 the reproduction curve falls even further below line $y = P$. The population-versus-time line then repeats itself in cycles, in this case every four years, never approaching an equilibrium level. A population of blowflies confined in a laboratory will exhibit such growth, as will certain predator–prey interactions.

We have drawn a line-segment graph of population versus time, since this is often done in practice.

2. a) Repeat the steps of Margin Exercise 1 for the reproduction curve shown below.

b) Is this an oscillating or a cyclic population?

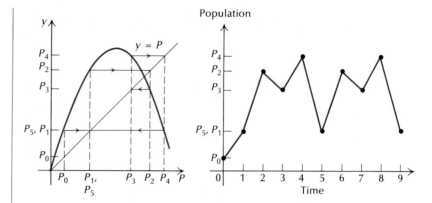

FIGURE 5

A cyclic population growth.

DO EXERCISE 2.

Suppose a certain population is growing in such a way that if P is the population at a particular time, then one year later there will be a 2% increase. The reproduction curve is given by

$$f(P) = P + 2\%P = 1 \cdot P + 0.02P = (1 + 0.02)P = 1.02P.$$

DO EXERCISE 3 ON THE FOLLOWING PAGE.

The reproduction curve $f(P) = 1.02P$ is shown in Fig. 6, along with the population-versus-time curve.

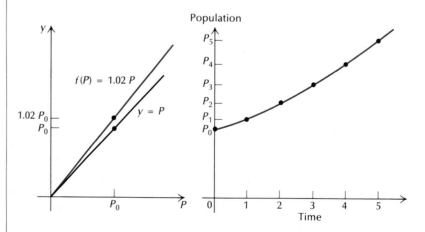

FIGURE 6

An uninhibited population growth.

3. ▦ In 1976 the population of the world was 4.0 billion. Suppose the reproduction curve is given by

$$f(P) = 1.019P.$$

Complete the following table.

Year	Population
1976	4.0 billion
1977	
1978	
1979	
1980	
1981	
1982	
1983	
1984	
1985	
1986	

4. An amount P is invested at $8\frac{1}{4}\%$ compounded annually. Find the reproduction curve.

5. A certain population is growing in such a way that if P is the population, then one year later it will have a 1% increase. Find the reproduction curve.

Note that the population increases indefinitely. We will study growth similar to this in Chapter 4. It is interesting that the figure above also models the growth of an amount, or "population," of money invested at 2% (simple interest) compounded annually.

Example 1   An amount P is invested at $7\frac{1}{2}\%$ compounded annually. Find the reproduction curve.

Solution

$$f(P) = P + 7\tfrac{1}{2}\%P = 1 \cdot P + 0.075P = (1 + 0.075)P = 1.075P$$

DO EXERCISES 4 AND 5.

Maximum Sustainable Harvest

We know that a population P will grow to $f(P)$ in one year. If this were a population of fur-bearing animals, then one could "harvest" the amount

$$f(P) - P$$

each year without depleting the initial population P. Now suppose we wanted the value of P_0 that would allow the harvest to be the largest. If we could determine that P_0, then we would let the population grow until it reached that level, and then would begin harvesting year after year the amount $f(P_0) - P_0$.

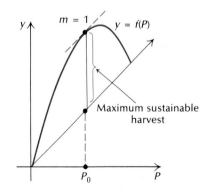

Let the harvest function H be given by

$$H(P) = f(P) - P.$$

Then

$$H'(P) = f'(P) - 1.$$

6. A certain population of fur-bearing animals has the reproduction curve

$$f(P) = P(8 - P),$$

where P is measured in thousands. Find the population at which the maximum sustainable harvest occurs. Find the maximum sustainable harvest.

Now if we assume that $H'(P)$ exists for all values of P and that there is only one critical point, it follows that the *maximum sustainable harvest* occurs at that value P_0 such that

$$H'(P_0) = f'(P_0) - 1 = 0 \quad \text{and} \quad H''(P_0) = f''(P_0) < 0.$$

Or, equivalently, we have the following.

THEOREM 9

The *maximum sustainable harvest* occurs at P_0 such that

$$f'(P_0) = 1 \quad \text{and} \quad f''(P_0) < 0,$$

and is given by

$$H(P_0) = f(P_0) - P_0.$$

Example 2   A certain population of fur-bearing animals has the reproduction curve

$$f(P) = P(10 - P),$$

where P is measured in thousands. Find the population at which the maximum sustainable harvest occurs. Find the maximum sustainable harvest.

Solution   Now

$$f(P) = 10P - P^2,$$

so

$$f'(P) = 10 - 2P \quad \text{and} \quad f''(P) = -2.$$

We set $f'(P) = 1$ and solve:

$$10 - 2P = 1$$
$$-2P = -9$$
$$P = 4.5.$$

There is a maximum since the second derivative is negative for all values of P. We find the maximum sustainable harvest by substituting 4.5 into the equation $H(P) = f(P) - P = (10P - P^2) - P$:

$$H(4.5) = [10 \cdot 4.5 - (4.5)^2] - 4.5 = 24.75 - 4.5 = 20.25.$$

Thus the maximum sustainable harvest is 20,250 at $P = 4500$.

DO EXERCISE 6.

EXERCISE SET 3.6

For each reproduction curve, (a) find the population at which the maximum sustainable harvest occurs, and (b) find the maximum sustainable harvest.

1. $f(P) = P(20 - P)$, where P is measured in thousands

2. $f(P) = P(6 - P)$, where P is measured in thousands

3. $f(P) = -0.025P^2 + 4P$, where P is measured in thousands. This is the reproduction curve in the Hudson Bay area for the *snowshoe hare*, a fur-bearing animal.

4. $f(P) = -0.01P^2 + 2P$, where P is measured in thousands. This is the reproduction curve in the Hudson Bay area for the *lynx*, a fur-bearing animal.

A prey: a snowshoe hare.

A predator of the snowshoe hare: a lynx.

5. $f(P) = 40\sqrt{P}$, where P is measured in thousands. Assume that this is the production curve for the brown-trout population in a large lake.

6. $f(P) = 30\sqrt{P}$, where P is measured in thousands. Assume that this is the production curve for the blue-gill population in a large lake.

7. $f(P) = 1.08P$

8. $f(P) = 1.075P$

3.7

OBJECTIVES

a) Given a function $y = f(x)$ and a value for Δx, find Δy.

b) Given a function $y = f(x)$, find dy, and compute dy given a value of Δx or dx.

c) Use differentials to make approximations of numbers like

$$\sqrt{27} \quad \text{or} \quad \sqrt[3]{10}.$$

DIFFERENTIALS

Delta Notation

Recall the difference quotient,

$$\frac{f(x + h) - f(x)}{h},$$

which is used to define the derivative of a function at x. The number h was considered to be a *change* in x. Another notation for such a change is Δx, read "delta x." The expression Δx is *not* the product of

Δ and x, but is its own entity; that is, it is a new type of variable that represents the *change* in the value of x from a *first* value to a *second*. Thus,

$$\Delta x = (x + h) - x = h.$$

If subscripts are used for the first and second values of x, we have

$$\Delta x = x_2 - x_1, \quad \text{or} \quad x_2 = x_1 + \Delta x.$$

Now Δx can be positive or negative.

Example 1

a) If $x_1 = 4$ and $\Delta x = 0.7$, then $x_2 = 4.7$.
b) If $x_1 = 4$ and $\Delta x = -0.7$, then $x_2 = 3.3$.

We usually omit the subscripts and use x and $x + \Delta x$.

Now suppose we have a function given by $y = f(x)$. A change in x from x to $x + \Delta x$ yields a change in y from $f(x)$ to $f(x + \Delta x)$. The change in y is given by

$$\Delta y = f(x + \Delta x) - f(x).$$

Example 2   For $y = x^2$, $x = 4$, and $\Delta x = 0.1$, find Δy.

Solution

$$\Delta y = (4 + 0.1)^2 - 4^2 = (4.1)^2 - 4^2 = 16.81 - 16 \doteq 0.81$$

Example 3   For $y = x^3$, $x = 2$, and $\Delta x = -0.1$, find Δy.

Solution

$$\Delta y = [2 + (-0.1)]^3 - 2^3 = (1.9)^3 - 2^3 = 6.859 - 8 = -1.141$$

1. For $y = x^2$, $x = 3$, and $\Delta x = -0.1$, find Δy.

DO EXERCISES 1 AND 2.

If delta notation is used, the difference quotient

$$\frac{f(x + h) - f(x)}{h}$$

becomes

$$\frac{f(x + \Delta x) - f(x)}{\Delta x} = \frac{\Delta y}{\Delta x}.$$

We can then express the derivative as

$$\frac{dy}{dx} = \lim_{\Delta x \to 0} \frac{\Delta y}{\Delta x}.$$

Note how the delta notation resembles the Leibniz notation. For values of Δx close to 0, we have the approximation

$$\frac{dy}{dx} \approx \frac{\Delta y}{\Delta x}, \quad \text{or} \quad f'(x) \approx \frac{\Delta y}{\Delta x}.$$

Multiplying both sides of the second expression by Δx, we get

2. For $y = x^3$, $x = 2$, and $\Delta x = 1$, find Δy.

$$\Delta y \approx f'(x)\,\Delta x.$$

We can see this in the following graph.

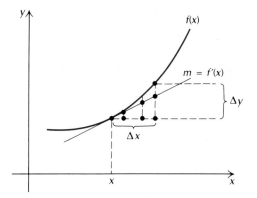

Recall that the derivative is a limit of slopes $\Delta y/\Delta x$ of secant lines. Thus as Δx gets smaller, the ratio $\Delta y/\Delta x$ gets closer to dy/dx. Note too that over small intervals the tangent line is a good approximation, linearly, to the function. Thus it is reasonable to assume that average rates of change $\Delta y/\Delta x$ of the function are approximately the same as the slope of the tangent line.

3. Approximate $\sqrt{67}$ using $\Delta y \approx f'(x)\,\Delta x$. See how close the approximation is by finding $\sqrt{67}$ directly on your calculator.

Let us use the fact that

$$\Delta y \approx f'(x)\,\Delta x$$

to make certain approximations, such as square roots.

Example 4   Approximate $\sqrt{27}$ using $\Delta y \approx f'(x)\,\Delta x$.

Solution   We first think of the number closest to 27 that is a perfect square. This is 25. What we will do is approximate how y, or $\sqrt{x}$, changes when 25 changes by $\Delta x = 2$. Let

$$y = f(x) = \sqrt{x}.$$

Then

$$\Delta y = \sqrt{x + \Delta x} - \sqrt{x} = \sqrt{x + \Delta x} - y,$$

so

$$y + \Delta y = \sqrt{x + \Delta x}.$$

Now

$$\Delta y \approx f'(x)\,\Delta x = \tfrac{1}{2}x^{-1/2}\,\Delta x = \frac{1}{2\sqrt{x}}\,\Delta x.$$

Let $x = 25$ and $\Delta x = 2$. Then

$$\Delta y \approx f'(x)\,\Delta x = \frac{1}{2\sqrt{25}} \cdot 2 = \frac{1}{\sqrt{25}} = \frac{1}{5} = 0.2.$$

So

$$\sqrt{27} = \sqrt{x + \Delta x} = y + \Delta y \approx \sqrt{25} + 0.2 = 5 + 0.2 = 5.2.$$

To five decimal places, $\sqrt{27} = 5.19615$. Thus our approximation is fairly accurate.

DO EXERCISE 3.

Suppose we have a total-cost function $C(x)$. When $\Delta x = 1$, we have

$$\Delta C \approx C'(x).$$

Whether this is a good approximation depends on the function and on the values of x. Let us consider an example.

Example 5   For the total-cost function

$$C(x) = 2x^3 - 12x^2 + 30x + 200:$$

4. Consider the total-cost function

$$C(x) = 0.01x^2 + 4x + 500.$$

a) Find ΔC and $C'(x)$ when $x = 5$ and $\Delta x = 1$.

b) Find ΔC and $C'(x)$ when $x = 100$ and $\Delta x = 1$.

a) find ΔC and $C'(x)$ when $x = 2$ and $\Delta x = 1$;

b) find ΔC and $C'(x)$ when $x = 100$ and $\Delta x = 1$.

Solution

a) $\Delta C = C(2 + 1) - C(2) = C(3) - C(2) = \$236 - \$228 = \$8.$

Recall that $C(2)$ is the total cost of producing 2 units and $C(3)$ is the total cost of producing 3 units, so $C(3) - C(2)$, or \$8, is the cost of the third unit. Now

$$C'(x) = 6x^2 - 24x + 30, \quad \text{so} \quad C'(2) = \$6.$$

b) $\Delta C = C(100 + 1) - C(100) = C(101) - C(100) = \$58,220.$

Note that this is the cost of the 101st unit. Now

$$C'(100) = \$57,630.$$

Note that in part (a) we might not consider the approximation between ΔC and $C'(x)$ to be too good, whereas in part (b) the approximation might be considered quite good since the numbers are so large. We have purposely used $\Delta x = 1$ to illustrate the following.

$$C'(x) \approx C(x + 1) - C(x)$$

Marginal cost is (approximately) the cost of the $(x + 1)$st, or next, unit.

This is the historical definition that economists have given to marginal cost.

Similarly, the following is true.

$$R'(x) \approx R(x + 1) - R(x)$$

Marginal revenue is (approximately) the revenue from the sale of the $(x + 1)$st, or next, unit.

And

$$P'(x) \approx P(x + 1) - P(x)$$

Marginal profit is (approximately) the profit from the production and sale of the $(x + 1)$st, or next, unit.

DO EXERCISE 4.

Differentials

Up to now we have not defined the symbols dy and dx as separate entities, and we have treated dy/dx as one symbol. We now define dy and dx. These symbols are called differentials.

DEFINITION

For $y = f(x)$, we define

$$dx, \quad \text{called the } \textit{differential of } x, \quad \text{by } dx = \Delta x$$

and

$$dy, \quad \text{called the } \textit{differential of } y, \quad \text{by } dy = f'(x) \, dx.$$

We can represent dx and dy pictorially as follows. Note that $dx = \Delta x$, but $dy \neq \Delta y$, though $dy \approx \Delta y$, for small values of dx.

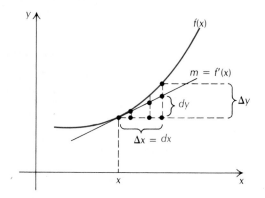

Example 6   For $y = x(4 - x)^3$, (a) find dy and (b) find dy when $x = 5$ and $dx = 0.2$.

Solution

a)   First, find dy/dx.

$$\frac{dy}{dx} = -x[3(4 - x)^2] + (4 - x)^3 \qquad \text{Using the Product Rule and the Extended Power Rule}$$

$$= (4 - x)^2[-3x + (4 - x)]$$

$$= (4 - x)^2[-4x + 4] \qquad \text{Simplifying}$$

$$= -4(4 - x)^2(x - 1)$$

5. For $y = x^3 - 5x^2 - 4x + 6$:

   a) find dy;

   b) find dy when $x = 3$ and $dx = 0.1$.

Then

$$dy = -4(4 - x)^2(x - 1)\, dx.$$

Note that the expression for dy contains *two* variables, x and dx.

   b) When $x = 5$ and $dx = 0.2$,

$$dy = -4(4 - 5)^2(5 - 1)0.2 = -4(-1)^2(4)(0.2) = -3.2.$$

DO EXERCISE 5.

EXERCISE SET 3.7

In Exercises 1–8, find Δy and $f'(x)\,\Delta x$.

1. For $y = f(x) = x^2$, $x = 2$, and $\Delta x = 0.01$.

2. For $y = x^3$, $x = 2$, and $\Delta x = 0.01$.

3. For $y = f(x) = x + x^2$, $x = 3$, and $\Delta x = 0.04$.

4. For $y = f(x) = x - x^2$, $x = 3$, and $\Delta x = 0.02$.

5. For $y = f(x) = 1/x^2$, $x = 1$, and $\Delta x = 0.5$.

6. For $y = f(x) = 1/x$, $x = 1$, and $\Delta x = 0.2$.

7. For $y = f(x) = 3x - 1$, $x = 4$, and $\Delta x = 2$.

8. For $y = f(x) = 2x - 3$, $x = 8$, and $\Delta x = 0.5$.

9. For the total-cost function

$$C(x) = 0.01x^2 + 0.6x + 30,$$

find ΔC and $C'(x)$ when $x = 70$ and $\Delta x = 1$.

10. For the total-cost function

$$C(x) = 0.01x^2 + 1.6x + 100,$$

find ΔC and $C'(x)$ when $x = 80$ and $\Delta x = 1$.

11. For the total-revenue function

$$R(x) = 2x,$$

find ΔR and $R'(x)$ when $x = 70$ and $\Delta x = 1$.

12. For the total-revenue function

$$R(x) = 3x,$$

find ΔR and $R'(x)$ when $x = 80$ and $\Delta x = 1$.

13. a) Using $C(x)$ of Exercise 9 and $R(x)$ of Exercise 11, find the total profit $P(x)$.

   b) Find ΔP and $P'(x)$ when $x = 70$ and $\Delta x = 1$.

14. a) Using $C(x)$ of Exercise 10 and $R(x)$ of Exercise 12, find the total profit $P(x)$.

   b) Find ΔP and $P'(x)$ when $x = 80$ and $\Delta x = 1$.

Approximate using $\Delta y \approx f'(x)\,\Delta x$.

15. $\sqrt{19}$ **16.** $\sqrt{10}$ **17.** $\sqrt{102}$ **18.** $\sqrt{103}$ **19.** $\sqrt[3]{10}$ **20.** $\sqrt[3]{28}$

21. The spherical volume of a tumor is given by

$$V = \tfrac{4}{3}\pi r^3,$$

where r is the radius in centimeters. By approximately how much does the volume increase when the radius is increased from 1 cm to 1.2 cm? Use 3.14 for π.

22. The circular area of a healing wound is given by

$$A = \pi r^2,$$

where r is the radius in centimeters. By approximately how much does the area decrease when the radius is decreased from 2 cm to 1.9 cm? Use 3.14 for π.

Find dy.

23. $y = (2x^3 + 1)^{3/2}$

24. $y = x^3(2x + 5)^2$

25. $y = \sqrt[5]{x + 27}$

26. $y = \dfrac{x^3 + x + 2}{x^2 + 3}$

27. $y = x^4 - 2x^3 + 5x^2 + 3x - 4$

28. $y = (7 - x)^8$

29. In Exercise 27, find dy when $x = 2$ and $dx = 0.1$.

30. In Exercise 28, find dy when $x = 1$ and $dx = 0.01$.

31. Suppose a rope surrounds the earth at the equator. The rope is lengthened by 10 ft. By about how much is the rope raised above the earth?

3.8

OBJECTIVES

a) Differentiate implicitly and find the slope of a curve at a given point.

b) Solve related-rate problems.

IMPLICIT DIFFERENTIATION AND RELATED RATES*

Implicit Differentiation

Consider the equation

$$y^3 = x.$$

This equation *implies* that y is a function of x, for if we solve for y, we get

$$y = \sqrt[3]{x}$$
$$= x^{1/3}.$$

We know from our work in this chapter that

$$\frac{dy}{dx} = \frac{1}{3} x^{-2/3}. \tag{1}$$

A method known as *implicit differentiation* allows us to find dy/dx *without* solving for y. We use the Chain Rule, treating y as a function of x. We then use the Extended Power Rule and differentiate both sides of

$$y^3 = x$$

with respect to x:

$$\frac{d}{dx} y^3 = \frac{d}{dx} x.$$

The derivative on the left side is found using the Extended Power Rule:

$$3y^2 \frac{dy}{dx} = 1.$$

---

* This section can be omitted without loss of continuity.

1. For $y^5 = 2x$, use implicit differen-
    tiation to find

    $$\frac{dy}{dx}.$$

    Leave the answer expressed in terms
    of y.

Then

$$\frac{dy}{dx} = \frac{1}{3y^2}, \quad \text{or} \quad \frac{1}{3}y^{-2}.$$

We can show that this indeed gives us the same answer as Eq. (1)
by replacing y by $x^{1/3}$:

$$\frac{dy}{dx} = \frac{1}{3}y^{-2} = \frac{1}{3}(x^{1/3})^{-2} = \frac{1}{3}x^{-2/3}.$$

DO EXERCISE 1.

Often, it is difficult or impossible to solve for y, obtaining an
explicit expression in terms of x. For example, the equation

$$y^3 + x^2y^5 - x^4 = 27$$

determines y as a function of x, but it would be difficult to solve for
y. We can nevertheless find a formula for the derivative of y *without*
solving for y. This involves computing dy^n/dx for various integers n,
and hence involves the Extended Power Rule in the form

$$\frac{d}{dx}y^n = ny^{n-1} \cdot \frac{dy}{dx}.$$

Example 1   For

$$y^3 + x^2y^5 - x^4 = 27:$$

a)   find dy/dx using implicit differentiation;

b)   find the slope of the tangent line to the curve at the point $(0, 3)$.

Solution

a)   We differentiate the term x^2y^5 using the Product Rule. Note that
     any time an expression involving y is differentiated, dy/dx must be
     a factor of the answer. When an expression involving just x is
     differentiated, there is no factor dy/dx.

$$\frac{d}{dx}(y^3 + x^2y^5 - x^4) = \frac{d}{dx}(27)$$

$$\frac{d}{dx}y^3 + \frac{d}{dx}x^2y^5 - \frac{d}{dx}x^4 = 0$$

$$3y^2 \cdot \frac{dy}{dx} + x^2 \cdot 5y^4 \cdot \frac{dy}{dx} + 2x \cdot y^5 - 4x^3 = 0$$

2. For

$$y^3 + x^2y^4 + x^3 = 8:$$

a) find

$$\frac{dy}{dx}$$

using implicit differentiation;

b) find the slope of the tangent line at $(0, 2)$.

Then

$$3y^2 \cdot \frac{dy}{dx} + 5x^2y^4 \cdot \frac{dy}{dx} = 4x^3 - 2xy^5 \qquad \begin{array}{l}\text{Getting all terms involving}\\ dy/dx \text{ alone on one side.}\end{array}$$

$$(3y^2 + 5x^2y^4) \frac{dy}{dx} = 4x^3 - 2xy^5$$

$$\frac{dy}{dx} = \frac{4x^3 - 2xy^5}{3y^2 + 5x^2y^4} \qquad \begin{array}{l}\text{Solving for } dy/dx. \text{ Leave the}\\ \text{answer in terms of } x \text{ and } y.\end{array}$$

b) To find the slope of the tangent line to the curve at $(0, 3)$, we replace x by 0 and y by 3:

$$\frac{dy}{dx} = \frac{4 \cdot 0^3 - 2 \cdot 0 \cdot 3^5}{3 \cdot 3^2 + 5 \cdot 0^2 \cdot 3^4}$$

$$= 0.$$

DO EXERCISE 2.

The demand function for a product (see Sections 1.5 and 2.6) is often given implicitly.

3. For the following equation, differentiate implicitly to find dp/dx:

$$100\sqrt{p} = 800 - x.$$

Example 2 For the following demand equation, differentiate implicitly to find dp/dx:

$$x = \sqrt{200 - p^3}.$$

Solution

$$\frac{d}{dx} x = \frac{d}{dx} (\sqrt{200 - p^3})$$

$$1 = \frac{1}{2} (200 - p^3)^{-1/2} \cdot (-3p^2) \cdot \frac{dp}{dx}$$

$$1 = \frac{-3p^2}{2\sqrt{200 - p^3}} \cdot \frac{dp}{dx}$$

$$\frac{2\sqrt{200 - p^3}}{-3p^2} = \frac{dp}{dx}$$

DO EXERCISE 3.

4. A stone is thrown into a pond. A circular ripple is spreading over the pond in such a way that its radius r is increasing at the rate of 3 feet per second at the moment when r goes through the value $r = 4$ ft. At that moment, how fast is the disturbed area increasing?

The radius of each ripple is a function of time.

Related Rates

Suppose y is a function of x, say

$$y = f(x),$$

and x varies with time t (as a function of time t). Since y depends on x and x depends on t, y also depends on t. That is, y is also a function of time t. The Chain Rule gives the following:

$$\frac{dy}{dt} = \frac{dy}{dx} \cdot \frac{dx}{dt}.$$

Thus the rate of change of y is *related* to the rate of change of x. Let us see how this comes up in problems. It helps to keep in mind that any variable can be thought of as a function of time t, even though a specific expression in terms of t may not be given.

Example 3 A restaurant supplier services the restaurants in a circular area in such a way that its radius r is increasing at the rate of 2 miles per year at the moment when r goes through the value $r = 5$ mi. At that moment, how fast is the area increasing?

Solution The area A and the radius r are always related by the equation for the area of a circle:

$$A = \pi r^2.$$

We take the derivative of both sides with respect to t:

$$\frac{dA}{dt} = 2\pi r \cdot \frac{dr}{dt}.$$

At the moment in question, $dr/dt = 2$ mi/yr (miles per year) and $r = 5$ mi, so

$$\frac{dA}{dt} = 2\pi (5 \text{ mi}) \left(2 \frac{\text{mi}}{\text{yr}} \right)$$

$$= 20\pi \frac{\text{mi}^2}{\text{yr}}$$

$$\approx 63 \text{ square miles per year.}$$

DO EXERCISE 4.

5. For a certain product, a company determines that total revenue from the sale of x units is given by

$$R(x) = 200x - x^2$$

and that total cost is given by

$$C(x) = 5000 + 8x.$$

Suppose the company is producing and selling x units at a rate of 8 per day at the moment the 100th unit is produced. At the same moment, what is the rate of change of (a) total revenue? (b) total cost? (c) total profit?

Example 4 *Business: Rate of change of revenue, cost, and profit.* For a company making stereos, total revenue from the sale of x stereos is given by

$$R(x) = 1000x - x^2,$$

and total cost is given by

$$C(x) = 3000 + 20x.$$

Suppose the company is producing and selling stereos at the rate of 10 stereos per day at the moment when the 400th stereo is produced. At that same moment, what is the rate of change of (a) total revenue? (b) total cost? (c) total profit?

Solution

a) $\dfrac{dR}{dt} = 1000 \cdot \dfrac{dx}{dt} - 2x \cdot \dfrac{dx}{dt}$ Differentiating with respect to time

      $= 1000 \cdot 10 - 2(400)10$ Substituting 10 for dx/dt and 400 for x

      $= \$2000$ per day

b) $\dfrac{dC}{dt} = 20 \cdot \dfrac{dx}{dt}$ Differentiating with respect to time

      $= 20(10)$

      $= \$200$ per day

c) Since $P = R - C$,

$$\frac{dP}{dt} = \frac{dR}{dt} - \frac{dC}{dt} = \$2000 \text{ per day} - \$200 \text{ per day}$$

$$= \$1800 \text{ per day.}$$

DO EXERCISE 5.

EXERCISE SET 3.8

Differentiate implicitly to find dy/dx. Then find the slope of the curve at the given point.

1. $xy - x + 2y = 3$; $(-5, \frac{2}{3})$

2. $xy + y^2 - 2x = 0$; $(1, -2)$

3. $x^2 + y^2 = 1$; $\left(\dfrac{1}{2}, \dfrac{\sqrt{3}}{2}\right)$

4. $x^2 - y^2 = 1$; $(\sqrt{3}, \sqrt{2})$

5. $x^2y - 2x^3 - y^3 + 1 = 0$; $(2, -3)$

6. $4x^3 - y^4 - 3y + 5x + 1 = 0$; $(1, -2)$

Differentiate implicitly to find dy/dx.

7. $2xy + 3 = 0$ 8. $x^2 + 2xy = 3y^2$ 9. $x^2 - y^2 = 16$ 10. $x^2 + y^2 = 25$

11. $y^5 = x^3$ 12. $y^3 = x^5$ 13. $x^2y^3 + x^3y^4 = 11$ 14. $x^3y^2 - x^5y^3 = -19$

For the following demand equations, differentiate implicitly to find dp/dx.

15. $p^2 + p + 2x = 40$ 16. $xp^3 = 24$ 17. $(p + 4)(x + 3) = 48$ 18. $1000 - 300p + 25p^2 = x$

APPLICATIONS

19. *Biomedical: Rate of change of a tumor.* The volume of a tumor is given by

    $$V = \tfrac{4}{3}\pi r^3.$$

    The radius is increasing at the rate of 0.03 centimeter per day (cm/day) at the moment when $r = 1.2$ cm. How fast is the volume changing at that moment?

20. *Biomedical: Rate of change of a healing wound.* The area of a healing wound is given by

    $$A = \pi r^2.$$

    The radius is decreasing at the rate of 1 millimeter per day $(-1$ mm/day) at the moment when $r = 25$ mm. How fast is the area decreasing at that moment?

Business: Rate of change of total revenue, cost, and profit. Find the rates of change of total revenue, cost, and profit for each of the following.

21. $R(x) = 50x - 0.5x^2$,

    $C(x) = 4x + 10$,

    when $x = 30$ and $dx/dt = 20$ units per day

22. $R(x) = 50x - 0.5x^2$,

    $C(x) = 10x + 3$,

    when $x = 10$ and $dx/dt = 5$ units per day

23. $R(x) = 2x$,

    $C(x) = 0.01x^2 + 0.6x + 30$,

    when $x = 20$ and $dx/dt = 8$ units per day

24. $R(x) = 280x - 0.4x^2$,

    $C(x) = 5000 + 0.6x^2$,

    when $x = 200$ and $dx/dt = 300$ units per day

25. Two cars start from the same point at the same time. One travels north at 25 miles per hour (mph), and the other travels east at 60 mph. How fast is the distance between them increasing at the end of 1 hr? [*Hint:* $D^2 = x^2 + y^2$. To find D after 1 hr, solve $D^2 = 25^2 + 60^2$.]

26. A ladder 26 ft long leans against a vertical wall. If the lower end is being moved away from the wall at the rate of 5 ft/sec, how fast is the height of the top decreasing (this will be a negative rate) when the lower end is 10 ft from the wall?

Biomedical: Poiseuille's Law. The flow of blood in a blood vessel is faster toward the center of the vessel and slower toward the outside. The speed of the blood V is given by

$$V = \frac{p}{4Lv}(R^2 - r^2),$$

where

R = the radius of the blood vessel,

r = the distance of the blood from the center of the vessel,

and p, L, and v are physical constants related to pressure, length, and viscosity of the blood vessels, respectively.

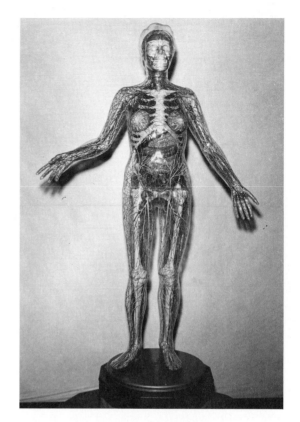

The flow of blood in a blood vessel can be modeled by Poiseuille's Law.

27. Assume that r is a constant as well as p, L, and v.

   a) Find the rate of change dV/dt in terms of R and dR/dt when $L = 1$ mm, $p = 100$, and $v = 0.05$.

   b) A person goes out into the cold to shovel snow. Cold air has the effect of contracting blood vessels far from the heart. Suppose a blood vessel contracts at a rate of

   $$\frac{dR}{dt} = -0.0015 \text{ mm/min}$$

   at a place in the blood vessel where the radius is $R = 0.0075$ mm. Find the rate of change dV/dt at that location.

28. Assume that r is a constant as well as p, L, and v.

   a) Find the rate of change dV/dt in terms of R and dR/dt when $L = 1$ mm, $p = 100$, and $v = 0.05$.

   b) When shoveling snow in cold air, a person with a history of heart trouble can develop angina (chest pains) due to contracting blood vessels. To counteract this, a person may take a nitroglycerin tablet, which dilates the blood vessels. Suppose that after a nitroglycerin tablet is taken, a blood vessel dilates at a rate of

   $$\frac{dR}{dt} = 0.0025 \text{ mm/min}$$

   at a place in the blood vessel where the radius is $R = 0.02$ mm. Find the rate of change dV/dt.

29. Two variable quantities A and B are found to be related by the equation

$$A^3 + B^3 = 9.$$

What is the rate of change dA/dt at the moment when $A = 2$ and $dB/dt = 3$?

30. Two variable quantities G and H, nonnegative, are found to be related by the equation

$$G^2 + H^2 = 25.$$

What is the rate of change dH/dt when $dG/dt = 3$ and $G = 0$? $G = 1$? $G = 3$?

EXTENSION EXERCISES

Differentiate implicitly to find dy/dx.

31. $\sqrt{x} + \sqrt{y} = 1$

32. $\dfrac{1}{x^2} + \dfrac{1}{y^2} = 5$

33. $y^3 = \dfrac{x-1}{x+1}$

34. $y^2 = \dfrac{x^2 - 1}{x^2 + 1}$

35. $x^{3/2} + y^{2/3} = 1$

36. $(x - y)^3 + (x + y)^3 = x^5 + y^5$

Differentiate implicitly to find dy/dx and d^2y/dx^2.

37. $xy + x - 2y = 4$

38. $y^2 - xy + x^2 = 5$

39. $x^2 - y^2 = 5$

40. $x^3 - y^3 = 8$

SUMMARY AND REVIEW: CHAPTER 3

The following contains a summary of what you should be able to do after completing this chapter. The review exercises are for practice. Answers are at the back of the book. If you miss an exercise, restudy the section indicated alongside the answers.

You should be able to:

Find maximum and minimum values of functions using Maximum–Minimum Principles 1 and/or 2.

Find the maximum and minimum values, if they exist, over the indicated interval. Where no interval is specified, use the real line.

1. $f(x) = \frac{1}{3}x^3 + 3x^2 + 9x + 2$

2. $f(x) = x^2 - 10x + 8; \ [-2, 6]$

3. $f(x) = 4x^3 - 6x^2 - 24x + 5; \ [-2, 3]$

4. $f(x) = 5x - 7; \ [-1, 1]$

5. $f(x) = x^2 - \dfrac{2}{x}, \ (-\infty, 0)$

6. $f(x) = 5x - 7$

7. $f(x) = 3x^4 + 2x^3 - 3x^2 + 1; \ [-3, 0]$

8. $f(x) = 5x^2 + \dfrac{5}{x^2}; \ (0, \infty)$

9. $f(x) = 5x^4 - x^5; \ [-1, 5]$

10. $f(x) = -x^2 + 5x + 7$

Solve maximum–minimum problems.

11. Of all the numbers whose sum is 60, find the two that have the maximum product.

12. Find the minimum value of $Q = x^2 - 2y^2$, where $x - 2y = 1$.

13. If $R(x) = 52x - 0.5x^2$ and $C(x) = 22x - 1$, find the maximum profit and the number of units that must be produced and sold in order to yield this maximum profit.

14. A rectangular box with a square base and a cover is to contain 2500 ft^3. If the cost per ft^2 for the bottom is $2, for the top is $3, and for the sides is $1, what should the area be in order to minimize the cost?

Given certain inventory costs, find how many times a year a store should reorder a product, and in what lot size, in order to minimize total inventory costs.

15. A store in California sells 360 multispeed bicycles per year. It costs $8 to store one bicycle for one year. To reorder bicycles there is a fixed cost of $10 plus $2 for each bicycle. How many times per year should the store order bicycles, and in what lot size, in order to minimize inventory costs?

Given a reproduction cycle described by a formula, find the population at which the maximum sustainable harvest occurs, and find the maximum sustainable harvest.

16. A certain population has a reproduction curve given by $f(P) = P(71 - P)$, where P is measured in thousands. Find the population at which the maximum sustainable harvest occurs, and find the maximum sustainable harvest.

Given a function $y = f(x)$ and a value for Δx, find Δy. Given a function $y = f(x)$, find dy, and compute dy given a value of Δx or dx. Use differentials to approximate roots of numbers.

Given $y = f(x) = x^3 - x$.

17. Find Δy and $f'(x)\,\Delta x$, given that $x = 3$ and $\Delta x = -0.5$.

18. a) Find dy.

    b) Find dy when $x = 2$ and $dx = 0.01$.

19. Approximate $\sqrt{69}$ using $\Delta y \approx f'(x)\,\Delta x$.

Use derivatives to sketch graphs of functions and determine relative maxima and minima.

Sketch the graph of the function. Determine the relative maxima and minima.

20. $y = \dfrac{1}{2}x + \dfrac{1}{x}$

21. $y = x^3 + x^2 - x + 3$

Differentiate implicitly and find the slope of a curve at a given point. Solve related-rate problems.

22. Differentiate implicitly to find dy/dx. Then find the slope of the curve at the given point.

$$2x^3 + 2y^3 = -9xy; \; (-1, -2)$$

23. A ladder 25 ft long leans against a vertical wall. If the lower end is being moved away from the wall at the rate of 6 ft/sec, how fast is the height of the top decreasing when the lower end is 7 ft from the wall?

24. Find the rates of change of total revenue, cost, and profit for

$$R(x) = 120x - 0.5x^2 \quad \text{and} \quad C(x) = 15x + 6,$$

when $x = 100$ and $dx/dt = 30$ units per day.

EXTENSION EXERCISES

25. Find the maximum and minimum values, if they exist, over the indicated interval.

$$f(x) = (x - 3)^{2/5}; \ (-\infty, \infty)$$

26. Differentiate implicitly to find dy/dx.

$$(x - y)^4 + (x + y)^4 = x^6 + y^6$$

27. Find the relative maxima and minima of $y = x^4 - 8x^3 - 270x^2$.

TEST: CHAPTER 3

Find the maximum and minimum values, if they exist, over the indicated interval. Where no interval is specified, use the real line.

1. $f(x) = x(6 - x)$

2. $f(x) = x^3 + x^2 - x + 1; \ [-2, \frac{1}{2}]$

3. $f(x) = -x^2 + 8.6x + 10$

4. $f(x) = -2x + 5; \ [-1, 1]$

5. $f(x) = -2x + 5$

6. $f(x) = 3x^2 - x - 1$

7. $f(x) = x^2 + \dfrac{128}{x}; \ (0, \infty)$

8. Of all numbers whose difference is 8, find the two that have the minimum product.

9. Minimize $Q = x^2 + y^2$, where $x - y = 10$.

10. Find the maximum profit and the number of units that must be produced and sold in order to yield the maximum profit.

$$R(x) = x^2 + 110x + 60,$$
$$C(x) = 1.1x^2 + 10x + 80$$

11. From a piece of cardboard 60 in. by 60 in., square corners are cut out so the sides can be folded up to make a box. What dimensions will yield a box of maximum volume? What is the maximum volume?

12. A sporting goods store sells 1225 tennis rackets per year. It costs $2 to store one tennis racket for one year. To reorder tennis rackets, there is a fixed cost of $1, plus $0.50 for each tennis racket. How many times per year should the sporting goods store order tennis rackets, and in what lot size, in order to minimize inventory costs?

13. Consider the reproduction curve $f(P) = P(100 - P)$, where P is measured in thousands. Find the population at which the maximum sustainable harvest occurs. Find the maximum sustainable harvest.

14. For $y = f(x) = x^2 - 3, x = 5$, and $\Delta x = 0.1$, find Δy and $f'(x) \, \Delta x$.

15. Approximate $\sqrt{104}$ using $\Delta y \approx f'(x) \, \Delta x$.

16. For $y = \sqrt{x^2 + 3}$, (a) find dy and (b) find dy when $x = 2$ and $dx = 0.01$.

17. Sketch the graph of

$$f(x) = 3x^4 + 4x^3 - 6x^2 - 12x + 1.$$

Determine the relative maxima and minima.

18. Differentiate implicitly to find dy/dx. Then find the slope of the curve at the given point.

$$x^3 + y^3 = 9; \; (1, 2)$$

19. A board 13 ft long leans against a vertical wall. If the lower end is being moved away from the wall at the rate of 0.4 ft/sec, how fast is the upper end coming down when the lower end is 12 ft from the wall?

EXTENSION EXERCISES

20. Find the maximum and minimum values, if they exist, over the indicated interval.

$$f(x) = \frac{x^2}{1 + x^3}; \; [0, \infty)$$

21. The total cost of producing x units of a product is given by

$$C(x) = 100x + 100\sqrt{x} + \frac{\sqrt{x^3}}{100}.$$

a) Find the average cost $A(x)$.

b) Find the minimum value of $A(x)$.

4

EXPONENTIAL AND LOGARITHMIC FUNCTIONS

In this chapter we will consider two kinds of functions that are closely related: *exponential functions* and *logarithmic functions*. We will learn to find derivatives of such functions. Both are rich in application, with exponential functions applying particularly to problems of population growth.

AN APPLICATION

Peter Minuit of the Dutch West India Company purchased Manhattan Island from the Indians in 1626 for $24 worth of merchandise. Assuming an exponential rate of inflation of 8%, how much will Manhattan be worth in 1988?

THE MATHEMATICS

The value V of Manhattan Island is given by

$$V(t) = \$24e^{0.08t},$$

This is an *exponential function*.

where t is the number of years since 1626.

4.1

EXPONENTIAL FUNCTIONS

Graphs of Exponential Functions

The following bar graph shows the cost of a 60-second commercial during the telecast of the Super Bowl. A curve drawn along the graph would approximate the graph of an *exponential function*. We now consider such graphs.

Year	Cost of a 60-sec TV commercial during the Super Bowl
1967	$80,000
1970	$200,000
1977	$324,000
1981	$550,000
1983	$800,000
1985	$1,100,000

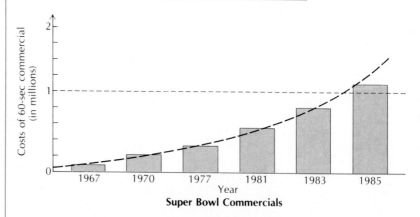

Super Bowl Commercials

In Chapter 1 we reviewed definitions of such expressions as a^x, where x is a rational number. For example,

$$a^{2.34}, \quad \text{or} \quad a^{234/100},$$

means "raise a to the 234th power and then take the 100th root."

What about expressions with irrational exponents, such as $2^{\sqrt{2}}$, 2^{π}, or $2^{-\sqrt{3}}$? An irrational number is a number named by an infinite, non-

1. ($\blacksquare$ with a $\boxed{y^x}$ key). Complete this table. Round to six decimal places.

r	2^r
3	
3.1	
3.14	
3.141	
3.1415	
3.14159	

Considering $2^\pi = \lim_{r \to \pi} 2^r$, for rational values of r, what seems to be the value of 2^π to two decimal places?

repeating decimal. Let us consider 2^π. We know that π is irrational with infinite, nonrepeating decimal expansion:

$$3.141592654\ldots.$$

This means that π is approached as a limit by the rational numbers

$$3, \quad 3.1, \quad 3.14, \quad 3.141, \quad 3.1415, \ldots,$$

so it seems reasonable that 2^π should be approached as a limit by the rational powers

$$2^3, \quad 2^{3.1}, \quad 2^{3.14}, \quad 2^{3.141}, \quad 2^{3.1415}, \ldots.$$

DO EXERCISE 1.

In general, a^x is approximated by the values of a^r for rational numbers r near x; a^x is the limit of a^r as r approaches x through rational values. In summary, for $a > 0$, the definition of a^x for rational numbers x can be extended to arbitrary real numbers x in such a way that the usual laws of exponents, such as

$$a^x \cdot a^y = a^{x+y}, \qquad a^x \div a^y = a^{x-y}, \qquad (a^x)^y = a^{xy}, \quad \text{and} \quad a^{-x} = \frac{1}{a^x},$$

still hold. Moreover, the function so obtained,

$$f(x) = a^x,$$

is continuous.

DEFINITION

An *exponential function* f is given by

$$f(x) = a^x, \qquad a > 0, \quad a \neq 1.$$

The number a is called the *base*.

The following are examples of exponential functions:

$$f(x) = 2^x, \qquad f(x) = (\tfrac{1}{2})^x, \qquad f(x) = (0.4)^x.$$

Note that in contrast to power functions like $y = x^2$ or $y = x^3$, the variable in an exponential function is in the exponent, not the base. Exponential functions have extensive application. Let us consider their graphs.

2. Consider $y = f(x) = 3^x$.   `CSS`

a) Complete this table of function values.

x	0	$\frac{1}{2}$	1	2	-1	-2
3^x						

b) Graph $f(x) = 3^x$.

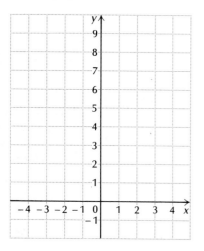

Example 1   Graph $y = f(x) = 2^x$.

Solution

a) First we find some function values.

x	0	$\frac{1}{2}$	1	2	3	-1	-2
$y = f(x)$ (or 2^x)	1	1.4	2	4	8	$\frac{1}{2}$	$\frac{1}{4}$

Note: For

$x = 0, y = 2^0 = 1;$

$x = \frac{1}{2}, y = 2^{1/2} = \sqrt{2} \approx 1.4;$

$x = 1, y = 2^1 = 2;$

$x = 2, y = 2^2 = 4;$

$x = 3, y = 2^3 = 8;$

$x = -1, y = 2^{-1} = \frac{1}{2};$

$x = -2, y = 2^{-2} = \dfrac{1}{2^2} = \dfrac{1}{4}.$

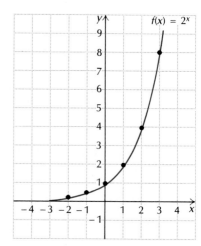

b) Next, we plot the points and connect them with a smooth curve as shown in the figure. The graph is continuous, increasing, and concave up.

DO EXERCISE 2.

Example 2   Graph $y = f(x) = (\frac{1}{2})^x$.

Solution

a) We first find some function values. Before we do this, note that

$$y = f(x) = (\tfrac{1}{2})^x = (2^{-1})^x = 2^{-x}.$$

This will ease our work.

3. Consider $y = f(x) = (\frac{1}{3})^x$. `CSS`

a) Complete this table of function values.

x	0	$\frac{1}{2}$	1	2	-1	-2
y						

b) Graph $f(x) = (\frac{1}{3})^x$.

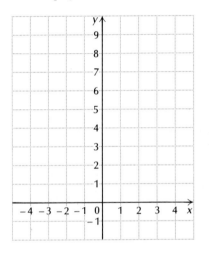

x	0	$\frac{1}{2}$	1	2	-1	-2	-3
y	1	0.7	$\frac{1}{2}$	$\frac{1}{4}$	2	4	8

Note: For

$$x = 0, y = 2^{-0} = 1;$$
$$x = \tfrac{1}{2}, y = 2^{-1/2}$$
$$= \frac{1}{\sqrt{2}} \approx \frac{1}{1.4} \approx 0.7;$$
$$x = 1, y = 2^{-1} = \tfrac{1}{2};$$
$$x = 2, y = 2^{-2} = \tfrac{1}{4};$$
$$x = -1, y = 2^{-(-1)} = 2;$$
$$x = -2, y = 2^{-(-2)} = 4;$$
$$x = -3, y = 2^{-(-3)} = 8.$$

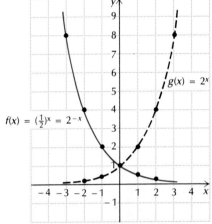

b) We plot these points and connect them with a smooth curve as shown by the solid curve in the figure. The graph is continuous, decreasing, and concave up. The dashed curve shows $g(x) = 2^x$ for comparison.

DO EXERCISE 3.

The following are some properties of the exponential function for various bases.

1. The function $f(x) = a^x$, where $a > 1$, is a positive, increasing, continuous function; and as x gets smaller, a^x approaches 0. The graph is concave up.

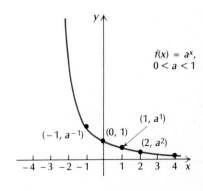

4. In order to investigate

$$\lim_{h \to 0} \frac{2^h - 1}{h},$$

we choose a sequence of numbers h, approaching 0, and compute

$$\frac{2^h - 1}{h}.$$

a) Complete this table.

h	$\dfrac{2^h - 1}{h}$
0.5	
0.25	
0.175	
0.0625	
0.03125	
0.00111	

b) What is the value of

$$\lim_{h \to 0} \frac{2^h - 1}{h},$$

to the nearest tenth?

5. a) Complete this table.

h	$\dfrac{3^h - 1}{h}$
0.5	
0.25	
0.175	
0.0625	
0.03125	
0.00111	

---

2. The function $f(x) = a^x$, where $0 < a < 1$, is a positive, decreasing, continuous function; and as x gets larger, a^x approaches 0. The graph is concave up.

When $a = 1$, $f(x) = a^x = 1^x = 1$, and is a constant function. This is why we do not allow 1 to be the base of an exponential function.

The Derivative of a^x, the Number e

Let us consider finding the derivative of the exponential function

$$f(x) = a^x.$$

The derivative is given by

$$f'(x) = \lim_{h \to 0} \frac{f(x + h) - f(x)}{h} \qquad \text{Definition of the derivative}$$

$$= \lim_{h \to 0} \frac{a^{x+h} - a^x}{h} \qquad \text{Substituting } a^{x+h} \text{ for } f(x + h) \text{ and } a^x \text{ for } f(x)$$

$$= \lim_{h \to 0} \frac{a^x \cdot a^h - a^x \cdot 1}{h}$$

$$= \lim_{h \to 0} a^x \cdot \left(\frac{a^h - 1}{h}\right)$$

$$= a^x \cdot \lim_{h \to 0} \frac{a^h - 1}{h}. \qquad \text{Since the variable is } h \text{ and } h \to 0, \text{ we treat } a^x \text{ as a constant, and the limit of a constant times a function is the constant times the limit.}$$

We get

$$f'(x) = a^x \cdot \lim_{h \to 0} \frac{a^h - 1}{h}. \qquad (1)$$

In particular, for $g(x) = 2^x$,

$$g'(x) = 2^x \cdot \lim_{h \to 0} \frac{2^h - 1}{h}.$$

Note that the limit does not depend on the value of x at which we are evaluating the derivative. In order for $g'(x)$ to exist, we must determine whether

$$\lim_{h \to 0} \frac{2^h - 1}{h} \text{ exists.}$$

Let us investigate this question.

DO EXERCISE 4.

b) What is the value of

$$\lim_{h \to 0} \frac{3^h - 1}{h},$$

to the nearest tenth?

6. a) Complete this table.

x	-3	-2	-1	0	1	2	3
2^x			0.5				
$(0.7)2^x$			0.35				

b) Using the same set of axes, graph $g(x) = 2^x$ with a solid line and $g'(x) \approx (0.7)2^x$ with a dashed line.

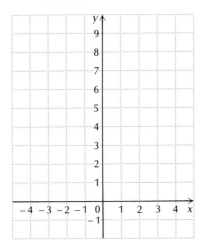

Margin Exercise 4 suggests that $(2^h - 1)/h$ has a limit as h approaches 0 and that its approximate value is 0.7, so that

$$g'(x) \approx (0.7)2^x.$$

In other words, the derivative is a constant times the function value 2^x. Similarly, for $t(x) = 3^x$,

$$t'(x) = 3^x \cdot \lim_{h \to 0} \frac{3^h - 1}{h}.$$

Again we can find an approximation for the limit that does not depend on the value of x at which we are evaluating the derivative.

DO EXERCISE 5. (EXERCISE 5 BEGINS ON THE PRECEDING PAGE.)

Margin Exercise 5 suggests that $(3^h - 1)/h$ has a limit as h approaches 0 and that its approximate value is 1.1, so that

$$t'(x) \approx (1.1)3^x.$$

In other words, the derivative is a constant times the function value 3^x.

DO EXERCISES 6 AND 7. (EXERCISE 7 IS ON THE FOLLOWING PAGE.)

In Fig. 1 we have graphed $g(x) = 2^x$ and $g'(x) \approx (0.7)2^x$. Note that the graph of g' lies *below* the graph of g.

In Fig. 2 we have graphed $t(x) = 3^x$ and $t'(x) \approx (1.1)3^x$. Note that the graph of t' lies *above* the graph of t.

FIGURE 1

FIGURE 2

7. a) Complete this table.

x	-2	-1	0	1	2
3^x					
$(1.1)3^x$					

b) Using the same set of axes, graph $h(x) = 3^x$ with a solid line and $h'(x) \approx (1.1)3^x$ with a dashed line.

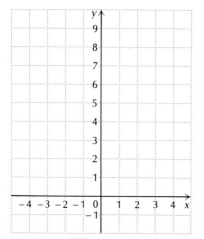

8. (🖩 with a y^x key). The compound-interest formula, which we developed in Chapter 1, is

$$A = P\left(1 + \frac{i}{n}\right)^{nt},$$

where A is the amount that an initial investment P will be worth after t years at interest rate i, compounded n times per year.

Suppose $1 is an initial investment at 100% interest for 1 year (no bank would pay this!). The formula above becomes

$$A = \left(1 + \frac{1}{n}\right)^n.$$

We might expect that there is exactly one base a between 2 and 3 for which a^x and its derivative have the same graph. This conjecture can be proved (though we will not do it here).

DEFINITION

We define the number e as the unique positive real number for which

$$\lim_{h \to 0} \frac{e^h - 1}{h} = 1.$$

It follows that for the exponential function $f(x) = e^x$,

$$f'(x) = e^x \cdot \lim_{h \to 0} \frac{e^h - 1}{h} = e^x \cdot 1 = e^x.$$

In Margin Exercise 8 you will not only consider an application of e, but also find decimal approximations.

DO EXERCISE 8.

Suppose we were to have the compounding periods n increase indefinitely. The amount in the investment of Margin Exercise 8 would be growing at interest compounded continuously, and would approach about $2.718. It can be shown that the number e can be described by a limit.

THEOREM 1

$$e = \lim_{n \to \infty} \left(1 + \frac{1}{n}\right)^n$$

That is, e is that number which

$$\left(1 + \frac{1}{n}\right)^n$$

CSS

approaches as n gets larger without bound. To ten decimal places, e is given by

$$e = 2.7182818284 \ldots.$$

We have established that for the function $f(x) = e^x$, we also have $f'(x) = e^x$, or, simply, the following.

Complete this table. Round to six decimal places.

n	$\left(1 + \dfrac{1}{n}\right)^n$
1 (compounding annually)	
2 (compounding semiannually)	
3	
4 (compounding quarterly)	
5	
100	
365 (compounding daily)	
8760 (compounding hourly)	
525,600 (compounding every minute)	

Differentiate.

9. $y = 6e^x$

10. $y = x^3 e^x$

11. $f(x) = \dfrac{e^x}{x^2}$

THEOREM 2

$$\frac{d}{dx} e^x = e^x$$

Note that this says that the derivative (the slope of the tangent line) at any x is the same as the function value. Let us find some other derivatives.

Example 3

$$\frac{d}{dx} 3e^x = 3e^x$$

Example 4

$$\frac{d}{dx} x^2 e^x = x^2 \cdot e^x + 2x \cdot e^x \qquad \text{By the Product Rule}$$

$$= e^x(x^2 + 2x), \quad \text{or} \quad xe^x(x + 2) \qquad \text{Factoring}$$

Example 5

$$\frac{d}{dx}\left(\frac{e^x}{x^3}\right) = \frac{x^3 \cdot e^x - e^x(3x^2)}{x^6} \qquad \text{By the Quotient Rule}$$

$$= \frac{x^2 e^x(x - 3)}{x^6} \qquad \text{Factoring}$$

$$= \frac{e^x(x - 3)}{x^4} \qquad \text{Simplifying}$$

DO EXERCISES 9–11.

Suppose we have a more complicated expression in the exponent, such as

$$h(x) = e^{x^2 - 5x}.$$

This is a composition of functions. Suppose, in general, that we have

$$h(x) = e^{f(x)} = g[f(x)], \quad \text{where } g(x) = e^x.$$

Now $g'(x) = e^x$. Then by the Chain Rule (Section 2.8) we have

$$h'(x) = g'[f(x)] \cdot f'(x) = e^{f(x)} \cdot f'(x).$$

For the case above, $f(x) = x^2 - 5x$, so $f'(x) = 2x - 5$. Then

$$h'(x) = f'[g(x)] \cdot f'(x) = e^{f(x)} \cdot f'(x) = e^{x^2 - 5x}(2x - 5).$$

Differentiate.

12. $f(x) = e^{-4x}$

13. $y = e^{x^3 + 8x}$

14. $f(x) = e^{\sqrt{x^2 + 5}}$

15. Graph $f(x) = 2e^{-x}$. Use your calculator. For example, for $x = 3$, $f(3) = 2e^{-3} = 2(0.0498) \approx 0.1$.

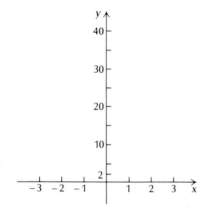

The next rule, which we have proven using the Chain Rule, allows us to find derivatives of functions like the one above.

THEOREM 3

$$\frac{d}{dx} e^{f(x)} = f'(x) e^{f(x)}$$

The following gives us a way to remember this rule.

$h(x) = e^{x^2 - 5x}$ ①   Take the derivative of the exponent.

$h'(x) = (2x - 5)e^{x^2 - 5x}$ ②   Multiply the derivative of the exponent by the original function.

Example 6

$$\frac{d}{dx} e^{3x} = 3e^{3x}$$

Example 7

$$\frac{d}{dx} e^{-x^2 + 4x - 7} = (-2x + 4)e^{-x^2 + 4x - 7}$$

Example 8

$$\frac{d}{dx} e^{\sqrt{x^2 - 3}} = \frac{1}{2} (x^2 - 3)^{-1/2} \cdot 2x \cdot e^{\sqrt{x^2 - 3}}$$

$$= x(x^2 - 3)^{-1/2} \cdot e^{\sqrt{x^2 - 3}}$$

$$= \frac{xe^{\sqrt{x^2 - 3}}}{\sqrt{x^2 - 3}}$$

DO EXERCISES 12–14.

Graphs of e^x, e^{-x}, and $1 - e^{-kx}$

We use a calculator with an $\boxed{e^x}$ key to find approximate values of e^x and e^{-x}. We can also use a power key $\boxed{y^x}$ with $y = 2.7183$ or some other approximation for e. With these we can draw graphs (Figs. 3 and 4) of the functions. Note that the graph of e^{-x} is a reflection, or "mirror image," of the graph of e^x across the y-axis.

DO EXERCISE 15.

16. a) Complete this table for

$$f(x) = 1 - e^{-x}.$$

x	0	$\frac{1}{2}$	1	2	3	4
$f(x)$						

b) Graph $f(x) = 1 - e^{-x}$.   CSS

FIGURE 3   FIGURE 4

Functions of the type $f(x) = 1 - e^{-kx}$ are also important.

Example 9   Graph $f(x) = 1 - e^{-2x}$, for nonnegative values of x.

Solution   We obtain these values using a calculator with an $\boxed{e^x}$ key.

x	$f(x)$
0	0
$\frac{1}{2}$	0.63
1	0.86
2	0.98
3	0.998

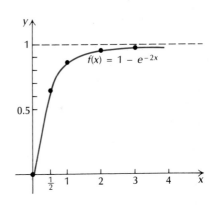

For example,

$$\begin{aligned} f(1) &= 1 - e^{-2(1)} \\ &= 1 - e^{-2} \\ &= 1 - 0.135335 \approx 0.86. \end{aligned}$$

DO EXERCISE 16.

In general, the graph of $f(x) = 1 - e^{-kx}$, for $k > 0$, increases from 0, since $f'(x) = ke^{-kx} > 0$, and approaches 1 as x gets larger; that is, $\lim_{x \to \infty} (1 - e^{-kx}) = 1$.

A word of caution! Functions of the type a^x (for example, 2^x, 3^x, and e^x) are different from functions of the type x^a (for example, x^2, x^3, $x^{1/2}$). For a^x the variable is in the exponent. For x^a the variable is in the base. The derivative of a^x is not xa^{x-1}. In particular, we have the following.

$$\frac{d}{dx} e^x \neq xe^{x-1}, \quad \text{but} \quad \frac{d}{dx} e^x = e^x$$

EXERCISE SET 4.1

Graph.

CSS

1. $y = 4^x$
2. $y = 5^x$
3. $y = (0.4)^x$
4. $y = (0.2)^x$
5. $x = 4^y$
6. $x = 5^y$

Differentiate.

7. $f(x) = e^{3x}$
8. $f(x) = e^{2x}$
9. $f(x) = 5e^{-2x}$
10. $f(x) = 4e^{-3x}$

11. $f(x) = 3 - e^{-x}$
12. $f(x) = 2 - e^{-x}$
13. $f(x) = -7e^x$
14. $f(x) = -4e^x$

15. $f(x) = \frac{1}{2}e^{2x}$
16. $f(x) = \frac{1}{4}e^{4x}$
17. $f(x) = x^4e^x$
18. $f(x) = x^5e^x$

19. $f(x) = \dfrac{e^x}{x^4}$
20. $f(x) = \dfrac{e^x}{x^5}$
21. $f(x) = e^{-x^2+7x}$
22. $f(x) = e^{-x^2+8x}$

23. $f(x) = e^{-x^2/2}$
24. $f(x) = e^{x^2/2}$
25. $y = e^{\sqrt{x-7}}$
26. $y = e^{\sqrt{x-4}}$

27. $y = \sqrt{e^x - 1}$
28. $y = \sqrt{e^x + 1}$
29. $y = xe^{-2x} + e^{-x} + x^3$
30. $y = e^x + x^3 - xe^x$

31. $y = 1 - e^{-x}$
32. $y = 1 - e^{-3x}$
33. $y = 1 - e^{-kx}$
34. $y = 1 - e^{-mx}$

Graph.

CSS

35. $f(x) = e^{2x}$
36. $f(x) = e^{(1/2)x}$
37. $f(x) = e^{-2x}$
38. $f(x) = e^{-(1/2)x}$

39. $f(x) = 1 - e^{-x}$, for nonnegative values of x
40. $f(x) = 2(1 - e^{-x})$, for nonnegative values of x

APPLICATIONS

41. *Business.* A company's total cost, in millions of dollars, is given by

$$C(t) = 100 - 50e^{-t},$$

where $t = $ time. Find each of the following.

a) The marginal cost $C'(t)$
b) $C'(0)$
c) $C'(4)$

42. *Business.* A company's total cost, in millions of dollars, is given by

$$C(t) = 200 - 40e^{-t},$$

where $t = $ time. Find each of the following.

a) The marginal cost $C'(t)$
b) $C'(0)$
c) $C'(5)$

EXTENSION EXERCISES

Differentiate.   `CSS`

43. $y = (e^{3x} + 1)^5$ **44.** $y = (e^{x^2} - 2)^4$ **45.** $y = \dfrac{e^{3t} - e^{7t}}{e^{4t}}$ **46.** $y = \sqrt[3]{e^{3t} + t}$

47. $y = \dfrac{e^x}{x^2 + 1}$ **48.** $y = \dfrac{e^x}{1 - e^x}$ **49.** $f(x) = e^{\sqrt{x}} + \sqrt{e^x}$ **50.** $f(x) = \dfrac{1}{e^x} + e^{1/x}$

51. $f(x) = e^{x/2} \cdot \sqrt{x - 1}$ **52.** $f(x) = \dfrac{xe^{-x}}{1 + x^2}$ **53.** $f(x) = \dfrac{e^x - e^{-x}}{e^x + e^{-x}}$ **54.** $f(x) = e^{e^x}$

($\blacksquare$ with $\boxed{y^x}$ key). Each of the following is an expression for e. Find the function values that are approximations for e. Round to five decimal places.

55. $e = \lim_{t \to 0} f(t)$; $f(t) = (1 + t)^{1/t}$. Find $f(1)$, $f(0.5)$, $f(0.2), f(0.1)$, and $f(0.001)$.

56. $e = \lim_{t \to 1} g(t)$; $g(t) = t^{1/(t-1)}$. Find $g(0.5)$, $g(0.9)$, $g(0.99), g(0.999)$, and $g(0.9998)$.

57. Find the maximum value of $f(x) = x^2 e^{-x}$ on $[0, 4]$.

58. Find the minimum value of $f(x) = xe^x$ on $[-2, 0]$.

59. Sketch the graph of $y = x^2 e^{-x}$.

60. Sketch the graph of $y = e^{-x^2}$.

4.2

OBJECTIVES

a) Differentiate functions involving natural logarithms.

b) Given an exponential equation, write an equivalent logarithmic equation.

c) Given a logarithmic equation, write an equivalent exponential equation.

d) Given $\log_a 3 = 1.099$ and $\log_a 5 = 1.609$, find logarithms like $\log_a 15$ and $\log_a 5a$.

e) Solve an equation like $e^t = 40$ for t.

f) Solve problems involving exponential and natural logarithmic functions.

LOGARITHMIC FUNCTIONS

Graphs of Logarithmic Functions

Suppose we want to solve the equation

$$10^x = 1000.$$

We are trying to find that power of 10 which will give 1000. We can see that the answer is 3. The number 3 is called the "logarithm, base 10, of 1000."

DEFINITION

The definition of a *logarithm* is as follows:

$$y = \log_a x \quad \text{means} \quad x = a^y, \quad a > 0, \quad a \neq 1.$$

The number a is called the *logarithmic base*.

Thus for logarithms base 10, $\log_{10} x$ is that number y such that $x = 10^y$. Therefore, a logarithm can be thought of as an exponent. We

1. Write equivalent exponential equations.
   a) $\log_b P = T$
   b) $\log_9 3 = \frac{1}{2}$
   c) $\log_{10} 1000 = 3$
   d) $\log_{10} 0.1 = -1$

can convert from a logarithmic equation to an exponential equation, and conversely, as follows.

Logarithmic equation	Exponential equation
$\log_a M = N$	$a^N = M$
$\log_{10} 100 = 2$	$10^2 = 100$
$\log_{10} 0.01 = -2$	$10^{-2} = 0.01$
$\log_{49} 7 = \frac{1}{2}$	$49^{1/2} = 7$

DO EXERCISES 1 AND 2.

In order to graph a logarithmic equation, we can graph its equivalent exponential equation.

Example 1 Graph $y = \log_2 x$.

Solution We first write the equivalent exponential equation:

$$x = 2^y.$$

We select values for y and find the corresponding values of 2^y. Then we plot points, remembering that x is still the first coordinate.

2. Write equivalent logarithmic equations.
   a) $e^k = T$
   b) $16^{1/4} = 2$
   c) $10^4 = 10,000$
   d) $10^{-3} = 0.001$

x, or 2^y	y
1	0
2	1
4	2
8	3
$\frac{1}{2}$	-1
$\frac{1}{4}$	-2

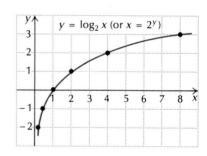

① —Select y.

② —Compute x.

3. Graph $y = \log_3 x$.

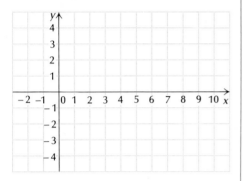

4. Consider

$$f(x) = 10^x$$

and

$$g(x) = \log_{10} x.$$

a) Find $f(3)$.

b) Find $g(1000)$.

c) Use ▦ to find $g(5)$.

d) Use ▦ to find $f(0.699)$.

DO EXERCISE 3.

The graphs of $f(x) = 2^x$ and $g(x) = \log_2 x$ are shown below using the same axes. Note that we can obtain the graph of g by reflecting the graph of f across the line $y = x$. Graphs obtained in this manner are known as *inverses* of each other.

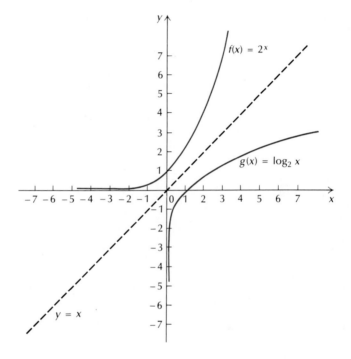

While we cannot develop inverses in detail, it is of interest to note that they "undo" each other. For example,

$$f(3) = 2^3 = 8 \qquad \text{The input 3 gives the output 8.}$$

and

$$g(8) = \log_2 8 = 3. \qquad \text{The input 8 gets us back to 3.}$$

DO EXERCISE 4.

Basic Properties of Logarithms

The following are some basic properties of logarithms. The proofs follow from the properties of exponents.

THEOREM 4

Property 1. $\log_a MN = \log_a M + \log_a N$

Property 2. $\log_a \dfrac{M}{N} = \log_a M - \log_a N$

Property 3. $\log_a M^k = k \cdot \log_a M$

Property 4. $\log_a a = 1$

Property 5. $\log_a a^k = k$

Property 6. $\log_a 1 = 0$

Proof of Properties 1 and 2.   Let $X = \log_a M$ and $Y = \log_a N$. Writing the equivalent exponential equations, we then have

$$M = a^X \quad \text{and} \quad N = a^Y.$$

Then by the properties of exponents (see Section 1.1), we have

$$MN = a^X \cdot a^Y = a^{X+Y},$$

so

$$\log_a MN = X + Y$$
$$= \log_a M + \log_a N;$$

and

$$\frac{M}{N} = a^X \div a^Y = a^{X-Y},$$

so

$$\log_a \frac{M}{N} = X - Y$$
$$= \log_a M - \log_a N.$$

Proof of Property 3.   Let $X = \log_a M$. Then

$$a^X = M,$$

so

$$(a^X)^k = M^k, \quad \text{or} \quad a^{Xk} = M^k.$$

Thus,

$$\log_a M^k = Xk = k \cdot \log_a M.$$

Proof of Property 4. $\log_a a = 1$ because $a^1 = a$.

Given

$$\log_a 2 = 0.301,$$
$$\log_a 5 = 0.699,$$

find each of the following.

5. $\log_a 4$

6. $\log_a 10$

7. $\log_a \frac{2}{5}$

8. $\log_a \frac{5}{2}$

9. $\log_a \frac{1}{5}$

10. $\log_a \sqrt{a^3}$

11. $\log_a 5a$

12. $\log_a 16$

Proof of Property 5. $\log_a a^k = k$ because $(a^k) = a^k$.

Proof of Property 6. $\log_a 1 = 0$ because $a^0 = 1$.

Let us illustrate these properties.

Example 2   Given

$$\log_a 2 = 0.301 \quad \text{and} \quad \log_a 3 = 0.477,$$

find each of the following.

a) $\log_a 6$

$\log_a 6 = \log_a (2 \cdot 3) = \log_a 2 + \log_a 3$ Property 1
$$= 0.301 + 0.477$$
$$= 0.778$$

b) $\log_a \frac{2}{3}$

$\log_a \frac{2}{3} = \log_a 2 - \log_a 3$ Property 2
$$= 0.301 - 0.477$$
$$= -0.176$$

c) $\log_a 81$

$\log_a 81 = \log_a 3^4 = 4 \log_a 3$ Property 3
$$= 4(0.477)$$
$$= 1.908$$

d) $\log_a \frac{1}{3}$

$\log_a \frac{1}{3} = \log_a 1 - \log_a 3$ Property 2
$$= 0 - 0.477$$ Property 6
$$= -0.477$$

e) $\log_a \sqrt{a}$

$\log_a \sqrt{a} = \log_a a^{1/2} = \frac{1}{2}$ Property 5

f) $\log_a 2a$

$\log_a 2a = \log_a 2 + \log_a a$ Property 1
$$= 0.301 + 1$$ Property 4
$$= 1.301$$

g) $\log_a 5$

No way to find using these properties.
($\log_a 5 \neq \log_a 2 + \log_a 3$)

h) $\dfrac{\log_a 3}{\log_a 2}$

$\dfrac{\log_a 3}{\log_a 2} = \dfrac{0.477}{0.301} = 1.58$

We simply divided and used none of the properties.

DO EXERCISES 5–12.

Common Logarithms

Logarithms to the base 10 are called *common logarithms*. When we write

$$\log M,$$

13. ▦ Find each logarithm. Round to four decimal places.

   a) log 31,456

   b) log 0.9080701

   c) log 78.6

   d) log 7.86

   e) log 0.786

   f) log 0.0786

   g) log 0.00786

with no base indicated, base 10 is understood. Note the following comparison of common logarithms and powers of 10.

$$1000 = 10^3$$
$$100 = 10^2$$
$$10 = 10^1$$
$$1 = 10^0$$
$$0.1 = 10^{-1}$$
$$0.01 = 10^{-2}$$
$$0.001 = 10^{-3}$$

The common logarithms at the right follow from the powers at the left.

$$\log 1000 = 3$$
$$\log 100 = 2$$
$$\log 10 = 1$$
$$\log 1 = 0$$
$$\log 0.1 = -1$$
$$\log 0.01 = -2$$
$$\log 0.001 = -3$$

Since $\log 100 = 2$ and $\log 1000 = 3$, it seems reasonable that $\log 500$ is somewhere between 2 and 3. Tables were generally used for such approximations, but with the advent of the calculator that method of finding logarithms is being used less and less. Using a calculator with a $\boxed{\log}$ key, we find that $\log 500 = 2.6990$, rounded to four decimal places.

Before calculators and computers became so readily available, common logarithms were used extensively to do certain kinds of computations. In fact, computation is the reason logarithms were developed. Since the standard notation we use for numbers is based on 10, it is logical that base-10, or common, logarithms were used for computations. Common logarithms can be found using a calculator. Today, computations with common logarithms are mainly of historical interest; the logarithmic functions, base e, are of modern importance.

DO EXERCISE 13.

Natural Logarithms

The number e, which is approximately 2.718282, was developed in Section 4.1, and has extensive application in many fields. The number $\log_e x$ is called the *natural logarithm* of x and is abbreviated $\ln x$; that is,

---

DEFINITION

$$\ln x = \log_e x.$$

The following is a restatement of the basic properties of logarithms in terms of natural logarithms.

Given

$$\ln 2 = 0.6931,$$
$$\ln 5 = 1.6094,$$

find each of the following.

14. $\ln 10$

15. $\ln \frac{2}{5}$

16. $\ln \frac{5}{2}$

17. $\ln 16$

18. $\ln 5e$

19. $\ln \sqrt{e}$

20. $\ln \frac{1}{5}$

THEOREM 5

Property 1.	$\ln MN = \ln M + \ln N$
Property 2.	$\ln \dfrac{M}{N} = \ln M - \ln N$
Property 3.	$\ln a^k = k \cdot \ln a$
Property 4.	$\ln e = 1$
Property 5.	$\ln e^k = k$
Property 6.	$\ln 1 = 0$

Let us illustrate these properties.

Example 3   Given

$$\ln 2 = 0.6931 \quad \text{and} \quad \ln 3 = 1.0986,$$

find each of the following.

a) $\ln 6$

$$\ln 6 = \ln (2 \cdot 3) = \ln 2 + \ln 3 \qquad \text{Property 1}$$
$$= 0.6931 + 1.0986$$
$$= 1.7917$$

b) $\ln 81$

$$\ln 81 = \ln (3^4)$$
$$= 4 \ln 3 \qquad \text{Property 3}$$
$$= 4(1.0986)$$
$$= 4.3944$$

c) $\ln \frac{2}{3}$

$$\ln \tfrac{2}{3} = \ln 2 - \ln 3 \qquad \text{Property 2}$$
$$= 0.6931 - 1.0986$$
$$= -0.4055$$

d) $\ln \frac{1}{3}$

$$\ln \tfrac{1}{3} = \ln 1 - \ln 3 \qquad \text{Property 2}$$
$$= 0 - 1.0986 \qquad \text{Property 6}$$
$$= -1.0986$$

e) $\ln 2e$

$$\ln 2e = \ln 2 + \ln e \qquad \text{Property 1}$$
$$= 0.6931 + 1 \qquad \text{Property 4}$$
$$= 1.6931$$

f) $\ln \sqrt{e^3}$

$$\ln \sqrt{e^3} = \ln e^{3/2}$$
$$= \tfrac{3}{2} \qquad \text{Property 5}$$

DO EXERCISES 14–20.

▦ Find each logarithm. Round to six decimal places.

21. ln 2

22. ln 20

23. ln 100

24. ln 0.07432

25. ln 1.08

26. ln 0.9999

Finding Natural Logarithms Using a Calculator

You should have a calculator with a $\boxed{\ln}$ key. You can find natural logarithms directly using this key.

Example 4 Find each logarithm on your calculator. Round to six decimal places.

a) $\ln 5.24 = 1.656321$

b) $\ln 0.00001277 = -11.268412$

DO EXERCISES 21–26.

Exponential Equations

If an equation contains a variable in an exponent, we call the equation *exponential*. We can use logarithms to manipulate or solve exponential equations.

Example 5 Solve $e^t = 40$ for t.

Solution

$$
\begin{aligned}
\ln e^t &= \ln 40 \qquad \text{Taking the natural logarithm on both sides} \\
t &= \ln 40 \qquad \text{Property 5} \\
t &\approx 3.688879 \qquad ▦ \\
t &\approx 3.7
\end{aligned}
$$

It should be noted that this is an approximation for t even though an equals sign is often used.

Example 6 Solve $e^{-0.04t} = 0.05$ for t.

Solution

$$
\begin{aligned}
\ln e^{-0.04t} &= \ln 0.05 \qquad \text{Taking the natural logarithm on both sides} \\
-0.04t &= \ln 0.05 \qquad \text{Property 5} \\
t &= \frac{\ln 0.05}{-0.04} \\
t &\approx \frac{-2.995732}{-0.04} \qquad ▦ \\
t &\approx 75
\end{aligned}
$$

Solve for t.

27. $e^t = 80$

28. $e^{-0.06t} = 0.07$

Calculator note. For purposes of space and explanation, we have rounded the value of ln 0.05 to -2.995732 in an intermediate step. When using your calculator you should find

$$\frac{\ln 0.05}{-0.04},$$

obtaining

$$\frac{-2.995732274}{-0.04}.$$

Then divide, and round at the end. Answers at the back of the book have been found in this manner. Remember, the number of places in a table or on a calculator may affect the accuracy of the answer. Usually, your answer should agree to at least three digits.

DO EXERCISES 27 AND 28.

Graphs of Natural Logarithmic Functions

There are two ways in which we might obtain the graph of $y = f(x) = \ln x$. One is by writing its equivalent equation $x = e^y$. Then we select values for y and use a calculator to find the corresponding values of e^y. We then plot points, remembering that x is still the first coordinate.

x, or e^y	y
0.1	-2
0.4	-1
1.0	0
2.7	1
7.4	2
20.1	3

①——Select y.

②——Compute x.

29. **a)** Complete using .

x	0.5	1	2	3	4
$\ln x$			0.7		

b) Graph $y = \ln x$.   CSS

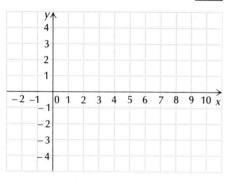

Note again that the functions are inverses of each other. That is, the graph of $y = \ln x$ is a reflection, or mirror image, across the line $y = x$, of the graph of $y = e^x$.

The second way of graphing $y = \ln x$ is to use a calculator to find function values. For example, $\ln 2 = 0.6931 \approx 0.7$.

DO EXERCISE 29.

These properties follow.

THEOREM 6

$\ln x$ exists only for positive numbers x.
$\ln x < 0$ for $0 < x < 1$.
$\ln x > 0$ for $x > 1$.

The Derivative of $\ln x$

Consider $f(x) = \ln x$. We can show that $f'(x) = 1/x$ (the slope of the tangent line at x is just the reciprocal of x). We are trying to find the derivative of

$$f(x) = \ln x. \tag{1}$$

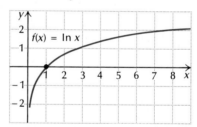

We first write its equivalent exponential equation:

$$e^{f(x)} = x. \qquad \ln x = \log_e x = f(x), \text{ so } e^{f(x)} = x, \tag{2}$$
$$\text{by the definition of logarithms}$$

Now we differentiate on both sides of this equation:

$$\frac{d}{dx} e^{f(x)} = \frac{d}{dx} x$$
$$f'(x) \cdot e^{f(x)} = 1$$
$$f'(x) \cdot x = 1 \qquad \text{Substituting } x \text{ for } e^{f(x)} \text{ from Eq. (2)}$$
$$f'(x) = \frac{1}{x}.$$

Differentiate.

30. $y = 5 \ln x$

31. $f(x) = x^3 \ln x + 4x$

32. $f(x) = \dfrac{\ln x}{x^2}$

Thus we have the following.

CSS

THEOREM 7

$$\frac{d}{dx} \ln x = \frac{1}{x}$$

This is true only for positive values of x, since $\ln x$ is defined only for positive numbers. (For negative numbers x, this derivative formula becomes

$$\frac{d}{dx} \ln |x| = \frac{1}{x},$$

but we will consider such a case very little in this text.)

Let us find some derivatives.

Example 7

$$\frac{d}{dx} 3 \ln x = \frac{3}{x}$$

Example 8

$$\frac{d}{dx} (x^2 \ln x + 5x) = x^2 \cdot \frac{1}{x} + 2x \cdot \ln x + 5 \qquad \text{Using the Product Rule on } x^2 \ln x$$

$$= x + 2x \cdot \ln x + 5 \qquad \text{Simplifying}$$

$$= x(1 + 2 \ln x) + 5$$

Example 9

$$\frac{d}{dx} \left(\frac{\ln x}{x^3} \right) = \frac{x^3 \cdot (1/x) - (\ln x)(3x^2)}{x^6} \qquad \text{By the Quotient Rule}$$

$$= \frac{x^2 - 3x^2 \ln x}{x^6}$$

$$= \frac{x^2(1 - 3 \ln x)}{x^6} \qquad \text{Factoring}$$

$$= \frac{1 - 3 \ln x}{x^4} \qquad \text{Simplifying}$$

DO EXERCISES 30–32.

Suppose we have a more complicated expression, such as

$$h(x) = \ln (x^2 - 8x).$$

This is a composition of functions. Suppose, in general, that we have

$$h(x) = \ln f(x) = g[f(x)], \quad \text{where } g(x) = \ln x.$$

Now $g'(x) = 1/x$. Then by the Chain Rule (Section 2.8), we have

$$h'(x) = g'[f(x)] \cdot f'(x) = g'[\ln f(x)] \cdot f'(x).$$

For the above case, $f(x) = x^2 - 8x$, so $f'(x) = 2x - 8$. Then

$$h'(x) = \frac{1}{x^2 - 8x} \cdot (2x - 8) = \frac{2x - 8}{x^2 - 8x}.$$

The following rule, which we have proven using the Chain Rule, allows us to find derivatives of functions like the one above.

THEOREM 8

$$\frac{d}{dx} \ln f(x) = f'(x) \cdot \frac{1}{f(x)}$$

The following gives us a way of remembering this rule.

$$h(x) = \ln (x^2 - 8x)$$

$$h'(x) = (2x - 8) \cdot \frac{1}{x^2 - 8x}$$

① Differentiate the "inside" function.

② Multiply by the reciprocal of the "inside" function.

Example 10

$$\frac{d}{dx} \ln 3x = 3 \cdot \frac{1}{3x} = \frac{1}{x}$$

Note that we could have done this another way, using Property 1:

$$\ln 3x = \ln 3 + \ln x;$$

then

$$\frac{d}{dx} \ln 3x = \frac{d}{dx} \ln 3 + \frac{d}{dx} \ln x = 0 + \frac{1}{x} = \frac{1}{x}.$$

Differentiate.

33. $f(x) = \ln 5x$

34. $f(x) = \ln (3x^2 + 4)$

35. $y = \ln (\ln 5x)$

36. $y = \ln \left(\dfrac{x^5 - 2}{x} \right)$

Example 11

$$\frac{d}{dx} \ln (x^2 - 5) = 2x \cdot \frac{1}{x^2 - 5} = \frac{2x}{x^2 - 5}$$

Example 12

$$\frac{d}{dx} \ln (\ln x) = \frac{1}{x} \cdot \frac{1}{\ln x} = \frac{1}{x \ln x}$$

Example 13

$$\frac{d}{dx} \ln \left(\frac{x^3 + 4}{x} \right) = \frac{d}{dx} [\ln (x^3 + 4) - \ln x] \qquad \text{By Property 2. This avoids using the Quotient Rule.}$$

$$= 3x^2 \cdot \frac{1}{x^3 + 4} - \frac{1}{x}$$

$$= \frac{3x^2}{x^3 + 4} - \frac{1}{x}$$

$$= \frac{3x^2}{x^3 + 4} \cdot \frac{x}{x} - \frac{1}{x} \cdot \frac{x^3 + 4}{x^3 + 4}$$

$$= \frac{(3x^2)x - (x^3 + 4)}{x(x^3 + 4)}$$

$$= \frac{3x^3 - x^3 - 4}{x(x^3 + 4)}$$

$$= \frac{2x^3 - 4}{x(x^3 + 4)}$$

DO EXERCISES 33–36.

Applications

Example 14 *Forgetting.* In a psychological experiment students were shown a set of nonsense syllables, such as POK, and asked to recall them every second thereafter. The percentage $R(t)$ who retained the syllables after t seconds was found to be given by

$$R(t) = 80 - 27 \ln t, \quad \text{for } t \geqslant 1.$$

Strictly speaking, the function is not continuous, but in order to use calculus, we "fill in" the graph with a smooth curve, considering $R(t)$ to be defined for any number $t \geqslant 1$. This is not unreasonable, since we are now able to find the percentage who retained the syllables after

Conduct your own memory experiment. Study this photograph carefully. Then put it aside and write down as many items as you can. Wait a half-hour and again write down as many as you can. Do this five more times. Make a graph of the number of items you remember versus the time. Does the graph appear to be logarithmic?

$t = 3.417$ seconds, instead of merely after integer values such as 1, 2, 3, 4, and so on.

a) What percentage retained the syllables after 1 second?

b) Find $R'(t)$, the rate of change of R with respect to t.

c) Find the maximum and minimum values, if they exist.

37. *Advertising.* A model for advertising response is given by

$$N(a) = 500 + 200 \ln a, \qquad a \geq 1,$$

where

$N(a) =$ the number of units sold, and

$a =$ the amount spent on advertising, in thousands of dollars.

a) How many units were sold after spending $1000? (Substitute 1 for a, not 1000.)

b) Find $N'(a)$.

c) Find the maximum and minimum values, if they exist. CSS

Solution

a) $R(1) = 80 - 27 \cdot \ln 1 = 80 - 27 \cdot 0 = 80\%$

b) $R'(t) = -27 \cdot \dfrac{1}{t} = -\dfrac{27}{t}$

c) Now $R'(t)$ exists for all values of t in the interval $[1, \infty)$. Note that for $t \geq 1$, $-27/t < 0$. Thus there are no critical points and R is decreasing. Then R has a maximum value at the endpoint 1. This maximum value is $R(1)$, or 80%. There is no minimum value.

DO EXERCISE 37.

Example 15 *Business.* A company begins a radio advertising campaign in New York City to market a new product. The percentage of the "target market" that buys a product is normally a function of the length of the advertising campaign. The radio station estimates this percentage as $(1 - e^{-0.04t})$ for this type of product, where $t =$ the number of days of the campaign. The target market is estimated to be 1,000,000 people and the price per unit is $0.50. The costs of advertising are $1000 per day. Find the length of the advertising campaign that will result in the maximum profit.

Solution That the percentage of the target market that buys the product can be modeled by $f(t) = 1 - e^{-0.04t}$ is justified if we look at its graph. The function increases from 0 (0%) toward 1 (100%). The longer the advertising campaign, the larger the percentage of the market that has bought the product. (See also the discussion in Chapter 6.)

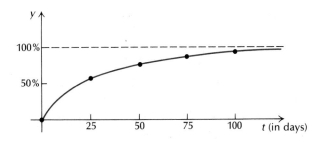

Recall the profit function (here expressed in terms of time, t):

$$\text{Profit} = \text{Revenue} - \text{Cost}$$
$$P(t) = R(t) - C(t).$$

38. Solve the problem in Example 15 when the price per unit is $0.80.

a) Find $R(t)$.

$R(t)$ = (Price per unit) · (Target market) · (Percentage buying)

$R(t) = 0.5(1{,}000{,}000)(1 - e^{-0.04t}) = 500{,}000 - 500{,}000e^{-0.04t}$

b) Find $C(t)$.

$C(t)$ = (Advertising costs per day) · (Number of days)

$C(t) = 1000t$

c) Find $P(t)$, and take its derivative.

$$P(t) = R(t) - C(t)$$
$$P(t) = 500{,}000 - 500{,}000e^{-0.04t} - 1000t$$
$$P'(t) = (-0.04)(-500{,}000e^{-0.04t}) - 1000$$
$$P'(t) = 20{,}000e^{-0.04t} - 1000$$

d) Set the first derivative equal to 0 and solve.

$$20{,}000e^{-0.04t} - 1000 = 0$$
$$20{,}000e^{-0.04t} = 1{,}000$$
$$e^{-0.04t} = \frac{1{,}000}{20{,}000} = 0.05 \qquad (1)$$
$$\ln e^{-0.04t} = \ln 0.05$$
$$-0.04t = \ln 0.05$$
$$t = \frac{\ln 0.05}{-0.04}$$
$$t = \frac{-2.995732}{-0.04} \quad \boxed{\boxminus}$$
$$t \approx 75$$

e) We have only one critical point, so we can use the second derivative to determine whether we have a maximum:

$$P''(t) = -0.04(20{,}000e^{-0.04t}) = -800e^{-0.04t}.$$

Since exponential functions are positive, $e^{-0.04t} > 0$ for all numbers t. Thus, since $-800e^{-0.04t} < 0$ for all numbers t, $P''(t)$ is less than 0 for $t = 75$ and we have a maximum.

The length of the advertising campaign must be 75 days in order to result in maximum profit.

DO EXERCISE 38.

EXERCISE SET 4.2

Write an equivalent exponential equation.

1. $\log_2 8 = 3$
2. $\log_3 81 = 4$
3. $\log_8 2 = \frac{1}{3}$
4. $\log_{27} 3 = \frac{1}{3}$
5. $\log_a K = J$
6. $\log_a J = K$
7. $\log_b T = v$
8. $\log_c Y = t$

Write an equivalent logarithmic equation.

9. $e^M = b$
10. $e^t = p$
11. $10^2 = 100$
12. $10^3 = 1000$
13. $10^{-1} = 0.1$
14. $10^{-2} = 0.01$
15. $M^p = V$
16. $Q^n = T$

Given $\log_b 3 = 1.099$ and $\log_b 5 = 1.609$, find each of the following.

17. $\log_b 15$
18. $\log_b \frac{3}{5}$
19. $\log_b \frac{5}{3}$
20. $\log_b \frac{1}{3}$
21. $\log_b \frac{1}{5}$
22. $\log_b \sqrt{b}$
23. $\log_b \sqrt{b^3}$
24. $\log_b 3b$
25. $\log_b 5b$
26. $\log_b 9$
27. $\log_b 25$
28. $\log_b 75$

Given $\ln 4 = 1.3863$ and $\ln 5 = 1.6094$, find each of the following. Do not use a calculator.

29. $\ln 20$
30. $\ln \frac{4}{5}$
31. $\ln \frac{5}{4}$
32. $\ln \frac{1}{5}$
33. $\ln \frac{1}{4}$
34. $\ln 5e$
35. $\ln 4e$
36. $\ln \sqrt{e^6}$
37. $\ln \sqrt{e^8}$
38. $\ln 25$
39. $\ln 16$
40. $\ln 100$

▦ Find the logarithm. Round to six decimal places.

41. $\ln 5894$
42. $\ln 99{,}999$
43. $\ln 0.0182$
44. $\ln 0.00087$
45. $\ln 1.88$
46. $\ln 18.8$
47. $\ln 0.0188$
48. $\ln 0.188$
49. $\ln 906$
50. $\ln 8100$
51. $\ln 0.011$
52. $\ln 0.00056$

Solve for t.

53. $e^t = 100$
54. $e^t = 1000$
55. $e^t = 60$
56. $e^t = 90$
57. $e^{-t} = 0.1$
58. $e^{-t} = 0.01$
59. $e^{-0.02t} = 0.06$
60. $e^{0.07t} = 2$

Differentiate.

61. $y = -6 \ln x$
62. $y = -4 \ln x$
63. $y = x^4 \ln x - \frac{1}{2}x^2$
64. $y = x^5 \ln x - \frac{1}{4}x^4$
65. $y = \dfrac{\ln x}{x^4}$
66. $y = \dfrac{\ln x}{x^5}$
67. $y = \ln \dfrac{x}{4}$
68. $y = \ln \dfrac{x}{2}$

$\left[Hint: \ln \dfrac{x}{4} = \ln x - \ln 4 \right]$

69. $f(x) = \ln (5x^2 - 7)$
70. $f(x) = \ln (7x^3 + 4)$
71. $f(x) = \ln (\ln 4x)$
72. $f(x) = \ln (\ln 3x)$
73. $f(x) = \ln \left(\dfrac{x^2 - 7}{x} \right)$
74. $f(x) = \ln \left(\dfrac{x^2 + 5}{x} \right)$
75. $f(x) = e^x \ln x$
76. $f(x) = e^{2x} \ln x$
77. $f(x) = \ln (e^x + 1)$
78. $f(x) = \ln (e^x - 2)$
79. $f(x) = (\ln x)^2$
    [*Hint:* Use the Extended Power Rule.]
80. $f(x) = (\ln x)^3$

APPLICATIONS   CSS

81. *Psychology: Forgetting.* Students in college botany took a final exam. They took equivalent forms of the exam in monthly intervals thereafter. The average score, $S(t)$ in percent, after t months was found to be given by

$$S(t) = 68 - 20 \ln (t + 1), \qquad t \geqslant 0.$$

a) What was the average score when they initially took the test, $t = 0$?

b) What was the average score after 4 months?

c) What was the average score after 24 months?

d) What percentage of the initial score did they retain after 2 years (24 months)?

e) Find $S'(t)$.

f) Find the maximum and minimum values, if they exist.

82. *Psychology: Forgetting.* Students in college zoology took a final exam. They took equivalent forms of the exam in monthly intervals thereafter. The average score, $S(t)$ in percent, after t months was found to be given by

$$S(t) = 78 - 15 \ln (t + 1), \qquad t \geqslant 0.$$

a) What was the average score when they initially took the test, $t = 0$?

b) What was the average score after 4 months?

c) What was the average score after 24 months?

d) What percentage of the initial score did they retain after 2 years (24 months)?

e) Find $S'(t)$.

f) Find the maximum and minimum values, if they exist.

83. *Business: Advertising.* A model for advertising response is given by

$$N(a) = 1000 + 200 \ln a, \qquad a \geqslant 1,$$

where

$N(a)$ = the number of units sold, and

a = the amount spent on advertising, in thousands of dollars.

a) How many units were sold after spending $1000 ($a = 1$) on advertising?

b) Find $N'(a)$ and $N'(10)$.

c) Find the maximum and minimum values, if they exist.

84. *Business: Advertising.* A model for advertising response is given by

$$N(a) = 2000 + 500 \ln a, \qquad a \geqslant 1,$$

where

$N(a)$ = the number of units sold, and

a = the amount spent on advertising, in thousands of dollars.

a) How many units were sold after spending $1000 ($a = 1$) on advertising?

b) ·Find $N'(a)$ and $N'(10)$.

c) Find the maximum and minimum values, if they exist.

85. *Psychology: Walking speed.* Bornstein and Bornstein found in a study that the average walking speed v of a person living in a city of population p, in thousands, is given by

$$v(p) = 0.37 \ln p + 0.05,$$

where v is in feet per second.

a) The population of Seattle is 531,000. What is the average walking speed of a person living in Seattle? Find $v(531)$.

b) The population of New York is 7,900,000. What is the average walking speed of a person living in New York?

c) Find $v'(p)$. Interpret $v'(p)$.

86. *Biomedical: The Reynolds number.* For many kinds of animals the Reynolds number R is given by

$$R = A \ln r - Br,$$

where A and B are positive constants and r is the radius of the aorta. Find the maximum value of R.

87. *Business.* Solve Example 15 given that the costs of advertising are $2000 per day.

88. *Business.* Solve Example 15 given that the costs of advertising are $4000 per day.

89. *Biomedical: Acceptance of a new medicine.* The percentage P of doctors who accept a new medicine is given by

$$P(t) = 1 - e^{-0.2t},$$

where $t = $ the time in months.

a) Find $P(1)$ and $P(6)$.

b) Find $P'(t)$.

c) How many months will it take for 90% of the doctors to accept the new medicine?

90. *Psychology: The Hullian learning model.* A typist learns to type W words per minute after t weeks of practice, where W is given by

$$W(t) = 100(1 - e^{-0.3t}).$$

a) Find $W(1)$ and $W(8)$.

b) Find $W'(t)$.

c) After how many weeks will the typist's speed be 95 words per minute?

91. *Business: Growth of a stock.* The value of a stock is modeled by

$$V(t) = \$58(1 - e^{-1.1t}) + \$20,$$

where V is the value of the stock after time t, in months.

a) Find $V(1)$ and $V(12)$.

b) Find $V'(t)$.

c) After how many months will the value of the stock be $75?

92. *Business: Marginal revenue.* The demand function for a certain product is given by

$$p = D(x) = 800e^{-0.125x}.$$

Recall that total revenue is given by $R(x) = xD(x)$.

a) Find $R(x)$.

b) Find the marginal revenue, $R'(x)$.

c) At what value of x will the revenue be maximum?

EXTENSION EXERCISES

Differentiate.

93. $y = (\ln x)^{-4}$

94. $y = (\ln x)^n$

95. $f(t) = \ln (t^3 + 1)^5$

96. $f(t) = \ln (t^2 + t)^3$

97. $f(x) = [\ln (x + 5)]^4$

98. $f(x) = \ln [\ln (\ln 3x)]$

99. $f(t) = \ln [(t^3 + 3)(t^2 - 1)]$

100. $f(t) = \ln \dfrac{1 - t}{1 + t}$

101. $y = \ln \dfrac{x^5}{(8x + 5)^2}$

102. $y = \ln \sqrt{5 + x^2}$

103. $f(t) = \dfrac{\ln t^2}{t^2}$

104. $f(x) = \frac{1}{5}x^5 (\ln x - \frac{1}{5})$

105. $y = \dfrac{x^{n+1}}{n+1}\left(\ln x - \dfrac{1}{n+1}\right)$

106. $y = \dfrac{x \ln x - x}{x^2 + 1}$

107. $y = \ln (t + \sqrt{1 + t^2})$

108. $f(x) = \ln \dfrac{1 + \sqrt{x}}{1 - \sqrt{x}}$

109. $f(x) = \ln [\ln x]^3$

110. $f(x) = \dfrac{\ln x}{1 + (\ln x)^2}$

111. Find $\lim\limits_{h \to 0} \dfrac{\ln (1 + h)}{h}$.

112. ▦ Which is larger, e^π or π^e?

113. ▦ Find $\sqrt[e]{e}$. Compare it to other expressions of the type $\sqrt[x]{x}, x > 0$. What can you conclude?

114. Find the minimum value of $f(x) = x \ln x$.   **CSS**

115. Find the minimum value of $f(x) = x^2 \ln x$.

116. Sketch the graph of $y = \dfrac{\ln x}{x^2}$.

Solve for t.

117. $P = P_0 e^{kt}$

118. $P = P_0 e^{-kt}$

Use input–output tables to find each limit.

119. $\lim_{x \to \infty} \ln x$

120. $\lim_{x \to 1} \ln x$

Verify each of the following.

121. $\ln x = \dfrac{\log x}{\log e} \approx 2.3026 \log x$

122. $\log x = \dfrac{\ln x}{\ln 10} \approx 0.4343 \ln x$

4.3

APPLICATIONS: THE UNINHIBITED GROWTH MODEL, $dP/dt = kP$

OBJECTIVES

a) State the solution of an equation

$$\frac{dP}{dt} = kP$$

as

$$P(t) = P_0 e^{kt}.$$

b) Given a growth rate, find the doubling time.

c) Given a doubling time, find the growth rate.

d) Solve applied problems involving exponential growth.

Exponential Growth

What will the world population be in 1996?

Consider the function

$$f(x) = 2e^{3x}.$$

1. **a)** Differentiate $y = 5e^{4x}$.

   b) Express dy/dx in terms of y.

Differentiating, we get

$$f'(x) = 3 \cdot 2e^{3x} = 3 \cdot f(x).$$

Graphically, this says that the derivative, or slope of the tangent line, is simply the constant 3 times the function value.

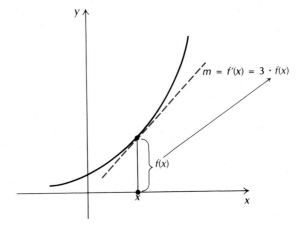

DO EXERCISE 1.

2. **a)** State the solution to

$$\frac{dN}{dt} = kN.$$

   b) State the solution to

$$f'(t) = k \cdot f(t).$$

In general, we have the following.

THEOREM 9

A function $y = f(x)$ satisfies the equation

$$\frac{dy}{dx} = ky \qquad [f'(x) = k \cdot f(x)]$$

if and only if

$$y = ce^{kx} \qquad [f(x) = ce^{kx}]$$

for some constant c.

No matter what the variables, you should be able to write the solution.

Example 1   The solution to $dA/dt = kA$ is $A = ce^{kt}$, or $A(t) = ce^{kt}$.

Example 2   The solution to $dP/dt = kP$ is $P = ce^{kt}$, or $P(t) = ce^{kt}$.

Example 3   The solution to $f'(Q) = k \cdot f(Q)$ is $f(Q) = ce^{kQ}$.

DO EXERCISE 2.

EXPONENTIAL AND LOGARITHMIC FUNCTIONS

3. *Exploratory exercises: Growth.* Use a sheet of $8\frac{1}{2}'' \times 11''$ paper. Cut it into two equal pieces. Then cut these into four equal pieces. Then cut these into eight equal pieces, and so on, performing five cutting steps.

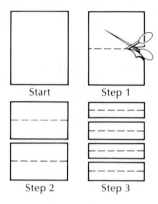

Start   Step 1

Step 2   Step 3

a) Place all the pieces in a stack and measure the thickness.

b) A piece of paper is typically 0.004 in. thick. Check the calculation in part (a) by completing this table.

	t	$0.004 \cdot 2^t$
Start	0	$0.004 \cdot 2^0$, or 0.004
Step 1	1	$0.004 \cdot 2^1$, or 0.008
Step 2	2	$0.004 \cdot 2^2$, or 0.016
Step 3	3	
Step 4	4	
Step 5	5	

c) Compute the thickness of the paper (in miles) after 25 steps.

The equation

$$\frac{dP}{dt} = kP, \quad k > 0 \qquad [P'(t) = k \cdot P(t), \quad k > 0]$$

is the basic model of uninhibited population growth, whether it be a population of humans, a bacteria culture, or money invested at interest compounded continuously. Neglecting special inhibiting and stimulating factors, we know that a population normally reproduces itself at a rate proportional to its size, and this is exactly what the equation $dP/dt = kP$ says. The solution to the equation is

$$P(t) = ce^{kt}, \tag{1}$$

where t = the time. At $t = 0$, we have some "initial" population $P(0)$ that we will represent by P_0. We can rewrite Eq. (1) in terms of P_0 as

$$P_0 = P(0) = ce^{k \cdot 0} = ce^0 = c \cdot 1 = c.$$

Thus $P_0 = c$, so we can express $P(t)$ as

$$P(t) = P_0 e^{kt}.$$

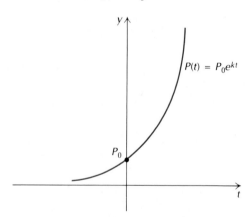

Its graph is this curve, which shows how uninhibited growth results in a "population explosion."

DO EXERCISE 3.

The constant k is called the *rate of exponential growth,* or simply the *growth rate.* This is not the rate of change of the population size, which is

$$\frac{dP}{dt} = kP,$$

266

4. *Business.* Suppose an amount P_0 is invested in a savings account where interest is compounded continuously at 13% per year. That is, the balance P grows at the rate given by

$$\frac{dP}{dt} = 0.13P.$$

a) Find the solution of the equation in terms of P_0 and 0.13.

b) Suppose $1000 is invested. What is the balance after 1 year?

c) After what period of time will an investment of $1000 double itself?   CSS

Under ideal conditions, the growth rate of this rabbit population might be 11.7% per day. When will this population of rabbits double?

but the constant that P must be multiplied by in order to get its rate of change. It is thus a different use of the word *rate*. It is like the *interest rate* paid by a bank. If the interest rate is 12%, or 0.12, we do not mean that your bank balance P is growing at the rate of 0.12 dollars per year, but at the rate of $0.12P$ dollars per year. We therefore express the rate as 12% per year, rather than 0.12 dollars per year. We could say that the rate is 0.12 dollars *per dollar* per year. When interest is compounded continuously, the interest rate is a true exponential growth rate.

Example 4   *Business: Interest compounded continuously.* Suppose an amount P_0 is invested in a savings account where interest is compounded continuously at 12% per year. That is, the balance P grows at the rate given by

$$\frac{dP}{dt} = 0.12P.$$

a) Find the solution of the equation in terms of P_0 and 0.12.

b) Suppose $100 is invested. What is the balance after 1 year?

c) After what period of time will an investment of $100 double itself?

Solution

a) $P(t) = P_0 e^{0.12t}$

b) $P(1) = 100e^{0.12(1)} = 100e^{0.12} = 100(1.127497)$
$$\approx \$112.75 \quad \blacksquare$$

c) We are looking for the time T at which $P(T) = \$200$. The number T is called the *doubling time*. To find T, we solve the equation:

$$200 = 100e^{0.12 \cdot T}$$
$$2 = e^{0.12T}. \qquad \text{Multiplying by } \tfrac{1}{100}$$

We use natural logarithms to solve this equation:

$$\ln 2 = \ln e^{0.12T}$$
$$\ln 2 = 0.12T \qquad \text{Property 5: } \ln e^k = k$$
$$\frac{\ln 2}{0.12} = T$$
$$\frac{0.693147}{0.12} = T, \quad \text{or} \quad 5.8 \approx T.$$

Thus $100 will double itself in 5.8 years.

DO EXERCISE 4.

Let us consider another development of the formula

$$P(t) = P_0 e^{kt}$$

for interest compounded continuously. First, consider the compound-interest formula

$$A = P_0 \left(1 + \frac{k}{n} \right)^{nt},$$

letting $P = P_0$ and $k = i$. We are interested in what happens as n gets very large, that is, as n approaches ∞. To determine this limit we first make a substitution:

$$h = \frac{n}{k}.$$

Then

$$hk = n \quad \text{and} \quad \frac{1}{h} = \frac{k}{n}.$$

Note that k is a positive constant. Thus, as n gets large so does h, and as h gets large so does n. To find a formula for continuously compounded interest, we evaluate the following limit:

$$P(t) = \lim_{h \to \infty} \left[P_0 \left(1 + \frac{k}{n} \right)^{nt} \right] \qquad \text{Letting the number of compounding periods become infinite}$$

$$= P_0 \lim_{h \to \infty} \left[\left(1 + \frac{1}{h} \right)^{hkt} \right] \qquad \text{The limit of a constant times a function is the constant times the limit. We also substitute } 1/h \text{ for } k/n \text{ and } hk \text{ for } n.$$

$$= P_0 \left[\lim_{h \to \infty} \left(1 + \frac{1}{h} \right)^{h} \right]^{kt} \qquad \text{The limit of a power is the power of the limit: a form of L2 and L3 in Section 2.1}$$

$$= P_0 [e]^{kt}. \qquad \text{The definition of } e$$

We can find a general expression relating the growth rate k and the doubling time T by solving the following equation:

$$2P_0 = P_0 e^{kT}$$

$$2 = e^{kT} \qquad \text{Multiplying by } 1/P_0$$

$$\ln 2 = \ln e^{kT}$$

$$\ln 2 = kT.$$

5. Complete this table relating growth rate k and doubling time T.

Growth rate k (% per year)	Doubling time T (in years)
2%	
	10
14%	
	15
1%	

THEOREM 10

The growth rate k and the doubling time T are related by

$$kT = \ln 2 = 0.693147,$$

or

$$k = \frac{\ln 2}{T} = \frac{0.693147}{T},$$

and

$$T = \frac{\ln 2}{k} = \frac{0.693147}{k}.$$

Note that this relationship between k and T does not depend on P_0.

Example 5 At one time in Canada, a bank advertised that it would double your money in 6.6 years. What is the interest rate on such an account, assuming interest were compounded continuously?

Solution

$$k = \frac{\ln 2}{T} = \frac{0.693147}{6.6} = 0.105 = 10.5\%$$

DO EXERCISE 5.

Example 6 *Ecology: World population growth.* The population of the world was 4.9 billion in 1986. On the basis of data available at that time, it was estimated that the population P was growing exponentially at the rate of 1.6% per year. That is, $dP/dt = 0.016P$, where $t =$ the time, in years, from 1986. (To facilitate computations we assume that the population was 4.9 billion at the beginning of 1986.)

a) Find the solution to the equation, assuming $P_0 = 4.9$ and $k = 0.016$.
b) Estimate the world population in 1996 ($t = 10$).
c) After what period of time will the population be double that in 1986?

Solution

a) $P(t) = 4.9e^{0.016t}$
b) $P(10) = 4.9e^{0.016(10)} = 4.9e^{0.16} = 4.9(1.173511)$ ▦
$$\approx 5.8 \text{ billion}$$

6. The population of the United States was 241 million in 1986. It was estimated that the population P was growing exponentially at the rate of 0.9% per year. That is,

$$\frac{dP}{dt} = 0.009P,$$

where t = the time in years.

a) Find the solution to the equation, assuming that $P_0 = 241$ and $k = 0.009$.

b) Estimate the population of the United States in 1989 ($t = 3$).

c) After what period of time will the population be double that in 1986?   CSS

c) $T = \dfrac{\ln 2}{k} = \dfrac{0.693147}{0.016} \approx 43.3 \text{ yr}$

Thus, according to this model, the 1986 population will double by the year 2029. (No wonder ecologists are alarmed!)

DO EXERCISE 6.

The Rule of 70

The relationship between doubling time T and interest rate k is the basis of a rule often used in the investment world, called the *Rule of 70*. To estimate how long it will take to double your money at varying rates of return, divide 70 by the rate of return. To see how this works, let the interest rate $k = r\%$. Then,

$$T = \frac{\ln 2}{k} = \frac{0.693147}{r\%} = \frac{0.693147}{r \times 0.01}$$

$$= \frac{0.693147}{r \times 0.01} \cdot \frac{100}{100} = \frac{69.3147}{r} \approx \frac{70}{r}.$$

Modeling Other Phenomena

Example 7 *Biomedical: Alcohol absorption and the risk of having an accident.* Extensive research has provided data relating the risk R (in percent) of having an automobile accident to the blood alcohol level b (in percent). Note that these data are not a perfect fit (see the part of the graph between $b = 0$ and $b = 0.05$), but we can approximate the data with an exponential function. The modeling assumption is that the rate of change of the risk R with respect to the blood alcohol level b is given by

$$\frac{dR}{db} = kR.$$

a) Find the solution of the equation, assuming $R_0 = 1\%$.

b) Find k using the data point $R(0.14) = 20$. (This is how one might fit the data to an exponential equation.)

c) Rewrite $R(b)$ in terms of k.

d) At what blood alcohol level will the risk of having an accident be 100%? Round to the nearest hundredth.

Some myths about alcohol (Indianapolis Alcohol Safety Action Project). It's a fact—the blood alcohol concentration (BAC) in the human body is measurable. And there's no cure for its effect on the central nervous system except time. It takes time for the body's metabolism to recover.

That means a cup of coffee, a cold shower, and fresh air can't erase the effect of several drinks.

There are variables, of course: a person's body weight, how many drinks have been consumed in a given time, how much has been eaten, and so on. These account for different BAC levels. But the myth that some people can "handle their liquor" better than others is a gross rationalization—especially when it comes to driving. Some people can act more sober than others. But an automobile doesn't act; it reacts.

7. *Business: Cost of a 60-second commercial during the Super Bowl.* Past data on the cost of a 60-sec commercial during the Super Bowl are given in the table and graph below.

Year	Cost of a 60-sec TV commercial during the Super Bowl
1967	$80,000
1970	$200,000
1977	$324,000
1981	$550,000
1983	$800,000
1985	$1,100.000

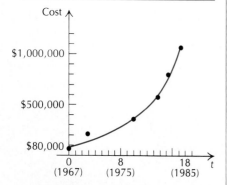

It appears that we can fit an exponential function to the data. We accept the modeling assumption that the rate of change of the cost of Super Bowl commercial C with respect to time t is given by $dC/dt = kC$.

a) Find the solution to the equation, assuming $C_0 = \$80,000$ [at $t = 0$ (1967), $C = \$80$ (in thousands)].

b) Find k using the data point $C(18) = \$1,100$ thousand. That is, in 1985, $\$1,100$ thousand, or $\$1,100,000$, was the cost of a 60-sec commercial.

c) Rewrite $C(t)$ in terms of k.

d) What will the cost of a 60-sec commercial be in 1995?

Solution

a) Since both R and b are percents, we omit the % symbol for ease of computation. The solution is

$$R(b) = e^{kb}, \quad \text{since } R_0 = 1.$$

b) We solve this equation for k, using natural logarithms:

$$20 = e^{k(0.14)} = e^{0.14k}.$$

$$\ln 20 = \ln e^{0.14k}$$

$$\ln 20 = 0.14k$$

$$\frac{\ln 20}{0.14} = k$$

$$\frac{2.995732}{0.14} = k \quad \boxdot$$

$$21.4 = k. \qquad \text{Rounding to the nearest tenth}$$

c) $R(b) = e^{21.4b}$

d) We solve this equation for b:

$$100 = e^{21.4b}$$

$$\ln 100 = \ln e^{21.4b}$$

$$\ln 100 = 21.4b$$

$$\frac{\ln 100}{21.4} = b$$

$$\frac{4.605170}{21.4} = b \quad \boxdot$$

$$0.22 = b. \qquad \text{Rounded to the nearest hundredth}$$

Calculator note: The calculations done in this problem can be performed more conveniently on your calculator by not stopping to round. For example, in part (b) we find $\ln 20$ and divide by 0.14, obtaining $21.39808767 \ldots$. We then use that value for k in part (d). Answers will be found that way in the exercises. You may note some variance in the last one or two decimal places if you round as you go.

Thus when the blood alcohol level is 0.22%, according to this model, the risk of an accident is 100%. From the graph, we see that this would occur after 12 1-oz drinks of 86-proof whiskey. "Theoretically," the model tells us that after 12 drinks of whiskey one is "sure" to have an accident. This might be questioned in reality, since a person who has had 12 drinks might not be able to drive at all.

DO EXERCISE 7.

EXERCISE SET 4.3

1. State the solution to $dQ/dt = kQ$ in terms of Q_0.

2. State the solution to $dR/dt = kR$ in terms of R_0.

APPLICATIONS

3. *Business: Compound interest.* Suppose P_0 is invested in a savings account where interest is compounded continuously at 9% per year. That is, the balance P grows at the rate given by $\quad$ CSS

$$\frac{dP}{dt} = 0.09P.$$

a) Find the solution to the equation in terms of P_0 and 0.09.

b) Suppose $1000 is invested. What is the balance after 1 year? after 2 years?

c) When will an investment of $1000 double itself?

4. *Business: Compound interest.* Suppose P_0 is invested in a savings account where interest is compounded continuously at 10% per year. That is, the balance P grows at the rate given by

$$\frac{dP}{dt} = 0.10P.$$

a) Find the solution to the equation in terms of P_0 and 0.10.

b) Suppose $20,000 is invested. What is the balance after 1 year? after 2 years?

c) When will an investment of $20,000 double itself?

5. *Ecology: Population growth.* The growth rate of the population of Central America is 3.5% per year (one of the highest in the world). What is the doubling time?

6. *Ecology: Population growth.* The growth rate of the population of Alaska is 2.8% per year (one of the highest of the fifty states). What is the doubling time?

7. *Business: Annual interest rate.* A bank advertises that it compounds interest continuously and that it will double your money in 10 years. What is its annual interest rate?
CSS

8. *Business: Annual interest rate.* A bank advertises that it compounds interest continuously and that it will double your money in 12 years. What is its annual interest rate?

9. *Ecology: Population growth.* The population of the USSR was 209 million in 1959. It was estimated that the population P was growing exponentially at the rate of 1% per year, that is,

$$\frac{dP}{dt} = 0.01P.$$

a) Find the solution to the equation, assuming $P_0 = 209$ and $k = 0.01$.

b) Estimate the population of the USSR in 1999.

c) After what period of time will the population be double that of 1959?

10. *Ecology: Population growth.* The population of Europe west of the USSR was 430 million in 1961. It was estimated that the population was growing exponentially at the rate of 1% per year, that is,

$$\frac{dP}{dt} = 0.01P.$$

a) Find the solution to the equation, assuming $P_0 = 430$ and $k = 0.01$.

b) Estimate the population of Europe in 1991.

c) After what period of time will the population be double that of 1961?

11. *Biomedical: Blood alcohol level.* In Example 7 (on alcohol absorption), at what blood alcohol level will the risk of an accident be 80%?

12. *Biomedical: Blood alcohol level.* In Example 7, at what blood alcohol level will the risk of an accident be 90%?

13. *Business: Franchise expansion.* A national hamburger firm is selling franchises throughout the country. The president estimates that the number of franchises N will increase at the rate of 10% per year, that is,

$$\frac{dN}{dt} = 0.10N.$$

a) Find the solution to the equation, assuming the number of franchises at $t = 0$ is 50.

b) How many franchises will there be in 20 years?

c) After what period of time will the initial number of 50 franchises double?

14. *Business: Franchise expansion.* Pizza, Unltd., a national pizza firm, is selling franchises throughout the country. The president estimates that the number of franchises N will increase at the rate of 15% per year, that is,

$$\frac{dN}{dt} = 0.15N.$$

a) Find the solution to the equation, assuming the number of franchises at $t = 0$ is 40.

b) How many franchises will there be in 20 years?

c) After what period of time will the initial number of 40 franchises double?

15. *Ecology: Oil demand.* The growth rate of the demand for oil in the United States is 10% per year. When will the demand be double that of 1980?

16. *Ecology: Coal demand.* The growth rate of the demand for coal in the world is 4% per year. When will the demand be double that of 1980?

17. *Ecology: Population growth.* The population of Tempe, Arizona, was 107 thousand in 1980. In 1984 it was 118 thousand. Assuming the exponential model:

a) Find the value k ($P_0 = 107$), and write the equation.

b) Estimate the population of Tempe in 1996.

18. *Ecology: Population growth.* The population of Austin, Texas, was 253 thousand in 1970. In 1980 it was 345 thousand. Assuming the exponential model:

a) Find the value k ($P_0 = 253$), and write the equation.

b) Estimate the population of Austin in 2000.

19. *Business: Cost of a movie ticket.* The average cost of a movie ticket in 1955 was $0.58. In 1985 the average cost was $3.54. Assuming the exponential model:

a) Find the value k ($P_0 = \$0.58$), and write the equation.

b) Estimate the average cost of a movie ticket in 1995.

c) After what period of time will the cost of a movie ticket be double that spent in 1985?

20. *Business: Cost of a double-dip ice cream cone.* In 1970 the cost of a double-dip ice cream cone was 52¢. In 1978 it was 66¢. Assuming the exponential model:

a) Find the value k ($P_0 = 52$), and write the equation.

b) Estimate the cost of a cone in 1990.

c) After what period of time will the cost of a cone be twice that of 1978?

21. *Business: Consumer price index.* The *consumer price index* compares the costs of goods and services over various years, where 1967 is used as a base (P_0). The same goods and services that cost $100 in 1967 cost $184.50 in 1977. Assuming the exponential model:

a) Find the value k ($P_0 = \$100$), and write the equation.

b) Estimate what the same goods and services will cost in 1987.

c) After what period of time did the same goods and services cost double that of 1967?

22. *Business: Job opportunities.* It is estimated that there were 714,000 accountants employed in 1972, and it is projected that there will be 935,000 accountants needed in 1985. Assuming the exponential model:

a) Find the value k ($P_0 = 714{,}000$), and write the equation.

b) Estimate the number of accountants needed in 1990.

c) After what period of time will the need for accountants be double that of 1972?

23. *Business: Value of Manhattan Island.* Peter Minuit of the Dutch West India Company purchased Manhattan Island from the Indians in 1626 for $24 worth of merchandise. Assuming an exponential rate of inflation of 8%, how much will Manhattan be worth in 1988?

24. *Ecology: Population growth in the Virgin Islands.* The U.S. Virgin Islands have one of the highest growth rates in the world, 9.6%. In 1970 the population was 75,150. The land area of the Virgin Islands is 3,097,600 square yards. Assuming this growth rate continues and is exponential, after what period of time will the population of the Virgin Islands be such that there is one person for every square yard of land?

25. *Ecology: Bicentennial growth of the United States.* The population of the United States in 1776 was about 2,508,000. In its bicentennial year the population was about 216,000,000. Assuming the exponential model, what was the growth rate of the United States through its bicentennial year?

26. *Business: Cost of a first-class postage stamp.* The cost of a first-class postage stamp in 1962 was 4¢. In 1986 it was 22¢. This was exponential growth. What was the growth rate? What will be the cost of a first-class postage stamp in 1989? in 1997?

27. *Business: Average salary of major-league baseball players.* The average salary of major-league baseball players in 1970 was $29,303. In 1985 the average salary was $363,000. This was exponential growth. What was the growth rate? What will the average salary be in 1989? in 1997?

28. ▦ *Business: Cost of a Hershey bar.* The cost of a Hershey bar in 1962 was $0.05, and was increasing at an exponential growth rate of 9.7%. What will the cost of a Hershey bar be in 1991? in 1997?

EXTENSION EXERCISES

Business: Effective annual yield. Suppose $100 is invested at 12% compounded continuously for 1 year. We know from Example 4 that the balance will be $112.75. This is the same as if $100 were invested at 12.75% and compounded once a year (simple interest). The 12.75% is called the *effective annual yield.* In general, if P_0 is invested at $k(\%)$ compounded continuously, then the effective annual yield is that number i satisfying $P_0(1 + i) = P_0 e^k$. Then $1 + i = e^k$, or

$$\text{Effective annual yield} = i = e^k - 1.$$

29. An amount is invested at 14% per year compounded continuously. What is the effective annual yield?

30. An amount is invested at 8% per year compounded continuously. What is the effective annual yield?

31. The effective annual yield on an investment compounded continuously is 9.42%. At what rate was it invested?

32. The effective annual yield on an investment compounded continuously is 10.52%. At what rate was it invested?

33. Find an expression relating the growth rate k and the *tripling time* T_3.

34. Find an expression relating the growth rate k and the *quadrupling time* T_4.

35. Gather data concerning population growth in your city. Estimate the population in 1990; in 2000.

36. A quantity Q_1 grows exponentially with a doubling time of 1 year. A quantity Q_2 grows exponentially with a doubling time of 2 years. If the initial amounts of Q_1 and Q_2 are the same, when will Q_1 be twice the size of Q_2?

37. A growth rate of 100% per day corresponds to what exponential growth rate per hour?

38. Show that any two measurements of an exponentially growing population will determine k. That is, show that if y has the values y_1 at t_1 and y_2 at t_2, then

$$k = \frac{\ln (y_2/y_1)}{t_2 - t_1}.$$

4.4

OBJECTIVES

a) State the solution to an equation

$$\frac{dP}{dt} = -kP$$

as

$$P(t) = P_0 e^{-kt}.$$

b) Given a decay rate, find the half-life.

c) Given a half-life, find the decay rate.

d) Solve applied problems involving decay.

e) Solve applied problems involving Newton's law of cooling.

1. Using the same set of axes, graph $y = e^{2x}$ and $y = e^{-2x}$.

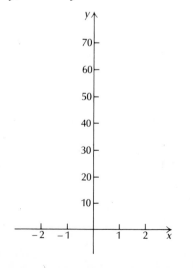

APPLICATIONS: DECAY

DO EXERCISE 1.

In the equation of population growth $dP/dt = kP$, the constant k is actually given by

$$k = (\text{Birth rate}) - (\text{Death rate}).$$

Thus a population "grows" only when the *birth rate* is greater than the *death rate*. When the birth rate is less than the death rate, k will be negative so the population will be decreasing, or "decaying," at a rate proportional to its size. For convenience in our computations we will express such a negative value as $-k$, where $k > 0$. The equation

$$\frac{dP}{dt} = -kP \quad (\text{where } k > 0)$$

shows P to be *decreasing* as a function of time, and the solution

$$P(t) = P_0 e^{-kt}$$

shows it to be decreasing exponentially. This is *exponential decay*. The amount present initially at $t = 0$ is again P_0.

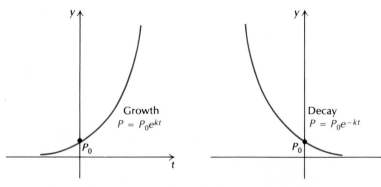

2. Xenon-133 has a decay rate of 14% per day. The rate of change of an amount N is given by

$$\frac{dN}{dt} = -0.14N.$$

a) Find the solution to the equation in terms of N_0.

b) Suppose 1000 grams of xenon-133 is present at $t = 0$. How much will remain after 10 days?

c) After what time will half of the 1000 grams remain?

Radioactive Decay

Radioactive elements decay exponentially; that is, they disintegrate at a rate that is proportional to the amount present.

Example 1 Strontium-90 has a decay rate of 2.8% per year. The rate of change of an amount N is given by

$$\frac{dN}{dt} = -0.028N.$$

a) Find the solution to the equation in terms of N_0 (the amount present at $t = 0$).

b) Suppose 1000 grams of strontium-90 is present at $t = 0$. How much will remain after 100 years?

c) After what amount of time will half of the 1000 grams remain?

CSS

Solution

a) $N(t) = N_0 e^{-0.028t}$

b) $N(100) = 1000e^{-0.028(100)} = 1000e^{-2.8}$

$$= 1000(0.060810) \qquad \text{▦}$$

$$\approx 60.8 \text{ grams}$$

c) We are asking, "At what time T will $N(T) = 500$?" The number T is called the *half-life*. To find T we solve this equation:

$$500 = 1000e^{-0.028T}$$

$$\tfrac{1}{2} = e^{-0.028T}$$

$$\ln \tfrac{1}{2} = \ln e^{-0.028T}$$

$$\ln 1 - \ln 2 = -0.028T$$

$$0 - \ln 2 = -0.028T$$

$$\frac{-\ln 2}{-0.028} = T$$

$$\frac{\ln 2}{0.028} = T$$

$$\frac{0.693147}{0.028} = T$$

$$25 \approx T.$$

Thus the half-life of strontium-90 is 25 years.

DO EXERCISE 2.

How can scientists determine that an animal bone has lost 30% of its carbon-14? The assumption is that the percentage of carbon-14 in the atmosphere and in living plants and animals is the same. When a plant or animal dies, the amount of carbon-14 decays exponentially. The scientist burns the animal bone and uses a Geiger counter to determine the percentage of the smoke that is carbon-14. It is the amount this varies from the percentage in the atmosphere that tells how much carbon-14 has been lost.

The process of carbon-14 dating was developed by the American chemist Willard F. Libby in 1952. It is known that the radioactivity in a living plant is 16 disintegrations per gram per minute. Since the half-life of carbon-14 is 5750 years, a dead plant with an activity of 8 disintegrations per gram per minute is 5750 years old, one with an activity of 4 disintegrations per gram per minute is 11,500 years old, and so on. Carbon-14 dating can be used to measure the age of objects from 30,000 to 40,000 years old. Beyond such an age it is too difficult to measure the radioactivity and some other method would have to be used.

We can find a general expression relating the decay rate k and the half-life T by solving this equation:

$$\tfrac{1}{2}P_0 = P_0 e^{-kT}$$
$$\tfrac{1}{2} = e^{-kT}$$
$$\ln \tfrac{1}{2} = \ln e^{-kT}$$
$$\ln 1 - \ln 2 = -kT$$
$$0 - \ln 2 = -kT$$
$$-\ln 2 = -kT$$
$$\ln 2 = kT.$$

Again, we have the following.

THEOREM 11

The *decay rate k* and the *half-life T* are related by

$$kT = \ln 2 = 0.693147,$$

or

$$k = \frac{\ln 2}{T} \quad \text{and} \quad T = \frac{\ln 2}{k}.$$

Thus the half-life T depends only on the decay rate k. In particular, it is independent of the initial population size.

The effect of half-life is shown in this radioactive decay curve. The exponential function gets close to, but never reaches, 0 as t gets larger. Thus, in theory, a radioactive substance never completely decays.

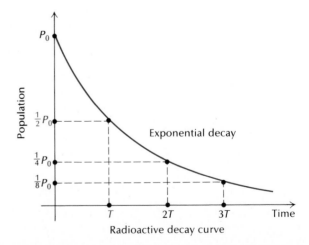

Radioactive decay curve

3. The decay rate of cesium-137 is 2.3% per year. What is its half-life?

Example 2   Plutonium, a common product and ingredient of nuclear reactors, is of great concern to those who are against the building of nuclear reactors. Its decay rate is 0.003% per year. What is its half-life?

Solution

$$T = \frac{\ln 2}{k} = \frac{0.693147}{0.00003} = 23{,}105 \text{ years}$$

DO EXERCISES 3 AND 4.

Example 3   *Carbon dating.* The radioactive element carbon-14 has a half-life of 5750 years. The percentage of carbon-14 present in the remains of plants and animals can be used to determine age. Archaeologists found that the linen wrapping from one of the Dead Sea Scrolls had lost 22.3% of its carbon-14. How old was the linen wrapping?

Solution

a) Find the decay rate k.

4. The half-life of barium-140 is 13 days. What is its decay rate?

$$k = \frac{\ln 2}{T} = \frac{0.693147}{5750} = 0.0001205, \quad \text{or } 0.01205\% \text{ per year}$$

b) Find the exponential equation for the amount $N(t)$ that remains from an initial amount N_0 after t years.

$$N(t) = N_0 e^{-0.0001205t}$$

(*Note:* This equation can be used for any subsequent carbon-dating problem.)

c) If an animal bone has lost 22.3% of its carbon-14 from an initial amount P_0, then 77.7% (P_0) is the amount present. To find the age t of the bone, we solve the following equation for t:

$$77.7\%P_0 = P_0 e^{-0.0001205t}$$
$$0.777 = e^{-0.0001205t}$$
$$\ln 0.777 = \ln e^{-0.0001205t}$$
$$\ln 0.777 = -0.0001205t$$
$$-0.2523149 = -0.0001205t$$
$$\frac{0.2523149}{0.0001205} = t$$
$$2093 \approx t.$$

5. How old is a skeleton that has lost 80% of its carbon-14?

Exploratory exercises: Cooling. Draw a glass of hot tap water. Place a thermometer in the glass and check the temperature. Check the temperature every 30 minutes thereafter. Plot your data on this graph, and connect the points with a smooth curve.

(Start)   Time (minutes)

a) What was the temperature at the start?

b) At what temperature does there seem to be a leveling off of the graph?

c) What is the difference between your answers to parts (a) and (b)?

d) How does the temperature in part (b) compare with the room temperature?

In 1947, a Bedouin youth looking for a stray goat climbed into a cave at Kirbet Qumran on the shores of the Dead Sea near Jericho and came upon earthenware jars containing an incalculable treasure of ancient manuscripts. Shown here are fragments of those so-called Dead Sea Scrolls, a portion of some 600 or so texts found so far and which concern the Jewish books of the Bible. Officials date them before 70 A.D., making them the oldest Biblical manuscripts by 1000 years.

Thus the linen wrapping of the Dead Sea Scrolls is about 2093 years old.

DO EXERCISE 5.

Newton's Law of Cooling

Before you study the following, do the exploratory exercises in the margin.

Newton's Law of Cooling

The temperature T of a cooling object drops at a rate that is proportional to the difference $T - C$, where C is the constant temperature of the surrounding medium. Thus,

$$\frac{dT}{dt} = -k(T - C). \qquad (1)$$

The solution to Eq. (1) is

$$T = T(t) = ae^{-kt} + C. \qquad (2)$$

We can check this by differentiating:

$$\frac{dT}{dt} = -kae^{-kt} = -k(ae^{-kt}) = -k(T - C).$$

Example 4 The temperature of a cup of freshly brewed coffee is 200°, and the room temperature is 70°. The temperature cools to 190° in 5 minutes.

a) What is the temperature after 10 minutes?
b) How long does it take the temperature to cool to 90°?

Solution

a) We first find the value of a in Eq. (2). At $t = 0$, $T = 200°$. We solve the following equation for a:

$$200 = ae^{-k \cdot 0} + 70$$
$$200 = a \cdot 1 + 70$$
$$130 = a.$$

Now we find k using the fact that at $t = 5$, $T = 190°$. We solve the following equation for k:

$$190 = 130e^{-k \cdot 5} + 70$$
$$120 = 130e^{-5k}$$
$$\tfrac{12}{13} = e^{-5k}$$
$$\ln \tfrac{12}{13} = \ln e^{-5k}$$
$$\ln \tfrac{12}{13} = -5k$$
$$-5k = \ln 12 - \ln 13$$
$$k = -\tfrac{1}{5}(\ln 12 - \ln 13)$$
$$= -\tfrac{1}{5}(2.484907 - 2.564949) \qquad \text{▦}$$
$$\approx 0.016.$$

Now we find the temperature at $t = 10$:

$$T(10) = 130e^{-0.016(10)} + 70 = 130e^{-0.16} + 70$$
$$= 130(0.852144) + 70 \approx 181°.$$

b) To find how long it will take the temperature to cool to 90°, we solve for t: CSS

$$90 = 130e^{-0.016t} + 70$$
$$20 = 130e^{-0.016t}$$
$$\tfrac{2}{13} = e^{-0.016t}$$

6. The temperature of a hot cup of soup is 200°. The room temperature is 70°. The temperature cools to 190° in 8 minutes.

   a) Find the value of the constant a in Newton's law of cooling.

   b) Find the value of the constant k.

   c) What is the temperature after 10 minutes?

   d) How long does it take the soup to cool to 80°?

7. Return to the data you found in the exploratory exercises. Find an equation that "fits" the data. Use this equation to check values of other data points. How do they compare? Is it ever "theoretically" possible for the temperature of the water to be the same as the room temperature?

Thus,

$$\ln \tfrac{2}{13} = \ln e^{-0.016t}$$

$$= -0.016t$$

$$-0.016t = \ln 2 - \ln 13$$

$$t = -\frac{1}{0.016}(\ln 2 - \ln 13)$$

$$t = -\frac{1}{0.016}(0.693147 - 2.564949)$$

$$t \approx 117 \text{ minutes.}$$

DO EXERCISES 6 AND 7.

The graph of $T(t) = ae^{-kt} + C$ is shown below. Note that $\lim_{t \to \infty} T(t) = C$. The temperature of the object decreases toward the temperature of the surrounding medium.

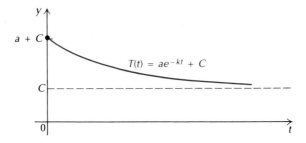

Mathematically, this model tells us that the temperature never reaches C, but in practice this happens eventually. At least, the temperature of the cooling object gets so close to that of the surrounding medium that no device could detect a difference. Let us now see how Newton's law of cooling can be used in solving a crime.

Example 5 *When was the murder committed?* The police discover the body of a calculus professor. Critical to solving the crime is determining when the murder was committed. The police call the coroner, who arrives at 12:00 P.M. He immediately takes the temperature of the body and finds it to be 94.6°. He waits 1 hour, takes the temperature again, and finds it to be 93.4°. He also notes that the temperature of the room is 70°. When was the murder committed?

8. A butcher is murdered and the body is thrown in a cooler where the temperature is 40°. The coroner, arriving at 3:00 P.M., takes the temperature of the body, finds it to be 73.8°, waits 1 hour, takes the temperature again, and finds it to be 70.3°. When was the murder committed?

Solution   We first find a in the equation $T(t) = ae^{-kt} + C$. Assuming the temperature of the body was normal when the murder occurred, we have $T = 98.6°$ at $t = 0$. Thus,

$$98.6° = ae^{-k \cdot 0} + 70°,$$

so

$$a = 28.6°.$$

Thus T is given by $T(t) = 28.6e^{-kt} + 70$.

We want to find the number of hours N since the murder was committed. To find N we must first determine k. From the two temperature readings the coroner made, we have

$$94.6 = 28.6e^{-kN} + 70, \qquad \text{or} \quad 24.6 = 28.6e^{-kN}; \qquad \textbf{(1)}$$
$$93.4 = 28.6e^{-k(N+1)} + 70, \quad \text{or} \quad 23.4 = 28.6e^{-k(N+1)}. \qquad \textbf{(2)}$$

Dividing Eq. (1) by Eq. (2), we get

$$\frac{24.6}{23.4} = \frac{28.6e^{-kN}}{28.6e^{-k(N+1)}}$$
$$= e^{-kN + k(N+1)} = e^{-kN + kN + k} = e^{k}.$$

We solve this equation for k:

$$\ln \frac{24.6}{23.4} = \ln e^{k} \qquad \text{Taking the natural logarithm on both sides}$$

$$\ln 24.6 - \ln 23.4 = k$$

$$3.202746 - 3.152736 \approx k$$

$$0.05 \approx k.$$

Now we substitute back into Eq. (1) and solve for N:

$$24.6 = 28.6e^{-0.05N}$$
$$\ln 24.6 = \ln 28.6e^{-0.05N}$$
$$\ln 24.6 = \ln 28.6 + \ln e^{-0.05N}$$
$$\ln 24.6 - \ln 28.6 = -0.05N$$
$$\frac{-0.150660}{-0.05} = N$$
$$3 \approx N.$$

The coroner arrived at 12:00 P.M., so the murder was committed at about 9:00 A.M.

DO EXERCISE 8.

EXERCISE SET 4.4

1. The decay rate of iodine-131 is 9.6% per day. What is its half-life?

2. The decay rate of krypton-85 is 6.3% per year. What is its half-life?

3. The half-life of polonium is 3 minutes. What is its decay rate?

4. The half-life of lead is 22 years. What is its decay rate?

5. Of an initial amount of 1000 grams of polonium, how much will remain after 20 minutes? See Exercise 3 for the value of k.

6. Of an initial amount of 1000 grams of lead, how much will remain after 100 years? See Exercise 4 for the value of k.

7. *Carbon dating.* How old is a piece of wood that has lost 90% of its carbon-14?

8. *Carbon dating.* How old is an ivory tusk that has lost 40% of its carbon-14?

9. *Carbon dating.* How old is a Chinese artifact that has lost 60% of its carbon-14?

10. *Carbon dating.* How old is a skeleton that has lost 50% of its carbon-14?

11. In a *chemical reaction* substance A decomposes at a rate proportional to the amount of A present.

    a) Write an equation relating A to the amount left of an initial amount A_0 after time t.

    b) It is found that 8 grams of A will reduce to 4 grams in 3 hours. After what time will there be only 1 gram left?

12. In a *chemical reaction* substance A decomposes at a rate proportional to the amount of A present.

    a) Write an equation relating A to the amount left of an initial amount A_0 after time t.

    b) It is found that 10 pounds of A will reduce to 5 pounds in 3.3 hours. After what time will there be only 1 pound left?

13. *Weight loss.* The initial weight of a starving animal is W_0. Its weight W after t days is given by

    $$W = W_0 e^{-0.008t}.$$

    a) What percentage of its weight does it lose each day?

    b) What percentage of its initial weight remains after 30 days?

14. *Weight loss.* The initial weight of a starving animal is W_0. Its weight W after t days is given by

    $$W = W_0 e^{-0.009t}.$$

    a) What percentage of its weight does it lose each day?

    b) What percentage of its initial weight remains after 30 days?

15. *Satellite power.* The power supply of a satellite is a radio-isotope. The power output P, in watts, decreases at a rate proportional to the amount present; P is given by

    $$P = 50e^{-0.004t},$$

    where t = the time in days.

    a) How much power will be available after 375 days?

    b) What is the half-life of the power supply?

    c) The satellite's equipment cannot operate on fewer than 10 watts of power. How long can the satellite stay in operation?

    d) How much power did the satellite have to begin with?

16. *Atmospheric pressure.* Atmospheric pressure P at altitude a is given by

    $$P = P_0 e^{-0.00005a},$$

    where P_0 = the pressure at sea level. Assume $P_0 = 14.7$ lb/in^2 (pounds per square inch).

    a) Find the pressure at an altitude of 1000 ft.

    b) Find the pressure at 20,000 ft.

    c) At what altitude is the pressure 1.47 lb/in^2?

17. *Salvage value.* A business estimates that the salvage value V of a piece of machinery after t years is given by

$$V(t) = \$40,000e^{-t}.$$

a) What did the machinery cost initially?

b) What is the salvage value after 2 years?

18. *Supply and demand.* The supply and demand for the sale of stereos by a sound company are given by

$$S(x) = e^x \quad \text{and} \quad D(x) = 163,000e^{-x},$$

where $S(x)$ = the price at which the company is willing to supply x stereos and $D(x)$ = the demand price for a quantity of x stereos. Find the equilibrium point. For reference, see Section 1.5.

THE BEER–LAMBERT LAW

A beam of light enters a medium such as water or smoky air with initial intensity I_0. Its intensity is decreased depending on the thickness (or concentration) of the medium. The intensity I at a depth (or concentration) of x units is given by

$$I = I_0 e^{-\mu x}.$$

The constant μ ("mu"), called the *coefficient of absorption*, varies with the medium.

19. *Light through sea water* has $\mu = 1.4$ when x is measured in meters (m).

a) What percentage of I_0 remains at a depth of sea water that is 1 m? 2 m? 3 m?

b) Plant life cannot exist below 10 m. What percentage of I_0 remains at 10 m?

20. *Light through smog.* Particulate concentrations of pollution reduce sunlight. In a smoggy area $\mu = 0.01$ and x = the concentration of particulates measured in micrograms per cubic meter. What percentage of an initial amount I_0 of sunlight passes through smog that has a concentration of 100 micrograms per cubic meter?

21. The temperature of a hot liquid is $100°$ and the room temperature is $75°$. The liquid cools to $90°$ in 10 minutes.

a) Find the value of the constant a in Newton's law of cooling.

b) Find the value of the constant k. Round to the nearest hundredth.

c) What is the temperature after 20 minutes?

d) How long does it take the liquid to cool to $80°$?

22. The temperature of a hot liquid is $100°$ and is placed in a refrigerator where the temperature is $40°$. The liquid cools to $90°$ in 5 minutes.

a) Find the value of the constant a in Newton's law of cooling.

b) Find the value of the constant k. Round to the nearest hundredth.

c) What is the temperature after 10 minutes?

d) How long does it take the liquid to cool to $41°$?

23. The coroner arrives at the scene of a murder at 11 P.M. She takes the temperature of the body and finds it to be $85.9°$. She waits 1 hour, takes the temperature again, and finds it to be $83.4°$. She notes that the room temperature is $60°$. When was the murder committed?

24. The coroner arrives at the scene of a murder at 2 A.M. He takes the temperature of the body and finds it to be $61.6°$. He waits 1 hour, takes the temperature again, and finds it to be $57.2°$. The body is in a meat freezer, where the temperature is $10°$. When was the murder committed?

25. *Ecology: Population decrease of Cincinnati.* The population of Cincinnati was 453,000 in 1970 and 385,000 in 1980. Assuming the population is decreasing according to the exponential-decay model:

a) Find the value k, and write the equation.

b) Estimate the population of Cincinnati in 1990.

c) After what period of time will Cincinnati have just 1 person?

26. *Ecology: Population decrease of Panama.* The population of Panama was 1,464,000 in 1970 and 1,260,000 in 1980. Assuming the population is decreasing according to the exponential-decay model:

   a) Find the value k, and write the equation.   b) Estimate the population of Panama in 2000.

   c) After what period of time will the population of Panama be 100,000?

27. *Business: Present value.* Following the birth of a child, a parent wants to make an initial investment of P_0 that will grow to $10,000 by the child's 20th birthday. Interest is compounded continuously at 8%. What should the initial investment be? Such an amount is called the *present value* of $10,000 due 20 years from now.

28. *Ecology: Population of the USSR in a preceding year.* The population of the USSR was 258 million in 1980 and was growing at the rate of 1% per year. What was the population in 1970? in 1940?

29. *Business: Consumer price index.* The consumer price index compares the costs of goods and services over various years, where 1967 is used as a base. The same goods and services that cost $100 in 1967 cost $42 in 1940. Assuming the exponential-decay model:

   a) Find the value k, and write the equation. Round to the nearest hundredth.

   b) Estimate what the same goods and services cost in 1900.

30. *Carbon dating.* Recently, while digging in Chaco Canyon, New Mexico, archeologists found corn pollen that was 4000 years old. This was evidence that Indians had begun cultivating crops in the Southwest centuries earlier than scientists had thought. What percent of the carbon-14 had been lost from the pollen?

4.5

OBJECTIVES

a) Express a power like 2^3 as a power of e.

b) Differentiate functions involving a^x.

c) Differentiate functions involving $\log_a x$.

THE DERIVATIVES OF a^x AND $\log_a x$ *

The Derivative of a^x

To find the derivative of a^x, for any base a, we first express it as a power of e. In order to do this, we first prove the following.

PROPERTY 7. $b^{\log_b x} = x$

To prove this, let

$$y = \log_b x.$$

Then, by the definition of a logarithm,

$$b^y = x.$$

---

* This section can be omitted without loss of continuity.

Express as a power of e.

1. 4^5

Substituting $\log_b x$ for y, we have

$$b^{\log_b x} = x.$$

We can now express a^x as a power of e. Using Property 7, where $b = e$ and $x = a$, we have

$$a = e^{\ln a}. \qquad \text{Remember: } \ln a = \log_e a.$$

Raising both sides to the power x, we get

$$a^x = (e^{\ln a})^x$$
$$= e^{x \cdot \ln a}. \qquad \text{Multiplying exponents}$$

Thus we have the following.

THEOREM 12

$$a^x = e^{x \cdot \ln a}$$

---

2. 2^x

Example 1   Express as a power of e.

a) 3^2
$$\begin{aligned}3^2 &= e^{2 \cdot \ln 3}\\ &\approx e^{2(1.098612)}\\ &= e^{2.1972}\end{aligned}$$

b) 10^x
$$\begin{aligned}10^x &= e^{x \cdot \ln 10}\\ &\approx e^{x(2.3026)}\\ &= e^{2.3026x}\end{aligned}$$

DO EXERCISES 1 AND 2.

Now we can differentiate.

Example 2

$$\begin{aligned}\frac{d}{dx}\, 2^x &= \frac{d}{dx}\, e^{x \cdot \ln 2}\\[2mm] &= \left[\frac{d}{dx}\, (x \cdot \ln 2)\right] \cdot e^{x \cdot \ln 2}\\[2mm] &= (\ln 2)\,(e^{\ln 2})^x\\[2mm] &= (\ln 2)\, 2^x\end{aligned}$$

3. Differentiate $y = 5^x$.

We completed this by taking the derivative of $x \ln 2$ and replacing $e^{x \cdot \ln 2}$ by 2^x. Note that $\ln 2 \approx 0.7$, so the above verifies our earlier approximation of the derivative of 2^x as $(0.7)2^x$ (see Section 4.1).

DO EXERCISE 3.

In general,

$$\frac{d}{dx} a^x = \frac{d}{dx} e^{x \cdot \ln a}$$

$$= \left[\frac{d}{dx} (x \cdot \ln a) \right] \cdot e^{x \ln a}$$

$$= (\ln a) \, a^x.$$

Thus we have the following.

THEOREM 13

$$\frac{d}{dx} a^x = (\ln a) \, a^x$$

Example 3

$$\frac{d}{dx} 3^x = (\ln 3) \, 3^x$$

Example 4

$$\frac{d}{dx} (1.4)^x = (\ln 1.4)(1.4)^x$$

Compare these formulas:

$$\frac{d}{dx} a^x = (\ln a) \, a^x; \qquad \frac{d}{dx} e^x = e^x.$$

It is the simplicity of the latter formula that is a reason for the use of the base e in calculus. The many applications of e in natural phenomena provide other reasons.

One other result also follows from what we have done. If

$$f(x) = a^x,$$

we know that

$$f'(x) = a^x(\ln a).$$

Differentiate.

4. $f(x) = 4^x$

In Section 4.1 we also showed that

$$f'(x) = a^x \left(\lim_{h \to 0} \frac{a^h - 1}{h} \right).$$

Since $a^x > 0$, we have the following.

THEOREM 14

$$\ln a = \lim_{h \to 0} \frac{a^h - 1}{h}$$

DO EXERCISES 4 AND 5.

The Derivative of $\log_a x$

Just as the derivative of a^x is expressed in terms of $\ln a$, so too is the derivative of $\log_a x$. To find this derivative we first express $\log_a x$ in terms of $\ln a$ using Property 7:

$$a^{\log_a x} = x.$$

5. $f(x) = (4.3)^x$

Then

$$\ln a^{\log_a x} = \ln x$$

$$(\log_a x) \cdot \ln a = \ln x \qquad \text{By Property 3, treating } \log_a x \text{ as an exponent}$$

and

$$\log_a x = \boxed{\frac{1}{\ln a}} \cdot \ln x.$$

$$\underline{\qquad} \text{ constant}$$

The derivative of $\log_a x$ follows.

THEOREM 15

$$\frac{d}{dx} \log_a x = \frac{1}{\ln a} \cdot \frac{1}{x}$$

Comparing this with

$$\frac{d}{dx} \ln x = \frac{1}{x},$$

we again see a reason for the use of the base e in calculus.

Differentiate.

6. $y = \log_2 x$

7. $y = -7 \log x$

8. $f(x) = x^6 \log x$

Example 5

$$\frac{d}{dx} \log_3 x = \frac{1}{\ln 3} \cdot \frac{1}{x}$$

Example 6

$$\frac{d}{dx} \log x = \frac{1}{\ln 10} \cdot \frac{1}{x} \qquad \log x = \log_{10} x$$

Example 7

$$\frac{d}{dx} x^2 \log x = x^2 \frac{1}{\ln 10} \cdot \frac{1}{x} + 2x \log x \qquad \text{By the Product Rule}$$

$$= \frac{x}{\ln 10} + 2x \log x = x\left(\frac{1}{\ln 10} + 2 \log x\right)$$

DO EXERCISES 6–8.

EXERCISE SET 4.5

Express as a power of e.

1. 5^4 **2.** 2^3 **3.** $(3.4)^{10}$ **4.** $(5.3)^{20}$

5. 4^k **6.** 5^R **7.** 8^{kT} **8.** 10^{kR}

Differentiate.

9. $y = 6^x$ **10.** $y = 7^x$ **11.** $f(x) = 10^x$ **12.** $f(x) = 100^x$

13. $f(x) = x(6.2)^x$ **14.** $f(x) = x(5.4)^x$ **15.** $y = x^3 10^x$ **16.** $y = x^4 5^x$

APPLICATIONS

Earthquake magnitude. The magnitude R (measured on the Richter scale) of an earthquake of intensity I is defined as

$$R = \log \frac{I}{I_0},$$

where I_0 is a minimum intensity used for comparison. When one earthquake is 10 times as intense as another, its magnitude on the Richter scale is 1 higher. If one earthquake is 100 times as intense as another, its magnitude on the Richter scale is 2 higher, and so on. Thus an earthquake whose magnitude is 7 on the Richter scale is 10 times as intense as an earthquake whose magnitude is 6. Earthquakes can be interpreted as multiples of the minimum intensity I_0.

17. In 1986 there was an earthquake near Cleveland, Ohio. It had an intensity of $10^5 \cdot I_0$. What was its magnitude on the Richter scale?

18. The Anchorage, Alaska, earthquake on March 27, 1964, had an intensity of $10^{8.4} \cdot I_0$. What was its magnitude on the Richter scale?

19. *Earthquake intensity.* The intensity I of an earthquake is given by

$$I = I_0 10^R,$$

where $R =$ the magnitude on the Richter scale and $I_0 =$ the minimum intensity, where $R = 0$, used for comparison.

a) Find I, in terms of I_0, for an earthquake of magnitude 7 on the Richter scale.

b) Find I, in terms of I_0, for an earthquake of magnitude 8 on the Richter scale.

c) Compare your answers to parts (a) and (b).

d) Find the rate of change dI/dR.

20. *Intensity of sound.* The intensity of a sound is given by

$$I = I_0 10^{0.1L},$$

where $L =$ the loudness of the sound as measured in decibels and $I_0 =$ the minimum intensity detectable by the human ear.

a) Find I, in terms of I_0, for the loudness of a power mower, which is 100 decibels.

b) Find I, in terms of I_0, for the loudness of just audible sound, which is 10 decibels.

c) Compare your answers to parts (a) and (b).

d) Find the rate of change dI/dL.

This photograph shows part of the damage of the earthquake in Anchorage, Alaska, in 1964.

Differentiate.

21. $y = \log_4 x$

22. $y = \log_5 x$

23. $f(x) = 2 \log x$

24. $f(x) = 5 \log x$

25. $f(x) = \log \dfrac{x}{3}$

26. $f(x) = \log \dfrac{x}{5}$

27. $y = x^3 \log_8 x$

28. $y = x \log_6 x$

29. *Earthquake magnitude.* The magnitude R (measured on the Richter scale) of an earthquake of intensity I is defined as

$$R = \log \frac{I}{I_0},$$

where $I_0 =$ the minimum intensity (used for comparison). (The exponential form of this definition is given in Exercise 19.) Find the rate of change dR/dI.

30. *Loudness of sound.* The loudness L of a sound of intensity I is defined as

$$L = 10 \log \frac{I}{I_0},$$

where $I_0 =$ the minimum intensity detectable by the human ear and $L =$ the loudness measured in decibels. (The exponential form of this definition is given in Exercise 20.) Find the rate of change dL/dI.

31. *Response to drug dosage.* The response y to a dosage x of a drug is given by

$$y = m \log x + b.$$

The response may be hard to measure with a number. The patient might sweat more, have an increase in temperature, or pass out. Find the rate of change dy/dx.

33. *Ecology: Recycling aluminum cans.* It is known that $\frac{1}{4}$ of all aluminum cans distributed will be recycled each year. A beverage company distributes 250,000 cans. The number still in use after time t, in years, is given by

$$N(t) = 250,000\left(\tfrac{1}{4}\right)^t.$$

Find $N'(t)$.

32. *Business: Double declining-balance depreciation.* An office machine is purchased for $5200. Under certain assumptions its salvage value V depreciates according to a method called double declining balance, basically 80% each year, and is given by

$$V(t) = \$5200(0.80)^t,$$

where t is the time, in years. Find $V'(t)$.

34. Find $\displaystyle\lim_{h \to 0} \frac{3^h - 1}{h}$.

EXTENSION EXERCISES

Use the Chain Rule and other formulas given in this section to differentiate each of the following. In some cases, it may help to take the logarithm on both sides of the equation before differentiating.

35. $f(x) = 3^{2x}$

36. $y = 2^{x^4}$

37. $y = x^x, x > 0$

38. $y = \log_3 (x^2 + 1)$

39. $f(x) = x^{e^x}, x > 0$

40. $y = a^{f(x)}$

41. $y = \log_a f(x), f(x)$ positive

42. $y = [f(x)]^{g(x)}, f(x)$ positive

4.6

OBJECTIVE

a) Given a demand function, find the elasticity and the value(s) of x for which the total revenue is maximized.

AN ECONOMIC APPLICATION: ELASTICITY OF DEMAND*

Suppose x represents the quantity of goods sold and p is the price per unit of the goods. Recall that x and p are related by the demand function

$$p = D(x).$$

Suppose there is a change Δx in the quantity sold. The percent change

---

* This section can be omitted without loss of continuity.

in quantity is

$$\frac{\Delta x}{x}.$$

A change in the quantity sold produces a change Δp in the price. The percent change in price is

$$\frac{\Delta p}{p}.$$

The ratio of these percents is given by

$$\frac{(\Delta x/x)}{(\Delta p/p)},$$

which can be expressed as

$$\frac{p}{x} \cdot \frac{1}{\Delta p/\Delta x}. \qquad (1)$$

For continuous functions

$$\lim_{\Delta x \to 0} \frac{\Delta p}{\Delta x} = \frac{dp}{dx},$$

and the limit as Δx approaches 0 of the expression in Eq. (1) becomes

$$\frac{p}{x} \cdot \frac{1}{dp/dx} = \frac{D(x)}{x} \cdot \frac{1}{D'(x)} = \frac{1}{x} \cdot \frac{D(x)}{D'(x)}.$$

DEFINITION

The *elasticity of demand E* is given by

$$E = -\frac{p}{x} \cdot \frac{1}{dp/dx} = -\frac{1}{x} \cdot \frac{D(x)}{D'(x)}.$$

The numbers x and p are always nonnegative. The slope of the demand curve dp/dx is always negative, since the demand curve is decreasing. The minus sign makes E nonnegative and easier, for our purposes, to work with. We will find the second expression for elasticity the most useful for computations.

Example 1   A company determines that the demand function for a certain product is given by

$$p = D(x) = 200 - x.$$

Find each of the following.

a) The elasticity as a function of x
b) The elasticity at $x = 70$ and at $x = 150$
c) The value of x for which $E = 1$
d) The total-revenue function
e) The value of x for which the revenue is a maximum

Solution

a) To find the elasticity, we first find

$$D'(x) = -1.$$

Then we substitute -1 for $D'(x)$ and $200 - x$ for $D(x)$ in the second expression for elasticity:

$$E(x) = -\frac{1}{x} \cdot \frac{D(x)}{D'(x)}$$

$$= -\frac{1}{x} \cdot \frac{200 - x}{-1} = \frac{200 - x}{x}.$$

b) $E(70) = \dfrac{200 - 70}{70} = \dfrac{130}{70} = \dfrac{13}{7},$

$E(150) = \dfrac{200 - 150}{150} = \dfrac{50}{150} = \dfrac{1}{3}$

c) We set $E(x) = 1$ and solve for x:

$$\frac{200 - x}{x} = 1$$

$200 - x = x$ We multiply by x assuming $x \neq 0$.

$200 = 2x$

$100 = x.$

d) Recall that the total revenue $R(x)$ is given by $x D(x)$. Then

$$\begin{aligned} R(x) &= x D(x) \\ &= x(200 - x) \\ &= 200x - x^2. \end{aligned}$$

e) To find the value of x that maximizes total revenue, we find $R'(x)$:

$$R'(x) = 200 - 2x.$$

Now $R'(x)$ exists for all values of x in the interval $[0, \infty)$. Thus we

1. A company determines that the demand function for a product is given by

    $$p = D(x) = 300 - x.$$

    Find each of the following.

    a)   The elasticity as a function of x

    b)   The elasticity at $x = 100$ and at $x = 200$

    c)   The value of x for which $E = 1$

    d)   The total-revenue function

    e)   The value of x for which the revenue is a maximum

solve:

$$R'(x) = 200 - 2x = 0$$
$$-2x = -200$$
$$x = 100.$$

Since there is only one critical point, we can try to use the second derivative to see if we have a maximum:

$$R''(x) = -2, \quad \text{a constant.}$$

Thus $R''(100)$ is negative, so $R(100)$ is a maximum.

DO EXERCISE 1.

Note in parts (c) and (e) of both Example 1 and Margin Exercise 1 that the value of x for which $E = 1$ is the same as the value of x for which the total revenue is a maximum. This is always true.

THEOREM 16

Total revenue is a maximum at the value(s) of x for which $E = 1$.

We can prove this as follows. We know that

$$R(x) = x\, D(x),$$

so

$$R'(x) = 1 \cdot D(x) + x\, D'(x)$$
$$= D(x)\left[1 + \frac{x\, D'(x)}{D(x)}\right] \qquad \text{Check this by multiplying.}$$
$$= D(x)\left[1 - \frac{1}{E}\right]$$

$$R'(x) = 0 \quad \text{when } 1 - \frac{1}{E} = 0, \text{ or } E = 1.$$

It helps to look at a typical demand curve in relation to elasticity and total revenue.

The demand curve is decreasing overall. For values of x for which $E > 1$, the total revenue is increasing. For values of x for which $E < 1$, the total revenue is decreasing. For the value of x for which $E = 1$, the total revenue is a maximum.

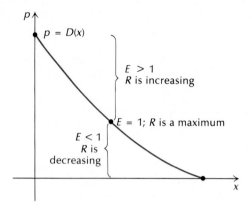

DEFINITION

For a particular value of x:

1. The demand is *inelastic* if $E < 1$;
2. The demand is *elastic* if $E > 1$;
3. The demand has *unit elasticity* if $E = 1$.

In summary, suppose a company puts more units of a product, say a new calculator, on the market and the total revenue increases. Then we say the demand is *elastic*. If the total revenue decreases, we say the demand is *inelastic*.

EXERCISE SET 4.6

For each demand function, find (a) the elasticity and (b) the value(s) of x for which the total revenue is a maximum.

1. $p = D(x) = 400 - x$

2. $p = D(x) = 500 - x$

3. $p = D(x) = 200 - 4x$

4. $p = D(x) = 500 - 2x$

5. $p = D(x) = \dfrac{400}{x}$

6. $p = D(x) = \dfrac{3000}{x}$

7. $p = D(x) = \sqrt{500 - x}$

8. $p = D(x) = \sqrt{300 - x}$

9. $p = D(x) = 100e^{-0.25x}$

10. $p = D(x) = 200e^{-0.05x}$

11. $p = D(x) = \dfrac{100}{(x + 3)^2}$

12. $p = D(x) = \dfrac{300}{(x + 8)^2}$

EXTENSION EXERCISES

13. *Constant elasticity curve*
  a) Find the elasticity of the demand function

$$p = D(x) = \frac{k}{x^n},$$

where k is a positive constant and n is an integer greater than 0.

  b) Is the value of the elasticity dependent on the quantity sold?

  c) Does the total revenue have a maximum? When?

15. Let

$$L(x) = \ln D(x).$$

Describe the elasticity in terms of $L'(x)$.

14. *Exponential demand curve*
  a) Find the elasticity of the demand function

$$p = D(x) = Ae^{-kx},$$

where A and k are positive constants.

  b) Is the value of the elasticity dependent on the quantity sold?

  c) Does the total revenue have a maximum? At what value of x?

SUMMARY AND REVIEW: CHAPTER 4

The following contains a summary of what you should be able to do after completing this chapter. The review exercises are for practice. Answers are at the back of the book. If you miss an exercise, restudy the section indicated alongside the answers.

A summary of the important formulas for this chapter is given on the inside front cover.

You should be able to:

Differentiate exponential and logarithmic functions.

Differentiate.

1. $y = \ln x$ **2.** $y = e^x$ **3.** $y = \ln (x^4 + 5)$ **4.** $y = e^{2\sqrt{x}}$

5. $f(x) = \ln x^6$ **6.** $f(x) = e^{4x} + x^4$ **7.** $f(x) = \dfrac{\ln x}{x^3}$ **8.** $f(x) = e^{x^2} \cdot \ln 4x$

9. $f(x) = e^{4x} - \ln \dfrac{x}{4}$ **10.** $g(x) = x^8 - 8 \ln x$ **11.** $y = \dfrac{\ln e^x}{e^x}$

Given certain logarithmic values, find other logarithmic values using properties of logarithms.

Given $\log_a 2 = 1.8301$ and $\log_a 7 = 5.0999$, find each of the following.

12. $\log_a 14$ **13.** $\log_a \frac{2}{7}$ **14.** $\log_a 28$ **15.** $\log_a 3.5$ **16.** $\log_a \sqrt{7}$ **17.** $\log_a \frac{1}{4}$

State the solution of $dP/dt = kP$ as $P(t) = P_0 e^{kt}$. Given a growth rate, find the doubling time, and vice versa. Solve applied problems involving exponential growth.

18. State the solution to $dQ/dt = kQ$, in terms of Q_0.

19. The population of a certain city doubled in 16 years. What was the growth rate of the city? Round to the nearest tenth of a percent.

20. $8300 is deposited in a savings and loan association where the interest rate is 8% compounded continuously. How long will it take the $8300 to double itself? Round to the nearest tenth of a year.

21. *Business: Cost of a prime-rib dinner.* The average cost C of a prime-rib dinner was $4.65 in 1962. In 1986 it was $15.81. Assuming that the exponential-growth model applies:

   a) Find the exponential-growth rate, and write the equation.

   b) What will the cost of such a dinner be in 1987? in 1997?

22. A clothing firm is selling franchises throughout the United States and Canada. It is estimated that the number of franchises will increase at the rate of 12% per year; that is,

$$\frac{dN}{dt} = 0.12N, \quad \text{where } t \text{ is the time in years.}$$

   a) Find the solution to the equation, assuming that the number of franchises in 1988 ($t = 0$) is 60.

   b) How many franchises will there be in 1992?

   c) When will the number of franchises be 120? Round to the nearest tenth of a year.

State the solution to an equation $dP/dt = -kP$ as $P(t) = P_0 e^{-kt}$. Given a decay rate, find the half-life, and vice versa. Solve applied problems involving decay.

23. The decay rate of a certain radioactive isotope is 13% per year. What is its half-life? Round to the nearest tenth of a year.

24. The half-life of radon-222 is 3.8 days. What is its decay rate? Round to the nearest tenth of a percent.

25. A certain radioactive element has a decay rate of 7% per day; that is,

$$\frac{dA}{dt} = -0.07A,$$

where $A =$ the amount of the element present at time t and $t =$ the time in days.

   a) Find a solution to the equation if the amount of the element present at $t = 0$ is 800 grams.

   b) After 20 days, how much of the 800 grams will remain? Round to the nearest gram.

   c) After what period of time does half the original amount remain?

Differentiate functions involving a^x and $\log_a x$.

Differentiate.

26. $y = 3^x$

27. $f(x) = \log_{15} x$

Given a demand function, find the elasticity and the value(s) of x for which the total revenue is maximized.

28. a) Find the elasticity of the demand function

$$p = D(x) = \frac{600}{(x+4)^2}.$$

b) Find the value of x for which the total revenue is a maximum.

EXTENSION EXERCISES

29. Differentiate $y = \dfrac{e^{2x} + e^{-2x}}{e^{2x} - e^{-2x}}$.

30. Find the minimum value of $f(x) = x^4 \ln 4x$.

TEST: CHAPTER 4

Differentiate.

1. $y = e^x$

2. $y = \ln x$

3. $f(x) = e^{-x^2}$

4. $f(x) = \ln \dfrac{x}{7}$

5. $f(x) = e^x - 5x^3$

6. $f(x) = 3e^x \ln x$

7. $y = \ln (e^x - x^3)$

8. $y = \dfrac{\ln x}{e^x}$

Given $\log_b 2 = 0.2560$ and $\log_b 9 = 0.8114$, find each of the following.

9. $\log_b 18$

10. $\log_b 4.5$

11. $\log_b 3$

12. State the solution to $dM/dt = kM$ in terms of M_0.

13. The doubling time of a certain bacteria culture is 4 hours. What is the growth rate? Round to the nearest tenth of a percent.

14. An investment is made at 6.931% per year compounded continuously. What is the doubling time? Round to the nearest year.

15. The cost C of a gallon of milk was $0.54 in 1941. In 1985 the cost of a gallon of milk was $2.31. Assuming the exponential-growth model applies:

a) Find the exponential-growth rate, and write the equation.

b) Find the cost of a gallon of milk in 1990; in 2000.

16. A dose of a drug is injected into the body of a patient. The drug amount in the body decreases at the rate of 10% per hour; that is,

$$\frac{dA}{dt} = -0.1A,$$

where A = the amount in the body and t = the time in hours.

a) A dose of 3 cubic centimeters (cc) is administered. Assuming $A_0 = 3$ and $k = 0.1$, find the solution to the equation.

b) How much of the initial dose of 3 cc will remain after 10 hours?

c) After what period of time does half the original dose remain?

17. The decay rate of zirconium is 1.1% per day. What is its half-life?

18. The half-life of tellurium is 1,000,000 years. What is its decay rate? As a percent, round to six decimal places.

Differentiate.

19. $f(x) = 20^x$

20. $y = \log_{20} x$

21. Find each of the following.

    a) The elasticity of the demand function

$$p = D(x) = 400e^{-0.2x}$$

    b) The value of x for which the total revenue is a maximum.

EXTENSION EXERCISES

22. Differentiate $y = x\,(\ln x)^2 - 2x \ln x + 2x$.

23. Find the maximum and minimum values of $f(x) = x^4 e^{-x}$ on $[0, 10]$.

5

INTEGRATION

Suppose we do the reverse of differentiating, that is, suppose we try to find a function whose derivative is a given function. This process is called *integration* and it is the main topic of this chapter. We will see that we can use integration to find the area under a curve over a closed interval as well as to find the accumulation of a certain quantity over an interval.

We first consider the meaning of integration and then we learn several techniques for integrating.

AN APPLICATION

Suppose that in a certain memory experiment the rate of memorizing is given by

$$M'(t) = -0.003t^2 + 0.2t,$$

where $M'(t)$ = the memory rate in words per minute. How many words are memorized in the first 10 minutes (from $t = 0$ to $t = 10$)?

THE MATHEMATICS

The number of words memorized in the first 10 minutes is given by

$$\int_0^{10} (-0.003t^2 + 0.2t) \, dt.$$

└── This is an *integral*.

This expression also gives the area under the graph of $M'(t) = -0.003t^2 + 0.2t$ over the interval $[0, 10]$.

5.1

THE ANTIDERIVATIVE

a) Find the indefinite integral (antiderivative) of a given function.

b) Find a function f with a given derivative and function value.

c) Solve applied problems involving antiderivatives.

In Chapters 2, 3, and 4 we considered several interpretations of the derivative, some of which are listed below.

Function	Derivative
Distance	Velocity
Revenue	Marginal revenue
Cost	Marginal cost
Population	Rate of growth of population

For population we actually considered the derivative first and then the function. Many problems can be solved by doing the reverse of differentiation, called *antidifferentiation*.

The Antiderivative

Suppose that y is a function of x and that the derivative is the constant 8. Can we find y? It is easy to see that one such function is $8x$. That is, $8x$ is a function whose derivative is 8. Are there other functions whose derivative is 8? Yes. Here are some examples:

$$8x + 3, \qquad 8x - 10, \qquad 8x + \sqrt{2}.$$

All these functions are $8x$ plus some constant. There are no other functions having a derivative of 8 other than those of the form $8x + C$. Another way of saying this is that any two functions having a derivative of 8 must differ by a constant. In general, any two having the same derivative differ by a constant.

THEOREM 1

If two functions F and G have the same derivative on an interval, then

$$F(x) = G(x) + C, \quad \text{where } C \text{ is a constant.}$$

The reverse of differentiating is *antidifferentiating*, and the result of antidifferentiating is an *antiderivative*. Above, we found antideriva-

Find three antiderivatives.

1. $\dfrac{dy}{dx} = 7$

2. $\dfrac{dy}{dx} = -2$

Find the general form of the antiderivative.

3. x

4. x^3

5. e^x

6. $\dfrac{1}{x}$

Evaluate. Don't forget the constant of integration!

7. $\int x^3\,dx$

8. $\int x\,dx$

9. $\int \dfrac{1}{x}\,dx, x > 0$

tives of the function 8. There are several of them, but they are all $8x$ plus some constant.

Example 1   Antidifferentiate (find the antiderivatives of) x^2.   CSS

Solution   One antiderivative is $x^3/3$. All other antiderivatives differ from this by a constant, so we can denote them as follows:

$$\frac{x^3}{3} + C.$$

This is the *general form* of the antiderivative of x^2.

DO EXERCISES 1–6.

Integrals and Integration

In some contexts, the process of antidifferentiation is called *integration,* and the general form of the antiderivative is referred to as an *indefinite integral*. A common notation for the indefinite integral, or antiderivative, from Leibniz, is

$$\int f(x)\,dx.$$

The symbol $\int$ is called an *integral sign*. The symbol dx plays no apparent role at this point in our development, but will be useful later. In this context, $f(x)$ is called the *integrand*. We illustrate this notation using the preceding example.

Example 2   Evaluate $\int x^2\,dx$.

Solution

$$\int x^2\,dx = \frac{x^3}{3} + C$$

The symbol on the left is read "the integral of x^2, dx." (The "dx" is often omitted in the reading.) In this case, the integrand is x^2. The constant C is called the *constant of integration*.

Example 3   Evaluate $\int e^x\,dx$.

Solution

$$\int e^x\,dx = e^x + C$$

DO EXERCISES 7–9.

To integrate (or antidifferentiate) we make use of differentiation formulas, in effect, reading them in reverse. Below are some of these, stated in reverse, as integration formulas. These can be checked by differentiating the right-hand side and noting that the result is, in each case, the integrand.

THEOREM 2

Basic Integration Formulas

1. $\int k\,dx = kx + C$ (k a constant)

2. $\int x^r\,dx = \dfrac{x^{r+1}}{r+1} + C,$ provided $r \neq -1$

(To integrate a power of x other than -1, increase the power by 1 and divide by the increased power.)

3. $\int x^{-1}\,dx = \int \dfrac{1}{x}\,dx = \ln x + C,$ $x > 0,$

$\int x^{-1}\,dx = \ln |x| + C,$ $x < 0$

(We will generally consider $x > 0$.)

4. $\int e^x\,dx = e^x + C$

The following rules allow us to find many other integrals. They can be obtained by reversing two familiar differentiation rules.

THEOREM 3

Rule A. $\int kf(x)\,dx = k \int f(x)\,dx$
(The integral of a constant times a function is the constant times the integral.)

Rule B. $\int [f(x) + g(x)]\,dx = \int f(x)\,dx + \int g(x)\,dx$
(The integral of a sum is the sum of the integrals.)

Example 4

$$\int (5x + 4x^3)\,dx = \int 5x\,dx + \int 4x^3\,dx \qquad \text{Rule B}$$

$$= 5 \int x\,dx + 4 \int x^3\,dx \qquad \text{Rule A}$$

Then,

$$\int (5x + 4x^3)\,dx = 5 \cdot \frac{x^2}{2} + 4 \cdot \frac{x^4}{4} + C = \frac{5}{2}x^2 + x^4 + C.$$

10. Evaluate. Don't forget the constant of integration!

$$\int (7x^4 + 2x)\, dx$$

Don't forget the constant of integration! It is not necessary to write two constants of integration. If we did, we could add them and consider C as the sum.

Note:

We can always check by differentiating.

Thus, in Example 4,

$$\frac{d}{dx}\left(\frac{5}{2}x^2 + x^4 + C\right) = 2 \cdot \frac{5}{2} \cdot x + 4x^3 = 5x + 4x^3.$$

DO EXERCISE 10.

Example 5

$$\int (e^x - \sqrt{x})\, dx = \int e^x\, dx - \int \sqrt{x}\, dx$$

$$= \int e^x\, dx - \int x^{1/2}\, dx$$

$$= e^x - \frac{x^{(1/2)+1}}{\frac{1}{2}+1} + C = e^x - \frac{x^{3/2}}{\frac{3}{2}} + C$$

$$= e^x - \frac{2}{3}x^{3/2} + C$$

Evaluate. Don't forget the constant of integration.

11. $\int (e^x - x^{2/5})\, dx$

Example 6

$$\int \left(1 - \frac{3}{x} + \frac{1}{x^4}\right) dx = \int 1\, dx - 3\int \frac{dx}{x} + \int x^{-4}\, dx$$

$$= x - 3\ln x + \frac{x^{-4+1}}{-4+1} + C$$

$$= x - 3\ln x - \frac{x^{-3}}{3} + C$$

12. $\int \left(\frac{5}{x} - 7 + \frac{1}{x^6}\right) dx$

DO EXERCISES 11 AND 12.

Another Look at Antiderivatives

The graphs of the antiderivatives of x^2 are the graphs of the functions

$$y = \int x^2\, dx = \frac{x^3}{3} + C$$

for the various values of the constant C.

13. Using the same set of axes, graph $y = \dfrac{x^3}{3}$, $y = \dfrac{x^3}{3} + 1$, and $y = \dfrac{x^3}{3} - 1$.

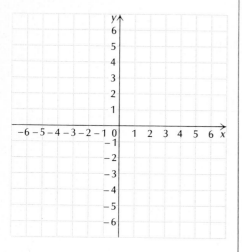

DO EXERCISE 13.

As we can see in the following graphs, x^2 is the derivative of each function. That is, the tangent line at the point

$$\left(a, \frac{a^3}{3} + C \right)$$

has slope a^2. The curves $(x^3/3) + C$ fill up the plane, with exactly one curve going through any given point (x_0, y_0).

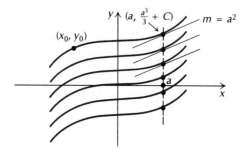

Suppose we look for an antiderivative of x^2 with a specified value at a certain point, say, $f(-1) = 2$. We find that there is only one such function.

Example 7 Find the function f such that

$$f'(x) = x^2 \quad \text{and} \quad f(-1) = 2.$$

Solution

a) We find $f(x)$ by integrating:

$$f(x) = \int x^2 \, dx = \frac{x^3}{3} + C.$$

14. Find f such that

$$f'(x) = x^2 \quad \text{and} \quad f(-2) = 5.$$

b) The condition $f(-1) = 2$ allows us to find C:

$$f(-1) = \frac{(-1)^3}{3} + C = 2,$$

and solving for C we get

$$-\tfrac{1}{3} + C = 2$$
$$C = 2 + \tfrac{1}{3}, \quad \text{or} \quad \tfrac{7}{3}.$$

Thus, $f(x) = (x^3/3) + (7/3)$.

DO EXERCISES 14 AND 15.

15. Find g such that

$$g'(x) = 2x - 4 \quad \text{and} \quad g(2) = 9.$$

Applications

Example 8   A company determines that the marginal cost C' of producing the xth unit of a certain product is given by

$$C'(x) = x^3 + 2x.$$

Find the total-cost function C, assuming that fixed costs (costs when 0 units are produced) are \$45.

Solution

a) We integrate to find $C(x)$, using K for the integration constant to avoid confusion with the cost function C:

$$C(x) = \int C'(x)\, dx = \int (x^3 + 2x)\, dx = \frac{x^4}{4} + x^2 + K.$$

16. A company determines that the marginal cost C' of producing the xth unit of a certain product is given by

$$C'(x) = x^2 + 5x.$$

Find the total-cost function C, assuming fixed costs to be \$35.

b) Fixed costs are \$45. This means that $C(0) = 45$. This allows us to determine the value of K:

$$C(0) = \frac{0^4}{4} + 0^2 + K = 45$$
$$K = 45.$$

Thus, $C(x) = x^4/4 + x^2 + 45$.

DO EXERCISE 16.

Recall that the position coordinate at time t of an object moving along a number line is $s(t)$. Then

$$s'(t) = v(t) = \text{the } \textit{velocity} \text{ at time } t,$$
$$v'(t) = a(t) = \text{the } \textit{acceleration} \text{ at time } t.$$

17. Suppose $v(t) = 4t^3$ and $s(0) = 13$. Find $s(t)$.

Example 9 Suppose $v(t) = 5t^4$ and $s(0) = 9$. Find $s(t)$.

Solution

a) We find $s(t)$ by integrating:

$$s(t) = \int v(t)\, dt$$
$$= \int 5t^4\, dt = t^5 + C.$$

b) The condition $s(0) = 9$ allows us to determine C:

$$s(0) = 0^5 + C = 9$$
$$C = 9.$$

Thus, $s(t) = t^5 + 9$.

DO EXERCISE 17.

Example 10 Suppose $a(t) = 12t^2 - 6$, $v(0) = $ initial velocity $= 5$, and $s(0) = 10$. Find $s(t)$.

18. Suppose $a(t) = 24t^2 - 12$, $v(0) = 7$, and $s(0) = 8$. Find $s(t)$.

Solution

a) We find $v(t)$ by integrating $a(t)$:

$$v(t) = \int a(t)\, dt = \int (12t^2 - 6)\, dt = 4t^3 - 6t + C_1.$$

b) The condition $v(0) = 5$ allows us to find C_1:

$$v(0) = 4 \cdot 0^3 - 6 \cdot 0 + C_1 = 5$$
$$C_1 = 5.$$

Thus, $v(t) = 4t^3 - 6t + 5$.

c) We find $s(t)$ by integrating $v(t)$:

$$s(t) = \int v(t)\, dt = \int (4t^3 - 6t + 5)\, dt = t^4 - 3t^2 + 5t + C_2.$$

d) The condition $s(0) = 10$ allows us to find C_2:

$$s(0) = 0^4 - 3 \cdot 0^2 + 5 \cdot 0 + C_2 = 10$$
$$C_2 = 10.$$

Thus, $s(t) = t^4 - 3t^2 + 5t + 10$.

DO EXERCISE 18.

EXERCISE SET 5.1

Evaluate.

1. $\int x^6 \, dx$

2. $\int x^7 \, dx$

3. $\int 2 \, dx$

4. $\int 4 \, dx$

5. $\int x^{1/4} \, dx$

6. $\int x^{1/3} \, dx$

7. $\int (x^2 + x - 1) \, dx$

8. $\int (x^2 - x + 2) \, dx$

9. $\int (t^2 - 2t + 3) \, dt$

10. $\int (3t^2 - 4t + 7) \, dt$

11. $\int 5e^x \, dx$

12. $\int 3e^x \, dx$

13. $\int (x^3 - x^{8/7}) \, dx$

14. $\int (x^4 - x^{6/5}) \, dx$

15. $\int \frac{1000}{x} \, dx$

16. $\int \frac{500}{x} \, dx$

17. $\int \frac{dx}{x^2} \left(\text{or} \int \frac{1}{x^2} \, dx \right)$

18. $\int \frac{dx}{x^3}$

Find f such that:

19. $f'(x) = x - 3, f(2) = 9$

20. $f'(x) = x - 5, f(1) = 6$

21. $f'(x) = x^2 - 4, f(0) = 7$

22. $f'(x) = x^2 + 1, f(0) = 8$

APPLICATIONS

23. A company determines that the marginal cost C' of producing the xth unit of a certain product is given by

$$C'(x) = x^3 - 2x.$$

Find the total-cost function C, assuming that fixed costs are $100.

24. A company determines that the marginal cost C' of producing the xth unit of a certain product is given by

$$C'(x) = x^3 - x.$$

Find the total-cost function C, assuming that fixed costs are $200.

25. A company determines that the marginal revenue R' from selling the xth unit of a certain product is given by

$$R'(x) = x^2 - 3.$$

a) Find the total-revenue function R, assuming $R(0) = 0$.

b) Why is $R(0) = 0$ a reasonable assumption?

26. A company determines that the marginal revenue R' from selling the xth unit of a certain product is given by

$$R'(x) = x^2 - 1.$$

a) Find the total-revenue function R, assuming $R(0) = 0$.

b) Why is $R(0) = 0$ a reasonable assumption?

Find $s(t)$.

27. $v(t) = 3t^2, s(0) = 4$

28. $v(t) = 2t, s(0) = 10$

Find $v(t)$.

29. $a(t) = 4t, v(0) = 20$

30. $a(t) = 6t, v(0) = 30$

Find $s(t)$.

31. $a(t) = -2t + 6$, $v(0) = 6$, $s(0) = 10$

32. $a(t) = -6t + 7$, $v(0) = 10$, $s(0) = 20$

33. For a freely falling object, $a(t) = -32$ ft/sec², $v(0) = $ initial velocity $= v_0$, and $s(0) = $ initial height $= s_0$. Find a general expression for $s(t)$ in terms of v_0 and s_0.

34. A ball is thrown from a height of 10 ft, $s(0) = 10$, at an initial velocity of 80 ft/sec, $v(0) = 80$. How long will it take the ball to hit the ground? (See Exercise 33.)

35. A car with constant acceleration goes from 0 to 60 mph in $\frac{1}{2}$ minute. How far does the car travel during that time?

36. *Efficiency of a machine operator.* The rate at which a machine operator's efficiency E (expressed as a percentage) changes with respect to time t is given by

$$\frac{dE}{dt} = 40 - 10t,$$

where $t = $ the number of hours that the operator has been at work.

a) Find $E(t)$, given that the operator's efficiency after working 2 hr is 72%; that is, $E(2) = 72$.

b) Use the answer to part (a) to find the operator's efficiency after 4 hr; after 8 hr.

37. *Efficiency of a machine operator.* The rate at which a machine operator's efficiency E (expressed as a percentage) changes with respect to time t is given by

$$\frac{dE}{dt} = 30 - 10t,$$

where $t = $ the number of hours that the operator has been at work.

a) Find $E(t)$, given that the operator's efficiency after working 2 hr is 72%; that is, $E(2) = 72$.

b) Use the answer to part (a) to find the operator's efficiency after 3 hr; after 5 hr.

38. *Psychology: Memory.* In a certain memory experiment the rate of memorizing is given by

$$M'(t) = 0.2t - 0.003t^2,$$

where $M(t)$ is the number of Spanish words memorized in t minutes.

a) Find $M(t)$ if it is known that $M(0) = 0$.

b) How many words are memorized in 8 minutes?

39. *Biomedical.* The area A of a healing wound is decreasing at the rate given by

$$A'(t) = -43.4t^{-2}, \qquad 1 \leq t \leq 7,$$

where t is the time in days and A is in square centimeters.

a) Find $A(t)$ if $A(1) = 39.7$.

b) Find the area of the wound after 7 days.

A machine operator's efficiency changes with respect to time.

EXTENSION EXERCISES

Find f.

40. $f'(t) = \sqrt{t} + \dfrac{1}{\sqrt{t}}$, $f(4) = 0$

41. $f'(t) = t^{\sqrt{3}}$, $f(0) = 8$

Evaluate. Each of the following can be integrated using the rules developed in this section, but some algebra may be required beforehand.

42. $\int (5t + 4)^2 \, dt$

43. $\int (x - 1)^2 x^3 \, dx$

44. $\int (1 - t) \sqrt{t} \, dt$

45. $\int \frac{(t + 3)^2}{\sqrt{t}} \, dt$

46. $\int \frac{x^4 - 6x^2 - 7}{x^3} \, dx$

47. $\int (t + 1)^3 \, dt$

48. $\int \frac{1}{\ln 10} \frac{dx}{x}$

49. $\int be^{ax} \, dx$

50. $\int (3x - 5)(2x + 1) \, dx$

51. $\int \sqrt[3]{64x^4} \, dx$

52. $\int \frac{x^2 - 1}{x + 1} \, dx$

53. $\int \frac{t^3 + 8}{t + 2} \, dt$

5.2

OBJECTIVES

a) **Find the area under a curve on a given closed interval.**

b) **Interpret the area under a curve in two other ways.**

AREA

We now consider the application of integration to finding areas of certain regions. Consider a function whose outputs are positive in an interval (the function might be 0 at one of the endpoints). We wish to find the area of the region between the graph of the function and the x-axis on that interval.

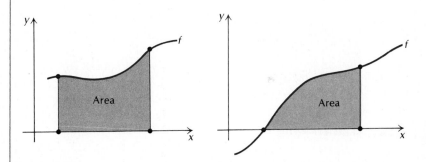

Let us first consider a constant function $f(x) = m$ on the interval from 0 to x, $[0, x]$.

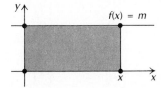

1. Consider the constant function $f(x) = 3$.

a) Find $A(x)$.
b) Find $A(1), A(2)$, and $A(5)$.
c) Graph $A(x)$.
d) How do $f(x)$ and $A(x)$ compare?

2. A clothing firm, Raggs, Ltd., determines that the marginal cost of each suit it produces is $50.

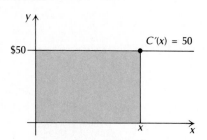

a) Find the total cost $C(x)$ of producing x suits, assuming that fixed costs are $0 (that is, ignore fixed costs).
b) Find the area of the shaded rectangle. Compare your answer with (a).
c) Graph $C(x)$. Why is this an increasing function?

The figure formed is a rectangle, and its area is mx. Suppose we allow x to vary, giving us rectangles of different areas. The area of each rectangle is still mx. We now have an area *function:*

$$A(x) = mx.$$

Its graph is shown below.

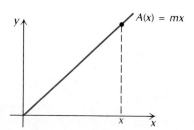

DO EXERCISES 1–3. (EXERCISE 3 IS ON THE FOLLOWING PAGE.)

Let us consider next the linear function $f(x) = mx$ on the interval from 0 to x, $[0, x]$.

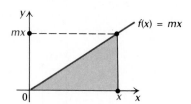

The figure formed this time is a triangle, and its area is $\frac{1}{2}$ the base times the height, $\frac{1}{2} \cdot x \cdot (mx)$, or $\frac{1}{2}mx^2$. If we allow x to vary, we again get an area function:

$$A(x) = \tfrac{1}{2}mx^2.$$

Its graph is as shown below.

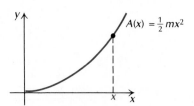

DO EXERCISE 4 ON THE FOLLOWING PAGE.

3. With better management Raggs, Ltd., of Margin Exercise 2, is able to decrease its production costs by $10 per suit for every hundred suits it produces. This is shown below.

a) Find the total cost of producing 400 suits.

b) Find the total area of the rectangles. Compare your answer with (a).

4. Consider the function $f(x) = 3x$.

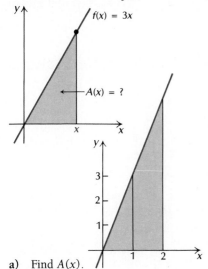

a) Find $A(x)$.

b) Find $A(1)$.

c) Find $A(2)$.

d) Find $A(3.5)$.

e) Graph $A(x)$.

f) How do $f(x)$ and $A(x)$ compare?

Now consider the linear function $f(x) = mx + b$ on the interval from 0 to x, $[0, x]$.

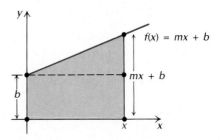

The figure formed this time is a trapezoid, and its area is $\frac{1}{2}$ the height times the sum of the lengths of its parallel sides (or, noting the dashed line, the area of the triangle plus the area of the rectangle):

$$\frac{1}{2} \cdot x \cdot [b + (mx + b)],$$

or

$$\frac{1}{2} \cdot x \cdot (mx + 2b),$$

or

$$\frac{1}{2}mx^2 + bx.$$

If we allow x to vary, we again get an area function:

$$A(x) = \frac{1}{2}mx^2 + bx.$$

Its graph is as shown below.

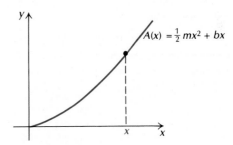

DO EXERCISE 5 ON THE FOLLOWING PAGE.

Now we consider the function $f(x) = x^2 + 1$ on the interval from 0 to x, $[0, x]$. The graph of the region in question is as shown, but it is not so easy this time to find the area function because the graph of

5. Raggs, Ltd., of Margin Exercise 3, installs new sewing machines. This allows the marginal cost per suit to decrease continually in such a way that

$$C'(x) = -0.1x + 50.$$

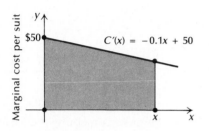

a) Find the total cost of producing x suits, ignoring fixed costs.

b) Find the area of the shaded trapezoid.

c) Find the total cost of producing 400 suits. Compare this answer with that of Margin Exercise 3.

$f(x)$ is not a straight line. Let us tabulate our previous results and look for a pattern.

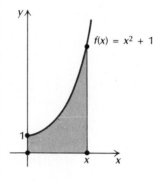

$f(x)$	$A(x)$
$f(x) = 3$	$A(x) = 3x$
$f(x) = m$	$A(x) = mx$
$f(x) = 3x$	$A(x) = \frac{3}{2}x^2$
$f(x) = mx$	$A(x) = \frac{1}{2}mx^2$
$f(x) = mx + b$	$A(x) = \frac{1}{2}mx^2 + bx$

You may have conjectured that the area function $A(x)$ is an antiderivative of $f(x)$. In the following exploratory exercises you will investigate further.

EXPLORATORY EXERCISES: FINDING AREAS

1. The region under the graph of $f(x) = x^2 + 1$, on the interval $[0, 2]$, is shown here.

a) Make a copy of the shaded region on thin paper.

b) Cut up the shaded region in any way you wish in order to fill up squares in the grid below. Make an estimate of the total area.

c) Using the antiderivative

$$F(x) = \frac{x^3}{3} + x,$$

find $F(2)$.

2. Repeat Exercises 1(a) and (b) for the shaded region of the graph shown here.

c)   Using the antiderivative

$$F(x) = \frac{x^3}{3} + x,$$

find $F(3)$.

d)   Compare your answers with parts (b) and (c).

The conjecture concerning areas and antiderivatives (or integrals) is true. It is expressed as follows.

THEOREM 4

Let f be a positive, continuous function on an interval $[a, b]$, and let $A(x)$ be the area of the region between the graph of f and the x-axis on the interval $[a, x]$. Then $A(x)$ is a differentiable function of x and

$$A'(x) = f(x).$$

Proof.   The situation described in the theorem is shown here.

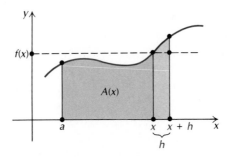

The derivative of $A(x)$ is, by definition of a derivative,

$$A'(x) = \lim_{h \to 0} \frac{A(x + h) - A(x)}{h}.$$

Note from the figure that $A(x + h) - A(x)$ is the area of the small, shaded, vertical strip. The area of this small strip is approximately that of a rectangle of base h and height $f(x)$, especially for small values of h. Thus we have

$$A(x + h) - A(x) \approx f(x) \cdot h.$$

Now

$$A'(x) = \lim_{h \to 0} \frac{A(x + h) - A(x)}{h} = \lim_{h \to 0} \frac{f(x) \cdot h}{h} = \lim_{h \to 0} f(x) = f(x),$$

since $f(x)$ does not involve h.

The theorem above also holds if $f(x) = 0$ at one or both endpoints of the interval $[a, b]$.

Since the area function A is an antiderivative of f, and since any two antiderivatives differ by a constant, we easily conclude that the area function and any antiderivative differ by a constant.

We can think of the function A as given by

$$A(x) = \text{the area on the interval } [a, x],$$

where a is some fixed point and x varies.

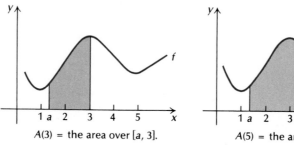

A(3) = the area over [a, 3].

A(5) = the area over [a, 5].

Now let us find some areas.

Example 1   Find the area under the graph of $y = x^2 + 1$ on the interval $[-1, 2]$.

Solution

a) We first make a drawing. This includes a graph of the function as well as the region in question.

b) Next we make a drawing showing a portion of the region from -1 to x. Now $A(x)$ is the area of this portion, that is, in the interval $[-1, x]$.

6. Find the area under the graph of $y = x^2 + 3$ on the interval [1, 2].

 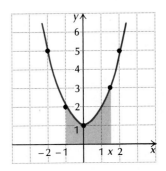

c)   Now

$$A(x) = \int (x^2 + 1)\, dx = \frac{x^3}{3} + x + C,$$

where C must be determined. Since we know that $A(-1) = 0$ (there is no area above the number -1), we can substitute for x in $A(x)$, as follows:

$$A(-1) = \frac{(-1)^3}{3} + (-1) + C = 0$$

$$-\frac{1}{3} - 1 + C = 0$$

$$C = \frac{4}{3}.$$

This determines that $C = \frac{4}{3}$, so we have

$$A(x) = \frac{x^3}{3} + x + \frac{4}{3}.$$

Then the area in the interval $[-1, 2]$ is $A(2)$. We compute $A(2)$ as follows:

$$A(2) = \frac{2^3}{3} + 2 + \frac{4}{3}$$

$$= \frac{8}{3} + 2 + \frac{4}{3}$$

$$= \frac{12}{3} + 2$$

$$= 6.$$

DO EXERCISE 6.

Example 2   Find the area under the graph of $y = x^3$ on the interval [0, 5].

7. Find the area under the graph of
   $y = x^2 + x$ on the interval $[0, 3]$.

Solution

a) We first make a drawing that includes a graph of the function
    and the region in question.

 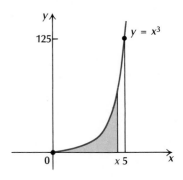

b) Next we make a drawing showing a portion of that region from
    0 to x. Now $A(x)$ is the area of this portion, that is, on the interval
    $[0, x]$.

c) Now

$$A(x) = \int x^3 \, dx = \frac{x^4}{4} + C,$$

where C must be determined. Since we know that $A(0) = 0$, we
can substitute 0 for x in $A(x)$, as follows:

$$A(0) = \frac{0^4}{4} + C = 0$$

$$C = 0.$$

This determines C. Thus,

$$A(x) = \frac{x^4}{4}.$$

Then the area in the interval $[0, 5]$ is $A(5)$. We can compute $A(5)$
as follows:

$$A(5) = \frac{5^4}{4} = \frac{625}{4} = 156\tfrac{1}{4}.$$

DO EXERCISE 7.

Since the area under a curve is an antiderivative, area can also be associated with various kinds of functions. For example, if we have a velocity function over an interval $[0, b]$, then the area under the curve in that interval is the total distance. Suppose the velocity function is

$$v(t) = t^3.$$

In 5 hours the total distance covered is $156\frac{1}{4}$. We can see this in Example 2, simply by changing the variable from x to t.

For a marginal-cost function over the interval $[0, x]$, the area under the curve is the total cost of producing x units, or the accumulated cost.

Example 3   Raggs, Ltd., goes even further to reduce production costs. In addition to purchasing new sewing machines, the president has air conditioning installed and takes a calculus course. These cause the marginal cost per suit to decrease rapidly in such a way that

$$C'(x) = 0.0003x^2 - 0.2x + 50.$$

Find the total cost of producing 400 suits (ignore fixed costs).

Solution

a)   First we make a drawing that includes a graph of the function and the region in question.

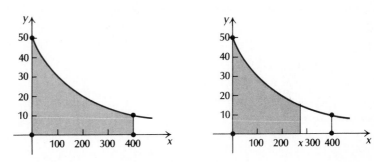

b)   Next we make a drawing showing a portion of that region from 0 to x. Now $A(x)$ is the area of that portion, that is, on the interval $[0, x]$.

c)   Now,

$$C(x) = A(x) = \int (0.0003x^2 - 0.2x + 50)\, dx$$
$$= 0.0001x^3 - 0.1x^2 + 50x + K,$$

8. Refer to Example 3.

    a) Compare $10,400 to your answer for Margin Exercise 5. Has the company reduced total costs?

    b) Find the total cost of producing 100 suits.

where K must be determined. We are ignoring fixed costs, $K = 0$, so we have

$$C(x) = 0.0001x^3 - 0.1x^2 + 50x.$$

Then the area in the interval $[0, 400]$ is $A(400)$, or $C(400)$. We can compute $C(400)$ as follows:

$$C(400) = 0.0001 \cdot 400^3 - 0.1 \cdot 400^2 + 50 \cdot 400, \quad \text{or} \quad \$10,400.$$

DO EXERCISE 8.

EXERCISE SET 5.2

Find the area under the given curve on the interval indicated.

1. $y = 4$; $[1, 3]$

2. $y = 5$; $[1, 3]$

3. $y = 2x$; $[1, 3]$

4. $y = x^2$; $[0, 3]$

5. $y = x^2$; $[0, 5]$

6. $y = x^3$; $[0, 2]$

7. $y = x^3$; $[0, 1]$

8. $y = 1 - x^2$; $[-1, 1]$

9. $y = 4 - x^2$; $[-2, 2]$

10. $y = e^x$; $[0, 2]$

11. $y = e^x$; $[0, 3]$

12. $y = \dfrac{1}{x}$; $[1, 2]$

13. $y = \dfrac{1}{x}$; $[1, 3]$

14. $y = x^2 - 4x$; $[-4, -2]$

15. $y = x^2 - 4x$; $[-4, -1]$

In each case, give two interpretations of the shaded region.

16.

17.

18.

19.

20.

21.

22.

23.

24.

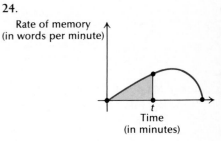

25. A particle starts out from the origin. Its velocity at time t is given by

$$v(t) = 3t^2 + 2t.$$

a) Find the distance the particle has traveled after t hours.

b) Find the distance the particle has traveled after 5 hours.

27. A sound company determines that the marginal cost of producing the xth stereo is given by

$$C'(x) = 100 - 0.2x, \qquad C(0) = 0.$$

It also determines that its marginal revenue from the sale of the xth stereo is given by

$$R'(x) = 100 + 0.2x, \qquad R(0) = 0.$$

a) Find the total cost of producing x stereos.

b) Find the total revenue from selling x stereos.

c) Find the total profit from the production and sale of x stereos.

d) Find the total profit from the production and sale of 1000 stereos.

26. A particle starts out from the origin. Its velocity at time t is given by

$$v(t) = 4t^3 + 2t.$$

a) Find the distance the particle has traveled after t hours.

b) Find the distance the particle has traveled after 2 hours.

28. A refrigeration company determines that the marginal cost of producing the xth refrigerator is given by

$$C'(x) = 50 - 0.4x, \qquad C(0) = 0.$$

It also determines that its marginal revenue from the sale of the xth refrigerator is given by

$$R'(x) = 50 + 0.4x, \qquad R(0) = 0.$$

a) Find the total cost of producing x refrigerators.

b) Find the total revenue from selling x refrigerators.

c) Find the total profit from the production and sale of x refrigerators.

d) Find the total profit from the production and sale of 1000 refrigerators.

EXTENSION EXERCISES

Find the area under the curve on the interval indicated.

29. $y = \dfrac{x^2 - 1}{x - 1}$; [2, 3]

30. $y = \dfrac{x^5 - x^{-1}}{x^2}$; [1, 5]

31. $y = (x - 1)\sqrt{x}$; [4, 16]

32. $y = (x + 2)^3$; [0, 1]

33. $y = \dfrac{\sqrt[3]{x^2} - 1}{\sqrt[3]{x}}$; [1, 8]

34. $y = \dfrac{x^3 + 8}{x + 2}$; [0, 1]

5.3

OBJECTIVES

a) Evaluate a definite integral.

b) Find the area under a graph on [a, b].

c) Solve applied problems involving definite integrals.

INTEGRATION ON AN INTERVAL: THE DEFINITE INTEGRAL

Let f be a positive, continuous function on an interval $[a, b]$. We know that f has an antiderivative, namely, $A(x)$. Let F and G be any two antiderivatives of f. Then

$$F(b) - F(a) = G(b) - G(a).$$

To understand this, recall that F and G differ by a constant; that is, $F(x) = G(x) + C$. Then

$$F(b) - F(a) = [G(b) + C] - [G(a) + C] = G(b) - G(a).$$

Thus the difference $F(b) - F(a)$ has the same value for all antiderivatives of f. It is called the *definite integral* of f from a to b.

Definite integrals are generally symbolized as follows:

$$\int_a^b f(x) \, dx.$$

This is read "the integral from a to b of $f(x)$ dx" (the "dx" is sometimes omitted from the reading). From the preceding development we see that to find a definite integral $\int_a^b f(x) \, dx$, we first find an antiderivative $F(x)$. The simplest one is the one for which the constant of integration is 0. We evaluate F at b and at a and subtract.

DEFINITION

Let f be any positive, continuous function on the interval $[a, b]$, and let F be any antiderivative of f. Then

$$\int_a^b f(x) \, dx \quad \text{is defined as} \quad F(b) - F(a).$$

Evaluating definite integrals is called *integrating*. The numbers a and b are known as the *limits of integration*.

Example 1 Evaluate $\int_a^b x^2 \, dx$.

Solution Using the antiderivative $F(x) = x^3/3$, we have

$$\int_a^b x^2 \, dx = \frac{b^3}{3} - \frac{a^3}{3}.$$

Evaluate.

1. $\int_a^b 2x\,dx$

2. $\int_a^b e^x\,dx$

DO EXERCISES 1 AND 2.

It is convenient to use an intermediate notation:

$$\int_a^b f(x)\,dx = [F(x)]_a^b$$
$$= F(b) - F(a).$$

We now evaluate several definite integrals.

Example 2

$$\int_{-1}^2 x^2\,dx = \left[\frac{x^3}{3}\right]_{-1}^2 = \frac{2^3}{3} - \frac{(-1)^3}{3}$$
$$= \frac{8}{3} - \left(-\frac{1}{3}\right) = \frac{8}{3} + \frac{1}{3} = 3$$

Example 3

$$\int_0^3 e^x\,dx = [e^x]_0^3 = e^3 - e^0 = e^3 - 1$$

Example 4

$$\int_1^4 (x^2 - x)\,dx = \left[\frac{x^3}{3} - \frac{x^2}{2}\right]_1^4 = \left(\frac{4^3}{3} - \frac{4^2}{2}\right) - \left(\frac{1^3}{3} - \frac{1^2}{2}\right)$$
$$= \left(\frac{64}{3} - \frac{16}{2}\right) - \left(\frac{1}{3} - \frac{1}{2}\right)$$
$$= \frac{64}{3} - 8 - \frac{1}{3} + \frac{1}{2} = 13\frac{1}{2}$$

Example 5

$$\int_1^e \left(1 + 2x - \frac{1}{x}\right)dx = [x + x^2 - \ln x]_1^e$$
$$= (e + e^2 - \ln e) - (1 + 1^2 - \ln 1)$$
$$= (e + e^2 - 1) - (1 + 1 - 0)$$
$$= e + e^2 - 1 - 1 - 1$$
$$= e + e^2 - 3$$

For some applications that we will soon consider it is important to note that in $\int_a^b f(x)\,dx$, $a < b$. That is, the larger number is on the top!

Evaluate.

3. $\int_1^3 2x\, dx$

4. $\int_{-2}^0 e^x\, dx$

5. $\int_0^1 (2x - x^2)\, dx$

6. $\int_1^e \left(1 + 3x^2 - \dfrac{1}{x}\right) dx$

DO EXERCISES 3–6.

The area under a curve can be expressed by a definite integral.

THEOREM 5

Let f be a positive, continuous function over the closed interval $[a, b]$. The area under the graph of f on the interval $[a, b]$ is

$$\int_a^b f(x)\, dx.$$

Proof

Let

$$A(x) = \text{the area of the region over } [0, x].$$

Then

$$A'(x) = f(x),$$

so $A(x)$ is an antiderivative of $f(x)$. Then

$$\int_a^b f(x)\, dx = A(b) - A(a).$$

But $A(b) - A(a)$ is the area over $[0, b]$ minus the area over $[0, a]$, which is the area over $[a, b]$.

Let us now find some areas.

Example 6   Find the area under $y = x^2 + 1$ on $[-1, 2]$.

Solution

$$\int_{-1}^2 (x^2 + 1)\, dx = \left[\frac{x^3}{3} + x\right]_{-1}^2$$

$$= \left(\frac{2^3}{3} + 2\right) - \left(\frac{(-1)^3}{3} + (-1)\right)$$

$$= \left(\frac{8}{3} + 2\right) - \left(-\frac{1}{3} - 1\right)$$

$$= \frac{8}{3} + 2 + \frac{1}{3} + 1 = 6$$

7. Find the area under $y = x^2 + 3$ on [1, 2]. Don't forget that it helps to draw the graph. (Compare the result with Margin Exercise 6 of Section 5.2.)

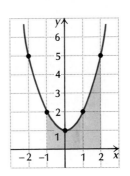

Compare this with Example 1 in Section 5.2.

DO EXERCISE 7.

Example 7 Find the area under $y = x^3$ on [0, 5].

Solution

$$\int_0^5 x^3\, dx = \left[\frac{x^4}{4}\right]_0^5 = \frac{5^4}{4} - \frac{0^4}{4} = \frac{625}{4}$$

8. Find the area under $y = x^2 + x$ on [0, 3]. (Compare the result with Margin Exercise 7 of Section 5.2.)

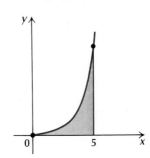

Compare this with Example 2 in Section 5.2.

DO EXERCISE 8.

Example 8 Find the area under $y = 1/x$ on [1, 4].

Solution

$$\int_1^4 \frac{dx}{x} = [\ln x]_1^4 = \ln 4 - \ln 1$$

$$= \ln 4 \approx 1.3863 \quad \blacksquare$$

9. Find the area under $y = 1/x$ on $[1, 7]$.

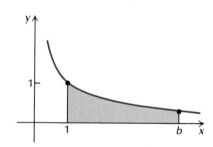

Example 9   Find the area under $y = 1/x^2$ on $[1, b]$.

Solution

$$\int_1^b \frac{dx}{x^2} = \int_1^b x^{-2}\, dx = \left[\frac{x^{-2+1}}{-2+1} \right]_1^b$$

$$= \left[\frac{x^{-1}}{-1} \right]_1^b = \left[-\frac{1}{x} \right]_1^b = \left(-\frac{1}{b} \right) - \left(-\frac{1}{1} \right)$$

$$= 1 - \frac{1}{b}$$

10. Find the area under $y = 1/x^4$ on $[1, b]$. $\boxed{\text{CSS}}$

DO EXERCISES 9 AND 10.

The following properties of definite integrals can be derived rather easily from the definition of a definite integral and from the properties of the indefinite integral.

PROPERTY 1

$$\int_a^b k \cdot f(x)\, dx = k \cdot \int_a^b f(x)\, dx$$

The integral of a constant times a function is the constant times the integral of the function. That is, we can "factor out" a constant from the integrand.

---

11. Evaluate $\int_1^2 20x^3 \, dx$.

Example 10

$$\int_0^5 100e^x \, dx = 100 \int_0^5 e^x \, dx$$
$$= 100[e^x]_0^5$$
$$= 100(e^5 - e^0)$$
$$= 100(e^5 - 1)$$

DO EXERCISE 11.

PROPERTY 2

$$\int_a^b [f(x) + g(x)] \, dx = \int_a^b f(x) \, dx + \int_a^b g(x) \, dx$$

The integral of a sum is the sum of the integrals.

PROPERTY 3

For $a < c < b$,

$$\int_a^b f(x) \, dx = \int_a^c f(x) \, dx + \int_c^b f(x) \, dx$$

For any number c between a and b, the integral from a to b is the integral from a to c plus the integral from c to b.

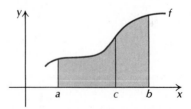

Property 3 has particular application when a function is defined in different ways over subintervals.

Example 11   Find the area under the graph of $y = f(x)$ from -4 to 5, where

$$f(x) = \begin{cases} 9, & \text{if } x < 3, \\ x^2, & \text{if } x \geqslant 3. \end{cases}$$

12. Find the area under the graph of $y = f(x)$ from -3 to 2, where

$$f(x) = \begin{cases} 4, & \text{if } x < 0, \\ 4 - x^2, & \text{if } x \geqslant 0. \end{cases}$$

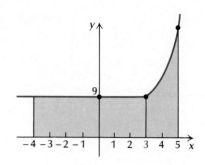

Solution

$$\int_{-4}^{5} f(x)\, dx = \int_{-4}^{} f(x)\, dx + \int^{5} f(x)\, dx$$

$$= \int_{-4}^{3} 9\, dx + \int_{3}^{5} x^2\, dx$$

$$= 9 \int_{-4}^{3} dx + \int_{3}^{5} x^2\, dx$$

$$= 9[x]^{3}_{-4} + \left[\frac{x^3}{3}\right]^{5}_{3}$$

$$= 9[3 - (-4)] + \left(\frac{5^3}{3}\right) - \left(\frac{3^3}{3}\right)$$

$$= 95\tfrac{2}{3}$$

DO EXERCISE 12.

An Application

Example 12 *Business: Accumulated sales.* The sales of a company are expected to grow continuously at a rate given by the function

$$S'(t) = 100e^t,$$

where $S'(t) =$ the sales rate in dollars per day at time t in days.

a) Find the accumulated sales for the first 7 days.

b) On what day will accumulated sales exceed $810,000?

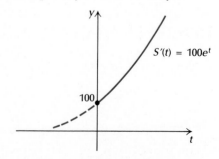

13. The sales of a company are expected to grow continuously at a rate given by the function

$$S'(t) = 200e^t,$$

where $S'(t)$ = the sales rate, in dollars per day, at time t, in days.

a) Find the accumulated sales for the first 8 days.

b) On what day will accumulated sales exceed $300,000?

c) The accumulated sales from the 8th through the 10th day is given by the integral from 7 to 10,

$$\int_7^{10} S'(t) \, dt.$$

Find this.

Solution

a) Accumulated sales through day 7 are

$$\int_0^7 S'(t) \, dt = \int_0^7 100e^t \, dt = 100 \int_0^7 e^t \, dt$$
$$= 100[e^t]_0^7 = 100(e^7 - e^0)$$
$$= 100(1096.633158 - 1)$$
$$= 100(1095.633158)$$
$$\approx \$109,563.32.$$

b) Accumulated sales through day k are

$$\int_0^k S'(t) \, dt = \int_0^k 100e^t \, dt$$
$$= 100 \int_0^k e^t \, dt$$
$$= 100[e^t]_0^k = 100(e^k - e^0)$$
$$= 100(e^k - 1).$$

We set this equal to $810,000 and solve for k:

$$100(e^k - 1) = 810{,}000$$
$$e^k - 1 = 8100$$
$$e^k = 8101.$$

We solve this equation for k using natural logarithms:

$$e^k = 8101$$
$$\ln e^k = \ln 8101$$
$$k = 8.999743 \qquad \blacksquare$$
$$k \approx 9.$$

Accumulated sales will exceed $810,000 approximately on day 9.

DO EXERCISE 13.

EXERCISE SET 5.3

Evaluate.

CSS

1. $\int_0^1 (x - x^2) \, dx$

2. $\int_1^2 (x^2 - x) \, dx$

3. $\int_{-1}^1 (x^2 - x^4) \, dx$

4. $\int_0^b e^x \, dx$

5. $\int_a^b e^t \, dt$

6. $\int_0^a (ax - x^2) \, dx$

7. $\int_a^b 3t^2 \, dt$

8. $\int_a^b 4t^3 \, dt$

9. $\int_1^e \left(x + \frac{1}{x}\right) dx$

10. $\int_1^e \left(x - \frac{1}{x}\right) dx$

11. $\int_0^1 \sqrt{x} \, dx$

12. $\int_0^1 3\sqrt{x} \, dx$

13. $\int_0^1 \frac{10}{17} t^3 \, dt$

14. $\int_0^1 \frac{12}{13} t^2 \, dt$

Find the area under the graph on the interval indicated.

15. $y = x^3$; [0, 2]

16. $y = x^4$; [0, 1]

17. $y = x^2 + x + 1$; [2, 3]

18. $y = 2 - x - x^2$; [-2, 1]

19. $y = 5 - x^2$; [-1, 2]

20. $y = e^x$; [-2, 3]

21. $y = e^x$; [-1, 5]

22. $y = 2x + \frac{1}{x^2}$; [1, 4]

23. $y = 2x - \frac{1}{x^2}$; [1, 3]

Find the area under the graph on [-2, 3], where:

24. $f(x) = \begin{cases} x^2, & \text{if } x < 1, \\ 1, & \text{if } x \geqslant 1. \end{cases}$

25. $f(x) = \begin{cases} 4 - x^2, & \text{if } x < 0, \\ 4, & \text{if } x \geqslant 0. \end{cases}$

APPLICATIONS

26. *Business: Accumulated sales.* Raggs, Ltd., estimates that its sales will grow continuously at a rate given by the function

$$S'(t) = 10e^t,$$

where $S'(t)$ = the sales rate, in dollars per day, at time t, in days.

a) Find the accumulated sales for the first 5 days.

b) Find the sales from the 2nd day through the 5th day. This is the integral from 1 to 5.

c) On what day will accumulated sales exceed $40,000?

27. *Business: Accumulated sales.* A company estimates that its sales will grow continuously at a rate given by the function

$$S'(t) = 20e^t,$$

where $S'(t)$ = the sales rate, in dollars per day, at time t, in days.

a) Find the accumulated sales for the first 5 days.

b) Find the sales from the 2nd day through the 5th day. This is the integral from 1 to 5.

c) On what day will accumulated sales exceed $20,000?

28. A particle starts out from the origin. Its velocity at time t is given by

$$v(t) = 3t^2 + 2t.$$

How far does it travel from the 2nd hour through the 5th hour (from $t = 1$ to $t = 5$)?

29. A particle starts out from the origin. Its velocity at time t is given by

$$v(t) = 4t^3 + 2t.$$

How far does it travel from the start through the 3rd hour (from $t = 0$ to $t = 3$)?

30. *Business.* Raggs, Ltd., determines that the marginal cost per suit is given by

$$C'(x) = 0.0003x^2 - 0.2x + 50.$$

Ignoring fixed costs, find the total cost of producing the 101st suit through the 400th suit (integrate from $x = 100$ to $x = 400$).

31. *Business.* In Exercise 30, find the cost of producing the 201st suit through the 400th suit (integrate from $x = 200$ to $x = 400$).

Psychology: Memorizing. In the psychological process of memorizing, the rate of memorizing (say, in words per minute) increases with respect to time, but eventually a maximum rate of memorizing is reached from which the memory rate begins to decrease.

32. Suppose in a certain memory experiment that the rate of memorizing is given by

$$M'(t) = -0.009t^2 + 0.2t,$$

where $M'(t)$ = the memory rate in words per minute. How many words are memorized in the first 10 minutes (from $t = 0$ to $t = 10$)?

CSS

33. Suppose in a certain memory experiment that the rate of memorizing is given by

$$M'(t) = -0.003t^2 + 0.2t,$$

where $M'(t)$ = the memory rate in words per minute. How many words are memorized in the first 10 minutes (from $t = 0$ to $t = 10$)?

34. *Psychology: Business.* A company is producing a new product. However, due to the nature of the product, it is felt that the time required to produce each unit will decrease as the workers become more familiar with production procedure. It is determined that the function for the learning process is

$$T(x) = ax^b,$$

where

$T(x)$ = the cumulative average time to produce x units,

x = the number of units produced,

a = the hours required to produce the 1st unit, and

b = the slope of the learning curve.

a) Find an expression for the total time required to produce 100 units.

b) Suppose $a = 100$ hr and $b = -0.322$. Find the total time required to produce 100 units.

EXTENSION EXERCISES

Evaluate.

CSS

35. $\int_1^2 (4x + 3)(5x - 2)\, dx$

36. $\int_2^5 (t + \sqrt{3})(t - \sqrt{3})\, dt$

37. $\int_0^1 (t + 1)^3\, dt$

38. $\int_1^3 \left(x - \frac{1}{x}\right)^2 dx$

39. $\int_1^3 \frac{t^5 - t}{t^3}\, dt$

40. $\int_4^9 \frac{t + 1}{\sqrt{t}}\, dt$

41. $\int_3^5 \frac{x^2 - 4}{x - 2}\, dx$

42. $\int_0^1 \frac{t^3 + 1}{t + 1}\, dt$

5.4

a) Evaluate the definite integral of a continuous function.

b) Find the area of a region bounded by two graphs.

c) Solve applied problems involving the area between two graphs.

THE DEFINITE INTEGRAL: THE AREA BETWEEN CURVES

We have considered the definite integral for functions that are positive on an interval $[a, b]$ (the function might be 0 at one or both endpoints). Now we will consider functions that have negative values. First, let us evaluate the integral of a function without negative values:

$$\int_0^2 x^2 \, dx = \left[\frac{x^3}{3} \right]_0^2 = \frac{2^3}{3} - \frac{0^3}{3} = \frac{8}{3}.$$

Thus the area of the shaded region is $\frac{8}{3}$.

Now let us consider the function $y = -x^2$ on the interval $[0, 2]$. Even though we have not defined the definite integral for functions with negative values, let us apply the evaluation procedures and see what we get:

$$\int_0^2 -x^2 \, dx = \left[-\frac{x^3}{3} \right]_0^2$$

$$= \left[-\frac{2^3}{3} - \left(-\frac{0^3}{3} \right) \right] = -\frac{8}{3}.$$

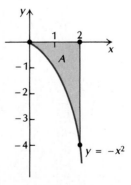

1. To find the area of the shaded region, evaluate

$$\int_0^2 x^3 \, dx.$$

2. To find the area of the shaded region, evaluate

$$\int_0^2 -x^3 \, dx.$$

The graphs of these two functions are reflections of each other across the x-axis. Thus the areas of the shaded regions are the same—that is, $\frac{8}{3}$. The evaluation procedure in the second case gave us $-\frac{8}{3}$. This illustrates that for negative-valued functions, the definite integral gives us the additive inverse of the area between the curve and the x-axis.

DO EXERCISES 1 AND 2.

Now let us consider the function $x^2 - 1$ on $[-1, 2]$. It has both positive and negative values. We will apply the preceding evaluation procedures, even though function values are not all nonnegative. We will do this in two ways.

First,

$$\int_{-1}^2 (x^2 - 1) \, dx = \int_{-1}^1 (x^2 - 1) \, dx + \int_1^2 (x^2 - 1) \, dx$$

$$= \left[\frac{x^3}{3} - x\right]_{-1}^1 + \left[\frac{x^3}{3} - x\right]_1^2$$

$$= \left[\left(\frac{1^3}{3} - 1\right) - \left(\frac{(-1)^3}{3} - (-1)\right)\right]$$

$$\quad + \left[\left(\frac{2^3}{3} - 2\right) - \left(\frac{1^3}{3} - 1\right)\right]$$

$$= \left[-\frac{4}{3}\right] + \left[\frac{4}{3}\right] = 0.$$

This shows that the area of the region under the x-axis is the same as the area of the region over the x-axis.

Now let us evaluate in another way:

$$\int_{-1}^2 (x^2 - 1) \, dx = \left[\frac{x^3}{3} - x\right]_{-1}^2 = \left(\frac{2^3}{3} - 2\right) - \left[\frac{(-1)^3}{3} - (-1)\right]$$

$$= \left(\frac{8}{3} - 2\right) - \left(-\frac{1}{3} + 1\right)$$

$$= \frac{8}{3} - 2 + \frac{1}{3} - 1 = 0.$$

3. Evaluate.

a) $\int_0^1 (x^2 - x)\, dx$

b) $\int_1^2 (x^2 - x)\, dx$

c) $\int_0^2 (x^2 - x)\, dx$

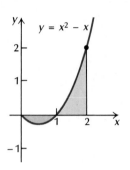

4. Evaluate.

a) $\int_{-2}^0 x^3\, dx$

b) $\int_0^2 x^3\, dx$

c) $\int_{-2}^2 x^3\, dx$

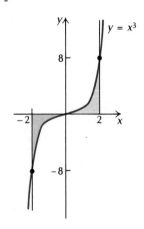

This result is consistent with the first. Thus we are motivated to extend our definition of definite integral to include any continuous function, having positive, negative, or zero values.

DEFINITION

For any function continuous on an interval $[a, b]$,

$$\int_a^b f(x)\, dx = F(b) - F(a),$$

where F is any antiderivative of f.

DO EXERCISES 3 AND 4.

The definite integral turns out to be the area above the x-axis minus the area below.

Example 1   Decide whether $\int_a^b f(x)\, dx$ is positive, negative, or zero.

a)   b)   c)

                    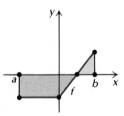

Solution

a) $\int_a^b f(x)\, dx = 0$

There is the same area above as below.

b) $\int_a^b f(x)\, dx > 0$

There is more area above.

c) $\int_a^b f(x)\, dx < 0$

There is more area below.

DO EXERCISES 5–8 ON THE FOLLOWING PAGE.

In each exercise,

a) decide whether $\int_a^b f(x)\,dx$ is positive, negative, or zero;

b) express $\int_a^b f(x)\,dx$ in terms of A.

5.

6.

7.

8.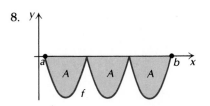

The Area of a Region Bounded by Two Graphs

Suppose we want to find the area of a region bounded by the graphs of two functions $y = f(x)$ and $y = g(x)$.

CSS

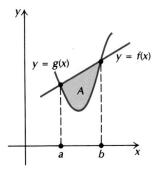

Note that the area of the desired region, A, is that of A_2 minus that of A_1.

Thus,

$$A = \int_a^b f(x)\,dx - \int_a^b g(x)\,dx \quad \text{or} \quad A = \int_a^b [f(x) - g(x)]\,dx.$$

In general, we have the following.

THEOREM 6

Let f and g be continuous functions and suppose $f(x) \geqslant g(x)$ over the interval $[a, b]$. Then the area of the region between the two curves, from $x = a$ to $x = b$, is

$$\int_a^b [f(x) - g(x)]\,dx.$$

Example 2   Find the area of the region bounded by the graphs of $y = 2x + 1$ and $y = x^2 + 1$.

Solution

a) First, make a reasonably accurate sketch to ensure that you have the right configuration. Note which is the *upper* graph. Here it is $2x + 1 \geqslant x^2 + 1$ over the interval $[0, 2]$.

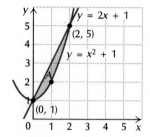

b) Second, if boundaries are not stated, determine the first coordinates of possible points of intersection. Occasionally you can do this just by looking at the graph. If not, you can solve the system of equations as follows. At the points of intersection, $y = x^2 + 1$ and $y = 2x + 1$, so

$$x^2 + 1 = 2x + 1$$
$$x^2 - 2x = 0$$
$$x(x - 2) = 0$$
$$x = 0 \quad \text{or} \quad x = 2.$$

Thus the interval with which we are concerned is $[0, 2]$.

c) Compute the area as follows:

$$\int_0^2 [(2x + 1) - (x^2 + 1)] \, dx = \int_0^2 (2x - x^2) \, dx$$
$$= \left[x^2 - \frac{x^3}{3} \right]_0^2$$
$$= \left(2^2 - \frac{2^3}{3} \right) - \left(0^2 - \frac{0^3}{3} \right)$$
$$= 4 - \frac{8}{3}$$
$$= \frac{4}{3}.$$

9. Find the area of the region bounded by the graphs of

$$y = x \quad \text{and} \quad y = x^2. \quad \boxed{\text{CSS}}$$

DO EXERCISE 9.

An Application

Example 3 *Emission control.* A clever college student develops an engine that is believed to meet federal standards for emission control. The engine's rate of emission is given by

$$E(t) = 2t^2,$$

where $E(t)$ = the emissions in billions of pollution particulates per year at time t, in years. The emission rate of a conventional engine is given by

$$C(t) = 9 + t^2.$$

The curves are shown here.

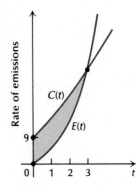

a) At what point in time will the emission rates be the same?

b) Before the time in part (a), what is the reduction in emissions resulting from using the student's engine?

Solution

a) The rates of emission will be the same with $E(t) = C(t)$, or

$$2t^2 = 9 + t^2$$
$$t^2 - 9 = 0$$
$$(t - 3)(t + 3) = 0$$
$$t = 3 \quad \text{or} \quad t = -3.$$

Since negative time has no meaning in this problem, the emission rates will be the same when $t = 3$ years.

10. A company determines that its marginal revenue per unit is given by

$$R'(x) = 14x, \qquad R(0) = 0.$$

Its marginal cost per unit is given by

$$C'(x) = 4x - 10, \qquad C(0) = 0.$$

Total profit from the production and sale of k units is given by

$$P(k) = R(k) - C(k)$$
$$= \int_0^k [R'(x) - C'(x)]\, dx.$$

a) Find $P(k)$.

b) Find $P(5)$.

b) The reduction in emissions is represented by the area of the shaded region in the graph above. It is the area between $y = 9 + t^2$ and $y = 2t^2$, from $t = 0$ to $t = 3$, and is computed as follows:

$$\int_0^3 [(9 + t^2) - 2t^2]\, dt = \int_0^3 (9 - t^2)\, dt$$
$$= \left[9t - \frac{t^3}{3} \right]_0^3$$
$$= \left(9 \cdot 3 - \frac{3^3}{3} \right) - \left(9 \cdot 0 - \frac{0^3}{3} \right)$$
$$= 27 - 9$$
$$= 18 \text{ billion pollution particulates.}$$

DO EXERCISE 10.

EXERCISE SET 5.4

Find the area of the region bounded by the given graphs.   CSS

1. $y = x, y = x^3, x = 0, x = 1$

2. $y = x, y = x^4$

3. $y = x + 2, y = x^2$

4. $y = x^2 - 2x, y = x$

5. $y = 6x - x^2, y = x$

6. $y = x^2 - 6x, y = -x$

7. $y = 2x - x^2, y = -x$

8. $y = x^2, y = \sqrt{x}$

9. $y = x, y = \sqrt{x}$

10. $y = 3, y = x, x = 0$

11. $y = 5, y = \sqrt{x}, x = 0$

12. $y = x^2, y = x^3$

13. $y = 4 - x^2, y = 4 - 4x$

14. $y = x^2 + 1, y = x^2, x = 1, x = 3$

15. $y = x^2 + 3, y = x^2, x = 1, x = 2$

APPLICATIONS

16. *Business: Profit.* A company determines that its marginal revenue per day is given by

$$R'(t) = 100e^t, \qquad R(0) = 0,$$

where $R(t) = $ the revenue, in dollars, on the tth day. Also, its marginal cost per day is given by

$$C'(t) = 100 - 0.2t, \qquad C(0) = 0,$$

where $C(t) = $ the cost, in dollars, on the tth day. Find the total profit from $t = 0$ to $t = 10$ (the first 10 days). *Note:*

$$P(T) = R(T) - C(T) = \int_0^T [R'(t) - C'(t)]\, dt.$$

17. *Psychology: Memorizing.* In a certain memory experiment, subject A is able to memorize words at the rate

$$m'(t) = -0.009t^2 + 0.2t \quad \text{(words per minute)}.$$

In the same memory experiment subject B is able to memorize at the rate

$$M'(t) = -0.003t^2 + 0.2t \quad \text{(words per minute)}.$$

a) Which subject has the higher rate of memorization?

b) How many more words does that subject memorize from $t = 0$ to $t = 10$ (during the first 10 minutes)?

EXTENSION EXERCISES

Find the area of the region bounded by the given graphs.

18. $y = x^2, y = x^{-2}, x = 1, x = 5$

19. $y = e^x, y = e^{-x}, x = 0, x = 1$

20. $y = x^2, y = \sqrt[3]{x^2}, x = 1, x = 8$

21. $y = x^2, y = x^3, x = -1, x = 1$

22. $x + 2y = 2, y - x = 1, 2x + y = 7$

23. $y = x + 6, y = -2x, y = x^3$

24. *Biomedical: Poiseuille's Law.* The flow of blood in a blood vessel is faster toward the center of the vessel and slower toward the outside. The speed of the blood is given by

$$V = \frac{p}{4Lv}(R^2 - r^2),$$

where R = the radius of the blood vessel, r = the distance of the blood from the center of the vessel, and p, v, and L are physical constants related to pressure and viscosity of the blood and the length of the blood vessel. If R is constant, we can think of V as a function of r:

$$V(r) = \frac{p}{4Lv}(R^2 - r^2).$$

Total blood flow Q is given by

$$Q = \int_0^R 2\pi \cdot V(r) \cdot r \cdot dr.$$

Find Q.

5.5

OBJECTIVES

a) Evaluate integrals using substitution.

b) Solve applied problems involving integration by substitution.

INTEGRATION TECHNIQUES: SUBSTITUTION

The following formulas provide a basis for an integration technique called *substitution*.

A. $\int u^r \, du = \dfrac{u^{r+1}}{r+1} + C$, provided $r \neq -1$

B. $\int e^u \, du = e^u + C$

C. $\int \dfrac{1}{u} \, du = \ln u + C, u > 0$; or $\int \dfrac{1}{u} \, du = \ln |u| + C, u < 0$

(We will generally consider $u > 0$.)

1. For $y = f(x) = 6x^2 + x$, find dy.

Recall the Leibniz notation, dy/dx, for a derivative. We gave specific definitions of the differentials dy and dx in Section 3.7. Recall that

$$\frac{dy}{dx} = f'(x)$$

and

$$dy = f'(x)\,dx$$

We will make extensive use of this notation in this section.

Example 1   For $y = f(x) = x^3$, find dy.

Solution   We have

$$\frac{dy}{dx} = f'(x) = 3x^2,$$

so

$$dy = f'(x)\,dx = 3x^2\,dx.$$

Example 2   For $u = g(x) = \ln x$, find du.

2. For $u = g(x) = x + 3$, find du.

Solution   We have

$$\frac{du}{dx} = g'(x) = \frac{1}{x},$$

so

$$du = g'(x)\,dx = \frac{1}{x}\,dx, \quad \text{or} \quad \frac{dx}{x}.$$

DO EXERCISES 1 AND 2.

So far the dx in

$$\int f(x)\,dx$$

has played no role in integrating other than to indicate the variable of integration. Now it will be convenient to make use of dx. Consider the integral

$$\int 2x \cdot e^{x^2}\,dx.$$

3. Evaluate $\int 3x^2 e^{x^3}\, dx$.

If we set

$$u = x^2,$$

then

$$du = 2x\, dx.$$

If we substitute u for x^2 and du for $2x\, dx$, then the integral takes on the form

$$\int e^u\, du.$$

Since

$$\int e^u\, du = e^u + C,$$

it follows that

$$\int 2x \cdot e^{x^2}\, dx = \int e^u\, du$$
$$= e^u + C$$
$$= e^{x^2} + C.$$

In effect, we have used the Chain Rule in reverse. We can check the result by differentiation. This procedure is referred to as *substitution,* or *change of variable.* It is a *trial-and-error* procedure; that is, if we try a substitution that doesn't result in an integrand that can be easily integrated, we try another. It will not always work! It will work if the integrand fits one of the rules A, B, or C.

DO EXERCISE 3.

Let us consider some additional examples.

Example 3

$$\int \frac{2x\, dx}{1 + x^2} = \int \frac{du}{u} \qquad \text{Substitution} \qquad \boxed{\begin{array}{l} \text{Let} \quad u = 1 + x^2; \\ \text{then } du = 2x\, dx. \end{array}}$$

$$= \ln u + C$$

$$= \ln (\boxed{1 + x^2}) + C$$

Remember that this is a trial-and-error process. Suppose we had made the substitution

$$u = x^2.$$

4. Evaluate $\int \dfrac{2x\,dx}{5 + x^2}$.

Then we would have

$$du = 2x\,dx,$$

and the integral would become

$$\int \frac{du}{1 + u}.$$

This is still not easily integrated, so we would try another substitution.

DO EXERCISE 4.

Example 4

$$\int \frac{2x\,dx}{(1 + x^2)^2} = \int \frac{du}{u^2} \quad \underline{\text{Substitution}} \quad \boxed{\begin{array}{l} u = 1 + x^2; \\ du = 2x\,dx. \end{array}}$$

$$= \int u^{-2}\,du = -u^{-1} + C$$

$$= -\frac{1}{u} + C = -\frac{1}{1 + x^2} + C$$

5. Evaluate $\int \dfrac{2x\,dx}{(3 + x^2)^2}$.

DO EXERCISE 5.

Example 5

$$\int \frac{\ln 3x\,dx}{x} = \int u\,du \quad \underline{\text{Substitution}} \quad \boxed{\begin{array}{l} u = \ln 3x; \\ du = \dfrac{1}{x}\,dx \end{array}}$$

$$= \frac{u^2}{2} + C$$

$$= \frac{(\ln 3x)^2}{2} + C$$

6. Evaluate $\int \dfrac{\ln x\,dx}{x}$.

DO EXERCISE 6.

Example 6   Evaluate $\int xe^{x^2}\,dx$.

Solution   Suppose we try

$$u = x^2;$$

then we have

$$du = 2x\,dx.$$

7. Evaluate $\int x^2 \cdot e^{x^3}\, dx$.

Evaluate.

8. $\int e^{5x}\, dx$

9. $\int e^{0.02x}\, dx$

10. $\int e^{-x}\, dx$

We don't quite have $2x\, dx$. We have only $x\, dx$ and will need to supply a 2. We do this by multiplying by 1, using $\frac{1}{2} \cdot 2$:

$$\frac{1}{2} \cdot 2 \cdot \int x e^{x^2}\, dx = \frac{1}{2} \int 2x e^{x^2}\, dx$$

$$= \frac{1}{2} \int e^{x^2} (2x\, dx)$$

$$= \frac{1}{2} \int e^{u}\, du$$

$$= \frac{1}{2}\, e^{u} + C$$

$$= \frac{1}{2}\, e^{x^2} + C.$$

DO EXERCISE 7.

Example 7

$$\left.\begin{aligned} \int b e^{ax}\, dx &= \frac{b}{a} \int a e^{ax}\, dx \\ &= \frac{b}{a} \int e^{u}\, du \end{aligned}\right\} \quad \underline{\text{Substitution}}\ \boxed{\begin{aligned} u &= ax; \\ du &= a\, dx \end{aligned}}$$

$$= \frac{b}{a}\, e^{u} + C$$

$$= \frac{b}{a}\, e^{ax} + C$$

Note that this gives us a formula for integrating $b e^{ax}$.

DO EXERCISES 8–10.

Example 8

$$\int \frac{dx}{x+3} = \int \frac{du}{u} \quad \underline{\text{Substitution}}\ \boxed{\begin{aligned} u &= x+3; \\ du &= dx \end{aligned}}$$

$$= \ln u + C = \ln (x+3) + C$$

With practice, you will be able to make certain substitutions mentally and just write down the answer. Examples 7 and 8 are good illustrations of this.

11. Evaluate $\int x^3(x^4 + 5)^{19}\, dx$.

Example 9

$\int x^2(x^3 + 1)^{10}\, dx$

$$= \frac{1}{3} \int 3x^2(x^3 + 1)^{10}\, dx \Bigg|$$

Substitution $\boxed{u = x^3 + 1; \\ du = 3x^2\, dx}$

$$= \frac{1}{3} \int u^{10}\, du$$

$$= \frac{1}{3} \cdot \frac{u^{11}}{11} + C$$

$$= \frac{1}{33}(x^3 + 1)^{11} + C$$

DO EXERCISE 11.

12. Evaluate $\int_1^e \frac{\ln x\, dx}{x}$. (See Margin Exercise 6.)

Example 10   Evaluate $\int_0^1 x^2(x^3 + 1)^{10}\, dx$.

Solution

a) First we find the indefinite integral (shown in Example 9).

b) Then we evaluate the definite integral on $[0, 1]$:

$$\int_0^1 x^2(x^3 + 1)^{10}\, dx = \left[\frac{1}{33}(x^3 + 1)^{11}\right]_0^1$$

$$= \frac{1}{33}[(1^3 + 1)^{11} - (0^3 + 1)^{11}]$$

$$= \frac{1}{33}(2^{11} - 1^{11})$$

$$= \frac{2^{11} - 1}{33}.$$

DO EXERCISE 12.

EXERCISE SET 5.5

Evaluate. (Be sure to check by differentiating!)

1. $\int \dfrac{3x^2\, dx}{7 + x^3}$

2. $\int \dfrac{3x^2\, dx}{1 + x^3}$

3. $\int e^{4x}\, dx$

4. $\int e^{3x}\,dx$

5. $\int e^{x/2}\,dx$

6. $\int e^{x/3}\,dx$

7. $\int x^3 e^{x^4}\,dx$

8. $\int x^4 e^{x^5}\,dx$

9. $\int t^2 e^{-t^3}\,dt$

10. $\int t e^{-t^2}\,dt$

11. $\int \dfrac{\ln 4x\,dx}{x}$

12. $\int \dfrac{\ln 5x\,dx}{x}$

13. $\int \dfrac{dx}{1+x}$

14. $\int \dfrac{dx}{5+x}$

15. $\int \dfrac{dx}{4-x}$

16. $\int \dfrac{dx}{1-x}$

17. $\int t^2(t^3-1)^7\,dt$

18. $\int t(t^2-1)^5\,dt$

19. $\int (x^4+x^3+x^2)^7(4x^3+3x^2+2x)\,dx$

20. $\int (x^3-x^2-x)^9(3x^2-2x-1)\,dx$

21. $\int \dfrac{e^x\,dx}{4+e^x}$

22. $\int \dfrac{e^t\,dt}{3+e^t}$

23. $\int \dfrac{\ln x^2}{x}\,dx$

24. $\int \dfrac{(\ln x)^2}{x}\,dx$

25. $\int \dfrac{dx}{x\ln x}$

26. $\int \dfrac{dx}{x\ln x^2}$

27. $\int \sqrt{ax+b}\,dx$

28. $\int x\sqrt{ax^2+b}\,dx$

29. $\int be^{ax}\,dx$

30. $\int P_0 e^{kt}\,dt$

Evaluate.

31. $\int_0^1 2x e^{x^2}\,dx$

32. $\int_0^1 3x^2 e^{x^3}\,dx$

33. $\int_0^1 x(x^2+1)^5\,dx$

34. $\int_1^2 x(x^2-1)^7\,dx$

35. $\int_1^3 \dfrac{dt}{1+t}$

36. $\int_1^3 e^{2x}\,dx$

37. $\int_1^4 \dfrac{2x+1}{x^2+x-1}\,dx$

38. $\int_1^3 \dfrac{2x+3}{x^2+3x}\,dx$

39. $\int_0^b e^{-x}\,dx$

40. $\int_0^b 2e^{-2x}\,dx$

41. $\int_0^b m e^{-mx}\,dx$

42. $\int_0^b k e^{-kx}\,dx$

43. $\int_0^4 (x-6)^2\,dx$

44. $\int_0^3 (x-5)^2\,dx$

APPLICATIONS

45. ▦ *Sociology: Divorce.* The U.S. divorce rate is approximated by

$$D(t) = 100{,}000 e^{0.025t},$$

where $D(t)$ = the number of divorces occurring at time t and t = the number of years measured from 1900. That is, $t = 0$ corresponds to 1900, $t = 88\frac{9}{365}$ corresponds to January 9, 1988, and so on.

a)   Find the total number of divorces from 1900 to 1988. Note that this is

$$\int_0^{88} D(t)\,dt.$$

b)   Find the total number of divorces from 1980 to 1988. Note that this is

$$\int_{80}^{88} D(t)\,dt.$$

46. ▦ *Business: Value of an investment.* A company buys a new machine for $250,000. The marginal revenue from the sale of products produced by the machine is projected to be

$$R'(t) = 4000t.$$

The salvage value of the machine decreases at the rate of

$$V(t) = 25{,}000e^{-0.1t}.$$

The total profit from the machine after T years is given by

$$P(T) = \left(\begin{array}{c}\text{Revenue from}\\\text{sale of product}\end{array}\right) + \left(\begin{array}{c}\text{Revenue from}\\\text{sale of machine}\end{array}\right) - \left(\begin{array}{c}\text{Cost of}\\\text{machine}\end{array}\right)$$

$$= \int_0^T R'(t)\,dt + \int_0^T V(t)\,dt - \$250{,}000.$$

a) Find $P(T)$.

b) Find $P(10)$.

EXTENSION EXERCISES

Evaluate.

47. $\displaystyle\int 5x\sqrt{1-4x^2}\,dx$

48. $\displaystyle\int \frac{dx}{ax+b}$

49. $\displaystyle\int \frac{x^2}{e^{x^3}}\,dx$

50. $\displaystyle\int \frac{e^{\sqrt{t}}}{\sqrt{t}}\,dt$

51. $\displaystyle\int \frac{e^{1/t}}{t^2}\,dt$

52. $\displaystyle\int \frac{(\ln x)^{99}}{x}\,dx$

53. $\displaystyle\int \frac{dx}{x(\ln x)^4}$

54. $\displaystyle\int (e^t + 2)e^t\,dt$

55. $\displaystyle\int x^2\sqrt{x^3+1}\,dx$

56. $\displaystyle\int \frac{t^2}{\sqrt[4]{2+t^3}}\,dt$

57. $\displaystyle\int \frac{x-3}{(x^2-6x)^{1/3}}\,dx$

58. $\displaystyle\int \frac{[(\ln x)^2 + 3(\ln x) + 4]}{x}\,dx$

59. $\displaystyle\int \frac{t^3\ln(t^4+8)}{t^4+8}\,dt$

60. $\displaystyle\int \frac{t^2+2t}{(t+1)^2}\,dt$

61. $\displaystyle\int \frac{x^2+6x}{(x+3)^2}\,dx$

Hint: $\dfrac{t^2+2t}{(t+1)^2} = \dfrac{t^2+2t+1-1}{t^2+2t+1} = 1 - \dfrac{1}{(t+1)^2}$

62. $\displaystyle\int \frac{x+3}{x+1}\,dx$

63. $\displaystyle\int \frac{t-5}{t-4}\,dt$

64. $\displaystyle\int \frac{dx}{x(\ln x)^n}$

Hint: Divide $\dfrac{x+3}{x+1} = 1 + \dfrac{2}{x+1}$.

65. $\displaystyle\int \frac{dx}{e^x+1}$

66. $\displaystyle\int \frac{e^x-e^{-x}}{e^x+e^{-x}}\,dx$

67. $\displaystyle\int \frac{(\ln x)^n}{x}\,dx$

Hint: $\dfrac{1}{e^x+1} = \dfrac{e^{-x}}{1+e^{-x}}$

68. $\displaystyle\int_0^b \frac{e^{-mx}}{1+ae^{-mx}}\,dx$

69. $\displaystyle\int \frac{dx}{x\ln x\,[\ln(\ln x)]}$

70. $\displaystyle\int 5x^2(2x^3-7)^n\,dx$

71. $\displaystyle\int 9x(7x^2+9)^n\,dx$

5.6

a) Evaluate integrals using integration by parts.

b) Solve applied problems involving integration by parts.

INTEGRATION TECHNIQUES: INTEGRATION BY PARTS

Recall the product rule for derivatives:

$$\frac{d}{dx}\,uv = \frac{du}{dx}\,v + \frac{dv}{dx}\,u = u\,\frac{dv}{dx} + v\,\frac{du}{dx}.$$

Integrating both sides, we get

$$uv = \int u\,\frac{dv}{dx}\,dx + \int v\,\frac{du}{dx}\,dx$$

$$= \int u\,dv + \int v\,du.$$

Solving for $\int u\,dv$, we get the following.

THEOREM 7

The Integration-by-Parts Formula

$$\int u\,dv = uv - \int v\,du$$

This equation can be used as a formula for integrating in certain situations—that is, situations in which an integrand is a product of two functions, and one of the functions can be integrated using the techniques we have already developed. For example,

$$\int xe^x\,dx$$

can be considered as

$$\int x(e^x\,dx) = \int u\,dv,$$

where $u = x$ and $dv = e^x\,dx$.

We already know how to integrate $e^x\,dx$, or dv. The simplest antiderivative is e^x. This is v. Now since $du = dx$, the formula gives us

$$\overset{u}{\int} \overset{dv}{(x)} \overset{}{(e^x\,dx)} = \overset{u}{(x)}\overset{v}{(e^x)} - \overset{}{\int} \overset{v}{(e^x)}\overset{du}{(dx)}$$

$$= xe^x - e^x + C.$$

This method of integrating is called *integration by parts*.

1. Evaluate $\int 3x \cdot e^{3x}\, dx$.

Note that integration by parts, like substitution, is a trial-and-error process. In the preceding example, suppose we had reversed the roles of x and e^x. We would have obtained

$$u = e^x, \qquad dv = x\, dx,$$

$$du = e^x\, dx, \qquad v = \frac{x^2}{2},$$

and

$$\int (e^x)(x\, dx) = (e^x)\left(\frac{x^2}{2}\right) - \int \left(\frac{x^2}{2}\right)(e^x\, dx).$$

Now the integrand on the right is more difficult to integrate than the one we started with. When we can integrate *both* factors of an integrand, and thus have a choice as to how to apply the integration-by-parts formula, it can happen that only one (and maybe none) of the possibilities will work.

DO EXERCISE 1.

Let us consider some additional examples.

2. Evaluate $\int x \cdot \ln x\, dx$.

Example 1   Evaluate $\int \ln x\, dx$.

Solution   Note that $\int (dx/x) = \ln x + C$, but we do not yet know how to find $\int \ln x\, dx$. Let

$$u = \ln x \quad \text{and} \quad dv = dx.$$

Then

$$du = \frac{1}{x}\, dx \quad \text{and} \quad v = x.$$

Using the integration-by-parts formula gives

$$\overset{u}{} \quad \overset{dv}{} \qquad \overset{u}{}\ \overset{v}{} \qquad \overset{v}{}\ \overset{du}{}$$

$$\int (\ln x)(dx) = (\ln x)x - \int x\left(\frac{1}{x}\, dx\right)$$

$$= x \ln x - \int dx = x \ln x - x + C.$$

DO EXERCISE 2.

Example 2   Evaluate $\int x\sqrt{x + 1}\, dx$.

3. Evaluate $\int x\sqrt{x+3}\,dx$

Solution We let

$$u = x \quad \text{and} \quad dv = (x+1)^{1/2}\,dx.$$

Then

$$du = dx \quad \text{and} \quad v = \tfrac{2}{3}(x+1)^{3/2}.$$

Note that we had to use substitution to integrate dv. Using the integration-by-parts formula gives us

$$\int x\sqrt{x+1}\,dx = x \cdot \tfrac{2}{3}(x+1)^{3/2} - \tfrac{2}{3}\int (x+1)^{3/2}\,dx$$
$$= \tfrac{2}{3}x(x+1)^{3/2} - \tfrac{2}{3}\cdot\tfrac{2}{5}(x+1)^{5/2} + C$$
$$= \tfrac{2}{3}x(x+1)^{3/2} - \tfrac{4}{15}(x+1)^{5/2} + C.$$

4. Evaluate $\int_1^2 x\ln x\,dx$. (See Margin Exercise 2.)

DO EXERCISE 3.

Example 3 Evaluate $\int_1^2 \ln x\,dx$.

Solution

a) First find the indefinite integral (Example 1).

b) Then evaluate the definite integral:

$$\int_1^2 \ln x\,dx = [x\ln x - x]_1^2$$
$$= (2\ln 2 - 2) - (1\cdot\ln 1 - 1)$$
$$= 2\ln 2 - 2 + 1$$
$$= 2\ln 2 - 1.$$

DO EXERCISE 4.

EXERCISE SET 5.6

Evaluate using integration by parts. Check by differentiating.

1. $\int 5xe^{5x}\,dx$

2. $\int 2xe^{2x}\,dx$

3. $\int x^3(3x^2\,dx)$

4. $\int x^2(2x\,dx)$

5. $\int xe^{2x}\,dx$

6. $\int xe^{3x}\,dx$

7. $\int xe^{-2x}\,dx$

8. $\int xe^{-x}\,dx$

9. $\int x^2\ln x\,dx$

10. $\int x^3\ln x\,dx$

11. $\int x\ln x^2\,dx$

12. $\int x^2\ln x^3\,dx$

13. $\int \ln(x+3)\,dx$

14. $\int \ln(x+1)\,dx$

15. $\int (x+2)\ln x\,dx$

16. $\int (x+1)\ln x\,dx$

17. $\int (x - 1) \ln x \, dx$

18. $\int (x - 2) \ln x \, dx$

19. $\int x\sqrt{x + 2} \, dx$

20. $\int x\sqrt{x + 4} \, dx$

21. $\int x^3 \ln 2x \, dx$

22. $\int x^2 \ln 5x \, dx$

23. $\int x^2 e^x \, dx$

24. $\int (\ln x)^2 \, dx$

25. $\int x^2 e^{2x} \, dx$

26. $\int x^{-5} \ln x \, dx$

Evaluate using integration by parts.

27. $\int_1^2 x^2 \ln x \, dx$

28. $\int_1^2 x^3 \ln x \, dx$

29. $\int_2^6 \ln (x + 3) \, dx$

30. $\int_0^5 \ln (x + 1) \, dx$

31. $\int_0^1 x e^x \, dx$

32. $\int_0^1 x e^{-x} \, dx$

APPLICATIONS

33. *Ecology: Electrical energy use.* The rate of electrical energy used by a family, in kilowatt hours per day, is given by

$$K(t) = 10te^{-t},$$

where t is the time, in hours. That is, t is in the interval $[0, 24]$.

a) How many kilowatt hours does the family use in the first T hours of a day ($t = 0$ to $t = T$)?

b) How many kilowatt hours does the family use in the first 4 hours of the day?

34. *Biomedical: Drug dosage.* Suppose an oral dose of a drug is taken. From that time, the drug is assimilated in the body and excreted through the urine. The total amount of the drug that has passed through the body in time T is given by

$$\int_0^T E(t) \, dt,$$

where E is the rate of excretion of the drug through the urine. A typical rate of excretion function is

$$E(t) = te^{-kt},$$

where $k > 0$ and t is time, in hours.

a) Use integration by parts to find a formula for

$$\int_0^T E(t) \, dt.$$

b) ▦ Find

$$\int_0^{10} E(t) \, dt, \quad \text{when } k = 0.2 \text{ mg/hr.}$$

EXTENSION EXERCISES

Evaluate using integration by parts.

35. $\int \sqrt{x} \ln x \, dx$

36. $\int x^n \ln x \, dx$

37. $\int \frac{te^t}{(t + 1)^2} \, dt$

38. $\int x^2 (\ln x)^2 \, dx$

39. $\int \frac{\ln x}{\sqrt{x}} \, dx$

40. $\int x^n (\ln x)^2 \, dx$

41. Verify that, for any positive integer n,

$$\int x^n e^x \, dx = x^n e^x - n \int x^{n-1} e^x \, dx.$$

42. Verify that, for any positive integer n,

$$\int (\ln x)^n \, dx = x(\ln x)^n - n \int (\ln x)^{n-1} \, dx.$$

5.7

INTEGRATION TECHNIQUES: USING TABLES

a) Evaluate integrals using a table of integration formulas.

1. Using Table 1, evaluate

$$\int \frac{1}{x(7 + 2x)} \, dx.$$

Tables of Integration Formulas

You have probably noticed that, generally speaking, integration is more difficult and "tricky" than differentiation. Because of this, integral formulas that are reasonable and/or important have been gathered into tables. Table 1 at the back of the book, though quite brief, is such an example. Entire books of integration formulas are available in libraries, and lengthy tables are also available in mathematics handbooks. Such tables are usually classified by the form of the integrand. The idea is to properly match the integral in question with a formula in the table. Sometimes some algebra or a technique such as integration by substitution or parts may be needed as well as a table.

Example 1 Evaluate

$$\int \frac{dx}{x(3 - x)}.$$

Solution This integral fits *Formula 20* in Table 1:

$$\int \frac{1}{x(ax + b)} \, dx = \frac{1}{b} \ln \left(\frac{x}{ax + b} \right) + C.$$

In our integral, $a = -1$ and $b = 3$, so we have, by the formula,

$$\int \frac{1}{x(3 - x)} \, dx = \int \frac{dx}{x(-1 \cdot x + 3)}$$

$$= \frac{1}{3} \ln \left(\frac{x}{-1 \cdot x + 3} \right) + C$$

$$= \frac{1}{3} \ln \left(\frac{x}{3 - x} \right) + C.$$

DO EXERCISE 1.

Example 2 Evaluate

$$\int \frac{5x}{7x - 8} \, dx.$$

Using Table 1, evaluate.

2. $\int \dfrac{3x}{5 - 2x}\, dx$

3. $\int \dfrac{x}{(5 - 2x)^2}\, dx$

4. Using Table 1, evaluate

$$\int \sqrt{25y^2 - 4}\, dy.$$

Solution   If we first factor 5 out of the integral, then the integral fits *Formula 18* in Table 1:

$$\int \frac{x}{ax + b}\, dx = \frac{b}{a^2} + \frac{x}{a} - \frac{b}{a^2} \ln{(ax + b)} + C.$$

In our integral, $a = 7$ and $b = -8$, so we have, by the formula,

$$\int \frac{5x}{7x - 8}\, dx = 5 \int \frac{x}{7x - 8}\, dx$$

$$= 5\left[\frac{-8}{7^2} + \frac{x}{7} - \frac{-8}{7^2} \ln{(7x - 8)}\right] + C$$

$$= 5\left[\frac{-8}{49} + \frac{x}{7} + \frac{8}{49} \ln{(7x - 8)}\right] + C$$

$$= -\frac{40}{49} + \frac{5x}{7} + \frac{40}{49} \ln{(7x - 8)} + C.$$

DO EXERCISES 2 AND 3.

Example 3   Evaluate $\int \sqrt{16x^2 + 3}\, dx$.

Solution   This formula almost fits *Formula 22* in Table 1:

$$\int \sqrt{x^2 \pm a^2}\, dx = \tfrac{1}{2}[x\sqrt{x^2 \pm a^2} \pm a^2 \ln{(x + \sqrt{x^2 \pm a^2})}] + C.$$

But the x^2 coefficient needs to be 1. To achieve this we first factor out 16, as follows. Then we apply *Formula 22* in Table 1:

$$\int \sqrt{16x^2 + 3}\, dx = \int \sqrt{16(x^2 + \tfrac{3}{16})}\, dx$$

$$= \int 4\sqrt{x^2 + \tfrac{3}{16}}\, dx$$

$$= 2[x\sqrt{x^2 + \tfrac{3}{16}} + \tfrac{3}{16} \ln{(x + \sqrt{x^2 + \tfrac{3}{16}})}] + C.$$

In our integral, $a^2 = 3/16$ and $a = \sqrt{3}/4$, though we did not need to use a in this form when applying the formula.

DO EXERCISE 4.

Example 4   Evaluate

$$\int \frac{dx}{x^2 - 25}.$$

5. Using Table 1, evaluate

$$\int \frac{2}{49 - x^2} \, dx.$$

Solution This integral fits *Formula 14* in Table 1:

$$\int \frac{1}{x^2 - a^2} \, dx = \frac{1}{2a} \ln \left(\frac{x - a}{x + a} \right) + C.$$

In our integral, $a = 5$, so we have, by the formula,

$$\int \frac{dx}{x^2 - 25} = \frac{1}{10} \ln \left(\frac{x - 5}{x + 5} \right) + C.$$

DO EXERCISE 5.

Example 5 Evaluate $\int (\ln x)^3 \, dx$.

Solution This integral fits *Formula 9* in Table 1:

$$\int (\ln x)^n \, dx = x(\ln x)^n - n \int (\ln x)^{n-1} \, dx + C, \qquad n \neq -1.$$

6. Using Table 1, evaluate

$$\int (\ln x)^2 \, dx.$$

We must apply the formula three times:

$$\int (\ln x)^3 \, dx$$

$$= x(\ln x)^3 - 3 \int (\ln x)^2 \, dx + C \qquad \text{Formula 9}$$

$$= x(\ln x)^3 - 3 \left[x(\ln x)^2 - 2 \int \ln x \, dx \right] + C \qquad \begin{array}{l} \text{Applying Formula 9} \\ \text{again} \end{array}$$

$$= x(\ln x)^3 - 3 \left[x(\ln x)^2 - 2 \left(x \ln x - \int dx \right) \right] + C \qquad \begin{array}{l} \text{Applying} \\ \text{Formula 9 for} \\ \text{the third time} \end{array}$$

$$= x(\ln x)^3 - 3x(\ln x)^2 + 6x \ln x - 6x + C.$$

DO EXERCISE 6.

EXERCISE SET 5.7

Evaluate using Table 1.

1. $\int xe^{-3x} \, dx$

2. $\int xe^{4x} \, dx$

3. $\int 5^x \, dx$

4. $\int \frac{1}{\sqrt{x^2 - 9}} \, dx$

5. $\int \frac{1}{16 - x^2} \, dx$

6. $\int \frac{1}{x\sqrt{4 + x^2}} \, dx$

7. $\int \frac{x}{5 - x} \, dx$

8. $\int \frac{x}{(1 - x)^2} \, dx$

9. $\int \dfrac{1}{x(5-x)^2}\,dx$ 　　10. $\int \sqrt{x^2+9}\,dx$ 　　11. $\int \ln 3x\,dx$ 　　12. $\int \ln \tfrac{4}{5}x\,dx$

13. $\int x^4 e^{5x}\,dx$ 　　14. $\int x^3 e^{-2x}\,dx$ 　　15. $\int x^3 \ln x\,dx$ 　　16. $\int 5x^4 \ln x\,dx$

17. $\int \dfrac{dx}{\sqrt{x^2+7}}$ 　　18. $\int \dfrac{3\,dx}{x\sqrt{1-x^2}}$ 　　19. $\int \dfrac{10\,dx}{x(5-7x)^2}$ 　　20. $\int \dfrac{2}{5x(7x+2)}\,dx$

21. $\int \dfrac{-5}{4x^2-1}\,dx$ 　　22. $\int \sqrt{9t^2-1}\,dt$ 　　23. $\int \sqrt{4m^2+16}\,dm$ 　　24. $\int \dfrac{3\ln x}{x^2}\,dx$

25. $\int \dfrac{-5\ln x}{x^3}\,dx$ 　　26. $\int (\ln x)^4\,dx$ 　　27. $\int \dfrac{e^x}{x^{-3}}\,dx$ 　　28. $\int \dfrac{3}{\sqrt{4x^2+100}}\,dx$

EXTENSION EXERCISES

Evaluate using Table 1.

29. $\int \dfrac{8}{3x^2-2x}\,dx$ 　　　　30. $\int \dfrac{x\,dx}{4x^2-12x+9}$ 　　　　31. $\int \dfrac{dx}{x^3-4x^2+4x}$

32. $\int e^x \sqrt{e^{2x}+1}\,dx$ 　　　　33. $\int \dfrac{-e^{-2x}\,dx}{9-6e^{-x}+e^{-2x}}$ 　　　　34. $\int \dfrac{\sqrt{(\ln x)^2+49}}{2x}\,dx$

5.8

OBJECTIVES

a) Approximate
$$\int_a^b f(x)\,dx$$
by adding the areas of rectangles.

b) Find the average value of a function over a given interval.

THE DEFINITE INTEGRAL AS A LIMIT OF SUMS*

We now consider approximating the area of a region by dividing it into subregions that are almost rectangles. In the figure below, $[a, b]$ has been divided into 4 subintervals, each having width Δx, or $(b-a)/4$.

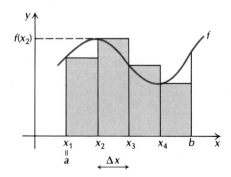

---

* This material does arise in some future sections. With some care, however, this topic can be considered optional.

The area under a curve can be approximated by a sum of rectangular areas.

The heights of the rectangles shown are

$$f(x_1), \qquad f(x_2), \qquad f(x_3), \quad \text{and} \quad f(x_4).$$

The area of the region under the curve is approximately the sum of the areas of the four rectangles:

$$f(x_1)\,\Delta x + f(x_2)\,\Delta x + f(x_3)\,\Delta x + f(x_4)\,\Delta x.$$

We can name this sum using *summation notation,* which utilizes the Greek capital letter sigma, Σ:

$$\sum_{i=1}^{4} f(x_i)\,\Delta x, \quad \text{or} \quad \sum_{i=1}^{4} f(x_i)\,\Delta x.$$

This is read "the sum of the numbers $f(x_i)\,\Delta x$ from $i = 1$ to $i = 4$." To recover the original expression, we substitute the numbers 1 through 4 successively into $f(x_i)\,\Delta x$ and write plus signs between the results.

Write the summation notation.

1. $1 + 4 + 9 + 16 + 25 + 36$

Example 1   Write summation notation for $2 + 4 + 6 + 8 + 10$.

Solution

$$2 + 4 + 6 + 8 + 10 = \sum_{i=1}^{5} 2i$$

Example 2   Write summation notation for

$$g(x_1)\,\Delta x + g(x_2)\,\Delta x + \cdots + g(x_{19})\,\Delta x.$$

Solution

2. $e + e^2 + e^3 + e^4$

$$g(x_1)\,\Delta x + g(x_2)\,\Delta x + \cdots + g(x_{19})\,\Delta x = \sum_{i=1}^{19} g(x_i)\,\Delta x$$

DO EXERCISES 1–3.

3. $P(x_1)\,\Delta x + P(x_2)\,\Delta x + \cdots + P(x_{38})\,\Delta x$

Example 3   Express $\sum_{i=1}^{4} 3^i$ without using summation notation.

Solution

$$\sum_{i=1}^{4} 3^i = 3^1 + 3^2 + 3^3 + 3^4, \quad \text{or} \quad 120$$

Example 4   Express $\sum_{i=1}^{30} h(x_i)\,\Delta x$ without using summation notation.

Express without using summation notation.

4. $\displaystyle\sum_{i=1}^{3} 4^i$

5. $\displaystyle\sum_{i=1}^{5} ie^i$

6. $\displaystyle\sum_{i=1}^{20} t(x_i)\,\Delta x$

Solution

$$\sum_{i=1}^{30} h(x_i)\,\Delta x = h(x_1)\,\Delta x + h(x_2)\,\Delta x + \cdots + h(x_{30})\,\Delta x$$

DO EXERCISES 4–6.

Approximation of area by rectangles becomes more accurate as we use more rectangles and smaller subintervals, as we show in the figures below.

8 subintervals

24 subintervals

In general, the interval $[a, b]$ is divided into n equal subintervals, each of width $\Delta x = (b - a)/n$. The heights of the rectangles are

$$f(x_1), f(x_2), \ldots, f(x_n).$$

The area of the region under the curve is approximated by the sum of the areas of the rectangles:

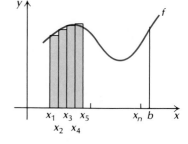

$$\sum_{i=1}^{n} f(x_i)\,\Delta x.$$

We now obtain the actual area by letting the number of intervals increase indefinitely and by then taking the limit. The area is thus given by

$$A = \lim_{n \to \infty} \sum_{i=1}^{n} f(x_i)\,\Delta x.$$

The area is also given by a definite integral:

$$\int_a^b f(x)\,dx = \lim_{n \to \infty} \sum_{i=1}^{n} f(x_i)\,\Delta x.$$

The fact that we can so express the integral of a function (positive or otherwise) as a limit of a sum or in terms of an antiderivative is so important that it has a name: *The Fundamental Theorem of Integral Calculus.*   CSS

The Fundamental Theorem of Integral Calculus

If a function f has an antiderivative F on $[a, b]$, then

$$\int_a^b f(x)\,dx = F(b) - F(a) = \lim_{n \to \infty} \sum_{i=1}^{n} f(x_i)\,\Delta x.$$

It is interesting to envision that, as we take the limit on the right, the summation sign stretches into something reminiscent of an S (the integral sign) and Δx is defined to be dx. This is also a motivation for the use of dx in the integral notation.

This result allows us to approximate the value of a definite integral by a sum, making it as good as we please by taking n sufficiently large.

Example 5   Raggs, Ltd., determines that the marginal cost per suit is given by

$$C'(x) = 0.0003x^2 - 0.2x + 50.$$

Approximate the total cost of producing 400 suits by computing the sum $\sum_{i=1}^{4} C'(x_i)\,\Delta x$.

Solution   The interval $[0, 400]$ is divided into 4 subintervals, each of length $\Delta x = (400 - 0)/4 = 100$. Now x_i is varying from $x_1 = 0$ to $x_4 = 300$.

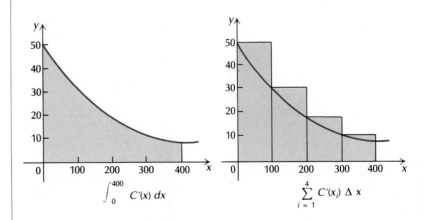

$$\int_0^{400} C'(x)\,dx \qquad\qquad \sum_{i=1}^{4} C'(x_i)\,\Delta x$$

7. Referring to Example 5, find

$$\sum_{i=1}^{8} C'(x_i)\,\Delta x,$$

where the interval [0, 400] is divided into 8 equal subintervals of length

$$\Delta x = \frac{400 - 0}{8} = 50.$$

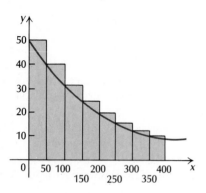

Thus we have

$$\sum_{i=1}^{4} C'(x_i)\,\Delta x = C'(0)\cdot 100 + C'(100)\cdot 100 + C'(200)\cdot 100$$
$$+ C'(300)\cdot 100$$
$$= 50\cdot 100 + 33\cdot 100 + 22\cdot 100 + 17\cdot 100$$
$$= \$12{,}200.$$

Now

$$\int_{0}^{400} C'(x)\,dx = \$10{,}400. \qquad \text{See Example 3 of Section 5.2.}$$

Thus the approximation is not too far off, even though the number of subintervals is small. In Margin Exercise 7 you will obtain a better approximation using 8 subintervals.

DO EXERCISES 7 AND 8. (EXERCISE 8 IS ON THE FOLLOWING PAGE.)

The fact that an integral can be approximated by a sum is useful when the antiderivative of a function does not have an elementary formula. For example, for the function $e^{-x^2/2}$, important in probability, there is no formula for the antiderivative. Thus tables of approximate values of its integral have been computed using summation methods.

The Average Value of a Function

Suppose that

$$T = f(t)$$

is the temperature at time t recorded at a weather station on a certain day. The station uses a 24-hour clock, so the domain of the temperature function is the interval [0, 24]. The function is continuous.

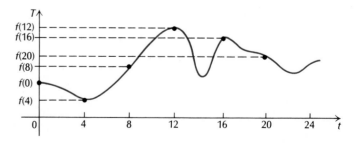

8. In graphs (a) and (b), compute the areas of each rectangle to four decimal places. Then add them to approximate the area under the curve $y = 1/x$ over $[1, 7]$.

a)

b)

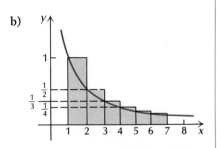

c) Evaluate

$$\int_1^7 \frac{1}{x}\, dx.$$

Find this answer using a calculator and compare it to (a) and (b).

To find the average temperature for the given day, we might take six temperature readings at 4-hour intervals, starting at midnight:

$$T_0 = f(0), \qquad T_1 = f(4),$$
$$T_2 = f(8), \qquad T_3 = f(12),$$
$$T_4 = f(16), \qquad T_5 = f(20).$$

The average reading would then be the sum of these six readings divided by 6:

$$T_{av} = \frac{T_0 + T_1 + T_2 + T_3 + T_4 + T_5}{6}.$$

This computation of the average temperature may not give the most useful answer. For example, suppose it is a hot summer day, and at 2:00 in the afternoon (hour 14 on the 24-hour clock) there is a short thunderstorm that cools the air for an hour between our readings. This temporary dip would not show up in the average computed above.

What can we do? We could take 48 readings at half-hour intervals. This should give a better result. In fact, the shorter the time between readings, the better the result should be. It seems reasonable that we might define the *average value of T* over the interval $[0, 24]$ to be the limit, as n approaches ∞, of the average of n values:

$$\text{Average value of } T = \lim_{n \to \infty} \frac{1}{n} \sum_{i=1}^n T_i$$

$$= \lim_{n \to \infty} \frac{1}{n} \sum_{i=1}^n f(t_i).$$

Note that this is not too far from our definition of an integral. All we would need is to get $24/n$, which is Δt, into the summation. We do this by expressing $1/n$ as $(1/24) \cdot (24/n)$. Then

$$\text{Average value of } T = \lim_{n \to \infty} \frac{1}{n} \sum_{i=1}^n f(t_i)$$

$$= \lim_{n \to \infty} \frac{1}{24} \cdot \frac{24}{n} \sum_{i=1}^n f(t_i)$$

$$= \frac{1}{24} \lim_{n \to \infty} \sum_{i=1}^n f(t_i) \cdot \frac{24}{n}$$

$$= \frac{1}{24} \lim_{n \to \infty} \sum_{i=1}^n f(t_i)\, \Delta t \qquad \Delta t = \frac{24}{n}$$

$$= \frac{1}{24} \int_0^{24} f(t)\, dt.$$

DEFINITION

Let f be a continuous function over a closed interval $[a, b]$. Its *average value*, y_{av}, is given by

$$y_{av} = \frac{1}{b - a} \int_a^b f(x)\, dx.$$

Let us consider average value in another way. If we multiply on both sides of

$$y_{av} = \frac{1}{b - a} \int_a^b f(x)\, dx$$

by $b - a$, we get

$$(b - a)y_{av} = \int_a^b f(x)\, dx.$$

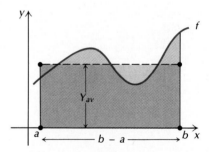

Now the expression on the left will give the area of a rectangle of length $b - a$ and height y_{av}. The area of such a rectangle is the same as the area bounded by $y = f(x)$ on the interval $[a, b]$.

Example 6 Find the average value of $f(x) = x^2$ over the interval $[0, 2]$.

Solution The average value is

$$\frac{1}{2} \int_0^2 x^2\, dx = \frac{1}{2} \left[\frac{x^3}{3} \right]_0^2$$

$$= \frac{1}{2} \left(\frac{2^3}{3} - \frac{0^3}{3} \right)$$

$$= \frac{1}{2} \cdot \frac{8}{3} = \frac{4}{3}, \quad \text{or} \quad 1\frac{1}{3}.$$

9. Find the average value of $f(x) = x^3$ over the interval $[0, 2]$.

10. The temperature over a 10-hour period is given by

$$f(t) = -t^2 + 5t + 40, \qquad 0 \leqslant t \leqslant 10.$$

   a) Find the average temperature.
   b) Find the minimum temperature.
   c) Find the maximum temperature.

11. The sales of a company are expected to grow according to the function

$$S(t) = 100t + t^2,$$

   where $S(t) =$ the sales, in dollars, on the tth day. Find the average sales from $t = 1$ to $t = 4$ (from the 1st to the 4th day).

Note that although the values of $f(x)$ increase from 0 to 4, we would not expect the average value to be 2, because we see from the graph that $f(x)$ is less than 2 over more than half the interval.

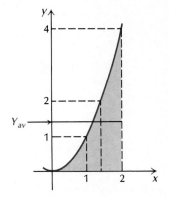

DO EXERCISES 9 AND 10.

Example 7   The emissions of an engine are given by

$$E(t) = 2t^2,$$

where $E(t) =$ the emissions in billions of pollution particulates at time t, in years. Find the average emissions from $t = 1$ to $t = 5$.

Solution   The average emissions are

$$\frac{1}{5-1} \int_1^5 2t^2\, dt = \frac{1}{4}\left[\frac{2}{3} t^3\right]_1^5$$

$$= \frac{1}{4} \cdot \frac{2}{3}\,(5^3 - 1^3)$$

$$= \frac{1}{6}\,(125 - 1)$$

$$= 20\tfrac{2}{3} \text{ billion pollution particulates.}$$

DO EXERCISE 11.

EXERCISE SET 5.8

1. **a)** Approximate

$$\int_1^7 (dx/x^2)$$

   by computing the area of each rectangle to four decimal places and then adding.

   b) Evaluate

$$\int_1^7 (dx/x^2).$$

   Compare the answer to (a).

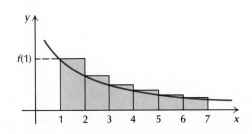

2. a) Approximate

$$\int_0^5 (x^2 + 1)\, dx$$

by computing the area of each rectangle and then adding.

b) Evaluate

$$\int_0^5 (x^2 + 1)\, dx.$$

Compare the answer to (a).

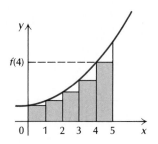

Find the average value over the given interval.

3. $y = 2x^3$; $[-1, 1]$

4. $y = 4 - x^2$; $[-2, 2]$

5. $y = e^x$; $[0, 1]$

6. $y = e^{-x}$; $[0, 1]$

7. $y = x^2 - x + 1$; $[0, 2]$

8. $y = x^2 + x - 2$; $[0, 4]$

9. $y = 3x + 1$; $[2, 6]$

10. $y = 4x + 1$; $[3, 7]$

11. $y = x^n$; $[0, 1]$

12. $y = x^n$; $[1, 2]$

APPLICATIONS

13. *Psychology: Results of studying.* A student's score on a test is a function

$$S(t) = t^2, \qquad t \text{ in } [0, 10],$$

where $S(t) =$ the score after t hours of study.

a) Find the maximum score the student can achieve and the number of hours of study required to attain it.

b) Find the average score over the 10-hour interval.

14. *Psychology: Results of practice.* A typist's speed over a 4-minute interval is given by

$$W(t) = -6t^2 + 12t + 90, \qquad t \text{ in } [0, 4],$$

where $W(t) =$ the speed in words per minute at time t.

a) Find the speed at the beginning of the interval.

b) Find the maximum speed and when it occurs.

c) Find the average speed over the 4-minute interval.

16. *Sociology: Average population of a city.* The population of a city increased and then decreased over an 8-year period according to the function

$$P(t) = -0.1t^2 + t + 3, \qquad 0 \leqslant t \leqslant 8,$$

where P is in millions and t is time.

a) Find the average population.

b) Find the minimum population.

c) Find the maximum population.

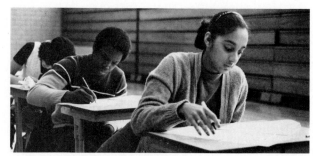

A student's test score is a function of the time spent studying.

15. *Sociology: Average population.* The population of the United States is given by

$$P(t) = 241e^{0.009t},$$

where P is in millions and t is the number of years since 1986. Find the average value of the population from 1986 to 1996.

17. *Biomedical: Average drug dosage.* The amount of a drug in the body at time t is given by

$$A(t) = 3e^{-0.1t},$$

where A is in cubic centimeters and t is time in hours.

a) What is the initial dosage of the drug?

b) What is the average amount in the body over a 2-hour period?

EXTENSION EXERCISES

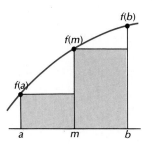

The Trapezoidal Rule. Another way to approximate an integral is to replace each rectangle in the sum (see the figure at the right) by a trapezoid (as shown in the second figure). The area of a trapezoid is $h(c_1 + c_2)/2$, where c_1 and c_2 are the lengths of the parallel sides. Thus, in the second figure,

$$\int_a^b f(x)\,dx = \int_a^m f(x)\,dx + \int_m^b f(x)\,dx$$

$$\approx \Delta x \frac{f(a) + f(m)}{2} + \Delta x \frac{f(m) + f(b)}{2}$$

$$\approx \Delta x \left[\frac{f(a)}{2} + f(m) + \frac{f(b)}{2}\right].$$

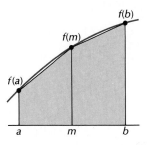

For an interval $[a, b]$ subdivided into n equal subintervals of length $\Delta x = (b - a)/n$, we get the approximation

$$\int_a^b f(x)\,dx \approx \Delta x \left[\frac{f(a)}{2} + f(x_2) + f(x_3) + \cdots + f(x_n) + \frac{f(b)}{2}\right],$$

where $x_1 = a$ and $x_n = x_{n-1} + \Delta x$ or $x_n = a + (n - 1)\,\Delta x$. This is called the *Trapezoidal Rule.*

18. Use the Trapezoidal Rule and the interval subdivision of Exercise 1 to approximate

$$\int_1^7 \frac{dx}{x^2}.$$

19. Use the Trapezoidal Rule and the interval subdivision of Exercise 2 to approximate

$$\int_0^5 (x^2 + 1)\,dx.$$

For those studying this chapter in the text *Applied Calculus,* 2nd ed. by Bittinger and Morrel (Reading, MA: Addison-Wesley, 1988), there is an extended treatment of the Trapezoidal Rule in Chapter 11.)

SUMMARY AND REVIEW: CHAPTER 5

The following contains a summary of what you should be able to do after completing this chapter. The review exercises are for practice. Answers are at the back of the book. If you miss an exercise, restudy the section indicated alongside the answers.

A summary of the important formulas for this chapter is given on the inside back cover of this book. Where directed, use Table 1 at the back of the book.

You should be able to:

Evaluate the indefinite integral of a function using the basic formulas given in Theorems 2 and 3. Evaluate a definite integral using the basic formulas. Find the area under a curve on a closed interval. Interpret the area under a curve in two other ways.

Evaluate.

1. $\int 8x^4 \, dx$

2. $\int (3e^x + 2) \, dx$

3. $\int \left(3t^2 + 7t + \frac{1}{t}\right) dt$

Find the area under the curve on the interval indicated.

4. $y = 3 - x^2$; $[-2, 1]$

5. $y = x^2 + 3x + 6$; $[0, 2]$

In each case, give two interpretations of the shaded region.

6.

7.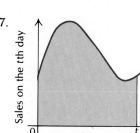

Evaluate.

8. $\int_a^b x^5 \, dx$

9. $\int_{-1}^1 (x^3 - x^4) \, dx$

10. $\int_0^1 (e^x + x) \, dx$

11. $\int_1^3 \frac{3}{x} \, dx$

Find the area of a region bounded by two curves.

Decide whether $\int_a^b f(x) \, dx$ is positive, negative, or zero.

12.

13.

14.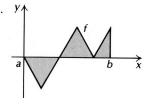

15. Find the area of the region bounded by $y = 3x^2$ and $y = 9x$.

Evaluate integrals using the integration techniques of substitution, integration by parts, and tables.

Evaluate using substitution. Do not use Table 1.

16. $\int x^3 e^{x^4} \, dx$

17. $\int \frac{24t^5}{4t^6 + 3} \, dt$

18. $\int \frac{\ln 4x}{2x} \, dx$

19. $\int 2e^{-3x} \, dx$

Evaluate using integration by parts. Do not use Table 1.

20. $\int 3xe^{3x} \, dx$

21. $\int \ln x^7 \, dx$

22. $\int 3x^2 \ln x \, dx$

Evaluate using Table 1.

23. $\int \frac{1}{49 - x^2} \, dx$

24. $\int x^2 e^{5x} \, dx$

25. $\int \frac{x}{7x + 1} \, dx$

26. $\int \frac{dx}{\sqrt{x^2 - 36}}$

27. $\int x^6 \ln x \, dx$

28. $\int xe^{8x} \, dx$

Approximate a definite integral by adding areas of rectangles. Find the average value of a function.

29. Approximate $\int_1^4 (2/x)\, dx$ by computing the area of each rectangle to three decimal places and adding.

30. Find the average value of $y = e^{-x} + 5$ over $[0, 2]$.

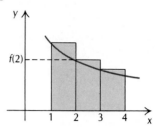

Solve applied problems involving integration.

31. A particle starts out from the origin. Its velocity at any time t, $t \geqslant 0$, is given by $v(t) = 3t^2 + 2t$. Find the distance that the particle travels during the first 4 hours (from $t = 0$ to $t = 4$).

32. A company estimates that its sales will grow continuously according to the function $S'(t) = 3e^{3t}$, where $S'(t) =$ sales, in dollars, on the tth day. Find the accumulated sales for the first 4 days.

EXTENSION EXERCISES

Evaluate.

33. $\int \dfrac{t^4 \ln (t^5 + 3)}{t^5 + 3}\, dt.$

34. $\int \dfrac{dx}{e^x + 2}$

35. $\int \dfrac{\ln \sqrt{x}}{x}\, dx$

36. $\int x^{91} \ln x\, dx$

37. $\int \ln\left(\dfrac{x-3}{x-4}\right) dx$

38. $\int \dfrac{dx}{x(\ln x)^4}$

TEST: CHAPTER 5

Evaluate.

1. $\int dx$

2. $\int 1000x^4\, dx$

3. $\int \left(e^x + \dfrac{1}{x} + x^{3/8}\right) dx$

Find the area under the curve on the interval indicated.

4. $y = x - x^2;\ [0, 1]$

5. $y = \dfrac{4}{x};\ [1, 3]$

6. Give two interpretations of the shaded area.

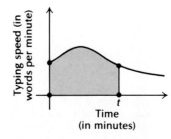

Evaluate.

7. $\int_{-1}^{2} (2x + 3x^2)\, dx$

8. $\int_{0}^{1} e^{-2x}\, dx$

9. $\int_{a}^{b} \dfrac{dx}{x}$

10. Decide whether $\int_{a}^{b} f(x)\, dx$ is positive, negative, or zero.

Evaluate using substitution. Do not use Table 1.

11. $\int \dfrac{dx}{x + 8}$

12. $\int e^{-0.5x}\, dx$

13. $\int t^3 (t^4 + 1)^9\, dt$

Evaluate using integration by parts. Do not use Table 1.

14. $\int xe^{5x}\, dx$

15. $\int x^3 \ln x^4\, dx$

Evaluate using Table 1.

16. $\int 2^x\, dx$

17. $\int \dfrac{dx}{x(7 - x)}$

18. Find the average value of $y = 4t^3 + 2t$ over $[-1, 2]$.

19. Find the area of the region bounded by $y = x$, $y = x^5$, $x = 0$, and $x = 1$.

20. Approximate

$$\int_{0}^{5} (25 - x^2)\, dx$$

by computing the area of each rectangle and adding.

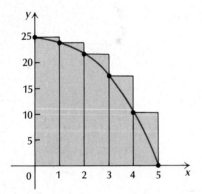

21. An air conditioning company determines that the marginal cost of the xth air conditioner is given by

$$C'(x) = -0.2x + 500, \qquad C(0) = 0.$$

Find the total cost of producing 100 air conditioners.

22. A typist's speed over a 4-minute interval is given by

$$W'(t) = -6t^2 + 12t + 90, \qquad t \text{ in } [0, 4],$$

where $W(t) = $ the speed in words per minute at time t. How many words are typed during the second minute (from $t = 1$ to $t = 2$)?

EXTENSION EXERCISES

Evaluate using any method.

23. $\int \dfrac{[(\ln x)^3 - 4(\ln x)^2 + 5]}{x}\, dx$

24. $\int \ln\left(\dfrac{x + 3}{x + 5}\right) dx$

6

APPLICATIONS OF INTEGRATION

In this chapter we study a wide variety of applications of integration. We first consider an application in economics to finding consumer's surplus and producer's surplus. Then we study the integration of functions involved in exponential growth and decay. Here we find not only applications to business, but also to such environmental concerns as the depletion of resources and the buildup of radioactivity in the atmosphere. Integration has extensive application to probability and statistics, to finding volume, and to many situations involving the solution of differential equations.

Topics in this chapter can be chosen to fit the needs of the student and the course.

AN APPLICATION

The depletion of oil. The world reserves of oil are 670,700 million barrels. In 1980 ($t = 0$) the world use of oil was 66,164 million barrels, and the growth rate for the use of oil was 10%. Assuming that this growth rate continues and that no new reserves are discovered, when will the world reserves of oil be exhausted?

THE MATHEMATICS

To find the time T at which the world reserves of oil will be exhausted, we solve the equation

$$670,700 = \int_0^T 66,164e^{0.11t}\, dt.$$

6.1

OBJECTIVE

a) Given a demand function $D(x)$ and a supply function $S(x)$, find the equilibrium point and the consumer's surplus and the producer's surplus at the equilibrium point.

ECONOMIC APPLICATION: CONSUMER'S SURPLUS AND PRODUCER'S SURPLUS

Recall that the consumer's demand curve $D(x)$ gives the demand price per unit that the consumer is willing to pay for x units. It is a decreasing function. The producer's supply curve $S(x)$ gives the price per unit at which the producer is willing to supply x units. It is an increasing function. The equilibrium point (x_E, p_E) is the intersection of the two curves. (See Section 1.5.)

Suppose the following figure represents the demand curve of college students for movies. Suppose we consider this curve for just one such student, Samantha.

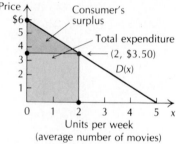

We want to examine the utility, or *benefit,* Samantha receives from going to movies. Samantha goes to 0 movies per week if the price is $6. She will go to 1 movie per week if the price is $4.80. Suppose she goes to 1 movie. Then the total expenditure to her is $4.80 · 1, or $4.80. The area of the gray shaded region represents her total expenditure, $4.80. Look at the area of the color region. It is $\frac{1}{2}(1)(\$1.20)$, or $0.60. The total area under the curve is a measure of the *total benefit,* or utility, of going to 1 movie, and is $5.40. The area of the color region, $0.60, is a measure of the benefit Samantha gets but does not have to pay for. Economists define this amount as *consumer's surplus.* It is the benefit of living in a society where the price consumers are willing to pay decreases when more units are purchased.

Suppose Samantha goes to 2 movies per week if the price is $3.50. The total area under the curve is $8.25 and represents the total benefit to Samantha of going to 2 movies. The total amount Samantha actually spends is $3.50(2), or $7, and is the area of the gray shaded region. It is her total expenditure. Look at the area of the color region. It (the consumer's surplus) is $\frac{1}{2}(2)$ (2.50), or $1.25. It is a measure of the benefit Samantha received but did not have to pay for.

Suppose the demand function is actually a curve, like the following.

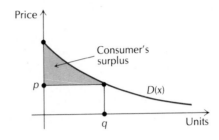

If Samantha goes to q movies when the price is p, then the total expenditure to Samantha is qp. The total area under the curve is the total utility, or total benefits received, and is

$$\int_0^q D(x)\,dx.$$

The *consumer's surplus* is the total area under the curve minus the total expenditure, qp, and is the total benefits minus the total costs, which is given by

$$\int_0^q D(x)\,dx - qp.$$

DEFINITION

Suppose $y = D(x)$ describes the demand function for a commodity. Then the consumer's surplus is defined for point (q, p) as

$$\int_0^q D(x)\,dx - qp.$$

Example 1   Find the consumer's surplus for the demand function $D(x) = (x - 5)^2$ when $x = 3$.

Solution   When $x = 3$, $D(3) = (3 - 5)^2 = (-2)^2 = 4$. Then

$$\text{Consumer's surplus} = \int_0^3 (x - 5)^2\,dx - 3 \cdot 4$$

$$= \int_0^3 (x^2 - 10x + 25)\,dx - 12$$

1. Find the consumer's surplus for the demand function

$$D(x) = (x - 6)^2$$

when $x = 4$.

and

$$= \left[\frac{x^3}{3} - 5x^2 + 25x \right]_0^3 - 12$$

$$= \left(\frac{3^3}{3} - 5(3)^2 + 25(3) \right)$$

$$- \left(\frac{0^3}{3} - 5(0)^2 + 25(0) \right) - 12$$

$$= (9 - 45 + 75) - 0 - 12 = \$27.$$

DO EXERCISE 1.

Suppose we now look at the supply curve for the movies. At a price of \$0 per movie the producer is willing to supply 0 movies. At a price of \$1.75 per movie, the producer is willing to supply 1 movie and makes total receipts of 1(\$1.75), or \$1.75. The area of the gray shaded triangle represents the total cost to the producer of producing 1 movie, and is $\frac{1}{2}(1)$ (\$1.75), or \$0.875. The area of the color triangle is also \$0.875, and represents the surplus over cost. It is a contribution to profit. Economists call this number the *producer's surplus*. It is the benefit to the producer of living in a society where a greater amount of a commodity will be supplied when the price increases.

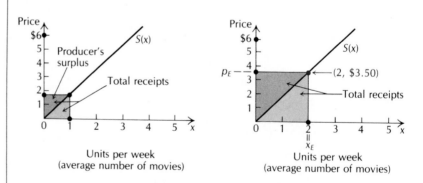

Units per week
(average number of movies)

At a price of \$3.50 the producer is willing to supply 2 movies and makes total receipts of 2(\$3.50), or \$7. The area of the gray shaded triangle represents the total cost to the producer of actually making the 2 movies, and is $\frac{1}{2}(2)$ (\$3.50), or \$3.50. The area of the color triangle is also \$3.50 and is the producer's surplus. It is a contribution to the profit of the producer.

2. Find the producer's surplus for the supply function

$$y = x^2 + x + 10$$

when $x = 1$.

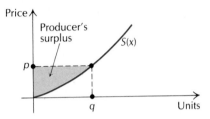

Suppose the supply function $S(x)$ is actually a curve. The producer will supply q movies if the price is p. The total receipts are qp. The producer's surplus is the total receipts minus the area under the curve and is given by

$$qp - \int_0^q S(x)\,dx.$$

DEFINITION

Suppose $y = S(x)$ is the supply function for a commodity. The *producer's surplus* is defined for the point (q, p) as

$$qp - \int_0^q S(x)\,dx.$$

Example 2 Find the producer's surplus for $S(x) = x^2 + x + 3$ when $x = 3$.

Solution When $x = 3$, $S(3) = 3^2 + 3 + 3 = 15$. Then

$$
\begin{aligned}
\text{Producer's surplus} &= 3 \cdot 15 - \int_0^3 (x^2 + x + 3)\,dx \\
&= 45 - \left[\frac{x^3}{3} + \frac{x^2}{2} + 3x\right]_0^3 \\
&= 45 - \left(\left[\frac{3^3}{3} + \frac{3^2}{2} + 3(3)\right] - \left[\frac{0^3}{3} + \frac{0^2}{2} + 3(0)\right]\right) \\
&= 45 - (9 + \tfrac{9}{2} + 9) + 0 = \$22.50.
\end{aligned}
$$

DO EXERCISE 2.

The equilibrium point (x_E, p_E) is the point at which the supply and demand curves intersect. It is that point where the sellers and buyers come together and purchases and sales actually occur. The following shows the equilibrium point and the consumer's and producer's surplus for the movie curves and for the general case.

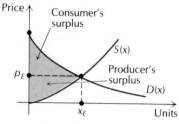

Example 3   Given

$$D(x) = (x - 5)^2,$$
$$S(x) = x^2 + x + 3,$$

find each of the following.

a) The equilibrium point

b) The consumer's surplus at the equilibrium point

c) The producer's surplus at the equilibrium point

Solution

a) To find the equilibrium point, we set $D(x) = S(x)$ and solve:

$$(x - 5)^2 = x^2 + x + 3$$
$$x^2 - 10x + 25 = x^2 + x + 3$$
$$-10x + 25 = x + 3$$
$$22 = 11x$$
$$\tfrac{22}{11} = x$$
$$2 = x.$$

Thus $x_E = 2$ units. To find p_E we substitute x_E into either $D(x)$ or $S(x)$. If we choose $D(x)$, then

$$p_E = D(x_E) = D(2) = (2 - 5)^2 = (-3)^2 = \$9 \text{ per unit.}$$

Thus the equilibrium point is $(2, \$9)$.

b) The consumer's surplus at the equilibrium point is

$$\int_0^{x_E} D(x)\, dx - x_E p_E,$$

3. Given

$$D(x) = (x - 6)^2,$$
$$S(x) = x^2 + x + 10,$$

find each of the following.

a) The equilibrium point

b) The consumer's surplus at the equilibrium point

c) The producer's surplus at the equilibrium point

or

$$\int_0^2 (x - 5)^2\, dx - 2 \cdot 9 = \left[\frac{(x - 5)^3}{3}\right]_0^2 - 18$$

$$= \left[\frac{(2 - 5)^3}{3} - \frac{(0 - 5)^3}{3}\right] - 18$$

$$= \frac{(-3)^3}{3} - \frac{(-5)^3}{3} - 18$$

$$= -\frac{27}{3} + \frac{125}{3} - \frac{54}{3}$$

$$= \frac{44}{3} = \$14.67.$$

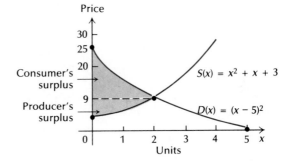

c) The producer's surplus at the equilibrium point is

$$x_E p_E - \int_0^{x_E} S(x)\, dx,$$

or

$$2 \cdot 9 - \int_0^2 (x^2 + x + 3)\, dx = 2 \cdot 9 - \left[\frac{1}{3}x^3 + \frac{1}{2}x^2 + 3x\right]_0^2$$

$$= 18 - \left[\left(\frac{1}{3} \cdot 2^3 + \frac{1}{2} \cdot 2^2 + 3 \cdot 2\right)\right.$$

$$\left. - \left(\frac{1}{3} \cdot 0^3 + \frac{1}{2} \cdot 0^2 + 3 \cdot 0\right)\right]$$

$$= 18 - \left(\frac{8}{3} + 2 + 6\right)$$

$$= \frac{22}{3} = \$7.33.$$

DO EXERCISE 3.

EXERCISE SET 6.1

In each exercise, find (a) the equilibrium point, (b) the consumer's surplus at the equilibrium point, and (c) the producer's surplus at the equilibrium point.

1. $D(x) = -\frac{5}{6}x + 10$, $S(x) = \frac{1}{2}x + 2$ 2. $D(x) = -2x + 8$, $S(x) = x + 2$
3. $D(x) = (x - 4)^2$, $S(x) = x^2 + 2x + 6$ 4. $D(x) = (x - 3)^2$, $S(x) = x^2 + 2x + 1$
5. $D(x) = (x - 6)^2$, $S(x) = x^2$ 6. $D(x) = (x - 8)^2$, $S(x) = x^2$

EXTENSION EXERCISES

7. ▦ $D(x) = e^{-x + 4.5}$, $S(x) = e^{x - 5.5}$ 8. ▦ $D(x) = \sqrt{56 - x}$, $S(x) = x$

6.2

OBJECTIVES

a) Find the balance in a savings account from an initial investment at a given interest rate, compounded continuously, for a given period of time t.

b) Find the amount of a continuous money flow over a period of time at an interest rate compounded continuously.

c) Given a continuous flow of money into an investment at the rate of P_0 dollars per year, find P_0 so that the amount of a continuous money flow over a specified time will be a given amount.

d) Find the total use of a natural resource over a given period of time.

APPLICATIONS OF THE MODEL $\int_0^T P_0 e^{kt} \, dt$

In this chapter we will make frequent use of the integration formula

$$\int b e^{ax} \, dx = \frac{b}{a} e^{ax} + C. \tag{1}$$

You should be able to use it quickly. It was derived by substitution in Example 7 of Section 5.5.

Recall the basic model of exponential growth (Section 4.3):

$$P'(t) = k \cdot P(t), \quad \text{or} \quad \frac{dP}{dt} = kP.$$

The solution to the equation is

$$P(t) = P_0 e^{kt}. \tag{2}$$

Thus $P(t)$ is an antiderivative of $kP_0 e^{kt}$, as we can see by using Eq. (1):

$$\int k P_0 e^{kt} \, dt = \frac{kP_0}{k} e^{kt} = P_0 e^{kt}.$$

One application of Eq. (2) is to compute the balance of a savings account after t years from an initial investment of P_0 at interest rate k, compounded continuously.

1. Find the balance in a savings account after 2 years from an initial investment of $1000 at an interest rate of 7.5% compounded continuously.

Example 1 Find the balance in a savings account after 3 years from an initial investment of $1000 at interest rate 8% compounded continuously.

Solution Using Eq. (2) with $k = 0.08$, $t = 3$, and $P_0 = \$1000$, we get

$$P(3) = 1000e^{0.08(3)} = 1000e^{0.24} \approx 1000(1.271249)$$

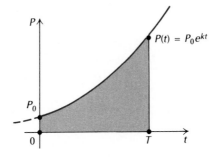

$$\approx \$1271.25.$$

DO EXERCISE 1.

The Integral $\int_0^T P_0 e^{kt}\, dt$

Consider the integral of $P_0 e^{kt}$ over the interval $[0, T]$:

$$\int_0^T P_0 e^{kt}\, dt = \left[\frac{P_0}{k} \cdot e^{kt} \right]_0^T = \frac{P_0}{k}\left(e^{kT} - e^{k \cdot 0} \right) = \frac{P_0}{k}\left(e^{kT} - 1 \right).$$

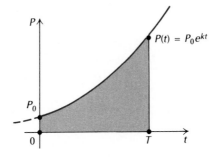

Thus,

$$\int_0^T P_0 e^{kt}\, dt = \frac{P_0}{k}\left(e^{kT} - 1 \right). \tag{3}$$

In the remainder of this section we consider two applications of this definite integral.

Continuous Money Flow

Suppose that money is flowing continuously into a savings account at an annual rate of $1000 per year at an interest rate of 8% compounded continuously. This means that over a small amount of time dt, the bank pays into the account a small part of the principal plus interest on all other money accumulated in the account up to this time (since $t = 0$).

2. Find the amount of a continuous money flow where $2000 per year is being invested at 7.5% compounded continuously for 10 years.   `CSS`

The amount that is paid in over time dt is

$$\$1000e^{0.08t}\, dt.$$

Suppose we want to find the accumulation of all these amounts over a 5-year period. That accumulation is given by the integral

$$\int_0^5 \$1000e^{0.08t}\, dt = \left[\frac{1000}{0.08} e^{0.08t} \right]_0^5 = 12{,}500(e^{0.08\cdot 5} - e^{0.08\cdot 0})$$

$$= 12{,}500(e^{0.4} - 1)$$

$$\approx 12{,}500(1.491825 - 1)$$

$$\approx \$6147.81.$$

Economists call $6147.81 the *amount of a continuous money flow*. In this case, the money is flowing according to a constant function $R(t) = \$1000$. Money could also be flowing according to some variable function, say, $R(t) = 2t - 7$ or $R(t) = t^2$.

DEFINITION

If the rate of flow of money into an investment is given by some function $R(t)$, then the *amount of a continuous money flow* at interest rate k, compounded continuously, over time T is given by

$$\int_0^T R(t)e^{kt}\, dt.$$

Example 2   Find the amount of a continuous money flow where $1000 per year is being invested at 8% compounded continuously for 15 years.

Solution

$$\int_0^{15} \$1000e^{0.08t}\, dt = \frac{1000}{0.08}[e^{0.08(15)} - 1] = 12{,}500(e^{1.2} - 1)$$

$$\approx 12{,}500(3.320117 - 1)$$

$$\approx \$29{,}001.46$$

DO EXERCISE 2.

Sometimes we might want to know how money should be flowing into an investment so that we end up with a specified amount.

3. Consider a continuous flow of money into an investment at the rate of P_0 dollars per year. What should P_0 be so that the amount of a continuous money flow over 20 years, at interest rate 7.5% compounded continuously, will be $10,000?

Example 3   Consider a continuous flow of money into an investment at the rate of P_0 dollars per year. What should P_0 be so that the amount of a continuous money flow over 20 years, at interest rate 8% compounded continuously, will be $10,000?

Solution   We find P_0 such that

$$10,000 = \int_0^{20} P_0 e^{0.08t}\, dt.$$

We solve the following equation:

$$10,000 = \frac{P_0}{0.08}\left(e^{0.08(20)} - 1\right)$$
$$800 = P_0(e^{1.6} - 1)$$
$$800 \approx P_0(4.953032 - 1)$$
$$800 = P_0(3.953032)$$
$$\$202.38 \approx P_0.$$

DO EXERCISE 3.

Ecology: Depletion of Natural Resources

Another application of the integral of exponential growth concerns

$$P(t) = P_0 e^{kt}$$

as a model of the demand for natural resources. Suppose that P_0 represents the amount of a natural resource (such as coal or oil) used at time $t = 0$ and that the growth rate for the use of this resource is k. Then, assuming exponential growth (which is the case for the use of many resources), the amount to be used at time t is $P(t)$, given by

$$P(t) = P_0 e^{kt}.$$

The total amount used during an interval $[0, T]$ is given by

$$\int_0^T P(t)\, dt = \int_0^T P_0 e^{kt}\, dt = \frac{P_0}{k}\left(e^{kT} - 1\right). \tag{4}$$

Example 4   *The demand for copper.* In 1980 ($t = 0$) the world use of copper was 21,350,000 tons, and the demand for it was growing exponentially at the rate of 15% per year. If the growth continues at this rate, how many tons of copper will the world use from 1980 to 1990?

4. *The demand for oil.* In 1980 ($t = 0$) the world use of oil was 66,164 million barrels, and the demand for it was growing exponentially at the rate of 10% per year. If the demand continues at this rate, how many barrels of oil will the world use from 1980 to 1990?

Solution   Using Eq. (4), we have

$$\int_0^{10} 21,350,000 e^{0.15t}\,dt = \frac{21,350,000}{0.15}\,(e^{0.15 \cdot 10} - 1)$$
$$= 142,333,333(e^{1.5} - 1) \quad \blacksquare$$
$$= 142,333,333(4.481689 - 1) \quad \blacksquare$$
$$= 142,333,333(3.481689)$$
$$\approx 495,560,400.$$

Thus from 1980 to 1990 the world will use 495,560,400 tons of copper.

DO EXERCISE 4.

Example 5   *The depletion of copper.* The world reserves of copper are 689,000,000 tons. Assuming the growth rate in Example 4 continues and that no new reserves are discovered, when will the world reserves of copper be exhausted?

Bingham Canyon mine in Utah has produced more copper than any other mine in history. The grade of ore, however, has dropped from 1.93 percent copper in 1906 to 0.6 percent today. At present it is planned that mining will cease when the percentage of copper reaches 0.4.

Solution   Using Eq. (4), we want to find T such that

$$689,000,000 = \frac{21,350,000}{0.15}\,(e^{0.15T} - 1).$$

5. *The depletion of oil.* The world reserves of oil are 670,700 million barrels. In 1980 $(t = 0)$ the world use of oil was 66,164 million barrels, and the growth rate for the use of oil was 10%. Assuming that this growth rate continues and that no new reserves are discovered, when will the world reserves of oil be exhausted?

We solve for T as follows:

$$689,000,000 = 142,333,333(e^{0.15T} - 1)$$

$$\frac{689,000,000}{142,333,333} = e^{0.15T} - 1 \qquad \text{▦}$$

$$4.840749 = e^{0.15T} - 1$$

$$5.840749 = e^{0.15T}$$

$$\ln 5.840749 = \ln e^{0.15T} \qquad \text{Taking the natural logarithm on both sides}$$

$$\ln 5.840749 = 0.15T \qquad \text{Recall: } \ln e^k = k.$$

$$\frac{\ln 5.840749}{0.15} = T$$

$$\frac{1.764859}{0.15} = T \qquad \text{▦}$$

$$12 \approx T. \qquad \text{Rounding to the nearest one}$$

Thus 12 years from 1980 (or by 1992), the world reserves of copper will be exhausted.

DO EXERCISE 5.

EXERCISE SET 6.2

APPLICATIONS

1. Find the amount in a savings account after 3 years from an initial investment of $100 at 9% compounded continuously.

2. Find the amount in a savings account after 4 years from an initial investment of $100 at 10% compounded continuously.

3. Find the amount of a continuous money flow where $100 per year is being invested at 9% compounded continuously for 20 years.

4. Find the amount of a continuous money flow where $100 per year is being invested at 10% compounded continuously for 20 years.

5. Find the amount of a continuous money flow where $1000 per year is being invested at 8.5% compounded continuously for 40 years.

6. Find the amount of a continuous money flow where $1000 per year is being invested at 7.5% compounded continuously for 40 years.

7. What should P_0 be so that the amount of a continuous money flow over 20 years, at interest rate 8.5% compounded continuously, will be $50,000?

8. What should P_0 be so that the amount of a continuous money flow over 20 years, at interest rate 7.5% compounded continuously, will be $50,000?

9. What should P_0 be so that the amount of a continuous money flow over 30 years, at interest rate 9% compounded continuously, will be $40,000?

10. What should P_0 be so that the amount of a continuous money flow over 30 years, at interest rate 10% compounded continuously, will be $40,000?

11. *The demand for aluminum ore.* In 1980 ($t = 0$) the world use of aluminum ore was 64,674,000 tons, and the demand for it was growing exponentially at the rate of 12% per year. If the demand continues to grow at this rate, how many tons of aluminum ore will the world use from 1980 to 1990?

13. *The depletion of aluminum ore.* The world reserves of aluminum ore are 22,670,000,000 tons. Assuming that the growth rate of Exercise 11 continues and that no new reserves are discovered, when will the world reserves of aluminum ore be exhausted?

12. *The demand for natural gas.* In 1980 ($t = 0$) the world use of natural gas was 52,360 billion cubic feet, and the demand for it was growing exponentially at the rate of 4% per year. If the demand continues to grow at this rate, how many cubic feet of natural gas will the world use from 1980 to 1990?

14. *The depletion of natural gas.* The world reserves of natural gas are 2,911,000 billion cubic feet. Assuming that the growth rate of Exercise 12 continues and that no new reserves are discovered, when will the world reserves of natural gas be exhausted?

EXTENSION EXERCISES

Stock dividends. The total dividends on stock $D(t)$ that a company pays in time T is given by

$$D(T) = \int_0^T d_0 e^{pkt}\, dt,$$

where d_0 = the instantaneous dividend payment at time 0, p = the percentage of the company's earnings that it retains, and k = the rate of return that a company can earn on its assets if it were to invest them.

15. Find the total dividends when $d_0 = \$10$, $p = 80\%$, $k = 15\%$, and $T = 50$ years.

16. Find a general formula for $D(T)$.

Find the amount of a continuous money flow when:

17. $R(t) = 2000t + 7$, $k = 8\%$, $T = 30$ years.

18. $R(t) = t^2$, $k = 7\%$, $T = 40$ years.

6.3

APPLICATIONS OF THE MODEL $\int_0^T Pe^{-kt}\, dt$

OBJECTIVES

a) Find the present value of an investment due t years later at a certain interest rate compounded continuously.

b) Find the accumulated present value of an investment.

A representative of a financial institution is often asked to solve a problem like the following.

Example 1 Following the birth of a child, a parent wants to make an initial investment of P_0 that will grow to $\$10,000$ by the child's 20th birthday. Interest is compounded continuously at 8%. What should the initial investment be?

Solution Using the equation $P = P_0 e^{kt}$, we find P_0 such that

$$10,000 = P_0 e^{0.08 \cdot 20}, \quad \text{or} \quad 10,000 = P_0 e^{1.6}.$$

1. Following the birth of a child, a parent wants to make an initial investment P_0 that will grow to $10,000 by the child's 20th birthday. Interest is compounded continuously at 7.5%. What should this initial investment be?

Now

$$\frac{10,000}{e^{1.6}} = P_0, \quad \text{or} \quad 10,000e^{-1.6} = P_0,$$

and, using a calculator, we have

$$P_0 = 10,000e^{-1.6} = 10,000(0.201897) \approx \$2018.97.$$

Thus the parent must deposit $2018.97, which will grow to $10,000 by the child's 20th birthday.

Economists call $2018.97 the *present value* of $10,000 due 20 years from now at 8% compounded continuously. The process of computing present value is called *discounting*.

DO EXERCISE 1.

In general, the present value P_0 of an amount P due t years later is found by solving the following equation for P_0:

$$P_0 e^{kt} = P$$

$$P_0 = \frac{P}{e^{kt}} = Pe^{-kt}.$$

2. Find the present value of $40,000 due 5 years later at 10% compounded continuously.   **CSS**

THEOREM 1

The *present value* P_0 of an amount P due t years later at interest rate k, compounded continuously, is given by

$$P_0 = Pe^{-kt}.$$

Note that this can be interpreted as exponential decay from the future back to the present.

DO EXERCISE 2.

Suppose we know that a continuous flow of money will go into an investment at the rate of P dollars per year, from now until some time T in the future. If an infinitesimal amount of time dt passes,

$$(P)\,dt$$

dollars will have accumulated. The present value of that amount is

$$[P \cdot dt]e^{-kt},$$

3. Find the accumulated present value of an investment over a 20-year period if there is a continuous money flow of $1800 per year and the current interest rate is 10% compounded continuously.

where k is the current interest rate compounded continuously. The accumulation of all the present values is given by the integral

$$\int_0^T Pe^{-kt}\,dt,$$

and is called the *accumulated present value*. Evaluating this integral using Eq. (1) of Section 6.2, we get

$$\int_0^T Pe^{-kt}\,dt = \frac{P}{-k}(e^{-kT} - e^{-k \cdot 0}) = \frac{P}{k}(1 - e^{-kT}).$$

In the preceding case money is flowing according to a constant function $R(t) = P$. Money could also be flowing according to some variable function such as $R(t) = 2t + 8$ or $R(t) = t^3$.

THEOREM 2

The *accumulated present value* of a continuous money flow into an investment at the rate of $R(t)$ dollars per year from now until some time T in the future is given by

$$\int_0^T R(t)e^{-kt}\,dt,$$

where k is the current interest rate compounded continuously.

Example 2 Find the accumulated present value of an investment over a 5-year period if there is a continuous money flow of $2400 per year and the current interest rate is 14% compounded continuously.

Solution The accumulated present value is

$$\int_0^5 \$2400e^{-0.14t}\,dt = \frac{2400}{0.14}(1 - e^{-0.14 \cdot 5})$$

$$= 17{,}142.86(1 - e^{-0.7})$$

$$= 17{,}142.86(1 - 0.496585)$$

$$\approx \$8629.97.$$

DO EXERCISE 3.

The preceding example is an application of the model

$$\int_0^T Pe^{-kt}\,dt = \frac{P}{k}(1 - e^{-kT}).$$

This model can be applied to a calculation of the buildup of a specific amount of radioactive material released into the atmosphere annually. Some of the material decays, but more continues to be released. The amount present at time T is given by the integral above.

EXERCISE SET 6.3

APPLICATIONS

1. Following the birth of a child, a parent wants to make an initial investment P_0 that will grow to $5000 by the child's 20th birthday. Interest is compounded continuously at 9%. What should the initial investment be?

2. Following the birth of a child, a parent wants to make an initial investment P_0 that will grow to $5000 by the child's 20th birthday. Interest is compounded continuously at 10%. What should the initial investment be?

3. Find the present value of $60,000 due 8 years later at 12% compounded continuously.

4. Find the present value of $50,000 due 16 years later at 14% compounded continuously.

5. Find the accumulated present value of an investment over a 10-year period if there is a continuous money flow of $2700 per year and the current interest rate is 9% compounded continuously.

6. Find the accumulated present value of an investment over a 10-year period if there is a continuous money flow of $2700 per year and the current interest rate is 10% compounded continuously.

7. An MBA accepts a position as president of a company at the age of 35. Assuming retirement at age 65 and an annual salary of $45,000 that is paid in a continuous money flow, what is the president's accumulated present value? The current interest rate is 8%, compounded continuously.

8. A college dropout takes a job as a truck driver at the age of 25. Assuming retirement at age 65 and an annual salary of $14,000 that is paid in a continuous money flow, what is the truck driver's accumulated present value? The current interest rate is 7%, compounded continuously.

9. *Radioactive buildup.* Plutonium has a decay rate of 0.003% per year. Suppose plutonium is released into the atmosphere each year for 20 years at the rate of 1 pound per year. What is the total amount of radioactive buildup?

10. *Radioactive buildup.* Cesium-137 has a decay rate of 2.3% per year. Suppose cesium-137 is released into the atmosphere each year for 20 years at the rate of 1 pound per year. What is the total amount of radioactive buildup?

EXTENSION EXERCISES

Find the accumulated present value when:

11. $R(t) = t$, $k = 8\%$, and $T = 20$ years.

12. $R(t) = e^t$, $k = 7\%$, and $T = 10$ years.

Accumulated present values of dividends. Suppose $d(t)$ represents the instantaneous dividend payment of a stock at time t. Then $d(t)e^{-mt}$ is the present value of that payment, where m is the current interest rate. The accumulation of all present values from time 0 to time T is given by

$$D_p(T) = \int_0^T d(t)e^{-mt}\,dt.$$

13. Find $D_p(T)$ when $d(t) = \$10$, $m = 8\%$, and $T = 10$ years.

14. Find $D_p(T)$ when $d(t) = t$, $m = 7\%$, and $T = 20$ years.

6.4

OBJECTIVES

a) Determine whether an improper integral is convergent or divergent, and calculate its value if it is convergent.

b) Solve applied problems involving improper integrals.

1. Complete.

b	$1 - \dfrac{1}{b}$
2	$1 - \frac{1}{2}$, or $\frac{1}{2}$
3	
10	
100	
200	

IMPROPER INTEGRALS

Let us try to find the area of the region under the graph of $y = 1/x^2$ on the interval $[1, \infty)$.

   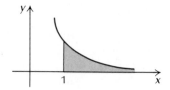

Note that this region is of infinite extent. We have not yet considered how to find the area of such a region. Let us find the area under the curve on the interval from 1 to b, and then see what happens as b gets very large. The area under the graph on $[1, b]$ is

$$\int_1^b \frac{dx}{x^2} = \left[-\frac{1}{x} \right]_1^b$$

$$= \left(-\frac{1}{b} \right) - \left(-\frac{1}{1} \right)$$

$$= -\frac{1}{b} + 1 = 1 - \frac{1}{b}.$$

Then

$$\lim_{b \to \infty} [\text{area from 1 to } b] = \lim_{b \to \infty} \left(1 - \frac{1}{b} \right).$$

Let us investigate this limit.

DO EXERCISE 1.

Note that as $b \to \infty$, $1/b \to 0$, so $[1 - 1/b] \to 1$. Thus,

$$\lim_{b \to \infty} [\text{area from 1 to } b] = \lim_{b \to \infty} \left(1 - \frac{1}{b} \right) = 1.$$

We *define* the area from 1 to ∞ to be this limit. Here we have an example of an infinitely long region with a finite area.

2. Find the area of the region under the graph of

$$y = \frac{1}{x^2}$$

on the interval $[2, \infty)$.

DO EXERCISE 2.

Such areas may not always be finite. Let us try to find the area of the region under the graph of $y = 1/x$ on the interval $[1, \infty)$.

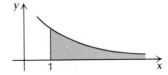

By definition, the area A from 1 to ∞ is the limit as $b \longrightarrow \infty$ of the area from 1 to b, so

$$A = \lim_{b \to \infty} \int_1^b \frac{dx}{x} = \lim_{b \to \infty} [\ln x]_1^b = \lim_{b \to \infty} (\ln b - \ln 1) = \lim_{b \to \infty} \ln b.$$

In Section 4.2 we learned that $\ln b$ increases indefinitely as b increases. Therefore, the limit does not exist.

Thus we have an infinitely long region with an infinite area. Note that the graphs of $y = 1/x^2$ and $y = 1/x$ have similar shapes, but the region under one of them has a finite area and the other does not.

3. Find the area under the graph of

$$y = \frac{1}{x}$$

from $x = 2$ to $x = \infty$.

DO EXERCISE 3.

An integral such as

$$\int_a^\infty f(x)\, dx,$$

with an upper limit of ∞, is called an *improper integral*. Its value is defined to be the following limit.

DEFINITION

$$\int_a^\infty f(x)\, dx = \lim_{b \to \infty} \int_a^b f(x)\, dx$$

If the limit exists, then we say that the improper integral *converges*, or is *convergent*. If the limit does not exist, then we say that the improper integral *diverges*, or is *divergent*. Thus,

$$\int_1^\infty \frac{dx}{x^2} = 1 \ \ converges; \quad \text{and} \quad \int_1^\infty \frac{dx}{x} \ \ diverges.$$

Determine whether the improper integral is convergent or divergent, and calculate its value if it is convergent.

4. $\int_0^\infty 5e^{-5x}\, dx$

Example 1

$$\int_0^\infty 2e^{-2x}\, dx = \lim_{b \to \infty} \int_0^b 2e^{-2x}\, dx$$

$$= \lim_{b \to \infty} \left[2\left(-\frac{1}{2}\right)e^{-2x} \right]_0^b$$

$$= \lim_{b \to \infty} \left[-e^{-2x} \right]_0^b$$

$$= \lim_{b \to \infty} \left[-e^{-2b} - (-e^{-2 \cdot 0}) \right]$$

$$= \lim_{b \to \infty} \left(-e^{-2b} + 1 \right)$$

$$= \lim_{b \to \infty} \left(1 - \frac{1}{e^{2b}} \right)$$

Now as $b \to \infty$, we know that $e^{2b} \to \infty$ (from Chapter 4), so

$$\frac{1}{e^{2b}} \to 0 \quad \text{and} \quad \left(1 - \frac{1}{e^{2b}} \right) \to 1.$$

Thus,

$$\int_0^\infty 2e^{-2x}\, dx = 1.$$

(The integral is convergent.)

5. $\int_0^\infty 2x\, dx$

DO EXERCISES 4 AND 5.

Following are definitions of two other types of improper integral.

DEFINITIONS

1. $\displaystyle\int_{-\infty}^b f(x)\, dx = \lim_{a \to -\infty} \int_a^b f(x)\, dx$

2. $\displaystyle\int_{-\infty}^\infty f(x)\, dx = \int_{-\infty}^c f(x)\, dx + \int_c^\infty f(x)\, dx$

In order for $\int_{-\infty}^\infty f(x)\, dx$ to converge, both integrals on the right above must converge.

Applications

In Section 6.3 we learned that the accumulated present value of a continuous money flow of P dollars per year from now until time T in the

6. Find the accumulated present value of an investment for which there is a perpetual continuous money flow of $3600 per year. The current interest rate is 10%, compounded continuously.

future is given by

$$\int_0^T Pe^{-kt}\,dt = \frac{P}{k}\left(1 - e^{-kT}\right),$$

where k is the current interest rate. Suppose the money flow is to continue perpetually. Under this assumption, the accumulated present value over this infinite time period would be

$$\int_0^\infty Pe^{-kt}\,dt = \lim_{T\to\infty}\int_0^T Pe^{-kt}\,dt$$

$$= \lim_{T\to\infty}\frac{P}{k}\left(1 - e^{-kT}\right)$$

$$= \lim_{T\to\infty}\frac{P}{k}\left(1 - \frac{1}{e^{kT}}\right)$$

$$= \frac{P}{k}.$$

THEOREM 3

The *accumulated present value* of a continuous money flow into an investment at the rate of P dollars per year perpetually is given by

$$\frac{P}{k},$$

where k is the current interest rate compounded continuously.

Example 2 Find the accumulated present value of an investment for which there is a perpetual continuous money flow of $2000 per year. The current interest rate is 8%, compounded continuously.

Solution The accumulated present value is 2000/0.08, or $25,000.

DO EXERCISE 6.

When an amount P of radioactive material is being released into the atmosphere annually, the amount present at time T is given by

$$\int_0^T Pe^{-kt}\,dt = \frac{P}{k}\left(1 - e^{-kT}\right).$$

As $T \longrightarrow \infty$ (the radioactive material is to be released forever), the buildup of radioactive material approaches a limiting value P/k.

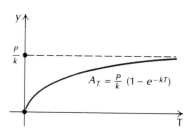

It is no wonder scientists and environmentalists become so concerned about continued nuclear detonations. Eventually the buildup is "here to stay."

EXERCISE SET 6.4

Determine whether the improper integral is convergent or divergent, and calculate its value if it is convergent.

1. $\int_3^\infty \dfrac{dx}{x^2}$

2. $\int_4^\infty \dfrac{dx}{x^2}$

3. $\int_3^\infty \dfrac{dx}{x}$

4. $\int_4^\infty \dfrac{dx}{x}$

5. $\int_0^\infty 3e^{-3x}\, dx$

6. $\int_0^\infty 4e^{-4x}\, dx$

7. $\int_1^\infty \dfrac{dx}{x^3}$

8. $\int_1^\infty \dfrac{dx}{x^4}$

9. $\int_0^\infty \dfrac{dx}{1+x}$

10. $\int_0^\infty \dfrac{4\, dx}{1+x}$

11. $\int_1^\infty 5x^{-2}\, dx$

12. $\int_1^\infty 7x^{-2}\, dx$

13. $\int_0^\infty e^x\, dx$

14. $\int_0^\infty e^{2x}\, dx$

15. $\int_3^\infty x^2\, dx$

16. $\int_5^\infty x^4\, dx$

17. $\int_0^\infty xe^x\, dx$

18. $\int_0^\infty \ln x\, dx$

19. $\int_0^\infty me^{-mx}\, dx,\, m > 0$

20. $\int_0^\infty Qe^{-kt}\, dt,\, k > 0$

APPLICATIONS

21. Find the accumulated present value of an investment for which there is a perpetual continuous money flow of $3600 per year. The current interest rate is 8%.

22. Find the accumulated present value of an investment for which there is a perpetual continuous money flow of $4500 per year. The current interest rate is 9%.

23. *Radioactive buildup.* Plutonium has a decay rate of 0.003% per year. Suppose plutonium is released into the atmosphere each year perpetually at the rate of 1 pound per year. What is the limiting value of the radioactive buildup?

24. *Radioactive buildup.* Cesium-137 has a decay rate of 2.3% per year. Suppose cesium-137 is released into the atmosphere each year perpetually at the rate of 1 pound per year. What is the limiting value of the radioactive buildup?

EXTENSION EXERCISES

Determine whether the improper integral is convergent or divergent, and calculate its value if it is convergent.

25. $\displaystyle\int_0^\infty \frac{dx}{x^{2/3}}$

26. $\displaystyle\int_1^\infty \frac{dx}{\sqrt{x}}$

27. $\displaystyle\int_0^\infty \frac{dx}{(x+1)^{3/2}}$

28. $\displaystyle\int_{-\infty}^0 e^{2x}\,dx$

29. $\displaystyle\int_0^\infty xe^{-x^2}\,dx$

30. $\displaystyle\int_{-\infty}^\infty xe^{-x^2}\,dx$

Accumulated present values of stock dividends paid perpetually. The accumulation of all present values of dividends that are assumed to be paid perpetually is given by

$$V = \int_0^\infty d(t)e^{-mt}\,dt,$$

where $d(t)$ is the instantaneous dividend payment and m is the current interest rate.

31. Find V when $d(t) = e^{-t}$ and $m = 7\%$.

32. Find V when $d(t) = \$1000$ and $m = 8\%$.

Biomedical: Drug dosage. Suppose an oral dose of a drug is taken. From that time, the drug is assimilated in the body and excreted through the urine. The total amount of the drug that has passed through the body in time T is given by

$$\int_0^T E(t)\,dt,$$

where E is the rate of excretion of the drug through the urine. A typical rate of excretion function is $E(t) = te^{-kt}$, where $k > 0$ and t is time in hours.

33. Find $\int_0^\infty E(t)\,dt$ and interpret the answer. That is, what does the integral represent?

34. A physician prescribes a dosage of 100 mg. Find k.

6.5

OBJECTIVES

a) Verify that a given function satisfies the property

$$\int_a^b f(x)\,dx = 1$$

for being a probability density function.

b) Find k such that a function like

$$f(x) = kx^2$$

is a probability density function over an interval $[a, b]$.

c) Solve applied problems involving probability density functions.

PROBABILITY

The definite integral plays a role in the theory of probability. Briefly, the *probability* of an event is a number from 0 to 1 that represents the chances of the event occurring. It is the "relative frequency" of occurrence—the percentage of times we expect an event to occur in a large number of trials.

Example 1 What is the probability of drawing an ace from a well-shuffled deck of cards?

Solution Since there are 52 possible outcomes, each card has the same chance of being drawn, and since there are 4 aces, the probability of drawing an ace is $\frac{4}{52}$ or $\frac{1}{13}$, or about 7.7%.

In practice we may not draw an ace 7.7% of the time, but in a large number of trials, after shuffling the cards and drawing one, replacing

1. In Example 2, what is the probability that the ball you are holding is:

   a)   black?

   b)   yellow?

   c)   green?

   d)   purple?

2. Consider this dartboard.

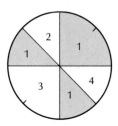

You throw a dart at the board without aiming at any particular region. What is the probability that you will score a:

   a)   1?

   b)   2?

   c)   3?

   d)   4?

A desire to calculate odds in games of chance gave rise to the theory of probability.

the card, shuffling the cards and drawing one, we would expect to draw an ace about 7.7% of the time. That is, the more draws we make, the closer we expect to get to 7.7%.

Example 2   A bag contains 7 black balls, 6 yellow balls, 4 green balls, and 3 red balls. The bag is shaken well and you remove 1 ball without looking. What is the probability that it is red? that it is white?

Solution   There are 20 balls altogether and of these 3 are red, so the probability of drawing a red ball is $\frac{3}{20}$. There are no white balls, so the probability of drawing a white one is $\frac{0}{20}$, or 0.

DO EXERCISES 1 AND 2.

Let us consider a table of probabilities from Example 2.

Color	Probability
Black (B)	$\frac{7}{20}$
Yellow (Y)	$\frac{6}{20}$
Green (G)	$\frac{4}{20}$
Red (R)	$\frac{3}{20}$

Note that the sum of these probabilities is 1. We are certain that we will draw either a black, yellow, green, or red ball. The probability of that event is 1. Let us arrange these data from the table into what is called a *frequency graph,* or *histogram,* which shows the fraction of times each event occurs (the probability of each event).

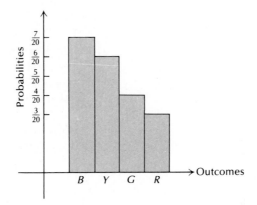

If we assign a width of 1 to each rectangle, then the sum of the areas of the rectangles is 1.

Continuous Random Variables

Suppose we throw a dart at a number line in such a way that it always lands in the interval [1, 3]. Let x be the number that the dart hits. Note that x is a quantity that can be observed (or measured) repeatedly and whose possible values consist of an entire interval of real numbers. Such a variable is called a *continuous random variable.*

Suppose we throw the dart a large number of times and it lands in the subinterval [1.6, 2.8] 43% of the time. The probability, then, that the dart lands in that interval is 0.43.

Let us consider some other examples of continuous random variables.

3. Suppose that dosage x of a drug is from 15 mg to 25 mg. What interval is determined?

Example 3   Suppose that x is the arrival time of buses at a bus stop in a three-hour period from 2:00 P.M. to 5:00 P.M. The interval is [2, 5]. Then x is a continuous random variable distributed over the interval [2, 5].

Example 4   Suppose that x is the corn acreage of each farm in the United States and Canada. The interval is [0, a], where a is the highest acreage. (Not knowing what the highest acreage might be, we could say that the interval is [0, ∞) to allow for all possibilities. *Note:* It might be argued that there is a value in [0, a] or [0, ∞) for which no farm has that acreage, but for practical purposes all values are included in our consideration.)

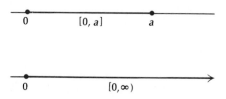

Then x is a continuous random variable distributed over the interval [0, a] or [0, ∞).

4. Suppose that distance x is the distance between successive cars on a highway. What interval is determined?

DO EXERCISES 3 AND 4.

Considering Example 3 on the arrival times of buses, suppose that we wanted to know the probability that a bus will arrive between 4:00 P.M. and 5:00 P.M., as represented by

$$P([4, 5]), \quad \text{or} \quad P(4 \leqslant x \leqslant 5).$$

In some cases it is possible to find a function over [2, 5] such that areas over subintervals give the probabilities that a bus will arrive during these subintervals. For example, suppose we have a constant function $f(x) = \frac{1}{3}$ that will give us these probabilities. Look at its graph.

5. Find the probability that a bus will arrive between 2:30 P.M. and 4:30 P.M.

The area under the curve is $3 \cdot \frac{1}{3}$, or 1. The probability that a bus will arrive between 4:00 P.M. and 5:00 P.M. is that fraction of the large area that lies over the interval $[4, 5]$. That is,

$$P([4, 5]) = \tfrac{1}{3} = 33\tfrac{1}{3}\%.$$

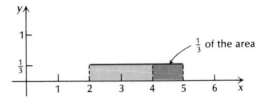

The probability that a bus will arrive between 2:00 P.M. and 4:30 P.M. is $\frac{5}{6}$, or $83\frac{1}{3}\%$.

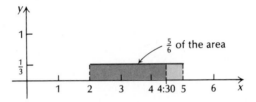

DO EXERCISE 5.

Note that any interval of length 1 has probability $\frac{1}{3}$. This may not always happen. Suppose we have a function

$$f(x) = \tfrac{3}{117}x^2$$

whose definite integral over the interval $[4, 5]$ would yield the probability that a bus will arrive between 4:00 P.M. and 5:00 P.M. Then

$$P([4, 5]) = \int_4^5 f(x)\, dx$$

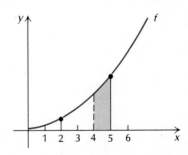

$$= \int_4^5 \tfrac{3}{117}x^2\, dx$$

$$= [\tfrac{3}{117} \cdot \tfrac{1}{3}x^3]_4^5$$

$$= \tfrac{1}{117}(5^3 - 4^3)$$

$$= \tfrac{61}{117} \approx 0.52.$$

Thus 52% of the time you will be able to catch a bus between 4:00 P.M. and 5:00 P.M. The function f is called a *probability density function.* Its

6. Verify Property 3 of the definition of a probability density function for

$$f(x) = \tfrac{2}{3}x \quad \text{over} \quad [1, 2].$$

integral over *any* subinterval gives the probability that x "lands" in that subinterval.

DEFINITION

Let x be a continuous random variable distributed over some interval $[a, b]$. A function f is said to be a *probability density function* for x if:

1. f is nonnegative over $[a, b]$, that is, $f(x) \geq 0$ for all x in $[a, b]$;

2. for any subinterval $[c, d]$ of $[a, b]$, the probability $P([c, d])$, or $P(c \leq x \leq d)$, that x lands in that subinterval is given by

$$P([c, d]) = \int_c^d f(x)\, dx;$$

3. the probability that x lands in $[a, b]$ is 1:

$$\int_a^b f(x)\, dx = 1;$$

that is, we are "certain" that x is in the interval $[a, b]$.

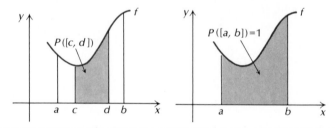

Example 5   Verify Property 3 of the definition above for

$$f(x) = \tfrac{3}{117}x^2.$$

Solution   The "big" interval under consideration is $[2, 5]$. So

$$\int_2^5 \tfrac{3}{117} x^2\, dx = [\tfrac{3}{117} \cdot \tfrac{1}{3} x^3]_2^5 = \tfrac{1}{117}(5^3 - 2^3) = \tfrac{117}{117} = 1.$$

DO EXERCISE 6.

Example 6   A company produces transistors. It determines that the life t of a transistor is from 3 to 6 years and that the probability density function for t is given by

$$f(t) = \frac{24}{t^3}, \quad \text{for } 3 \leq t \leq 6.$$

7. Referring to Example 6:

  a) Verify Property 3 of the definition of a probability density function.

  b) Find the probability that a transistor will last no more than 5 years.

  c) Find the probability that a transistor will last from 4 to 6 years.

a) Find the probability that a transistor will last no more than 4 years.

b) Find the probability that a transistor will last from 4 to 5 years.

Solution

a) The probability that a transistor will last no more than 4 years is

$$P(3 \leqslant t \leqslant 4) = \int_3^4 \frac{24}{t^3}\, dt$$

$$= \left[24\left(-\frac{1}{2}t^{-2}\right)\right]_3^4$$

$$= \left[-\frac{12}{t^2}\right]_3^4$$

$$= -12\left(\frac{1}{4^2} - \frac{1}{3^2}\right)$$

$$= -12\left(\frac{1}{16} - \frac{1}{9}\right)$$

$$= -12\left(-\frac{7}{144}\right)$$

$$= \frac{7}{12} \approx 0.58.$$

b) The probability that a transistor will last from 4 to 5 years is

$$P(4 \leqslant t \leqslant 5) = \int_4^5 \frac{24}{t^3}\, dt$$

$$= \left[24\left(-\frac{1}{2}t^{-2}\right)\right]_4^5$$

$$= \left[-\frac{12}{t^2}\right]_4^5$$

$$= -12\left(\frac{1}{5^2} - \frac{1}{4^2}\right)$$

$$= -12\left(\frac{1}{25} - \frac{1}{16}\right)$$

$$= -12\left(-\frac{9}{400}\right)$$

$$= \frac{27}{100} = 0.27.$$

DO EXERCISE 7.

8. Find k such that

$$f(x) = kx^2$$

is a probability density function over the interval $[1, 3]$.

Constructing Probability Density Functions

Suppose you have an arbitrary nonnegative function $f(x)$ whose definite integral over some interval $[a, b]$ is K. Then

$$\int_a^b f(x)\, dx = K.$$

Multiply on both sides by $1/K$:

$$\frac{1}{K}\int_a^b f(x)\, dx = \frac{1}{K}\cdot K = 1 \quad \text{or} \quad \int_a^b \frac{1}{K}\cdot f(x)\, dx = 1.$$

Thus when we multiply the function $f(x)$ by $1/K$, we have a function whose area over the given interval is 1.

Example 7   Find k such that

$$f(x) = kx^2$$

is a probability density function over the interval $[2, 5]$.

Solution   We have

$$\int_2^5 x^2\, dx = \left[\frac{x^3}{3}\right]_2^5 = \frac{5^3}{3} - \frac{2^3}{3} = \frac{125}{3} - \frac{8}{3} = \frac{117}{3}.$$

9. Find k such that

$$f(x) = kx^3$$

is a probability density function over the interval $[0, 1]$.

Thus,

$$k = \frac{1}{\left(\frac{117}{3}\right)} = \frac{3}{117} \quad \text{and} \quad f(x) = \frac{3}{117}x^2.$$

DO EXERCISES 8 AND 9.

Uniform Distributions

Suppose the probability density function of a continuous random variable is constant. How is it described? Consider the following graph.

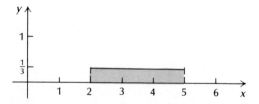

The length of the shaded rectangle is the length of the interval $[2, 5]$, which is 3. In order for the shaded area to be 1, the height of the rectangle must be $\frac{1}{3}$. Thus, $f(x) = \frac{1}{3}$.

10. A number x is selected at random from the interval $[7, 15]$. The probability density function for x is given by

$$f(x) = \tfrac{1}{8}, \quad \text{for } 7 \leqslant x \leqslant 15.$$

Find the probability that a number selected is in the subinterval $[11, 13]$.

For the general case, consider this graph.

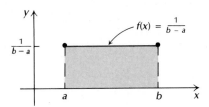

The length of the shaded rectangle is the length of the interval $[a, b]$, which is $b - a$. In order for the shaded area to be 1, the height of the rectangle must be $1/(b - a)$. Thus, $f(x) = 1/(b - a)$.

DEFINITION

A continuous random variable x is said to be *uniformly distributed* over an interval $[a, b]$ if it has a probability density function f given by

$$f(x) = \frac{1}{b - a}, \quad \text{for } a \leqslant x \leqslant b.$$

Example 8 A number x is selected at random from the interval $[40, 50]$. The probability density function for x is given by

$$f(x) = \tfrac{1}{10}, \quad \text{for } 40 \leqslant x \leqslant 50.$$

Find the probability that a number selected is in the subinterval $[42, 48]$.

Solution The probability is

$$P(42 \leqslant x \leqslant 48) = \int_{42}^{48} \tfrac{1}{10}\, dx = \tfrac{1}{10}[x]_{42}^{48}$$

$$= \tfrac{1}{10}(48 - 42) = \tfrac{6}{10} = 0.6.$$

DO EXERCISE 10.

Example 9 A company produces sirens for tornado warnings. The maximum loudness L of the sirens ranges from 70 to 100 decibels. The probability density function for L is

$$f(L) = \tfrac{1}{30}, \quad \text{for } 70 \leqslant L \leqslant 100.$$

A siren is selected at random off the assembly line. Find the probability that its maximum loudness is from 70 to 92 decibels.

11. A person arrives at a bus stop. The waiting time t for a bus is 0 to 20 minutes. The probability density function for t is

$$f(t) = \tfrac{1}{20}, \quad \text{for } 0 \leqslant t \leqslant 20.$$

What is the probability that the person will have to wait no more than 5 minutes for a bus?

Solution   The probability is

$$P(70 \leqslant L \leqslant 92) = \int_{70}^{92} \tfrac{1}{30}\, dL$$

$$= \tfrac{1}{30}[L]_{70}^{92}$$

$$= \tfrac{1}{30}(92 - 70)$$

$$= \tfrac{22}{30} = \tfrac{11}{15} \approx 0.73.$$

DO EXERCISE 11.

Exponential Distributions

The duration of a phone call, the distance between successive cars on a highway, and the amount of time required to learn a task are all examples of exponentially distributed random variables. That is, their probability density functions are exponential.

DEFINITION

A continuous random variable is *exponentially distributed* if it has a probability density function given by

$$f(x) = ke^{-kx}, \quad \text{over the interval } [0, \infty).$$

---

12. Verify that

$$\int_0^\infty ke^{-kx}\, dx = 1.$$

The function $f(x) = 2e^{-2x}$ is such a probability density function. That

$$\int_0^\infty 2e^{-2x}\, dx = 1$$

is shown in Section 6.4. The general case

$$\int_0^\infty ke^{-kx}\, dx = 1$$

can be verified in a similar way.

DO EXERCISE 12.

A transportation planner can determine probabilities that cars are certain distances apart.

Why is it reasonable to assume that distance between cars is exponentially distributed? Part of the reason is that there are many more cases in which distances are small, though we can find other distributions that are "skewed" in this manner. The same argument holds for the duration of a phone call. That is, there are more short calls than long ones. The rest of the reason might lie in analyzing data involving such distances or phone calls.

Example 10 *Transportation planning.* The distance x, in feet, between successive cars on a certain stretch of highway has a probability density function

$$f(x) = ke^{-kx}, \quad \text{for } 0 \leqslant x < \infty,$$

where $k = 1/a$ and $a =$ the average distance between successive cars over some period of time.

A transportation planner determines that the average distance between cars on a certain stretch of highway is 166 ft. What is the probability that the distance between cars is 50 ft or less?

Solution We first determine k:

$$k = \tfrac{1}{166}$$
$$\approx 0.006024.$$

The probability density function for x is

$$f(x) = 0.006024e^{-0.006024x}, \quad \text{for } 0 \leqslant x < \infty.$$

The probability that the distance between cars is 50 ft or less is

$$
\begin{aligned}
P(0 \leqslant x \leqslant 50) &= \int_0^{50} 0.006024e^{-0.006024x}\, dx \\
&= \left[\frac{0.006024}{-0.006024} e^{-0.006024x} \right]_0^{50} \\
&= [-e^{-0.006024x}]_0^{50} \\
&= (-e^{-0.006024 \cdot 50}) - (-e^{-0.006024 \cdot 0}) \\
&= -e^{-0.301200} + 1 \\
&= 1 - e^{-0.301200} \\
&= 1 - 0.739930 \\
&\approx 0.260.
\end{aligned}
$$

13. A transportation planner determines that the average distance between cars on a certain stretch of highway is 125 ft. What is the probability that the distance between cars is 50 ft or less?

DO EXERCISE 13.

EXERCISE SET 6.5

Verify Property 3 of the definition of a probability density function over the given interval.

1. $f(x) = 2x$, $[0, 1]$
2. $f(x) = \frac{1}{4}x$, $[1, 3]$
3. $f(x) = \frac{1}{3}$, $[4, 7]$
4. $f(x) = \frac{1}{4}$, $[9, 13]$

5. $f(x) = \frac{3}{26}x^2$, $[1, 3]$
6. $f(x) = \frac{3}{64}x^2$, $[0, 4]$
7. $f(x) = \frac{1}{x}$, $[1, e]$
8. $f(x) = \frac{1}{e-1}e^x$, $[0, 1]$

9. $f(x) = \frac{3}{2}x^2$, $[-1, 1]$
10. $f(x) = \frac{1}{3}x^2$, $[-2, 1]$
11. $f(x) = 3e^{-3x}$, $[0, \infty)$
12. $f(x) = 4e^{-4x}$, $[0, \infty)$

Find k such that the function is a probability density function over the given interval.

13. $f(x) = kx$, $[1, 3]$
14. $f(x) = kx$, $[1, 4]$
15. $f(x) = kx^2$, $[-1, 1]$

16. $f(x) = kx^2$, $[-2, 2]$
17. $f(x) = k$, $[2, 7]$
18. $f(x) = k$, $[3, 9]$

19. $f(x) = k(2 - x)$, $[0, 2]$
20. $f(x) = k(4 - x)$, $[0, 4]$
21. $f(x) = \frac{k}{x}$, $[1, 3]$

22. $f(x) = \frac{k}{x}$, $[1, 2]$
23. $f(x) = ke^x$, $[0, 3]$
24. $f(x) = ke^x$, $[0, 2]$

APPLICATIONS

25. A dart is thrown at a number line in such a way that it always lands in the interval $[0, 10]$. Let $x =$ the number the dart hits. Suppose the probability density function for x is given by

    $$f(x) = \tfrac{1}{50}x, \quad \text{for } 0 \leqslant x \leqslant 10.$$

    Find $P(2 \leqslant x \leqslant 6)$, the probability that it lands in $[2, 6]$.

26. Suppose in the situation of Exercise 25 that the dart always lands in the interval $[0, 5]$, and that the probability density function for x is given by

    $$f(x) = \tfrac{3}{125}x^2, \quad \text{for } 0 \leqslant x \leqslant 5.$$

    Find $P(1 \leqslant x \leqslant 4)$, the probability that it lands in $[1, 4]$.

27. A number x is selected at random from the interval $[4, 20]$. The probability density function for x is given by

    $$f(x) = \tfrac{1}{16}, \quad \text{for } 4 \leqslant x \leqslant 20.$$

    Find the probability that a number selected is in the subinterval $[9, 17]$.

28. A number x is selected at random from the interval $[5, 29]$. The probability density function for x is given by

    $$f(x) = \tfrac{1}{24}, \quad \text{for } 5 \leqslant x \leqslant 29.$$

    Find the probability that a number selected is in the subinterval $[13, 29]$.

29. A transportation planner determines that the average distance between cars on a certain highway is 100 ft. What is the probability that the distance between cars is 40 ft or less?

30. A transportation planner determines that the average distance between cars on a certain highway is 200 ft. What is the probability that the distance between cars is 10 ft or less?

31. A telephone company determines that the duration t of a phone call is an exponentially distributed random variable with probability density function

    $$f(t) = 2e^{-2t}, \quad 0 \leqslant t < \infty.$$

    Find the probability that a phone call will last no more than 5 minutes.

32. Referring to the data in Exercise 31, find the probability that a phone call will last no more than 2 minutes.

33. In a psychology experiment, the time t, in seconds, that it takes a rat to learn its way through a maze is an exponentially distributed random variable with probability density function

$$f(t) = 0.02e^{-0.02t}, \quad 0 \leqslant t < \infty.$$

Find the probability that a rat will learn its way through a maze in 150 seconds or less.

34. Assume the situation and equation in Exercise 33, but find the probability that a rat will learn its way through a maze in 50 seconds or less.

35. The *time to failure t*, in hours, of a certain machine can often be assumed to be exponentially distributed with probability density function

$$f(t) = ke^{-kt}, \quad 0 \leqslant t < \infty,$$

where $k = 1/a$ and $a =$ the average time that will pass before a failure occurs. Suppose the average time that will pass before a failure occurs is 100 hours. What is the probability that a failure will occur in 50 hours or less?

The time it takes a rat to learn its way through a maze is an exponentially distributed random variable.

36. The *reliability* of the machine (the probability that it will work) in Exercise 35 is defined as

$$R(T) = 1 - \int_0^T 0.01e^{-0.01t}\, dt,$$

where $R(T)$ is the reliability at time T. Find $R(T)$.

EXTENSION EXERCISES

37. The function $f(x) = x^3$ is a probability density on $[0, b]$. What is b?

38. The function $f(x) = 12x^2$ is a probability density on $[-a, a]$. What is a?

6.6

OBJECTIVES

a) Given a probability density function, find $E(x)$, $E(x^2)$, the mean, the variance, and the standard deviation.

b) Use Table 2 at the back of the book to evaluate probabilities involving a normal distribution.

PROBABILITY: EXPECTED VALUE; THE NORMAL DISTRIBUTION

Expected Value

Let us again consider throwing a dart at a number line in such a way that it always lands in the interval $[1, 3]$.

Suppose we throw the dart at the line 100 times and keep track of the numbers it hits. Then suppose we calculate the arithmetic mean (or average) $\bar{x}$ of all these numbers:

$$\bar{x} = \frac{x_1 + x_2 + x_3 + \cdots + x_{100}}{100} = \frac{\sum_{i=1}^{100} x_i}{100} = \sum_{i=1}^{100} x_i \cdot \frac{1}{100}.$$

The expression

$$\sum_{i=1}^{n} x_i \cdot \frac{1}{n}$$

is analogous to the integral

$$\int_{1}^{3} x \cdot f(x) \, dx,$$

where f is the probability density function for x and where, in this case, $f(x)$ is the constant function 1. Now $1/n$ can be considered to be Δx and $f(x)$ gives a "weight" to x. We add all the $x_i \cdot 1 \cdot (1/n)$ values when we find $\sum_{i=1}^{n} x_i \cdot 1 \cdot (1/n)$; and, similarly, we add all the values $x \cdot f(x)$ values when we find $\int_{1}^{3} x \cdot f(x) \, dx$.

Suppose we have the probability density function $f(x) = \frac{1}{4}x$. This function gives more "weight" to the right side of the interval than to the left. Perhaps more points are attained when the dart lands on the right. Then

$$\int_{1}^{3} x \cdot f(x) \, dx = \int_{1}^{3} x \cdot \frac{1}{4} x \, dx = \left[\frac{1}{4} \cdot \frac{x^3}{3} \right]_{1}^{3} = \left[\frac{x^3}{12} \right]_{1}^{3}$$

$$= \frac{1}{12} (3^3 - 1^3) = \frac{26}{12} \approx 2.17.$$

Suppose we continue to throw the dart and compute averages. The more times we throw the dart, the closer we expect the averages to come to 2.17.

Let x be a continuous random variable over the interval $[a, b]$ with probability density function f.

DEFINITION

The expected value of x is defined by

$$E(x) = \int_{a}^{b} x \cdot f(x) \, dx,$$

where f is a probability density function for x.

---

1. Given the probability density function

$$f(x) = 2x \quad \text{over } [0, 1],$$

find $E(x)$ and $E(x^2)$.

The concept of expected value generalizes to other functions of x. Suppose $y = g(x)$. Then we have the following.

DEFINITION

The expected value of $g(x)$ is defined by

$$E(g(x)) = \int_a^b g(x) \cdot f(x)\, dx,$$

where f is a probability density function for x.

For example,

$$E(x) = \int_a^b x f(x)\, dx,$$

$$E(x^2) = \int_a^b x^2 f(x)\, dx,$$

$$E(e^x) = \int_a^b e^x f(x)\, dx,$$

and

$$E(2x + 3) = \int_a^b (2x + 3) f(x)\, dx.$$

Example 1 Given the probability density function

$$f(x) = \tfrac{1}{2}x \quad \text{over } [0, 2],$$

find $E(x)$ and $E(x^2)$.

Solution

$$E(x) = \int_0^2 x \cdot \frac{1}{2} x\, dx = \int_0^2 \frac{1}{2} x^2\, dx$$

$$= \frac{1}{2} \left[\frac{x^3}{3} \right]_0^2$$

$$= \frac{1}{2} \left(\frac{2^3}{3} - \frac{0^3}{3} \right) = \frac{1}{2} \cdot \frac{8}{3} = \frac{4}{3};$$

$$E(x^2) = \int_0^2 x^2 \cdot \frac{1}{2} x\, dx = \int_0^2 \frac{1}{2} x^3\, dx = \frac{1}{2} \left[\frac{x^4}{4} \right]_0^2$$

$$= \frac{1}{2} \left(\frac{2^4}{4} - \frac{0^4}{4} \right) = \frac{1}{2} \cdot \frac{16}{4} = 2$$

DO EXERCISE 1.

DEFINITION

The *mean* μ of a continuous random variable is defined to be $E(x)$. That is,

$$\mu = E(x) = \int_a^b x f(x)\, dx,$$

where f is a probability density function for x.

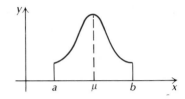

If we could imagine cutting out the region shown under the curve, we would find that the mean is the balance point. (The symbol μ is the lower-case Greek letter "mu.") Note that the mean can be thought of as an "average" on the x-axis in contrast to the "average value of a function" that lies on the y-axis.

DEFINITION

The *variance* σ^2 of a continuous random variable is defined as

$$\sigma^2 = E(x^2) - \mu^2 = E(x^2) - [E(x)]^2,$$

$$= \int_a^b x^2 f(x)\, dx - \left[\int_a^b x f(x)\, dx\right]^2.$$

The *standard deviation* σ of a continuous random variable is defined as

$$\sigma = \sqrt{\text{variance}}.$$

The symbol σ is the lower-case Greek letter "sigma."

Example 2   Given the probability density function

$$f(x) = \tfrac{1}{2}x \quad \text{over } [0, 2],$$

find the mean, the variance, and the standard deviation.

Solution   From Example 1, we have

$$E(x) = \tfrac{4}{3} \quad \text{and} \quad E(x^2) = 2.$$

2. Given the probability density function

$$f(x) = 2x \quad \text{over } [0, 1],$$

find the mean, the variance, and the standard deviation.

Then

$$\text{the mean} = \mu = E(x) = \tfrac{4}{3};$$
$$\text{the variance} = \sigma^2 = E(x^2) - [E(x)]^2$$
$$= 2 - (\tfrac{4}{3})^2 = 2 - \tfrac{16}{9}$$
$$= \tfrac{18}{9} - \tfrac{16}{9} = \tfrac{2}{9};$$
$$\text{the standard deviation} = \sigma = \sqrt{\tfrac{2}{9}} = \tfrac{1}{3}\sqrt{2} \approx 0.47.$$

Loosely speaking, we say that the standard deviation is a measure of how close the graph of f is to the mean. Note these examples.

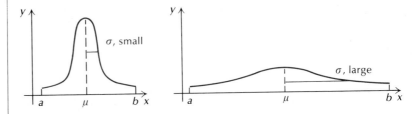

DO EXERCISE 2.

The Normal Distribution

Suppose the average on a test is 70. Usually there are about as many scores above the average as there are below; and the further away from the average a particular score is, the fewer people there are who get that score. In this example, it is probable that more people would score in the 80s than in the 90s; and more people would score in the 60s than in the 50s. Test scores, heights of human beings, and weights of human beings are all examples of random variables that may be *normally* distributed.

Consider the function

$$g(x) = e^{-x^2/2} \quad \text{over the interval } (-\infty, \infty).$$

This function has the entire set of real numbers as its domain. Its graph is the bell-shaped curve shown at the top of the following page. We can find function values by using a calculator:

$$y = e^{-x^2/2}.$$

x	0	1	2	3	-1	-2	-3
y	1	0.6	0.1	0.01	0.6	0.1	0.01

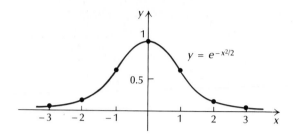

This function has an antiderivative, but that antiderivative has no elementary formula. Nevertheless, it has been shown that its improper integral converges over the interval $(-\infty, \infty)$ and

$$\int_{-\infty}^{\infty} e^{-x^2/2}\, dx = \sqrt{2\pi}.$$

That is, although an expression for the antiderivative cannot be found, there is a numerical value for the improper integral evaluated over the set of real numbers. Note that since the area is not 1, the function g is not a probability density function, but the following is:

$$\frac{1}{\sqrt{2\pi}}\, e^{-x^2/2}.$$

DEFINITION

A continuous random variable x has a *standard normal distribution* if its probability density function is

$$f(x) = \frac{1}{\sqrt{2\pi}}\, e^{-x^2/2} \quad \text{over } (-\infty, \infty).$$

This distribution has a mean of 0 and a standard deviation of 1. Its graph is shown below.

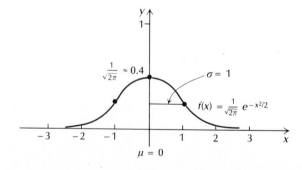

The general case is defined as follows.

DEFINITION

A continuous random variable x is *normally distributed* with mean μ and standard deviation σ if its probability density function is given by

$$f(x) = \frac{1}{\sigma\sqrt{2\pi}} \cdot e^{-(1/2)[(x-\mu)/\sigma]^2} \quad \text{over } (-\infty, \infty).$$

The graph is a transformation of the graph of the standard density. This can be shown by translating the graph along the x-axis and changing the way the graph is clustered about the mean. Some examples are shown below.

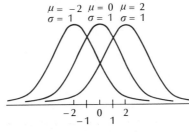

Normal distributions with same standard deviations but different means

Normal distributions with same means but different standard deviations

The normal distribution is extremely important in statistics; it underlies much of the research in the behavioral and social sciences. Because of this, tables of approximate values of the definite integral of the standard density functions have been prepared. Table 2 at the back of the book is such a table. It contains values of

$$P(0 \leqslant x \leqslant t) = \int_0^t \frac{1}{\sqrt{2\pi}} e^{-x^2/2} \, dx.$$

The symmetry of the graph about the mean allows many types of probabilities to be computed from the table.

Example 3   Let x be a continuous random variable with standard normal density. Using Table 2, find each of the following.

a) $P(0 \leqslant x \leqslant 1.68)$ b) $P(-0.97 \leqslant x \leqslant 0)$

c) $P(-2.43 \leqslant x \leqslant 1.01)$ d) $P(1.90 \leqslant x \leqslant 2.74)$

e) $P(-2.98 \leqslant x \leqslant -0.42)$ f) $P(x \geqslant 0.61)$

Solution

a) $P(0 \leqslant x \leqslant 1.68)$ is the area bounded by the standard normal curve and the lines $x = 0$ and $x = 1.68$. We look this up in Table 2 by going down the left column to 1.6, then moving to the right to the column headed 0.08. There we read 0.4535. Thus,

$$P(0 \leqslant x \leqslant 1.68) = 0.4535.$$

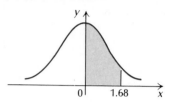

b) Because of the symmetry of the graph,

$$P(-0.97 \leqslant x \leqslant 0)$$
$$= P(0 \leqslant x \leqslant 0.97)$$
$$= 0.3340.$$

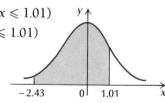

c) $P(-2.43 \leqslant x \leqslant 1.01)$
$$= P(-2.43 \leqslant x \leqslant 0) + P(0 \leqslant x \leqslant 1.01)$$
$$= P(0 \leqslant x \leqslant 2.43) + P(0 \leqslant x \leqslant 1.01)$$
$$= 0.4925 + 0.3438$$
$$= 0.8363$$

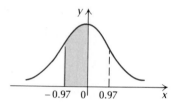

d) $P(1.90 \leqslant x \leqslant 2.74)$
$$= P(0 \leqslant x \leqslant 2.74) - P(0 \leqslant x \leqslant 1.90)$$
$$= 0.4969 - 0.4713$$
$$= 0.0256$$

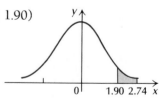

e) $P(-2.98 \leqslant x \leqslant -0.42)$
$$= P(0.42 \leqslant x \leqslant 2.98)$$
$$= P(0 \leqslant x \leqslant 2.98) - P(0 \leqslant x \leqslant 0.42)$$
$$= 0.4986 - 0.1628$$
$$= 0.3358$$

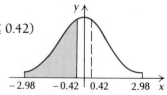

3. Let x be a continuous random variable with standard normal density. Using Table 2, find each of the following.

a) $P(0 \leqslant x \leqslant 2.17)$

b) $P(-1.76 \leqslant x \leqslant 0)$

c) $P(-1.77 \leqslant x \leqslant 2.53)$

d) $P(0.49 \leqslant x \leqslant 1.75)$

e) $P(-1.66 \leqslant x \leqslant -1.00)$

f) $P(x \geqslant 1.87)$

f) $P(x \geqslant 0.61)$

$$= P(x \geqslant 0) - P(0 \leqslant x \leqslant 0.61)$$

$$= 0.5000 - 0.2291$$

$$= 0.2709$$

(Because of the symmetry about the line $x = 0$, half the area is on each side of the line, and since the entire area is 1, $P(x \geqslant 0) = 0.5000$.)

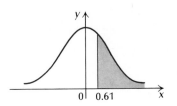

DO EXERCISE 3.

In many applications, a normal distribution is not standard. It would be a hopeless task to make tables for all values of the mean μ and the standard deviation σ. In such cases, the transformation

$$X = \frac{x - \mu}{\sigma}$$

standardizes the distribution, permitting the use of Table 2 at the back of the book. That is,

$$P(a \leqslant x \leqslant b) = P\left(\frac{a - \mu}{\sigma} \leqslant X \leqslant \frac{b - \mu}{\sigma}\right),$$

and the probability on the right can be found using Table 2. To see this, consider

$$P(a \leqslant x \leqslant b) = \int_a^b \frac{1}{\sigma\sqrt{2\pi}} e^{-(1/2)[(x-\mu)/\sigma]^2} \, dx,$$

and make the substitution

$$X = \frac{x - \mu}{\sigma} = \frac{x}{\sigma} - \frac{\mu}{\sigma}.$$

Then

$$dX = \frac{1}{\sigma} \, dx.$$

When $x = a$, $X = (a - \mu)/\sigma$; and when $x = b$, $X = (b - \mu)/\sigma$. Then

$$P(a \leqslant x \leqslant b) = \int_a^b \frac{1}{\sigma\sqrt{2\pi}} e^{-(1/2)[(x-\mu)/\sigma]^2} \, dx$$

4. The daily profits p of a small business firm are normally distributed with mean $200 and standard deviation $40. Find the probability that the daily profit will be from $230 to $250.

and

$$P(a \leqslant x \leqslant b) = \int_{(a-\mu)/\sigma}^{(b-\mu)/\sigma} \frac{1}{\sqrt{2\pi}} e^{-(1/2)X^2} dX$$

The integrand is now in the form of the standard density.

$$= P\left(\frac{a-\mu}{\sigma} \leqslant X \leqslant \frac{b-\mu}{\sigma}\right).$$

We can look this up in Table 2.

Example 4   The weights w of the students in a calculus class are normally distributed with mean 150 lb and standard deviation 25 lb. Find the probability that a student's weight is from 160 lb to 180 lb.

Solution   We first standardize the weights:

$$180 \text{ is standardized to } \frac{b-\mu}{\sigma} = \frac{180-150}{25} = 1.2;$$

$$160 \text{ is standardized to } \frac{a-\mu}{\sigma} = \frac{160-150}{25} = 0.4.$$

Then

$$\begin{aligned} P(160 \leqslant w \leqslant 180) &= P(0.4 \leqslant X \leqslant 1.2) \qquad \text{Now we can use Table 2.} \\ &= P(0 \leqslant X \leqslant 1.2) - P(0 \leqslant X \leqslant 0.4) \\ &= 0.3849 - 0.1554 \\ &= 0.2295. \end{aligned}$$

Thus the probability that a student's weight is from 160 lb to 180 lb is 0.2295. That is, about 23% of the students have weights from 160 lb to 180 lb.

DO EXERCISE 4.

EXERCISE SET 6.6

For each probability density function, over the given interval, find $E(x)$, $E(x^2)$, the mean, the variance, and the standard deviation.

1. $f(x) = \frac{1}{3}$, $[2, 5]$

2. $f(x) = \frac{1}{4}$, $[3, 7]$

3. $f(x) = \frac{2}{9}x$, $[0, 3]$

4. $f(x) = \frac{1}{8}x$, $[0, 4]$

5. $f(x) = \frac{2}{3}x$, $[1, 2]$

6. $f(x) = \frac{1}{4}x$, $[1, 3]$

7. $f(x) = \frac{1}{3}x^2$, $[-2, 1]$

8. $f(x) = \frac{3}{2}x^2$, $[-1, 1]$

9. $f(x) = \frac{1}{\ln 3} \cdot \frac{1}{x}$, $[1, 3]$

10. $f(x) = \frac{1}{\ln 2} \cdot \frac{1}{x}$, $[1, 2]$

Let x be a continuous random variable with standard normal density. Using Table 2, find each of the following.

11. $P(0 \leqslant x \leqslant 2.69)$

12. $P(0 \leqslant x \leqslant 0.04)$

13. $P(-1.11 \leqslant x \leqslant 0)$

14. $P(-2.61 \leqslant x \leqslant 0)$

15. $P(-1.89 \leqslant x \leqslant 0.45)$

16. $P(-2.94 \leqslant x \leqslant 2.00)$

17. $P(1.76 \leqslant x \leqslant 1.86)$

18. $P(0.76 \leqslant x \leqslant 1.45)$

19. $P(-1.45 \leqslant x \leqslant -0.69)$

20. $P(-2.45 \leqslant x \leqslant -1.69)$

21. $P(x \geqslant 3.01)$

22. $P(x \geqslant 1.01)$

23. a) $P(-1 \leqslant x \leqslant 1)$

  b) What percentage of the area is from -1 to 1?

24. a) $P(-2 \leqslant x \leqslant 2)$

  b) What percentage of the area is from -2 to 2?

Let x be a continuous random variable that is normally distributed with mean $\mu = 22$ and standard deviation $\sigma = 5$. Using Table 2, find each of the following.

25. $P(24 \leqslant x \leqslant 30)$

26. $P(22 \leqslant x \leqslant 27)$

27. $P(19 \leqslant x \leqslant 25)$

28. $P(18 \leqslant x \leqslant 26)$

APPLICATIONS

29. At the time of this writing, the bowling scores S of the author of this text were normally distributed with mean 190 and standard deviation 23.

  a) Find the probability that a score is from 190 to 213.

  b) Find the probability that a score is from 160 to 175.

  c) Find the probability that a score is greater than 200.

30. The daily production N of stereos by a recording company is normally distributed with mean 1000 and standard deviation 50. The company promises to pay bonuses to its employees on those days when the production of stereos is 1100 or more. What percentage of the days will the company have to pay a bonus?

31. The number of daily orders N received by a mail-order firm is normally distributed with mean 250 and standard deviation 20. The company has to hire extra help or pay overtime on those days when the number of orders received is 300 or higher. What percentage of the days will the company have to hire extra help or pay overtime?

32. The scores S on a psychology test are normally distributed with mean 65 and standard deviation 20. A score of 80 to 89 is a B. What is the probability of getting a B?

EXTENSION EXERCISES

For each probability density function over the given interval, find $E(x)$, $E(x^2)$, the mean, the variance, and the standard deviation.

33. The uniform probability density

$$f(x) = \frac{1}{b - a} \quad \text{over } [a, b]$$

34. The exponential probability density

$$f(x) = ke^{-kx} \quad \text{over } [0, \infty)$$

Consider the transportation planning problem in Example 10 of Section 6.5. Explain why the result of this problem justifies $k = 1/a$, where a is the average distance between successive cars over some period of time.

Median. Let x be a continuous random variable over $[a, b]$ with probability density function f. Then the *median* of x is that number m for which

$$\int_0^m f(x)\, dx = \frac{1}{2}.$$

Find the median.

35. $f(x) = \frac{1}{2}x,\ [0, 2]$

36. $f(x) = \frac{3}{2}x^2,\ [-1, 1]$

37. $f(x) = ke^{-kx},\ [0, \infty)$

6.7

OBJECTIVE

a) Use integration to find the volume of a solid of revolution.

VOLUME

Consider the graph of $y = f(x)$. If the upper half-plane is rotated about the x-axis, then each point on the graph has a circular path, and the whole graph sweeps out a certain surface, called a *surface of revolution*.

The plane region between the graph, the x-axis, and the interval $[a, b]$ sweeps out a *solid of revolution*. To calculate the volume of this solid, we first approximate it by a finite sum of thin right circular cylinders. We divide the interval $[a, b]$ into equal subintervals, each of length Δx. Thus the height of each cylinder is Δx. The radius of each cylinder is $f(x_i)$, where x_i is the right-hand endpoint of the subinterval that delineates that cylinder.

Since the volume of a right circular cylinder is given by

$$V = \pi r^2 h,$$

each of the approximating cylinders has volume

$$\pi[f(x_i)]^2 \, \Delta x.$$

The volume of the solid of revolution is approximated by the sum of the volumes of all the cylinders:

$$V \approx \sum_{i=1}^{n} \pi[f(x_i)]^2 \, \Delta x.$$

The actual volume is the limit as the thickness of the cylinders approaches zero or the number of them approaches infinity:

$$V = \lim_{n \to \infty} \sum_{i=1}^{n} \pi[f(x_i)]^2 \, \Delta x.$$

This is just the definite integral of the function $y = \pi[f(x)]^2$.

THEOREM 4

$$V = \int_a^b \pi[f(x)]^2 \, dx \qquad \text{\textit{Volume of a solid of revolution rotated about the x-axis}}$$

Example 1   Find the volume of the solid of revolution generated by rotating the region under the graph of

$$y = \sqrt{x}$$

from $x = 0$ to $x = 1$ about the x-axis.

Solution

$$V = \int_0^1 \pi[f(x)]^2 \, dx$$

$$= \int_0^1 \pi[\sqrt{x}]^2 \, dx$$

$$= \int_0^1 \pi x \, dx$$

$$= \left[\frac{\pi x^2}{2} \right]_0^1$$

$$= \pi \left(\frac{1^2}{2} - \frac{0^2}{2} \right) = \frac{\pi}{2}$$

Find a curve that can be rotated to form the solid of revolution.

1. Find the volume of the solid of revolution generated by rotating the region under the graph of

$$y = x$$

from $x = 0$ to $x = 1$ about the x-axis.

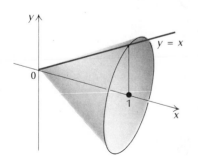

DO EXERCISE 1.

2. Find the volume of the solid of revolution generated by rotating the region under the graph of

$$y = e^x$$

from $x = -2$ to $x = 1$ about the x-axis.

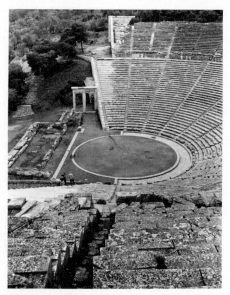

Explain how this could be interpreted as a solid of revolution.

Example 2   Find the volume of the solid of revolution generated by rotating the region under the graph of

$$y = e^x$$

from $x = -1$ to $x = 2$ about the x-axis.

Solution

$$V = \int_{-1}^{2} \pi[f(x)]^2 \, dx$$

$$= \int_{-1}^{2} \pi[e^x]^2 \, dx$$

$$= \int_{-1}^{2} \pi e^{2x} \, dx$$

$$= \left[\frac{\pi}{2} e^{2x}\right]_{-1}^{2}$$

$$= \frac{\pi}{2} (e^{2 \cdot 2} - e^{2(-1)})$$

$$= \frac{\pi}{2} (e^4 - e^{-2})$$

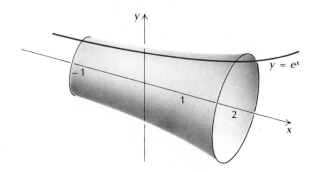

DO EXERCISE 2.

EXERCISE SET 6.7

Find the volume generated by revolving about the x-axis the regions bounded by the following graphs.

1. $y = \sqrt{x}, x = 0, x = 3$

2. $y = \sqrt{x}, x = 0, x = 2$

3. $y = x, x = 1, x = 2$

4. $y = x, x = 1, x = 3$

5. $y = e^x, x = -2, x = 5$

6. $y = e^x, x = -3, x = 2$

7. $y = 1/x, x = 1, x = 3$

8. $y = 1/x, x = 1, x = 4$

9. $y = 1/\sqrt{x}, x = 1, x = 3$

10. $y = 1/\sqrt{x}, x = 1, x = 4$

11. $y = 4, x = 1, x = 3$

12. $y = 5, x = 1, x = 3$

13. $y = x^2, x = 0, x = 2$

14. $y = x + 1, x = -1, x = 2$

15. $y = \sqrt{1 + x}, x = 2, x = 10$

16. $y = 2\sqrt{x}, x = 1, x = 2$ **17.** $y = \sqrt{4 - x^2}, x = -2, x = 2$ **18.** $y = \sqrt{r^2 - x^2}, x = -r, x = r$
Make a drawing for Exercise 18. Here you will derive a general formula for the volume of a sphere.

EXTENSION EXERCISES

Find the volume generated by revolving about the x-axis the regions bounded by the following graphs.

19. $y = \sqrt{\ln x}, x = e, x = e^3$ **20.** $y = \sqrt{xe^{-x}}, x = 1, x = 2$

21. Consider $y = 1/x$.

a) Find $\int_1^\infty \dfrac{1}{x} dx$.

b) Find the volume generated by revolving about the x-axis the region under the graph of $y = 1/x$ for $x \geqslant 1$.

Note that the area under the graph does not exist, although the volume does. This is like a can of paint that has a finite volume, but not enough paint to cover a cross section of the can.

SUMMARY AND REVIEW: CHAPTER 6

The following contains a summary of what you should be able to do after completing this chapter. The review exercises are for practice. Answers are at the back of the book. If you miss an exercise, restudy the section indicated alongside the answers.

You should be able to:

Given a demand function and a supply function, find the equilibrium point and the consumer's surplus and producer's surplus at the equilibrium point.

Given $D(x) = (x - 6)^2$ and $S(x) = x^2 + 12$, find each of the following.

1. The equilibrium point   2. The consumer's surplus at the equilibrium point

3. The producer's surplus at the equilibrium point

Find the amount of a continuous money flow over a period of time at an interest rate compounded continuously. Given a continuous flow of money into an investment at the rate of P_0 dollars per year, find P_0 so that the amount of a continuous money flow over a specified time will be a given amount. Find the total use of natural resources over a given period of time.

4. Find the amount of a continuous money flow where $2400 per year is being invested at 10%, compounded continuously, for 15 years.

5. Consider a continuous money flow into an investment at the rate of P_0 dollars per year. What should P_0 be so that the amount of a continuous money flow over 25 years at 12%, compounded continuously, will be $40,000?

6. *The demand for potash.* In 1978 $(t = 0)$ the world use of potash was 30,500 thousand tons, and the demand for it was growing at the rate of 3% per year. If the demand continues to grow at this rate, how many tons of potash will the world use from 1978 to 1998?

7. *The depletion of potash.* The world reserve of potash is 144,500,000 thousand tons. Assuming the growth rate in Exercise 6 continues and that no new reserves are discovered, when will the world reserve of potash be exhausted?

Find the present value of an investment due t years later at a certain interest rate, compounded continuously. Find the accumulated present value of an investment.

8. Find the present value of $100,000, due 50 years later at 12%, compounded continuously.

9. Find the accumulated present value of an investment over a 20-year period if there is a continuous money flow of $4800 per year and the current interest rate is 9%.

10. Find the accumulated present value of an investment for which the continuous money flow in Exercise 9 is perpetual.

Determine whether an improper integral is convergent or divergent, and calculate its value if it is convergent.

11. $\int_1^\infty \frac{1}{x^2}\, dx$

12. $\int_1^\infty e^{4x}\, dx$

13. $\int_0^\infty e^{-2x}\, dx$

Find k such that a function like $f(x) = kx^2$ is a probability density function over an interval. Solve applied problems involving probability density functions.

14. Find k such that $f(x) = k/x^3$ is a probability density function over the interval $[1, 2]$.

15. A person randomly arrives at a bus stop where the waiting time t for a bus is 25 minutes. The probability density function for t is $f(t) = \frac{1}{25}$, for $0 \leqslant t \leqslant 25$. Find the probability that a person will have to wait no more than 15 minutes for a bus.

Given a probability density function, find $E(x)$, $E(x^2)$, the mean, the variance, and the standard deviation. Use Table 2 at the back of the book to evaluate probabilities involving a normal distribution.

Given the probability density function $f(x) = 3x^2$ over $[0, 1]$, find each of the following.

16. $E(x^2)$

17. $E(x)$

18. The mean

19. The variance

20. The standard deviation

Let x be a continuous random variable with standard normal density. Using Table 2, find each of the following.

21. $P(0 \leqslant x \leqslant 1.85)$

22. $P(-1.74 \leqslant x \leqslant 1.43)$

23. $P(-2.08 \leqslant x \leqslant -1.18)$

24. $P(x \geqslant 0)$

25. The number of pizzas sold daily at Benito's Pizzeria is normally distributed with mean 400 and standard deviation 60. What is the probability that the number sold during one day is 480 or more?

Use integration to find the volume of a solid of revolution.

Find the volume generated by revolving about the x-axis the region bounded by each of the following graphs.

26. $y = x^3$, $x = 1$, $x = 2$

27. $y = \dfrac{1}{x + 2}$, $x = 0$, $x = 1$

EXTENSION EXERCISES

28. The function $f(x) = x^8$ is a probability density function on the interval $[-c, c]$. Find c.

Determine whether the improper integral is convergent or divergent, and calculate its value if it is convergent.

29. $\int_{-\infty}^{0} x^4 e^{-x^5} dx$

30. $\int_{0}^{\infty} \frac{dx}{(x + 1)^{4/3}}$

TEST: CHAPTER 6

Given the demand and supply functions $D(x) = (x - 7)^2$ and $S(x) = x^2 + x + 4$, find each of the following.

1. The equilibrium point

2. The consumer's surplus at the equilibrium point

3. The producer's surplus at the equilibrium point

4. Find the amount of a continuous money flow if $1200 per year is being invested at 6%, compounded continuously, for 15 years.

5. Consider a continuous money flow into an investment at the rate of P_0 dollars per year. What should P_0 be so that the amount of a continuous money flow over 25 years at 6%, compounded continuously, will be $20,000?

6. *The demand for iron ore.* In 1980 $(t = 0)$ the world use of iron ore was 1,017,500 thousand tons, and the demand for it was growing exponentially at the rate of 6% per year. If the demand continues to grow at this rate, how many tons of iron ore will the world use from 1980 to 1990?

7. *The depletion of iron ore.* The world reserves of iron ore are 103,000,000 thousand tons. Assuming that the growth rate of 6% per year continues and that no new reserves are discovered, when will the world reserves of iron ore be exhausted?

8. Following the birth of a child, a parent wants to make an initial investment P_0 that will grow to $10,000 by the child's 20th birthday. Interest is compounded continuously at 7%. What should the initial investment be?

9. Find the accumulated present value of an investment over a 20-year period where there is a continuous money flow of $3800 per year and the current interest rate is 11%.

10. Find the accumulated present value of an investment for which the continuous money flow in Question 9 is perpetual.

Determine whether the improper integral is convergent or divergent, and calculate its value if it is convergent.

11. $\int_{1}^{\infty} \frac{dx}{x^5}$

12. $\int_{0}^{\infty} \frac{3}{1 + x} dx$

13. Find k such that $f(x) = kx^3$ is a probability density function over the interval $[0, 2]$.

14. A telephone company determines that the length of time t of a phone call is an exponentially distributed random variable with probability density function

$$f(t) = 2e^{-2t}, \qquad 0 \leqslant t < \infty.$$

Find the probability that a phone call will last no more than 1 minute.

Given the probability density function $f(x) = \frac{1}{4}x$ over $[1, 3]$, find each of the following.

15. $E(x)$ 16. $E(x^2)$ 17. The mean

18. The variance   19. The standard deviation

Let x be a continuous random variable with standard normal density. Using Table 2, find each of the following.

20. $P(0 \leqslant x \leqslant 1.5)$ 21. $P(0.12 \leqslant x \leqslant 2.32)$ 22. $P(-1.61 \leqslant x \leqslant 1.76)$

23. The price per pound p of a T-bone steak at various stores in a certain city is normally distributed with mean \$4.75 and standard deviation \$0.25. What is the probability that the price per pound is \$4.80 or more?

Find the volume generated by revolving about the x-axis the regions bounded by the following graphs.

24. $y = \dfrac{1}{\sqrt{x}}, x = 1, x = 5$ 25. $y = \sqrt{2 + x}, x = 0, x = 1$

EXTENSION EXERCISES

26. The function $f(x) = x^5$ is a probability density on $[0, b]$. What is b?

27. Determine whether the following improper integral is convergent or divergent, and calculate its value if it is convergent.

$$\int_{-\infty}^{0} x^3 e^{-x^4} \, dx$$

7

FUNCTIONS OF SEVERAL VARIABLES

Functions that have more than one input are called *functions of several variables*. We introduce these functions in this chapter and learn to differentiate them to find what are called *partial derivatives*. Then we use these functions and their partial derivatives to solve maximum–minimum problems. Finally, we consider the integration of such functions.

AN APPLICATION

Wind speed of a tornado. Under certain conditions the *wind speed* of a tornado at a distance d from its center can be approximated by the function

$$S = \frac{aV}{0.51d^2},$$

where a is an atmospheric constant that depends on certain atmospheric conditions and V is the approximate volume of the tornado in cubic feet.

Approximate the wind speed 100 ft from the center of a tornado when its volume is 1,600,000 ft^3 and $a = 0.78$.

THE MATHEMATICS

To solve the problem we substitute 100 for d, 1,600,000 for V, and 0.78 for a in the equation:

$$S = \frac{aV}{0.51d^2},$$

This is a ⎤
function of several variables.

7.1

OBJECTIVE

a) Find a function value for a function of several variables.

FUNCTIONS OF SEVERAL VARIABLES

Suppose a one-product firm produces x items of its product at a profit of \$4 per item. Then its total profit $P(x)$ is given by

$$P(x) = 4x.$$

This is a function of one variable.

Suppose a two-product firm produces x items of one product at a profit of \$4 per item and y items of a second at a profit of \$6 per item. Then its total profit P is a function of the *two* variables x and y, and is given by

$$P(x, y) = 4x + 6y.$$

This function assigns to the input pair (x, y) a unique output number $4x + 6y$.

DEFINITION

A *function of two variables* assigns to each input pair (x, y) a unique number $f(x, y)$.

We can also think of a function of two variables as a machine that has two inputs. The domain of a function of two variables is a set of pairs (x, y) in the plane. When such a function is given by a formula, the domain consists of all ordered pairs (x, y) that are sensible replacements in the formula.

1. For $P(x, y) = 4x + 6y$:

    a) find $P(14, 12)$ and interpret its
        meaning;

    b) find $P(0, 8)$ and interpret its
        meaning.

2. A pitcher's *earned-run average* is
    given by

    $$A(n, i) = 9 \cdot \left(\frac{n}{i}\right),$$

    where n is the total number of
    earned runs given up in i innings of
    pitching. Find each of the following.

    a) $A(4, 5)$

    b) $A(7, \frac{2}{3})$

    c) $A(1, 15)$

3. For $V(x, y, z) = xyz$, find $V(5, 10, 40)$.

4. For the function C of Example 2,
    find each of the following.

    a) $C(9, 5, 3, 0)$

    b) $C(1, 2, 3, 4)$

Example 1   For $P(x, y) = 4x + 6y$, find $P(25, 10)$.

Solution $P(25, 10)$ is defined to be the value of the function found
by substituting 25 for x and 10 for y:

$$P(25, 10) = 4 \cdot 25 + 6 \cdot 10$$
$$= 100 + 60$$
$$= 160.$$

This means that by selling 25 items of the first product and 10 of the
second, the two-product firm will make a profit of $160.

DO EXERCISES 1–3.

The following are examples of *functions of several variables,* that
is, functions of two or more variables. If there are n variables, then
there are n inputs for such a function.

Example 2   The total cost of a company, in thousands of dollars, is
given by

$$C(x, y, z, w) = 4x^2 + 5y + z - \ln (w + 1),$$

where x dollars is spent for labor, y dollars for raw materials, z dollars
for advertising, and w dollars for machinery. This is a function of four
variables. Find $C(3, 2, 0, 10)$.

Solution   We substitute 3 for x, 2 for y, 0 for z, and 10 for w:

$$C(3, 2, 0, 10) = 4 \cdot 3^2 + 5 \cdot 2 + 0 - \ln (10 + 1)$$
$$\approx 4 \cdot 9 + 10 + 0 - 2.4$$
$$= \$43.6 \text{ thousand.}$$

DO EXERCISE 4.

Example 3   *Business: The cost of storage equipment.* A business purchases a piece of storage equipment that costs C_1 dollars and has
capacity V_1. Later it wishes to replace the original with a new piece
of equipment that costs C_2 dollars and has capacity V_2. The ratio of
the new capacity to the original is

$$\frac{V_2}{V_1} = k,$$

so

$$V_2 = kV_1.$$

5. Repeat Example 3 given that the original tank cost $65,000 and the new tank is triple the capacity of the original.

It has been found in industrial economics that the cost of the new piece of equipment can be estimated by the function of three variables:

$$C_2 = \left(\frac{V_2}{V_1}\right)^{0.6} C_1 = k^{0.6}C_1. \tag{1}$$

For $45,000, a beverage company buys a manufacturing tank that has a capacity of 10,000 gallons. Later it decides to buy a tank with double the capacity of the original. Estimate the cost of the new tank.

Solution   We substitute 20,000 for V_2, 10,000 for V_1, and 45,000 for C_1 in Eq. (1):

$$C_2 = \left(\frac{20,000}{10,000}\right)^{0.6} \cdot (45,000) = 2^{0.6}(45,000)$$
$$\approx (1.515712)(45,000)$$
$$= \$68,207.25.$$

Note that a 100% increase in capacity was achieved by about a 52% increase in cost. This is independent of any increase in labor, management, or the cost of other equipment resulting from the purchase of the tank.

DO EXERCISE 5.

Example 4   *The gravity model.* The number of telephone calls between two cities is given by

$$N(d, P_1, P_2) = \frac{2.8P_1P_2}{d^{2.4}},$$

where d is the distance between the cities and P_1 and P_2 are their populations.

Sociologists say that as two cities merge, the communication between them increases.

6. The constant function g is given by

$$g(x, y) = 4$$

for all inputs x and y. Find each of the following.

a) $g(-9, 10)$

b) $g(560, 43)$

A constant can also be thought of as a function of several variables.

Example 5 The constant function f is given by

$$f(x, y) = -3 \quad \text{for all inputs } x \text{ and } y.$$

Find $f(5, 7)$ and $f(-2, 0)$.

Solution Since this is a constant function, it has the value -3 for any x and y. Thus,

$$f(5, 7) = -3 \quad \text{and} \quad f(-2, 0) = -3.$$

DO EXERCISE 6.

Geometric Interpretations

Consider a function of two variables

$$z = f(x, y).$$

Recall the mapping interpretation of function that we considered in Chapter 1. As a mapping, a function of two variables can be thought of as mapping a point (x_1, y_1) in an xy-plane onto a point z_1 on a number line.

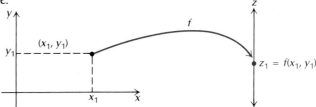

To graph a function of two variables, we need a three-dimensional coordinate system. The axes are generally placed as shown at right. The line z, called the z-axis, is placed perpendicular to the xy-plane at the origin.

To help visualize this, think of looking into the corner of a room, where the floor is the xy-plane and the z-axis is the intersection of the two walls.

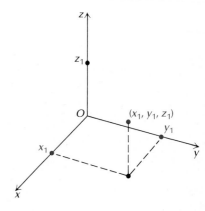

7. Using the axes shown below, graph $P_1(3, 2, 5)$, $P_2(2, 3, 1)$, and $P_3(-3, 2, 0)$.

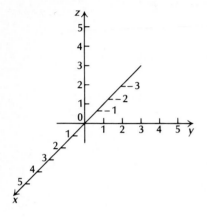

To plot a point (x_1, y_1, z_1) we locate the point (x_1, y_1) in the xy-plane and move up or down in space according to the value of z_1.

Example 6   Plot these points:

$$P_1(2, 3, 5) \ P_2(2, -2, -4), P_3(0, 5, 2), \text{ and } P_4(2, 3, 0).$$

Solution

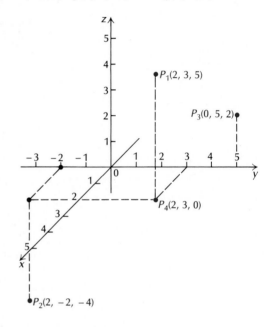

DO EXERCISE 7.

The *graph* of a function of two variables

$$z = f(x, y)$$

consists of ordered triples (x_1, y_1, z_1), where $z_1 = f(x_1, y_1)$. The domain of f is a region D in the xy-plane, and the graph of f is a surface S, as shown at the bottom of the preceding page.

Here are some equations and their graphs.

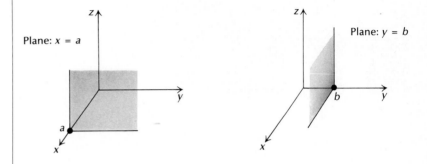

Elliptic paraboloid: $z = x^2 + y^2$

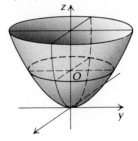

Hyperbolic paraboloid: $z = x^2 - y^2$

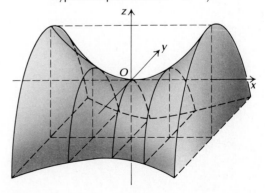

The following graphs have been generated by computer graphics.

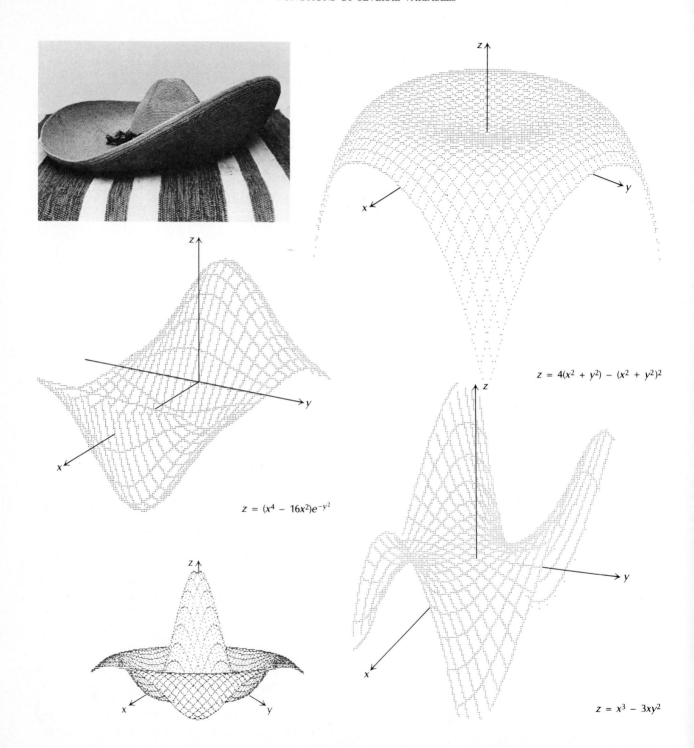

$z = 4(x^2 + y^2) - (x^2 + y^2)^2$

$z = (x^4 - 16x^2)e^{-y^2}$

$z = x^3 - 3xy^2$

EXERCISE SET 7.1

1. For $f(x, y) = x^2 - 2xy$, find $f(0, -2)$, $f(2, 3)$, and $f(10, -5)$.

2. For $f(x, y) = (y^2 + 3xy)^3$, find $f(-2, 0)$, $f(3, 2)$, and $f(-5, 10)$.

3. For $f(x, y) = 3^x + 7xy$, find $f(0, -2)$, $f(-2, 1)$, and $f(2, 1)$.

4. For $f(x, y) = \log_{10} x - 5y^2$, find $f(10, 2)$, $f(1, -3)$, and $f(100, 4)$.

5. For $f(x, y) = \ln x + y^3$, find $f(e, 2)$, $f(e^2, 4)$, and $f(e^3, 5)$.

6. For $f(x, y) = 2^x - 3^y$, find $f(0, 0)$, $f(1, 1)$, and $f(2, 2)$.

7. For $f(x, y, z) = x^2 - y^2 + z^2$, find $f(-1, 2, 3)$ and $f(2, -1, 3)$.

8. For $f(x, y, z) = 2^x + 5zy - x$, find $f(0, 1, -3)$ and $f(1, 0, -3)$.

APPLICATIONS

9. *Psychology: Intelligence quotient.* The intelligence quotient in psychology is given by

$$Q(m, c) = 100 \cdot \frac{m}{c},$$

where m is a person's mental age and c is his or her chronological, or actual age. Find $Q(21, 20)$ and $Q(19, 20)$.

10. *Business: Price-earnings ratio.* The *price-earnings ratio* of a stock is given by

$$R(P, E) = \frac{P}{E},$$

where P is the price per share of the stock and E is the earnings per share. Recently, the price per share of IBM stock was $\$287\frac{3}{8}$ and the earnings per share were $\$23.30$. Find the price-earnings ratio. Give decimal notation to the nearest tenth.

11. *Business: Yield of a stock.* The yield of a stock is given by

$$Y(D, P) = \frac{D}{P},$$

where D is the dividends per share of a stock and P is the price per share. Recently, the price per share of Goodyear stock was $\$16\frac{7}{8}$ and the dividends per share were $\$1.30$. Find the yield. Give percent notation to the nearest tenth of a percent.

12. *Biomedical: Poiseuille's Law.* The speed of blood in a vessel is given by

$$V(L, p, R, r, v) = \frac{p}{4Lv}(R^2 - r^2),$$

where R is the radius of the vessel, r is the distance of the blood from the center of the vessel, L is the length of the blood vessel, p is pressure, and v is viscosity. Find $V(1, 100, 0.0075, 0.0025, 0.05)$.

13. Consider the storage model in Example 3. For $\$100,000$, a company buys a storage tank that has a capacity of 80,000 gallons. Later it replaces the tank with a new tank with double the capacity of the original. Estimate the cost of the new tank.

14. *Wind speed of a tornado.* Under certain conditions the *wind speed* of a tornado at a distance d from its center can be approximated by the function

$$S = \frac{aV}{0.51d^2},$$

where a is an atmospheric constant that depends on certain atmospheric conditions and V is the approximate volume of the tornado in cubic feet. Approximate the wind speed 100 ft from the center of a tornado when its volume is 1,600,000 ft^3 and $a = 0.78$.

EXTENSION EXERCISES

Wind chill temperature. Wind speed affects the actual temperature, making a person colder due to extra heat loss from the skin. The *wind chill temperature* is what the temperature would have to be with no wind in order to give the same chilling effect. The wind chill temperature W is given by

$$W(v, T) = 91.4 - \frac{(10.45 + 6.68\sqrt{v} - 0.447v)(457 - 5T)}{110},$$

where *T* is the actual temperature as given by a thermometer, in degrees Fahrenheit, and *v* is the speed of the wind, in miles per hour. Find the wind chill temperature in each case. Round to the nearest one degree.   CSS

15. $T = 30°F, v = 25$ mph   **16.** $T = 20°F, v = 20$ mph   **17.** $T = 20°F, v = 40$ mph   **18.** $T = -10°F, v = 30$ mph

---

7.2

OBJECTIVES

PARTIAL DERIVATIVES

a) Find the partial derivatives of a given function.

b) Evaluate the partial derivatives of a function at a given point.

Finding Partial Derivatives

Consider the function f given by

$$z = f(x, y) = x^2y^3 + xy + 4y^2.$$

Suppose for the moment that we fix y at 3. Then

$$f(x, 3) = x^2(3^3) + x(3) + 4(3^2) = 27x^2 + 3x + 36.$$

Note that we now have a function of only one variable. Taking the first derivative with respect to x, we have

$$54x + 3.$$

DO EXERCISE 1.

1. Consider

$$f(x, y) = 1 - x^2 - y^2.$$

a) Fix y at 4 and find $f(x, 4)$.

b) The answer to part (a) could be interpreted as a function of one variable x. Find the first derivative.

Now, without replacing y by a specific number, let us consider y fixed. Then f becomes a function of x alone and we can calculate its derivative with respect to x. This derivative is called the *partial derivative of f with respect to x*. Notation for this partial derivative is

$$\frac{\partial f}{\partial x} \quad \text{or} \quad \frac{\partial z}{\partial x}.$$

Thus let us again consider the function

$$z = f(x, y) = \boxed{x}^2 y^3 + \boxed{x} y + 4y^2.$$

2. For $f(x, y) = 1 - x^2 - y^2$, find $\partial f/\partial x$.

3. For $z = 3x^2y + 5x^3$, find each of the following.

a) $\dfrac{\partial z}{\partial x}$

b) $\dfrac{\partial z}{\partial y}$

The color screens indicate the variable x when we fix y and treat it as a constant. The expressions y^3, y, and y^2 are then constants. We have

$$\frac{\partial f}{\partial x} = \frac{\partial z}{\partial x}$$
$$= 2xy^3 + y.$$

DO EXERCISE 2.

Similarly, we find $\partial f/\partial y$ or $\partial z/\partial y$ by fixing x (treating it as a constant) and calculating the derivative with respect to y. From

$$z = f(x, y) = x^2\,\boxed{y}\,^3 + x\,\boxed{y} + 4\,\boxed{y}\,^2, \qquad \text{The screens indicate the variable.}$$

we get

$$\frac{\partial f}{\partial y} = \frac{\partial z}{\partial y}$$
$$= 3x^2y^2 + x + 8y.$$

DO EXERCISE 3.

A definition of partial derivatives is as follows.

DEFINITION

For $z = f(x, y)$,

$$\frac{\partial z}{\partial x} = \lim_{h \to 0} \frac{f(x + h, y) - f(x, y)}{h},$$

$$\frac{\partial z}{\partial y} = \lim_{h \to 0} \frac{f(x, y + h) - f(x, y)}{h}.$$

Partial differentiation can be done for any number of variables.

Example 1 For $w = x^2 - xy + y^2 + 2yz + 2z^2 + z$, find

$$\frac{\partial w}{\partial x}, \qquad \frac{\partial w}{\partial y}, \qquad \text{and} \qquad \frac{\partial w}{\partial z}.$$

Solution In order to find $\partial w/\partial x$, we consider x the variable and the other letters the constants. From

$$w = \boxed{x}\,^2 - \boxed{x}\,y + y^2 + 2yz + 2z^2 + z,$$

4. For $t = xy + xz + x^2 + y^3$, find each of the following.

a) $\dfrac{\partial t}{\partial x}$

b) $\dfrac{\partial t}{\partial y}$

c) $\dfrac{\partial t}{\partial z}$

we get

$$\frac{\partial w}{\partial x} = 2x - y.$$

From

$$w = x^2 - \boxed{x}\,y\ +\ \boxed{y}^{\,2} + 2\,\boxed{y}\,z + 2z^2 + z,$$

we get

$$\frac{\partial w}{\partial y} = -x + 2y + 2z;$$

and from

$$w = x^2 - xy + y^2 + 2y\,\boxed{z}\ + 2\,\boxed{z}^{\,2} + \boxed{z}\,,$$

we get

$$\frac{\partial w}{\partial z} = 2y + 4z + 1.$$

DO EXERCISE 4.

5. For $f(x, y) = 3x^3y + 2xy$, find each of the following.

a) f_x

b) $f_x(-4, 1)$

c) f_y

d) $f_y(2, 6)$.

We will often make use of a simpler notation f_x for the partial derivative of f with respect to x and f_y for the partial derivative of f with respect to y.

Example 2   For $f(x, y) = 3x^2y + xy$, find f_x and f_y.

Solution   We have

$$f_x = 6xy + y,$$
$$f_y = 3x^2 + x.$$

For the function in the preceding example, let us evaluate f_x at $(2, -3)$:

$$f_x(2, -3) = 6 \cdot 2 \cdot (-3) + (-3) = -39.$$

If we use the notation $\partial f / \partial x = 6xy + y$, where $f = 3x^2y + xy$, the value of the partial derivative at $(2, -3)$ is given by

$$\left.\frac{\partial f}{\partial x}\right|_{(2,\,-3)} = 6 \cdot 2 \cdot (-3) + (-3) = -39,$$

but this notation is not as convenient as $f_x(2, -3)$.

DO EXERCISE 5.

6. For $f(x, y) = \ln (xy) + ye^x$, find f_x and f_y.

Example 3 For $f(x, y) = e^{xy} + y \ln x$, find f_x and f_y.

Solution

$$f_x = y \cdot e^{xy} + y \cdot \frac{1}{x} = ye^{xy} + \frac{y}{x},$$

$$f_y = x \cdot e^{xy} + 1 \cdot \ln x = xe^{xy} + \ln x$$

DO EXERCISE 6.

The Geometric Interpretation of Partial Derivatives

The *graph* of a function of two variables $z = f(x, y)$ is a surface S, as shown here.

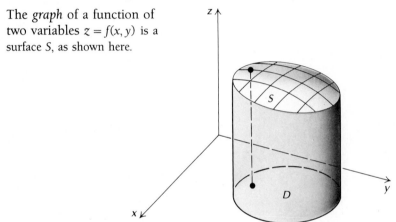

Now suppose we hold x fixed, say, at the value a. The set of all points for which $x = a$ is a plane parallel to the yz-plane, so when x is fixed at a, y and z vary along the plane, as shown in the following figure. The plane shown cuts the surface in some curve C. The partial derivative f_y gives the slope of tangent lines to this curve.

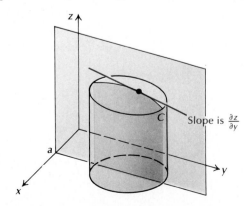

Similarly, if we hold y fixed, say at the value b, we obtain a curve C', as shown in the following figure. The partial derivative f_x gives the slope of tangent lines to this curve.

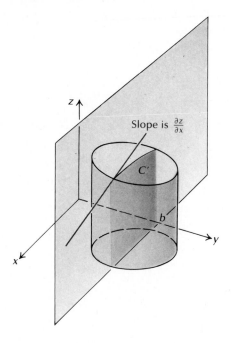

An Economic Application: The Cobb–Douglas Production Function

One model of production that is frequently considered in business and economics is the *Cobb–Douglas production function*:

$$p(x, y) = Ax^a y^{1-a}, \qquad A > 0 \quad \text{and} \quad 0 < a < 1,$$

where p is the number of units produced with x units of labor and y units of capital. (Capital is the cost of machinery, buildings, tools, and other supplies.) The partial derivatives

$$\frac{\partial p}{\partial x} \quad \text{and} \quad \frac{\partial p}{\partial y}$$

are called, respectively, the *marginal productivity of labor* and the *marginal productivity of capital*.

Example 4 A company has the following production function for a certain product:

$$p(x, y) = 50x^{2/3} y^{1/3}.$$

7. A company has the following production function for a certain product:

$$p(x, y) = 800x^{3/4}y^{1/4}.$$

a) Find the production from 81 units of labor and 625 units of capital.

b) Find the marginal productivities.

c) Evaluate the marginal productivities at $x = 81$ and $y = 625$.

a) Find the production from 125 units of labor and 64 units of capital.

b) Find the marginal productivities.

c) Evaluate the marginal productivities at $x = 125$ and $y = 64$.

Solution

a) $p(125, 64) = 50(125)^{2/3}(64)^{1/3} = 50(25)(4) = 5000$ units

b) $\dfrac{\partial p}{\partial x} = 50\left(\dfrac{2}{3}\right)x^{-1/3}y^{1/3} = \dfrac{100y^{1/3}}{3x^{1/3}}$, or $\dfrac{100}{3}\left(\dfrac{y}{x}\right)^{1/3}$;

$\dfrac{\partial p}{\partial y} = 50\left(\dfrac{1}{3}\right)x^{2/3}y^{-2/3} = \dfrac{50x^{2/3}}{3y^{2/3}}$, or $\dfrac{50}{3}\left(\dfrac{x}{y}\right)^{2/3}$

c) $\dfrac{\partial p}{\partial x}\bigg|_{(125, 64)} = \dfrac{100(64)^{1/3}}{3(125)^{1/3}} = \dfrac{100(4)}{3(5)} = 26\dfrac{2}{3}$;

$\dfrac{\partial p}{\partial y}\bigg|_{(125, 64)} = \dfrac{50(125)^{2/3}}{3(64)^{2/3}} = \dfrac{50(25)}{3(16)} = 26\dfrac{1}{24}$

How can we interpret marginal productivities? Suppose the amount spent on capital is fixed at, say, $y = 64$. Then if the amount of labor changes by a small amount, production will change by about 27 units. Suppose the amount of labor is held fixed at, say, $x = 125$. Then if the amount of capital spent changes slightly, this will produce a change of about 26 units of production.

A Cobb–Douglas production function is consistent with the law of diminishing returns. That is, if one input (of either labor or capital) is held fixed while the other increases infinitely, then production will eventually increase at a decreasing rate. With such functions it also turns out that if a certain maximum production is possible, then the expense of more labor, for example, will not prevent that maximum output from still being attainable.

DO EXERCISE 7.

EXERCISE SET 7.2

Find $\dfrac{\partial z}{\partial x}, \dfrac{\partial z}{\partial y}, \dfrac{\partial z}{\partial x}\bigg|_{(-2, -3)}$ and $\dfrac{\partial z}{\partial y}\bigg|_{(0, -5)}$

1. $z = 2x - 3xy$ 2. $z = (x - y)^3$ 3. $z = 3x^2 - 2xy + y$ 4. $z = 2x^3 + 3xy - x$

Find $f_x, f_y, f_x(-2, 4)$, and $f_y(4, -3)$.

5. $f(x, y) = 2x - 3y$ 6. $f(x, y) = 5x + 7y$

Find f_x, f_y, $f_x(-2, 1)$, and $f_y(-3, -2)$.

7. $f(x, y) = \sqrt{x^2 + y^2}$

8. $f(x, y) = \sqrt{x^2 - y^2}$

Find f_x and f_y.

9. $f(x, y) = e^{2x+3y}$

10. $f(x, y) = e^{3x-2y}$

11. $f(x, y) = e^{xy}$

12. $f(x, y) = e^{2xy}$

13. $f(x, y) = y \ln (x + y)$

14. $f(x, y) = x \ln (x + y)$

15. $f(x, y) = x \ln (xy)$

16. $f(x, y) = y \ln (xy)$

17. $f(x, y) = \dfrac{x}{y} - \dfrac{y}{x}$

18. $f(x, y) = \dfrac{x}{y} + \dfrac{y}{x}$

19. $f(x, y) = 3(2x + y - 5)^2$

20. $f(x, y) = 4(3x + y - 8)^2$

Find $\dfrac{\partial f}{\partial b}$ and $\dfrac{\partial f}{\partial m}$.

21. $f(b, m) = (m + b - 4)^2 + (2m + b - 5)^2 + (3m + b - 6)^2$

22. $f(b, m) = (m + b - 6)^2 + (2m + b - 8)^2 + (3m + b - 9)^2$

Find f_x, f_y, and f_λ.

23. $f(x, y, \lambda) = 3xy - \lambda(2x + y - 8)$

24. $f(x, y, \lambda) = 4xy - \lambda(3x - y + 7)$

25. $f(x, y, \lambda) = x^2 + y^2 - \lambda(10x + 2y - 4)$

26. $f(x, y, \lambda) = x^2 - y^2 - \lambda(4x - 7y - 10)$

APPLICATIONS

27. A company has the following production function for a certain product:

$$p(x, y) = 1800x^{0.621}y^{0.379},$$

where p is the number of units produced with x units of labor and y units of capital.

a) Find the production from 2500 units of labor and 1700 units of capital.

b) Find the marginal productivities.

c) Evaluate the marginal productivities at $x = 2500$ and $y = 1700$.

28. A company has the following production function for a certain product:

$$p(x, y) = 2400x^{2/5}y^{3/5},$$

where p is the number of units produced with x units of labor and y units of capital.

a) Find the production from 32 units of labor and 1024 units of capital.

b) Find the marginal productivities.

c) Evaluate the marginal productivities at $x = 32$ and $y = 1024$.

EXTENSION EXERCISES

Find f_x and f_t.

29. $f(x, t) = \dfrac{x^2 + t^2}{x^2 - t^2}$

30. $f(x, t) = \dfrac{x^2 - t}{x^3 + t}$

31. $f(x, t) = \dfrac{2\sqrt{x} - 2\sqrt{t}}{1 + 2\sqrt{t}}$

32. $f(x, t) = \sqrt[4]{x^3 t^5}$

33. $f(x, t) = 6x^{2/3} - 8x^{1/4}t^{1/2} - 12x^{-1/2}t^{3/2}$

34. $f(x, t) = \left(\dfrac{x^2 + t^2}{x^2 - t^2}\right)^5$

7.3

OBJECTIVE

a) **Find the four second-order partial derivatives of a function.**

1. Consider

$$z = 3xy^2 + 2xy + x^2.$$

   a) Find $\partial z/\partial y$.
   b) For the function in part (a), find the first partial derivative with respect to x.
   c) For the function in part (a), find the first partial derivative with respect to y; that is, differentiate "twice" with respect to y.

2. Consider

$$f(x, y) = 3xy^2 + 2xy + x^2.$$

   a) Find f_y.
   b) For the function in part (a), find the first partial derivative with respect to x. Denote this f_{yx}.
   c) For the function in part (a), find the first partial derivative with respect to y. Denote this f_{yy}.

HIGHER-ORDER PARTIAL DERIVATIVES

Consider

$$z = f(x, y) = 3xy^2 + 2xy + x^2. \tag{1}$$

Then

$$\frac{\partial z}{\partial x} = \frac{\partial f}{\partial x} = 3y^2 + 2y + 2x. \tag{2}$$

Suppose we find the first partial derivative of function (2) with respect to y. This will be a *second-order partial derivative*. Notation for it is as follows:

$$\frac{\partial}{\partial y}\left(\frac{\partial z}{\partial x}\right) = \frac{\partial}{\partial y}\left(\frac{\partial f}{\partial x}\right)$$

$$= \frac{\partial^2 z}{\partial y\,\partial x}$$

$$= \frac{\partial^2 f}{\partial y\,\partial x} = 6y + 2.$$

DO EXERCISE 1.

   We could also denote the preceding partial derivative using the notation f_{xy}. Then

$$f_{xy} = 6y + 2.$$

Note that in the notation f_{xy}, x and y are in the order (left to right) in which the differentiation is done. In the other symbolisms, that order is reversed, but the meaning is not.

DO EXERCISE 2.

   Notation for the four second-order partial derivatives is as follows.

3. For

$$z = f(x, y)$$
$$= 3xy^2 + 2xy + x^2 + x \ln y,$$

find the four second-order partial derivatives.

DEFINITION

1. $\dfrac{\partial^2 z}{\partial x\, \partial x} = \dfrac{\partial^2 f}{\partial x\, \partial x} = \dfrac{\partial^2 z}{\partial x^2} = \dfrac{\partial^2 f}{\partial x^2} = f_{xx}$

Take the partial with respect to x, and then with respect to x again.

2. $\dfrac{\partial^2 z}{\partial y\, \partial x} = \dfrac{\partial^2 f}{\partial y\, \partial x} = f_{xy}$

Take the partial with respect to x, and then with respect to y.

3. $\dfrac{\partial^2 z}{\partial x\, \partial y} = \dfrac{\partial^2 f}{\partial x\, \partial y} = f_{yx}$

Take the partial with respect to y, and then with respect to x.

4. $\dfrac{\partial^2 z}{\partial y\, \partial y} = \dfrac{\partial^2 f}{\partial y\, \partial y} = \dfrac{\partial^2 z}{\partial y^2} = \dfrac{\partial^2 f}{\partial y^2} = f_{yy}$

Take the partial with respect to y, and then with respect to y again.

Example 1 For

$$z = f(x, y) = x^2 y^3 + x^4 y + x e^y,$$

find the four second-order partial derivatives.

Solution

a) $\dfrac{\partial^2 f}{\partial x^2} = f_{xx} = \dfrac{\partial}{\partial x}\, (2xy^3 + 4x^3 y + e^y)$

   $\qquad = 2y^3 + 12x^2 y$

Differentiate twice with respect to x.

b) $\dfrac{\partial^2 f}{\partial y\, \partial x} = f_{xy} = \dfrac{\partial}{\partial y}\, (2xy^3 + 4x^3 y + e^y)$

   $\qquad = 6xy^2 + 4x^3 + e^y$

Differentiate with respect to x, and then with respect to y.

c) $\dfrac{\partial^2 f}{\partial x\, \partial y} = f_{yx} = \dfrac{\partial}{\partial x}\, (3x^2 y^2 + x^4 + x e^y)$

   $\qquad = 6xy^2 + 4x^3 + e^y$

Differentiate with respect to y, and then with respect to x.

d) $\dfrac{\partial^2 f}{\partial y^2} = f_{yy} = \dfrac{\partial}{\partial y}\, (3x^2 y^2 + x^4 + x e^y)$

   $\qquad = 6x^2 y + x e^y$

Differentiate twice with respect to y.

DO EXERCISE 3.

Note by comparing (b) and (c) above that

$$\frac{\partial^2 f}{\partial y\, \partial x} = \frac{\partial^2 f}{\partial x\, \partial y}. \qquad \text{And, similarly, } f_{xy} = f_{yx}.$$

Although this will be true for virtually all functions that we consider in this text, it is *not* true for all functions. Such an example is given in Exercise 19 of Exercise Set 7.3.

EXERCISE SET 7.3

Find the four second-order partial derivatives.

1. $f(x, y) = 3x^2 - xy + y$ 2. $f(x, y) = 5x^2 + xy - x$ 3. $f(x, y) = 3xy$

4. $f(x, y) = 4xy$ 5. $f(x, y) = x^5y^4 + x^3y^2$ 6. $f(x, y) = x^4y^3 - x^2y^3$

Find f_{xx}, f_{xy}, f_{yx}, and f_{yy}. (Remember, f_{yx} means to differentiate with respect to y, and then to x.)

7. $f(x, y) = 2x - 3y$ 8. $f(x, y) = 3x + 5y$

9. $f(x, y) = e^{2xy}$ 10. $f(x, y) = e^{xy}$

11. $f(x, y) = x + e^y$ 12. $f(x, y) = y - e^x$

13. $f(x, y) = y \ln x$ 14. $f(x, y) = x \ln y$

EXTENSION EXERCISES

Find f_{xx}, f_{xy}, f_{yx}, and f_{yy}.

15. $f(x, y) = \dfrac{x}{y^2} - \dfrac{y}{x^2}$

16. $f(x, y) = \dfrac{xy}{x - y}$

17. Consider $f(x, y) = \ln (x^2 + y^2)$. Show that f is a solution to the partial differential equation

$$\frac{\partial^2 f}{\partial x^2} + \frac{\partial^2 f}{\partial y^2} = 0.$$

18. Consider $f(x, y) = x^3 - 5xy^2$. Show that f is a solution to the partial differential equation

$$xf_{xy} - f_y = 0.$$

19. Consider the function f defined as follows:

$$f(x, y) = \begin{cases} \dfrac{xy(x^2 - y^2)}{x^2 + y^2}, & \text{for } (x, y) \neq (0, 0), \\ 0, & \text{for } (x, y) = (0, 0). \end{cases}$$

a) Find $f_x(0, y)$ by evaluating the limit

$$\lim_{h \to 0} \frac{f(h, y) - f(0, y)}{h}.$$

b) Find $f_y(x, 0)$ by evaluating the limit

$$\lim_{h \to 0} \frac{f(x, h) - f(x, 0)}{h}.$$

c) Then find and compare $f_{yx}(0, 0)$ and $f_{xy}(0, 0)$.

20. For $f = [\ln (x^3 + e^y)]^5$, find f_{xx}, f_{xy}, f_{yx}, and f_{yy}.

7.4

a) Find the relative maximum and minimum values of a function of two variables.

MAXIMUM–MINIMUM PROBLEMS

We will now find maximum and minimum values of functions of two variables.

DEFINITION

A function f of two variables:

i) has a *relative maximum* at (a, b) if

$$f(x, y) \leqslant f(a, b)$$

for all points in a circular region containing (a, b);

ii) has a *relative minimum* at (a, b) if

$$f(x, y) \geqslant f(a, b)$$

for all points in a circular region containing (a, b).

This definition is illustrated in Fig. 1.

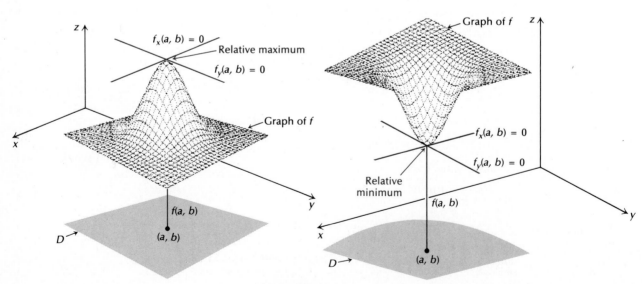

FIGURE 1

A relative maximum (minimum) may not be an "absolute" maximum (minimum), as illustrated below in Fig. 2.

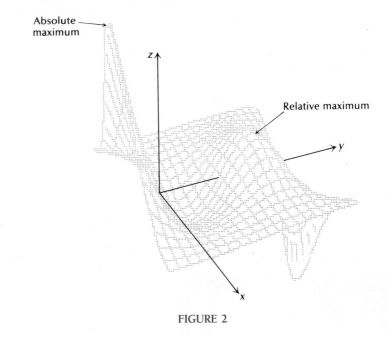

Absolute maximum

Relative maximum

FIGURE 2

Determining Maximum and Minimum Values

Suppose a function f assumes a relative maximum (or minimum) value at some point (a, b) inside its domain. If we hold y constant at the value b, then $f(x, b)$ is a function of one variable x having its relative maximum value at $x = a$, so its derivative must be 0 there. That is, $f_x = 0$ at the point (a, b). Similarly, $f_y = 0$ at (a, b). The equations

$$f_x = 0, \qquad f_y = 0$$

are thus satisfied by the point (a, b) at which the relative maximum occurs. We call a point (a, b) where both partial derivatives are 0 a *critical point*. This is comparable to the earlier definition for functions of one variable. Thus one strategy for finding relative maximum or minimum values is to solve the system of equations above to find critical points. Just as for functions of one variable, this strategy does *not* guarantee that we will have a relative maximum or minimum value. We have argued only that *if* f has a maximum or minimum value at (a, b), *then* both its partial derivatives must be 0 at that point. Look at Fig. 1. Then note Fig. 3, which illustrates a case in which the partial derivatives are 0 but the function does not have a maximum or minimum value.

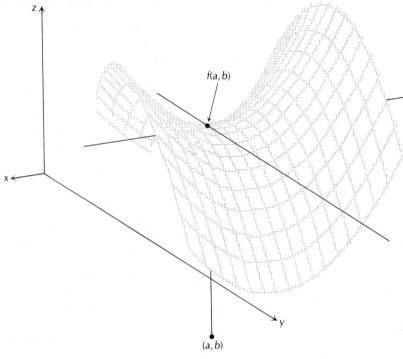

$f(a, b)$

(a, b)

FIGURE 3

Where is the saddle point?

In Fig. 3, suppose we fix y at a point b. Then $f(x, b)$, considered as a function of one variable, has a minimum at a, but f does not. Similarly, if we fix x at a, then $f(a, y)$, considered as a function of one variable, has a maximum at b, but f does not. The point $f(a, b)$ is called a *saddle point*. In other words, $f_x(a, b) = 0$ and $f_y(a, b) = 0$ (the point (a, b) is a critical point), but f does not attain a relative maximum or minimum value at (a, b). A saddle point for a function of two variables is comparable to a point of inflection for a function of one variable.

A test for finding relative maximum and minimum values that involves the use of first- and second-order partial derivatives is stated below. We will not prove this theorem.

THEOREM 1

The *D*-test

To find the relative maximum and minimum values of f, we do the following.

a) Find f_x, f_y, f_{xx}, and f_{xy}.

b) Solve the system of equations $f_x = 0, f_y = 0$. Let (a, b) represent a solution.

c) Evaluate D, where $D = f_{xx}(a, b) \cdot f_{yy}(a, b) - [f_{xy}(a, b)]^2$.

d) Then:

    i) f has a maximum at (a, b) if $D > 0$ and $f_{xx}(a, b) < 0$;

    ii) f has a minimum at (a, b) if $D > 0$ and $f_{xx}(a, b) > 0$;

    iii) f has neither a maximum nor a minimum at (a, b) if $D < 0$. The function has a *saddle point* at (a, b). See Fig. 3.

    iv) This test is not applicable if $D = 0$.

A relative maximum or minimum value *may not be an absolute maximum or minimum value.* Tests for absolute maximum or minimum values are rather complicated. We will restrict our attention to finding *relative* maximum or minimum values. Fortunately, in most applications, relative maximum and minimum values turn out to be absolute as well.

Example 1   Find the relative maximum and minimum values of

$$f(x, y) = x^2 + xy + y^2 - 3x.$$

Solution

a) Find f_x, f_y, f_{xx}, f_{yy}, and f_{xy}:

$$f_x = 2x + y - 3, \qquad f_y = x + 2y,$$
$$f_{xx} = 2; \qquad\qquad f_{yy} = 2;$$
$$f_{xy} = 1.$$

b) Solve the system of equations $f_x = 0, f_y = 0$:

$$2x + y - 3 = 0, \tag{1}$$
$$x + 2y = 0. \tag{2}$$

Solving Eq. (2) for x, we get $x = -2y$. Substituting $-2y$ for x in

The shape of a perfect tent. To give a tent roof the maximum strength possible, designers draw the fabric into a series of three-dimensional shapes that, viewed in profile, resemble a horse's saddle and that mathematicians call an anticlastic curve. Two people with a stretchy piece of fabric such as Lycra Spandex can duplicate the shape, as shown above. One person pulls up and out on two diagonal corners; the other person pulls down and out on the other two corners. The opposing tensions draw each point of the fabric's surface into rigid equilibrium. The more pronounced the curve, the stiffer the surface.

1. Find the relative maximum and minimum values of

$$f(x, y) = x^2 + 2xy + 2y^2 - 6y.$$

Eq. (1) and solving, we get

$$2(-2y) + y - 3 = 0$$
$$-4y + y - 3 = 0$$
$$-3y - 3 = 0$$
$$y = -1.$$

To find x when $y = -1$, we substitute -1 for y in either Eq. (1) or Eq. (2). We use Eq. (2):

$$x + 2(-1) = 0$$
$$x = 2.$$

Thus $(2, -1)$ is our candidate for a maximum or minimum value.

c) We must check to see whether $f(2, -1)$ is a maximum or minimum value:

$$D = f_{xx}(2, -1) \cdot f_{yy}(2, -1) - [f_{xy}(2, -1)]^2$$
$$= 2 \cdot 2 - [1]^2$$
$$= 3.$$

d) Thus $D = 3$ and $f_{xx}(2, -1) = 2$. Since $D > 0$ and $f_{xx}(2, -1) > 0$, it follows that f has a relative minimum at $(2, -1)$ and that minimum is found as follows:

$$f(2, -1) = 2^2 + 2(-1) + (-1)^2 - 3 \cdot 2$$
$$= 4 - 2 + 1 - 6 = -3.$$

DO EXERCISE 1.

Example 2   Find the relative maximum and minimum values of

$$f(x, y) = xy - x^3 - y^2.$$

Solution

a) Find f_x, f_y, f_{xx}, f_{yy}, and f_{xy}:

$$f_x = y - 3x^2, \qquad f_y = x - 2y,$$
$$f_{xx} = -6x; \qquad f_{yy} = -2;$$
$$f_{xy} = 1.$$

b) Solve the system of equations $f_x = 0, f_y = 0$:

$$y - 3x^2 = 0, \qquad (1)$$
$$x - 2y = 0. \qquad (2)$$

Solving Eq. (1) for y, we get $y = 3x^2$. Substituting $3x^2$ for y in Eq. (2) and solving, we get

$$x - 2(3x^2) = 0$$
$$x - 6x^2 = 0$$
$$x(1 - 6x) = 0. \quad \textbf{Factoring}$$

Setting each factor equal to 0 and solving, we have

$$x = 0 \quad \text{or} \quad 1 - 6x = 0$$
$$x = 0 \quad \text{or} \quad\quad x = \tfrac{1}{6}.$$

To find y when $x = 0$, we substitute 0 for x in either Eq. (1) or Eq. (2). We use Eq. (2):

$$0 - 2y = 0$$
$$-2y = 0$$
$$y = 0.$$

Thus $(0, 0)$ is one critical value (a candidate for a maximum or minimum). To find the other critical value, we substitute $\tfrac{1}{6}$ for x in either Eq. (1) or Eq. (2). We use Eq. (2):

$$\tfrac{1}{6} - 2y = 0$$
$$-2y = -\tfrac{1}{6}$$
$$y = \tfrac{1}{12}.$$

Thus $(\tfrac{1}{6}, \tfrac{1}{12})$ is another critical point.

c) We must check both $(0, 0)$ and $(\tfrac{1}{6}, \tfrac{1}{12})$ to see whether they yield maximum or minimum values:

For $(0, 0)$:

$$D = f_{xx}(0, 0) \cdot f_{yy}(0, 0) - [f_{xy}(0, 0)]^2$$
$$= [-6 \cdot 0] \cdot [-2] - [1]^2$$
$$= -1.$$

Since $D < 0$, it follows that $f(0, 0)$ is neither a maximum nor a minimum value, but a saddle point.

For $(\tfrac{1}{6}, \tfrac{1}{12})$:

$$D = f_{xx}(\tfrac{1}{6}, \tfrac{1}{12}) \cdot f_{yy}(\tfrac{1}{6}, \tfrac{1}{12}) - [f_{xy}(\tfrac{1}{6}, \tfrac{1}{12})]^2$$
$$= [-6 \cdot \tfrac{1}{6}] \cdot [-2] - [1]^2$$
$$= -1(-2) - 1$$
$$= 1.$$

2. Find the relative maximum and minimum values of

$$f(x, y) = 2xy - 4x^3 - y^2.$$

d) Thus $D = 1$ and $f_{xx}(\frac{1}{6}, \frac{1}{12}) = -1$. Since $D > 0$ and $f_{xx}(\frac{1}{6}, \frac{1}{12}) < 0$, it follows that f has a relative maximum at $(\frac{1}{6}, \frac{1}{12})$ and that maximum is found as follows:

$$f(\tfrac{1}{6}, \tfrac{1}{12}) = \tfrac{1}{6} \cdot \tfrac{1}{12} - (\tfrac{1}{6})^3 - (\tfrac{1}{12})^2$$
$$= \tfrac{1}{72} - \tfrac{1}{216} - \tfrac{1}{144} = \tfrac{1}{432}.$$

DO EXERCISE 2.

Example 3 *Business: Maximizing profit.* A firm produces two kinds of golf ball, one that sells for $3 each and the other for $2 each. The total revenue from the sale of x thousand balls at $3 each and y thousand at $2 each is given by

$$R(x, y) = 3x + 2y.$$

The company determines that the total cost, in thousands of dollars, of producing x thousand of the $3 ball and y thousand of the $2 ball is given by

$$C(x, y) = 2x^2 - 2xy + y^2 - 9x + 6y + 7.$$

Find the amount of each type of ball that must be produced and sold in order to maximize profit.

Solution Total profit $P(x, y)$ is given by

$$P(x, y) = R(x, y) - C(x, y)$$
$$= 3x + 2y - (2x^2 - 2xy + y^2 - 9x + 6y + 7)$$
$$P(x, y) = -2x^2 + 2xy - y^2 + 12x - 4y - 7.$$

a) Find P_x, P_y, P_{xx}, P_{yy}, and P_{xy}:

$$P_x = -4x + 2y + 12,$$
$$P_{xx} = -4;$$
$$P_y = 2x - 2y - 4,$$
$$P_{yy} = -2;$$
$$P_{xy} = 2.$$

b) Solve the system of equations $P_x = 0$, $P_y = 0$:

$$-4x + 2y + 12 = 0, \tag{1}$$
$$2x - 2y - 4 = 0. \tag{2}$$

3. *Business: Maximizing profit.* A firm produces two kinds of calculator, one that sells for $15 each and the other for $20 each. The total revenue from the sale of x thousand calculators at $15 each and y thousand at $20 each is given by

$$R(x, y) = 15x + 20y.$$

The company determines that the total cost, in thousands of dollars, of producing x thousand of the $15 calculator and y thousand of the $20 calculator is given by

$$C(x, y) = 3x^2 - 3xy + \tfrac{3}{2}y^2 + 6x + 14y - 50.$$

Find the amount of each type of calculator that must be produced and sold in order to maximize profit.

Adding these equations, we get

$$-2x + 8 = 0.$$

Then

$$-2x = -8$$
$$x = 4.$$

To find y when $x = 4$, we substitute 4 for x in either Eq. (1) or Eq. (2). We use Eq. (2):

$$2 \cdot 4 - 2y - 4 = 0$$
$$-2y + 4 = 0$$
$$-2y = -4$$
$$y = 2.$$

Thus $(4, 2)$ is our candidate for a maximum or minimum value.

c) We must check to see whether $P(4, 2)$ is a maximum or minimum:

$$D = P_{xx}(4, 2) \cdot P_{yy}(4, 2) - [P_{xy}(4, 2)]^2$$
$$= (-4)(-2) - 2^2$$
$$= 4.$$

d) Thus $D = 4$ and $P_{xx} = -4$. Since $D > 0$ and $P_{xx}(4, 2) < 0$, it follows that P has a relative maximum at $(4, 2)$. Thus in order to maximize profit, the company must produce and sell 4 thousand of the $3 golf balls and 2 thousand of the $2 golf balls.

DO EXERCISE 3.

EXERCISE SET 7.4

Find the relative maximum and minimum values.

1. $f(x, y) = x^2 + xy + y^2 - y$

2. $f(x, y) = x^2 + xy + y^2 - 5y$

3. $f(x, y) = 2xy - x^3 - y^2$

4. $f(x, y) = 4xy - x^3 - y^2$

5. $f(x, y) = x^3 + y^3 - 3xy$

6. $f(x, y) = x^3 + y^3 - 6xy$

7. $f(x, y) = x^2 + y^2 - 2x + 4y - 2$

8. $f(x, y) = x^2 + 2xy + 2y^2 - 6y + 2$

9. $f(x, y) = x^2 + y^2 + 2x - 4y$

10. $f(x, y) = 4y + 6x - x^2 - y^2$

11. $f(x, y) = 4x^2 - y^2$

12. $f(x, y) = x^2 - y^2$

APPLICATIONS

In these problems assume that relative maximum and minimum values are absolute maximum and minimum values.

13. *Business: Maximizing profit.* A firm produces two kinds of radio, one that sells for $17 each and the other for $21 each. The total revenue from the sale of x thousand radios at $17 each and y thousand at $21 each is given by

$$R(x, y) = 17x + 21y.$$

The company determines that the total cost, in thousands of dollars, of producing x thousand of the $17 radio and y thousand of the $21 radio is given by

$$C(x, y) = 4x^2 - 4xy + 2y^2 - 11x + 25y - 3.$$

Find the amount of each type of radio that must be produced and sold in order to maximize profit.

14. *Business: Maximizing profit.* A firm produces two kinds of baseball glove, one that sells for $18 each and the other for $25 each. The total revenue from the sale of x thousand gloves at $18 each and y thousand at $25 each is given by

$$R(x, y) = 18x + 25y.$$

The company determines that the total cost, in thousands of dollars, of producing x thousand of the $18 glove and y thousand of the $25 glove is given by

$$C(x, y) = 4x^2 - 6xy + 3y^2 + 20x + 19y - 12.$$

Find the amount of each type of glove that must be produced and sold in order to maximize profit.

15. A one-product company found that its profit, in millions of dollars, is a function P given by

$$P(a, p) = 2ap + 80p - 15p^2 - \tfrac{1}{10}a^2p - 100,$$

where a = the amount spent on advertising, in millions of dollars, and p = the price charged per item of the product, in dollars. Find the maximum value of P and the values of a and p at which it is attained.

16. A one-product company finds that its profit, in millions of dollars, is a function P given by

$$P(a, n) = -5a^2 - 3n^2 + 48a - 4n + 2an + 300,$$

where a = the amount spent on advertising, in millions of dollars, and n = the number of items sold, in thousands. Find the maximum value of P and the values of a and n at which it is attained.

EXTENSION EXERCISES

17. *Two-variable revenue maximization.* Boxowitz, Inc., a computer firm, markets two kinds of electronic calculator that compete with one another. Their demand functions are expressed by the following relationships:

$$q_1 = 78 - 6p_1 - 3p_2, \tag{1}$$
$$q_2 = 66 - 3p_1 - 6p_2, \tag{2}$$

where p_1 and p_2 = the price of each calculator, in multiples of $10, and q_1 and q_2 = the quantity of each calculator demanded, in hundreds of units.

a) Find a formula for the total-revenue function R in terms of the variables p_1 and p_2. [*Hint:* $R = p_1q_1 + p_2q_2$; then substitute expressions from Eqs. (1) and (2) to find $R(p_1, p_2)$.]

b) What prices p_1 and p_2 should be charged for each product in order to maximize total revenue?

c) How many units will be demanded?

d) What is the maximum total revenue?

18. Repeat Exercise 17, where

$$q_1 = 64 - 4p_1 - 2p_2 \quad \text{and} \quad q_2 = 56 - 2p_1 - 4p_2.$$

Find the relative maximum and minimum values and the saddle points.

19. $f(x, y) = e^x + e^y - e^{x+y}$ **20.** $f(x, y) = xy + \dfrac{2}{x} + \dfrac{4}{y}$

21. $f(x, y) = 2y^2 + x^2 - x^2y$

22. $S(b, m) = (m + b - 72)^2 + (2m + b - 73)^2 + (3m + b - 75)^2$

7.5

TOTAL DIFFERENTIAL

OBJECTIVES

a) Compute the total differential of a function of two or more variables.

b) Use the differential dw of a function w to approximate the actual change Δw in that function.

1. Find dz if $z = xe^y$.

If $y = f(x)$, then the *differential of y, dy*, is defined by

$$dy = f'(x)\, dx,$$

where dx is the *differential of x*. Note that dy is a function of two variables, x and dx.

In Section 3.7, we saw that linear approximation is a useful tool for certain problems involving functions of one variable, and so we'd like to extend the techniques developed there to functions of two or more variables. To do so, we'll use differentials, beginning with the following definition.

DEFINITION

Let $z = f(x, y)$. Then the *total differential dz* is given by

$$dz = f_x\, dx + f_y\, dy.$$

Example 1 Find dz if $z = x^2y^3$.

Solution Here, $f(x, y) = x^2y^3$, so that $f_x = 2xy^3$ and $f_y = 3x^2y^2$. According to the definition,

$$dz = f_x\, dx + f_y\, dy$$
$$= 2xy^3\, dx + 3x^2y^2\, dy.$$

DO EXERCISES 1 AND 2.

2. Find dz if $z = \ln{(xy)}$.

Example 2 Evaluate dz for $x = 2$, $y = -1$ if $z = xy - 2x^2$.

Solution Since $f(x, y) = xy - 2x^2$, it follows that $f_x = y - 4x$ and $f_y = x$, so

$$dz = f_x\, dx + f_y\, dy = (y - 4x)\, dx + x\, dy$$

3. Evaluate dz for $x = 2$, $y = 3$ if $z = xy^2$.

is valid for all values of x and y. If we substitute in $x = 2$ and $y = -1$, we find that

$$dz = (-1 - 4 \cdot 2)\, dx + 2\, dy$$
$$= (-1 - 8)\, dx + 2\, dy$$
$$= -9\, dx + 2\, dy.$$

DO EXERCISE 3.

Note in Example 2 that after specific values have been substituted for x and y, dz is still a function of the variables dx and dy. In general, if $z = f(x, y)$, then dz is a function of four variables: x, y, dx, and dy.

Example 3   Evaluate dz for $z = x^2 - y^2$ if $x = 2$, $y = 1$, $dx = 0.1$, and $dy = 0.15$.

Solution   Since $f(x, y) = x^2 - y^2$, we have $f_x = 2x$ and $f_y = -2y$. From the definition,

$$dz = 2x\, dx + -2y\, dy.$$

If we let $x = 2$, $y = 1$, $dx = 0.1$, and $dy = 0.15$ in the last equation, we get

$$dz = 2 \cdot 2 \cdot (0.1) - 2 \cdot 1 \cdot (0.15)$$
$$= 4(0.1) - 2(0.15)$$
$$= 0.4 - 0.3 = 0.1.$$

DO EXERCISE 4.

4. Evaluate dz for $z = x^3y + 5y^2$ if $x = 1$, $y = -1$, $dx = 0.25$, and $dy = -0.01$.

Our definition can be extended easily to functions of three or more variables.

DEFINITION

Let $w = f(x, y, z)$. Then the *total differential dw* is given by
$$dw = f_x\, dx + f_y\, dy + f_z\, dz.$$
If w is a function of n variables, say $w = f(x_1, x_2, \ldots, x_n)$, then we define the *total differential dw* as
$$dw = f_{x_1}\, dx_1 + f_{x_2}\, dx_2 + \cdots + f_{x_n}\, dx_n$$
$$= \sum_{i=1}^{n} f_{x_i}\, dx_i.$$

5. Find dw if $w = \dfrac{1}{x} - xy + 2z$.

Example 4 Find dw given that $w = x^2 + y^3 + z^4$.

Solution Here, $f(x, y, z) = x^2 + y^3 + z^4$. Thus $f_x = 2x$, $f_y = 3y^2$, and $f_z = 4z^3$. Substituting these quantities into the definition gives

$$dw = f_x \, dx + f_y \, dy + f_z \, dz$$
$$= 2x \, dx + 3y^2 \, dy + 4z^3 \, dz.$$

DO EXERCISES 5 AND 6.

6. Find dw if $w = (x^2 + y^2 + z^2)^{1/2}$.

Example 5 Find dw if $w = x_1^2 + 2x_2x_3 - x_4$.

Solution Here, $f(x_1, x_2, x_3, x_4) = x_1^2 + 2x_2x_3 - x_4$, so that $f_{x_1} = 2x_1$, $f_{x_2} = 2x_3$, $f_{x_3} = 2x_2$, and $f_{x_4} = -1$. Substitution yields

$$dw = f_{x_1} \, dx_1 + f_{x_2} \, dx_2 + f_{x_3} \, dx_3 + f_{x_4} \, dx_4$$
$$= 2x_1 \, dx_1 + 2x_3 \, dx_2 + 2x_2 \, dx_3 - 1 \, dx_4$$
$$= 2x_1 \, dx_1 + 2x_3 \, dx_2 + 2x_2 \, dx_3 - dx_4.$$

DO EXERCISE 7.

7. Find dw if $w = x_1^2 + x_2^2 + x_3^2 + x_4^2$.

Example 6 Evaluate dw for $x = 1, y = 2, z = 3$, if $w = xy - xz + 2yz$.

Solution Since $f(x, y, z) = xy - xz + 2yz$ and $f_x = y - z$, $f_y = x + 2z$, and $f_z = -x + 2y$, we get

$$dw = (y - z) \, dx + (x + 2z) \, dy + (2y - x) \, dz.$$

So, letting $x = 1$, $y = 2$, and $z = 3$, we have

$$dw = (2 - 3) \, dx + (1 + 2 \cdot 3) \, dy + (2 \cdot 2 - 1) \, dz$$

8. Evaluate dw for $x = -1, y = 1,$ and $z = 1$ if $w = x^2y^3 - z^2$.

or

$$dw = -dx + 7 \, dy + 3 \, dz.$$

$$w = xy - xz + 2yz$$
$$\frac{dw}{dx} + \frac{dw}{dy} + \frac{dw}{dz}$$
$$= df_x \quad df_y + f_z$$

DO EXERCISE 8.

Note that after the three variables x, y, and z have been assigned specific values in Example 6 above, w is still a function of the three variables dx, dy, and dz. If x, y, and z are allowed to vary, then we see that dw is actually a function of the six independent variables $x, y, z, dx, dy,$ and dz. If w is a function of n variables, say $w = f(x_1, x_2, \ldots, x_n)$, then dw is a function of the $2n$ variables $x_1, x_2, \ldots, x_n, dx_1, dx_2, \ldots, dx_n$.

9. Evaluate dw for $x = 1$, $y = -1$, $z = 3$, $dx = 0.1$, $dy = -0.2$, and $dz = 0$ if $w = \ln (xyz)$.

Example 7   Evaluate dw for $x = 2$, $y = 5$, $z = -3$, $dx = -0.01$, $dy = 0.02$, and $dz = -0.03$ if $w = 4x^2 - 2y^2 + 3z^2$.

Solution   Since $w = f(x, y, z) = 4x^2 - 2y^2 + 3z^2$, we have $f_x = 8x$, $f_y = -4y$, and $f_z = 6z$, so that $dw = 8x\,dx - 4y\,dy + 6z\,dz$. Substituting gives us

$$dw = 8 \cdot 2 \cdot (-0.01) - 4 \cdot 5 \cdot (0.02) + 6 \cdot (-3) \cdot (-0.03)$$
$$= 16(-0.01) - 20(0.02) - 18(-0.03)$$
$$= -0.16 - 0.40 + 0.54 = -0.02.$$

DO EXERCISE 9.

If $z = f(x, y)$, then the change Δz in z corresponding to changes Δx and Δy in x and y, respectively, is defined by

$$\Delta z = f(x + \Delta x, y + \Delta y) - f(x, y).$$

Example 8   Calculate Δz for $f(x, y) = x^2 - y^2$ if $x = 1$, $y = 2$, $\Delta x = 0.01$, and $\Delta y = -0.03$.

10. Calculate Δz for

$$f(x, y) = 3x^2 + 4y$$

if $x = 2$, $y = 1$, $\Delta x = 0.07$, and $\Delta y = -0.015$.

Solution   Since $x = 1$, $y = 2$, $\Delta x = 0.01$, and $\Delta y = -0.03$, we have $x + \Delta x = 1 + 0.01 = 1.01$ and $y + \Delta y = 2 + -0.03 = 1.97$. So,

$$\Delta z = f(x + \Delta x, y + \Delta y) - f(x, y)$$
$$= f(1.01, 1.97) - f(1, 2)$$
$$= ((1.01)^2 - (1.97)^2) - (1^2 - 2^2)$$
$$= (1.0201 - 3.8809) - (1 - 4)$$
$$= (-2.8608) - (-3) = 0.1392.$$

DO EXERCISE 10.

In Example 8 above, we calculated the exact change in $z = f(x, y)$ for certain values of x, y, Δx, and Δy. Just as we used the differential dy to approximate the exact change Δy when y is a function of x, we now want to use the differential dz to approximate the exact change Δz in $z = f(x, y)$.

Example 9   Use the differential dz to approximate Δz in Example 8 above.

Solution   We had $z = f(x, y) = x^2 - y^2$ in Example 8, so that $f_x = 2x$, $f_y = -2y$, and $dz = 2x\,dx - 2y\,dy$. If we use the values that were

11. Use the differential dz to approximate the exact change Δz in Margin Exercise 10.

given in Example 8, setting $\Delta x = dx = 0.01$ and $\Delta y = dy = -0.03$, then we get

$$dz = 2 \cdot 1 \cdot (0.01) - 2 \cdot (2) \cdot (-0.03)$$
$$= 2(0.01) - 4(-0.03)$$
$$= 0.02 + 0.12$$
$$= 0.14.$$

Comparing this with the exact value $\Delta z = 0.1392$ from Example 8, we see that $dz \approx \Delta z$.

DO EXERCISE 11.

These ideas extend immediately to functions of three or more variables.

Example 10   If $w = f(x, y, z) = 2x - y^2 + xz$, calculate both the exact change Δw and the approximate change dw for $x = 1$, $y = 1$, $z = 2$, $\Delta x = 0.01$, $\Delta y = -0.02$, and $\Delta z = 0.04$.

12. If $w = x^2 + y^2 + z^2$, calculate both the exact change Δw and the approximate change dw for $x = 4$, $y = -1$, $z = 5$, $\Delta x = 0.03$, $\Delta y = -0.04$, and $\Delta z = -0.01$.

Solution   Note that $x + \Delta x = 1.01$, $y + \Delta y = 0.98$, and $z + \Delta z = 2.04$, so that the exact change Δw is given by

$$\Delta w = f(x + \Delta x, y + \Delta y, z + \Delta z) - f(x, y, z)$$
$$= f(1.01, 0.98, 2.04) - f(1, 1, 2)$$
$$= [2(1.01) - (0.98)^2 + (1.01)(2.04)] - [2 \cdot 1 - 1^2 + 1 \cdot 2]$$
$$= [2.02 - 0.9604 + 2.0604] - [2 - 1 + 2]$$

or

$$\Delta w = 3.12 - 3 = 0.12.$$

To compute dw, note that $f_x = 2 + z$, $f_y = -2y$, and $f_z = x$, so that

$$dw = (2 + z)\, dx - 2y\, dy + x\, dz.$$

Setting $dx = \Delta x = 0.01$, $dy = \Delta y = -0.02$, and $dz = \Delta z = 0.04$, we get

$$dw = (2 + 2) \cdot (0.01) - 2 \cdot 1 \cdot (-0.02) + 1 \cdot (0.04)$$
$$= 4(0.01) - 2(-0.02) + 0.04$$
$$= 0.04 + 0.04 + 0.04$$
$$= 0.12.$$

In this case, coincidentally, $dw = \Delta w$.

DO EXERCISE 12.

13. If $z = \ln(xy)$, use differentials to calculate an approximate value for $\ln((1.01)(0.97))$.

14. If $z = \ln(x/y)$, use differentials to calculate an approximate value for $\ln(1.01/0.97)$.

If $z = f(x, y)$, then we had defined $\Delta z = f(x + \Delta x, y + \Delta y) - f(x, y)$. Rewriting, we get

$$f(x + \Delta x, y + \Delta y) = f(x, y) + \Delta z.$$

Now, if we use the approximation $dz \approx \Delta z$, we obtain

$$f(x + \Delta x, y + \Delta y) \approx f(x, y) + dz.$$

In a similar way, if $w = f(x, y, z)$, then we can show that

$$f(x + \Delta x, y + \Delta y, z + \Delta z) \approx f(x, y, z) + dw.$$

Example 11   If $z = f(x, y) = x^2 y^3$, use differentials to calculate an approximate value for $(1.01)^2(0.98)^3$.

Solution   We are asked to calculate $f(1.01, 0.98)$. If we set $x = 1$, $y = 1$, $\Delta x = 0.01$, and $\Delta y = -0.02$, then we must calculate

$$f(x + \Delta x, y + \Delta y).$$

From above, we already know that

$$f(x + \Delta x, y + \Delta y) \approx f(x, y) + dz.$$

But

$$f(x, y) = f(1, 1) = 1^2 (1)^3 = 1,$$

while

$$\begin{aligned}
dz &= f_x \, dx + f_y \, dy \\
&= 2xy^3 \, dx + 3x^2 y^2 \, dy \\
&= 2 \cdot 1 \cdot (1)^3 \, dx + 3(1)^2(1)^2 \, dy \\
&= 2 \, dx + 3 \, dy.
\end{aligned}$$

Setting $dx = \Delta x = 0.01$ and $dy = \Delta y = -0.02$, we find that

$$\begin{aligned}
dz &= 2(0.01) + 3(-0.02) \\
&= 0.02 - 0.06 = -0.04.
\end{aligned}$$

So $f(1.01, 0.98) \approx f(1, 1) + dz$ and

$$\approx 1 - 0.04 = 0.96.$$

In this case, we can calculate the actual value of $f(1.01, 0.98)$ with a calculator. We find that

$$f(1.01, 0.98) = (1.01)^2(0.98)^3 \approx 0.9601.$$

In this case, our approximation was rather good.

DO EXERCISES 13 AND 14.

15. The pressure P at the point (x, y, z) in a certain region of space is given (in pounds per square inch) by

$$P(x, y, z) = x^2 + 3xy + z^2.$$

Use differentials to calculate an approximate value for the pressure at $(1.99, 1.52, 3.98)$.

Example 12   The temperature T at the point (x, y, z) of a certain region of space is given by $T(x, y, z) = (x^2 + y^2 + z^2)^{3/2}$. Use differentials to calculate the approximate temperature at $(2.014, 2.15, 1.06)$.

Solution   We use

$$T(x + \Delta x, y + \Delta y, z + \Delta z) \approx T(x, y, z) + dT.$$

Since $T(2, 2, 1)$ is easy to evaluate, we let $x = 2$, $y = 2$, and $z = 1$; this forces us to choose $\Delta x = 0.014$, $\Delta y = 0.15$, and $\Delta z = 0.06$. Proceeding, we get

$$T(2, 2, 1) = (2^2 + 2^2 + 1^2)^{3/2}$$
$$= (4 + 4 + 1)^{3/2} = 9^{3/2} = 27.$$

Now,

$$dT = T_x\, dx + T_y\, dy + T_z\, dz$$
$$= \tfrac{3}{2}(x^2 + y^2 + z^2)^{1/2}\, 2x\, dx + \tfrac{3}{2}(x^2 + y^2 + z^2)^{1/2}\, 2y\, dy$$
$$+ \tfrac{3}{2}(x^2 + y^2 + z^2)^{1/2}\, 2z\, dz$$
$$= (x^2 + y^2 + z^2)^{1/2}(3x\, dx + 3y\, dy + 3z\, dz).$$

If we let $dx = \Delta x = 0.014$, $dy = \Delta y = 0.15$, and $dz = \Delta z = 0.06$, then we get

$$dT = (2^2 + 2^2 + 1^2)^{1/2}[3 \cdot 2 \cdot (0.014) + 3 \cdot 2 \cdot (0.15) + 3 \cdot 1 \cdot (0.06)]$$
$$= (9)^{1/2}[6(0.014) + 6(0.15) + 3(0.06)]$$
$$= 3[1.164] = 3.492.$$

Thus,

$$T(2.014, 2.15, 1.06) \approx T(2, 2, 1) + dT$$
$$= 27 + 3.492$$
$$= 30.492.$$

DO EXERCISE 15.

EXERCISE SET 7.5

Find dz.

1. $z = x^2 + 4y^2$

2. $z = 3x - 4y + xy$

3. $z = xe^{4y}$

4. $z = e^{xy}$

5. $z = x^4 + \dfrac{x}{y} + y^3$

6. $z = x \ln y$

Find dw.

7. $w = x^2 - y^2 + 3z^2$

8. $w = xy + 2xz + yz$

9. $w = xy^2z^4$

10. $w = \ln(xyz)$

11. $w = x_1^2 + x_2x_3 - x_4^2$

12. $w = x_1x_2x_3x_4$

Evaluate both Δz and dz.

13. $z = x^2 + y^2$, $x = 5$, $y = 12$; $dx = \Delta x = 0.01$, $dy = \Delta y = 0.02$

14. $z = 4x + 3y$, $x = 1$, $y = -1$; $dx = \Delta x = 0.03$, $dy = \Delta y = -0.02$

15. $z = e^{x+y}$, $x = y = 0$; $dx = \Delta x = 0.01$, $dy = \Delta y = 0.0005$

16. $z = \ln(xy)$, $x = 2$, $y = 0.5$; $dx = \Delta x = 0.02$, $dy = \Delta y = 0.01$

Evaluate both Δw and dw.

17. $w = x^2 + y^2 - z^2$, $x = 2$, $y = 2$, $z = 1$; $dx = \Delta x = 0.01$, $dy = \Delta y = 0.02$, $dz = \Delta z = 0.02$

18. $w = 4x - 2y + 3z$, $x = 1$, $y = 1$, $z = 2$; $dx = \Delta x = 0.06$, $dy = \Delta y = 0.13$, $dz = \Delta z = -0.04$

19. $w = xyz$, $x = 1$, $y = 0$, $z = 3$; $dx = \Delta x = 0.04$, $dy = \Delta y = -0.02$, $dz = \Delta z = 0.01$

20. $w = \ln(xz/y)$, $x = 2$, $y = 6$, $z = 3$; $dx = \Delta x = 0.05$, $dy = \Delta y = 0.08$, $dz = \Delta z = 0.01$

Use differentials to calculate an approximate value.

21. $(5.01)^2 + (11.94)^2$

22. $(1.06)(0.99)^3 + (2.03)^2$

23. $(1.04)^2 + (0.97)^2 + (1.02)^2 + (1.015)^2$

24. $(1.05)^3 - (0.93)^2$

25. Use differentials to find an approximate value for the area of a right triangle whose base and altitude are 2.013 ft and 4.501 ft, respectively.

26. A rectangular box has dimensions 1.54 ft by 1.01 ft by 7.05 ft. Use differentials to approximate its volume.

27. Use differentials to approximate the surface area of the box in Exercise 26.

28. Calculate dz if $z = x^y$, $x > 0$.

APPLICATIONS

29. A firm estimates that its profit from selling gold medallions at a price of g dollars each and silver medallions at a price of s dollars each is given by

    $$P(g, s) = -50,000 + 300s + 1000g - 3s^2 - g^2 + sg.$$

    a) Show that if the firm sells the silver medallions at $100 each and the gold ones at $500 each, its estimated profit is $250,000.

    b) Use the total differential to approximate the effect on estimated profit of a $1 increase in the price of both medallions.

30. Referring to Exercise 10 in Exercise Set 7.1, use differentials to estimate the change in the price-earnings ratio $R(P, E)$ of IBM common stock if the price per share increases from $287\frac{3}{8}$ to $288\frac{5}{8}$ while the earnings per share decreases from $23.30 to $22.80.

31. *Biomedical: Poiseuille's law.* Referring to Exercise 12 in Exercise Set 7.1, use differentials to find the approximate change in the speed of blood if p increases from 100 to 103, v decreases from 0.05 to 0.049, and R, L, and r are held constant.

32. *Business.* Referring to Exercise 40 in Exercise Set 7.1, use differentials to calculate the approximate increase in production for the company if the number of units of labor is increased from 32 to 35 and the number of units of capital is increased from 1024 to 1040. What is the exact increase in production?

33. If $w = x_1 x_2 \ldots x_n$, show that

$$dw = w\left[\frac{1}{x_1}\,dx_1 + \frac{1}{x_2}\,dx_2 + \cdots + \frac{1}{x_n}\,dx_n\right].$$

34. *Investments.* If the price P of a certain stock is given by

$$P = D\left(1 + \frac{1}{i}\right),$$

where D is the annual dividend per share and i is the annual interest rate, use differentials to estimate what happens to the price of the stock on the day its dividend is cut from \$5 to \$4 per share if the interest rate i increases from 6% to 6.5% on the same day.

7.6

OBJECTIVE

AN APPLICATION: THE LEAST-SQUARES TECHNIQUE

a) Find the regression line for a given set of data points and use the regression line to make predictions regarding further data.

The problem of fitting an equation to a set of data occurs frequently. We considered one procedure for doing this in Section 1.6. Such an equation provides a model of the phenomena from which predictions can be made. For example, in business one might want to predict future sales on the basis of past data. In ecology, one might want to predict future demands for natural gas on the basis of past need. Suppose we are trying to determine a linear equation

$$y = mx + b$$

to fit the data. To determine this equation is to determine the values of m and b. But how? Let us consider some factual data.

The graph shown in Fig. 4 appeared in a newspaper advertisement for the Indianapolis Life Insurance Company. It pertains to the total amount of life insurance in force in various years. The same data are compiled in the following table.

Total Individual Life
Insurance in Force
$ BILLIONS

FIGURE 4

1. **a)** Use Fig. 5 to predict the total
        amount of life insurance in force
        in 1988.

    b) Use Fig. 6 to predict the total
        amount of life insurance in force
        in 1988.

    c) Compare your answers.

Year, x	1. 1968	2. 1973	3. 1978	4. 1983	5. 1988
Total individual life insurance in force (in billions), y	$1.2	$2.2	$4.3	$5.2	?

Suppose we plot these points and try to draw a line through them
that fits. Note that there are several ways this might be done (see Figs.
5 and 6). Each would give a different estimate of the total insurance in
force in 1988.

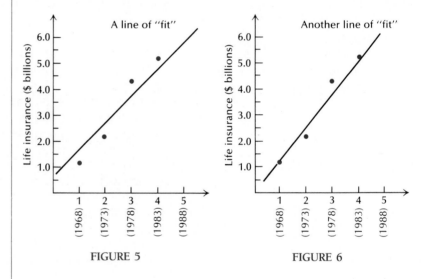

FIGURE 5   FIGURE 6

DO EXERCISE 1.

Note that time is incremented in fives of years, making computations easier. Consider the data points $(1, 1.2)$, $(2, 2.2)$, $(3, 4.3)$, and
$(4, 5.2)$ as plotted in Fig. 7.

We will try to fit these data with a line

$$y = mx + b$$

by determining the values of m and b. Note the y-errors, or y-deviations,
$y_1 - 1.2, y_2 - 2.2, y_3 - 4.3$, and $y_4 - 5.2$ between the observed points
$(1, 1.2)$, $(2, 2.2)$, $(3, 4.3)$, and $(4, 5.2)$ and the points $(1, y_1)$, $(2, y_2)$,
$(3, y_3)$, and $(4, y_4)$ on the line. We would like, somehow, to minimize

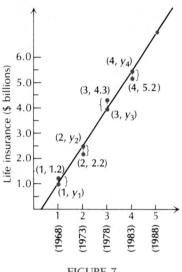

FIGURE 7

these deviations in order to have a good fit. One way of minimizing the deviations is based on the *least-squares assumption*.

The Least-Squares Assumption

The line of best fit is the line for which the sum of the squares of the *y*-deviations is a minimum. This is called the *regression line*.

Using the least-squares assumption for the life insurance data, we would minimize

$$(y_1 - 1.2)^2 + (y_2 - 2.2)^2 + (y_3 - 4.3)^2 + (y_4 - 5.2)^2, \quad \textbf{(1)}$$

and since the points $(1, y_1)$, $(2, y_2)$, $(3, y_3)$, and $(4, y_4)$ must be solutions of $y = mx + b$, it follows that

$$y_1 = m(1) + b = m + b,$$
$$y_2 = m(2) + b = 2m + b,$$
$$y_3 = m(3) + b = 3m + b,$$
$$y_4 = m(4) + b = 4m + b.$$

Substituting $m + b$ for y_1, $2m + b$ for y_2, $3m + b$ for y_3, and $4m + b$ for y_4 in Eq. (1), we have

$$(m+b-1.2)^2 + (2m+b-2.2)^2 + (3m+b-4.3)^2 + (4m+b-5.2)^2.$$

$$\textbf{(2)}$$

2. Consider the following factual data on the cost of a 30-second commercial during the broadcast of the Super Bowl.

Year, x	Cost of a 30-second commercial during the Super Bowl, y
1. 1979	$180,000
2. 1980	$234,000
3. 1981	$275,000
4. 1982	$345,000

a) Find the regression line.

b) Use the regression line to predict the cost in 1988; in 1995.

Thus, to find the regression line for the given set of data, we must find the values of m and b that minimize the function S given by the sum in Eq. (2).

To apply the D-test, we first find the partial derivatives $\partial S/\partial b$ and $\partial S/\partial m$:

$$\frac{\partial S}{\partial b} = 2(m + b - 1.2) + 2(2m + b - 2.2) + 2(3m + b - 4.3)$$
$$+ 2(4m + b - 5.2)$$
$$= 20m + 8b - 25.8;$$

$$\frac{\partial S}{\partial m} = 2(m + b - 1.2) + 2(2m + b - 2.2)2 + 2(3m + b - 4.3)3$$
$$+ 2(4m + b - 5.2)4$$
$$= 60m + 20b - 78.6.$$

We set these derivatives equal to 0 and solve the resulting system:

$$20m + 8b - 25.8 = 0, \qquad 5m + 2b = 6.45,$$
$$\text{or}$$
$$60m + 20b - 78.6 = 0; \qquad 15m + 5b = 19.65.$$

The solution to this system is

$$b = -0.30, \qquad m = 1.41.$$

We leave it to the reader to complete the D-test to verify that $(-0.30, 1.41)$ does, in fact, yield the minimum of S. We need not bother to compute $S(-0.30, 1.41)$.

The values of m and b are all we need to determine $y = mx + b$. The regression line is

$$y = 1.41x - 0.30.$$

We can extrapolate from the data to predict the total amount of life insurance in force in 1988:

$$y = 1.41(5) - 0.30 = 6.75.$$

Thus the total amount of life insurance in force in 1988 will be about $6.75 billion.

The method of least squares is a statistical process illustrated here with only four data points in order to simplify the explanation. Most statistical researchers would warn that many more than four data points should be used to get a "good" regression line. Furthermore, making predictions too far in the future from any linear model may be suspect. It can be done, but the further into the future the prediction is made, the more dubious one should be about the prediction.

DO EXERCISE 2.

*The Regression Line for an Arbitrary Collection of Data Points (c_1, d_1), (c_2, d_2), . . . , (c_n, d_n)

Look again at the regression line

$$y = 1.41x - 0.30$$

for the data points $(1, 1.2)$, $(2, 2.2)$, $(3, 4.3)$, and $(4, 5.2)$. Let us consider the arithmetic averages, or means, of the x-coordinates, denoted $\bar{x}$, and the y-coordinates, denoted $\bar{y}$:

$$\bar{x} = \frac{1 + 2 + 3 + 4}{4} = 2.5,$$

$$\bar{y} = \frac{1.2 + 2.2 + 4.3 + 5.2}{4} = 3.225.$$

It turns out that the point $(\bar{x}, \bar{y})$, or $(2.5, 3.225)$, is on the regression line for

$$3.225 = 1.41(2.5) - 0.30.$$

Thus the regression line is

$$y - \bar{y} = m(x - \bar{x}), \quad \text{or} \quad y - 3.225 = m(x - 2.5).$$

All that remains, in general, is to determine m.

Suppose we wanted to find the regression line for an arbitrary number of points (c_1, d_1), (c_2, d_2), . . . , (c_n, d_n). To do so, we find the values m and b that minimize the function S given by

$$S(b, m) = (y_1 - d_1)^2 + (y_2 - d_2)^2 + \cdots + (y_n - d_n)^2$$
$$= \sum_{i=1}^{n} (y_i - d_i)^2,$$

where $y_i = mc_i + b$.

Using a procedure like the one we used earlier to minimize S, we can show that $y = mx + b$ takes the form

$$y - \bar{y} = m(x - \bar{x}),$$

where

$$\bar{x} = \frac{\sum_{i=1}^{n} c_i}{n}, \quad \bar{y} = \frac{\sum_{i=1}^{n} d_i}{n}, \quad \text{and} \quad m = \frac{\sum_{i=1}^{n} (c_i - \bar{x})(d_i - \bar{y})}{\sum_{i=1}^{n} (c_i - \bar{x})^2}.$$

Let us see how this works out for the individual life-insurance example used previously.

---

* This part is considered optional and can be omitted without loss of continuity.

3. Repeat Margin Exercise 2(a) using the procedure just outlined in the optional part of this section.

c_i	d_i	$c_i - \bar{x}$	$(c_i - \bar{x})^2$	$(d_i - \bar{y})$	$(c_i - \bar{x})(d_i - \bar{y})$
1	1.2	-1.5	2.25	-2.025	3.0375
2	2.2	-0.5	0.25	-1.025	0.5125
3	4.3	0.5	0.25	1.075	0.5375
4	5.2	1.5	2.25	1.975	2.9625

$$\sum_{i=1}^{4} c_i = 10 \qquad \sum_{i=1}^{4} d_i = 12.9 \qquad \sum_{i=1}^{4} (c_i - \bar{x})^2 = 5 \qquad \sum_{i=1}^{4} (c_i - \bar{x})(d_i - \bar{y}) = 7.05$$

$$\bar{x} = 2.5 \qquad \bar{y} = 3.225$$

$$m = \frac{7.05}{5} = 1.41$$

Thus the regression line is

$$y - 3.225 = 1.41(x - 2.5),$$

which simplifies to

$$y = 1.41x - 0.30.$$

DO EXERCISE 3.

*Nonlinear Regression

It can happen that data do not seem to fit a linear equation, but when logarithms of either the x-values or the y-values (or both) are taken, a linear relationship will exist. Indeed, on considering the graph in Fig. 4 it is not unreasonable to expect these data to fit an exponential function.

Example 1 Use logarithms and regression to find an equation

$$y = Be^{kx}$$

that fits the data. Then estimate the total amount of life insurance in force in 1988.

Year, x	1. 1968	2. 1973	3. 1978	4. 1983
Total individual life insurance in force (in billions), y	$1.2	$2.2	$4.3	$5.2

---

* This part is considered optional and can be omitted without loss of continuity.

Solution   If we take the natural logarithm of both sides of

$$y = Be^{kx},$$

we get

$$\ln y = \ln B + kx.$$

Note that $\ln B$ and k are constants. So, if we replace $\ln y$ by a new variable Y, the equation takes the form of a linear function

$$Y = mx + b,$$

where $m = k$ and $b = \ln B$.

We are going to find this regression line, but before starting we need to find the logarithms of the y-values.

x	1	2	3	4
$Y = \ln y$	0.1823	0.7885	1.4586	1.6487

To find the regression line, we use the abbreviated procedure described in the preceding part of this section.

c_i	d_i	$c_i - \bar{x}$	$(c_i - \bar{x})^2$	$d_i - \bar{Y}$	$(c_i - \bar{x})(d_i - \bar{Y})$
1	0.1823	-1.5	2.25	-0.837225	1.255838
2	0.7885	-0.5	0.25	-0.231025	0.115513
3	1.4586	0.5	0.25	0.439075	0.219538
4	1.6487	1.5	2.25	0.629175	0.943763

$$\sum_{i=1}^{4} c_i = 10 \qquad \sum_{i=1}^{4} d_i \qquad \sum_{i=1}^{4} (c_i - \bar{x})^2 \qquad \sum_{i=1}^{4} (c_i - \bar{x})(d_i - \bar{Y})$$

$$\bar{x} = \frac{10}{4} = 2.5 \qquad = 4.0781 \qquad\qquad = 5 \qquad\qquad = 2.534652$$

$$\bar{Y} = \frac{4.0781}{4} \qquad\qquad m = \frac{2.534652}{5}$$

$$= 1.019525 \qquad\qquad = 0.5069304$$

Thus the regression line is

$$Y - 1.019525 = 0.5069304(x - 2.5),$$

4. a) Use natural logarithms and regression to find an equation

$$y = Be^{kx}$$

that fits the data in Margin Exercise 2 regarding the cost of Super Bowl commercials.

b) Estimate the cost of a Super Bowl commercial in 1988; in 1995.

which simplifies to

$$Y = 0.5069304x - 0.247801.$$

Recall that we were to find k and B. From this equation we know that

$$m = k = 0.5069304$$

and

$$b = \ln B = -0.247801.$$

To find B we use the definition of logarithms (or take the antilog) and get

$$B = e^{-0.247801} = 0.7805.$$

Then the desired equation is

$$y = 0.7805e^{0.5069304x}.$$

Using this equation, we then find that the total amount of life insurance in force in 1988 will be

$$y = 0.7805e^{0.5069304(5)} = \$9.8 \text{ billion}.$$

DO EXERCISE 4.

In conclusion, there are other kinds of nonlinear regression besides logarithmic. For example, a set of data might fit a quadratic equation

$$y = ax^2 + bx + c.$$

In such a case, one can still use regression to find the numbers a, b, and c that minimize the sums of squares of deviations.

EXERCISE SET 7.6

APPLICATIONS

1. *Cost of a ticket to a Broadway musical.* The factual data in the table at right show the average price of a ticket to a Broadway musical over several years.

   a) Find the regression line $y = mx + b$.

   b) Use the regression line to predict the average price of a ticket in 1988; in 1995.

Year, x	Average price of a Broadway musical
1. 1980	$30
2. 1981	45
3. 1982	40
4. 1983	45
5. 1984	45
6. 1985	45

2. Consider the factual data on natural gas demand shown at right.

   a) Find the regression line $y = mx + b$.

   b) Use the regression line to predict gas demand in 1990; in 2000.

Year, x	Demand (in quadrillion BTU)
1. 1950	19
2. 1960	21
3. 1970	22

3. A professor wanted to predict students' final examination scores on the basis of their midterm test scores. An equation was determined from data (see table at right) on the scores of three students who took the same course with the same instructor the previous semester.

   a) Find the regression line $y = mx + b$. [*Hint:* The y-deviations are $70m + b - 75$, $60m + b - 62$, and so on.]

   b) The midterm score of a student was 81. Use the regression line to predict the student's final exam score.

Midterm score (%), x	Final exam score (%), y
70	75
60	62
85	89

4. Consider the factual data at right showing the book value of the common stock of Eli Lilly and Co., a pharmaceutical firm, during four recent years of operation.

   a) Find the regression line $y = mx + b$.

   b) Use the regression line to predict the value of the stock in 1989; in 2000.

Year, x	Book value of stock, y
1. 1982	$13.55
2. 1983	14.43
3. 1984	14.91
4. 1985	16.44

EXTENSION EXERCISES

5. a) ▨ Find the regression line $y = mx + b$ that fits the set of data in the table.

   b) Use the regression line to predict the world record in the mile in 1985; in 2000.

   c) In August 1985 Steve Cram set a new world record of 3:46.31 for the mile. How does this compare with what can be predicted by the regression? [*Hint:* Convert each time to decimal notation; for example,

   $$4{:}24.5 = 4\tfrac{24.5}{60} = 4.4083.$$

Year, x (Use the actual year for x.)	World record in mile, y (min:sec)
1875 (Walter Slade)	4:24.5
1894 (Fred Bacon)	4:18.2
1923 (Paavo Nurmi)	4:10.4
1937 (Sidney Wooderson)	4:06.4
1942 (Gunder Haegg)	4:06.2
1945 (Gunder Haegg)	4:01.4
1954 (Roger Bannister)	3:59.4
1964 (Peter Snell)	3:54.4
1967 (Jim Ryun)	3:51.1
1975 (John Walker)	3:49.4
1979 (Sebastian Coe)	3:49.0
1980 (Steve Ovett)	3:48.8

6. **a)** Use logarithms and regression to find an equation

$$y = Be^{kx}$$

that fits this set of data.

b) Use the regression equation to find the population of the United States in 1990.

Year (from 1976), x	Population of U.S. (in millions), y
0	216
1	218
2	219
3	221

7. **a)** ▦ Use logarithms and regression to find an equation

$$y = Be^{kx}$$

that fits the set of data in Exercise 1.

b) Use the regression equation to predict average ticket price in 1988; in 1995.

c) Compare your answers with those of Exercise 1.

8. **a)** Use regression (but not logarithms) to find an equation

$$y = ax^2 + bx + c$$

that fits this set of data.*

b) Find the death rate of those who average 4 hours of sleep; 10 hours of sleep; 7.5 hours of sleep.

Average number of hours of sleep, x	Death rate in one year (per 100,000 males), y
5	1121
6	805
7	626
8	813
9	967

---

* The set of data in Exercise 8 comes from a study by Dr. Harold J. Morowitz.

7.7

OBJECTIVES

a) Find a maximum or minimum value of a given function subject to a given constant, using the method of LaGrange multipliers.

b) Solve applied problems involving LaGrange multipliers.

CONSTRAINED MAXIMUM AND MINIMUM VALUES: LAGRANGE MULTIPLIERS

Before we proceed in detail, let us return to a problem we considered in Chapter 3.

Example 1 A hobby store has 20 ft of fencing to fence off a rectangular electric-train area in one corner of its display room. The two sides up against the wall require no fence. What dimensions of the rectangle will maximize the area?

We maximize the function

$$A = xy$$

subject to the condition, or *constraint,* $x + y = 20$. Note that A is a function of two variables.

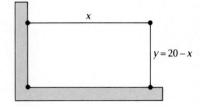

When we solved this earlier, we first solved the constraint for y:

$$y = 20 - x.$$

We then substituted $20 - x$ for y to obtain

$$A(x, 20 - x) = x(20 - x) = 20x - x^2,$$

which is a function of one variable. Next, we found a maximum value using Maximum–Minimum Principle 1 (see Section 3.4). By itself, the function of two variables

$$A(x, y) = xy$$

has no maximum value. This can be checked using the D-test. With the constraint $x + y = 20$, however, the function does have a maximum. We see this in the following figure.

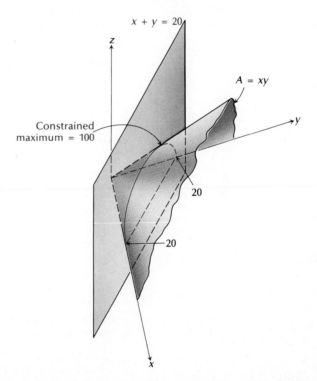

It may be quite difficult to solve a constraint for one variable. The procedure outlined below allows us to proceed without doing so.

THEOREM 2

The Method of LaGrange Multipliers

To find a maximum or minimum value of a function $f(x, y)$ subject to the constraint $g(x, y) = 0$, we do the following.

a) Form a new function:

$$F(x, y, \lambda) = f(x, y) - \lambda g(x, y).$$

b) Find the partial derivatives F_x, F_y, and F_λ.

c) Solve the system

$$F_x = 0, \qquad F_y = 0, \quad \text{and} \quad F_\lambda = 0.$$

Let (a, b) represent a solution of this system. We still must determine whether (a, b) yields a maximum or minimum, but we will assume one or the other in the problems considered here.

The variable λ (lambda) is called a *LaGrange multiplier*. We first illustrate the method of LaGrange multipliers by resolving the problem in Example 1.

Example 2 Find the maximum value of

$$A(x, y) = xy$$

subject to the constraint $x + y = 20$.

Solution

a) We form the new function F given by

$$F(x, y, \lambda) = xy - \lambda \cdot (x + y - 20).$$

Note that we first had to express $x + y = 20$ as $x + y - 20 = 0$.

b) We find the first partial derivatives:

$$F_x = y - \lambda,$$
$$F_y = x - \lambda,$$
$$F_\lambda = -(x + y - 20).$$

c) We set these derivatives equal to 0 and solve the resulting system:

$$y - \lambda = 0, \tag{1}$$
$$x - \lambda = 0, \tag{2}$$
$$x + y - 20 = 0. \tag{3}$$

[If $-(x + y - 20) = 0$, then $x + y - 20 = 0$.] From Eqs. (1) and (2)

1. A rancher has 50 ft of fencing to fence off a rectangular animal pen in the corner of a barn. What dimensions of the rectangle will yield the maximum area?

   a) Express the area as a function of two variables with a constraint.

   b) Find the maximum value of the function in part (a) using the method of LaGrange multipliers.

it follows that

$$x = y = \lambda.$$

Substituting λ for x and y in Eq. (3), we get

$$\lambda + \lambda - 20 = 0$$
$$2\lambda = 20$$
$$\lambda = 10.$$

Thus $x = \lambda = 10$ and $y = \lambda = 10$. The maximum occurs at $(10, 10)$ and is

$$A(10, 10) = 10 \cdot 10 = 100.$$

DO EXERCISE 1.

Example 3   Find the maximum value of

$$f(x, y) = 3xy$$

subject to the constraint

$$2x + y = 8.$$

[*Note:* f could be interpreted as a production function with budget constraint $2x + y = 8$.]

Solution

a) We form the new function F given by

$$F(x, y, \lambda) = 3xy - \lambda(2x + y - 8).$$

Note that we first had to express $2x + y = 8$ as $2x + y - 8 = 0$.

b) We find the first partial derivatives:

$$F_x = 3y - 2\lambda,$$
$$F_y = 3x - \lambda,$$
$$F_\lambda = -(2x + y - 8).$$

c) We set these derivatives equal to 0 and solve the resulting system:

$$3y - 2\lambda = 0, \qquad \textbf{(1)}$$
$$3x - \lambda = 0, \qquad \textbf{(2)}$$
$$-(2x + y - 8) = 0, \quad \text{or} \quad 2x + y - 8 = 0. \qquad \textbf{(3)}$$

Solving Eq. (1) for y, we get

$$y = \tfrac{2}{3}\lambda.$$

2. Find the maximum value of
$$f(x, y) = 5xy$$
subject to the constraint
$$4x + y = 20.$$

Solving Eq. (2) for x, we get
$$x = \frac{\lambda}{3}.$$

Substituting $(\frac{2}{3})\lambda$ for y and $(\lambda/3)$ for x in Eq. (3), we get
$$2\left(\frac{\lambda}{3}\right) + \left(\frac{2}{3}\lambda\right) - 8 = 0$$
$$\frac{4}{3}\lambda = 8$$
$$\lambda = \frac{3}{4} \cdot 8 = 6.$$

Then
$$x = \frac{\lambda}{3} = \frac{6}{3} = 2 \quad \text{and} \quad y = \frac{2}{3}\lambda = \frac{2}{3} \cdot 6 = 4.$$

The maximum of f subject to the constraint occurs at $(2, 4)$ and is
$$f(2, 4) = 3 \cdot 2 \cdot 4 = 24.$$

DO EXERCISE 2.

Example 4   *The beverage can problem.* The standard beverage can has a volume of 12 oz, or 26 in³. What dimensions yield the minimum surface area? Find the minimum surface area.

Solution   We want to minimize the function s given by
$$s(h, r) = 2\pi rh + 2\pi r^2$$
subject to the volume constraint
$$\pi r^2 h = 26, \quad \text{or} \quad \pi r^2 h - 26 = 0.$$

Note that s does not have a minimum without the constraint.

a)   We form the new function S given by

$$S(h, r, \lambda) = 2\pi rh + 2\pi r^2 - \lambda(\pi r^2 h - 26).$$

b)   We find the first partial derivatives:

$$\frac{\partial S}{\partial h} = 2\pi r - \lambda \pi r^2,$$

$$\frac{\partial S}{\partial r} = 2\pi h + 4\pi r - 2\lambda \pi rh,$$

$$\frac{\partial S}{\partial \lambda} = -(\pi r^2 h - 26).$$

c)   We set these derivatives equal to 0 and solve the resulting system:

$$2\pi r - \lambda \pi r^2 = 0, \qquad \textbf{(1)}$$
$$2\pi h + 4\pi r - 2\lambda \pi rh = 0, \qquad \textbf{(2)}$$
$$-(\pi r^2 h - 26) = 0, \quad \text{or} \quad \pi r^2 h - 26 = 0. \qquad \textbf{(3)}$$

Note that we can solve Eq. (1) for r:

$$\pi r(2 - \lambda r) = 0$$
$$\pi r = 0 \quad \text{or} \quad 2 - \lambda r = 0$$
$$r = 0 \quad \text{or} \qquad r = \frac{2}{\lambda}.$$

Since $r = 0$ cannot be a solution to the original problem, we continue by substituting $2/\lambda$ for r in Eq. (2):

$$2\pi h + 4\pi \cdot \frac{2}{\lambda} - 2\lambda \pi \cdot \frac{2}{\lambda} \cdot h = 0$$

$$2\pi h + \frac{8\pi}{\lambda} - 4\pi h = 0$$

$$\frac{8\pi}{\lambda} - 2\pi h = 0$$

$$-2\pi h = -\frac{8\pi}{\lambda},$$

so

$$h = \frac{4}{\lambda}.$$

Since $h = 4/\lambda$ and $r = 2/\lambda$, it follows that $h = 2r$. Substituting $2r$ for h

3. Repeat Example 4 for a can of 16 oz, or 35 in³.

in Eq. (3) yields

$$\pi r^2 (2r) - 26 = 0$$

$$2\pi r^3 - 26 = 0$$

$$2\pi r^3 = 26$$

$$\pi r^3 = 13$$

$$r^3 = \frac{13}{\pi}$$

$$r = \sqrt[3]{\frac{13}{\pi}} \approx 1.6 \text{ in.}$$

Thus when $r = 1.6$ in., $h = 3.2$ in., and the surface area is a minimum and is about

$$2\pi(1.6)(3.2) + 2\pi(1.6)^2, \quad \text{or} \quad 48.3 \text{ in}^2.$$

DO EXERCISE 3.

The actual dimensions of a standard-sized 12-oz beverage can are $r = 1.25$ in. and $h = 4.876$ in. A natural question after studying Example 4 is, "Why don't beverage companies make cans using the dimensions found in that example?" To do this at this time would mean a monumental cost in retooling. New can-making machines would have to be purchased at a cost of millions. New beverage-filling machines would have to be purchased. Vending machines would no longer be the correct size. A partial response to the desire to save aluminum has been found in recycling and in manufacturing cans with a rippled effect at the top. These cans require less aluminum. As a result of many engineering ideas, the amount of aluminum required to make 1000 cans has been reduced from 36.5 lb to 28.1 lb. The consumer is actually a very important factor in the shape of the can. Market research has shown that a can with the dimensions found in Example 4 is not as comfortable to hold and might not be accepted by consumers.*

_____

* Many thanks to Don Hauser of the Pepsi-Cola Co. and Bobby Ryals of the Continental Can Co. for the ideas in this paragraph.

EXERCISE SET 7.7

Find the maximum value of f subject to the given constraint.

1. $f(x, y) = xy;\ 2x + y = 8$

2. $f(x, y) = 2xy;\ 4x + y = 16$

3. $f(x, y) = 4 - x^2 - y^2;\ x + 2y = 10$

4. $f(x, y) = 3 - x^2 - y^2;\ x + 6y = 37$

Find the minimum value of f subject to the given constraint.

5. $f(x, y) = x^2 + y^2;\ 2x + y = 10$

6. $f(x, y) = x^2 + y^2;\ x + 4y = 17$

7. $f(x, y) = 2y^2 - 6x^2;\ 2x + y = 4$

8. $f(x, y) = 2x^2 + y^2 - xy;\ x + y = 8$

9. $f(x, y, z) = x^2 + y^2 + z^2;\ y + 2x - z = 3$

10. $f(x, y, z) = x^2 + y^2 + z^2;\ x + y + z = 1$

Use the method of LaGrange multipliers to solve these problems.

11. Of all numbers whose sum is 70, find the two that have the maximum product.

12. Of all numbers whose sum is 50, find the two that have the maximum product.

13. Of all numbers whose difference is 6, find the two that have the minimum product.

14. Of all numbers whose difference is 4, find the two that have the minimum product.

APPLICATIONS

15. A standard piece of typing paper has a perimeter of 39 in. Find the dimensions of the paper that will give the most typing area, subject to the perimeter constraint of 39 in. What is its area? Does the standard $8\frac{1}{2}$ in. × 11 in. paper have maximum area?

16. A carpenter is building a rectangular room with a fixed perimeter of 80 ft. What are the dimensions of the largest room that can be built? What is its area?

17. An oil drum of standard size has a volume of 200 gal, or 27 ft³. What dimensions yield the minimum surface area? Find the minimum surface area.

Do these drums appear to be made in such a way as to minimize surface area?

18. A juice can of standard size has a volume of 99 in³. What dimensions yield the minimum surface area? Find the minimum surface area.

19. The total sales S of a one-product firm are given by

$$S(L, M) = ML - L^2,$$

where $M =$ the cost of materials and $L =$ the cost of labor. Find the maximum value of this function subject to the budget constraint

$$M + L = 80.$$

20. The total sales S of a one-product firm are given by

$$S(L, M) = 2ML - L^2,$$

where $M =$ the cost of materials and $L =$ the cost of labor. Find the maximum value of this function subject to the budget constraint

$$M + L = 60.$$

21. A company is planning to construct a warehouse whose cubic footage is to be 252,000 ft³. Construction costs per square foot are estimated to be as follows.

Walls: $3.00

Floor: $4.00

Ceiling: $3.00

a) The total cost of the building is a function $C(x, y, z)$, where x is the length, y is the width, and z is the height. Find a formula for $C(x, y, z)$.

b) What dimensions of the building will minimize the total cost? What is the minimum cost?

22. A container company is going to construct a shipping container of volume 12 ft³ with a square bottom and top. The cost of the top and sides is $2 per square foot and for the bottom is $3 per square foot. What dimensions will minimize the cost of the container?

23. A product can be made entirely on machine A or machine B, or it can be made on both. The nature of the machines makes their cost functions differ:

$$\text{Machine A: } C(x) = 10 + \frac{x^2}{6},$$

$$\text{Machine B: } C(y) = 200 + \frac{y^3}{9}.$$

Total cost is given by $C(x, y) = C(x) + C(y)$. How many units should be made on each machine in order to minimize total costs if $x + y = 10{,}100$ units are required?

EXTENSION EXERCISES

Find the indicated maximum or minimum values of f subject to the given constraint.

24. Minimum: $f(x, y) = xy; x^2 + y^2 = 4$

25. Minimum: $f(x, y) = 2x^2 + y^2 + 2xy + 3x + 2y;$
    $y^2 = x + 1$

26. Maximum: $f(x, y, z) = x + y + z; x^2 + y^2 + z^2 = 1$

27. Maximum: $f(x, y, z) = x^2y^2z^2; x^2 + y^2 + z^2 = 1$

28. Maximum: $f(x, y, z) = x + 2y - 2z; x^2 + y^2 + z^2 = 4$

29. Maximum: $f(x, y, z, t) = x + y + z + t;$
    $x^2 + y^2 + z^2 + t^2 = 1$

30. Minimum: $f(x, y, z) = x^2 + y^2 + z^2; x - 2y + 5z = 1$

31. Suppose $p(x, y)$ represents the production of a two-product firm. We give no formula for p. The company produces x items of the first product at a cost of c_1 each and y items of the second product at a cost of c_2 each. The budget constraint B is given by

$$B = c_1 x + c_2 y.$$

Find the value of λ in the LaGrange multiplier method in terms of p_x, p_y, c_1, and c_2. The resulting equation is called the *Law of Equimarginal Productivity*.

32. A company has the following Cobb–Douglas production function for a certain product:

$$p(x, y) = 800x^{3/4}y^{1/4},$$

where $x =$ labor, measured in dollars, and $y =$ capital, measured in dollars. Suppose a company can make a total investment in labor and capital of $1,000,000. How should it allocate the investment between labor and capital in order to maximize production?

7.8

OBJECTIVE

a) Find the dimensions of a building of fixed floor area, with a square base, that will minimize travel time between the most remote points of the building.

AN APPLICATION: MINIMIZING TRAVEL TIME IN A BUILDING

In multilevel building design, one consideration is travel time between the most remote points in a rectangular building with a square base. Let us suppose each floor of a 12-story building has a square grid of hallways, as shown below.

This is the type of building we are considering.

Suppose you are standing at the most remote point P in the top northeast corner of this building. How long will it take you to reach the southwest corner on the first floor? (You will be going from point P to point Q in the illustration.)

Let us call the time t. We find a formula for t in two steps:

1. You are to go from the twelfth floor to the first floor. This is a move in a *vertical direction*; and

2. You must go across the first floor. This is a move in a *horizontal direction.*

The vertical time is h, the height of the top floor from the ground, divided by a, the speed at which you can travel in a vertical direction (elevator speed). Thus vertical time is h/a.

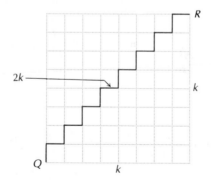

The horizontal time is the time it takes to go across the first level, using the square grid of hallways (from R to Q above). If each floor is a square with sides of length k, then the distance from R to Q is $2k$. If the walking speed is b, then the horizontal time is $2k/b$. Thus the time it will take to go from P to Q is a function of two variables, h and k, given by

$$t(h, k) = \text{Vertical time} + \text{Horizontal time} = \frac{h}{a} + \frac{2k}{b}.$$

Now, what happens if we must choose between two (or more) building plans with the same total floor area, but with different dimensions? Will the travel time be the same? Or will it be different for the two buildings?

First of all, what is the total floor area of a given building? Suppose the building has n floors, each a square of side k. Then the total floor area is given by

$$A = nk^2.$$

Note that the area of the roof is not included.

If h is the height of the top floor from the ground and c is the height of each floor, then $n = h/c$. Thus,

$$A = \frac{h}{c} k^2.$$

By working the exploratory exercises on the following page, let us return to the problem of the two buildings with the same total floor area, but with different dimensions, and see what happens to $t(h, k)$.

Now let us use calculus to solve a building planning problem.

Example 1 *Minimizing travel time.* The objective is to find the dimensions of a rectangular building with a square base that will minimize travel time t between the most remote points in the building. Each floor has a square grid of hallways. The height of the top floor from the ground is h, and the length of a side of each floor is k. The elevator speed is 10 ft/sec, and the average speed of a person walking is 4 ft/sec. The total floor area of the building is 40,000 ft². The height of each floor is 8 ft.

Solution We want to find the values of h and k that will minimize the function t given by

$$t(h, k) = \frac{h}{10} + \frac{2k}{4} = \frac{1}{10} h + \frac{1}{2} k$$

subject to the constraint

$$\frac{h}{8} k^2 = 40,000, \quad \text{or} \quad hk^2 = 320,000.$$

We first form the new function T given by

$$T(h, k, \lambda) = \tfrac{1}{10} h + \tfrac{1}{2} k - \lambda(hk^2 - 320,000).$$

EXPLORATORY EXERCISES

For each case, let the elevator speed $a = 10$ ft/sec, the walking speed $b = 4$ ft/sec, and the height of each floor $c = 15$ ft. Complete the following table.

Case	Building	Number of levels, n	Floor length (in feet), k	Total area (in square feet), A	Height (in feet), $h = cn$	Travel time (in seconds), $t(h, k)$
1	B_1	2	40.0	3200	30	23.00
	B_2	3	32.66	3200	45	20.83
2	B_1	2	60.0			
	B_2	3	48.99			
3	B_1	4	40.0			
	B_2	5	35.777			
4	B_1	5	60.0			
	B_2	10	42.246			
5	B_1	5	150.0			
	B_2	10	106.066			
6	B_1	10	40.0			
	B_2	17	30.679			
7	B_1	10	80.0			
	B_2	17	61.357			
8	B_1	17	40.0			
	B_2	26	32.344			
9	B_1	17	50.0			
	B_2	26	40.43			
10	B_1	26	77.0			
	B_2	50	55.525			

a) What did you notice about $t(h, k)$ when h and k are very nearly the same?
b) What about when h and k were very large?
c) Do you think there are values of h and k for a building with a given floor area that will minimize travel time?

We take the first derivatives and set them equal to 0:

$$T_h = \tfrac{1}{10} - \lambda k^2 = 0, \tag{1}$$

$$T_k = \tfrac{1}{2} - 2\lambda hk = 0, \tag{2}$$

$$T_\lambda = -(hk^2 - 320{,}000) = 0, \quad \text{or} \quad hk^2 - 320{,}000 = 0. \tag{3}$$

To clear of fractions, we multiply the first equation by 10 and the second by 2:

$$1 - 10\lambda k^2 = 0, \tag{4}$$

$$1 - 4\lambda hk = 0, \tag{5}$$

$$hk^2 - 320{,}000 = 0. \tag{6}$$

Solving Eq. (4) for λ, we get

$$\lambda = \frac{1}{10k^2}.$$

Solving Eq. (5) for λ, we get

$$\lambda = \frac{1}{4hk}.$$

Thus

$$\frac{1}{10k^2} = \frac{1}{4hk}. \tag{7}$$

Now k must be nonzero in the original problem. Assuming $k \neq 0$, we multiply Eq. (7) by k to simplify it and get

$$\frac{1}{10k} = \frac{1}{4h}.$$

Solving this equation for k, we get

$$k = \tfrac{2}{5}h. \tag{8}$$

Substituting $\tfrac{2}{5}h$ for k in Eq. (6), we get

$$h(\tfrac{2}{5}h)^2 - 320{,}000 = 0$$

$$\tfrac{4}{25}h^3 - 320{,}000 = 0$$

$$\tfrac{4}{25}h^3 = 320{,}000$$

$$h^3 = 320{,}000 \cdot \tfrac{25}{4} = 2{,}000{,}000$$

$$h = \sqrt[3]{2{,}000{,}000}$$

$$h \approx 126 \text{ ft.} \quad \blacksquare$$

1. Solve the example when the total floor area is 100,000 ft^2 and the height of each floor is 10 ft.

Then from Eq. (8) we find k:

$$k = \tfrac{2}{5} \cdot 126 \approx 50 \text{ ft.}$$

The height of the building is 126 ft plus 8 ft. The height of the top floor is added on. Thus the dimensions of the building are 50 ft by 50 ft by 134 ft.

DO EXERCISE 1.

EXERCISE SET 7.8

Given the conditions in each of Exercises 1 and 2, find the values of h and k that minimize travel time

$$t(h, k) = \frac{h}{a} + \frac{2k}{b}$$

subject to the floor-area constraint A given by

$$A = \frac{h}{c} k^2,$$

where h = the height of the top floor from the ground (ft), k = the length of the side of the base (ft), a = the elevator speed (ft/sec), b = the average speed of humans walking in the building (ft/sec), and c = the height of each floor (ft). Then find the dimensions of the building.

1. $a = 20, b = 5, c = 8, A = 80{,}000 \text{ ft}^2$

2. $a = 20, b = 5, c = 10, A = 60{,}000 \text{ ft}^2$

EXTENSION EXERCISE

3.   Find a general solution in terms of a, b, c, and A.

7.9

OBJECTIVE

a) Evaluate the double integral of $f(x, y)$ over a rectangle R whose sides are parallel to the axes by evaluating two appropriately chosen iterated integrals.

DOUBLE INTEGRATION: RECTANGULAR REGIONS

Consider a rectangle R in the plane whose sides are parallel to the x- and y-axes. Such a rectangle is shown on the following page.

Now, let $z = f(x, y)$ be a continuous function* whose domain contains R For the moment, assume that $f(x, y)$ is a positive function on

_____

* Intuitively speaking, a function of two variables is continuous if the surface formed by its graph has no "breaks." Formally, a function f is continuous at (a, b) if (1) $f(a, b)$ exists; (2) $\lim_{(x,y) \to (a,b)} f(x, y)$ exists; and (3) $\lim_{(x,y) \to (a,b)} f(x, y) = f(a, b)$.

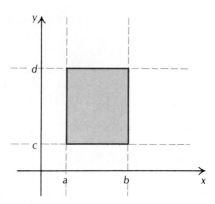

R and consider the problem of calculating the volume of the solid figure (see the figure below) whose base is R and whose "top" is the portion of $z = f(x, y)$ that lies directly above R. Since the top is a curved surface, the volume cannot be computed using the standard formula $V = A \cdot h$, where V is the volume, A is the area of the base, and h is the height.

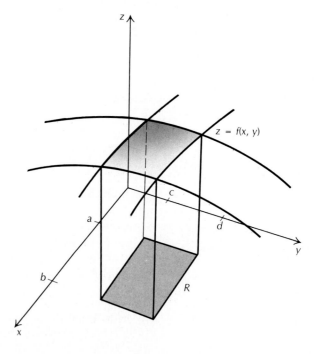

To calculate the volume, we proceed as follows. First, we divide the interval $[a, b]$ along the x-axis into n subintervals, all with the same length Δx. This means that $\Delta x = (b - a)/n$. It also means that the endpoints of the subintervals, $x_1 = a, x_2, x_3, \ldots, x_{n+1} = b$ are given by $x_{k+1} = a + k \cdot \Delta x$ for $k = 0, 1, 2, \ldots, n$.

Next, we divide the interval $[c, d]$ along the y-axis into m subintervals, all having the same length Δy. Then $\Delta y = (d - c)/m$ and $y_{k+1} = c + k \, \Delta y$ for $k = 0, 1, 2, \ldots, m$.

Note in the figure below that if we draw the lines passing through $x_2, x_3, \ldots, x_n$ that are parallel to the y-axis and the lines through $y_2, y_3, \ldots, y_m$ that are parallel to the x-axis, then these lines will divide R into small rectangles. There are exactly $m \cdot n$ of these smaller rectangles, each having area $\Delta x \cdot \Delta y$. To keep track of the rectangles, we label the rectangle that has x_i and y_j for its "smallest" x- and y-endpoints R_{ij}. So, the light-colored rectangle whose x-endpoints are x_1 and x_2 and whose y-endpoints are y_1 and y_2 is called R_{11}, since x_1 and y_1 are the smaller x- and y-endpoints. Similarly, the dark-colored rectangle is R_{22}.

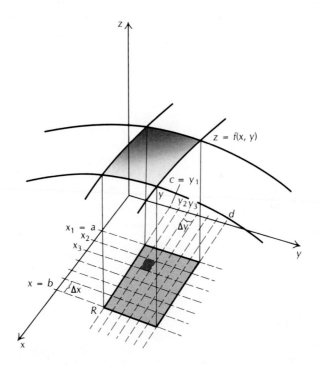

Continuing our calculation, look at the following figure.

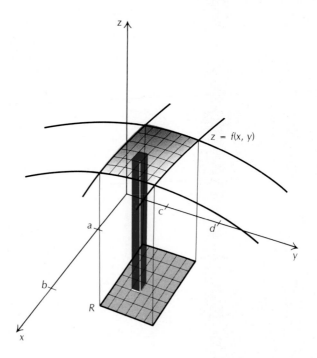

The portion of the volume that is colored in has the rectangle R_{43} for its base. The pictured volume looks like a tall, thin box, but it is not a (rectangular) box, since its top is slightly curved. Nevertheless, there is so little variation in its altitude that we can approximate its volume using the volume of the box whose base is R_{43} and whose altitude or height is $f(x_4, y_3)$, the value of $f(x, y)$ at the corner point of R_{43} nearest the origin in the figure. Thus the volume of the colored portion in the figure is approximately $f(x_4, y_3) \cdot \Delta x \cdot \Delta y$ since the area of the base rectangle B_{43} is $\Delta x \cdot \Delta y$. We can obtain a good approximation to the volume V of the original solid by constructing rectangular boxes for each base rectangle R_{ij} and then adding the volumes of these boxes. The volume of the box whose base is R_{ij} will be approximately equal to $f(x_i, y_j) \cdot \Delta x \cdot \Delta y$ for all $i = 1, \ldots, n$ and all $j = 1, \ldots, m$. Adding these numbers gives

$$V \approx \sum_{i=1}^{n} \sum_{j=1}^{m} f(x_i, y_j) \, \Delta x \, \Delta y.$$

We can increase the accuracy of the approximation by increasing m or n (or both). We expect that if we take the limit of these sums as m and n approach infinity, then we would obtain the exact volume. That is true.

DEFINITION

Let $z = f(x, y)$ be a continuous function on the rectangle R, where the sides of R are parallel to the x- and y-axes. Then the *double integral of $f(x, y)$ over R*, denoted

$$\iint_R f(x, y) \, dx \, dy,$$

is the limit

$$\lim_{m,n \to \infty} \sum_{i=1}^{n} \sum_{j=1}^{m} f(x_i, y_j) \, \Delta x \, \Delta y.$$

If $f(x, y)$ is continuous, we can show that the limit above always exists. If $f(x, y) \geqslant 0$ on R, then we define the volume of the solid whose base is R and whose altitude at each point (x, y) of R is $f(x, y)$ to be the double integral defined above.

Some double integrals can be calculated directly from the definition, but the calculations are tedious. We'd like to have a method of evaluating double integrals that is fairly easy to apply. Recall that although the integral

$$\int_a^b f(x) \, dx$$

can be defined as a limit of sums (see Section 5.7), it is usually evaluated using antiderivatives. There is a corresponding method for handling double integrals. The next few examples lay the groundwork for this method.

Example 1   Evaluate $\int_1^3 xy \, dx$.

Solution   First, we consider the key to the evaluation. The dx in the integral tells us that we are to carry out the integration with respect to x. This means that any other independent variables (like y above) that occur in the integral are to be treated as constants (much as you do when you calculate partial derivatives). Here, if we think of y as being a fixed number, then the antiderivative of xy is $y \cdot (x^2/2)$. Proceeding as usual, we get

$$\int_1^3 xy \, dx = \left[\frac{y \cdot x^2}{2} \right]_1^3 = \frac{y \cdot (3)^2}{2} - \frac{y \cdot (1)^2}{2}$$

$$= \frac{y \cdot 9}{2} - \frac{y \cdot 1}{2} = \frac{8y}{2} = 4y.$$

Note that the answer is a function of y alone.

1. Evaluate $\int_1^3 xy\, dx$ by factoring out the y-term and evaluating the remaining integral in the usual way.

DO EXERCISES 1 AND 2.

Example 2 Evaluate $\int_0^1 (xy + t)\, dy$.

Solution Here, the dy tells us that we are to integrate with respect to y, so that x and t are to be regarded as constants. In other words, we integrate from $y = 0$ to $y = 1$. Just as the antiderivative of $3y + 5$ is $3 \cdot (y^2/2) + 5y$, the antiderivative of $xy + t$ is $x \cdot (y^2/2) + ty$. Using the limits of integration, we get

$$\left[x \cdot \frac{y^2}{2} + ty \right]_0^1 = \left(x \cdot \frac{(1)^2}{2} + t \cdot 1 \right) - \left(x \cdot \frac{0^2}{2} + t \cdot 0 \right)$$

$$= \left(\frac{x}{2} + t \right) - (0 + 0)$$

$$= \frac{x}{2} + t.$$

2. Evaluate $\int_0^1 xy^2\, dx$.

Note again that there is no y in the answer. The expression is a function of x and t only.

DO EXERCISES 3–5.

3. Evaluate $\int_0^1 xy^2\, dy$.

Example 3 Evaluate $\int_0^2 \left\{ \int_1^3 xy\, dx \right\} dy$.

Solution Note that there are two integral signs here. This is our first example of a *repeated*, or *iterated*, *integral*. In order to evaluate it, we work from the inside out. We first attack the integral in brackets,

$$\int_1^3 xy\, dx,$$

4. Evaluate $\int_{-1}^2 (xy + x^2)\, dy$.

which we recognize as the integral from Example 1 above. Using the answer from that exercise, we get

$$\int_0^2 \left\{ \int_1^3 xy\, dx \right\} dy = \int_0^2 4y\, dy.$$

Since the antiderivative of $4y$ (we are integrating with respect to y here) is $4 \cdot (y^2/2) = 2y^2$, we get

5. Evaluate $\int_0^2 (rx^2 + ys)\, ds$.

$$\int_0^2 \left\{ \int_1^3 xy\, dx \right\} dy = \int_0^2 4y\, dy$$

$$= [2y^2]_0^2$$

$$= 2(2)^2 - 2(0)^2$$

$$= 2 \cdot 4 - 2 \cdot 0 = 8.$$

6. Evaluate

$$\int_0^1 \left\{ \int_0^1 xy^2 \, dx \right\} dy.$$

(See Margin Exercise 2.)

7. Evaluate

$$\int_0^5 \left\{ \int_0^1 xy^2 \, dy \right\} dx.$$

8. Evaluate

$$\int_0^1 \left\{ \int_0^1 xy^2 \, dy \right\} dx$$

and compare your answer with that of Margin Exercise 6.

9. Evaluate

$$\int_0^1 \left\{ \int_0^5 xy^2 \, dx \right\} dy$$

and compare your answer with that of Margin Exercise 7.

DO EXERCISES 6 AND 7.

Example 4 Evaluate

$$\int_1^3 \left\{ \int_0^2 xy \, dy \right\} dx.$$

Solution As before, we work from the inside out. So, we look first at

$$\int_0^2 xy \, dy.$$

The *dy* tells us to integrate with respect to *y*. The antiderivative will then be $x \cdot (y^2/2)$. So,

$$\int_0^2 xy \, dy = \left[x \cdot \frac{y^2}{2} \right]_0^2 = x \cdot \frac{(2)^2}{2} - x \cdot \frac{(0)^2}{2} = 2x.$$

Then,

$$\int_1^3 \left\{ \int_0^2 xy \, dy \right\} dx = \int_1^3 2x \, dx$$
$$= [x^2]_1^3$$
$$= (3)^2 - (1)^2$$
$$= 8.$$

That is,

$$\int_1^3 \left\{ \int_0^2 xy \, dy \right\} dx = 8.$$

Note now that the integrals in Examples 3 and 4 are the same except that the order of integration has been reversed.

DO EXERCISES 8 AND 9.

The relationship between the double integral

$$\iint_R f(x, y) \, dx \, dy$$

and the *iterated* (or repeated) *integral*, such as

$$\int_a^b \left\{ \int_c^d f(x, y) \, dy \right\} dx,$$

is given in the following theorem.

THEOREM 3

Let $f(x, y)$ be continuous on the rectangle R (R as in the definition of double integral above). Then

$$\iint_R f(x, y)\, dx\, dy = \int_c^d \left[\int_a^b f(x, y)\, dx \right] dy$$

$$= \int_a^b \left[\int_c^d f(x, y)\, dy \right] dx.$$

That is, the double integral of $f(x, y)$ over R can be evaluated as an *iterated integral* (in two different ways).

Example 5 Evaluate the double integral of $f(x, y) = x^2 y$ over the rectangle shown below.

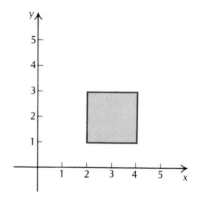

Solution We apply the theorem ($x^2 y$ is certainly continuous) to obtain

$$I = \iint_R x^2 y\, dx\, dy = \int_1^3 \left\{ \int_2^4 x^2 y\, dx \right\} dy.$$

We evaluate the inside integral first, getting

$$I = \int_1^3 \left\{ \left[\frac{x^3}{3} \cdot y \right]_2^4 \right\} dy.$$

As before, we treat y as a constant in finding the antiderivative above, since we are integrating with respect to x. So,

$$I = \int_1^3 \left\{ \frac{4^3}{3} y - \frac{2^3}{3} y \right\} dy = \int_1^3 \left\{ \frac{64}{3} y - \frac{8}{3} y \right\} dy$$

$$= \int_1^3 \frac{56}{3} y\, dy = \frac{56}{3} \int_1^3 y\, dy$$

10. a)   Evaluate $\int_1^3 x^2 y\, dy$.

 b)   Evaluate the double integral I from Example 5 as

$$\int_2^4 \left\{ \int_1^3 x^2 y\, dy \right\} dx.$$

and

$$I = \frac{56}{3}\left[\frac{y^2}{2}\right]_1^3 = \frac{56}{3}\left[\frac{3^2}{2} - \frac{1^2}{2}\right]$$

$$= \frac{56}{3}\left[\frac{9}{2} - \frac{1}{2}\right] = \frac{56}{3}\cdot 4 = \frac{224}{3}.$$

DO EXERCISES 10 AND 11.

Example 6   Evaluate

$$I = \iint_R \frac{x}{y}\, dx\, dy$$

(over the rectangle $R = \{(x, y) \mid 0 \leqslant x \leqslant 1, 1 \leqslant y \leqslant 2\}$ shown below) in *two* ways. This notation means: R is the set of all points (x, y) such that $0 \leqslant x \leqslant 1$ and $1 \leqslant y \leqslant 2$.

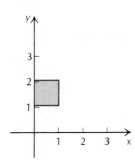

11.   Evaluate $\iint_R 3xy\, dx\, dy$, where R is the rectangular region in Example 5.

Solution   If we integrate first with respect to x and then with respect to y, we get

$$I = \int_1^2 \left\{ \int_0^1 \frac{x}{y}\, dx \right\} dy = \int_1^2 \frac{1}{y}\left\{ \int_0^1 x\, dx \right\} dy$$

$$= \int_1^2 \frac{1}{y}\cdot\left\{ \left[\frac{x^2}{2}\right]_0^1 \right\} dy = \int_1^2 \frac{1}{y}\left\{ \frac{1^2}{2} - \frac{0^2}{2} \right\} dy$$

$$= \int_1^2 \frac{1}{y}\cdot\frac{1}{2}\, dy = \left[\frac{1}{2}\ln|y|\right]_1^2$$

$$= \frac{1}{2}\ln|2| - \frac{1}{2}\ln|1| = \frac{1}{2}\ln 2 - 0 = \frac{1}{2}\ln 2.$$

12. Evaluate $I = \iint_R ye^x \, dx \, dy$ in two ways, where R is the rectangle in Example 6.

If we integrate first with respect to y and then with respect to x, we get

$$I = \int_0^1 \left\{ \int_1^2 \frac{x}{y} \, dy \right\} dx$$

$$= \int_0^1 x \left\{ \int_1^2 \frac{1}{y} \, dy \right\} dx$$

$$= \int_0^1 x \{ [\ln |y|]_1^2 \} \, dx$$

$$= \int_0^1 x \{ \ln 2 \} \, dx$$

$$= \ln 2 \cdot \left[\frac{x^2}{2} \right]_0^1$$

$$= \frac{1}{2} \ln 2.$$

DO EXERCISES 12 AND 13.

13. Evaluate $I = \iint_R xy^3 \, dx \, dy$ in two ways, where R is the rectangle in Example 6.

Example 7   Find the volume of the solid (shown below) whose base is the rectangle $R = \{ (x, y) \,|\, 0 \leqslant x \leqslant 1, \, 0 \leqslant y \leqslant 1 \}$ and whose height at (x, y) is given by $f(x, y) = x + y$.

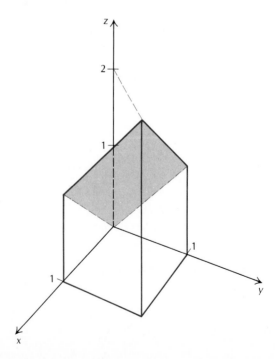

14. Find the volume of the solid whose base is the rectangle R in Example 7 and whose height at (x, y) is $f(x, y) = x^2 + y$.

Solution   According to the definition, the volume V is given by the double integral

$$V = \iint_R (x + y)\, dx\, dy.$$

Writing the double integral as two iterated integrals, we get

$$V = \int_0^1 \left\{ \int_0^1 (x + y)\, dx \right\} dy$$

$$= \int_0^1 \left\{ \left[\frac{x^2}{2} + xy \right]_0^1 \right\} dy$$

$$= \int_0^1 \left\{ \left(\frac{1^2}{2} + 1 \cdot y \right) - \left(\frac{(0)^2}{2} + 0 \cdot y \right) \right\} dy$$

$$= \int_0^1 \left(\frac{1}{2} + y \right) dy$$

$$= \left[\frac{1}{2} y + \frac{y^2}{2} \right]_0^1 = \left(\frac{1}{2} \cdot 1 + \frac{1^2}{2} \right) - \left(\frac{1}{2} \cdot 0 - \frac{(0)^2}{2} \right)$$

$$= \left(\frac{1}{2} + \frac{1}{2} \right) - (0 - 0) = 1.$$

Geometrically, we can see that this is correct by noting that our solid comprises one half of the rectangular box whose base is R and whose altitude is 2. This box has volume $1 \cdot 1 \cdot 2 = 2$. Thus, $1 = \frac{1}{2} \cdot 2$ is correct.

DO EXERCISE 14.

EXERCISE SET 7.9

Evaluate each of the following integrals.

1. $\int_0^1 x^2 y\, dx$

2. $\int_{-1}^1 (x^2 + xy + 1)\, dy$

3. $\int_{-1}^1 (x^2 + xy + 1)\, dx$

4. $\int_0^3 (ax^2 + bx + y)\, dx$

Evaluate each of the following iterated integrals.

5. $\int_0^1 \left\{ \int_0^2 x^2 y^3\, dy \right\} dx$

6. $\int_{-1}^0 \left\{ \int_0^1 (3y^2 - 1 + 2x)\, dx \right\} dy$

7. $\int_0^1 \left\{ \int_0^{\ln 2} 2xe^y\, dy \right\} dx$

8. $\int_0^1 \left\{ \int_0^1 (2axy + x^2)\, dy \right\} dx$

Evaluate the double integral $\iint_R f(x, y)\, dx\, dy$ over the indicated rectangle by reducing to an appropriate iterated integral.

9. $f(x, y) = x + y$, $R = \{(x, y) \,|\, 0 \leq x \leq 2, 1 \leq y \leq 3\}$

10. $f(x, y) = x$, $R = \{(x, y) \,|\, -1 \leq x \leq 1, 0 \leq y \leq 1\}$

11. $f(x, y) = \dfrac{x^2}{y}$, $R = \{(x, y) \,|\, 0 \leq x \leq 1, 2 \leq y \leq 4\}$

12. $f(x, y) = 1$, $R = \{(x, y) \,|\, a \leq x \leq b, c \leq y \leq d\}$

13. Find the volume of the solid whose base is the rectangle
$$R = \{(x, y) \,|\, 0 \leq x \leq 1, 0 \leq y \leq 1\}$$
and whose height at (x, y) is $f(x, y) = 3x^2 + 3y^2$.

14. Find the volume of the solid whose base is the rectangle
$$R = \{(x, y) \,|\, 0 \leq x \leq 1, 0 \leq y \leq 1\}$$
and whose height at (x, y) is $f(x, y) = e^x$.

15. Find the volume of the solid whose base is the rectangle
$$\{(x, y) \,|\, 1 \leq x \leq 2, 0 \leq y \leq 4\}$$
and whose height at (x, y) is $f(x, y) = \dfrac{y}{x}$.

DEFINITION

If $f(x, y)$ is a continuous function on the rectangle R, then f_{av}, *the average value of $f(x, y)$ on R*, is
$$\frac{1}{\text{area } R} \iint_R f(x, y)\, dx\, dy.$$

Calculate f_{av} for each of the following choices of $f(x, y)$ and R.

16. $f(x, y) = x^2 + y^2$, $R = \{(x, y) \,|\, 0 \leq x \leq 1, 0 \leq y \leq 1\}$

17. $f(x, y) = xy$, $R = \{(x, y) \,|\, 0 \leq x \leq 2, 0 \leq y \leq 2\}$

18. $f(x, y) = xy$, $R = \{(x, y) \,|\, -1 \leq x \leq 1, 0 \leq y \leq 1\}$

EXTENSION EXERCISES

19. Let $R = \{(x, y) \,|\, a \leq x \leq b, c \leq y \leq d\}$ and let $g(x)$ be continuous on $[a, b]$ and $h(y)$ be continuous on $[c, d]$. Let $f(x, y) = g(x)h(y)$ on R. Show that
$$f_{av} = g_{av}h_{av},$$
where f_{av} is the average value of $f(x, y)$ on R and g_{av} and h_{av} are the average values of $g(x)$ on $[a, b]$ and $h(y)$ on $[c, d]$, respectively.

7.10

OBJECTIVE

a) Evaluate the double integral $f(x, y)$ over a suitable region G by evaluating two appropriately chosen iterated integrals.

DOUBLE INTEGRATION: MORE GENERAL REGIONS

In Section 7.9, we saw how to define and evaluate the double integral of $f(x, y)$ over R, where R is a rectangle in the plane whose sides are parallel to the x- and y-axes. It is useful, however, to be able to define and evaluate the double integral of $f(x, y)$ over more general regions.

To avoid technical problems, we will restrict ourselves for the time being to integration over regions G that look like those shown below. In these cases, the region G lies between the graphs of $y = g(x)$, $y = h(x)$, $x = a$, and $x = b$, where $g(x)$ and $h(x)$ are smooth functions with $g(x) < h(x)$ for $a < x < b$.

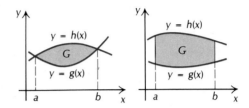

For such a region G, we can cover G with rectangles as shown below, form the sum $\sum_{i=1}^{n} \sum_{j=1}^{m} f(x_i, y_j)\, \Delta x\, \Delta y$ as before, and take the limit as m and n both tend to infinity. If $f(x, y)$ is continuous on G and if both $g(x)$ and $h(x)$ have a continuous derivative on $[a, b]$ and are as in the figure above, then this limit always exists. We denote it by

$$\iint_G f(x, y)\, dx\, dy.$$

We will soon see that we can evaluate the double integral by an appropriate iterated integral.

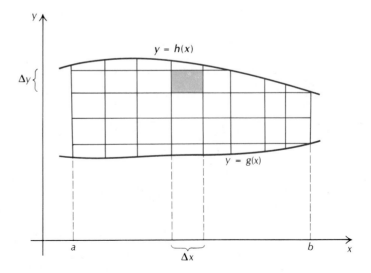

Example 1   Evaluate

$$\int_{x^2}^{x} xy\, dy.$$

1. Evaluate $\int_x^1 3y^2 \, dy$.

Solution   We proceed just as we did in Section 7.9. The dy indicates that we are to integrate with respect to y. The antiderivative of $f(x, y) = xy$ (with x regarded as a constant) is $x \cdot (y^2/2)$. So,

$$\int_{x^2}^x xy \, dy = \left[x \cdot \frac{y^2}{2} \right]_{x^2}^x$$

$$= x \cdot \frac{(x)^2}{2} - \frac{x \cdot (x^2)^2}{2}$$

$$= \frac{x^3}{2} - \frac{x^5}{2}$$

$$= \frac{x^3 - x^5}{2}.$$

2. Evaluate $\int_{-x}^x (2y + 4) \, dy$.

Note that we use the upper and lower limits of integration (to evaluate the integral) just as we would if they were constants.

DO EXERCISES 1 AND 2.

3. Evaluate the iterated integral

$$\int_0^1 \left\{ \int_x^1 3y^2 \, dy \right\} dx.$$

Example 2   Evaluate the iterated integral

$$\int_0^1 \left\{ \int_{x^2}^x xy \, dy \right\} dx.$$

Solution   We proceed as usual, evaluating the inside integral first. Using the result of Example 1, we get

$$\int_0^1 \left\{ \int_{x^2}^x xy \, dy \right\} dx = \int_0^1 \frac{x^3 - x^5}{2} \, dx$$

4. Evaluate the iterated integral

$$\int_0^1 \left\{ \int_{-x}^x (2y + 4) \, dy \right\} dx.$$

$$= \int_0^1 \left(\frac{1}{2} x^3 - \frac{1}{2} x^5 \right) dx$$

$$= \left[\frac{1}{8} x^4 - \frac{1}{12} x^6 \right]_0^1$$

$$= \left[\frac{1}{8} (1)^4 - \frac{1}{12} (1)^6 \right] - \left[\frac{1}{8} (0)^4 - \frac{1}{12} (0)^6 \right]$$

$$= \left(\frac{1}{8} - \frac{1}{12} \right) - (0) = \frac{3}{24} - \frac{2}{24} = \frac{1}{24}.$$

$$= \frac{1}{24}.$$

DO EXERCISES 3 AND 4.

To see why we can evaluate the double integral by iterated integrals, we consider the volume of the solid based on G and capped above by the piece of the surface $z = f(x, y)$ lying over G, where f is a positive (continuous) function of two variables.

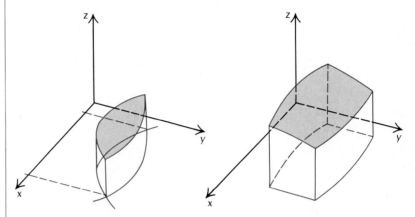

The cross section of this solid in the plane $x = x_0$ is the plane region under the graph of $z = f(x_0, y)$ from $y_1 = g(x_0)$ to $y_2 = h(x_0)$. Its area is

$$A(x_0) = \int_{y_1}^{y_2} f(x_0, y) \, dy = \int_{g(x_0)}^{h(x_0)} f(x_0, y) \, dy.$$

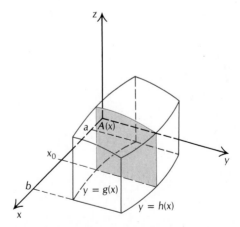

As we vary the slicing plane, the cross section also changes. It follows that the volume in question is given by

$$V = \int_a^b A(x) \, dx$$

$$V = \int_a^b \left\{ \int_{g(x)}^{h(x)} f(x, y) \, dy \right\} dx.$$

We give the relationship between the double integral $\iint_G f(x, y)\, dx\, dy$ and an iterated integral in the following theorem.

THEOREM 4

Let $g(x)$ and $h(x)$ have continuous derivatives and satisfy $g(x) < h(x)$ for $a \leqslant x \leqslant b$. Let G be either of the regions shown at the top of page 489, so that

$$G = \{(x, y)\,|\,a \leqslant x \leqslant b \quad \text{and} \quad g(x) \leqslant y \leqslant h(x)\}.$$

Let $f(x, y)$ be continuous at every point of G. Then the double integral

$$\iint_G f(x, y)\, dx\, dy$$

exists and is equal to the iterated integral

$$\int_a^b \left\{ \int_{g(x)}^{h(x)} f(x, y)\, dy \right\} dx.$$

Note that the iterated integral is taken from the lower boundary of G, $y = g(x)$, to the upper boundary of G, $y = h(x)$. In particular, the first (inside) integration takes place with respect to y. The second (outside) integration is taken from the leftmost value of x in G, $x = a$, to the rightmost value of x in G, $x = b$. The second integration, then, is taken with respect to the variable x.

Example 3   Evaluate $\iint_G f(x, y)\, dx\, dy$, where

$$G = \{(x, y)\,|\,0 \leqslant x \leqslant 1 \text{ and } x^2 \leqslant y \leqslant x\} \quad \text{and} \quad f(x, y) = xy.$$

Solution   We apply Theorem 4 to reduce the double integral to an iterated integral. The region G is shown below. As in the last section, the notation in the statement of the example is read: G is the set of all points (x, y) such that $0 \leqslant x \leqslant 1$ and $x^2 \leqslant y \leqslant x$. The first, or inside, integral will extend from the lower boundary $y = x^2$ to the upper boundary $y = x$. So it will be

$$\int_{x^2}^{x} f(x, y)\, dy = \int_{x^2}^{x} xy\, dy.$$

The second, or outside, integral will be taken from $x = 0$ to $x = 1$. So

$$\iint_G xy\, dx\, dy = \int_0^1 \left\{ \int_{x^2}^{x} xy\, dy \right\} dx.$$

5. **a)**   Sketch the region
$G = \{(x, y) \,|\, 0 \leqslant x \leqslant 1$ and
$x \leqslant y \leqslant 1\}$.

b)   Then evaluate

$$\iint_G 3y^2 \, dx \, dy.$$

6. **a)**   Sketch the region
$G = \{(x, y) \,|\, 0 \leqslant x \leqslant 1$ and
$-x \leqslant y \leqslant x\}$.

b)   Evaluate

$$\iint_G (2y + 4) \, dx \, dy.$$

7. **a)**   Sketch the region
$G = \{(x, y) \,|\, 0 \leqslant x \leqslant 2$ and
$-x \leqslant y \leqslant 1\}$.

b)   Evaluate

$$\iint_G 6x \, dx \, dy.$$

For the record, note that the order of integration is important in this case.

To complete the problem, note that the iterated integral above is the integral from Example 2. Using the answer from that example, we get

$$\iint_G xy \, dy \, dx = \tfrac{1}{24}.$$

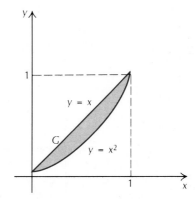

DO EXERCISES 5–7.

Example 4   Find the volume of the solid whose base is the triangular region $G = \{(x, y) \,|\, 0 \leqslant x \leqslant 1, 0 \leqslant y \leqslant x\}$ and whose height at each point of G is $z = f(x, y) = x - y$.

Solution   The solid is shown in the figure on the following page. As in Section 7.9, we define the volume V of the solid to be the double integral of $z = f(x, y) = x - y$ over G. Note that $x - y \geqslant 0$ on G. So

$$V = \iint_G f(x, y) \, dx \, dy.$$

We can evaluate the double integral by reducing it to an iterated integral. The first, or inside, integral will extend from the x-axis ($y = 0$) to the upper boundary $y = x$. The first integral will be

$$\int_0^x (x - y) \, dy.$$

The outer integral is taken from $x = 0$ to $x = 1$. So

$$V = \int_0^1 \left\{ \int_0^x (x - y) \, dy \right\} dx.$$

8. Find the volume of the solid whose base is the triangular region

$$G = \{(x, y) \,|\, 0 \leqslant x \leqslant 1, x \leqslant y \leqslant 1\}$$

and whose height at (x, y) is $z = f(x, y) = 3y^2$.

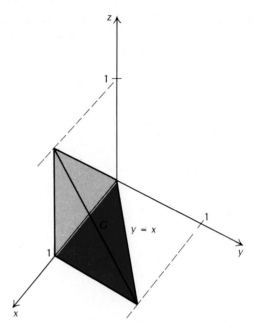

Evaluating, we get

$$V = \int_0^1 \left\{ \left[xy - \frac{y^2}{2} \right]_0^x \right\} dx$$

$$= \int_0^1 \left\{ \left(x \cdot x - \frac{(x)^2}{2} \right) - \left(x \cdot 0 - \frac{(0)^2}{2} \right) \right\} dx$$

$$= \int_0^1 \left\{ \left(x^2 - \frac{x^2}{2} \right) - 0 \right\} dx = \int_0^1 \frac{1}{2} x^2 \, dx$$

$$= \left[\frac{1}{2} \cdot \frac{x^3}{3} \right]_0^1 = \left(\frac{1}{2} \cdot \frac{1^3}{3} \right) - \left(\frac{1}{2} \cdot \frac{0^3}{3} \right)$$

$$= \frac{1}{2} \cdot \frac{1}{3} - \frac{1}{2} \cdot 0 = \frac{1}{6}.$$

DO EXERCISES 8 AND 9.

9. Find the volume of the solid whose base is the region

$$G = \{(x, y) \,|\, 0 \leqslant x \leqslant 1, 0 \leqslant y \leqslant e^x\}$$

and whose height is given by $z = 2y$.

DEFINITION

Let $f(x, y)$ be continuous on G, where G satisfies the condition of the previous theorem. Then *the average value of $f(x, y)$ on G*, denoted f_{av}, is defined by

$$f_{av} = \frac{1}{\text{area } (G)} \iint_G f(x, y) \, dx \, dy.$$

10. Calculate the average value of
    $f(x, y) = 3y^2$ over

    $G = \{(x, y) \mid 0 \leqslant x \leqslant 1, x \leqslant y \leqslant 1\}.$

11. Calculate the average value of
    $f(x, y) = x$ over

    $G = \{(x, y) \mid 0 \leqslant x \leqslant 1, x^2 \leqslant y \leqslant 1\}.$

Example 5   Calculate the average value of $f(x, y) = x - y$ over the region $G = \{(x, y) \mid 0 \leqslant x \leqslant 1, 0 \leqslant y \leqslant x\}$.

Solution   The region G is triangular. The triangle has base and height 1 so that its area is $\frac{1}{2}$. So

$$f_{av} = \frac{1}{\text{area }(G)} \iint_G f(x, y)\, dx\, dy$$

$$= \frac{1}{\frac{1}{2}} \iint_G (x - y)\, dx\, dy.$$

The double integral that appears here was evaluated in Example 4. Using that result, we have

$$f_{av} = 2 \cdot \frac{1}{6}$$

$$= \frac{1}{3}.$$

Note that in most cases, we need to use calculus to compute the area of G.

DO EXERCISES 10 AND 11.

Example 6   Evaluate $\iint_G x\, dx\, dy$, where G is the portion of the plane above $y = x^2$ and below $y = x + 2$.

Solution   This problem differs from previous ones only in that the x-limits of integration are not given. The region G is shown below.

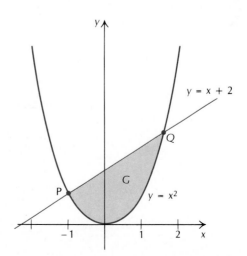

12. Calculate $\iint_G 2y\,dx\,dy$, where G is the portion of the plane above the parabola $y = x^2 + x$ and below the parabola $y = 1 - x^2$.

We must find the points P and Q of intersection. Setting $x^2 = x + 2$ and solving gives

$$x^2 - x - 2 = 0$$

or

$$(x - 2)(x + 1) = 0.$$

We see that P is $(-1, 1)$ and Q is $(2, 4)$. So we obtain

$$\iint_G x\,dx\,dy = \int_{-1}^{2} \left\{ \int_{x^2}^{x+2} x\,dy \right\} dx$$

$$= \int_{-1}^{2} \left\{ [xy]_{x^2}^{x+2} \right\} dx$$

$$= \int_{-1}^{2} (x^2 + 2x - x^3)\,dx$$

$$= \left[\frac{x^3}{3} + x^2 - \frac{x^4}{4} \right]_{-1}^{2}$$

$$= \left(\frac{8}{3} + 4 - \frac{16}{4} \right) - \left(\frac{-1}{3} + 1 - \frac{1}{4} \right)$$

$$= \frac{9}{3} + 3 - \frac{15}{4}$$

$$= \frac{9}{4}.$$

DO EXERCISE 12.

In the iterated integrals that have appeared so far in this section, the first, or inside, integral has always been taken with respect to y. If the region G looks like those below, however, it is more convenient to integrate first with respect to x.

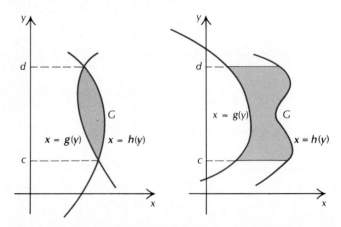

In the case shown, we integrate from the left-hand boundary $x = g(y)$ to the right-hand boundary $x = h(y)$. So

$$\int_{g(y)}^{h(y)} f(x, y)\, dx$$

is the inside, or first, integral. Then we integrate with respect to y, integrating from the smallest y-value in G, $y = c$, to the largest, $y = d$. The result is the equality

$$\iint_G f(x, y)\, dx\, dy = \int_c^d \left\{ \int_{g(y)}^{h(y)} f(x, y)\, dx \right\} dy.$$

Example 7   Evaluate $\iint_G 6xy\, dx\, dy$, where G is the region shown between $x = y^2 - 2$ and $y = x$.

Solution   Solving as in Example 6, we find that the region G looks like the following figure.

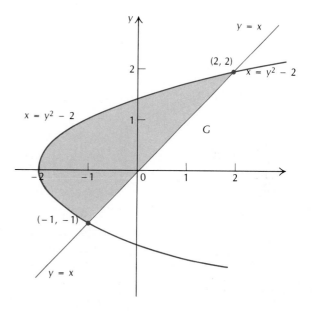

In this case, we integrate from the left-hand boundary, $x = y^2 - 2$, to the right-hand boundary, $x = y$, getting

$$\int_{y^2-2}^{y} 6xy\, dx.$$

Then we integrate from the smallest y-value in G, -1, to the largest,

13. Evaluate $\iint_G 2y\, dx\, dy$, where G is the region in the first quadrant to the left of $x = 2y - y^2$.

2, obtaining

$$\iint_G 6xy\, dx\, dy = \int_{-1}^{2} \left\{ \int_{y^2-2}^{y} 6xy\, dx \right\} dy.$$

Evaluating, we get

$$\int_{-1}^{2} \left\{ [3x^2y]_{y^2-2}^{y} \right\} dy = \int_{-1}^{2} \left\{ 3(y)^2 y - 3(y^2-2)^2 y \right\} dy$$

$$= \int_{-1}^{2} \left\{ 3y^3 - 3(y^4 - 4y^2 + 4)y \right\} dy$$

$$= \int_{-1}^{2} \left\{ 3y^3 - 3y^5 + 12y^3 - 12y \right\} dy$$

$$= \int_{-1}^{2} \left\{ -3y^5 + 15y^3 - 12y \right\} dy$$

$$= \left[-\frac{1}{2} y^6 + \frac{15}{4} y^4 - 6y^2 \right]_{-1}^{2}$$

$$= \left\{ -\frac{1}{2}(2)^6 + \frac{15}{4}(2)^4 - 6(2)^2 \right\}$$

$$\quad - \left\{ -\frac{1}{2}(-1)^6 + \frac{15}{4}(-1)^4 - 6(-1)^2 \right\}$$

$$= \{-32 + 60 - 24\} - \left\{ -\frac{1}{2} + \frac{15}{4} - 6 \right\}$$

$$= 4 + \frac{1}{2} - \frac{15}{4} + 6 = 10 + \frac{2}{4} - \frac{15}{4} = \frac{27}{4}.$$

DO EXERCISE 13.

In Example 7 and Margin Exercise 13, it is much more convenient to integrate first with respect to x. In some cases, we can apply both methods. In these cases, we obtain the same answer, as the following example shows.

Example 8   Rework Example 3, taking the inside integral with respect to x.

Solution   We integrate from the left-hand boundary, $x = y$, to the right-hand boundary, $x = \sqrt{y}$, getting

$$\int_{y}^{\sqrt{y}} xy\, dx.$$

14. Rework Example 4, integrating first with respect to x.

Integrating next from $y = 0$ to $y = 1$, we find that

$$\iint_G xy \, dx \, dy = \int_0^1 \left\{ \int_y^{\sqrt{y}} xy \, dx \right\} dy = \int_0^1 \left\{ \left[\frac{1}{2} x^2 y \right]_y^{\sqrt{y}} \right\} dy$$

$$= \int_0^1 \left\{ \frac{1}{2} (\sqrt{y})^2 y - \frac{1}{2} (y)^2 y \right\} dy = \int_0^1 \left\{ \frac{1}{2} y^2 - \frac{1}{2} y^3 \right\} dy$$

$$= \left[\frac{1}{6} y^3 - \frac{1}{8} y^4 \right]_0^1 = \frac{1}{6} - \frac{1}{8} = \frac{1}{24}.$$

DO EXERCISES 14 AND 15.

15. Rework Example 6 by first breaking up the given region into two smaller regions, integrating first with respect to x in both cases.

Application to Probability

Suppose we throw a dart at a region G in a plane. It lands on a point (x, y). We can think of (x, y) as a continuous random variable that assumes all values in some region G. A function f is said to be a *joint probability density* if

$$f(x, y) \geq 0 \quad \text{for all } (x, y) \text{ in } G$$

and

$$\iint_G f(x, y) \, dx \, dy = 1.$$

Now if G_0 is a subregion of G and is the sort of region we discussed earlier, then the probability that the dart "lands in G_0" is given by

$$\iint_{G_0} f(x, y) \, dx \, dy$$

For appropriate regions, we can evaluate this double integral as an iterated integral.

EXERCISE SET 7.10

Evaluate each of the following.

1. $\int_0^1 \int_0^1 2y \, dx \, dy$

2. $\int_0^1 \int_0^1 2x \, dx \, dy$

3. $\int_{-1}^1 \int_x^1 xy \, dy \, dx$

4. $\int_{-1}^1 \int_x^2 (x + y) \, dy \, dx$

5. $\int_0^1 \int_{-1}^3 (x + y) \, dy \, dx$

6. $\int_0^1 \int_{-1}^1 (x + y) \, dy \, dx$

7. $\int_0^1 \int_{x^2}^x (x + y) \, dy \, dx$

8. $\int_0^1 \int_{-1}^x (x^2 + y^2) \, dy \, dx$

9. $\int_0^2 \int_0^x (x + y^2) \, dy \, dx$

10. $\int_1^3 \int_0^x 2e^{x^2} \, dy \, dx$

Evaluate $\iint_G f(x, y)\, dx\, dy$ for the given function $f(x, y)$ and region G. Sketch G.

11. $f(x, y) = 1$, $G = \{(x, y) \mid 0 \leqslant x \leqslant 2, 0 \leqslant y \leqslant x\}$

12. $f(x, y) = x$, $G = \{(x, y) \mid 0 \leqslant y \leqslant 1 - x^2\}$

13. $f(x, y) = 1 + 2x$, $G = \{(x, y) \mid \frac{1}{2}y \leqslant x \leqslant y, y \leqslant 1\}$

14. $f(x, y) = 4x$, $G = \{(x, y) \mid y \geqslant 0, x^3 \leqslant y \leqslant x\}$

15. Find the volume of the solid capped by the surface $z = 1 - y - x^2$ over the region bounded above and below by $y = 0$ and $y = 1 - x^2$, and left and right by $x = 0$ and $x = 1$, by evaluating the integral

$$\int_0^1 \int_0^{1-x^2} (1 - y - x^2)\, dy\, dx.$$

16. Find the volume of the solid capped by the surface $z = x + y$ over the region bounded above and below by $y = 0$ and $y = 1 - x$, and left and right by $x = 0$ and $x = 1$, by evaluating the integral

$$\int_0^1 \int_0^{1-x} (x + y)\, dy\, dx.$$

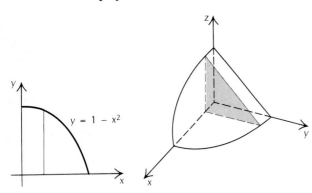

17. Find the volume of the solid whose base is the region in the xy-plane that is bounded by the parabola $y = 4 - x^2$ and the line $y = 3x$, while the top of the solid is capped by $z = x + 6$.

18. Find the volume of the solid whose base is the region in the xy-plane above the x-axis and below $y = 2 - x^2$ and which is capped by $z = 3y^2$.

Find the average value f_{av} of the given function over the region R.

19. $f(x, y) = x^2 + y^2$, $G = \{(x, y) \mid 0 \leqslant x \leqslant 2, 0 \leqslant y \leqslant 2x\}$

20. $f(x, y) = e^{x+y}$, $G = \{(x, y) \mid 0 \leqslant y \leqslant x, 0 \leqslant x \leqslant 1\}$

In each of the following, write an equivalent iterated integral with the order of integration reversed. As a check, evaluate *both* iterated integrals.

21. $\int_0^2 \int_1^{e^x} dy\, dx$

22. $\int_0^2 \int_0^y dx\, dy$

In Exercises 23–26, suppose that a continuous random variable has a joint probability function given by

$$f(x, y) = x^2 + \tfrac{1}{3}xy, \quad 0 \leqslant x \leqslant 1, \quad 0 \leqslant y \leqslant 2.$$

23. Find $\int_0^2 \int_0^1 f(x, y)\, dx\, dy$.

24. Find the probability that a point (x, y) is in the region bounded by $0 \leqslant x \leqslant \frac{1}{2}$, $1 \leqslant y \leqslant 2$, by evaluating the integral

$$\int_1^2 \int_0^{1/2} f(x, y)\, dx\, dy.$$

25. Find the probability that a point (x, y) is in the region $\{(x, y) \mid 0 \leqslant x \leqslant 1, 0 \leqslant y \leqslant 1\}$, that is, the lower half of the rectangle.

26. Find the probability that a point (x, y) is in the part of the rectangle above the line $y = 2 - 2x$.

EXTENSION EXERCISES

A *triple integral* of a continuous function $f(x, y, z)$ over an appropriate region G in three dimensions, say

$$\iiint_G f(x, y, z) \, dx \, dy \, dz,$$

can be defined in a manner similar to that in which the double integral was defined. Triple integrals can frequently be evaluated by the iterated integral

$$\int_a^b \left\{ \int_{G(x)}^{H(x)} \left\{ \int_{g(x,y)}^{h(x,y)} f(x, y, z) \, dz \right\} dy \right\} dz,$$

where the region G is bounded above and below by the surfaces $z = g(x, y)$ and $z = h(x, y)$ and where the y- and x-limits of integration are obtained from the boundaries of the region bounded by $g(x, y) = h(x, y)$ in the xy-plane. Evaluate each of the following triple integrals.

27. $\int_0^1 \int_1^3 \int_{-1}^2 (2x + 3y - z) \, dx \, dy \, dz$

28. $\int_0^2 \int_1^4 \int_{-1}^6 (8x - 2y + z) \, dx \, dy \, dz$

29. $\int_0^1 \int_0^{1-x} \int_0^{2-x} xyz \, dz \, dy \, dx$

30. $\int_0^2 \int_{2-y}^{6-2y} \int_0^{\sqrt{4-y^2}} z \, dz \, dx \, dy$

If $f(x, y, z)$ is continuous on a region G in three dimensions, we define *the average value* f_{av} of $f(x, y, z)$ *on* G by

$$f_{av} = \frac{1}{\text{vol } (G)} \iiint_G f(x, y, z) \, dx \, dy \, dz,$$

where vol (G) is the volume of G.

31. Find the average value of $f(x, y, z) = xyz$ on the unit cube

$$\{(x, y, z) \,|\, 0 \leqslant x \leqslant 1, 0 \leqslant y \leqslant 1 \text{ and } 0 \leqslant z \leqslant 1\}.$$

SUMMARY AND REVIEW: CHAPTER 7

The following contains a summary of what you should be able to do after completing this chapter. The review exercises are for practice. Answers are at the back of the book. If you miss an exercise, restudy the section indicated alongside the answer.

You should be able to:

Find the partial derivatives of a given function and the four second partial derivatives of a function.

Given $f(x, y) = e^y + 3xy^3 + 2y$, find each of the following.

1. f_x　　　　2. f_y　　　　3. f_{xy}　　　　4. f_{yx}　　　　5. f_{xx}　　　　6. f_{yy}

Given $z = x^2 \ln y + y^4$, find each of the following.

7. $\dfrac{\partial z}{\partial x}$　　8. $\dfrac{\partial z}{\partial y}$　　9. $\dfrac{\partial^2 z}{\partial y \, \partial x}$　　10. $\dfrac{\partial^2 z}{\partial x \, \partial y}$　　11. $\dfrac{\partial^2 z}{\partial x^2}$　　12. $\dfrac{\partial^2 z}{\partial y^2}$

Find the relative maximum and minimum values of a function of two variables.

Find the relative maximum and minimum values.

13. $f(x, y) = x^3 - 6xy + y^2 + 6x + 3y - \frac{1}{5}$

14. $f(x, y) = x^2 - xy + y^2 - 2x + 4y$

15. $f(x, y) = 3x - 6y - x^2 - y^2$

16. $f(x, y) = x^4 + y^4 + 4x - 32y + 80$

Find the regression line for a given set of data points and use the regression line to make predictions regarding further data.

17. Consider these data regarding the average cost of a ticket to a Broadway show over a recent five-year period.

Year, x	Average cost of a ticket to a Broadway show, y
1. 1980	$22.50
2. 1981	$27.50
3. 1982	$30.00
4. 1983	$32.50
5. 1984	$35.00
6. 1985	$35.00

a) Find the regression line $y = mx + b$.

b) Use the regression line to predict the average cost of a ticket in 1989; in 2000.

18. Consider these data regarding enrollment in colleges and universities during a recent three-year period.

Year, x	Enrollment (in millions), y
1	7.2
2	8.0
3	8.4

a) Find the regression line $y = mx + b$.

b) Use the regression line to predict enrollment in the fourth year.

Find a maximum or minimum value of a given function subject to a given constraint, using the method of Lagrange multipliers.

19. Find the minimum value of

$$f(x, y) = x^2 - 2xy + 2y^2 + 20$$

subject to the constraint $2x - 6y = 15$.

20. Find the maximum value of

$$f(x, y) = 6xy$$

subject to the constraint $2x + y = 20$.

Evaluate multiple integrals.

Evaluate.

21. $\int_0^1 \int_{x^2}^{3x} (x^3 + 2y)\, dy\, dx$

22. $\int_0^1 \int_{x^2}^{x} (x - y)\, dy\, dx$

Calculate the total differential dz, given z as a function of x and y.

23. Calculate the total differential dz given that

$$z = 5x^2 - e^{2y}.$$

24. Calculate the total differential dz given that

$$z = 2x^2y.$$

Use the total differential to approximate the actual change Δz in $z = f(x, y)$ resulting from changes Δx and Δy in x and y, respectively.

25. The legs of a right triangle are measured to be 5 ft and 12 ft. Find the maximum possible error in measuring the area of the triangle if errors of measurement of 0.1 ft and 0.4 ft, respectively, are possible in the measurement of the legs.

Use the total differential to approximate the actual change Δz in $z = f(x, y)$ resulting from changes Δx and Δy in x and y, respectively.

26. A rectangular box is found to measure 2 ft by 4 ft by 5 ft. If errors of 0.01 ft, 0.02 ft, and 0.04 ft are possible in the three measurements, respectively, use differentials to calculate the maximum possible error in measuring the volume of the box.

Evaluate an iterated integral by two integrations.

Evaluate.

27. $\int_0^1 \int_{x^2}^x (x - y)\, dy\, dx$

Evaluate the double integral of $f(x, y)$ over a rectangular region whose sides are parallel to the coordinate axes by evaluating two appropriately chosen iterated integrals.

28. Find the average value f_{av} of $f(x, y) = 4xy$ on the rectangle in the plane whose vertices are the four points $(0, 0)$, $(1, 0)$, $(0, 2)$, and $(1, 2)$.

Evaluate the double integral of $f(x, y)$ over a general region in the plane by evaluating two appropriately chosen iterated integrals.

29. Evaluate $\iint_G 12x^2\, dx\, dy$, where G is the region in the plane bounded by the positive x-axis, the positive y-axis, and the line $x + y = 2$.

30. Find the volume of the solid whose base is the semi-circular region $\{(x, y)\,|\,y \geq 0,\ x^2 + y^2 \leq 1\}$ and which is capped by the plane $z = 4y$.

EXTENSION EXERCISES

31. Evaluate $\int_0^2 \int_{1-2x}^{1-x} \int_0^{\sqrt{2-x^2}} z\, dz\, dy\, dx$.

32. Suppose beverages could be packaged in either a cylindrical container or a rectangular container with a square top and bottom. If we assume a volume of 26 in³, which container would have the minimum surface area?

TEST: CHAPTER 7

Given $f(x, y) = e^x + 2x^3y + y$, find each of the following.

1. $\dfrac{\partial f}{\partial x}$
2. $\dfrac{\partial f}{\partial y}$
3. $\dfrac{\partial^2 f}{\partial x^2}$
4. $\dfrac{\partial^2 f}{\partial x\, \partial y}$
5. $\dfrac{\partial^2 f}{\partial y\, \partial x}$
6. $\dfrac{\partial^2 f}{\partial y^2}$

7. Find the relative maximum and minimum values of
$$f(x, y) = x^2 - xy + y^3 - x.$$

8. Find the relative maximum and minimum values of
$$f(x, y) = y^2 - x^2.$$

9. Consider these data regarding total sales of a company during its first three years of operation.

Year, x	1	2	3
Sales (in millions), y	$10	$15	$19

a) Find the regression line $y = mx + b$.

b) Use the regression line to predict sales in the fourth year.

10. Find the maximum value of
$$f(x, y) = 6xy - 4x^2 - 3y^2$$
subject to the constraint $x + 3y = 19$.

11. Evaluate
$$\int_0^2 \int_1^x (x^2 - y) \, dy \, dx.$$

12. Calculate the total differential dz given that $z = x^3 + e^y$.

13. Calculate the total differential dz given that $z = 3xy^2$.

14. A rectangular box is found to measure 2 ft by 3 ft by 5 ft. If errors of 0.01 ft, 0.02 ft, and 0.04 ft are possible in the three measurements, respectively, use differentials to calculate the maximum possible error in measuring the volume of the box.

15. Find the volume of the solid whose base is the semicircular region
$$\{(x, y) \,|\, 0 \leqslant y, x^2 + y^2 \leqslant 1\}$$
and which is capped by the plane $z = 2y$.

16. Find the average value f_{av} of $f(x, y) = xy$ on the region of the plane bounded by the positive x- and y-axes and the curve $y = 1 - x^3$.

EXTENSION EXERCISES

17. A company has the following Cobb–Douglas production function for a certain product,
$$p(x, y) = 50x^{2/3}y^{1/3},$$
where $x =$ labor, measured in dollars, and $y =$ capital, measured in dollars. Suppose a company can make a total investment in labor and capital of $600,000. How should it allocate the investment between labor and capital in order to maximize production?

18. Find f_x and f_t:
$$f(x, t) = \frac{x^2 - 2t}{x^3 + 2t}.$$

8

TRIGONOMETRIC FUNCTIONS

Functions for which the graphs repeat themselves are called *periodic*. Of special importance in mathematics are the periodic *trigonometric functions*. These functions originated as a means of indirect measurement. For example, a surveyor might use trigonometry to measure the distance across a river without actually crossing the river. We will learn to differentiate and integrate these functions and to solve related problems.

AN APPLICATION

A company in a northern climate has sales of skis as given by

$$S(t) = 7\left(1 - \cos\frac{\pi}{6}t\right),$$

where S is sales in thousands of dollars during the tth month. Find the total sales for the first year.

THE MATHEMATICS

Total sales for the first year are given by

$$\int_0^{12} S(t)\,dt = \int_0^{12} 7\left(1 - \cos\frac{\pi}{6}t\right)dt.$$

This is a *trigonometric function*.

8.1

INTRODUCTION TO TRIGONOMETRY

Angles and Rotations

We now introduce the trigonometric functions together with their derivatives and integrals.

We first consider a rotating ray, with its endpoint at the origin of an xy-plane. The ray starts in position along the positive half of the x-axis. A counterclockwise rotation is called *positive,* and a clockwise rotation is called *negative.*

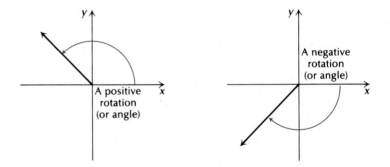

Note that the rotating ray and the positive half of the x-axis form an angle. Thus we often speak of "rotations" and "angles" interchangeably. The rotating ray is often called the *terminal side* of the angle, and the positive half of the x-axis is called the *initial side.*

Measures of Rotations or Angles

The size, or *measure,* of an angle, or rotation, may be given in degrees. Thus a complete revolution has a measure of 360°, half a revolution has a measure of 180°, and so on. We can also speak of an *angle* of 90° or 720° or −240°.

An angle between 0° and 90° has its terminal side in the first quadrant. An angle between 90° and 180° has its terminal side in the second quadrant. An angle between 180° and 270° has its terminal side in the third quadrant. An angle between 0° and −90° has its terminal side in the fourth quadrant.

Note that angles with measure 0°, 360°, and 720° have the same terminal side, as do angles with measure 270° and −90°.

1. In which quadrant does the terminal side of each angle lie?

   a) 47°

   b) 212°

   c) −43°

   d) −135°

   e) 365°

   f) −365°

   g) 740°

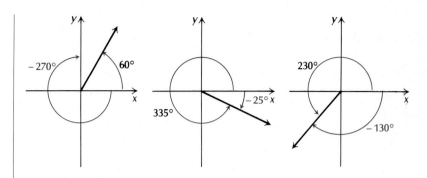

DO EXERCISE 1.

Measurement in Radians

A unit of angle or rotation measure other than the degree is very useful for many purposes. This unit is called the *radian*. Consider a circle with radius of length 1, centered at the origin. The measure of an angle, or rotation, in *radians* is the distance around this circle from the initial side of the angle to the terminal side.

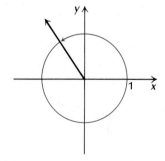

Since the circumference of the circle is $2\pi \cdot 1$, or 2π, a complete revolution (360°) has a measure of 2π radians. Half of this (180°) is π radians, and a fourth of this (90°) is $\pi/2$ radians. In general, we can convert from one measure to the other using this proportion.

THEOREM 1

$$\frac{\text{Radian measure}}{\pi} = \frac{\text{Degree measure}}{180}$$

Example 1   Convert 270° to radians.

2. Convert to radian measure. Leave answers in terms of π.

a) $135°$

b) $315°$

c) $-90°$

d) $720°$

e) $-225°$

f) $-315°$

g) $405°$

h) $480°$

Solution

$$\frac{\text{Radian measure}}{\pi} = \frac{270}{180}$$

$$\text{Radian measure} = \frac{270}{180} \cdot \pi, \quad \text{or} \quad \frac{3}{2}\pi$$

When no unit is specified for an angle measure, it is understood to be given in radians.

Example 2   Convert $\pi/4$ radians to degrees.

Solution

$$\frac{\pi/4}{\pi} = \frac{\text{Degree measure}}{180}$$

$$\text{Degree measure} = 180 \cdot \frac{\pi/4}{\pi}, \quad \text{or} \quad 45°$$

DO EXERCISES 2 AND 3.

Trigonometric Functions

The concept of rotation or angle is important to functions called *trigonometric,* or *circular,* functions.

Consider an angle t, measured in radians, shown here on a circle with radius 1.

3. Convert to degree measure.

a) $\dfrac{\pi}{3}$

b) $\dfrac{3}{4}\pi$

c) $\dfrac{5}{2}\pi$

d) 10π

e) $-\dfrac{7\pi}{6}$

f) 300π

g) -270π

h) $\dfrac{25\pi}{4}$

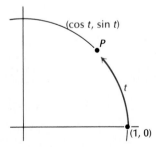

The terminal side of the angle intersects this circle at point P. The distance *around the circle* from $(1, 0)$ to P is t. We define $\cos t$ (cosine t) and $\sin t$ (sine t) as the first and second coordinates of P, respectively:

$$\cos t = \text{the first coordinate of } P,$$
$$\sin t = \text{the second coordinate of } P.$$

Certain values of these functions are easy to determine. When $t = \pi$, for example, the terminal side of the angle is on the horizontal axis and P is one unit to the left; hence the first coordinate is -1 and the second coordinate is 0. Thus,

$$\cos \pi = -1 \quad \text{and} \quad \sin \pi = 0.$$

Similarly, when $t = \pi/2$, the point P is one unit up on the vertical axis; hence the first coordinate is 0 and the second coordinate is 1. Thus,

$$\cos \frac{\pi}{2} = 0 \quad \text{and} \quad \sin \frac{\pi}{2} = 1.$$

When $t = 0$, the terminal side is on the horizontal axis and P is one unit to the right; hence the first coordinate is 1 and the second coordinate is 0. Thus,

$$\cos 0 = 1 \quad \text{and} \quad \sin 0 = 0.$$

We can also define the trigonometric functions in terms of right triangles. Suppose t is an angle of a right triangle, measured in degrees. Then

$$\sin t = \frac{\text{length of side opposite } t}{\text{length of hypotenuse}}$$

and

$$\cos t = \frac{\text{length of side adjacent to } t}{\text{length of hypotenuse}}.$$

Using properties of right triangles, we can develop these other values of the sine and cosine functions.

For $t = \pi/4 = 45°$,

$$\cos \frac{\pi}{4} = \cos 45° = \frac{\sqrt{2}}{2} \approx 0.707$$

and

$$\sin \frac{\pi}{4} = \sin 45° = \frac{\sqrt{2}}{2} \approx 0.707.$$

For $t = \pi/3 = 60°$,

$$\cos \frac{\pi}{3} = \cos 60° = \frac{1}{2} = 0.5$$

and

$$\sin \frac{\pi}{3} = \sin 60° = \frac{\sqrt{3}}{2} \approx 0.866.$$

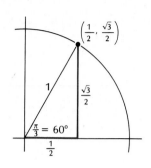

For $t = \pi/6 = 30°$,

$$\cos \frac{\pi}{6} = \cos 30° = \frac{\sqrt{3}}{2} \approx 0.866$$

and

$$\sin \frac{\pi}{6} = \sin 30° = \frac{1}{2} = 0.5.$$

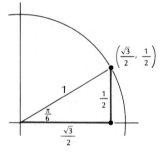

The following table summarizes these important values of the sine and cosine functions. It should be memorized.

t	t	$\sin t$	$\cos t$
0°	0	0	1
30°	$\dfrac{\pi}{6}$	$\dfrac{1}{2}$	$\dfrac{\sqrt{3}}{2}$
45°	$\dfrac{\pi}{4}$	$\dfrac{\sqrt{2}}{2}$	$\dfrac{\sqrt{2}}{2}$
60°	$\dfrac{\pi}{3}$	$\dfrac{\sqrt{3}}{2}$	$\dfrac{1}{2}$

Other function values follow from certain symmetries on the unit circle. Some are shown below.

4. Using this figure, find each of the following.

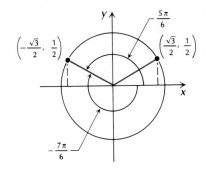

a) $\cos\left(\dfrac{5\pi}{6}\right)$

b) $\sin\left(\dfrac{5\pi}{6}\right)$

c) $\cos\left(-\dfrac{7\pi}{6}\right)$

d) $\sin\left(-\dfrac{7\pi}{6}\right)$

(a)

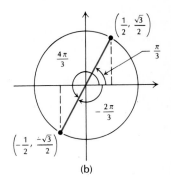

(b)

From (a) we get

$$\cos(-45°) = \cos\left(-\frac{\pi}{4}\right) = \frac{\sqrt{2}}{2} \approx 0.707,$$

$$\sin(-45°) = \sin\left(-\frac{\pi}{4}\right) = -\frac{\sqrt{2}}{2} \approx -0.707,$$

and

$$\cos 315° = \cos\left(\frac{7\pi}{4}\right) = \frac{\sqrt{2}}{2} \approx 0.707,$$

$$\sin 315° = \sin\left(\frac{7\pi}{4}\right) = -\frac{\sqrt{2}}{2} \approx -0.707.$$

From (b) we get

$$\cos(-120°) = \cos\left(-\frac{2\pi}{3}\right) = -\frac{1}{2} = -0.5,$$

$$\sin(-120°) = \sin\left(-\frac{2\pi}{3}\right) = -\frac{\sqrt{3}}{2} \approx -0.866,$$

and

$$\cos 240° = \cos\left(\frac{4\pi}{3}\right) = -\frac{1}{2} = -0.5,$$

$$\sin 240° = \sin\left(\frac{4\pi}{3}\right) = -\frac{\sqrt{3}}{2} \approx -0.866.$$

Approximations of the trigonometric values can be found on your calculator. Be sure to check whether it is using degrees or radians. Usually you can adapt to consider either.

DO EXERCISE 4.

Graphs of cos t and sin t CSS

Note that cos t and sin t are functions of t defined for all real numbers t. For very large $|t|$, we may "wrap around" the circle several times before coming to the terminal point P. Nevertheless, P still has one first coordinate, cos t, and one second coordinate, sin t. For example,

$$\cos 3\pi = \cos \pi = -1 \quad \text{and} \quad \sin 3\pi = \sin \pi = 0.$$

Also,

$$\cos\left(\frac{15\pi}{4}\right) = \cos\left(\frac{7\pi}{4}\right) = \frac{\sqrt{2}}{2} \quad \text{and} \quad \sin\left(\frac{15\pi}{4}\right) = \sin\left(\frac{7\pi}{4}\right) = -\frac{\sqrt{2}}{2}$$

Plotting points previously obtained, we graph the cosine and sine functions as follows.

The cosine function

The sine function

At the origin, t is, of course, 0. Thus the point P is on the horizontal axis, so cos 0 = 1 and sin 0 = 0. Moving to the right on the graphs corresponds to rotating the terminal side of the angle counterclockwise. Moving to the left corresponds to rotating the terminal side clockwise. Note in particular that sin $t = 0$ has solutions $t = 0, \pm\pi, \pm 2\pi, \ldots$, and cos $t = 0$ has solutions $t = \pm(\pi/2), \pm(3\pi/2), \pm(5\pi/2), \ldots$. Note that each curve repeats itself, as the terminal side makes successive revolutions. From 0 to 2π is one complete revolution, or *cycle*. The cycle repeats itself from there on. We say that the *period* of each function is 2π. Algebraically, this means that for any t,

$$\cos (t + 2\pi) = \cos t \quad \text{and} \quad \sin (t + 2\pi) = \sin t.$$ CSS

DEFINITION

A function f is *periodic* if there exists a positive number p such that

$$f(x + p) = f(x).$$

This means that adding p to an input does not change the output. The smallest such number p is called the *period*.

A printout from an electrocardiogram may form a periodic function.

Other Trigonometric Functions

The functions $\sin x$ and $\cos x$ are the basic trigonometric functions, but there are four others—the tangent, cotangent, secant, and cosecant functions—defined as follows.   `CSS`

DEFINITION

$$\tan x = \frac{\sin x}{\cos x}, \qquad x$$

$$\cot x = \frac{\cos x}{\sin x} = \frac{1}{\tan x},$$

$$\sec x = \frac{1}{\cos x},$$

$$\csc x = \frac{1}{\sin x},$$

provided the denominators are not equal to 0.

Let us find some values of the tangent function.

Example 3   Find $\tan \pi/6$ and $\tan \pi/2$.

5. Find each of the following.

   a) $\tan \dfrac{\pi}{4}$

   b) $\tan \dfrac{\pi}{3}$

   c) $\tan 0$

   d) $\sec \dfrac{\pi}{4}$

   e) $\sec \dfrac{\pi}{3}$

   f) $\sec 0$

6. Find each of the following.

   a) $\cot \dfrac{\pi}{6}$

   b) $\cot 0$

   c) $\cot \dfrac{\pi}{3}$

   d) $\csc \dfrac{\pi}{6}$

   e) $\csc 0$

   f) $\csc \dfrac{\pi}{4}$

Solution

$$\tan \frac{\pi}{6} = \frac{\sin (\pi/6)}{\cos (\pi/6)} = \frac{\frac{1}{2}}{\frac{\sqrt{3}}{2}} = \frac{1}{\sqrt{3}} = \frac{1}{\sqrt{3}} \cdot \frac{\sqrt{3}}{\sqrt{3}} = \frac{\sqrt{3}}{3} \approx 0.577,$$

$$\tan \frac{\pi}{2} = \frac{\sin (\pi/2)}{\cos (\pi/2)} = \frac{1}{0} = \text{undefined}$$

DO EXERCISES 5 AND 6.

The graph of $\tan x$ is shown below.

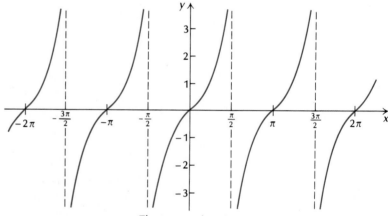

The tangent function

Identities

An *identity* is an equation that holds for all sensible replacements of its variables by real numbers. For example, $x + 3 = 3 + x$ is an identity. It is true for all real numbers. Another example of an identity is $(x^2 - 4)/(x - 2) = x + 2$. It is true for all real numbers except 2, which is *not* a sensible replacement in the expression on the left.

    The properties

$$\cos (x + 2\pi) = \cos x \quad \text{and} \quad \sin (x + 2\pi) = \sin x$$

hold for all real numbers x. They are examples of *trigonometric identities*. Another identity that holds for all real numbers is the following:

THEOREM 2

$$\sin^2 t + \cos^2 t = 1, \tag{1}$$

---

7. Multiply Identity (1) by

$$\frac{1}{\sin^2 t}$$

to develop another identity.

where $\sin^2 t$ means $(\sin t)^2$ and $\cos^2 t$ means $(\cos t)^2$. To see why this holds, note the right triangle inside the unit circle shown below.

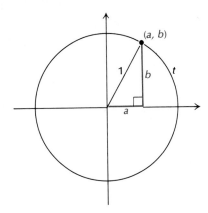

Since the length of the radius is 1, we know from the Pythagorean theorem that

$$a^2 + b^2 = 1^2, \quad \text{or} \quad a^2 + b^2 = 1;$$

and, since for any point (a, b) on the unit circle, $\cos t = a$ and $\sin t = b$, the identity follows.

If we multiply Identity (1) by $1/(\cos^2 t)$, we get another identity:

$$\frac{\sin^2 t}{\cos^2 t} + \frac{\cos^2 t}{\cos^2 t} = \frac{1}{\cos^2 t}.$$

THEOREM 3

$$\tan^2 t + 1 = \sec^2 t \qquad (2)$$

DO EXERCISE 7.

The following are *sum* and *difference identities.*

THEOREM 4

$$\cos (u + v) = \cos u \cos v - \sin u \sin v, \qquad (3)$$
$$\cos (u - v) = \cos u \cos v + \sin u \sin v, \qquad (4)$$
$$\sin (u + v) = \sin u \cos v + \cos u \sin v, \qquad (5)$$
$$\sin (u - v) = \sin u \cos v - \cos u \sin v \qquad (6)$$

8. Use Identity (6) to find sin 15°.

9. Use Identity (5) to find $\sin\left(\dfrac{\pi}{4}+\dfrac{\pi}{3}\right)$.

10. Let $u = v$ in Identity (5) to find an identity for sin 2u.

Example 4 Use Identity (4) to find cos 15°.

Solution If we think of 15° as 45° − 30°, then

$$\cos 15° = \cos (45° - 30°)$$
$$= \cos 45° \cos 30° + \sin 45° \sin 30°$$
$$= \frac{\sqrt{2}}{2}\cdot\frac{\sqrt{3}}{2} + \frac{\sqrt{2}}{2}\cdot\frac{1}{2}$$
$$= \frac{\sqrt{6} + \sqrt{2}}{4}.$$

DO EXERCISES 8 AND 9.

If we let $u = v$ in Identity (3), we obtain a *double-angle identity*:

$$\cos 2u = \cos (u + u)$$
$$= \cos u \cos u - \sin u \sin u,$$

or the following:

CSS

THEOREM 5

$$\cos 2u = \cos^2 u - \sin^2 u.$$

DO EXERCISE 10.

EXERCISE SET 8.1

In what quadrant does the terminal side of the angle lie?

1. 34°   **2.** 320°   **3.** −120°   **4.** −205°

Convert to radian measure. Leave answers in terms of π.

5. 30°   **6.** 15°   **7.** 60°   **8.** 200°   **9.** 75°   **10.** 300°

Convert to degree measure.

11. $\dfrac{3}{2}\pi$ **12.** $\dfrac{5}{4}\pi$ **13.** $-\dfrac{\pi}{4}$ **14.** $-\dfrac{\pi}{6}$

15. 8π **16.** -12π **17.** 1 radian   **18.** 2 radians

Find each of the following.

19. $\sin\dfrac{\pi}{3}$ **20.** $\cos\dfrac{\pi}{4}$ **21.** $\cos\dfrac{3\pi}{2}$ **22.** $\sin\dfrac{5\pi}{4}$

23. $\sin \dfrac{\pi}{6}$

24. $\cos \pi$

25. $\tan \dfrac{\pi}{2}$

26. $\tan \dfrac{\pi}{6}$

27. $\cot \dfrac{\pi}{3}$

28. $\cot \dfrac{\pi}{6}$

29. $\sec \pi$

30. $\csc \dfrac{\pi}{4}$

Verify the following identities.

31. $\tan (u + v) = \dfrac{\tan u + \tan v}{1 - \tan u \tan v}$

32. $\tan 2u = \dfrac{2 \tan u}{1 - \tan^2 u}$

🖩 If your calculator has trigonometric keys, find answers to the following problems to four decimal places. Check to see if your calculator is using degree measure or radian measure for angles.

33. $\sin 31.4°$

34. $\cos 1.07°$

35. $\tan 139.2°$

36. $\cot 153.5°$

37. $\cos (-1.91)$

38. $\sin (-11.2\pi)$

39. $\cot 49\pi$

40. $\tan (-17.4)$

APPLICATIONS

41. 🖩 *Biomedical: Temperature during an illness.* The temperature of a patient during a 12-day illness is given by

$$T(t) = 101.6° + 3 \sin\left(\dfrac{\pi}{8} t\right).$$

The graph is shown here. Find $T(0)$, $T(1)$, $T(2)$, $T(4)$, and $T(12)$. Round to the nearest tenth of a degree.

CSS

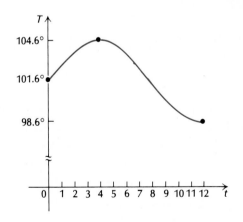

42. 🖩 *Business: Sales.* Sales of certain products fluctuate in cycles, as shown in this graph. A company in a northern climate has total sales of skis as given by

$$S(t) = 7\left(1 - \cos\dfrac{\pi}{6} t\right),$$

where S is sales in thousands of dollars during the tth month. Find $S(0)$, $S(1)$, $S(2)$, $S(3)$, $S(6)$, $S(12)$, and $S(15)$. Round to the nearest tenth.

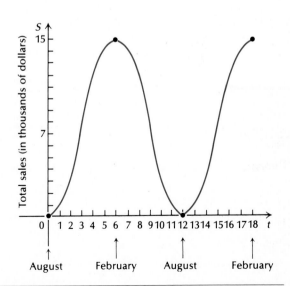

8.2

DERIVATIVES OF THE TRIGONOMETRIC FUNCTIONS

The development of the derivatives of $\sin x$ and $\cos x$ is comparable to the development of the derivative of e^x. We first find the value of the derivative at 0, and then extend this to the general formula. The derivatives at 0 are given by the following limits:

$$\sin' (0) = \lim_{h \to 0} \frac{\sin (0 + h) - \sin 0}{h} = \lim_{h \to 0} \frac{\sin h}{h};$$

$$\cos' (0) = \lim_{h \to 0} \frac{\cos (0 + h) - \cos 0}{h} = \lim_{h \to 0} \frac{\cos h - 1}{h}.$$

We find the limits using the following input–output table.

h (in radians) with decimal approximation	$\sin h$	$\cos h$	$\dfrac{\sin h}{h}$	$\dfrac{\cos h - 1}{h}$
$\dfrac{\pi}{2}$ (1.5708)	1	0	0.6366	−0.6366
$\dfrac{\pi}{3}$ (1.0472)	$\dfrac{\sqrt{3}}{2}$ (0.8660)	$\dfrac{1}{2}$ (0.5000)	0.8270	−0.4775
$\dfrac{\pi}{4}$ (0.7854)	$\dfrac{\sqrt{2}}{2}$ (0.7071)	$\dfrac{\sqrt{2}}{2}$ (0.7071)	0.9003	−0.3729
$\dfrac{\pi}{6}$ (0.5236)	$\dfrac{1}{2}$ (0.5000)	$\dfrac{\sqrt{3}}{2}$ (0.8660)	0.9549	−0.2559
$\dfrac{\pi}{20}$ (0.1571)	0.1564	0.9877	0.9955	−0.0783
$\dfrac{\pi}{50}$ (0.0628)	0.0628	0.9980	1.0000	−0.0318
$\dfrac{\pi}{4000}$ (0.0008)	0.0008	0.99997	1.0000	−0.0038

Values approaching 0 from the left can also be checked. The following values of the limits can be conjectured from the table:

$$\lim_{h \to 0} \frac{\sin h}{h} = 1 \quad \text{and} \quad \lim_{h \to 0} \frac{\cos h - 1}{h} = 0.$$

The results can be proved more formally, but we will not do so here.

Now let us consider the general derivatives:

$$\frac{d}{dx} \sin x = \lim_{h \to 0} \frac{\sin (x + h) - \sin x}{h}.$$

Using Identity (5), we get

$$\frac{\sin (x + h) - \sin x}{h} = \frac{\sin x \cos h + \cos x \sin h - \sin x}{h}$$

$$= \frac{\sin x \cos h - \sin x}{h} + \frac{\cos x \sin h}{h}$$

$$= \sin x \left(\frac{\cos h - 1}{h} \right) + \cos x \left(\frac{\sin h}{h} \right).$$

Then using the limits just developed, we have

$$\frac{d}{dx} \sin x = \lim_{h \to 0} \frac{\sin (x + h) - \sin x}{h}$$

$$= (\sin x) \cdot 0 + (\cos x) \cdot 1$$

$$= \cos x.$$

A development for the derivative of $\cos x$ is similar but uses Identity (3). The result is $-\sin x$.

In summary, we have the following. $\boxed{\text{CSS}}$

THEOREM 6

$$\frac{d}{dx} \sin x = \cos x \quad \text{and} \quad \frac{d}{dx} \cos x = -\sin x$$

The derivatives of the remaining trigonometric functions are computed from their definitions in terms of $\sin x$ and $\cos x$, together with the Quotient Rule and/or the Extended Power Rule. The remaining formulas are as follows.

THEOREM 7

$$\frac{d}{dx} \tan x = \sec^2 x, \qquad \frac{d}{dx} \cot x = -\csc^2 x,$$

$$\frac{d}{dx} \sec x = \tan x \sec x, \qquad \frac{d}{dx} \csc x = -\cot x \csc x$$

1. Prove that $\dfrac{d}{dx} \cot x = -\csc^2 x.$

Example 1   Prove that

$$\frac{d}{dx} \tan x = \sec^2 x.$$

Solution   By definition,

$$\tan x = \frac{\sin x}{\cos x}.$$

Thus we can find its derivative using the Quotient Rule:

$$\frac{d}{dx} \tan x = \frac{\cos x\,(\cos x) - \sin x\,(-\sin x)}{\cos^2 x}$$

$$= \frac{\cos^2 x + \sin^2 x}{\cos^2 x}$$

$$= \frac{1}{\cos^2 x}$$

$$= \sec^2 x.$$

DO EXERCISE 1.

2. Differentiate $y = \tan^3 x.$

Example 2   Find the derivative of $y = \sec^3 x.$

Solution   We use the Chain Rule:

$$\frac{dy}{dx} = 3 \sec^2 x \cdot \left(\frac{d}{dx} \sec x \right)$$

$$= 3 \sec^2 x \cdot \tan x \cdot \sec x.$$

Replacing the factors by the definitions in terms of $\sin x$ and $\cos x$, we can simplify this as follows:

$$3 \sec^2 x \cdot \tan x \cdot \sec x = 3 \cdot \frac{1}{\cos^2 x} \cdot \frac{\sin x}{\cos x} \cdot \frac{1}{\cos x}$$

$$= \frac{3 \sin x}{\cos^4 x}.$$

DO EXERCISE 2.

Using the Chain Rule, we can find other derivatives.

Example 3   Differentiate $f(x) = \sin (x^3 - 5x).$

Differentiate.

3. $f(x) = \sin (x^4 + 3x^2)$

Solution

$$f'(x) = \cos (x^3 - 5x) \cdot (3x^2 - 5)$$
$$= (3x^2 - 5) \cos (x^3 - 5x)$$

Example 4   Differentiate $y = \cos (e^{4x}) \sin x^2$.

Solution   We use the Product Rule and the Chain Rule:

$$\frac{dy}{dx} = -\sin (e^{4x}) \cdot (4e^{4x}) \sin x^2 + \cos (e^{4x}) \cdot 2x \cdot \cos x^2$$
$$= -4e^{4x} \sin (e^{4x}) \sin x^2 + 2x \cos (e^{4x}) \cos x^2.$$

DO EXERCISES 3 AND 4.

Applications

Equations of the type

$$y = A \sin (Bx - C) + D$$

have many applications. We can also express this equation as

$$y = A \sin \left[B\left(x - \frac{C}{B}\right) \right] + D.$$

4. $y = \cos (e^{x^2}) \sin x^3$

The numbers A, B, C, and D play an important role in graphing such an equation. The number A corresponds to a vertical stretching or shrinking of the graph of $y = \sin x$, B corresponds to a horizontal stretching or shrinking, C/B corresponds to a shift of the entire graph to the right or left, and D corresponds to a vertical translation. The following names are attached to the particular expressions involving some of the numbers:

$$\textit{Amplitude} = |A|,$$
$$\textit{Period} = \frac{2\pi}{B},$$
$$\textit{Phase shift} = \frac{C}{B}.$$

Example 5   A weight is attached to the end of a spring. When the weight is disturbed, it bobs up and down with a definite frequency. If the motion were to occur in a perfect vacuum and if the spring were perfectly elastic, then the oscillatory motion would continue undiminished forever. Suppose a spring oscillates in such a way that its vertical position at time t, from its position at rest, is given by

$$y = 3 \sin\left(2t + \frac{\pi}{2}\right) = 3 \sin\left[2\left(t - \left(-\frac{\pi}{4}\right)\right)\right].$$

a) Find the amplitude, the period, and the phase shift.
b) Graph the equation.
c) Find dy/dt.

Extended   Compressed   At rest

Solution

a) The amplitude $= |3| = 3$. The period $= 2\pi/2 = \pi$. The phase shift $= -(\pi/4)$.

b) We plot the various equations needed to get the final graph.

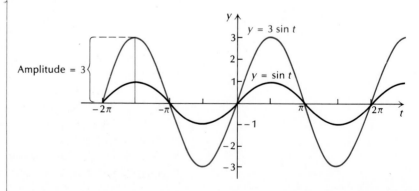

5. Suppose a spring oscillates in such a way that its vertical position at time t, from its position at rest, is given by

$$y = 4 \sin\left(2t - \frac{\pi}{2}\right).$$

a) Find the amplitude, the period, and the phase shift.

b) Graph the equation.

c) Find dy/dt.   CSS

6. A sound wave is given by

$$y = 0.03 \sin 1.198\pi x.$$

Find dy/dx.

7. A light wave is given by

$$L = A \sin\left(2\pi f T - \frac{d}{\omega}\right).$$

Find dL/dT.

The graph of $y = 3 \sin t$ is a vertical stretching, by a factor of 3, of the graph of $y = \sin t$. The period of $y = 3 \sin t$ is still 2π. The graph of $y = 3 \sin 2t$ is a horizontal shrinking, by a factor of $\frac{1}{2}$, of the graph of $y = 3 \sin t$. The period of $y = 3 \sin 2t$ is π.

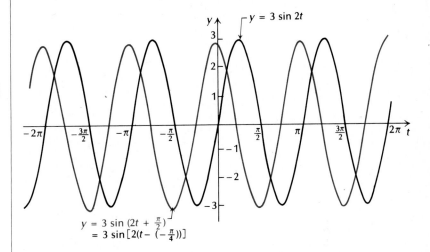

$$y = 3 \sin \left(2t + \frac{\pi}{2}\right)$$
$$= 3 \sin \left[2\left(t - \left(-\frac{\pi}{4}\right)\right)\right]$$

The graph of

$$y = 3 \sin\left[2\left(t - \left(-\frac{\pi}{4}\right)\right)\right]$$

is shifted $\pi/4$ units to the left of the graph of $y = 3 \sin 2t$. If the phase shift were positive, the shift would be to the right.

c) $\dfrac{dy}{dt} = 3 \cos\left(2t + \dfrac{\pi}{2}\right) \cdot 2 = 6 \cos\left(2t + \dfrac{\pi}{2}\right)$

DO EXERCISE 5.

The motion of the spring in the preceding example is called *simple harmonic motion*. Other types of simple harmonic motion are sound waves and light waves. For sound waves the amplitude is the loudness. For light waves the amplitude is the brightness. The *frequency*, which is the reciprocal of the period, is the tone of a sound wave and the color of a light wave.

DO EXERCISES 6 AND 7.

A Biological Application: Biorhythms

Some people conjecture that a person's life has cycles of good and bad days that begin at birth. There are supposedly three such cycles, or *biorhythms*.

1. *Physical.* A cycle represented by a sine function with a period of 23 days:

$$y = \sin \frac{2\pi}{23} t.$$

   This cycle is related to physical qualities such as vitality, strength, and energy.

2. *Emotional (sensitivity).* A cycle represented by a sine function with a period of 28 days:

$$y = \sin \frac{2\pi}{28} t.$$

   This cycle is related to creativity, moodiness, intuition, and cheerfulness.

3. *Mental (intellectual).* A cycle represented by a sine function with a period of 33 days:

$$y = \sin \frac{2\pi}{33} t.$$

   This cycle is related to the ability to study, think, react, and remember. (The positive part of *this* cycle would be a good time to study calculus.) Because these cycles have periods of different lengths, there are varying combinations of highs and lows throughout one's life. This is illustrated in the following graph.*

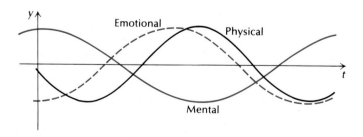

---

* For more on these "biorhythms," write to Biorhythm Computers, Inc., 298 Fifth Ave, New York, NY 10001.

Example 6 *Biology: Minimizing the surface area of a bee's cell.* Did you know that a honey bee constructs the cells in its comb in such a way that the minimum amount of wax is used?* One cell of a honeycomb is shown here. It is a prism whose base, the open part at the top, is a regular hexagon. The bottom comes together at point A. The surface area is given by

$$S(\theta) = 6ab + \frac{3}{2}\,a^2\left(\frac{\sqrt{3} - \cos\theta}{\sin\theta}\right),$$

where θ is the measure of what is called the angle of inclination. To maintain the prism shape of the cell, we can vary the values of θ between $0°$ and $90°$. Thus we want to minimize S on the open interval $(0, 90)$.

A honeycomb.

Solution   We first find $dS/d\theta$:

$$\frac{dS}{d\theta} = \frac{3}{2}\,a^2\left[\frac{\sin\theta\,(\sin\theta) - \cos\theta\,(\sqrt{3} - \cos\theta)}{\sin^2\theta}\right]$$

$$= \frac{3}{2}\,a^2\left[\frac{\sin^2\theta - \sqrt{3}\,\cos\theta + \cos^2\theta}{\sin^2\theta}\right]$$

$$= \frac{3}{2}\,a^2\left[\frac{1 - \sqrt{3}\,\cos\theta}{\sin^2\theta}\right].$$

---

* This discovery was published in a study by Sir D'Arcy Wentworth Thompson, *On Growth and Form* (Cambridge University Press, 1917).

8. A corridor of width a meets a corridor of width b at right angles. Workers wish to push a heavy beam of length L on dollies around the corner. Before starting, however, they want to be sure they can make the turn. How long a beam will go around the corner? (Disregard the width of the beam.)

The length L of the beam is given by

$$L = \frac{a}{\sin \theta} + \frac{b}{\cos \theta},$$

where θ is the angle shown above.

a) Find L in terms of a and b, such that L is of maximum length.

b) Use the formula in part (a) to find how long a beam will go around the corner when $a = 8$ and $b = 5\sqrt{5}$.   [CSS]

Critical points occur where $\sin \theta$ is 0; but if we check such values, we see that they are multiples of 90°, and that at such values the bottom of the cell would be flat, which is not the way it is to be constructed. Thus we set $dS/d\theta = 0$ and solve for θ with the assumption that $\sin \theta \neq 0$:

$$\frac{dS}{d\theta} = \frac{3}{2}\, a^2 \left[\frac{1 - \sqrt{3}\, \cos \theta}{\sin^2 \theta} \right] = 0$$

$$\frac{1 - \sqrt{3}\, \cos \theta}{\sin^2 \theta} = 0$$

$$1 - \sqrt{3}\, \cos \theta = 0 \qquad \text{Multiplying by } \sin^2 \theta$$

$$\cos \theta = \frac{1}{\sqrt{3}}$$

$$\approx 0.5774.$$

We need to find θ such that $\cos \theta \approx 0.5774$. We can work backward on a calculator using the [Arc] [cos] key(s). It follows that

$$\theta \approx 54.7°.$$

The study showed that bees tend to use this angle. That they do may be explained by a genetic theory that those who survive to pass on their genetic traits are those who are more "successful" at living. Perhaps bees that waste less time and wax by making cells with the minimum amount of wax were, at one time, such successful genetic survivors.

DO EXERCISE 8.

EXERCISE SET 8.2

Prove the following derivative formulas.

1. $\dfrac{d}{dx} \sec x = \tan x \sec x$

2. $\dfrac{d}{dx} \csc x = -\cot x \csc x$

Differentiate.

3. $y = x \sin x$

4. $y = x \cos x$

5. $f(x) = e^x \sin x$

6. $f(x) = e^x \cos x$

7. $y = \dfrac{\sin x}{x}$

8. $y = \dfrac{\cos x}{x}$

9. $f(x) = \sin^2 x$

10. $f(x) = \cos^2 x$

11. $y = \sin x \cos x$

12. $y = \cos^2 x + \sin^2 x$

13. $f(x) = \dfrac{\sin x}{1 + \cos x}$

14. $f(x) = \dfrac{1 - \cos x}{\sin x}$

15. $y = \tan^2 x$

16. $y = \sec^2 x$

17. $f(x) = \sqrt{1 + \cos x}$

18. $f(x) = \sqrt{1 - \sin x}$

19. $y = x^2 \cos x - 2x \sin x - 2 \cos x$

20. $y = x^2 \sin x - 2x \cos x + 2 \sin x$

21. $y = e^{\sin x}$

22. $y = e^{\cos x}$

23. Find d^2y/dx^2 if $y = \sin x$.

24. Find d^2y/dx^2 if $y = \cos x$.

25. $y = \sin (x^2 + x^3)$

26. $y = \sin (x^5 - x^4)$

27. $f(x) = \cos (x^5 - x^4)$

28. $f(x) = \cos (x^2 + x^3)$

29. $f(x) = \cos \sqrt{x}$

30. $f(x) = \sin \sqrt{x}$

31. $y = \sin (\cos x)$

32. $y = \cos (\sin x)$

33. $y = \sqrt{\cos 4x}$

34. $y = \sin^2 5x$

35. $f(x) = \cot \sqrt[3]{5 - 2x}$

36. $f(x) = \sec (\tan 7x)$

37. $y = \tan^4 3x - \sec^4 3x$

38. $y = \sqrt[5]{\cot 5x - \cos 5x}$

Differentiate. Use the formula $\dfrac{d}{du} \ln |u| = \dfrac{1}{u} \, du$.

39. $y = \ln |\sin x|$

40. $f(x) = \ln |x - \cos x|$

APPLICATIONS

41. *Biomedical: Temperature during an illness.* The temperature of a patient during a 12-day illness is given by

$$T(t) = 101.6° + 3 \sin \frac{\pi}{8} t.$$

Find $T'(t)$.

42. *Business: Sales.* A company in a northern climate has sales of skis as given by

$$S(t) = 7 \left(1 - \cos \frac{\pi}{6} t \right).$$

Find $S'(t)$.

43. *Satellite location.* A satellite circles the earth in such a manner that it is y miles from the equator (north or south, height not considered) t minutes after its launch, where

$$y = 5000 \left[\cos \frac{\pi}{45} (t - 10) \right].$$

Find dy/dt.

44. *Rollercoaster layout.* A rollercoaster is constructed in such a way that it is y meters above the ground x meters from the starting point, where

$$y = 15 + 15 \sin \frac{\pi}{50} x.$$

Find dy/dx.

45. A spring oscillates in such a way that its vertical position at time t, from its position at rest, is given by

$$y = 5 \sin (4t + \pi).$$

a)   Find the amplitude, the period, and the phase shift.

b)   Find dy/dt.

46. The current i at time t of a wire passing through a magnetic field is given by

$$i = I \sin (\omega t + a).$$

a)   Find the amplitude, the period, and the phase shift.

b)   Find di/dt.

47. A company determines that sales during the tth month are given by

$$S(t) = 40{,}000(\sin t + \cos t).$$

The sales are seasonal and fluctuate. Find $S'(t)$.

49. Referring to Margin Exercise 8, determine how long a beam will go around the corner when $a = 8$ and $b = 8$.

50. Referring to Margin Exercise 8, determine how long a beam will go around the corner when $a = 3\sqrt{3}$ and $b = 5\sqrt{5}$.

51. A V-shaped water trough is to be made with sides that are 10 in. wide. At what angle θ will the trough be able to carry the greatest volume of water?

52. The illumination from a light source is inversely proportional to the square of the distance from the light and directly proportional to the sine of the angle formed by the line from the light source and the horizontal. How high should a light be placed on a pole in order to maximize the illumination on the ground along the circumference of a circle of radius 25 ft?

48. A piston connected to a crankshaft (see the figure) moves up and down in such a way that its second coordinate after time t is given by

$$y(t) = \sin t + \sqrt{25 - \cos^2 t}.$$

Find the rate of change $y'(t)$.

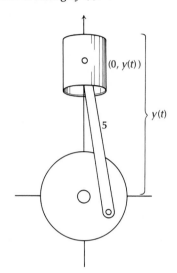

EXTENSION EXERCISES

Find the partial derivatives f_x, f_y, f_{xx}, f_{yx}, f_{xy}, and f_{yy}.

53. $f(x, y) = \sin 2y - ye^x$ **54.** $f(x, y) = e^{-x} \cos y$ **55.** $f(x, y) = \cos (2x + 3y)$

56. $f(x, y) = y \ln (\sin x)$ **57.** $f(x, y) = x^3 \tan (5xy)$ **58.** $f(x, y) = \sin x \cos y$

Differentiate implicitly to find y'.

59. $y = x \cos y$ **60.** $xy = e^x \sin y$

8.3

OBJECTIVE

a) Integrate trigonometric functions.

INTEGRATION OF THE TRIGONOMETRIC FUNCTIONS

Each of the previously developed differentiation formulas yields an integration formula. For example, we have the following.

1. Find the area under the graph of $y = \sin x$ on the interval $[0, \pi]$.

THEOREM 8

$$\int \sin x \, dx = -\cos x + C,$$

$$\int \cos x = \sin x + C,$$

$$\int \sec^2 x \, dx = \tan x + C,$$

$$\int \csc^2 x \, dx = -\cot x + C$$

Example 1   Find the area under the graph of $y = \cos x$ on the interval $[0, \pi/2]$.

Solution

$$\int_0^{\pi/2} \cos x \, dx = [\sin x]_0^{\pi/2} = \left(\sin \frac{\pi}{2}\right) - (\sin 0) = 1 \qquad \boxed{\text{CSS}}$$

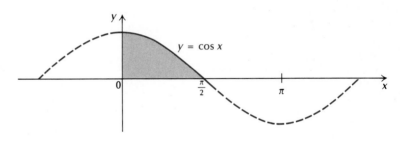

2. Integrate $\int (\sin x)^3 \cos x \, dx$.

DO EXERCISE 1.

We use substitution in the following examples.

Example 2

$$\int (\sin x)^2 \cos x \, dx = \int u^2 \, du \qquad \text{Substitution} \quad \boxed{\begin{array}{l} u = \sin x, \\ du = \cos x \, dx \end{array}}$$

$$= \frac{u^3}{3} + C$$

$$= \frac{(\sin x)^3}{3} + C$$

DO EXERCISE 2.

3. Integrate $\int \cos 4x \, dx$.

Example 3

$$\int \sin 3x \, dx = \frac{1}{3} \int \sin u \, du \qquad \underline{\text{Substitution}} \quad \boxed{\begin{array}{l} u = 3x, \\ du = 3 \, dx \end{array}}$$

$$= -\frac{1}{3} \cos u + C$$

$$= -\frac{1}{3} \cos 3x + C$$

DO EXERCISE 3.

In an expression in which natural logarithms of trigonometric functions are involved, absolute-value signs are usually necessary because the trigonometric functions alternate periodically between positive and negative values. Accordingly, we use the integration formula

$$\int \frac{du}{u} = \ln |u| + C.$$

Example 4

4. Integrate $\int \dfrac{2x - \sin x}{x^2 + \cos x} \, dx$.

$$\int \frac{1 + \cos x}{x + \sin x} \, dx = \int \frac{du}{u} \qquad \underline{\text{Substitution}} \quad \boxed{\begin{array}{l} u = x + \sin x, \\ du = (1 + \cos x) \, dx \end{array}}$$

$$= \ln |u| + C$$

$$= \ln |x + \sin x| + C$$

DO EXERCISE 4.

We use integration by parts in the following example.

Example 5   Integrate $\int x \cos x \, dx$.

Solution   Let

$$u = x \quad \text{and} \quad dv = \cos x \, dx.$$

Then

$$du = dx \quad \text{and} \quad v = \sin x.$$

Using the integration-by-parts formula, we get

$$\overset{u}{\int (x)} \overset{dv}{(\cos x \, dx)} = \overset{u}{(x)} \overset{v}{(\sin x)} - \int \overset{v}{(\sin x)} \overset{du}{(dx)}$$

$$= x \sin x - (-\cos x) + C = x \sin x + \cos x + C.$$

5. Integrate $\int x \sin x \, dx$.

6. Integrate $\int \csc x \, dx.$ $\left[\text{Hint: Multiply} \right.$ by 1 using $\left. \dfrac{\csc x + \cot x}{\csc x + \cot x}. \right]$

DO EXERCISE 5.

To find an integral such as $\int \sec x \, dx$, we must first use some algebra to rename the expression. Then we use substitution.

Example 6   Integrate $\int \sec x \, dx$.

Solution

$$\int \sec x \, dx = \int (\sec x) \cdot 1 \, dx \qquad \text{Multiplying by 1}$$

$$= \int \sec x \cdot \frac{\sec x + \tan x}{\sec x + \tan x} \, dx \qquad \text{Substituting } \frac{\sec x + \tan x}{\sec x + \tan x} \text{ for 1}$$

$$= \int \frac{\sec x \, (\sec x + \tan x)}{\sec x + \tan x} \, dx$$

$$= \int \frac{\sec x \tan x + \sec^2 x}{\sec x + \tan x} \, dx$$

$$= \int \frac{du}{u} \qquad \text{Substitution} \quad \boxed{\begin{array}{l} u = \sec x + \tan x, \\ du = (\sec x \tan x + \sec^2 x) \, dx \end{array}}$$

$$= \ln |u| + C$$

$$= \ln |\sec x + \tan x| + C$$

DO EXERCISE 6.

EXERCISE SET 8.3

1. Find the area under the graph of $y = \sin x$ on the interval $[0, \pi/3]$.

2. Find the area under the graph of $y = \cos x$ on the interval $[0, \pi/3]$.

Integrate using substitution.

3. $\int (\sin x)^4 \cos x \, dx$

4. $\int (\sin x)^5 \cos x \, dx$

5. $\int -(\cos x)^2 \sin x \, dx$

6. $\int (\cos x)^3 (-\sin x) \, dx$

7. $\int \cos (x + 3) \, dx$

8. $\int \sin (x + 4) \, dx$

9. $\int \sin 2x \, dx$

10. $\int \cos 3x \, dx$

11. $\int x \cos x^2 \, dx$

12. $\int x \sin x^2 \, dx$

13. $\int e^x \sin (e^x) \, dx$

14. $\int e^x \cos (e^x) \, dx$

15. $\int \tan x \, dx$ $\left[\text{Hint: } \tan x = \dfrac{\sin x}{\cos x}. \right]$

16. $\int \cot x \, dx$ $\left[\text{Hint: } \cot x = \dfrac{\cos x}{\sin x}. \right]$

Integrate by parts.

17. $\int x \cos 4x\,dx$

18. $\int x \sin 3x\,dx$

19. $\int 3x \cos x\,dx$

20. $\int 2x \sin x\,dx$

21. $\int x^2 \sin x\,dx$ [*Hint*: Let $u = x$ and $dv = x \sin x\,dx$ and use the result of Margin Exercise 5.]

22. $\int x^2 \cos x\,dx$ [*Hint*: Let $u = x$ and $dv = x \cos x\,dx$ and use the result of Example 5.]

23. $\int \tan^2 x\,dx$

24. $\int \cot^2 x\,dx$

APPLICATIONS

25. *Business: Total sales.* A company in a northern climate has sales of skis as given by

$$S(t) = 7\left(1 - \cos\frac{\pi}{6}t\right),$$

where S is sales in thousands of dollars during the tth month. Total sales for the first year are given by

$$\int_0^{12} S(t)\,dt.$$

Find the total sales.

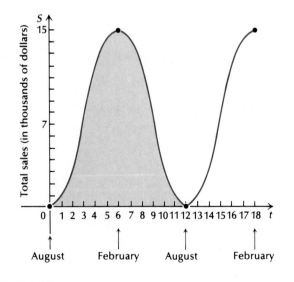

26. *Total area under a rollercoaster.* A rollercoaster is made in such a way that it is y meters above the ground x meters from the starting point, where

$$y = 15 + 15 \sin\frac{\pi}{50}x.$$

The area under the rollercoaster, from the starting point to a point 100 meters away, is given by

$$\int_0^{100} y\,dx.$$

Find this area.

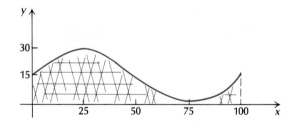

EXTENSION EXERCISES

Integrate by parts.

27. $\int \sin(\ln x)\,dx$

28. $\int \cos(\ln x)\,dx$

29. $\int e^x \cos x\,dx$

30. $\int e^x \sin x\,dx$

Integrate.

31. $\int \sec^2 7x\, dx$

32. $\int e^x \csc^2 (e^x)\, dx$

33. $\int \sec x\, (\sec x + \tan x)\, dx$

34. $\int \sec u \tan u\, du$

35. $\int \csc u \cot u\, du$

36. $\int e^{2x} \sec (e^{2x})\, dx$

37. $\int \dfrac{\cos^2 x}{\sin x}\, dx$

38. $\int \cot 7x \sin 7x\, dx$

39. $\int (1 + \sec x)^2\, dx$

40. $\int (1 - \csc x)^2\, dx$

8.4

OBJECTIVES

a) Find certain values of inverse trig-
    onometric functions.

b) Integrate certain functions whose
    antiderivatives involve inverse trig-
    onometric functions.

Find each of the following.   CSS

1. $\sin^{-1}\left(\dfrac{1}{2}\right)$

2. $\sin^{-1}(0)$

3. $\sin^{-1}\left(-\dfrac{\sqrt{3}}{2}\right)$

INVERSE TRIGONOMETRIC FUNCTIONS*

Look back at the graph of $y = \sin x$. Note that outputs ranged from -1 to 1. Suppose we wanted to work backward from an output to an input. More specifically, suppose x is some number such that $-1 \leqslant x \leqslant 1$ and that we wanted to find a number y such that

$$-\frac{\pi}{2} \leqslant y \leqslant \frac{\pi}{2} \quad \text{and} \quad \sin y = x.$$

This determines a function, called the *inverse sine function*, given by

$$y = \sin^{-1} x \quad \text{or} \quad \arcsin x.$$

Caution! The -1 is *not* an exponent in this context and $\sin^{-1} x$ is *not* the reciprocal of $\sin x$. Such a function is an *inverse trigonometric function.* Let us find a function value.

Example 1   Find $\sin^{-1} (\sqrt{3}/2)$.

Solution   *Think:* What number between $-(\pi/2)$ and $\pi/2$ is such that its sine is $\sqrt{3}/2$? That number is $\pi/3$. Thus,

$$\sin^{-1}\left(\frac{\sqrt{3}}{2}\right) = \frac{\pi}{3}.$$

DO EXERCISES 1–3.

_____

* This section can be omitted without loss of continuity.

Find each of the following.

4. $\tan^{-1}(\sqrt{3})$

The *inverse cosine function* is given by

$$y = \cos^{-1} x \quad \text{or} \quad y = \arccos x,$$

where

$$x = \cos y \quad \text{and} \quad 0 \leqslant y \leqslant \pi.$$

The *inverse tangent function* is given by

$$y = \tan^{-1} x \quad \text{or} \quad y = \arctan x,$$

where

$$x = \tan y \quad \text{and} \quad -\frac{\pi}{2} < y < \frac{\pi}{2}.$$

DO EXERCISES 4–6.

5. $\cos^{-1}\left(\dfrac{\sqrt{3}}{2}\right)$

Graphs of these functions are as follows.   CSS

6. $\cos^{-1}\left(-\dfrac{1}{2}\right)$

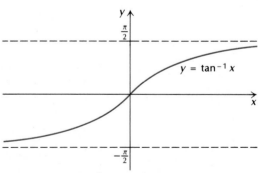

Inverse trigonometric functions are integrals of certain functions. You will often see them in tables of integrals.

7. Find

$$\int \frac{1}{1 + 25x^2} \, dx$$

using the substitution $u = 5x$ in the integral formula

$$\int \frac{1}{1 + u^2} \, du = \tan^{-1} u + C.$$

Example 2   Prove that

$$\int \frac{1}{1 + x^2} \, dx = \tan^{-1} x + C.$$

Solution   We use a method of integration known as *trigonometric substitution*. We simplify the integral using the identity $1 + \tan^2 u = \sec^2 u$.

$$\int \frac{1}{1 + x^2} \, dx$$

$$= \int \frac{1}{1 + \tan^2 u} \sec^2 u \, du \qquad \text{Substitution} \qquad \boxed{\begin{array}{l} u = \tan^{-1} x, \\ x = \tan u, \\ dx = \sec^2 u \, du \end{array}}$$

$$= \int \frac{1}{\sec^2 u} \sec^2 u \, du \qquad \text{By Identity (2)}$$

$$= \int du = u + C = \tan^{-1} x + C$$

DO EXERCISE 7.

Example 3   Prove that

$$\int -\frac{1}{\sqrt{1 - x^2}} \, dx = \cos^{-1} x + C.$$

Solution   Again we use substitution:

$$\int -\frac{1}{\sqrt{1 - x^2}} \, dx$$

$$= \int \frac{1}{\sqrt{1 - \cos^2 u}} \sin u \, du. \qquad \text{Substitution} \qquad \boxed{\begin{array}{l} u = \cos^{-1} x, \\ x = \cos u, \\ dx = -\sin u \, du \end{array}}$$

By Identity (1), $\sin^2 u + \cos^2 u = 1$, so

$$1 - \cos^2 u = \sin^2 u.$$

8. Find

$$\int -\frac{1}{\sqrt{1 - 4x^2}} \, dx$$

using the substitution $u = 2x$ in the integral

$$\int -\frac{1}{\sqrt{1 + u^2}} \, du = \cos^{-1} u + C.$$

Then, since u is such that $0 \leqslant u \leqslant \pi$, $\sin u \geqslant 0$, so $\sqrt{\sin^2 u} = \sin u$ and the integral becomes

$$\int \frac{1}{\sqrt{\sin^2 u}} \cdot \sin u \, du \qquad \text{By Identity (1)}$$

$$= \int \frac{1}{\sin u} \cdot \sin u \, du = \int du = u + C = \cos^{-1} x + C.$$

DO EXERCISE 8.

EXERCISE SET 8.4

Find each of the following.

1. $\sin^{-1}\left(\dfrac{\sqrt{2}}{2}\right)$

2. $\sin^{-1}\left(-\dfrac{\sqrt{2}}{2}\right)$

3. $\cos^{-1}(0)$

4. $\cos^{-1}\left(\dfrac{\sqrt{2}}{2}\right)$

5. $\tan^{-1}(1)$

6. $\tan^{-1}(-1)$

7. $\sin^{-1}\left(-\dfrac{1}{2}\right)$

8. $\cos^{-1}\left(-\dfrac{\sqrt{3}}{2}\right)$

Integrate using substitution.

9. $\displaystyle\int \frac{e^t}{1+e^{2t}}\,dt$

10. $\displaystyle\int \frac{-1}{\sqrt{1-25x^2}}\,dx$

11. Show that

$$\int \frac{1}{\sqrt{1-x^2}}\,dx = \sin^{-1}x + C.$$

12. Integrate using the formula in Exercise 11:

$$\int \frac{1}{\sqrt{1-49x^2}}\,dx.$$

▦ If your calculator has an inverse function key, find each of the following.

13. $\sin^{-1}(0.9874)$

14. $\cos^{-1}(-0.3487)$

15. $\cos^{-1}(0.9988)$

16. $\tan^{-1}(2000)$

EXTENSION EXERCISE

17. Since $\displaystyle\int \frac{1}{1+x^2}\,dx = \tan^{-1}x + C$, it follows that if $u = \tan^{-1}x$, then $\dfrac{du}{dx} = \dfrac{1}{1+x^2}$. Use this result and integration by parts to find $\displaystyle\int \tan^{-1}x\,dx$.

SUMMARY AND REVIEW: CHAPTER 8

The following contains a summary of what you should be able to do after completing this chapter. The review exercises are for practice. Answers are at the back of the book. If you miss an exercise, restudy the section indicated alongside the answers.

A summary of the important formulas for this chapter is given on the inside back cover of this book.

You should be able to:

Convert from radian measure to degree measure and from degree measure to radian measure. Find values of the trigonometric functions.

Convert to radian measure.

1. $240°$

2. $315°$

Convert to degree measure.

3. $\dfrac{5\pi}{6}$

4. $\dfrac{7\pi}{3}$

Find each of the following.

5. $\sin \dfrac{2\pi}{3}$

6. $\cos\left(-\dfrac{\pi}{2}\right)$

7. $\tan \dfrac{\pi}{4}$

8. $\csc \pi$

Differentiate trigonometric functions. Solve problems involving derivatives of the trigonometric functions.

Differentiate.

9. $y = \tan (3x^2 - x)$

10. $y = \sin^6 x$

11. $y = \ln (\tan x)$

12. $f(x) = e^{\sin 3x}$

13. $f(x) = \dfrac{1 + \cos 2x}{\sin 2x}$

14. $f(x) = \cos \sqrt{t} + \sqrt{\cos t}$

15. $y = \sqrt{\tan x}$

16. $y = e^x \sin x$

17. $f(x) = \sin 3x \cos 3x$

18. $y = \sin x - x \cos x$

19. A pendulum oscillates in such a way that its horizontal position at time t, from its position at rest, is given by

$$y = 3 \cos (3t - 2\pi).$$

Find dy/dt.

20. The temperature of a patient during an 8-day illness is given by

$$T(t) = 99.7° + 5 \sin \dfrac{\pi}{3} t.$$

Find $T'(t)$.

Integrate trigonometric functions. Integrate certain functions whose antiderivatives involve inverse trigonometric functions.

21. Find the area under the graph of $y = \cos x$ on the interval $[0, \pi/6]$.

Integrate using substitution.

22. $\displaystyle\int e^x \cos e^x \, dx$

23. $\displaystyle\int \sin (x - 1) \, dx$

24. $\displaystyle\int \cos 8x \, dx$

Integrate by parts.

25. $\displaystyle\int 3x \cos 2x \, dx$

26. $\displaystyle\int x \sin 3x \, dx$

Integrate.

27. $\displaystyle\int \dfrac{1}{\sqrt{1 - 64t^2}} \, dt$

28. $\displaystyle\int \dfrac{5 - \cos x}{5x - \sin x} \, dx$

EXTENSION EXERCISES

29. Find $\dfrac{d^3y}{dx^3}$: $y = e^{\sin x}$.

30. Find f_x and f_{xy}: $f(x, y) = \sin y \cos x$.

TEST: CHAPTER 8

1. Convert $120°$ to radian measure. (Leave the answer in terms of π.)

2. Convert $5\pi/6$ to degree measure.

Find each of the following.

3. $\sin \dfrac{\pi}{6}$

4. $\cos \pi$

Differentiate.

5. $y = \cos t$

6. $y = \sin (3x^2 - 5x)$

7. $f(x) = \dfrac{x}{\sin x}$

8. $f(t) = \tan^2 t$

9. $f(x) = \sqrt{\sin x + \cos x}$

10. $y = \dfrac{\sin x + \cos x}{\sin x - \cos x}$

11. A company has sales as given by

$$S(t) = 20\left(1 + \sin \frac{\pi}{8} t\right).$$

Find $S'(t)$.

12. Find the area under the graph of $y = \sin x$ on the interval $[0, \pi/6]$.

Integrate using substitution.

13. $\displaystyle\int (\sin x)^9 \cos x \, dx$

14. $\displaystyle\int \cos 5t \, dt$

15. Integrate by parts:

$$\int 5x \sin 5x \, dx.$$

16. Integrate using substitution:

$$\int \frac{1}{1 + 16t^2} \, dt.$$

17. Integrate:

$$\int \frac{2 - \cos x}{2x - \sin x} \, dx.$$

EXTENSION EXERCISES

18. Find

$$\frac{dy}{dx} \quad \text{and} \quad \frac{d^2y}{dx^2},$$

if $y = 2e^{\cos x}$.

19. Find f_x and f_{xy}:

$$f(x, y) = \frac{\cos x}{\cos y}.$$

9

DIFFERENTIAL EQUATIONS

A *differential equation* is an equation that involves derivatives or differentials. In this chapter we learn many ways to solve differential equations and we apply these methods to the solution of many kinds of problems.

AN APPLICATION

The population of this island is probably very close to its limiting value. How do we use differential equations to analyze population growth?

THE MATHEMATICS

A type of *differential equation* used to analyze population growth is

$$\frac{dP}{dt} = kP(L - P),$$

where L is the limiting value.

OBJECTIVES

a) Solve certain differential equations, giving both general and particular solutions.

b) Solve certain differential equations given a condition $f(a) = b$.

c) Verify that a given function is a solution of a given differential equation.

1. Solve $y' = 3x^2$.

9.1

DIFFERENTIAL EQUATIONS

A *differential equation* is an equation that involves derivatives or differentials. In Chapter 4 we studied one very important differential equation,

$$\frac{dP}{dt} = kP,$$

where P, or $P(t)$, is the population at time t. This equation is a model of uninhibited population growth. Its solution is

$$P = P_0 e^{kt},$$

where the constant P_0 is the size of the initial population; that is, at $t = 0$. As this one equation illustrated, differential equations are rich in application.

Solving Certain Differential Equations

In this chapter we will frequently use the notation y' for a derivative—mainly because it is simple. Thus, if $y = f(x)$, then

$$y' = \frac{dy}{dx} = f'(x).$$

We have already found solutions of certain differential equations when we found antiderivatives or indefinite integrals. The differential equation

$$\frac{dy}{dx} = g(x) \quad \text{or} \quad y' = g(x)$$

has the solution

$$y = \int g(x) \, dx.$$

Example 1   Solve $y' = 2x$.

Solution

$$y = \int 2x \, dx = x^2 + C$$

DO EXERCISE 1.

Look again at the solution to Example 1. Note the constant of integration. This solution is called a *general solution* because taking all values of C gives *all* the solutions. Taking specific values of C gives particular solutions. For example, the following are particular solutions of $y' = 2x$:

$$y = x^2 + 3, \qquad y = x^2, \qquad y = x^2 - 3.$$

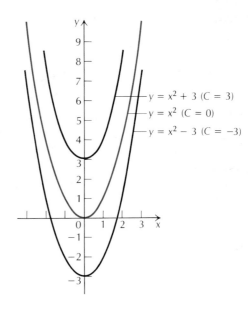

This graph shows the curves of a few particular solutions. The general solution can be envisioned as the set of all particular solutions, a *family* of curves.

DO EXERCISE 2.

Knowing the value of a function at a particular point may allow us to select a particular solution from the general solution.

Example 2   Solve $f'(x) = e^x + 5x - x^{1/2}$, given that $f(0) = 8$.

Solution

a)   First find the general solution:

$$f(x) = \int f'(x)\ dx = e^x + \tfrac{5}{2}x^2 - \tfrac{2}{3}x^{3/2} + C.$$

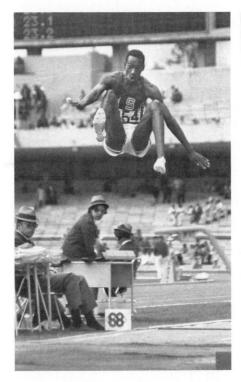

At the 1968 Olympic Games in Mexico City, Bob Beamon made what was believed to be a miracle long jump of 29 ft, $2\frac{1}{2}$ in. Many believed this was due to the altitude, which was 7400 ft. Using differential equations for analysis, M. N. Bearley refuted the altitude theory in "The Long Jump Miracle of Mexico City" (*Mathematics Magazine,* vol. 45, November 1972, pp. 241–246). Bearley argues that the world record jump was a result of Beamon's exceptional speed (9.5 sec in the 100-yd dash) and the fact that he hit the take-off board in perfect position.

2.   Given $y' = 3x^2$.

    a)   Write the general solution.

    b)   Write three particular solutions.

3. Solve

$$y' = 2x$$

given that $y = 7$ when $x = 1$.

4. Solve

$$f'(x) = \frac{1}{x} - 2x + x^{1/2}$$

given that $f(1) = 4$.

5. Show that

$$y = 2e^x - 7e^{3x}$$

is a solution of

$$y'' - 4y' + 3y = 0.$$

6. Show that

$$y = xe^{2x}$$

is a solution of

$$\frac{dy}{dx} - 2y = e^{2x}.$$

b) Since $f(0) = 8$, we substitute to find C:

$$8 = e^0 + \tfrac{5}{2} \cdot 0^2 - \tfrac{2}{3} \cdot 0^{3/2} + C$$
$$8 = 1 + C$$
$$7 = C.$$

Thus the solution is $f(x) = e^x + \tfrac{5}{2}x^2 - \tfrac{2}{3}x^{3/2} + 7$.

DO EXERCISES 3 AND 4.

Verifying Solutions

To verify that a function is a solution of a differential equation, we find the necessary derivatives and substitute.

Example 3 Show that $y = 4e^x + 5e^{3x}$ is a solution of

$$y'' - 4y' + 3y = 0.$$

Solution

a) We first find y' and y'':

$$y' = 4e^x + 15e^{3x},$$
$$y'' = 4e^x + 45e^{3x}.$$

b) Then we substitute in the differential equation, as follows.

$$
\begin{array}{c|c}
y'' - 4y' + 3y = 0 \\
\hline
(4e^x + 45e^{3x}) - 4(4e^x + 15e^{3x}) + 3(4e^x + 5e^{3x}) & 0 \\
4e^x + 45e^{3x} - 16e^x - 60e^{3x} + 12e^x + 15e^{3x} & \\
 & 0
\end{array}
$$

DO EXERCISES 5 AND 6.

EXERCISE SET 9.1

Find the general solution and three particular solutions.

1. $y' = 4x^3$

2. $y' = 6x^5$

3. $y' = e^{2x} + x$

4. $y' = e^{3x} - x$

5. $y' = \dfrac{3}{x} - x^2 + x^5$

6. $y' = \dfrac{5}{x} + x^2 - x^4$

Find the particular solution determined by the given condition.

7. $y' = x^2 + 2x - 3; y = 4$ when $x = 0$

8. $y' = 3x^2 - x + 5; y = 6$ when $x = 0$

9. $f'(x) = x^{2/3} - x; f(1) = -6$

10. $f'(x) = x^{2/5} + x; f(1) = -7$

11. Show that $y = x \ln x + 3x - 2$ is a solution of

$$y'' - \frac{1}{x} = 0 \quad \text{for } x > 0.$$

12. Show that $y = x \ln x - 5x + 7$ is a solution of

$$y'' - \frac{1}{x} = 0 \quad \text{for } x > 0.$$

13. Show that $y = e^x + 3xe^x$ is a solution of

$$y'' - 2y' + y = 0.$$

14. Show that $y = -2e^x + xe^x$ is a solution of

$$y'' - 2y' + y = 0.$$

15. Marginal cost for a certain product is $C'(x) = 2.6 - 0.02x$. Find the total-cost function $C(x)$ and the average cost $A(x)$, assuming fixed costs are \$120; that is, $C(0) = \$120$.

16. Marginal revenue for a certain product is $R'(x) = 300 - 2x$. Find the total-revenue function $R(x)$ assuming $R(0) = 0$.

EXTENSION EXERCISE

17. A firm's marginal profit P as a function of total cost C is given by

$$\frac{dP}{dC} = \frac{-200}{(C + 3)^{3/2}}.$$

a) Find the profit function $P(C)$ if $P = \$10$ when $C = \$61$.

b) At what cost will the firm break even $(P = 0)$?

9.2

OBJECTIVE

a) Solve certain differential equations using separation of variables.

SEPARATION OF VARIABLES

Consider the differential equation

$$\frac{dy}{dx} = 2xy. \tag{1}$$

We treat dy/dx as a quotient, as we did in Chapter 5. We multiply Eq. (1) by dx and then by $1/y$ to get

$$\frac{dy}{y} = 2x\,dx. \tag{2}$$

We now say that we have *separated the variables,* meaning that all the

1. Use separation of variables to solve

$$\frac{dy}{dx} = 3x^2y.$$

expressions involving y are on one side and all those involving x are on the other. We then integrate both sides of Eq. (2):

$$\int \frac{dy}{y} = \int 2x\, dx + C.$$

We use only one constant because any two antiderivatives differ by a constant:

$$\ln y = x^2 + C, \qquad y > 0.$$

Recall that the definition of logarithms says that if $\log_a b = t$, then $b = a^t$. Now, $\ln y = \log_e y = x^2 + C$, so by the definition of logarithms, we have

$$y = e^{x^2 + C}$$
$$= e^{x^2} \cdot e^C.$$

Thus the solution to differential equation (1) is

$$y = C_1 e^{x^2}, \quad \text{where } C_1 = e^C.$$

DO EXERCISE 1.

Example 1   Solve

$$3y^2 \frac{dy}{dx} + x = 0; \quad y = 5 \text{ when } x = 0.$$

Solution

a) We first separate the variables:

$$3y^2 \frac{dy}{dx} = -x$$
$$3y^2\, dy = -x\, dx.$$

We then integrate both sides:

$$\int 3y^2\, dy = \int -x\, dx + C$$
$$y^3 = -\frac{x^2}{2} + C$$
$$= C - \frac{x^2}{2}$$
$$y = \sqrt[3]{C - \frac{x^2}{2}}. \qquad \text{Taking the cube root}$$

2. Solve

$$3y^2 \frac{dy}{dx} - 2x = 0; \quad y = 7 \text{ when } x = 2.$$

3. Solve $\dfrac{dy}{dx} = \dfrac{5}{y}$.

b) Since $y = 5$ when $x = 0$, we substitute to find C:

$$5 = \sqrt[3]{C - \frac{0^2}{2}} \qquad \text{Substituting 5 for } y \text{ and 0 for } x$$

$$5 = \sqrt[3]{C}$$

$$125 = C \qquad \text{Cubing both sides}$$

The particular solution is

$$y = \sqrt[3]{125 - \frac{x^2}{2}}.$$

DO EXERCISE 2.

Example 2 Solve

$$\frac{dy}{dx} = \frac{x}{y}.$$

Solution We first separate variables:

$$y \frac{dy}{dx} = x$$

$$y \, dy = x \, dx.$$

We then integrate both sides:

$$\int y \, dy = \int x \, dx$$

$$\frac{y^2}{2} = \frac{x^2}{2} + C$$

$$y^2 = x^2 + 2C$$

$$y^2 = x^2 + C_1,$$

where $C_1 = 2C$. We make this substitution to simplify the equation. We then obtain the solutions

$$y = \sqrt{x^2 + C_1}$$

and

$$y = -\sqrt{x^2 + C_1}.$$

DO EXERCISE 3.

Example 3 Solve $y' = x - xy$.

4. Solve $y' = 2x + xy$.

Solution Before we separate variables, we replace y' by dy/dx:

$$\frac{dy}{dx} = x - xy.$$

Now we separate variables:

$$dy = (x - xy)\, dx$$
$$dy = x(1 - y)\, dx$$
$$\frac{dy}{1 - y} = x\, dx.$$

Then we integrate both sides:

$$\int \frac{dy}{1 - y} = \int x\, dx$$

$$-\ln(1 - y) = \frac{x^2}{2} + C, \qquad 1 - y > 0$$

$$\ln(1 - y) = -\frac{x^2}{2} - C$$

$$1 - y = e^{-x^2/2 - C}$$

$$-y = e^{-x^2/2 - C} - 1$$

$$y = -e^{-x^2/2 - C} + 1$$

$$= -e^{-x^2/2} \cdot e^{-C} + 1$$

Thus,

$$y = 1 + C_1 e^{-x^2/2}, \qquad \text{where } C_1 = -e^{-C}.$$

DO EXERCISE 4.

APPLICATION: ELASTICITY

Example 4 The elasticity of demand for a product is 1 for all $x > 0$; that is, $E(x) = 1$ for all $x > 0$. Find the demand function $p = D(x)$. (See Section 4.6.)

Solution Since $E(x) = 1$ for all $x > 0$,

$$1 = E(x) = -\frac{p}{x} \cdot \frac{1}{dp/dx}. \qquad (3)$$

Then

$$\frac{dp}{dx} = -\frac{p}{x}.$$

5. Find the demand function $p = D(x)$ given the elasticity condition

$$E(x) = 3 \quad \text{for all } x > 0.$$

Separating variables, we get

$$\frac{dp}{p} = -\frac{dx}{x}.$$

Now we integrate both sides:

$$\int \frac{dp}{p} = -\int \frac{dx}{x}$$

$$\ln p = -\ln x + C \qquad p > 0 \text{ since } E(x) = 1, x > 0; \text{ and } dp/dx < 0$$

since demand functions are decreasing. See Eq. (3).

We can express $C = \ln C_1$, since any real number C is the natural logarithm of some number C_1. Then

$$\ln p = \ln C_1 - \ln x$$

$$= \ln \frac{C_1}{x},$$

so

$$p = \frac{C_1}{x}.$$

This characterizes those demand functions for which the elasticity is always 1.

DO EXERCISE 5.

A Psychological Application: Reaction to a Stimulus

THE WEBER–FECHNER LAW

In psychology, one model of stimulus–response asserts that the rate of change dR/dS of the reaction R with respect to a stimulus S is inversely proportional to the stimulus. That is,

$$\frac{dR}{dS} = \frac{k}{S},$$

where k is some positive constant.

To solve this equation we first separate the variables:

$$dR = k \cdot \frac{dS}{S}.$$

We then integrate both sides:

$$\int dR = \int k \cdot \frac{dS}{S}$$

$$R = k \ln S + C. \tag{4}$$

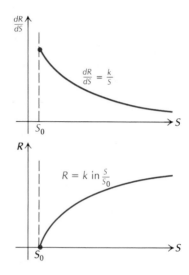

Now suppose we let S_0 be the lowest level of the stimulus that can be detected consistently. This is the *threshold value*, or *detection threshold*. For example, the lowest level of sound that can be consistently detected is the tick of a watch at 20 feet, under very quiet conditions. If S_0 is the lowest level of sound that can be detected, it seems reasonable that the reaction to it would be 0; that is, $R(S_0) = 0$. Substituting this condition in Eq. (4) we get

$$0 = k \ln S_0 + C,$$

or

$$-k \ln S_0 = C.$$

Replacing C in Eq. (4) by $-k \ln S_0$, we get

$$R = k \cdot \ln S - k \cdot \ln S_0$$
$$R = k(\ln S - \ln S_0).$$

Using a property of logarithms, we have

$$R = k \cdot \ln \frac{S}{S_0}.$$

6. *The Brentano–Stevens Law.* The Weber–Fechner Law has been the subject of great debate among psychologists as to its validity. The model

$$\frac{dR}{dS} = k \cdot \frac{R}{S},$$

where k is a positive constant, has also been conjectured and experimented with. Find the general solution to this equation. (This has also been referred to as the *Power Law of Stimulus–Response.*)

Look at the graphs of dR/dS and R shown earlier. Note that as the stimulus gets larger, the rate of change decreases; that is, the reaction becomes smaller as the stimulation you receive gets stronger. For example, suppose you are in a room with one lamp and that lamp has a 50-watt bulb in it. If the bulb were suddenly changed to 100 watts, you would probably be very aware of the difference. That is, your reaction would be strong. If the bulb were then changed to 150 watts, your reaction would not be as great as it was to the change from 50 to 100 watts. A change from a 150- to a 200-watt bulb would cause even less reaction, and so on.

For your interest, here are some other detection thresholds.

Stimulus	Detection threshold
Light	The flame of a candle 30 miles away on a dark night
Taste	Water diluted with sugar in the ratio of 1 teaspoon to 2 gallons
Smell	One drop of perfume diffused into the volume of three average-sized rooms
Touch	The wing of a bee dropped on your cheek at a distance of 1 centimeter (about $\frac{3}{8}$ of an inch)

DO EXERCISE 6.

EXERCISE SET 9.2

Solve.

1. $\dfrac{dy}{dx} = 4x^3y$

2. $\dfrac{dy}{dx} = 5x^4y$

3. $3y^2 \dfrac{dy}{dx} = 5x$

4. $3y^2 \dfrac{dy}{dx} = 7x$

5. $\dfrac{dy}{dx} = \dfrac{2x}{y}$

6. $\dfrac{dy}{dx} = \dfrac{x}{2y}$

7. $\dfrac{dy}{dx} = \dfrac{3}{y}$

8. $\dfrac{dy}{dx} = \dfrac{4}{y}$

9. $y' = 3x + xy,$
   $y = 5$ when $x = 0$

10. $y' = 2x - xy,$
    $y = 9$ when $x = 0$

11. $y' = 5y^{-2},$
    $y = 3$ when $x = 2$

12. $y' = 7y^{-2},$
    $y = 3$ when $x = 1$

13. $\dfrac{dy}{dx} = 3y$

14. $\dfrac{dy}{dx} = 4y$

15. $\dfrac{dP}{dt} = 2P$

16. $\dfrac{dP}{dt} = 4P$

17. **a)** Use separation of variables to solve the differential-equation model of uninhibited growth,

$$\frac{dP}{dt} = kP.$$

b) Rewrite the solution in terms of the condition $P_0 = P(0)$.

18. *Domar's capital expansion model is*

$$\frac{dI}{dt} = hkI,$$

where I = investment, h = investment productivity (constant), k = marginal productivity to consume (constant), and t = time.

a) Use separation of variables to solve the differential equation.

b) Rewrite the solution in terms of the condition $I_0 = I(0)$.

19. *Economics: Utility.* The reaction R, in pleasure units, by a consumer receiving S units of a product can be modeled by the differential equation

$$\frac{dR}{dS} = \frac{k}{S+1},$$

where k is a positive constant.

a) Use separation of variables to solve the differential equation.

b) Rewrite the solution in terms of the initial condition $R(0) = 0$.

c) Explain why the condition $R(0) = 0$ is reasonable.

20. *Newton's Law of Cooling.* The temperature T of a cooling object drops at a rate that is proportional to the difference $T - M$, where M is the constant temperature of the surrounding medium. Thus,

$$\frac{dT}{dt} = -k(T - M), \qquad \boxed{\text{CSS}}$$

where k is a positive constant and t is time.

a) Solve the differential equation.

b) Rewrite the solution in terms of the condition $T(0) = 200°$.

Elasticity. Find the demand function $p = D(x)$ given the following elasticity conditions.

21. $E(x) = \dfrac{200 - x}{x}$; $p = 190$ when $x = 10$

22. $E(x) = \dfrac{4}{x}$; $p = e^{-1}$ when $x = 4$

23. $E(x) = 2$ for all $x > 0$

24. $E(x) = k$ for some constant k and all $x > 0$

9.3

APPLICATIONS: THE INHIBITED GROWTH MODEL $dP/dt = kP(L - P)$

OBJECTIVES

a) State the differential equation

$$\frac{dP}{dt} = kP(L - P)$$

for the inhibited growth model.

b) State the solution to the inhibited growth model

$$P(t) = \frac{P_0 L}{P_0 + e^{-Lkt}(L - P_0)}.$$

(continued)

Recall the model of uninhibited growth

$$\frac{dP}{dt} = kP,$$

with solution (by separation of variables)

$$P = P_0 e^{kt}.$$

A more realistic model of population growth will take into account

c) Given the point of inflection, find the limiting value.

d) Given the values of the constants for the equation in part (b):

   i) write the equation for $P(t)$ in terms of the given constants;

   ii) sketch a graph of $P(t)$, given the time t at which the point of inflection occurs;

   iii) find t such that $P(t) = M$, for some number M.

e) Given certain constants and function values, find the value of k in the formula in part (b).

factors other than the size of the population. For example, there may be some compelling reason why the population P can never grow beyond a limiting value L—perhaps a limitation on food, living space, or other natural resources. In such cases we expect *growth rate* to lessen as the population size increases, and to approach 0 as P approaches L.

Example 1 Buy an ant colony at a pet shop. Place a fixed amount of food and water in the colony (enough to last a long period of time) and close it up. The growth rate of the ants is inhibited by the fixed amount of food and water.

The simplest model for such an inhibition to express itself is for the growth rate dP/dt to be directly proportional to *both* the population size P and to its remaining possible room for growth $L - P$. Then we have the following.

DEFINITION

The model of inhibited growth is

$$\frac{dP}{dt} = kP(L - P),$$

where $L = $ the *limiting value* and $k > 0$.

Work through the exploratory exercise set on the following page, either on your own or in class with your instructor.

Now let us solve the differential equation

$$\frac{dP}{dt} = kP(L - P).$$

Note that this equation defines P as a function of t implicitly. That is, it is assumed that P is a function of t even though t does not appear on the right-hand side of the equation. We first separate variables:

$$dP = kP(L - P)\, dt$$

$$\frac{dP}{P(L - P)} = k\, dt.$$

Now we integrate both sides. We use Table 1 to integrate the left side.

EXPLORATORY EXERCISES: SHIPWRECKED ON AN ISLAND

A ship carrying 100 passengers wrecks on an island, never to be rescued. The population grows over the next 120 years as indicated in the table.

Time, t (in years)	Population, P	Growth rate, $\Delta P/\Delta t$	Time, t (in years)	Population, P	Growth rate, $\Delta P/\Delta t$
0	100	0 (assumed)	70	470	
10	120	2	80	500	
20	150	3	90	520	
30	200		100	530	
40	270		110	535	
50	350		120	538	
60	420				

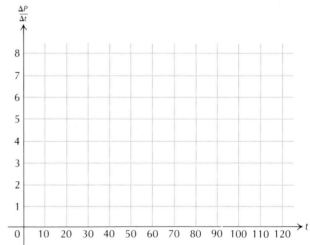

a) Complete the table.

b) Plot P versus t. Use solid dots and connect them with a smooth curve.

c) Plot $\Delta P/\Delta t$ versus t. Use open dots and connect them with a smooth curve.

d) Does P resemble the graph of an exponential function of the type $P(t) = P_0 e^{kt}$?

e) Between what values of t is $\Delta P/\Delta t$ increasing?

f) Between what values of t is $\Delta P/\Delta t$ decreasing?

g) Is there a point of inflection? Where?

h) Does there appear to be a limiting value to the population growth? If so, what is it?

1. Let $C = 1$, $L = 1$, and $k = 1$. Then Eq. (2) becomes

$$P(t) = \frac{e^t}{1 + e^t}.$$

We want to graph $P(t)$, but to ease computations we multiply by 1 as follows:

$$P(t) = \frac{e^t}{1 + e^t} \cdot \frac{e^{-t}}{e^{-t}}$$

$$= \frac{e^t \cdot e^{-t}}{1 \cdot e^{-t} + e^t \cdot e^{-t}}$$

$$P(t) = \frac{1}{e^{-t} + 1}.$$

a) 🖩 Complete this table of function values.

t	-3	-2	-1	0	1	2	3
$P(t)$							0.95

For example,

$$P(3) = \frac{1}{e^{-3} + 1} \approx \frac{1}{0.05 + 1}$$

$$= \frac{1}{1.05} \approx 0.95.$$

b) Graph $P(t)$.

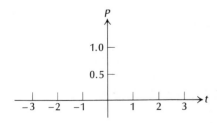

A similar example appears in Example 1 of Section 5.7.

$$\int \frac{dP}{P(L - P)} = \int k \, dt$$

$$\frac{1}{L} \ln \frac{P}{L - P} = kt + C$$

To solve for P we first find an expression for $P/(L - P)$:

$$\ln \frac{P}{L - P} = Lkt + LC$$

$$\frac{P}{L - P} = e^{Lkt + LC} = e^{Lkt} \cdot e^{LC}.$$

Now, letting $C_1 = e^{LC}$, we have

$$\frac{P}{L - P} = C_1 e^{Lkt}.$$

We complete the solution for P as follows, dropping the subscript from C_1 for simplicity:

$$P = Ce^{Lkt}(L - P)$$

$$P = CLe^{Lkt} - PCe^{Lkt}$$

$$P + PCe^{Lkt} = CLe^{Lkt}$$

$$P(1 + Ce^{Lkt}) = CLe^{Lkt}$$

$$P = \frac{CLe^{Lkt}}{1 + Ce^{Lkt}}.$$

THEOREM 1

The *model of inhibited growth*

$$\frac{dP}{dt} = kP(L - P) \qquad (1)$$

has the solution

$$P(t) = \frac{CLe^{Lkt}}{1 + Ce^{Lkt}}. \qquad (2)$$

DO EXERCISE 1.

For each value of the constant C we get an S-shaped curve. These curves fill up the strip $0 < P < L$.

2. Given

$$P = \frac{CLe^{Lkt}}{1 + Ce^{Lkt}} \qquad (1)$$

and

$$C = \frac{P_0}{L - P_0}. \qquad (2)$$

a) Replace C in Eq. (1) by the right-hand side of Eq. (2) and show your work in simplifying to get

$$P = \frac{P_0 L e^{Lkt}}{P_0 e^{Lkt} + (L - P_0)}. \qquad (3)$$

b) Simplify Eq. (3) by multiplying by e^{-Lkt}/e^{-Lkt}.

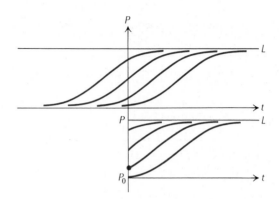

If we don't project these curves into the past (that is, if we don't use *negative time*), then the solution curves are confined to the part of the strip for which $t \geqslant 0$. For any initial population P_0, there is a growth curve emanating from $(0, P_0)$, showing the evolution of that population. That is, $P = P_0$ at $t = 0$ is a condition that picks out a unique solution. We can eliminate the constant C and express P in terms of P_0 as follows:

$$P_0 = P(0) = \frac{CLe^{Lk0}}{1 + Ce^{Lk0}} = \frac{CL}{1 + C}. \qquad (3)$$

Solving for C, we get

$$C = \frac{P_0}{L - P_0},$$

$$P(t) = \frac{P_0 L e^{Lkt}}{P_0 e^{Lkt} + (L - P_0)}. \qquad (4)$$

We can simplify this further by multiplying by e^{-Lkt}/e^{-Lkt}.

DO EXERCISE 2.

THEOREM 2

The *model of inhibited growth* (often called the *logistic function*)

$$\frac{dP}{dt} = kP(L - P)$$

has the solution

$$P(t) = \frac{P_0 L}{P_0 + e^{-Lkt}(L - P_0)}.$$

Recall that an inflection point is a point across which a curve changes concavity (see p. 180). The inflection point on the graph of $P(t)$ is important in application. It gives the time and population size at which the *rate* of population growth reaches a maximum and changes from an increasing function to a decreasing function. The following tells us where the inflection point of P occurs.

THEOREM 3

P **has an inflection point at that value of** t **where**

$$P = \tfrac{1}{2}L.$$

The following example is based on an actual experiment.*

Example 2   Before 1937 there were no pheasants on Protection Island off the coast of Washington State. In 1937 eight pheasants, two cocks and six hens, were released on the island. The following curve shows the growth of the pheasant population over the next six years.

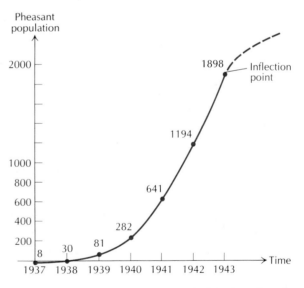

In 1943 the army arrived and started shooting the pheasants. It was theorized on the basis of the data that the population had reached

A ring-necked pheasant.

---

* A. S. Einarsen, "Some Factors Affecting Ring-necked Pheasant Population Density," *Murrelet*, **26** (1945): 2–9, 39–44.

3. The following shows the growth of an
ant colony with a point of inflection
at P_1. What is the limiting value?

its point of inflection. That is, the growth would stop turning upward
and start turning downward as it approached its limiting value. What
was the limiting value of the pheasant population?

Solution At the inflection point

$$P = \tfrac{1}{2}L.$$

Now $P = 1898$ from the graph, so

$$1898 = \tfrac{1}{2}L.$$
$$3796 = L.$$

DO EXERCISE 3.

The spread of an epidemic can also be described by the model of
inhibited growth.

Example 3 *Spread of an epidemic.* In a town whose total population
is 2000, the disease *Rottenich* creates an epidemic. The initial number
of people infected is 10. The rate of spread of the infection follows
the inhibited growth model

$$\frac{dP}{dt} = kP(L - P),$$

where

$P =$ the number of people infected after time t (in weeks),
$L = 2000$, the total population of the town,
$k = 0.003$, and
$P_0 =$ the number of people initially infected $= 10$.

a) Write the equation for $P(t)$ in terms of the given constants.
b) Sketch a graph of $P(t)$ given that the point of inflection occurs
at $t = 0.882$ week.
c) At what time t are 1600 people affected?

Solution

a) $P(t) = \dfrac{P_0 L}{P_0 + e^{-Lkt}(L - P_0)} = \dfrac{10 \cdot 2000}{10 + e^{-2000(.003)t}(2000 - 10)}$

$P(t) = \dfrac{20{,}000}{10 + 1990e^{-6t}}$

b) We first mark the limiting value 2000 on the vertical axis in the
figure and draw a horizontal line through it. Then we mark P_0,

4. A town whose population is 4000 is struck by the disease *Andromedamama*. The initial number of people infected is 20. The rate of spread of the infection follows the inhibited growth model

$$\frac{dP}{dt} = kP(L - P), \qquad \boxed{\text{CSS}}$$

where

P = the number infected after t (months),

L = 4000, the total population of the town,

k = 0.0005, and

P_0 = 20, the number of people initially infected.

a) Write the equation for $P(t)$ in terms of the given constants.

b) Sketch a graph of $P(t)$, given that the point of inflection occurs at $t = 2.9$ months. Use graph paper.

c) At what time t are 800 people infected?

or 10, on the vertical axis and mark the inflection point (0.882, 1000). Finally, we draw an S-shaped curve from P_0 through the inflection point and upward, approaching the limiting value.

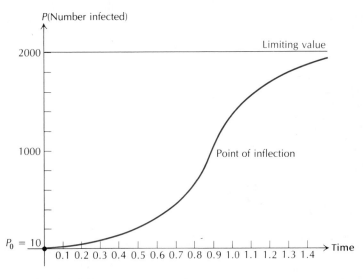

c) We solve the following equation for t:

$$1600 = \frac{20{,}000}{10 + 1990e^{-6t}}$$

$$1600(10 + 1990e^{-6t}) = 20{,}000$$

$$10 + 1990e^{-6t} = \frac{20{,}000}{1600} = 12.5$$

$$1990e^{-6t} = 2.5$$

$$e^{-6t} = \frac{2.5}{1990}.$$

We use natural logarithms to solve this equation:

$$\ln e^{-6t} = \ln \frac{2.5}{1990}$$

$$-6t = \ln 2.5 - \ln 1990$$

$$-6t = 0.916291 - 7.595890 \qquad \boxed{\text{▦}}$$

$$t = \frac{0.916291 - 7.595890}{-6}$$

$$t \approx 1.113 \text{ weeks.}$$

DO EXERCISE 4.

5. State the solution, $Q(y)$, to

$$\frac{dQ}{dy} = rQ(M - Q).$$

As you will see better after you have done the exercises in this section, and as you have already seen in the preceding example, the differential equation $dP/dt = kP(L - P)$ has many applications. The solution to the equation, however, can be used in all the applications. Of course, different variables or constants are used in different applications. For example, the solution to

$$\frac{dN}{dw} = cN(K - N)$$

is

$$N(w) = \frac{N_0 K}{N_0 + e^{-Kcw}(K - N_0)}.$$

DO EXERCISE 5.

Population Growth: Some Comments

Can a J-shaped curve (the graph of the equation $P = P_0 e^{kt}$) or an S-shaped curve be used to model world population growth? To find out let us look at a log–log graph* of population growth over the last one million years.

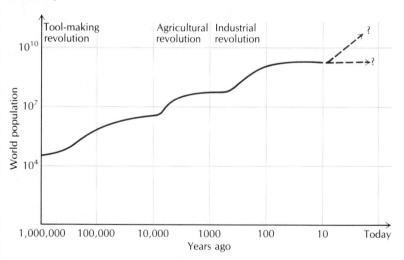

What happens is that we get a series of S-shaped curves. At three intervals there are surges of growth. These smaller intervals could be modeled by J-shaped curves. Following each surge there seems to be

---

* The axes are scaled using logarithms of the numbers in question. This condenses the graph.

a leveling off, but an advance in technology causes another surge. One comes after the *tool-making revolution*. Another comes after the *agricultural revolution,* and another after the *industrial revolution.* Judging from the graph, and from the current concerns about availability of food, energy, and other resources, we might speculate that we are in the midst of a leveling-off period when birth rate and death rate are about the same, or will be by the year 2000.

EXERCISE SET 9.3

For the inhibited growth model:

1. $P = 450$ at the point of inflection. What is the limiting value?

2. $P = 670$ at the point of inflection. What is the limiting value?

3. In an experiment* it was found that the rate of growth of yeast cells in a laboratory satisfied the inhibited growth model

$$\frac{dP}{dt} = kP(L - P),$$ CSS

where

P = the number of cells present after time t (in hours),

$L = 700$, the limiting value,

$k = 0.0008$, and

$P_0 = 10$.

a) Write the equation for $P(t)$ in terms of the given constants.

b) Sketch a graph of $P(t)$ given that the point of inflection occurs at $t = 8$.

c) At what time t are 500 cells present?

4. A company introduced a new product on a trial run in a city. They advertised the product on television and gathered data concerning the percentage P of people who bought the product in relation to the number of times the product was advertised on television. The graph of the data resembled an S-shaped curve, so they hypothesized that it satisfied the inhibited growth model

$$\frac{dP}{dt} = kP(L - P),$$

where

P = the percentage of people who buy the product,

t = the number of times the product is advertised on TV,

$L = 100\%$, or 1, the highest percentage of people who could buy the product,

$k = 0.13$, and

$P_0 = 2\%$, or 0.02, the percentage who bought the product without having seen an ad.

a) Write the equation for $P(t)$ in terms of the given constants.

b) Sketch a graph of $P(t)$ given that the point of inflection (50%) occurs at $t = 30$.

c) How many times t would the company have to advertise its product so that 80% of the people would buy?

---

* G. F. Gause, *The Struggle For Existence* (Baltimore: Williams and Wilkins, 1934).

In Exercises 5 and 6 you will study two problems concerning the *diffusion of information*. This is a sociological term for the spread of information such as a rumor, or news of some new fashion rage, or news of some new product or medicine. In a limited population L the rate of diffusion of the information is proportional to the number who have heard the information, N, and to the number who have not heard the information, $L - N$.

5. *Spread of a rumor.* In a college with a student population of 800, a group of 6 students spread the rumor "Men go for women who study calculus, and women go for men who study calculus!" The rate at which the rumor spreads satisfies the differential equation

$$\frac{dN}{dt} = kN(L - N),$$ `CSS`

where

   N = the number of people who have heard the rumor after t minutes,

   $L = 800$, and

   $N_0 = 6$, the number of people who started the rumor.

   a) Write the equation for $N(t)$, the solution to the differential equation, in terms of the given constants. Note that we do not yet know k.

   b) The point of inflection occurs at $t = 12.2$; that is, $N(12.2) = 400$. Use this to find k.

   c) After what time t have 700 students heard the rumor?

6. *Acceptance of hybrid corn.* The rate of acceptance of hybrid corn in Kentucky* followed the differential equation

$$\frac{dC}{dt} = kC(L - C),$$

where

   C = the percentage of corn acreage in hybrid at time t in years (after 1938),

   $L = 100\%$, or 1, and

   $C_0 = 3\%$, the percent of corn acreage in hybrid at $t = 0$ (1938).

   a) Write the equation for $C(t)$, the solution to the differential equation, in terms of the given constants. Note again that we do not yet know k.

   b) The point of inflection occurs at $t = 7$; that is, $C(7) = 50\%$, or 0.5. Use this to find k.

   c) After what time t is 80% of the acreage in hybrid corn?

Agricultural Statistics (U.S. Department of Agriculture, Washington, D.C., 1961).

The acceptance of hybrid corn in Kentucky followed the uninhibited growth model.

EXPLORATORY EXERCISES: A PSYCHOLOGY EXPERIMENT ON PERCEPTION

PREPARATION

a) Find a 100-gram ("gram" is abbreviated "g") weight to be used as a *standard*.

b) Find rocks (or other objects) that weigh, respectively, about 85 g, 90 g, 95 g, 105 g, 110 g, and 115 g.

CARRYING OUT THE EXPERIMENT

The experiment can be performed by the instructor or a student on the rest of the class.

a) Arrange the rocks in random order. Pick one. Hand the standard and the rock to each member of the class and ask the question, "Is the rock heavier than the standard?"

b) Repeat part (a) with each rock and keep track of the responses in a table such as the one shown here.

c) Plot the data points using the axes below and connect them with a smooth curve.

Rock weight (g)	Students						% yes
	1	2	3	4	5	...	
105	Yes	No	Yes				80
85	No	No	No				5
110							
100							
90							
95							
115							

d) Answer the following discussion questions.

1. Does the graph resemble an S-shaped curve?

2. Does a small percent of the class respond "yes" when handed the 85-g rock? Does this seem reasonable?

3. Does a large percent of the class respond "yes" when handed the 115-g rock? Does this seem reasonable?

4. What percent responds "yes" to the 100-g rock? Does this seem reasonable? Why?

5. How could the experiment be repeated to improve the shape of the curve?

7. A *perception* experiment in psychology is done as in the exploratory exercises on the preceding page. If you have not done the experiment, you should read it over. The percentage P who respond "yes" to a rock of weight w satisfies the differential equation

$$\frac{dP}{dw} = kP(L - P),$$

where

P = % who respond "yes" to a rock of weight w,

L = 100%, or 1, and

P_{85} = 4%, percent who respond "yes" to the weight of w = 85 g (gram).

Look the graph over. Doesn't it seem reasonable that only 4% would have judged the 85-g rock to be heavier than the 100-g rock? As the weight of the comparison rocks gets to the standard, the percent increases. When the comparison weighs 100 g, the same as the standard, 50% say it is heavier. That is, half would say it is heavier and half would say it isn't. Then as the comparison rock gets heavier and heavier, the percent of "yes" responses increases toward 100%.

a)   Write the equation for $P(w)$, the solution to the differential equation, in terms of the given constants. Note that $w = 0$ means 85 g, $w = 1$ means 90 g,

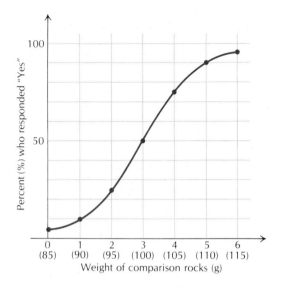

$w = 2$ means 95 g, and so on. Note that we have not yet determined k.

b)   Use the fact that $P = 50\% = 0.5$ at $w = 3$ (100 g) to find k.

c)   Rewrite the equation for $P(w)$ in terms of k and the other constants.

d)   At what weight w do 80% respond "yes"?

9.4

OBJECTIVES

a) State the solution to the growth model

$$\frac{dP}{dt} = k(L - P).$$

b) Use a calculator to graph equations of the type

$$P(t) = L(1 - e^{-kt}).$$

(continued)

APPLICATIONS: THE GROWTH MODEL $dP/dt = k(L - P)$

We have considered two models of growth:

$$\frac{dP}{dt} = kP, \qquad \text{Uninhibited growth model}$$

$$\frac{dP}{dt} = kP(L - P). \qquad \text{Inhibited growth model}$$

A third kind of growth assumes a limited population L, but assumes that the growth rate dP/dt is directly proportional only to its remaining possible room for growth $L - P$, independent of the number who al-

c) Given L and a function value for the equation in part (b), find k.

d) Given L and k for the equation in part (b), find a certain function value $P(t)$.

e) Given L and k for the equation in part (b), find t such that $P(t) = M$, for some given number M.

1. *Exploratory exercises: Advertising.* A company introduces a new product on a trial run in a city. It advertises the product on TV and gathers data concerning the percentage P of the people in the city who bought the product after it was advertised a certain number of times.

    a) The following table contains the data. Complete the calculations in the table.

t (Number of times ad ran)	P (% who bought product)	$\Delta P/\Delta t$ (% rate of change)
0	0	undefined
10	50	5
20	75	
30	87	
40	94	
50	97	

    b) Plot P versus t. Connect the points with a smooth curve.

(Exercise continued on next page.)

ready exist. The model is

$$\frac{dP}{dt} = k(L - P), \qquad \text{where } P = 0 \text{ when } t = 0.$$

This model has not proved suitable to describe biological population growth. Thus we will not consider that here. It is applicable, however, to other areas. Before solving the differential equation we consider one such application in Margin Exercise 1.

DO EXERCISE 1.

We have considered a problem similar to this before in Exercise 4 of Exercise Set 9.3 where we encountered an S-shaped curve. In Margin Exercise 1 the graph did not resemble an S-shaped curve. In most advertising applications a curve like that in Margin Exercise 1 is what occurs. Models can vary, of course.

Now let us solve the differential equation

$$\frac{dP}{dt} = k(L - P).$$

We first separate the variables:

$$\frac{dP}{L - P} = k \, dt.$$

Now we integrate both sides:

$$\int \frac{dP}{L - P} = \int k \, dt$$

$$-\ln (L - P) = kt + C$$

$$\ln (L - P) = -kt - C$$

$$L - P = e^{-kt - C} = e^{-kt} \cdot e^{-C}.$$

Letting $C_1 = e^{-C}$, we have

$$L - P = C_1 e^{-kt}$$

$$-P = -L + C_1 e^{-kt}$$

$$P = L - C_1 e^{-kt}.$$

Now $P = 0$ when $t = 0$, so

$$0 = L - C_1 e^{-k \cdot 0}$$

$$= L - C_1 \cdot 1 = L - C_1.$$

c) Plot $\Delta P/\Delta t$ versus t. Connect the points with a smooth curve.

d) Does P resemble an S-shaped curve?

e) Does there seem to be a point of inflection for P?

f) What appears to be the limiting value for P?

g) Does the slope of the graph in part (b) appear to be increasing? decreasing? neither?

2. Let $L = 3$ and $k = 1$, so   **CSS**

$$P(t) = 3(1 - e^{-t}).$$

a) 📱 Complete this table of function values.

t	0	1	2	3	4	5
$P(t)$			2.6			

For example,

$$P(2) = 3(1 - e^{-2})$$
$$\approx 3(1 - 0.135335)$$
$$\approx 2.6.$$

(Exercise continued on next page.)

Thus $L = C_1$, and

$$P = L - Le^{-kt}$$
$$= L(1 - e^{-kt}).$$

THEOREM 4

The solution to

$$\frac{dP}{dt} = k(L - P), \qquad P(0) = 0 \tag{1}$$

is

$$P(t) = L(1 - e^{-kt}). \tag{2}$$

DO EXERCISE 2.

As Margin Exercise 2 illustrates, the following is a graph of $P(t)$. Note that the graph starts at $(0, 0)$ and increases upward toward the limiting value L. The slope is decreasing and there is no point of inflection.

Now let us reconsider the problem on advertising in Margin Exercise 1.

Example 1   Reread Margin Exercise 1.

a) Suppose we want to "fit" Eq. (2) to these data. We know that $L = 100\%$, or 1, so

$$P(t) = 1 - e^{-kt}.$$

b)   Graph $P(t)$.

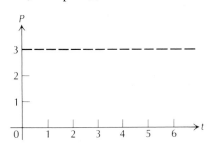

3. a)   In another advertising experiment it was determined that $L = 100\%$, so

$$P(t) = 1 - e^{-kt}.$$

One of the data points was

$$P(10) = 40\%, \quad \text{or} \quad 0.4.$$

Use this value to determine k. Round to the nearest hundredth.

b)   Rewrite $P(t)$ in terms of k.

c)   Use the equation in part (b) to find $P(60)$.

d)   How many times must the product be advertised so that 99% of the people will buy it?

Thus we must determine a value for k. A way to do this is to arbitrarily pick a data point, substitute, and solve for k. We use $(10, 50\%)$; that is, $P(10) = 50\%$, or 0.50. Substituting and solving, we have

$$0.5 = 1 - e^{-k \cdot 10}$$
$$-0.5 = e^{-10k}$$
$$0.5 = e^{-10k}.$$

We use natural logarithms to solve this equation:

$$\ln 0.5 = \ln e^{-10k}$$
$$-0.693147 = -10k$$
$$0.0693147 = k$$
$$0.07 \approx k.$$

b)   We then rewrite $P(t)$ in terms of k:

$$P(t) = 1 - e^{-0.07t}.$$

c)   Use the equation in part (b) to find $P(50)$:

$$P(50) = 1 - e^{-0.07 \cdot 50}$$
$$= 1 - e^{-3.5}$$
$$= 1 - 0.030197$$
$$\approx 0.9698.$$

Note that this is close to the data value of 97%. This may not always happen in practice.

d)   How many times must the product be advertised so that 99% of the people will buy it? We solve

$$0.99 = 1 - e^{-0.07t}$$

for t and get $t \approx 66$.

DO EXERCISE 3.

EXERCISE SET 9.4

Graph.

CSS

1. $P(t) = 4(1 - e^{-t})$ 2. $P(t) = 2(1 - e^{-t})$ 3. $P(t) = 1 - e^{-2t}$ 4. $P(t) = 1 - e^{-3t}$

APPLICATIONS

5. In a different advertising experiment it was determined that $L = 50\%$, or 0.5; that is, no matter how much a company advertised, the percentage of people who bought the product would never reach or exceed 50%.

   a) Express $P(t)$ in terms of $L = 0.5$.

   b) It was determined that $P(10) = 25\%$, or 0.25. Use this to determine k. Round to the nearest hundredth.

   c) Rewrite $P(t)$ in terms of k.

   d) Use the equation in part (c) to find $P(30)$.

   e) How many times must the product be advertised so that 49% of the people will buy it?

6. *Biomedical: Acceptance of a new medicine.* A different model for the diffusion of information—for example, awareness of a new medicine by doctors—is the differential equation

$$\frac{dP}{dt} = k(L - P),$$

where

$P =$ the percentage of doctors who are aware of the new medicine after time t (in months),

$L = 100\%$, or 1, and

$k =$ a constant.

   a) Write the equation for $P(t)$, the solution to the differential equation, in terms of L and k.

   b) Suppose $P(4) = 50\%$, or 0.5. Use this to determine k. Round to the nearest tenth.

   c) Rewrite $P(t)$ in terms of k.

   d) Use the equation in part (c) to find $P(6)$.

   e) How many months will it take for 90% of the doctors to become aware of the new medicine?

7. *Psychology: Hullian model of learning.* This model is expressed by

$$\frac{dP}{dt} = k(L - P),$$

where

$P =$ the probability of mastery of a certain concept after t learning trials,

$L = 1$, the limiting value of mastery of the learning, and

$k =$ a constant.

   a) Write the solution $P(t)$ in terms of L and k.

   b) Suppose $P(5) = 0.6$. Use this to determine k. Round to the nearest hundredth.

   c) Rewrite $P(t)$ in terms of k.

   d) Use the equation in part (c) to find $P(10)$.

   e) Find t such that $P(t) = 0.90$. That is, how many trials are necessary in order for the probability of mastery of learning to be 0.90?

8. *Biology: Yield due to the spread of fertilizer.* A farmer is growing soybeans in a field. The more fertilizer he spreads, the greater his yield, up to some limiting value L. A model for this, applied to any given specific farm, is

$$\frac{dY}{dn} = k(L - Y),$$

where

Y = the yield, in bushels per acre, upon spreading n pounds of fertilizer per acre,

$L = 60$ bushels per acre, and

k = a constant.

a) Write the solution $Y(n)$ in terms of L and k.

b) Suppose $Y(10) = 21$. Use this to determine k. Round to the nearest hundredth.

c) Rewrite $Y(n)$ in terms of k.

d) Use the equation in part (c) to find $Y(5)$.

e) How many pounds of fertilizer, per acre, must be spread to yield 42 bushels per acre?

(*Note:* This model is not completely valid in the sense that there usually is a certain amount of fertilizer that will kill the crop.)

9. Inside nuclear reactors, radioactive material is produced that satisfies

$$\frac{dM}{dt} = k\left(\frac{p}{k} - M\right), \qquad M = 0 \text{ when } t = 0,$$

where

M = the mass of radioactive material inside the reactor after time t (in years),

p = the rate at which the radioactive material is produced, and

k = the annual decay rate of the radioactive material.

Find the solution, $M(t)$, to the differential equation in terms of p and k.

The yield due to fertilizer can be modeled by $Y' = k(L - Y)$.

Nuclear power station at Rowe, Massachusetts, in operation since 1960.

10. The *growth rate of a certain stock* is modeled by

$$\frac{dV}{dt} = k(L - V), \qquad V = \$20 \text{ when } t = 0,$$

where

V = the value of the stock, per share, after time t (in months),

L = \$24.81, the limiting value of the stock, and

k = a constant.

Find the solution to the differential equation in terms of L and k.

11. *Fick's law in biology* is given by

$$\frac{dC}{dt} = \frac{kA}{V}(C_T - C),$$

where

C = the amount of a substance that passes through a tube with cross-sectional area A and volume V at time t,

k = a constant, and

C_T = the total amount of substance.

Find the solution to the differential equation assuming $C(0) = C_0$.

9.5

OBJECTIVES

a) Determine whether or not a given first-order differential equation is exact.

b) Solve an exact first-order differential equation.

EXACT EQUATIONS

Consider the differential equation

$$y' = \frac{dy}{dx} = \frac{-3x^2 - y}{x + 1}.$$

Although this equation appears to be a likely candidate for separation of variables, a little experimentation will convince you that this cannot be done.

Instead, we try the following approach. We write the equation in *differential form*. This means that we clear the equation of "fractions" by formally multiplying both sides by $(x + 1)\, dx$. This gives us

$$(x + 1)\, dy = (-3x^2 - y)\, dx,$$

or, rewriting,

$$(3x^2 + y)\, dx + (x + 1)\, dy = 0.$$

Now consider the function $F(x, y) = x^3 + xy + y$. Its partial derivatives are $F_x = 3x^2 + y$ and $F_y = x + 1$. The equation above is just

$$F_x\, dx + F_y\, dy = 0.$$

If this last expression looks familiar, it's because we studied the total differential in Chapter 7. We can rewrite this equation as

$$dF = 0,$$

1. Check by substitution that

$$y = \frac{C - x^3}{x + 1}$$

satisfies the differential equation

$$y' = \frac{dy}{dx} = \frac{-3x^2 - y}{x + 1}.$$

where we have used the fact that

$$dF = F_x \, dx + F_y \, dy.$$

Since $dF = 0$ when $F(x, y) = C$, the last equation suggests that the solution is $F(x, y) = C$. That turns out to be correct. We get

$$x^3 + xy + y = C$$

or, solving for y,

$$y(x + 1) = C - x^3$$

$$y = \frac{C - x^3}{x + 1}.$$

Now that we have the solution, we can check it by direct substitution.

DO EXERCISES 1 AND 2.

Unfortunately, it's not always possible to take a first-order differential equation, write it in differential form, and then find a function F such that the equation becomes $dF = 0$.

2. Write the differential equation

$$y' = \frac{dy}{dx} = \frac{-x}{y}$$

in differential form.

DEFINITION

The first-order differential equation

$$P(x, y) \, dx + Q(x, y) \, dy = 0$$

is *exact* if there is a function $F(x, y)$ such that

$$F_x = \frac{\partial F}{\partial x} = P(x, y) \quad \text{and} \quad F_y = \frac{\partial F}{\partial y} = Q(x, y).$$

In other words, $P(x, y) \, dx + Q(x, y) \, dy$, or simply $P \, dx + Q \, dy = 0$, is exact if we can solve it just as we solved the example above. The following theorem tells us when an equation is exact.

THEOREM 5

If $P(x, y)$ and $Q(x, y)$ have continuous partial derivatives for all (x, y) in a rectangle R, then

$$P(x, y) \, dx + Q(x, y) \, dy = 0$$

is *exact* if and only if

$$P_y = Q_x.$$

3. Show that

$$x\, dx + y\, dy = 0$$

is exact.

Example 1   Show that the equation

$$2xy\, dx + x^2\, dy = 0$$

is exact.

Solution   Here $P(x, y) = 2xy$ and $Q(x, y) = x^2$. So,

$$\frac{\partial P}{\partial y} = P_y = 2x \qquad \text{and} \qquad \frac{\partial Q}{\partial x} = Q_x = 2x.$$

Since $P_y = Q_x$, the equation is exact.

DO EXERCISES 3–5.

4. Show that

$$(6xy + 2y)\, dx + (3x^2 + 2x + 1)\, dy = 0$$

is exact.

Example 2   Solve $x\, dx + y\, dy = 0$.

Solution   We know from Margin Exercise 3 that this equation is exact. The problem here is to find $F(x, y)$. Now, $F(x, y)$ must satisfy $F_x = x$ and $F_y = y$. Consider the first equation. We can antidifferentiate it to get $F(x, y) = \frac{1}{2}x^2 + C$. To be correct, however, recall that since F is a function of two variables and since we are integrating with respect to x (and thinking of y as a constant), the constant C above may involve y-terms. We write

$$F(x, y) = \tfrac{1}{2}x^2 + g(y),$$

where g is an unknown function. To find g, recall that F must also satisfy $F_y = y$. Here, $F_y = \partial F/\partial y = g'(y)$. Thus $g'(y) = y$, and, anti-differentiating, we have $g(y) = \frac{1}{2}y^2 + C_1$, where C_1 is a constant. So we find that

5. **a)**   "Simplify" the equation

$$2xy\, dx + x^2\, dy = 0$$

by dividing by x.

b)   Show that the resulting equation is *not* exact.

$$F(x, y) = \tfrac{1}{2}x^2 + \tfrac{1}{2}y^2 + C_1.$$

The solution to

$$x\, dx + y\, dy = 0,$$

then, is given by $\frac{1}{2}x^2 + \frac{1}{2}y^2 + C_1 = C$, where C is a constant. We can simplify this result by subtracting C_1 from both sides and multiplying by 2. We then get

$$x^2 + y^2 = K,$$

where $K = 2(C - C_1)$ is a constant. The solution curves are a family of concentric circles, each centered at $(0, 0)$.

Note, in retrospect, that the constant C_1 was irrelevant and could have been left out. We will do so in the future.

6. Solve the exact equation in Example 1.

7. Solve $y\,dx + x\,dy = 0$.

8. Solve the equation in Margin Exercise 4.

9. Use the quadratic formula to solve
$$x^2y^2 - xy + 5x = 7$$
for y.

10. Solve the initial-value problem
$$2xy^3\,dx + 3x^2y^2\,dy = 0, \qquad y(1) = 2.$$

11. Solve the initial-value problem
$$(1 + 2y)\,dx + 2x\,dy = 0, \quad y(1) = -1.$$

DO EXERCISES 6–8.

Example 3 Solve the initial-value problem
$$(2xy^2 - y + 5)\,dx + (2x^2y - x)\,dy = 0, \qquad y(1) = 2.$$

Solution We first solve the equation just as we would if $y(1) = 2$ were not given. After solving the equation, we'll use the initial condition to eliminate the constant of integration.

Here, $P(x, y) = 2xy^2 - y + 5$ and $Q(x, y) = 2x^2y - x$. Testing for exactness, we find that $P_y = 4xy - 1$ and $Q_x = 4xy - 1$. The equation is exact, since $P_y = Q_x$.

To find $F(x, y)$, we recall that $F_x = P = 2xy^2 - y + 5$ and integrate with respect to x. We get

$$\begin{aligned} F(x, y) &= \int P(x, y)\,dx \\ &= \int (2xy^2 - y + 5)\,dx \\ &= x^2y^2 - xy + 5x + g(y). \end{aligned}$$

So $\partial F/\partial y = F_y = 2x^2y - x + g'(y)$. Since we also must have $F_y = Q = 2x^2y - x$, we see that $g'(y) = 0$. So, $g(y) = C_0$, a constant, and may be taken to be 0. Thus

$$F(x, y) = x^2y^2 - xy + 5x \quad \text{and} \quad x^2y^2 - xy + 5x = C$$

gives the solution.

To evaluate C, note that $y = 2$ when $x = 1$. Direct substitution above yields

$$(1)^2(2)^2 - 1 \cdot 2 + 5 \cdot 1 = C$$

or

$$\begin{aligned} C &= 1 \cdot 4 - 2 + 5 \\ &= 9 - 2 \\ &= 7. \end{aligned}$$

Therefore,

$$x^2y^2 - xy + 5x = 7.$$

It isn't necessary to do so, but you may solve the last equation for y by regarding $a = x^2$, $b = -x$, and $c = 5x - 7$ and applying the quadratic formula to $ay^2 + by + c = 0$.

DO EXERCISES 9–11.

EXERCISE SET 9.5

Determine whether the given equation is exact.

1. $(2x + 3) \, dx + 6y \, dy = 0$

2. $(2y + 3) \, dx + 5x \, dy = 0$

3. $(6y + 3) \, dx + 6x \, dy = 0$

4. $\sin y \, dx + \cos x \, dy = 0$

5. $\sin y \, dx + x \cos y \, dy = 0$

6. $\sin x \, dx + y \cos x \, dy = 0$

7. $\left(y^3 e^x + \dfrac{y}{x} \right) dx + (3y^2 e^x + 2y + \ln x) \, dy = 0$

8. $(4y^5 x^3 - 6xy^2) \, dx + (5x^4 y^4 - 6x^3 y) \, dy = 0$

9. Solve the equation in Exercise 1.

10. Solve the equation in Exercise 3.

11. Solve the equation in Exercise 5.

12. Solve the equation in Exercise 7.

Show that the following equations are exact. Then solve the initial-value problem.

13. $(3x^2 + 3y) \, dx + (3x + 3y^2) \, dy = 0, y(1) = 1$

14. $2xy \, dx + (1 + x^2) \, dy = 0, y(1) = \frac{1}{2}$

15. $(\cos y - y \sin x) \, dx + (\cos x - x \sin y) \, dy = 0, y(0) = 1$

16. $(x^3 + 2xy) \, dx + (x^2 + y^3) \, dy = 0, y(2) = 2$

17. Find the equation of the curve that satisfies the differential equation

$$\frac{dy}{dx} = \frac{y - x}{2y - x}$$

and passes through $(1, 1)$.

18. *Business.* A company finds that if p is the price per unit (in dollars) of its goods and if x represents the quantity of goods sold, then the elasticity of demand is

$$E(x) = \frac{1 + x^2}{1 + 3x^2}.$$

Use the fact that

$$E = -\frac{p}{x} \cdot \frac{dx}{dp}$$

(see Section 4.6) to set up and solve a differential equation for p, given that $p(2) = \$0.10$.

19. *Business.* A company discovers that the price P of one of its products and the number of units u of the product demanded by consumers at that price are related (for a certain range of values) by the equation

$$\frac{dP}{du} = \frac{-3P}{2P + 3u}.$$

If $P = 5$ when $u = 5$, find P when $u = 16$.

20. *Business.* A firm's average cost per unit A (in dollars) of producing x thousands of units of a certain product satisfies the equation

$$\frac{dA}{dx} = \frac{4x - 3A}{3x + 7}.$$

Find A as a function of x if $A(1) = 1$.

EXTENSION EXERCISE

21. Show that the substitution $y = vx$ in the differential equation in Exercise 17 changes it to a problem that can be solved by separation of the variables.

9.6

FIRST-ORDER LINEAR EQUATIONS

OBJECTIVE

a) Solve any first-order linear differential equation.

We begin this section with the following definition.

DEFINITION

A first-order differential equation is called *linear* if it can be written in the form

$$a_0(x)y' + a_1(x)y = b(x). \qquad (1)$$

Note that $a_0(x)$ and $a_1(x)$, which are called the *coefficients* of the equation, are functions of x alone. So too is the function $b(x)$ on the right-hand side of the equation. So, for example, the equation

$$(x^2 + 1)y' - xy = e^x$$

is a first-order linear differential equation with $a_0(x) = x^2 + 1$, $a_1(x) = -x$, and $b(x) = e^x$. On the other hand, $(y')^3(y) - e^y = 0$ is *not* a linear equation.

In this text, we will always assume that the coefficients $a_0(x)$ and $a_1(x)$ and the function $b(x)$ (sometimes called the *forcing function*) are all continuous on some closed interval $[c, d]$. This guarantees that Eq. (1) has a solution on $[c, d]$.

Example 1 Solve $xy' - 3x^3y = 0$.

Solution Comparing this equation with the definition above, we have $a_0(x) = x$, $a_1(x) = -3x^3$, and $b(x) = 0$. (Equations in which $b(x) = 0$ are called *homogeneous*.) Dividing through by x, we have

$$y' - 3x^2y = 0.$$

Using $y' = dy/dx$, we have

$$\frac{dy}{dx} = 3x^2y.$$

We see now that we can separate the variables in this equation. Doing so, we obtain

$$\frac{dy}{y} = 3x^2 \, dx.$$

1. Solve $xy' - y = 0$.

Integrating both sides, we have

$$\int \frac{dy}{y} = \int 3x^2 \, dx$$

$$\ln y = x^3 + C.$$

Applying the definition of logarithms to $\ln y = \log_e y = x^3 + C$ gives us

$$y = e^{x^3 + C}$$

or

$$y = e^{x^3} e^C$$

or

$$y = C_1 e^{x^3}, \quad \text{where } C_1 = e^C.$$

DO EXERCISE 1.

It turns out that solving an arbitrary *homogeneous* first-order differential equation is no harder than solving Example 1 or Margin Exercise 1. That is, every homogeneous linear first-order equation can be solved by separating the variables (see Exercise Set 9.6). Unfortunately, though, separation of variables usually doesn't work in the nonhomogeneous case. Nor is such an equation exact. Consider, for example,

$$a_0(x)y' + a_1(x)y = b(x).$$

Dividing through by $a_0(x)$ yields

$$y' + \frac{a_1(x)}{a_0(x)} y = \frac{b(x)}{a_0(x)},$$

where we assume that $a_0(x) \neq 0$. For simplicity, let

$$p(x) = \frac{a_1(x)}{a_0(x)} \quad \text{and} \quad q(x) = \frac{b(x)}{a_0(x)}.$$

Then our equation becomes

$$y' + p(x)y = q(x).$$

This is sometimes called the *standard form* of the equation. Now, letting $y' = dy/dx$ and rearranging gives

$$p(x)y - q(x) + \frac{dy}{dx} = 0$$

4. Solve $x^2 y' + 4xy = 8x$.

Now the left-hand side is just the derivative of $x^2 y$. That is, we have

$$(x^2 y)' = 5x^4.$$

Integrating, we have $x^2 y = x^5 + C$, where C is a constant of integration. So, solving for y, we now have

$$y = x^3 + \frac{C}{x^2} = x^3 + Cx^{-2},$$

valid for $x > 0$. Note that the solution is undefined at $x = 0$. We knew that our solution might not be valid there, since we had to divide through by x in the first step of the solution.

DO EXERCISES 4 AND 5.

Example 4   *Mixing*. A tank initially contains a brine solution consisting of 400 pounds of salt dissolved in 1200 gallons of water. Salt water containing 1 pound of salt per gallon of water is pumped into the tank at the rate of 5 gallons per minute. The brine is pumped out at the rate of 5 gallons per minute. Assuming that the mixture is kept uniform by stirring, find the amount of salt in the tank after 5 hours.

5. Solve $y' + 2y = e^{3x}$.

Solution   Let $A(t)$ be the amount of salt (in pounds) that is in the tank after t minutes. We are given that $A(0) = 400$ and we want to find $A(300)$. We will find this by setting up and then solving a differential equation for $A(t)$. Note that the expression $A'(t) = dA/dt$ represents the instantaneous rate of change (the rate of increase or decrease) of the amount of salt in the tank at time t.

Since salt water containing 1 pound per gallon is being pumped into the tank at a rate of 5 gallons per minute, we see that salt is entering the tank at the rate of 5 pounds per minute.

The mixture is being pumped out of the tank at the rate of 5 gallons per minute. How much salt does each gallon of the mixture contain? At time t, there are $A(t)$ pounds of salt dissolved in 1200 gallons of water. The concentration of salt (in pounds per gallon) at time t is $A(t)/1200$. Since 5 gallons per minute are being pumped out, this means that

$$\frac{A(t)}{1200} \cdot 5 = \frac{A(t)}{240}$$

pounds per minute are being pumped out.

Now the rate of change of salt in the tank, $dA/dt = A'(t)$, is just the difference between the rate at which salt enters the tank and the rate at which it leaves the tank at time t. That is,

$$\frac{dA}{dt} = \text{(Rate entering)} - \text{(Rate leaving)}$$

$$\frac{dA}{dt} = 5 - \frac{A(t)}{240}.$$

Rearranging, we have

$$A'(t) + \frac{A(t)}{240} = 5,$$

a first-order linear equation. To solve it, note that the integrating factor is $e^{\int (1/240)\, dt} = e^{t/240}$. The equation becomes

$$(e^{t/240} A)' = 5e^{t/240},$$

or, after integrating,

$$e^{t/240} A = 1200 e^{t/240} + C$$

or

$$A(t) = 1200 + Ce^{-t/240}.$$

Recall that $A(0) = 400$. Letting $t = 0$ and $A = 400$ above, we get

$$400 = 1200 + Ce^{-0/240}$$

or

$$400 = 1200 + Ce^{0}$$

and

$$C = -800.$$

6. Show that the equation

$$A'(t) = 5 - \frac{A(t)}{240}$$

from Example 4 can be solved by separating the variables.

7. Find the answer to the question in Example 4 if the solution pumped into the tanks contains only 0.6 pound of salt per gallon.

So

$$A(t) = 1200 - 800e^{-t/240}.$$

In particular,

$$A(300) = 1200 - 800e^{-300/240}$$
$$= 1200 - 800e^{-1.25}$$
$$= 1200 - 800(0.2865048)$$
$$\approx 1200 - 229.204$$
$$\approx 970.796 \text{ pounds.}$$

DO EXERCISES 6 AND 7.

EXERCISE SET 9.6

Solve by separating the variables.

1. $y' - 3y = 0$

2. $y' - 2xy = 0$

3. $y' + \frac{3}{x}y = 0$

4. $y' - 2xy = 2x$

In Exercises 5–8, show that the given equation is not exact. Then find an integrating factor $m(x)$ for the equation. Don't actually solve the equation.

5. $y' - 3y = e^{5x}$

6. $y' + \frac{4}{x}y = \sin x$

7. $y' - \frac{5}{x^2}y = x^3$

8. $y' + \frac{2x}{x^2+1}y = 3x + 2$

Solve by finding an appropriate integrating factor.

9. $y' - 3y = 0$

10. $y' + 4y = 0$

11. $y' - 2xy = 0$

12. $y' + \frac{3}{x}y = 0$

13. $y' + 5y = 0, y(0) = 7$

14. $xy' + y = 0, y(1) = 3$

15. $y' - 2xy = 2x$

16. $y' + \frac{3}{x}y = 6x^2 - 4$

17. $y' + \frac{2x}{1+x^2}y = x$

18. $y' - \frac{\sin x}{\cos x}y = \frac{1}{\cos x}$

19. $x^2y' + xy = x^4, y(2) = 5$

20. $y' + \frac{3x^2}{(1+x^3)}y = 1, y(0) = 2$

21. A tank initially contains 180 pounds of salt dissolved in 500 gallons of water. Salt water containing 0.5 pound of salt per gallon enters the tank at the rate of 3 gallons per minute. The mixture (kept uniform by stirring) is removed at the same rate. How many pounds of salt are in the tank after an hour? a day?

22. A 2000-gallon tank initially contains 200 pounds of salt dissolved in 500 gallons of water. Pure water is pumped into the tank at the rate of 5 gallons per minute, while the (well-stirred) mixture is drawn off at the rate of 2 gallons per minute. The process stops when the tank is full. How much salt is left in the tank when the tank is full?

23. *Newton's Law of Cooling.* The temperature T of a cooling object drops at a rate that is proportional to the difference $T - M$, where M is the constant temperature of the surrounding medium. Thus,

$$\frac{dT}{dt} = -k(T - M),$$

where k is a positive constant and t is time.

a) Use the technique of this section to solve the differential equation.

b) If a metal sphere that has been heated to 143°F is observed to cool to 117° in 30 minutes in a room whose temperature is kept constant at 70°, find the temperature of the sphere at any later time t. How long will it take the sphere to cool to 90°?

24. *Revenue growth.* The annual revenue R (in millions of dollars) of a company satisfies the equation

$$\frac{dR}{dt} = 6 - (0.1)R + (0.05)t$$

over a certain period of time, where t is the time in years. If the revenues are $4 million after two years, find the annual revenue after five years.

EXTENSION EXERCISES

25. Let $f(x)$ satisfy $f'(x) = p(x)$, where $p(x)$ is continuous on $[a, b]$.

a) Show that if $v = ye^{f(x)}$, then $v' = e^{f(x)}y' + p(x)e^{f(x)}y$.

b) Then show that under the substitution $v = ye^{f(x)}$, the equation

$$y' + y \cdot p(x) = q(x)$$

becomes

$$v' = e^{-\int p(x)\,dx} \cdot q(x),$$

which may be solved by an integration.

26. In each of the following, show that the given equation is not exact. Then show that the given function $m(x, y)$ is an integrating factor for the equation, and solve the equation.

a) $3y^4\,dx + 4xy^3\,dy = 0$, $m(x, y) = x^2$

b) $5xy\,dx + 3x^2\,dy = 0$; $m(x, y) = x^3y^2$

c) $(2y - 15xy^2)\,dx + (3x - 20x^2y)\,dy = 0$; $m(x, y) = xy^2$

d) $2xy\,dx + 5x^2\,dy = 0$; $m(x, y) = y^4$

27. Consider the first-order homogeneous equation $a_0y' + a_1y = 0$, where the coefficients are assumed to be constant. Show that $y = Ce^{-a_1x/a_0}$ is the solution to this equation.

28. Use the techniques of this section to solve $dP/dt = k(L - P)$ with the condition $P(0) = 0$.

9.7

OBJECTIVES

a) Solve a linear homogeneous differential equation with constant coefficients.

(continued)

HIGHER-ORDER DIFFERENTIAL EQUATIONS

In the preceding section, we saw that we could solve any first-order linear differential equation. Unfortunately, not every higher-order linear differential equation can be solved explicitly. For the moment, we'll consider second-order linear homogeneous equations with con-

b) Solve a linear (nonhomogeneous) differential equation with constant coefficients if the forcing function is a polynomial.

stant coefficients, that is, equations of the form

$$a_0 y'' + a_1 y' + a_2 y = 0, \tag{1}$$

where a_0, a_1, and a_2, the coefficients of the equation, are constants.

The corresponding first-order equation was solved explicitly in Exercise 27 of Exercise Set 9.6. The solution is an exponential function. This suggests that the solution to Eq. (1) may be an exponential function, which turns out to be correct. If we let $y = e^{rx}$ (r a constant) in Eq. (1) above, then, since $y' = re^{rx}$ and $y'' = r^2 e^{rx}$, we'll get

$$a_0(r^2 e^{rx}) + a_1(re^{rx}) + a_2(e^{rx}) = 0.$$

Noting that each term in the equation above has the common factor e^{rx}, we write

$$e^{rx}(a_0 r^2 + a_1 r + a_2) = 0.$$

Now e^{rx} is a positive number for all choices of real numbers r and x. Dividing by e^{rx}, we get

$$a_0 r^2 + a_1 r + a_2 = 0.$$

The quadratic formula yields two values of r.

Example 1 Solve $y'' - 5y' + 6y = 0$.

Solution We try $y = e^{rx}$. Then $y' = re^{rx}$ and $y'' = r^2 e^{rx}$. Substituting these quantities into the equation gives

$$r^2 e^{rx} - 5re^{rx} + 6e^{rx} = 0$$

or

$$e^{rx}(r^2 - 5r + 6) = 0$$

or

$$r^2 - 5r + 6 = 0.$$

This last equation factors as $(r - 2)(r - 3) = 0$, and we get solutions, or roots, 2 and 3. This gives us the two solutions e^{2x} and e^{3x}.

It turns out that the general solution to the equation is given by $y = C_1 e^{2x} + C_2 e^{3x}$, where C_1 and C_2 are arbitrary constants. We can verify this by finding y' and y'' and substituting y, y', and y'' into the original equation. We get

$$y = C_1 e^{2x} + C_2 e^{3x},$$
$$y' = 2C_1 e^{2x} + 3C_2 e^{3x},$$
$$y'' = 4C_1 e^{2x} + 9C_2 e^{3x}.$$

1. a) By substituting directly, show that $y = e^{2x}$ is a solution to

$$y'' - 5y' + 6y = 0.$$

b) Repeat part (a) for $y = e^{3x}$.

c) Repeat part (a) for $y = 2e^{3x} - 7e^{2x}$.

2. Solve $y'' - 6y' + 8y = 0$.

3. Solve $2y'' - 7y' + 3y = 0$.

Then, substituting, we get

$$
\begin{aligned}
y'' &- 5y' + 6y \\
&= (4C_1e^{2x} + 9C_2e^{3x}) - 5(2C_1e^{2x} + 3C_2e^{3x}) + 6(C_1e^{2x} + C_2e^{3x}) \\
&= 4C_1e^{2x} + 9C_2e^{3x} - 10C_1e^{2x} - 15C_2e^{3x} + 6C_1e^{2x} + 6C_2e^{3x} \\
&= 0.
\end{aligned}
$$

DO EXERCISES 1–3.

In Example 1 we used the fact that if $f(x)$ and $g(x)$ are solutions to a linear homogeneous differential equation, then so too is $C_1f(x) + C_2g(x)$. We state this important principle in the form of a theorem.

THEOREM 6

If $f_1(x), f_2(x), \ldots, f_n(x)$ are solutions to a linear *homogeneous* differential equation, then so too is

$$\sum_{i=1}^{n} C_if_i(x) = C_1f_1(x) + C_2f_2(x) + \cdots + C_nf_n(x),$$

where $C_1, C_2, \ldots, C_n$ are arbitrary constants.

Note that the equation need not have constant coefficients. Nor must it be a second-order equation. The only requirements are that it be linear and homogeneous. We leave the proof of this theorem for the exercises.

In Example 1, we obtained the expression $y = C_1e^{2x} + C_2e^{3x}$ for the solution to $y'' - 5y' + 6y = 0$. This expression is called the *general solution* to the equation. This means that every function of this form is a solution to the equation and that every solution to the equation can be written in this form.

For the equation $y'' - 5y' + 6y = 0$, the general solution contained two arbitrary constants. In all of the equations that we'll consider, the number of constants in the general solution will be equal to the order of the equation.

Example 2 Find the general solution to $y''' + y'' - 6y' = 0$.

Solution As in Example 1, we let $y = e^{rx}$. We obtain $y' = re^{rx}$, $y'' = r^2e^{rx}$, and $y''' = r^3e^{rx}$. Substituting into the equation, we obtain

$$r^3e^{rx} + r^2e^{rx} - 6re^{rx} = 0,$$

4. Solve $y''' + y'' - 2y' = 0$.

or

$$e^{rx}(r^3 + r^2 - 6r) = 0.$$

Since e^{rx} is never 0, we get $r^3 + r^2 - 6r = 0$, or

$$r(r^2 + r - 6) = 0.$$

Factoring yields $r(r + 3)(r - 2) = 0$, so that 0, -3, and 2 are its roots. It is easy to check that e^{-3x}, e^{2x}, and e^{0x}, or 1, are solutions to $y''' + y'' - 6y' = 0$. The general solution is

$$y = C_1 e^{-3x} + C_2 e^{2x} + C_3 \cdot 1,$$

5. Find the auxiliary equation of $y'' - 7y' + 12y = 0$.

or

$$y = C_1 e^{-3x} + C_2 e^{2x} + C_3.$$

DO EXERCISE 4.

In each of Examples 1 and 2, we made the substitution $y = e^{rx}$, reduced the given differential equation to a polynomial equation in the variable r, and then solved this polynomial equation for r. The resulting values of r determined the solution to the differential equation.

But there's no need to repeat the substitution $y = e^{rx}$ and the subsequent algebra each time. We can bypass this work, going directly from the given differential equation to the associated polynomial equation (called the *characteristic equation* or the *auxiliary equation*).

6. Find the auxiliary equation of $y''' - 6y'' + 11y' - 6y = 0$.

Example 3 Find the auxiliary equation of $y'' - 7y' + 10y = 0$.

Solution Replacing y'' by r^2, y' by r, and $y(= y^{(0)})$ by $r^0 = 1$ gives

$$r^2 - 7r + 10 = 0.$$

DO EXERCISES 5 AND 6.

7. Solve $y'' - 7y' + 12y = 0$.

Example 4 Solve $y'' - 7y' + 10y = 0$.

Solution From Example 3, the auxiliary equation is $r^2 - 7r + 10 = 0$. Factoring, we get $(r - 2)(r - 5) = 0$, so that its roots are 2 and 5. The general solution to the differential equation is

$$y = C_1 e^{2x} + C_2 e^{5x}.$$

DO EXERCISE 7.

8. Solve $y'' - 4y' + 4y = 0$.

The technique we have developed works well so long as the auxiliary equation has real roots, all of which are distinct. It must be modified if the real roots are repeated or if the roots are complex.

Example 5 Solve $y'' - 2y' + y = 0$.

Solution The auxiliary equation is $r^2 - 2r + 1 = 0$. Factoring yields $(r - 1)^2 = 0$. The roots are 1 and 1, or simply 1. (This is the case of repeated real roots.) Sure enough, $r = 1$ yields the solution $y = e^x$. We might try $y = C_1 e^x + C_2 e^x$ as the general solution. But

$$C_1 e^x + C_2 e^x = (C_1 + C_2)e^x = Ke^x,$$

where $K = C_1 + C_2$. In other words, simply repeating the e^x gets us nowhere.

Let's substitute $y = ve^x$ into the original equation. Then, by the Product Rule,

$$y' = ve^x + v'e^x$$

and

$$y'' = (ve^x + v'e^x) + (v'e^x + v''e^x) = ve^x + 2v'e^x + v''e^x$$

and substituting all of this into

$$y'' - 2y' + y = 0$$

gives

$$(ve^x + 2v'e^x + v''e^x) - 2(ve^x + v'e^x) + ve^x = 0.$$

Simplifying, we get $v''e^x = 0$, or, since e^x is never 0, $v'' = 0$. Integrating twice yields $v = C_1 + C_2 x$. Replacing v in the expression $y = ve^x$ gives

$$y = (C_1 + C_2 x)e^x$$

or

$$y = C_1 e^x + C_2 xe^x.$$

This is the general solution to the original equation and can be checked in the original equation.

DO EXERCISE 8.

The calculations in Example 5 weren't easy. Rather than repeat them each time repeated roots occur, we simply remember that if the root r is repeated exactly once, the corresponding solutions of the differential equation are e^{rx} and xe^{rx}.

9. Solve $y'' - 6y' + 9y = 0$.

Example 6 Solve $y'' + 4y' + 4y = 0$.

Solution The auxiliary equation is $r^2 + 4r + 4 = 0$, or $(r + 2)^2 = 0$. So -2 and -2 are the roots. The corresponding solutions to the differential equation are e^{-2x} and xe^{-2x}. The general solution is

$$y = C_1 e^{-2x} + C_2 xe^{-2x} + C_3 \cdot 1,$$

DO EXERCISES 9 AND 10.

If a given number, say r, appears as a root three times, we use e^{rx}, xe^{rx}, and $x^2 e^{rx}$ for the three solutions.

10. Solve $4y'' + 4y' + y = 0$.

Example 7 Solve $y''' - 3y'' + 3y' - y = 0$.

Solution The auxiliary equation is $r^3 - 3r^2 + 3r - 1 = 0$, or $(r - 1)^3 = 0$. So the roots are 1, 1, and 1. Three solutions are e^x, xe^x, and $x^2 e^x$. The general solution is

$$y = C_1 e^x + C_2 xe^x + C_3 x^2 e^x.$$

DO EXERCISE 11.

We have considered the cases in which the roots of the auxiliary equation are real numbers. Consider the equation $y'' + y = 0$. The auxiliary equation is $r^2 + 1 = 0$. The roots of this equation are $r = \pm\sqrt{-1}$, or $r = \pm i$, where i denotes $\sqrt{-1}$ (which doesn't exist as a real number). On the other hand, it's easy to solve $y'' + y = 0$. If $y = \sin x$, for example, then $y' = \cos x$ and $y'' = -\sin x$. In other words, $y = \sin x$ is one solution. A similar calculation (do it!) reveals that $y = \cos x$ is another solution. The general solution is $y = C_1 \sin x + C_2 \cos x$.

11. Solve $y''' + 3y'' + 3y' + y = 0$.

This example illustrates the general case. If the roots of the auxiliary equation are the complex numbers $r_1 \pm r_2 i$, then the two corresponding solutions are $e^{r_1 x} \cos (r_2 x)$ and $e^{r_1 x} \sin (r_2 x)$.

Example 8 Solve $y'' + 16y = 0$.

Solution The auxiliary equation is $r^2 + 16 = 0$. Solving directly yields $r^2 = -16$, or $r = \pm\sqrt{-16}$, or $r = \pm 4\sqrt{-1} = \pm 4i = 0 \pm 4i$. This is the case where $r_1 = 0$ and $r_2 = 4$. Two solutions are $e^{0x} \sin 4x = \sin 4x$ and $e^{0x} \cos (4x) = \cos 4x$. The general solution is $y = C_1 \sin 4x + C_2 \cos 4x$.

12. Solve $y'' + 9y = 0$.

DO EXERCISE 12.

Example 9   Solve $y'' - 2y' + 10y = 0$.

Solution   The auxiliary equation is $r^2 - 2r + 10 = 0$. Applying the quadratic formula yields

$$r = \frac{-(-2) \pm \sqrt{(-2)^2 - 4(1)(10)}}{2 \cdot 1}$$

13. Solve $y'' - 4y' + 29y = 0$.

$$= \frac{2 \pm \sqrt{4 - 40}}{2} = \frac{2 \pm \sqrt{-36}}{2}$$

$$= \frac{2 \pm 6\sqrt{-1}}{2} = 1 \pm 3i.$$

So $r_1 = 1$ and $r_2 = 3$. The corresponding solutions are $e^x \sin 3x$ and $e^x \cos 3x$ and the general solution is $y = C_1 e^x \sin 3x + C_2 e^x \cos 3x$.

DO EXERCISES 13 AND 14.

If some of the roots are real and some are complex, we simply combine our earlier techniques.

14. Solve $y'' + 2y' + 5y = 0$.

Example 10   Solve $y''' - 4y'' + 8y' - 8y = 0$.

Solution   The auxiliary equation is $r^3 - 4r^2 + 8r - 8 = 0$. This factors as $(r - 2)(r^2 - 2r + 4) = 0$. The roots turn out to be $r = 2$, $1 \pm \sqrt{3}i$. The corresponding solutions are e^{2x}, $e^x \sin (\sqrt{3}x)$, and $e^x \cos (\sqrt{3}x)$. The general solution is

15. Solve $y''' - 2y'' + 8y' = 0$.

$$y = C_1 e^{2x} + C_2 e^x \sin (\sqrt{3}x) + C_3 e^x \cos (\sqrt{3}x).$$

DO EXERCISE 15.

If complex roots are repeated, then we modify the solutions just as we did in the case of repeated real roots.

Example 11   Solve $y'''' + 2y'' + y = 0$.

Solution   The auxiliary equation is $r^4 + 2r^2 + 1 = 0$, or $(r^2 + 1)^2 = 0$. The roots are $\pm i$ and $\pm i$. Two solutions corresponding to $\pm i$ are $\sin x$ and $\cos x$. Two more corresponding to the second pair $\pm i$ are

16. Solve $y'''' + 8y'' + 16y = 0$.

$x \sin x$ and $x \cos x$. The general solution is

$$y = C_1 \sin x + C_2 \cos x + C_3 x \sin x + C_4 x \cos x.$$

DO EXERCISE 16.

Nonhomogeneous Equations

So far in this section, the equations considered have been linear homogeneous equations with constant coefficients. We'll relax that now, considering equations of the form

$$a_0 y^{(n)} + a_1 y^{(n-1)} + \cdots + a_{n-1} y' + a_n y = b(x); \qquad \text{(2)}$$

17. Check by direct calculation that $y = C_1 \sin x + C_2 \cos x + 5$ is a solution to $y'' + y = 5$ for all choices of C_1 and C_2.   `CSS`

that is, we no longer require that the equation be homogeneous. We'll consider only the case in which $b(x)$ is a polynomial.

It turns out to be easy to solve such an equation. To do so, we first find the general solution (call it y_h) to the *associated homogeneous equation,* that is, to Eq. (2) with $b(x) = 0$. Next, we find a *particular solution* (call it y_p) to Eq. (2). Finally, we add y_p to y_h to obtain the general solution to Eq. (2).

Example 12   Solve $y'' + y = 5$.

Solution   We must first solve the associated homogeneous equation, $y'' + y = 0$. From above, the general solution, y_h, is $y_h = C_1 \sin x + C_2 \cos x$. It remains to find a particular solution y_p. To do so, we make an educated guess. Here, we guess that $y_p = C$ must be a constant (more about how to make this "guess" later). Since $y_p' = 0$, we get $y_p'' = 0$, and, since y_p must satisfy $y'' + y = 5$, we find that

18. Solve $4y'' + 4y' + y = 3$ (see Margin Exercise 10).

$$(0) + (C) = 5,$$

or

$$C = 5.$$

In other words, $y_p = 5$ is a particular solution to $y'' + y = 5$. In retrospect, we see that this is correct.

To complete the problem, we add y_p and y_h, getting

$$y = C_1 \sin x + C_2 \cos x + 5.$$

DO EXERCISES 17 AND 18.

In Example 12 and Margin Exercise 18, it turned out that y_p was equal to the forcing term $b(x)$. That was a coincidence.

19. Solve $y'' - 9y = x^2 + 2x$.

Example 13   Solve $y'' - 6y' + 8y = 2x^2 - 1$.

Solution   We first solve $y'' - 6y' + 8y = 0$. From Margin Exercise 2, we know that its solution is $y_h = C_1 e^{4x} + C_2 e^{2x}$. It remains only to find a particular solution y_p. To do this, we try $y_p = Ax^2 + Bx + C$, where A, B, and C are numbers to be determined. Then $y_p' = 2Ax + B$ and $y_p'' = 2A$. Since y_p must satisfy the given equation, we substitute, getting

$$(2A) - 6(2Ax + B) + 8(Ax^2 + Bx + C) = 2x^2 - 1,$$

or

$$2A - 12Ax - 6B + 8Ax^2 + 8Bx + 8C = 2x^2 - 1.$$

Regrouping yields

$$8Ax^2 + (-12A + 8B)x + (2A - 6B + 8C) = 2x^2 - 1.$$

In order for these two expressions to be equal, the coefficients of corresponding terms must be equal. So,

$$8A = 2, \qquad x^2\text{-coefficients are equal}$$
$$8B - 12A = 0, \qquad x\text{-coefficients are equal}$$
$$2A - 6B + 8C = -1. \qquad \text{Constant terms are equal}$$

The second equation gives $B = \frac{3}{2}A$. Since $8A = 2$, we get $A = \frac{1}{4}$ and $B = \frac{3}{2} \cdot \frac{1}{4} = \frac{3}{8}$. Solving the third equation for C yields

$$8C = -2A + 6B - 1 = -2(\tfrac{1}{4}) + 6(\tfrac{3}{8}) - 1,$$

or

$$8C = -\tfrac{1}{2} + \tfrac{18}{8} - 1 = -\tfrac{2}{4} + \tfrac{9}{4} - \tfrac{4}{4} = \tfrac{3}{4},$$

and

$$C = \tfrac{3}{32}.$$

So,

$$y_p = \tfrac{1}{4}x^2 + \tfrac{3}{8}x + \tfrac{3}{32}.$$

The general solution to the given equation is

$$y = y_h + y_p,$$

or

$$y = C_1 e^{4x} + C_2 e^{2x} + \tfrac{1}{4}x^2 + \tfrac{3}{8}x + \tfrac{3}{32}.$$

DO EXERCISE 19.

20. Solve $y'' + y' = 7$.

The two examples above show that making the correct guess for the form of the particular solution y_p is the crucial step in solving these equations. If the forcing term is a polynomial, then we will usually let y_p be a polynomial of the same degree. This works in many cases. But consider the equation below.

Example 14   Solve $y'' - 2y' = 5$.

Solution   We first solve $y'' - 2y' = 0$. The auxiliary equation is $r^2 - 2r = 0$, or $r(r - 2) = 0$. So the roots are 0 and 2 and

$$y_h = C_1 e^{0 \cdot x} + C_2 e^{2x} = C_1 + C_2 e^{2x}.$$

To complete the problem, we need only find a particular solution y_p. Since the forcing term is a constant, 5, we might try $y_p = C$. Then $y_p' = 0$ and $y_p'' = 0$. Substitution into the original equation gives $(0) - 2(0) = 5$, or $0 = 5$, which is not true. We conclude that $y_p = C$ was the wrong trial solution. The trouble is that 0 was a root of the auxiliary equation. Whenever 0 is a root of the auxiliary equation, we must remember to multiply the trial solution we would generally use by x^m (m is the number of times that 0 occurs as a root of the auxiliary equation). Above, we would ordinarily have used $y_p = C$. Since 0 is a root of the auxiliary equation, we must multiply by x. So we use $y_p = Cx$. Then $y_p' = C$ and $y_p'' = 0$. Substituting, we obtain

$$(0) - 2(C) = 5,$$

or

$$-2C = 5,$$

or

$$C = -\tfrac{5}{2}.$$

Thus $y_p = -\tfrac{5}{2}x$ is a particular solution to $y'' - 2y' = 5$. The general solution is then

$$y = y_h + y_p$$
$$y = C_1 + C_2 e^{2x} - \tfrac{5}{2}x.$$

DO EXERCISE 20.

Example 15   Solve

$$y''' - y'' = 2x - 3,$$

$$y(0) = 1, \qquad y'(0) = 0, \qquad y''(0) = 3.$$

Solution   As before, we first find the general solution to the differential equation. Then we use the initial conditions $y(0) = 1$, $y'(0) = 0$, and $y''(0) = 3$ to eliminate the constants.

To solve $y''' - y'' = 2x - 3$, we first solve the associated homogeneous equation $y''' - y'' = 0$. The auxiliary equation is $r^3 - r^2 = 0$, or $r^2(r - 1) = 0$. The roots are 0, 0, and 1. So $y_h = C_1 + C_2 x + C_3 e^x$.

Next, we must find y_p. Since the forcing term is $2x - 3$, a polynomial of degree one, we would ordinarily use $y_p = Ax + B$. In this case, that won't work (as you would discover by trying it), since 0 is a root (here, a double root) of the auxiliary equation. Since 0 occurs twice as a root of $r^3 - r^2 = 0$, we multiply our usual trial solution, $Ax + B$, by x^2, obtaining $y_p = Ax^3 + Bx^2$. Then $y_p' = 3Ax^2 + 2Bx$ and $y_p'' = 6Ax + 2B$. Finally, $y_p''' = 6A$. Substituting into the original equation, we get

$$(6A) - (6Ax + 2B) = 2x - 3, \qquad \boxed{\text{CSS}}$$

or

$$-6Ax + (6A - 2B) = 2x - 3.$$

Equating corresponding coefficients yields $-6A = 2$ and $6A - 2B = -3$. Solving the first one gives $A = -\frac{1}{3}$. Substituting this into the second equation yields $6(-\frac{1}{3}) - 2B = -3$, or $-2 - 2B = -3$. Solving gives $-2B = -1$, or $B = \frac{1}{2}$. Thus $y_p = -\frac{1}{3}x^3 + \frac{1}{2}x^2$. The general solution is then

$$y = y_h + y_p = C_1 + C_2 x + C_3 e^x - \tfrac{1}{3}x^3 + \tfrac{1}{2}x^2.$$

It remains to use the initial conditions to eliminate C_1, C_2, and C_3. Letting $x = 0$ above, we get

$$y(0) = C_1 + C_2(0) + C_3 e^0 - \tfrac{1}{3}(0)^3 + \tfrac{1}{2}(0)^2,$$

or

$$y(0) = C_1 + C_3.$$

But $y(0) = 1$, so $C_1 + C_3 = 1$. Now, differentiating the solution gives

$$y' = C_2 + C_3 e^x - x^2 + x.$$

Letting $x = 0$, we have

$$y'(0) = C_2 + C_3 e^0 - (0)^2 + 0,$$

or

$$y'(0) = C_2 + C_3.$$

So, since $y'(0) = 0$, we get $C_2 + C_3 = 0$. Finally, differentiating again

21. Solve $y''' + 2y'' = 6x - 4$

22. Solve $y'' - y = -5, y(0) = 1$,
$y'(0) = 13$.

gives

$$y'' = C_3 e^x - 2x + 1,$$

so that

$$y''(0) = C_3 e^0 - 2(0) + 1 = C_3 + 1.$$

Since $y''(0) = 3$, we have $C_3 + 1 = 3$, or $C_3 = 2$. This enables us to find C_1 and C_2, for $C_2 + C_3 = 0$ gives $C_2 = -C_3 = -2$ and $C_1 + C_3 = 1$ yields $C_1 = 1 - C_3 = 1 - 2 = -1$. So,

$$y = -1 - 2x + 2e^x - \tfrac{1}{3}x^3 + \tfrac{1}{2}x^2.$$

Observe that three initial conditions are needed to eliminate the three constants in the general solution.

DO EXERCISES 21 AND 22.

EXERCISE SET 9.7

Solve.

1. $y'' - 6y' + 5y = 0$

2. $y'' - 8y' + 15y = 0$

3. $y'' - y' - 2y = 0$

4. $y'' + 2y' - 3y = 0$

5. $y'' + 3y' + 2y = 0$

6. $y'' + 5y' + 6y = 0$

7. $2y'' - 5y' + 2y = 0$

8. $2y'' - y' - y = 0$

9. $y'' - 9y = 0$

10. $y'' - 3y = 0$

11. $y'' + 10y' + 25y = 0$

12. $y'' - 8y' + 16y = 0$

13. $4y'' + 12y' + 9y = 0$

14. $y''' + 8y'' + 16y' = 0$

15. $y''' - y'' - 8y' + 12y = 0$

16. $y''' - 7y'' + 11y' - 5y = 0$

17. $y''' + 3y'' + 3y' + y = 0$

18. $y'''' - 4y''' = 0$

19. $y'''' - 5y''' + 4y'' = 0$

20. $y'''' + 3y''' - y'' - 3y' = 0$

21. $y'' + 36y = 0$

22. $y'' + 4y' + 13y = 0$

23. $y'' + 8y' + 41y = 0$

24. $y'' + y' + y = 0$

25. $y''' + 2y'' + 5y' = 0$

26. $y''' - 2y'' + 4y' - 8y = 0$

27. $y''' + y'' + 15y' - 17y = 0$

28. $y'''' + 4y'' + 4y = 0$

29. $y'' + y = 7$

30. $y'' - 3y' + 2y = 4$

31. $y'' - 2y' + y = 3$

32. $y'' - 7y' + 10y = 10x - 27$

33. $y'' + 4y' + 4y = 8 - 12x$

34. $y'' - 4y' + 5y = 5x + 1$

35. $y'' + 4y' + 3y = 6x^2 - 4$

36. $y'' + 4y = 8x^2 - 12x$

37. $y'' - 3y' = 4$

38. $y''' + y' = -2$

39. $y''' + y'' = -2$

40. $y'' - 2y' = -4x$

41. $y''' + 4y'' + 20y' = 40x - 12$

42. $y'' - y = 3x^2 - 8x + 5$

43. $y'' - y' = 0, y(0) = 0, y'(0) = -1$

44. $y'' + y = 0, y(0) = 4, y'(0) = 3$

45. $y'' - y = 0, y(0) = 1, y'(0) = 3$

46. $y'' - 2y' + y = 0, y(0) = 1, y'(0) = -2$

47. $y''' - y' = 0, y(0) = 1, y'(0) = 8, y''(0) = 4$

48. $y'' - y' - 2y = 2x - 1, y(0) = 6, y'(0) = 0$

49. $y'' + 9y = 9 - 9x, y(0) = 3, y'(0) = 2$

50. $y''' + 3y'' = 18x - 18, y(0) = 3, y'(0) = -1, y''(0) = 10$

EXTENSION EXERCISES

51. *Commodities.* Over a certain period of time, a trader notes that the price P of gold satisfies the differential equation

$$P'' + 120P' - 575P = 290,$$

where P is measured in dollars per ounce and t is measured in trading days. If the initial price is $304 per ounce and the initial value of P' is $\frac{1}{10}$, find the price of gold after 50 trading days as well as the highest trading price per ounce over the next 50 trading days.

53. Let y_1 and y_2 both be solutions to the second-order linear homogeneous differential equation

$$a_0(x)y'' + a_1(x)y' + a_2(x)y = 0.$$

(Note that we are *not* assuming that the coefficients are constants.) Show that $C_1 y_1 + C_2 y_2$ is also a solution to the equation, where C_1 and C_2 are arbitrary constants.

55. **a)** Let y_p be a solution to

$$a_0(x)y'' + a_1(x)y' + a_2(x)y = b(x),$$

and let y_h be a solution to

$$a_0(x)y'' + a_1(x)y' + a_2(x)y = 0.$$

Show that $y_p + y_h$ is a solution to the nonhomogeneous equation.

b) Let y_1 and y_2 *both* be solutions to the nonhomogeneous equation

$$a_0(x)y'' + a_1(x)y' + a_2(x)y = b(x).$$

Show that $y_1 - y_2$ is a solution to the homogeneous equation

$$a_0(x)y'' + a_1(x)y' + a_2(x)y = 0.$$

c) Use part (b) to show that if y_1 is a given solution to

$$a_0(x)y'' + a_1(x)y' + a_2(x)y = b(x),$$

then every other solution to it is of the form $y_1 + y_2$, where y_2 is a solution to the associated homogeneous equation.

57. Repeat Exercise 56 if the roots are 7, 9, $2 \pm 4i$, and $2 \pm 4i$.

52. *Population growth.* The estimated number N of a certain bacteria (in thousands) present t hours after the beginning of an experiment satisfies the differential equation

$$N'' - 3N' + 2N = 0.$$

Find the approximate number of bacteria present after 4 hours if $N(0) = 5$ and $N'(0) = 15$.

54. Show that the conclusion of Exercise 53 holds for the nth-order homogeneous linear equation

$$a_0(x)y^{(n)} + a_1(x)y^{(n-1)} + \cdots + a_{n-1}(x)y' + a_n(x)y = 0.$$

56. You are told that the roots of the auxiliary equation of a certain sixth-order linear homogeneous differential equation with constant coefficients are 2, 2, 3, 3, 3, and 3. Find the general solution to the equation.

9.8

1. Solve

$$\frac{dx}{dt} = 4x,$$

$$\frac{dy}{dt} = y.$$

SYSTEMS OF DIFFERENTIAL EQUATIONS

In many applications we are interested in finding the values of two (or more) quantities, say x and y, that are functions of yet another variable t (usually thought of as being time), whose rates of change are given as a system of equations.

Example 1 Solve

$$\frac{dx}{dt} = 5x,$$

$$\frac{dy}{dt} = 3y.$$

Solution We want to find x and y in terms of t. In this case, that's particularly easy to do because the equations in the system aren't interrelated. That is, the first equation doesn't involve y and the second one doesn't involve x.

The first equation is a first-order linear equation (with variables separable) whose solution is $x = C_1 e^{5t}$. Similarly, the solution to the second equation is $y = C_2 e^{3t}$.

Note that there will be two constants in the solution.

DO EXERCISE 1.

Example 2 Solve

$$\frac{dx}{dt} = y,$$

$$\frac{dy}{dt} = 4x.$$

Solution In this system the variables *are* interrelated, or coupled. That is, x and y both occur in the first equation as well as the second equation. We can "uncouple" them as follows. We differentiate the first equation with respect to t to get

$$\frac{d^2x}{dt^2} = x'' = \frac{dy}{dt}.$$

2. Solve

$$\frac{dx}{dt} = 2y,$$

$$\frac{dy}{dt} = 8x.$$

From the second equation, we have

$$\frac{dy}{dt} = 4x.$$

Combining the equations gives us

$$x'' = \frac{dy}{dt} = 4x,$$

or

$$x'' - 4x = 0.$$

This is a second-order linear homogeneous equation with constant co-efficients. Its auxiliary equation is $r^2 - 4 = 0$, or $(r + 2)(r - 2) = 0$. Since 2 and -2 are the roots, we find that $x(t) = C_1 e^{2t} + C_2 e^{-2t}$. Note that since $y = dx/dt$, we know immediately that we have

$$y(t) = 2C_1 e^{2t} - 2C_2 e^{-2t} = \frac{dx}{dt}.$$

DO EXERCISES 2 AND 3.

3. Solve

$$\frac{dx}{dt} = y,$$

$$\frac{dy}{dt} = -4x.$$

We can solve any first-order linear system of the form

$$\frac{dx}{dt} = ax + by,$$

$$\frac{dy}{dt} = cx + dy,$$

where a, b, c, and d are constants, by combining the technique shown in Example 2 with a little algebra. To see how, consider the following example.

Example 3   Solve

$$\frac{dx}{dt} = x + 2y,$$

$$\frac{dy}{dt} = 4x - y.$$

Solution   Our objective is to combine the two equations so as to eliminate one of the dependent variables, say y. We write the first equation as $x' = x + 2y$, and then differentiate it to get

$$x'' = x' + 2y'.$$

4. Check the solutions obtained in Example 3 by direct substitution into the original system of equations.

5. Solve

$$\frac{dx}{dt} = 4x + 3y,$$

$$\frac{dy}{dt} = -4x - 4y.$$

Now, we'll use the second equation, $y' = 4x - y$, to eliminate y' above. We get

$$x'' = x' + 2(4x - y),$$

or

$$x'' = x' + 8x - 2y.$$

Unfortunately, a y-term remains. But we can eliminate it by using the original version of the first equation. Solving $x' = x + 2y$ for $2y$ yields $2y = x' - x$. Substituting this into the last equation gives us

$$x'' = x' + 8x - (x' - x),$$

or

$$x'' = x' + 8x - x' + x,$$

or

$$x'' - 9x = 0.$$

We have eliminated y and what remains is a second-order linear homogeneous equation in x. Its auxiliary equation is $r^2 - 9 = 0$. The roots here are 3 and -3, so that $x = C_1 e^{3t} + C_2 e^{-3t}$. It still remains to find y. Recall though that in the process of eliminating y, we had obtained $2y = x' - x$. Now using our solution for x gives

$$2y = (3C_1 e^{3t} - 3C_2 e^{-3t}) - (C_1 e^{3t} + C_2 e^{-3t}),$$

or

$$2y = 2C_1 e^{3t} - 4C_2 e^{-3t},$$

or

$$y = C_1 e^{3t} - 2C_2 e^{-3t}.$$

So the solution is

$$x = C_1 e^{3t} + C_2 e^{-3t},$$
$$y = C_1 e^{3t} - 2C_2 e^{-3t}.$$

DO EXERCISES 4 AND 5.

Example 4 Solve

$$\frac{dx}{dt} = y, \qquad x(0) = 1,$$

$$\frac{dy}{dt} = -x + 2y, \qquad y(0) = 3.$$

6. Solve

$$\frac{dx}{dt} = 3x + y, \qquad x(0) = 1,$$

$$\frac{dy}{dt} = -x + y, \qquad y(0) = 3.$$

Solution   Just as in previous sections, we'll first find the general solution to the problem. Then we'll use the initial conditions to evaluate the constants in the general solution.

Differentiating the first equation gives $x'' = y'$. Using the second equation to replace y', we get $x'' = -x + 2y$. From the first equation, we have $y = x'$. Combining the last two equations gives $x'' = -x + 2x'$, or $x'' - 2x' + x = 0$.

Again, we have a second-order linear equation with constant coefficients. The auxiliary equation is $r^2 - 2r + 1 = 0$, or $(r-1)^2 = 0$, so that the roots are 1 and 1. Since the roots are repeated, we have

$$x = C_1 e^t + C_2 t e^t.$$

Since $y = dx/dt = x'$, we get

$$y = C_1 e^t + C_2 t e^t + C_2 e^t,$$

or

$$y = (C_1 + C_2) e^t + C_2 t e^t.$$

Letting $t = 0$ in the equation for x gives

$$x(0) = C_1 e^0 + C_2 \cdot 0 \cdot e^0 = C_1.$$

Since $x(0) = 1$ was one of the initial conditions, we have $x(0) = C_1 = 1$. Letting $t = 0$ in the equation for y gives $y(0) = C_1 + C_2$. Since $y(0) = 3$, we have $C_1 + C_2 = 3$. Since $C_1 = 1$, we must have $C_2 = 2$. So, $x = e^t + 2te^t$ and $y = 3e^t + 2te^t$.

It is easy to check that these solutions satisfy both the original equations and the initial conditions.

7. Solve

$$\frac{dx}{dt} = 3x - y, \qquad x(0) = 3,$$

$$\frac{dy}{dt} = 2x, \qquad y(0) = 5.$$

DO EXERCISES 6 AND 7.

If $f(t)$ and $g(t)$ are polynomials in t, then the preceding techniques can be used to solve systems of the form

$$\frac{dx}{dt} = ax + by + f(t),$$

$$\frac{dy}{dt} = cx + dy + g(t),$$

where, as above, a, b, c, and d are constants.

Example 5   Solve

$$\frac{dx}{dt} = x + y + 4 \quad \text{and} \quad \frac{dy}{dt} = 4x + y + 3.$$

8. Solve

$$\frac{dx}{dt} = 4x + y,$$

$$\frac{dy}{dt} = 2x + 3y + 50.$$

Solution We differentiate the first equation to obtain $x'' = x' + y'$. Using the second equation to replace y' yields

$$x'' = x' + (4x + y + 3),$$

or

$$x'' = x' + 4x + y + 3.$$

Solving the first equation for y yields $y = x' - x - 4$, so that combining the last two equations gives us

$$x'' = x' + 4x + (x' - x - 4) + 3,$$

or

$$x'' = 2x' + 3x - 1,$$

or

$$x'' - 2x' - 3x = -1.$$

Note that the only difference between this example and our previous ones is that we end up with a nonhomogeneous second-order equation for x rather than a homogeneous one.

Here, the auxiliary equation of the associated homogeneous equation $x'' - 2x' - 3x = 0$ is $r^2 - 2r - 3 = 0$, which factors as $(r - 3)(r + 1) = 0$. The roots are 3 and -1, and $x_h = C_1 e^{3t} + C_2 e^{-t}$. Note that the constant function $x_p = \frac{1}{3}$ is a particular solution, so that

$$x = C_1 e^{3t} + C_2 e^{-t} + \tfrac{1}{3}$$

is the general solution for x. Since $y = x' - x - 4$, we get

$$y = (3C_1 e^{3t} - C_2 e^{-t}) - (C_1 e^{3t} + C_2 e^{-t} + \tfrac{1}{3}) - 4$$
$$y = 3C_1 e^{3t} - C_2 e^{-t} - C_1 e^{3t} - C_2 e^{-t} - \tfrac{1}{3} - \tfrac{12}{3},$$

or

$$y = 2C_1 e^{3t} - 2C_2 e^{-t} - \tfrac{13}{3}.$$

DO EXERCISE 8.

Example 6 Tank A contains 2000 pounds of salt dissolved in 1000 gallons of water. Tank B contains 1000 pounds of salt dissolved in 1000 gallons of water. The mixture from tank A is pumped to tank B at the rate of 500 gallons per day, while that from tank B is pumped to tank A at the same rate. Assuming that the mixture in each tank is kept uniform by stirring, how much salt is in each tank after t days?

Solution   We know that the initial concentration of salt in tank A is 2000 lb/1000 gal = 2 lb/gal. The initial concentration in tank B is 1000 lb/1000 gal = 1 lb/gal. As we pump the saltier mixture from tank A to tank B, we'd expect the mixture in tank B to get saltier. At the same time, the mixture being pumped from tank B to tank A should gradually dilute the mixture in tank A. We might guess that the concentrations in each tank might approach an equilibrium concentration.

To see what actually happens, let

$A(t)$ = the number of pounds of salt in tank A after t days, and

$B(t)$ = the number of pounds of salt in tank B after t days.

Note that $A(0) = 2000$ and $B(0) = 1000$. The concentration of salt in tank A after t days is $(A(t)/1000)$ lb/gal, since tank A always contains exactly 1000 gallons of water. Similarly, the concentration in tank B after t days is $B(t)/1000$ lb/gal. To set up our differential equations we observe that

Rate of change of salt in tank A = (Rate of flow in)

$-$ (Rate of flow out).

The rate at which salt flows into tank A is $(500) \cdot B(t)/1000$ pounds per day; that is, the product of the flow into tank A, 500 gallons per day, and the concentration in tank B, $B(t)/1000$ pounds per gallon.

Similarly, the flow out of tank A is $500 \cdot A(t)/1000$. Thus the rate of change of the amount of salt in tank A, dA/dt, is

$$\frac{dA}{dt} = \frac{500B(t)}{1000} - \frac{500A(t)}{1000},$$

or

$$\frac{dA}{dt} = \frac{1}{2} B(t) - \frac{1}{2} A(t).$$

The amount of salt flowing into tank B is just the negative of that flowing into tank A (equivalently, what flows out of tank A goes into tank B and vice versa). So

$$\frac{dB}{dt} = -\frac{dA}{dt} = \frac{1}{2} A(t) - \frac{1}{2} B(t).$$

We have a system of equations, namely,

$$\frac{dA}{dt} = -\frac{1}{2} A + \frac{1}{2} B \quad \text{and} \quad \frac{dB}{dt} = \frac{1}{2} A - \frac{1}{2} B$$

with the initial conditions $A(0) = 2000$ and $B(0) = 1000$.

Differentiating the first equation, $A' = -\frac{1}{2}A + \frac{1}{2}B$, gives $A'' = -\frac{1}{2}A' + \frac{1}{2}B'$. Using the second equation, we get

$$A'' = -\tfrac{1}{2}A' + \tfrac{1}{2} \cdot (\tfrac{1}{2}A - \tfrac{1}{2}B),$$

or

$$A'' = -\tfrac{1}{2}A' + \tfrac{1}{4}A - \tfrac{1}{4}B.$$

From the first equation, we have $\frac{1}{2}B = A' + \frac{1}{2}A$ or, substituting above,

$$A'' = -\tfrac{1}{2}A' + \tfrac{1}{4}A - \tfrac{1}{4}(2A' + A),$$

or

$$A'' = -A'.$$

The auxiliary equation of $A'' + A' = 0$ is $r^2 + r = 0$, or $r(r + 1) = 0$. Since its roots are 0 and -1, we get

$$A(t) = C_1 + C_2 e^{-t}.$$

Since $B = A + 2A'$ (solving the first equation for B), we get

$$B(t) = (C_1 + C_2 e^{-t}) + 2(-C_2 e^{-t}),$$

or

$$B(t) = C_1 - C_2 e^{-t}.$$

To evaluate the constants, recall that $A(0) = 2000$, whereas if we let $t = 0$ in our formula for $A(t)$, we get

$$A(0) = C_1 + C_2 e^{-0} = C_1 + C_2.$$

So $C_1 + C_2 = 2000$. Similarly, $B(0) = 1000$ while $B(0) = C_1 - C_2 e^{-0} = C_1 - C_2$. Thus $C_1 - C_2 = 1000$. Adding, we get

$$\begin{aligned} C_1 + C_2 &= 2000 \\ C_1 - C_2 &= 1000 \\ \hline 2C_1 \phantom{{}- C_2} &= 3000 \end{aligned}$$

or $C_1 = 1500$. From the first equation, $1500 + C_2 = 2000$, or $C_2 = 500$. We have

$$A(t) = 1500 + 500e^{-t}, \qquad \boxed{\text{CSS}}$$
$$B(t) = 1500 - 500e^{-t}.$$

Note that $A(t) + B(t) = 3000$, which is correct, since salt neither enters nor leaves the system.

As for our intuitive feeling that the concentration of salt in each tank should approach an equilibrium concentration, note that if t is a large positive number, then e^{-t} is a very small (positive) number. If

9. Find the answer to Example 6 if tank A initially contains 1600 pounds of salt and tank B contains 800 pounds.

you have a calculator handy, you can check that

$$e^{-3} \approx 0.04978707,$$
$$e^{-14} \approx 0.00000083,$$

and

$$e^{-30} \approx 1 \times 10^{-13} \qquad (0.0000000000001).$$

Then we can see that as t increases, the amount of salt in each tank will approach 1500 pounds, as we might have expected.

DO EXERCISES 9 AND 10.

Example 7 *Commodities prices.* The prices of two commodities after t days are $P(t)$ and $Q(t)$, respectively. Traders note that over a period of time the prices obey the system

$$\frac{dP}{dt} = 3000 - 5Q$$

and

10. Find the answer to Example 6 if both tanks initially contain 1100 pounds of salt. (*Warning:* Before computing, stop and *think!*)

$$\frac{dQ}{dt} = 20P - 5000.$$

If the initial prices are $P(0) = \$260$ and $Q(0) = \$560$, find the prices at any later time (assume that t is in days and P and Q are in dollars per unit).

Solution   Differentiating the first equation yields $P'' = -5Q'$. Since $Q' = 20P - 5000$ (the second equation), we get

$$P'' = -5(20P - 5000),$$

or

$$P'' = -100P + 25,000,$$

or

$$P'' + 100P = 25,000.$$

We have a nonhomogeneous second-order equation with constant co-efficients. Note that the constant function $P_p = 250$ is a particular solution to the equation. Since the auxiliary equation of the associated homogeneous equation $(P'' + 100P = 0)$ is $r^2 + 100 = 0$, the roots are $r = \pm 10i$ and we get

$$P_h = C_1 \sin (10t) + C_2 \cos (10t).$$

So

$$P = C_1 \sin (10t) + C_2 \cos (10t) + 250.$$

Solving the first equation for Q gives $5Q = -dP/dt + 3000$, or

$$Q = -\frac{1}{5} \cdot \frac{dP}{dt} + 600.$$

So

$$Q = -\tfrac{1}{5}(10C_1 \cos (10t) - 10C_2 \sin (10t)) + 600$$
$$Q = -2C_1 \cos (10t) + 2C_2 \sin (10t) + 600,$$

or

$$Q = 2C_2 \sin (10t) - 2C_1 \cos (10t) + 600.$$

We were given that $P(0) = 260$. From our solution,

$$P(0) = C_1 \sin (10 \cdot 0) + C_2 \cos (10 \cdot 0) + 250$$
$$P(0) = C_1 \sin (0) + C_2 \cos (0) + 250,$$

or

$$P(0) = C_2 + 250.$$

So $260 = C_2 + 250$, or $C_2 = 10$. Similarly, $Q(0) = 560$, while, from the solution,

$$Q(0) = 2C_2 \sin (10 \cdot 0) - 2C_1 \cos (10 \cdot 0) + 600,$$

or

$$Q(0) = 2C_2 \sin 0 - 2C_1 \cos 0 + 600,$$

or

$$Q(0) = 600 - 2C_1.$$

So $560 = 600 - 2C_1$ and we get $C_1 = 20$. Thus after t days, the prices of the commodities are

$$P(t) = 20 \sin (10t) + 10 \cos (10t) + 250$$

and

$$Q(t) = 20 \sin (10t) - 40 \cos (10t) + 600.$$

11. If $P(0) = \$24$, $Q(0) = \$38$, and P and Q are related by

$$\frac{dP}{dt} = 160 - 4Q,$$

$$\frac{dQ}{dt} = 16P - 320,$$

find $P(t)$ and $Q(t)$.

The following table gives the values of P and Q after the first few days.

t	P	Q
0	260.00	560.00
1	230.73	622.68
2	272.34	601.94
3	231.78	574.07
4	258.23	641.58
5	254.40	556.15

The prices fluctuate back and forth about the values \$250 and \$600, respectively. It is *not* true, however, that the greatest price for P coincides with the least for Q (there is a time lag involved).

DO EXERCISE 11.

Example 8   *Population growth.* The populations of two species at time t are $X(t)$ and $Y(t)$, respectively. If it is assumed that the species interact according to

$$\frac{dX}{dt} = 3X - 5Y,$$

$$\frac{dY}{dt} = 5X - 3Y - 480,$$

and if $X(0) = 160$ and $Y(0) = 92$, find $X(t)$ and $Y(t)$. Describe what happens to the populations over a long period of time.

Solution   The usual sequence of steps (we omit the details) leads to the second-order equation $X'' + 16X = 2400$. The general solution to this is $X = C_1 \sin 4t + C_2 \cos 4t + 150$. Solving for Y in the first equation yields

$$Y = (\tfrac{3}{5}C_1 + \tfrac{4}{5}C_2) \sin 4t + (-\tfrac{4}{5}C_1 + \tfrac{3}{5}C_2) \cos 4t + 90.$$

Using the initial conditions yields $160 = C_2 + 150$, or $C_2 = 10$, and $92 = -\tfrac{4}{5}C_1 + \tfrac{3}{5}C_2 + 90$, or $2 = -\tfrac{4}{5}C_1 + 6$, or $C_1 = 5$. Substitution and simplification result in

$$X = 5 \sin 4t + 10 \cos 4t + 150,$$
$$Y = 11 \sin 4t + 2 \cos 4t + 90.$$

CSS

The populations fluctuate. If we use the fact that $A \sin (\lambda t) + B \cos (\lambda t)$ always remains between $\pm \sqrt{A^2 + B^2}$, then we see that the X-population stays between $150 \pm \sqrt{5^2 + 10^2}$, or between 139 and 161, while the Y-population remains between 79 and 101 (approximately).

EXERCISE SET 9.8

Find the general solution to each of the following systems of equations.

1. $x' = 2x, y' = 3y$
2. $x' = -x, y' = 4y$
3. $x' = y, y' = -2x + 3y$
4. $x' = x + 2y, y' = x + 2y$
5. $x' = 2x + y, y' = 3x + 4y$
6. $x' = 3x - 4y, y' = 2x - 3y$
7. $x' = -\frac{1}{2}x + y, y' = \frac{1}{2}x$
8. $x' = -x - 3y, y' = -y$
9. $x' = 4x + y, y' = -x + 2y$
10. $x' = 2x - y, y' = 4x - 2y$
11. $x' = 2y, y' = -18x$
12. $x' = x - y, y' = 5x - y$
13. $x' = 3x - 5y, y' = x - y$
14. $x' = -3x + 5y, y' = -x + y$

Find the functions that satisfy both the system of equations and the given initial conditions.

15. $x' = 2x - y, y' = 2x + 5y, x(0) = 3, y(0) = -5$
16. $x' = x + y, y' = x + y, x(0) = 2, y(0) = 0$
17. $x' = 2x + 3y, y' = -3x + 8y, x(0) = 1, y(0) = 1$
18. $x' = 2x + 3y, y' = -3x + 8y, x(0) = 1, y(0) = 2$
19. $x' = y, y' = -4x, x(0) = 1, y(0) = 2$
20. $x' = x + 5y, y' = -x - 3y, x(0) = 10, y(0) = -3$

Find the solution to each of the following systems of equations, finding the particular solution if initial conditions are given.

21. $x' = 5x - y + 5, y' = 2x + 2y + 2$
22. $x' = 2x + y + 1, y' = -2y - 22$
23. $x' = y + 2t + 3, y' = -x + 4t - 2$
24. $x' = -x + y + 2t + 2, y' = 2x + 2t - 1$
25. $x' = 2x + y + 3, y' = 5x - 2y + 12,$ $x(0) = 0, y(0) = -2$
26. $x' = y + 5, y' = -x + 2y + 10, x(0) = 1, y(0) = -2$

APPLICATIONS

27. *Mixing chemicals.* Tank A initially contains 200 pounds of salt dissolved in 500 gallons of water, while tank B contains 100 pounds of salt dissolved in 500 gallons of water. The mixture in tank A is pumped to tank B at the rate of 100 gallons per day and the mixture from B is pumped to A at the same rate. Set up the system of differential equations for $A(t)$ and $B(t)$, the number of pounds of salt in tanks A and B, after t days, respectively. Then solve them. What is the "limiting concentration" of salt in tank A? in tank B? `CSS`

28. *Population growth.* The numbers $X(t)$ and $Y(t)$ of two species in a certain area (in hundreds) satisfy the system of equations

$$\frac{dX}{dt} = X + Y - 100, \qquad X(0) = 40,$$

$$\frac{dY}{dt} = -8X - 3Y + 450, \qquad Y(0) = 60,$$

where t is measured in months.

a) Find $X(t)$ and $Y(t)$.
b) Find $X(1), Y(1), X(10),$ and $Y(10)$.
c) What are $X(\infty)$ and $Y(\infty)$?

29. *Commodities prices.* The prices P and Q of two inter-related commodities are determined by the following system of equations:

$$P' = P + 13Q - 400, \qquad P(0) = 49,$$
$$Q' = -2P - Q + 100, \qquad Q(0) = 32.$$

Find $P(t)$ and $Q(t)$.

30. *Projected sales.* A securities analyst finds that the projected sales $X(t)$ and $Y(t)$ of two competing firms (in millions of dollars) are given by

$$X' = 4Y - 100t, \qquad X(0) = 377,$$
$$Y' = -X + 40t + 400, \qquad Y(0) = 12.$$

Find $X(t)$ and $Y(t)$ (where t is in years).

EXTENSION EXERCISES

31. Solve

$$x' = -2x + 2y + 2e^t,$$
$$y' = 2x + y.$$

32. Solve

$$\frac{dx}{dt} = -2x + y - z,$$
$$\frac{dy}{dt} = y + z,$$
$$\frac{dz}{dt} = 3z.$$

SUMMARY AND REVIEW: CHAPTER 9

The following contains a summary of what you should be able to do after completing this chapter. The review exercises are for practice. Answers are at the back of the book. If you miss an exercise, restudy the section indicated alongside the answer.

You should be able to:

Solve certain differential equations, giving both general and particular solutions.

Solve.

1. $\frac{dy}{dx} = 6x^2$

2. $\frac{dy}{dx} = 6x^2y$

3. $y' = 5x^2 - x^2y$

4. $\frac{dy}{dx} = 2x^2 - 4x + 3$, given that $y = \frac{2}{3}$ when $x = 1$

5. $\frac{dv}{dt} = 2v^{-3}$

Solve certain differential equations given a condition $f(a) = b$.

6. Solve $dy/dx = 4x^3 - 2x, y(0) = 1$.

Verify that a given function is a solution of a given differential equation.

7. Verify that $y = x^2$ is a solution to $x(dy/dx) - 2y = 0$.

State the differential equation for the inhibited growth model, state the solution to it, and sketch the graph of the solution, including the point of inflection and the limiting value.

8. Graph $P(t) = 1 - e^{-4t}$.

9. A group of 8 students at State University start the rumor, "All final examinations for the year have been cancelled." The rate at which this rumor spreads satisfies the equation

$$\frac{dN}{dt} = kN(L - N),$$

where

N = the number of students that have heard this rumor after t minutes,

$L = 1200$, the total enrollment of the university,

$N_0 = 8$, the number of students who started the rumor, and

$k = 0.0002$.

a) Write the equation for $N(t)$.

b) Sketch a graph of $N(t)$ given that the point of inflection occurs at $t = 21$ minutes.

c) After how many minutes t will 800 students have heard the rumor?

State the solution to the growth model $dP/dt = k(L - P)$ and sketch the solution, given enough information to determine L and k.

10. The Hullian model of learning is $dP/dt = k(L - P)$, where

P = the probability of mastery of the process of differentiating quotients after t learning trials,

$L = 1$, the limiting value of the probability of mastery of the process, and

k = a constant.

a) Write the solution $P(t)$ in terms of L and k.

b) Suppose that the probability of mastery of differentiating quotients is 0.80 after 10 trials. Use this information to determine k to the nearest hundredth.

c) Rewrite $P(t)$ in terms of k.

d) Find $P(12)$.

e) Using the value of k determined in part (b), find the number of trials required so that $P(t) = 0.95$.

11. For the inhibited growth model, $P = 375$ at the point of inflection. What is the limiting value?

Determine whether a first-order differential equation is exact, and, if it is, solve it.

Solve.

12. $2x\,dx + 3y^2\,dy = 0, y(0) = 2$

13. $(4x^3 + 6xy)\,dx + (3x^2 - 4y^3)\,dy = 0, y(0) = 1$

Solve a first-order linear differential equation.

Solve.

14. $xy' + 4y = 0, y(1) = 3$ 15. $y' - 2y = e^x$

Solve a second-order linear differential equation with constant coefficients provided that the forcing function is either 0 (the homogeneous case) or a (nonzero) polynomial.

Solve.

16. $y'' - 7y' + 10y = 0$ 17. $y'' - 6y' + 9y = 0$ 18. $y'' + 25y = 100$

Solve a system of linear differential equations if the forcing functions are polynomials.

Solve.

19. $x' = 2x + 6y, y' = 4x$ 20. $x' = 18y, y' = -2x$
21. $x' = -4x + y, y' = -4x$ 22. $x' = 3x + y, y' = x + 3y, x(0) = 2, y(0) = 0$
23. $x' = y, y' = 4x, x(0) = 3, y(0) = 6$

TEST: CHAPTER 9

1. Solve the differential equation $y' = x^2 + 3x - 5$, given the condition $y = 7$ when $x = 0$.

Solve the following differential equations.

2. $\dfrac{dy}{dx} = 8x^7 y$ 3. $\dfrac{dy}{dx} = \dfrac{9}{y}$ 4. $\dfrac{dy}{dt} = 6y$

5. For the inhibited growth model, $P = 680$ at the point of inflection. What is the limiting value?
6. The growth rate for a certain population of bacteria cells is given by

$$\frac{dP}{dt} = kP(L - P),$$

where

$P =$ the number of cells present after time t (in hours),
$L = 800$, the limiting value,
$k = 0.0009$, and
$P_0 = 20$.

a) Write the solution $P(t)$ in terms of the given constraints.
b) Sketch a graph of $P(t)$ given that the point of inflection occurs at $t = 5$.
c) At what time t are 500 cells present?

7. Graph $P(t) = 5(1 - e^{-t})$.

8. The growth rate of a company is modeled by

$$\frac{dS}{dt} = k(L - S), \qquad S = \$0 \text{ when } t = 0;$$

where

$S =$ the total sales, in millions of dollars, in the tth year of operation,

$L = \$20$ million, the limiting value, and

$k =$ a constant.

a) Write the solution $S(t)$ in terms of L and k.

b) Suppose $S(2) = \$4$ million; that is, sales during the second year are $4 million. Use this to determine k. Round to the nearest hundredth.

c) Rewrite $S(t)$ in terms of k.

d) Use the equation in part (c) to find $S(10)$, the sales in the tenth year.

e) In what year will the sales be $15 million?

Solve each of the following.

9. $3x^2\, dx + 2y\, dy = 0, y(0) = 1$

10. $(3x^2 + 4xy)\, dx + (2x^2 - 5y^4)\, dy = 0, y(0) = 1$

11. $xy' + 3y = 0, y(1) = 4$

12. $y' - 3y = e^{4x}$

13. $y'' - 5y' + 6y = 0$

14. $y'' - 12y' + 36y = 0$

15. $y'' + 36y = 72$

16. $x' = 2x + 5y, y' = 3x$

17. $x' = 6y, y' = -24x$

18. $x' = -6x + 3y, y' = -3x$

19. $x' = -6x + y, y' = -2x - 3y, x(0) = 0, y(0) = -1$

10

TAYLOR SERIES AND l'HÔPITAL'S RULE

In this chapter we will learn about infinite series of numbers and infinite series of functions as well as how to calculate with these series. These calculations will, in turn, enable us to evaluate limits and to evaluate certain integrals that cannot be calculated easily in any other way.

AN APPLICATION

A corporation deposits $1000 in a bank. The bank may, and does, loan out 90% of the money deposited. That money, in turn, is deposited in another bank, which, in turn, lends out 90% of it. Assuming the process continues indefinitely, find the total economic effect of the initial $1000 deposit.

THE MATHEMATICS

The total effect of the initial $1000 deposit can be modeled as the sum of the infinite series

$$\$1000 + \$1000(0.90) \\ + \$1000(0.90)^2 \\ + \$1000(0.90)^3 + \cdots.$$

This series is a *geometric series* whose sum is given by

$$s = \$1000/[1 - 0.90] = \$10,000.$$

The sum $10,000 is the result of what is referred to in economics as the *multiplier effect*.

10.1

TAYLOR'S FORMULA FOR POLYNOMIALS

OBJECTIVE

a)   Expand a given polynomial in powers of $x - a$ (for any given real number a).

The function f given by

$$f(x) = 2x^2 - 16x + 35$$

can also be expressed in the form

$$f(x) = 2(x - 3)^2 - 4(x - 3) + 5.$$ CSS

This can be checked by multiplying and simplifying the right-hand side of the second equation. In the second form, $f(x)$ is expressed as a sum of powers of $x - 3$. In general, the formulas for many functions can be written in the form*

$$f(x) = \sum_{n=0}^{\infty} c_n(x - a)^n,$$

where a is a fixed number and $c_1, c_2, \ldots, c_n, \ldots$ is a sequence of real numbers. In this chapter we will develop expansions like this. Each is called a *Taylor series expansion*. This tool is useful in many areas of mathematics.

To introduce the subject, we first consider the Taylor expansion for polynomial functions. Recall the following definition from Section 1.5.

DEFINITION

A *polynomial function (in x)* is a function that can be written as

$$f(x) = c_n x^n + c_{n-1} x^{n-1} + \cdots + c_1 x + c_0,$$

where $c_n, c_{n-1}, \ldots c_1, c_0$ are real numbers, with $c_n \neq 0$, called the *coefficients* of the polynomial and where n is a nonnegative integer called the *degree* of the polynomial.

Thus each of the following is a polynomial function:

$$f(x) = -5 \qquad \text{(a constant function)},$$
$$f(x) = x + 3 \qquad \text{(a linear function)},$$

and

$$f(x) = 2x^3 - 4x^2 + x + 1.$$

---

* You may need to review summation notation in Section 5.8.

1. Simplify

$$\frac{2(x-3)^2 + 4(x-3) + 1}{x-3}$$

for $x \neq 3$.

The polynomial $2x^3 - 4x^2 + x + 1$ has degree 3. The polynomial $x + 3$ has degree 1 (since $x^1 = x$). The polynomial $0 \cdot x^5 + 3x^4$ has degree 4. The polynomial 6, or $6x^0$, has degree 0.

Consider the following example.

Example 1 Simplify

$$\frac{x^2 - 3x + 1}{x}, \qquad x \neq 0.$$

Solution We write this expression in the form

$$\frac{x^2}{x} - \frac{3x}{x} + \frac{1}{x},$$

and, performing the indicated divisions, we obtain

$$x - 3 + \frac{1}{x} \quad \text{for } x \neq 0.$$

DO EXERCISE 1.

2. Use the identity

$$x^3 - 6x^2 + 14x - 8$$
$$= (x-2)^3 + 2(x-2) + 4$$

to simplify

$$\frac{x^3 - 6x^2 + 14x - 8}{(x-2)^2}.$$

Example 2 Simplify

$$\frac{x^3 - 3x^2 + 3x - 4}{x-1}, \qquad x \neq 1.$$

Solution This looks harder. We can write this equation as

$$\frac{x^3}{x-1} - \frac{3x^2}{x-1} + \frac{3x}{x-1} - \frac{4}{x-1},$$

but this doesn't simplify as in Example 1. Suppose, however, that we know that

$$x^3 - 3x^2 + 3x - 4 = (x-1)^3 - 3.$$

This is true, although you would not know it offhand. Then we can simplify as follows:

$$\frac{x^3 - 3x^2 + 3x - 4}{x-1} = \frac{(x-1)^3 - 3}{x-1} = \frac{(x-1)^3}{x-1} - \frac{3}{x-1}$$

$$= (x-1)^2 - \frac{3}{x-1},$$

whenever $x \neq 1$.

DO EXERCISE 2.

In Example 2, the key to the solution was the identity

$$x^3 - 3x^2 + 3x - 4 = (x - 1)^3 - 3.$$

We were given the expansion of $x^3 - 3x^2 + 3x - 4$ in powers of $x - 1$. We want to be able to write down identities of this kind quickly and easily. Suppose that we want to write the polynomial $2x^2 - 16x + 35$ in powers of $x - 3$. That is, we want to find real numbers c_2, c_1, and c_0 so that

$$2x^2 - 16x + 35 = c_2(x - 3)^2 + c_1(x - 3) + c_0.$$

One possible solution is to multiply out the right-hand side of the equation, grouping the x^2-, x-, and constant terms together. Then we can equate corresponding coefficients to obtain three linear equations in the three unknowns c_2, c_1, and c_0, each of which can then be found. But here's another way.

Example 3   Let $f(x) = 2x^2 - 16x + 35$. Find c_2, c_1, and c_0 so that

$$f(x) = c_2(x - 3)^2 + c_1(x - 3) + c_0. \tag{1}$$

Solution   The equation above must hold for all real numbers x. So it must be true if we let $x = 3$. We obtain

$$\begin{aligned}
f(3) &= c_2(3 - 3)^2 + c_1(3 - 3) + c_0 \\
&= c_2 \cdot 0 + c_1 \cdot 0 + c_0 \\
&= c_0.
\end{aligned}$$

Since $f(3) = 2(3)^2 - 16(3) + 35 = 18 - 48 + 35 = 5$, we get

$$c_0 = f(3) = 5.$$

Next, we apply some calculus. We differentiate both sides of Eq. (1) with respect to x to get

$$\begin{aligned}
f'(x) &= 2c_2(x - 3) + c_1 + 0 \\
&= 2c_2(x - 3) + c_1.
\end{aligned} \tag{2}$$

Letting $x = 3$, we get

$$f'(3) = 2c_2(3 - 3) + c_1,$$

or

$$c_1 = f'(3).$$

Since $f(x) = 2x^2 - 16x + 35$, then

$$f'(x) = 4x - 16 \quad \text{and} \quad f'(3) = 4(3) - 16 = -4.$$

3. Let $f(x) = x^2 - 4x + 2$. Find c_2, c_1, and c_0 so that

$$f(x) = c_2(x - 3)^2 + c_1(x - 3) + c_0.$$

So we get

$$c_1 = f'(3)$$
$$= -4.$$

To find c_2, we differentiate Eq. (2). We get $f''(x) = 2c_2$, a constant function. Since $f'(x) = 4x - 16$, we also have $f''(x) = 4$, so that $2c_2 = 4$, or $c_2 = 2$. Thus $c_2 = 2$, $c_1 = -4$, $c_0 = 5$, and

$$2x^2 - 16x + 35 = 2(x - 3)^2 - 4(x - 3) + 5.$$

DO EXERCISE 3.

Example 4 Write $f(x) = x^3 - 7x + 3$ in powers of $x - 2$.

Solution We want to write

$$f(x) = c_3(x - 2)^3 + c_2(x - 2)^2 + c_1(x - 2) + c_0. \tag{3}$$

We let $x = 2$ in Eq. (3) to get

$$f(2) = c_3(2 - 2)^3 + c_2(2 - 2)^2 + c_1(2 - 2) + c_0$$
$$= c_0.$$

Differentiating, we obtain

$$f'(x) = 3c_3(x - 2)^2 + 2c_2(x - 2) + c_1. \tag{4}$$

Letting $x = 2$, we get

$$f'(2) = c_1.$$

Differentiating Eq. (4) yields

$$f''(x) = 6c_3(x - 2) + 2c_2,$$

and, letting $x = 2$, we get

$$f''(2) = 2c_2.$$

Another differentiation yields

$$f'''(x) = 6c_3.$$

Since this equation is true for all values of x—that is, since $f'''(x)$ is a constant function—we have, in particular,

$$f'''(2) = 6c_3.$$

Before we compute c_3, c_2, c_1, and c_0, note that there seems to be a pattern to the c_n's. That is,

$$c_0 = f(2),$$
$$c_1 = f'(2),$$
$$c_2 = \tfrac{1}{2}f''(2),$$
$$c_3 = \tfrac{1}{6}f'''(2).$$

Returning to the problem, since $f(x) = x^3 - 7x + 3$, we have

$$c_0 = f(2) = 2^3 - 7 \cdot 2 + 3$$
$$= 8 - 14 + 3 = -3.$$

Differentiating, we have $f'(x) = 3x^2 - 7$, so that

$$c_1 = f'(2) = 3(2)^2 - 7$$
$$= 12 - 7 = 5.$$

Also, $f''(x) = 6x$, so that $f''(2) = 6 \cdot 2 = 12$, and $c_2 = \tfrac{1}{2}f''(2) = \tfrac{1}{2} \cdot 12 = 6$. Finally, $f'''(x) = 6$, so that $c_3 = \tfrac{1}{6}f'''(2) = \tfrac{1}{6} \cdot 6 = 1$. Thus,

$$x^3 - 7x + 3 = (x - 2)^3 + 6(x - 2)^2 + 5(x - 2) - 3.$$

Factorial Notation

Before continuing, we introduce some notation. Products like $3 \cdot 2 \cdot 1$ or $6 \cdot 5 \cdot 4 \cdot 3 \cdot 2 \cdot 1$ will occur so often that it is convenient to adopt a notation for them. We define $5 \cdot 4 \cdot 3 \cdot 2 \cdot 1 = 5!$, read "5 factorial."

DEFINITION

We define $n!$, read "n factorial," as follows:

$$n! = n(n - 1)(n - 2) \cdots 3 \cdot 2 \cdot 1,$$

for any integer n, $n > 1$. For convenience in later notation, we define $1!$ (*one factorial*) to be 1 and $0!$ (*zero factorial*) to be 1.

Thus we have

$$3! = 3 \cdot 2 \cdot 1 = 6,$$
$$5! = 5 \cdot 4 \cdot 3 \cdot 2 \cdot 1 = 120,$$

and

$$6! = 6 \cdot 5 \cdot 4 \cdot 3 \cdot 2 \cdot 1$$
$$= 6 \cdot (5 \cdot 4 \cdot 3 \cdot 2 \cdot 1)$$
$$= 6 \cdot 5! = 6 \cdot 120 = 720.$$

4. Compute.

a) $4!$

b) $\dfrac{16}{2!}$

c) $\dfrac{-5}{0!}$

d) $\dfrac{720}{6!}$

DO EXERCISE 4.

For a moment, look back at Example 4, where we listed the coefficients $c_3, c_2, c_1,$ and c_0 in the expansion of $f(x) = x^3 - 7x + 3$ in powers of $x - 2$. Note now that we can rewrite them as follows:

$$c_0 = f(2) = \frac{f(2)}{0!} \qquad \text{since } 0! = 1,$$

$$c_1 = f'(2) = \frac{f'(2)}{1!} \qquad \text{since } 1! = 1,$$

$$c_2 = \frac{1}{2}f''(2) = \frac{f''(2)}{2}$$

$$= \frac{f''(2)}{2!},$$

and

$$c_3 = \frac{1}{6}f'''(2) = \frac{f'''(2)}{6}$$

$$= \frac{f'''(2)}{3!}.$$

In words, c_3 is the value of the third derivative of $f(x)$ at $x = 2$ (since we were expanding in powers of $x - 2$) divided by $3!$.

To make the pattern stand out more clearly, recall (Section 2.9) that $f^{(3)}(x)$ is sometimes used for the third derivative of $f(x)$. Thus $f^{(1)}(x) = f'(x)$ and $f^{(6)}(x)$ denotes the sixth derivative of $f(x)$. Finally, we let $f^{(0)}(x) = f(x)$. Then we can rewrite the preceding formulas as follows:

$$c_0 = \frac{f(2)}{0!} = \frac{f^{(0)}(2)}{0!},$$

$$c_1 = \frac{f'(2)}{1!} = \frac{f^{(1)}(2)}{1!},$$

$$c_2 = \frac{f''(2)}{2!} = \frac{f^{(2)}(2)}{2!},$$

and

$$c_3 = \frac{f'''(2)}{3!} = \frac{f^{(3)}(2)}{3!}.$$

Examples 3 and 4 illustrate the following theorem.

THEOREM 1

Taylor's Formula For Polynomials

Let $f(x)$ be a polynomial of degree n and let a be any number. Then $f(x)$ has a unique expansion in powers of $x - a$:

$$f(x) = c_0 + c_1(x - a) + c_2(x - a)^2 + \cdots + c_{n-1}(x - a)^{n-1} + c_n(x - a)^n.$$

Here, the jth coefficient, c_j, can be computed from the formula

$$c_j = \frac{f^{(j)}(a)}{j!}, \qquad j = 0, 1, \ldots, n.$$

Example 5   Write $f(x) = x^3 - 7x^2 + 2x + 3$ in powers of $x - 1$.

Solution   We apply Taylor's formula with $a = 1$. Thus,

$$c_0 = \frac{f^{(0)}(1)}{0!} = \frac{(1)^3 - 7(1)^2 + 2 \cdot 1 + 3}{1}$$

$$= 1 - 7 + 2 + 3$$

$$= -1.$$

To compute c_1, we need a formula for $f'(x)$. Differentiating $f(x) = x^3 - 7x^2 + 2x + 3$, we get

$$f^{(1)}(x) = f'(x)$$
$$= 3x^2 - 14x + 2.$$

Hence,

$$c_1 = \frac{f^{(1)}(1)}{1!} = \frac{3(1)^2 - 14 \cdot 1 + 2}{1}$$

$$= -9.$$

Continuing, we have $f^{(2)}(x) = f''(x) = 6x - 14$, so that

$$c_2 = \frac{f^{(2)}(1)}{2!} = \frac{6 \cdot 1 - 14}{2} = \frac{-8}{2} = -4.$$

Finally, $f^{(3)}(x) = f'''(x) = 6$, so that

$$c_3 = \frac{f^{(3)}(1)}{3!} = \frac{6}{3!}$$

$$= \frac{6}{6} = 1.$$

5. Write $f(x) = 2x^2 - x + 3$ in powers of $x - 1$.

Note that $f^{(4)}(x) = 0$ for all x. In particular, c_4 must be 0. Since the same is true for all higher derivatives of $f(x)$, we can stop here—that is, we have found all the nonzero c_j's. So the expansion of $f(x) = x^3 - 7x^2 + 2x + 3$ in powers of $x - 1$ is

$$f(x) = -1 - 9(x - 1) - 4(x - 1)^2 + (x - 1)^3,$$

or, equivalently,

$$f(x) = (x - 1)^3 - 4(x - 1)^2 - 9(x - 1) - 1.$$

DO EXERCISES 5 AND 6.

Example 6   Expand $f(x) = x^3 - 7x^2 + 2x + 3$ in powers of $x - 2$.

Solution   We have

$$c_0 = \frac{f^{(0)}(2)}{0!} = \frac{f(2)}{0!}$$
$$= (2)^3 - 7(2)^2 + 2 \cdot 2 + 3$$
$$= 8 - 28 + 4 + 3$$
$$= -13.$$

6. Write $f(x) = x^3 - x$ in powers of $x - 3$.

Since $f'(x) = 3x^2 - 14x + 2$ (from Example 5), then

$$c_1 = \frac{f^{(1)}(2)}{1!} = \frac{3 \cdot 2^2 - 14 \cdot 2 + 2}{1}$$
$$= \frac{12 - 28 + 2}{1}$$
$$= -14.$$

Continuing, we have $f^{(2)}(x) = 6x - 14$ and

$$c_2 = \frac{f^{(2)}(2)}{2!} = \frac{6 \cdot 2 - 14}{2} = \frac{-2}{2} = -1.$$

Since $f^{(3)}(x) = 6$,

$$c_3 = \frac{f^{(3)}(2)}{3!} = \frac{6}{6} = 1.$$

So

$$f(x) = (x - 2)^3 - (x - 2)^2 - 14(x - 2) - 13.$$

Note that even though the function being expanded is the same in Examples 5 and 6, the coefficients of the two expansions are different.

7. Write $f(x) = x^2 - 3x + 1$ in powers of $x - 4$.

DO EXERCISES 7 AND 8.

Example 7   Expand $f(x) = x^2 + 3x + 1$ in powers of $x + 1$.

Solution   Since $x + 1 = x - (-1)$, we have $a = -1$. So

$$c_0 = \frac{f^{(0)}(-1)}{0!} = \frac{(-1)^2 + 3(-1) + 1}{1}$$
$$= 2 - 3 = -1.$$

Next, $f^{(1)}(x) = f'(x) = 2x + 3$, so that

$$c_1 = \frac{f^{(1)}(-1)}{1!} = \frac{2(-1) + 3}{1} = 1.$$

Finally, $f^{(2)}(x) = 2$ and

$$c_2 = \frac{f^{(2)}(-1)}{2!} = \frac{2}{2} = 1.$$

So

$$f(x) = 1 \cdot (x + 1)^2 + 1 \cdot (x + 1) - 1$$

8. Write $f(x) = x^2 - 3x + 1$ in powers of $x - 2$.

or

$$f(x) = (x + 1)^2 + (x + 1) - 1. \tag{5}$$

Note that if we multiply this last expression out, we obtain

$$f(x) = (x^2 + 2x + 1) + (x + 1) - 1$$
$$= x^2 + (2x + x) + (1 + 1 - 1)$$
$$= x^2 + 3x + 1.$$

Our check shows that formula (5) is correct.

So far, we have regarded Taylor's theorem for polynomials only as a way to write a given polynomial $f(x)$ in powers of $x - a$ (a given). There is a slightly different point of view that will be helpful to us later in the chapter. According to this view, Taylor's theorem says that if $f(x)$ is a polynomial of degree n, and if we know the value of $f(x)$ and its first n derivatives at $x = a$, then we can find the value of $f(x)$ at any point. To illustrate this, we consider the following example.

Example 8   Find a polynomial of degree 3 such that $f(5) = 2$, $f'(5) = 0$, $f''(5) = -3$, and $f'''(5) = 12$.

9. Find a polynomial $f(x)$ of degree 2 that satisfies $f(1) = 2, f'(1) = -1$, and $f''(1) = 2$.

10. Find a polynomial $f(x)$ of degree 3 such that $f(-1) = 1, f'(-1) = -\frac{1}{2}$, $f''(-1) = 4$, and $f'''(-1) = 18$.

11. Find a polynomial of degree 5 that satisfies $f(0) = 1, f'(0) = 2$, $f''(0) = -1, f^{(3)}(0) = 42$, $f^{(4)}(0) = 12$, and $f^{(5)}(0) = -60$.

Solution   If we use the given data to expand $f(x)$ about $x = 5$ (that is, in powers of $x - 5$), we obtain

$$c_0 = \frac{f^{(0)}(5)}{0!} = \frac{f(5)}{1} = 2,$$

$$c_1 = \frac{f^{(1)}(5)}{1!} = \frac{0}{1} = 0,$$

$$c_2 = \frac{f^{(2)}(5)}{2!} = \frac{-3}{2} = -\frac{3}{2},$$

and

$$c_3 = \frac{f^{(3)}(5)}{3!} = \frac{12}{6} = 2.$$

Thus,

$$f(x) = 2(x - 5)^3 - \tfrac{3}{2}(x - 5)^2 + 0(x - 5) + 2$$
$$= 2(x - 5)^3 - \tfrac{3}{2}(x - 5)^2 + 2.$$

From Theorem 1, we know that $f(x)$ is the *only* polynomial of degree 3 that satisfies the given conditions.

DO EXERCISES 9–11.

EXERCISE SET 10.1

Simplify.

1. $\dfrac{x^2 - 4x + 3}{x}, x \neq 0$

2. $\dfrac{(x + 4)^2 - 3(x + 4) + 1}{x + 4}, x \neq -4$

3. $\dfrac{2(x - 3)^2 - 6(x - 3) + 7}{x - 3}, x \neq 3$

4. $\dfrac{x^3 - 5x^2 + 7x - 3}{x}, x \neq 0$

5. $\dfrac{(x - 2)^3 + 7(x - 2) + 4}{x - 2}, x \neq 2$

6. $\dfrac{3(x + 4)^3 + 3(x + 4)^2 + 5(x + 4)}{x + 4}, x \neq -4$

Expand each of the given polynomials in powers of $x - a$.

7. $x^2 - 2x + 3, a = 2$

8. $x^2 - 2x + 3, a = 1$

9. $x^2 - 5x + 7, a = 2$

10. $x^2 - 5x + 7, a = 1$

11. $x^2 - 5x + 7, a = -1$

12. $x^2 - 5x + 7, a = 0$

13. $x^2 - 3x + 4, a = -2$

14. $x^2 - 3x + 4, a = 2$

15. $x^3 - 2x^2 + 4x + 1, a = 1$

16. $x^3 - 2x^2 + 4x + 1, a = 2$

17. $x^3 - 2x^2 + 4x + 1, a = -1$

18. $x^3 + 5, a = 1$

19. $2x^2 - 6x + 3, a = 3$

20. $x^3 - x - 1, a = 2$

21. $3x^4 - 7x^2 + 1, a = 1$

22. $x^4 - 5x^3 + 3, a = 1$

23. $x^5 - 5x^4 + 5x - 1, a = 1$

24. $x^5 - 3x + 1, a = 1$

25. Check the answers to Exercises 11 and 13 by multiplying them out and comparing your results with the given polynomial.

26. Repeat Exercise 25 using the solutions to Exercise 15 and 21.

Simplify using Taylor's formula.

27. $\dfrac{x^2 - 5}{x - 1}, x \neq 1$

28. $\dfrac{x^2 - 3x + 1}{x - 2}, x \neq 2$

29. $\dfrac{x^2 + 1}{x + 1}, x \neq -1$

30. $\dfrac{x^2 - 7x - 1}{x + 3}, x \neq -3$

EXTENSION EXERCISES

31. Calculate $f'(1)$ and $f''(1)$ for

$$f(x) = x^4 - 4x^3 + 6x^2 - 4x + 4.$$

Conclude that $x = 1$ is a critical point for $f(x)$ but that the Second Derivative Test (Section 3.3) fails. Next, write $f(x)$ in powers of $x - 1$. Use your answers to show that $f(x)$ has a minimum point at $x = 1$ and that $x = 1$ is the only maximum or minimum point for $f(x)$.

32. Calculate $f'(1)$ and $f''(1)$ for

$$f(x) = x^3 - 3x^2 + 3x - 3.$$

Conclude, as in Exercise 31, that the Second Derivative Test fails. Write $f(x)$ in powers of $x - 1$. Use your answers to conclude that $f(x)$ has no maximum or minimum points and that $f(x)$ has exactly one point of inflection, at $x = 1$.

10.2

OBJECTIVES

a) Compute the Taylor polynomial of degree n for $f(x)$ at the point $x = a$.

b) Find a Taylor polynomial and use it to approximate a function value.

TAYLOR POLYNOMIALS

Finding Taylor Polynomials

We now extend the ideas discussed in Section 10.1 to a wide class of functions.

DEFINITION

Let $f(x)$ have derivatives of all orders. Then the *Taylor polynomial* $p_n(x)$ *of degree n at x = a* is defined to be the polynomial given by

$$p_n(x) = c_0 + c_1(x - a) + \cdots + c_n(x - a)^n$$

$$= \sum_{j=0}^{n} c_j(x - a)^j,$$

where

$$c_j = \frac{f^{(j)}(a)}{j!}, \quad \text{for } j = 0, 1, \ldots, n.$$

1. Find the Taylor polynomial of degree 2 at $x = 0$ if $f(x) = 2e^{-3x}$.

Compare this with the formula in Theorem 1 (Section 10.1).

Example 1 Find the Taylor polynomial of degree 2 at $x = 0$ if $f(x) = e^x$.

Solution We must apply the definition with $n = 2$ and $a = 0$. Doing so, we obtain $p_2(x) = c_0 + c_1 x + c_2 x^2$, where

$$c_j = \frac{f^{(j)}(0)}{j!}.$$

But $f^{(0)}(x) = f(x) = e^x$, so that

$$c_0 = \frac{f^{(0)}(0)}{0!} = \frac{e^0}{0!} = 1.$$

Since $f^{(1)}(x) = f'(x) = e^x$ and $f^{(2)}(x) = e^x$, we have

$$c_1 = \frac{f^{(1)}(0)}{1!} = \frac{e^0}{1} = 1$$

and

$$c_2 = \frac{f^{(2)}(0)}{2!} = \frac{e^0}{2} = \frac{1}{2}.$$

So

$$p_2(x) = 1 + x + \tfrac{1}{2}x^2.$$

DO EXERCISE 1.

Example 2 Find the Taylor polynomial of degree 2 at $x = -1$ if $f(x) = x^2 + 3x + 1$.

Solution We have $p_2(x) = c_0 + c_1(x + 1) + c_2(x + 1)^2$, where

$$c_j = \frac{f^{(j)}(-1)}{j!}.$$

Since $f(x) = f^{(0)}(x) = x^2 + 3x + 1$, we get

$$c_0 = \frac{f^{(0)}(-1)}{0!}$$

$$= \frac{(-1)^2 + 3(-1) + 1}{1}$$

$$= \frac{1 - 3 + 1}{1} = -1.$$

2. a) Find the Taylor polynomial of
       degree 2 at $x = 4$ if $f(x) = x^2 - 3x + 1$.

   b) Compare Example 2 and Margin
       Exercise 2 with Example 7 and
       Margin Exercise 7 of Section 10.1.

Similarly, $f'(x) = f^{(1)}(x) = 2x + 3$ and

$$c_1 = \frac{f^{(1)}(-1)}{1!} = \frac{2(-1) + 3}{1} = \frac{-2 + 3}{1} = 1,$$

and since $f''(x) = f^{(2)}(x) = 2$, we have

$$c_2 = \frac{f^{(2)}(-1)}{2!} = \frac{2}{2} = 1.$$

So

$$p_2(x) = -1 + (x + 1) + (x + 1)^2.$$

DO EXERCISE 2.

Example 3   Find the Taylor polynomials of degrees 2 and 3 at $x = 0$
if $f(x) = \cos x$.

Solution   Recalling that $f'(x) = f^{(1)}(x) = -\sin x$ and that $f''(x) = f^{(2)}(x) = -\cos x$, we have

$$c_0 = \frac{f^{(0)}(0)}{0!} = \frac{\cos (0)}{1} = 1,$$

$$c_1 = \frac{f^{(1)}(0)}{1!} = \frac{-\sin (0)}{1} = 0,$$

and

$$c_2 = \frac{f^{(2)}(0)}{2!} = \frac{-\cos (0)}{2!} = \frac{-1}{2} = -\frac{1}{2}.$$

We find that

$$p_2(x) = 1 + 0 \cdot x - \tfrac{1}{2}x^2 = 1 - \tfrac{1}{2}x^2.$$

To compute $p_3(x)$, we need only calculate

$$c_3 = \frac{f^{(3)}(0)}{3!}.$$

Since $f''(x) = -\cos x$, differentiation yields $f'''(x) = f^{(3)}(x) = \sin x$
and $f^{(3)}(0) = \sin (0) = 0$. So

$$c_3 = \frac{f^{(3)}(0)}{3!} = 0,$$

and we have

$$p_3(x) = 1 + 0 \cdot x - \tfrac{1}{2}x^2 + 0 \cdot x^3 = 1 - \tfrac{1}{2}x^2.$$

3. Find the Taylor polynomials of degrees 2 and 3 at $x = 0$ if $f(x) = \sin x$.

DO EXERCISE 3.

Example 4 Compute the Taylor polynomial of degree 2 at $x = 1$ for

$$f(x) = \frac{1}{1 + x^2}.$$

Solution Since $f^{(0)}(1) = \frac{1}{2}$, we have $c_0 = \frac{1}{2}$. Applying the Quotient Rule, we get

$$f'(x) = \frac{(1 + x^2) \cdot 0 - 1(2x)}{(1 + x^2)^2} = \frac{-2x}{(1 + x^2)^2}.$$

So

$$f'(1) = f^{(1)}(1) = \frac{-2(1)}{(1 + 1^2)^2} = \frac{-2}{4} = -\frac{1}{2}$$

and

$$c_1 = \frac{f^{(1)}(1)}{1!} = -\frac{1}{2}.$$

Finally, differentiating $f'(x)$ yields

$$f''(x) = \frac{(1 + x^2)^2(-2) - (-2x) \cdot 2(1 + x^2) \cdot 2x}{(1 + x^2)^4}$$

$$= \frac{-2(1 + x^2)^2 + 8x^2(1 + x^2)}{(1 + x^2)^4}$$

$$= \frac{-2(1 + x^2) + 8x^2}{(1 + x^2)^3}$$

$$= \frac{6x^2 - 2}{(1 + x^2)^3}.$$

4. Find the Taylor polynomial of degree 2 at $x = 0$ if $f(x) = 1/(1 - x^2)$.

So

$$c_2 = \frac{f^{(2)}(1)}{2!} = \frac{1}{2!} \cdot \frac{6(1)^2 - 2}{(1 + 1^2)^3}$$

$$= \frac{1}{2} \cdot \frac{4}{8} = \frac{1}{4}.$$

Thus,

$$p_2(x) = \frac{1}{2} - \frac{1}{2}(x - 1) + \frac{1}{4}(x - 1)^2.$$

DO EXERCISE 4.

Approximating Using Taylor Polynomials

We turn now to a geometrical or graphical interpretation of Taylor polynomials. Let $f(x) = e^x$, the graph that appears in color in the figure below. We'll consider the Taylor polynomials for e^x at $x = 0$.

The Taylor polynomial of degree 0 at $x = 0$ is the constant function $p_0(x) = f(0) = e^0 = 1$. This function is the horizontal (black) line passing through $(0, 1)$ in the figure below.

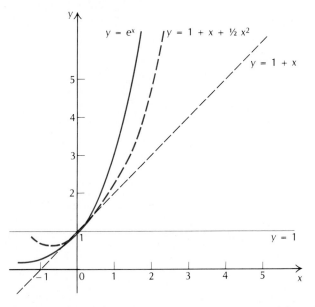

The Taylor polynomial of degree 1 at $x = 0$ is $p_1(x) = f(0) + f'(0)x = 1 + x$. This function is also linear and its graph is the dashed line in the figure above. Note that it, too, passes through $(0, 1)$. Since the slope of $1 + x$ is 1 and the slope of e^x at $(0, 1)$ is also $e^0 = 1$, we see that $1 + x$ is the tangent line to the graph of $y = e^x$ at $(0, 1)$.

In general, the Taylor polynomial of degree 1 for $f(x)$ at $x = a$ is always the tangent line to the graph of $y = f(x)$ at the point $(a, f(a))$ on the graph.

Next, consider the Taylor polynomial of degree 2 for e^x at $x = 0$. From Example 1, we have $p_2(x) = 1 + x + \frac{1}{2}x^2$. This polynomial is the color dashed curve in the figure above. Since $p_2(0) = 1$, it passes through $(0, 1)$. By the way in which it was constructed, $p_2'(0) = 1 = f'(0)$ and $p_2''(0) = 1 = f''(0)$. We might, therefore, think of $p_2(x)$ as being the "tangent quadratic" to the graph of $f(x) = e^x$ at $x = 0$.

Note that although $p_2(x)$ agrees with e^x only at $x = 0$, it is "close to e^x" for values of x that are close to 0. This suggests that we might use the Taylor polynomials of e^x to approximate e^x.

5. Find the Taylor polynomial of degree 3 for e^x at $x = 0$ and use it to calculate an approximate value for $e^{0.1}$.

CSS

Example 5 Use $p_2(x) = 1 + x + \frac{1}{2}x^2$ to calculate an approximate value for $e^{0.1}$.

Solution We know that

$$p_2(0.1) = 1 + 0.1 + \tfrac{1}{2}(0.1)^2 = 1.105.$$

Thus $e^{0.1}$ is approximately equal to 1.105. A hand-held calculator gives $e^{0.1} = 1.1051709$, so that the error involved here is about 0.0001709. We can obtain a more accurate answer by using Taylor polynomials of higher degree.

DO EXERCISE 5.

Example 6 Use the Taylor polynomial of degree 3 to calculate $\sin\left(\frac{1}{5}\right)$.

Solution We calculate $p_3(x)$ for $f(x) = \sin x$ at $x = 0$ and then approximate $\sin\left(\frac{1}{5}\right)$ by $p_3\left(\frac{1}{5}\right)$. Since $f'(x) = \cos x$, $f''(x) = -\sin x$, and $f'''(x) = -\cos x$, we have

$$c_0 = \frac{f^{(0)}(0)}{0!} = \frac{\sin(0)}{1} = 0,$$

$$c_1 = \frac{f^{(1)}(0)}{1!} = \frac{\cos(0)}{1} = 1,$$

$$c_2 = \frac{f^{(2)}(0)}{2!} = \frac{-\sin(0)}{2} = 0,$$

6. Use a Taylor polynomial to approximate $\tan\left(\frac{1}{10}\right)$.

and

$$c_3 = \frac{f^{(3)}(0)}{3!} = \frac{-\cos(0)}{6} = \frac{-1}{6},$$

so that $p_3(x) = x - \frac{1}{6}x^3$. Now

CSS

$$p_3\left(\tfrac{1}{5}\right) = \tfrac{1}{5} - \tfrac{1}{6}\left(\tfrac{1}{5}\right)^3 = 0.2 - 0.0013333 = 0.1986667.$$

Thus,

$$\sin \tfrac{1}{5} \approx 0.1986667.$$

(Recall that $\approx$ means *approximately equal to*.) A hand-held calculator (make sure that yours is set to handle radians before you compute) yields

$$\sin\left(\tfrac{1}{5}\right) \approx 0.19866933.$$

DO EXERCISE 6.

EXERCISE SET 10.2

Find $p_2(x)$ for the given function and the specified value of x.

CSS

1. $y = e^{-x}; x = 0$

2. $y = \dfrac{1}{x}; x = 1$

3. $y = \sin(x^2); x = 0$

4. $y = 5x^2 - 6x + 3; x = 0$

5. $f(x) = 2x^3 - 4x^2 + x - 1; x = 0$

6. $f(x) = \sqrt{x}; x = 1$

7. $f(x) = \sec x; x = 0$

8. $f(x) = \ln x; x = 1$

9. $y = x \sin x; x = 0$

10. $y = x^2 \cos x; x = 0$

Find $p_3(x)$ for the given function and the specified value of x.

11. $y = e^{(1/2)x}; x = 0$

12. $y = xe^x; x = 0$

13. $f(x) = x^3; x = 1$

14. $f(x) = \ln(1 + x); x = 0$

15. $y = \ln(1 - x); x = 0$

16. $y = \dfrac{1}{1 - x}; x = 0$

Find both $p_4(x)$ and $p_5(x)$ for the given function and the specified value of x.

17. $f(x) = \sin x; x = 0$

18. $f(x) = e^x; x = 0$

19. $y = \cos x; x = 0$

20. $y = \ln(1 + x); x = 0$

Sketch the graphs of $f(x)$, $p_0(x)$, $p_1(x)$, and $p_2(x)$ near the given value of x for each of the following.

21. $f(x) = \sin x; x = 0$

22. $f(x) = \cos x; x = 0$

23. $f(x) = e^{-x}; x = 0$

24. $f(x) = \sqrt{x}; x = 1$

Use $p_3(x)$ from above to calculate an approximate value for each of the following.

25. $\sin(0.13)$

26. $\cos\left(\frac{1}{8}\right)$

27. $e^{-0.05}$

28. $\ln(1.05)$

29. $\ln(0.95)$

30. $\sin\left(\frac{1}{5}\right)$

APPLICATIONS

31. *Psychology: The Hullian model of learning.* It is found that the probability P of mastery of a certain concept after t trials is given by

$$P(t) = 1 - e^{-0.18t}.$$

a) Compute $P(1)$, $P(2)$, and $P(3)$.

b) Find the Taylor polynomial $p_3(t)$ of order 3 for $P(t)$ at $t = 0$.

c) Use the polynomial in (b) to calculate approximate values for $P(1)$, $P(2)$, and $P(3)$.

32. *Newton's Law of Cooling.* An object heated to 200°F is placed in a room whose temperature is kept constant at 50°F. The temperature of the object after t hours in the room is determined to be given by

$$T(t) = 50 + 150e^{-0.5t}.$$

a) Find the temperature of the object after 1 and 2 hours.

b) Use the Taylor polynomial $p_3(t)$ of degree 3 at $t = 0$ to compute approximate values for $T(1)$ and $T(2)$.

33. *Ecology: Population growth.* The population of fish (in thousands) in a certain lake is given by

$$P(t) = \sqrt{22 + t}$$

over a five-year period beginning at $t = 0$.

a) Calculate the Taylor polynomial $p_2(t)$ of order 2 at $t = 3$ for $P(t)$.

b) Use $p_2(3.5)$ to find an approximate value for $P(3.5)$. Do the same for $P(4)$.

34. *Science: Motion of a spring.* The displacement y of a spring after t seconds is given by

$$y(t) = 6.5 \sin\left(t + \frac{\pi}{4}\right).$$

a) Find the Taylor polynomial $p_2(t)$ for $y(t)$ at $t = \pi/4$.

b) Use $p_2(\pi/3)$ to compute an approximate value for the displacement after $\pi/3$ seconds.

10.3

INFINITE SERIES

OBJECTIVES

a) Tell whether a given geometric series converges or diverges by calculating its ratio. If it converges, compute its sum.

b) Tell whether an alternating series converges or diverges. If it converges, approximate its sum by the nth partial sum and determine the error involved in this approximation.

An *infinite series* is a special kind of sum

$$a_1 + a_2 + \cdots + a_n + \cdots, \tag{1}$$

where each number of the *sequence* $a_1, a_2, \ldots a_n$ is a real number. The numbers $a_1, a_2, \ldots, a_n$ are called the *terms* of the infinite series.

Two examples of infinite series are

$$\frac{1}{2} + \frac{1}{4} + \cdots + \frac{1}{2^n} + \cdots \tag{2}$$

and

$$1 + 2 + \cdots + n + \cdots.$$

In the first series, the nth term of the series is $1/2^n$. The fourth term, for instance, is

$$\frac{1}{2^4} = \frac{1}{16}$$

and the first term is

$$\frac{1}{2^1} = \frac{1}{2}.$$

In the second series, the nth term is n.

In ordinary arithmetic, numbers are added two at a time. To add $2 + 4 + 5 + 8$, for example, requires three additions. But an infinite sum like

$$\frac{1}{2} + \frac{1}{4} + \cdots + \frac{1}{2^n} + \cdots$$

would require an infinite number of additions, so it cannot be carried out in a finite length of time. Thus in order to give meaning to an expression like (1) or (2), we use the limit concept.

If the series (1) is given, then

$$s_n = a_1 + a_2 + \cdots + a_n,$$

is called the nth *partial sum* of the series.

DEFINITION

The infinite series (1) *converges to s*, or *has the sum s*, written

$$s = a_1 + a_2 + \cdots + a_n + \cdots,$$

if the limit of the partial sums s_n of the series is s. Here s is a (finite) real number. In other words,

$$s = \lim_{n \to \infty} (a_1 + a_2 + \cdots + a_n)$$

$$= \lim_{n \to \infty} s_n.$$

If the partial sums do not have a limit, then we say that the series *diverges*.

Example 1   Find the sum of Eq. (2), that is, of

$$\frac{1}{2} + \frac{1}{4} + \frac{1}{8} + \cdots + \frac{1}{2^n} + \cdots.$$

Solution   Let's consider s_n for a few values of n:

$$s_1 = \tfrac{1}{2} \qquad\qquad\qquad = \tfrac{1}{2} = 0.5,$$
$$s_2 = \tfrac{1}{2} + \tfrac{1}{4} \qquad\qquad = \tfrac{3}{4} = 0.75,$$
$$s_3 = \tfrac{1}{2} + \tfrac{1}{4} + \tfrac{1}{8} \qquad = \tfrac{7}{8} = 0.875,$$
$$s_4 = \tfrac{1}{2} + \tfrac{1}{4} + \tfrac{1}{8} + \tfrac{1}{16} \qquad = \tfrac{15}{16} = 0.9375,$$
$$s_5 = \tfrac{1}{2} + \tfrac{1}{4} + \tfrac{1}{8} + \tfrac{1}{16} + \tfrac{1}{32} = \tfrac{31}{32} = 0.96875.$$

Note that when s_n is written as a fraction, as in the next-to-last column, the numerator is one less than the denominator. Since the denominator is 2^n, this suggests that

$$s_n = \frac{2^n - 1}{2^n} = 1 - \frac{1}{2^n}$$

1. **a)** Find an expression for the *n*th term a_n of the infinite series

$$\frac{2}{3} + \frac{2}{9} + \frac{2}{27} + \frac{2}{81} + \cdots.$$

  b) Calculate the first six partial sums of the series above. (Round your answers to seven decimal places.)

2. **a)** Find an expression for the *n*th term of

$$1 + 3 + 5 + 7 + 9 + \cdots.$$

  b) Calculate the first five partial sums of this series.

  c) Find an expression for the *n*th partial sum s_n of this series.

  d) Does the series converge?

for all *n*. This formula is true. Note that as *n* increases, the sequence $1/2^n$ decreases rapidly toward 0. So

$$\lim_{n \to \infty} \left(1 - \frac{1}{2^n}\right) = 1.$$

Then

$$s = \lim_{n \to \infty} s_n = 1.$$

DO EXERCISE 1.

Example 2   Find the sum of $1 + 1 + \cdots + 1 + \cdots$.

Solution   The *n*th partial sum of this series is

$$\underbrace{1 + 1 + \cdots + 1}_{n \text{ times}} = n.$$

So the partial sums tend to infinity as *n* increases. Since $+\infty$ is not a real number, the series diverges.

DO EXERCISE 2.

A special case in which we can always determine convergence or divergence is that of a geometric series.

DEFINITION

The **infinite series**

$$a + ar + ar^2 + \cdots + ar^{n-1} + \cdots \tag{3}$$

is called a *geometric series with initial term a and ratio r*. Here $a \neq 0$.

The series in Example 1 is a geometric series with initial term $\frac{1}{2}$ and ratio $\frac{1}{2}$.

If $a + ar + \cdots + ar^{n-1} + \cdots$ is a geometric series, then the ratio of the second term, ar, to the first term, a, is

$$\frac{ar}{a} = r.$$

The ratio of the third term, ar^2, to the second term, ar, is

$$\frac{ar^2}{ar} = r.$$

3. a) Find an expression for the nth term a_n of the geometric series

$$2 + \tfrac{2}{5} + \tfrac{2}{25} + \cdots.$$

b) Find the fifth and sixth terms of the series above.

In general, the ratio of any term of a geometric series to the preceding term is r. In other words, $a_{n+1}/a_n = r$, or $a_{n+1} = ra_n$ for all n.

Example 3 Find the fifth and sixth terms of the geometric series $2 + 6 + 18 + \cdots$.

Solution The initial term a is 2. The ratio of the second term to the first is $\tfrac{6}{2} = 3$. So $r = 3$. This means that the nth term of the series is $ar^{n-1} = 2 \cdot 3^{n-1}$. Thus the fifth term is $2 \cdot 3^{5-1} = 2 \cdot 3^4 = 2 \cdot 81 = 162$. The sixth term is three times the fifth term, or $a_6 = 3 \cdot 162 = 486$.

DO EXERCISES 3 AND 4.

The Partial Sums of a Geometric Series

We want to find a formula for the nth partial sum s_n of a geometric series:

$$s_n = a + ar + ar^2 + \cdots + ar^{n-1}.$$

If we multiply both sides of this equation by the ratio r, we get

$$rs_n = ar + ar^2 + \cdots + ar^{n-1} + ar^n.$$

4. Find the initial term and the ratio of the geometric series

$$\tfrac{1}{5} + \tfrac{1}{15} + \tfrac{1}{45} + \cdots.$$

If we multiply the formula for s_n by -1, we have

$$-s_n = -a - ar - ar^2 - \cdots - ar^{n-1}.$$

Adding these expressions, we get

$$rs_n - s_n = ar^n - a,$$

or

$$(r - 1)s_n = a(r^n - 1).$$

To solve this equation for s_n, we can divide both sides by $r - 1$ (if $r \neq 1$), getting the following formula.

THEOREM 2

The nth partial sum of a geometric series is given by

$$s_n = \frac{a(r^n - 1)}{r - 1} \qquad (4)$$

for any $r \neq 1$.

---

5. Find the nth partial sum s_n of the geometric series

$$2 + \tfrac{2}{3} + \tfrac{2}{9} + \tfrac{2}{27} + \cdots.$$

Example 4   Find the sum of the first n terms of the geometric series

$$\frac{1}{2} + \frac{1}{4} + \cdots + \frac{1}{2^n} + \cdots.$$

CSS

Solution   This is a geometric series with $a = \tfrac{1}{2}$ and $r = \tfrac{1}{2}$. Using Eq. (4), we have

$$s_n = \frac{\dfrac{1}{2}\left[\left(\dfrac{1}{2}\right)^n - 1\right]}{\dfrac{1}{2} - 1} = \frac{\dfrac{1}{2}\left[\left(\dfrac{1}{2}\right)^n - 1\right]}{-\dfrac{1}{2}}$$

$$s_n = -\left[\left(\frac{1}{2}\right)^n - 1\right]$$

or

$$s_n = 1 - \frac{1}{2^n}.$$

Note that this justifies our use of this formula in Example 1.

DO EXERCISE 5.

The Sum of a Geometric Series

We want to be able to tell when a given geometric series converges and when it diverges. If it converges, we want to be able to find its sum. Formula (4) enables us to do this. For if r is a number between -1 and 1, then it follows that r^n gets closer and closer to 0 as n gets larger. (Pick a number r between -1 and 1 and check this by finding larger and larger powers on your calculator.) As r^n gets closer and closer to 0, so does ar^n. Since

$$s_n = \frac{a(r^n - 1)}{r - 1} = \frac{ar^n - a}{r - 1},$$

we see that s_n gets closer to

$$\frac{-a}{r - 1} = \frac{a}{1 - r}$$

as n increases, if r is between -1 and 1.

If $|r|$ is greater than 1, then $|r|^n$ gets larger and larger as n increases. (▦ Check this on your calculator.) In this case, s_n does not have a limit.

6. Find the sum of the geometric series

$$2 + \tfrac{2}{5} + \tfrac{2}{25} + \cdots .$$

If $|r| = 1$, the series also diverges. This case is considered in Exercises 53 and 54.

THEOREM 3

The geometric series $a + ar + ar^2 + \cdots + ar^n + \cdots$ converges to

$$s = \frac{a}{1 - r}$$

if $|r| < 1$, that is, if r is between -1 and 1. Otherwise, it diverges.

Example 5   Find the sum of

$$2 + \frac{2}{3} + \frac{2}{9} + \frac{2}{27} + \cdots + \frac{2}{3^{n-1}} + \cdots .$$

Solution   This is a geometric series with initial term $a = 2$ and ratio $r = \tfrac{1}{3}$. Since r is between -1 and 1, the series converges and the sum is

$$s = \frac{a}{1 - r} = \frac{2}{1 - \tfrac{1}{3}}$$

$$= \frac{2}{\tfrac{2}{3}} = 2 \cdot \frac{3}{2} = 3.$$

DO EXERCISES 6 AND 7.

7. Find the sum of the geometric series

$$\tfrac{1}{5} + \tfrac{1}{15} + \tfrac{1}{45} + \cdots .$$

Example 6   Find the sum of

$$1 - 1 + 1 - 1 + \cdots + (-1)^{n+1} + \cdots .$$

Solution   This is a geometric series with initial term $a = 1$ and ratio $r = -1$. Since r is not between -1 and 1, this series diverges. Note that the partial sums of this series are the sequence $1, 0, 1, 0, 1, 0, 1, \ldots$, which has no limit.

DO EXERCISE 8.

8. Find the sum of

$$1 - 2 + 4 - 8 + \cdots .$$

Example 7   Find the sum of

$$1 - \frac{1}{2} + \frac{1}{4} - \frac{1}{8} + \cdots + \left(-\frac{1}{2}\right)^{n-1} + \cdots .$$

9. Find the sum of

$$1 - \frac{2}{5} + \frac{4}{25} - \frac{8}{125} + \cdots.$$

Solution   This is a geometric series with $a = 1$ and $r = -\frac{1}{2}$. Thus it converges and its sum is

$$\frac{a}{1 - r} = \frac{1}{1 - (-\frac{1}{2})} = \frac{1}{1 + \frac{1}{2}} = \frac{1}{\frac{3}{2}} = \frac{2}{3}.$$

DO EXERCISE 9.

Example 8   *Economic multiplier.* The United States banking laws require banks to maintain a reserve equivalent to a certain proportion of their outstanding deposits. This enables such banks, when they wish and when they can find borrowers, to loan out a certain proportion of the funds that have been deposited in them. Let us assume that this proportion is 0.90 (or 90%). Now suppose a corporation deposits $1000 in a bank that subsequently is able to loan the maximum amount legally possible, and this loan is redeposited elsewhere, and so on. What is the total effect of the $1000 on the economy?

10. Find s if the proportion in Example 8 is changed to 0.95 (or 95%).

Solution   The total effect can be modeled as the sum of the infinite geometric series

$$\$1000 + \$1000(0.90) + \$1000(0.90)^2 + \$1000(0.90)^3 + \cdots,$$

which is given by

$$s = \frac{\$1000}{1 - 0.90}$$
$$= \$10,000.$$

11. Find s if the proportion in Example 8 is changed to 85%.

The sum $10,000 is the result of what is referred to in economics as the *multiplier effect*.

DO EXERCISES 10 AND 11.

Example 9   Write $0.63636363\ldots$, or $0.\overline{63}$, as a fraction. (The bar above indicates the portion that repeats.)

Solution   The key to the problem is to note that we can write this decimal as a geometric series. Thus,

$$0.63636363\ldots = 0.63 + 0.0063 + 0.000063 + \cdots$$
$$= \frac{63}{100} + \frac{63}{10,000} + \frac{63}{1,000,000} + \cdots.$$

12. Write $0.4444\ldots$ as a fraction.

The right-hand side is a geometric series with initial term $a = \frac{63}{100} = 0.63$ and ratio $r = \frac{1}{100} = 0.01$. Since the ratio is between -1 and 1, the sum of the series can be expressed as

$$\frac{a}{1-r} = \frac{\frac{63}{100}}{1 - \frac{1}{100}} = \frac{\frac{63}{100}}{\frac{99}{100}}$$

$$= \frac{63}{99} = \frac{7}{11}.$$

Thus,

$$\frac{7}{11} = 0.63636363\ldots, \quad \text{or} \quad 0.\overline{63}.$$

13. Write $0.4594594\ldots$ as a fraction.

DO EXERCISES 12 AND 13.

Alternating Series

We reconsider the series

$$1 - \frac{1}{2} + \frac{1}{4} - \frac{1}{8} + \cdots + \left(-\frac{1}{2}\right)^{n-1} + \cdots = \frac{2}{3}.$$

The first six partial sums are as follows:

$$
\begin{aligned}
s_1 &= 1 & &= 1 = 1, \\
s_2 &= 1 - \tfrac{1}{2} & &= \tfrac{1}{2} = 0.5, \\
s_3 &= 1 - \tfrac{1}{2} + \tfrac{1}{4} & &= \tfrac{3}{4} = 0.75, \\
s_4 &= 1 - \tfrac{1}{2} + \tfrac{1}{4} - \tfrac{1}{8} & &= \tfrac{5}{8} = 0.625, \\
s_5 &= 1 - \tfrac{1}{2} + \tfrac{1}{4} - \tfrac{1}{8} + \tfrac{1}{16} & &= \tfrac{11}{16} = 0.6875, \\
s_6 &= 1 - \tfrac{1}{2} + \tfrac{1}{4} - \tfrac{1}{8} + \tfrac{1}{16} - \tfrac{1}{32} & &= \tfrac{21}{32} = 0.65625.
\end{aligned}
$$

14. Calculate the first six partial sums of
$$2 - \tfrac{2}{3} + \tfrac{2}{9} - \tfrac{2}{27} + \tfrac{2}{81} - \cdots.$$

Note that the odd-numbered partial sums $1, 0.75, 0.6875, \ldots$ are all greater than $\frac{2}{3}$, the sum of the given series. The even-numbered partial sums are all less than $\frac{2}{3}$.

Furthermore, the odd-numbered partial sums $1, 0.75, 0.6875, \ldots$ are decreasing. The even-numbered partial sums $0.5, 0.625, 0.65625, \ldots$ are increasing.

DO EXERCISE 14.

The discussion above and Margin Exercise 14 illustrate the following theorem.

15. Determine whether or not

$$1 - \frac{1}{4} + \frac{1}{9} - \frac{1}{16} + \cdots$$

$$+ (-1)^{n+1} \frac{1}{n^2} + \cdots$$

converges or diverges.

THEOREM 4

The Alternating Series Test

Let

$$a_1 \geqslant a_2 \geqslant \cdots \geqslant a_n \geqslant \cdots$$

be positive numbers with $\lim_{n \to \infty} a_n = 0$. Then the *alternating series*

$$a_1 - a_2 + a_3 - a_4 + \cdots + (-1)^{n-1} a_n + \cdots$$

converges. Furthermore, the odd-numbered partial sums are all greater than s, the sum of the series. In fact, they decrease to s. The even-numbered partial sums are all less than s (and increase to s).

Example 10   Determine whether the series

$$1 - \frac{1}{2} + \frac{1}{3} - \frac{1}{4} + \cdots + (-1)^{n-1} \frac{1}{n} + \cdots$$

converges or diverges.

Solution   The given series is an alternating series with $a_n = 1/n$ for all positive integers n. Since these numbers decrease to 0, the series converges.

DO EXERCISE 15.

Example 11   Show that

$$1 - \frac{1}{8} + \frac{1}{27} + \cdots + (-1)^{n-1} \frac{1}{n^3} + \cdots$$

converges, and use the fourth and fifth partial sums to estimate the sum s of the series.

Solution   The series is an alternating series with $a_n = 1/n^3$ for each positive integer n. Since these numbers decrease to 0, the series converges. The fourth partial sum is

$$1 - \tfrac{1}{8} + \tfrac{1}{27} - \tfrac{1}{64} \approx 0.896412.$$

The fifth partial sum is

$$1 - \tfrac{1}{8} + \tfrac{1}{27} - \tfrac{1}{64} + \tfrac{1}{125} \approx 0.904412.$$

According to the test, the sum s of the series is between these two

16. Use the third and fourth partial sums to estimate the sum of

$$1 - \frac{1}{16} + \frac{1}{81} - \frac{1}{256} + \cdots$$

$$+ (-1)^{n+1} \frac{1}{n^4} + \cdots.$$

numbers. Thus,

$$0.896412 < s < 0.904412.$$

DO EXERCISE 16.

In Example 11 we saw that the sum s of the series

$$1 - \frac{1}{8} + \frac{1}{27} + \cdots + (-1)^{n-1} \frac{1}{n^3} + \cdots$$

was between 0.896412 and 0.904412, the fourth and fifth partial sums of the series. The difference between the fifth partial sum and the fourth partial sum of this series is the fifth term of the series, $\frac{1}{125}$. If we want to use 0.896412 as an approximate value for s, then the error involved is no more than $\frac{1}{125} = 0.008$. This fact is so important that we state it separately.

THEOREM 5

If $a_1 \geqslant a_2 \geqslant \cdots \geqslant a_n \geqslant \cdots$ are positive numbers with $\lim\limits_{n \to \infty} a_n = 0$ and if s is the sum of the (convergent) alternating series

$$a_1 - a_2 + a_3 + \cdots + (-1)^{n-1} a_n + \cdots,$$

then the absolute value of the difference between s and the nth partial sum is less than or equal to a_{n+1}. That is,

$$|s - s_n| \leqslant a_{n+1}.$$

Example 12   Let

$$s = 1 - \frac{1}{7} + \frac{1}{49} + \cdots + \frac{(-1)^{n-1}}{7^{n-1}} + \cdots.$$

Find the error involved in approximating s by the fourth partial sum

$$s_4 = 1 - \tfrac{1}{7} + \tfrac{1}{49} - \tfrac{1}{343}$$

$$\approx 0.87463557.$$

Check by finding s explicitly.

Solution   By the alternating series test, we know that the series converges. According to Theorem 5 above, the error involved in approximating s by 0.87463557 is less than $1/7^4 = \frac{1}{2401} \approx 0.0004165$. Since the given series happens to be a geometric series with $a = 1$ and

17. a) Use the third and fourth partial sums to estimate the sum of

$$1 - \frac{1}{9} + \frac{1}{81} - \frac{1}{729} + \cdots.$$

b) Then check your answer by finding the sum of this geometric series.

$r = -\frac{1}{7}$, we see that the sum is

$$s = \frac{a}{1 - r} = \frac{1}{1 - (-\frac{1}{7})} = \frac{1}{\frac{8}{7}} = \frac{7}{8} = 0.875.$$

Thus the error involved in approximating s by 0.87463557 is 0.00036443, so that our error estimate above is correct.

DO EXERCISE 17.

It is important to keep in mind that the alternating-series test applies only if the series is indeed alternating.

Example 13 Use the alternating-series test to determine whether

$$1 + \frac{1}{2} + \frac{1}{3} + \cdots + \frac{1}{n} + \cdots$$

converges or diverges.

Solution This series is not an alternating series! Thus the alternating series test cannot be applied. Contrary to what you might expect, this series *diverges*, as we will show in Exercise Set 10.3.

EXERCISE SET 10.3

Find an expression for the nth term a_n of the series.

1. $\frac{1}{5} + \frac{1}{25} + \frac{1}{125} + \cdots$

2. $1 + 4 + 9 + 16 + 25 + \cdots$

3. $\frac{1}{2} + \frac{2}{3} + \frac{3}{4} + \frac{4}{5} + \frac{5}{6} + \cdots$

4. $1 - \frac{1}{2} + \frac{1}{3} - \frac{1}{4} + \cdots$

5. $\frac{2}{3} + \frac{4}{9} + \frac{8}{27} + \frac{16}{81} + \frac{32}{243} + \cdots$

6. $\frac{1}{4} - \frac{1}{16} + \frac{1}{64} - \frac{1}{256} + \cdots$

7. $1 - 2 + 3 - 4 + \cdots$

8. $1 + \frac{1}{6} + \frac{1}{36} + \frac{1}{216} + \cdots$

Find the initial term, the ratio, and the fifteenth partial sum for the geometric series.

9. $1 + \frac{1}{3} + \frac{1}{9} + \cdots$

10. $1 + \frac{1}{5} + \frac{1}{25} + \cdots$

11. $1 - \frac{1}{3} + \frac{1}{9} - \cdots$

12. $1 - \frac{1}{5} + \frac{1}{25} - \cdots$

13. $\frac{1}{8} - \frac{1}{12} + \frac{1}{18} - \cdots$

14. $\frac{1}{2} + \frac{2}{3} + \frac{8}{9} + \cdots$

15. $\frac{5}{3} - \frac{2}{3} + \frac{4}{15} - \cdots$

16. $\frac{1}{3} - \frac{1}{2} + \frac{3}{4} - \frac{9}{8} + \cdots$

Determine whether the geometric series converges or diverges. If it converges, find its sum.

17. $1 + \frac{1}{8} + \frac{1}{64} + \cdots$

18. $1 + 3 + 9 + 27 + \cdots$

19. $1 + \frac{1}{9} + \frac{1}{81} + \frac{1}{729} + \cdots$

20. $1 - 2 + 4 - 8 + \cdots$

21. $1 - \frac{2}{3} + \frac{4}{9} - \cdots$

22. $\frac{1}{2} - \frac{1}{2} + \frac{1}{2} - \frac{1}{2} + \cdots$

23. $\frac{1}{4} + \frac{1}{4} + \frac{1}{4} + \cdots$

24. $\frac{1}{2} - \frac{1}{6} + \frac{1}{18} - \cdots$

Write the decimal as a rational fraction (the bar above indicates the portion that repeats).

25. $0.777\overline{7}$

26. $0.4141\overline{41}$

27. $2.3838\overline{38}$

28. $0.03666\overline{6}$

29. $0.142857\overline{142857}$

30. $0.1234567890\overline{1234567890}$

Determine whether the infinite series converges or diverges.

31. $\frac{1}{7} + \frac{2}{7} + \frac{4}{7} + \frac{8}{7} + \cdots$

32. $1 - \frac{1}{3} + \frac{1}{5} - \frac{1}{7} + \cdots$

33. $1 - \frac{9}{10} + \frac{81}{100} - \frac{729}{1000} + \cdots$

34. $1 - \frac{1}{9} + \frac{1}{25} - \frac{1}{49} + \frac{1}{81} - \cdots$

35. $1 - \frac{1}{4} + \frac{1}{7} - \frac{1}{10} + \cdots$

36. $\frac{1}{9} - \frac{1}{6} + \frac{1}{4} - \frac{3}{8} + \cdots$

37. $1 - \frac{3}{4} + \frac{9}{16} - \frac{27}{64} + \cdots$

38. $-\frac{1}{5} + \frac{1}{6} - \frac{1}{7} + \frac{1}{8} - \cdots$

39. $\frac{1}{5} - \frac{1}{7} + \frac{1}{9} - \frac{1}{11} + \cdots$

40. $\frac{1}{36} + \frac{1}{24} + \frac{1}{16} + \frac{3}{32} + \cdots$

Use the fourth and fifth partial sums of the (convergent) series to estimate the sum of the series.

41. $1 - \frac{1}{3} + \frac{1}{9} - \frac{1}{27} + \cdots$

42. $1 - \frac{1}{6} + \frac{1}{36} - \frac{1}{216} + \cdots$

43. $1 - \dfrac{1}{16} + \dfrac{1}{81} - \dfrac{1}{256} + \cdots + (-1)^{n+1} \cdot \dfrac{1}{n^4} + \cdots$

44. $\dfrac{1}{2} - \dfrac{1}{8} + \dfrac{1}{24} - \dfrac{1}{64} + \cdots + (-1)^{n-1} \cdot \dfrac{1}{n2^n} + \cdots$

45. $\dfrac{1}{9} - \dfrac{1}{25} + \dfrac{1}{49} - \dfrac{1}{81} + \cdots + (-1)^{n+1} \cdot \dfrac{1}{(2n+1)^2} + \cdots$

46. $-\dfrac{1}{4} + \dfrac{1}{16} - \dfrac{1}{36} + \dfrac{1}{64} - \dfrac{1}{100} + \cdots$

APPLICATIONS

47. *Economics.* The government makes an $8,000,000,000 expenditure for a new type of aircraft. If 75% of this gets spent again, and 75% of that gets spent, and so on, what is the total effect on the economy?

48. *Medicine.* A patient is to receive 1 mg (milligram) of a certain drug once every day. By the time the next dose is administered, it is expected that 50% of the drug will have been eliminated from the patient's system.

   a) Find the amount of the drug in the patient's system immediately after the first four doses have been administered.

   b) If we assume that the pattern in part (a) holds, how much of the drug will be present after the fifteenth dose is administered?

   c) When will the amount of the drug in the patient's system exceed 3 mg?

49. If, in Problem 48, the daily dosage is 5 mg and the percentage eliminated daily is 60%, find each of the following.

   a) The amount of the drug in the patient's system immediately after the first, second, and third doses have been administered

   b) The amount of the drug in the patient's system immediately after the ninth dose has been administered

   c) The "limiting value" of the amount of the drug in the patient's system, that is, the sum of the geometric series in question

50. *Finance.* A certain *perpetuity* is to pay $100 each year (starting today). If the current interest rate is i, then the present value of a payment of $100 n$ years from today is defined to be $100(1 + i)^{-n}$, so that the *total* present value P of the payments is given by

$$P = 100 + \frac{100}{1+i} + \frac{100}{(1+i)^2} + \cdots .$$

   a) Find the present value of the perpetuity if $i = 0.10$ (10% interest).

   b) What happens to the present value if the interest rate drops from 10% to 8%?

51. *Investments.* A certain stock pays a dividend of $5 per year at present. If the dividend grows 2% each year and if the interest rate is 12.5%, then the present value V of the next twelve payments (beginning today) is given by

$$V = 5 + 5\left(\frac{1.02}{1.125}\right) + 5\left(\frac{1.02}{1.125}\right)^2 + \cdots + 5\left(\frac{1.02}{1.125}\right)^{11}.$$

Find the sum.

52. *Chemistry.* A factory deposits waste material containing 1 kg (kilogram) of a certain radioactive isotope on a slag heap each day. It is known that after one day, 3% of the isotope will decay.

a) How much of the isotope will be present after the first, second, and third deposits?

b) How much of the isotope will be present after the fiftieth deposit?

c) Find the limiting value of the amount of isotope in the slag heap.

EXTENSION EXERCISES

53. By computing the nth partial sum s_n, show that a geometric series with initial term $a \neq 0$ and ratio $r = 1$ diverges.

54. By computing the nth partial sum s_n, show that a geometric series with initial term $a \neq 0$ and ratio $r = -1$ diverges.

55. A patient is able to eliminate p ($0 < p < 1$) of the concentration of a certain drug between successive doses. If the daily dosage of the drug is d mg, find the limiting value of the amount of the drug in the patient's system.

56. A perpetuity pays $\$r$ annually beginning today. If the interest rate is i, then the present value of the annuity, denoted P, is given by

$$P = r + \frac{r}{1 + i} + \frac{r}{(1 + i)^2} + \cdots.$$

Find the sum of the series.

57. Find the value of the perpetuity in Exercise 56 if the first payment is to be made one year from today.

58. We consider the harmonic series

$$1 + \frac{1}{2} + \frac{1}{3} + \frac{1}{4} + \frac{1}{5} + \cdots + \frac{1}{n} + \cdots.$$

Note that

$$\frac{1}{3} + \frac{1}{4} > \frac{1}{4} + \frac{1}{4} = \frac{1}{2},$$

that

$$\frac{1}{5} + \frac{1}{6} + \frac{1}{7} + \frac{1}{8} > \frac{1}{8} + \frac{1}{8} + \frac{1}{8} + \frac{1}{8} = \frac{1}{2},$$

and so forth.

a) Show that

$$\frac{1}{2^n + 1} + \frac{1}{2^n + 2} + \cdots + \frac{1}{2^{n+1}} > \frac{1}{2}.$$

b) Use the result from part (a) to show that $s_4 > 2$, $s_8 > \frac{5}{2}$, $s_{16} > 3$; then show that $s_{2^{n+1}} > (n + 3)/2$ for all integers n.

c) Conclude that the harmonic series diverges.

10.4

TAYLOR SERIES

a) Compute the Taylor series expansion for a given function.

b) Use Taylor series and the alternating-series test to approximate certain integrals.

In Section 10.2, we saw that we can approximate many functions by using an appropriately constructed polynomial, called the Taylor polynomial (of degree n) of $f(x)$ at $x = a$. Although exact equality occurs only at $x = a$, the approximation is excellent near $x = a$ and the accuracy can be improved by using a polynomial of increased degree. This raises the question of whether we can obtain exact equality by taking the limit (as n tends to infinity) of the Taylor polynomial of degree n. When this occurs, we say that $f(x)$ is *represented by its Taylor series at* $x = a$.

DEFINITION

If $f(x)$ has derivatives of all orders at $x = a$, then *the Taylor series of $f(x)$ at $x = a$* is the series

$$\sum_{j=0}^{\infty} c_j(x - a)^j,$$

where

$$c_j = \frac{f^{(j)}(a)}{j!} \quad \text{for } j = 0, 1, \ldots.$$

Note that the nth partial sum of the series is given by

$$p_n(x) = \sum_{j=0}^{n} c_j(x - a)^j,$$

the Taylor polynomial of $f(x)$ (of degree n) at $x = a$. We say that $f(x)$ is *represented by its Taylor series at $x = a$* if

$$f(x) = \sum_{j=0}^{\infty} c_j(x - a)^j$$

$$= \lim_{n \to \infty} \sum_{j=0}^{n} c_j(x - a)^j$$

$$= \lim_{n \to \infty} p_n(x).$$

Example 1   Compute the Taylor series of $f(x) = e^x$ at $x = 0$.

Solution   In this problem, $a = 0$. We must compute the coefficients, given by $c_j = f^{(j)}(0)/j!$. Since $f(x) = e^x$, the required derivatives are easy to compute in this case. We have $f^{(j)}(x) = e^x$ for all j. So $f^{(j)}(0) =$

1. Compute the Taylor series of

$$f(x) = e^{-x} \quad \text{at } x = 0.$$

$e^0 = 1$ for all values of j and $c_j = f^{(j)}(0)/j! = 1/j!$ for all j. Thus the Taylor series for e^x at $x = 0$ is

$$1 + x + \frac{x^2}{2!} + \frac{x^3}{3!} + \cdots + \frac{x^n}{n!} + \cdots = \sum_{j=0}^{\infty} \frac{x^j}{j!}.$$

DO EXERCISE 1.

Example 2   Compute the Taylor series of $f(x) = \sin x$ at $x = 0$.

Solution   We have $f'(x) = \cos x$ and $f''(x) = f^{(2)}(x) = -\sin x$. Differentiating again, we obtain $f^{(3)}(x) = -\cos x$, and, differentiating again, $f^{(4)}(x) = \sin x$. Note that $f^{(4)}(x) = f(x)$, the function with which we started. This means that $f^{(5)}(x) = f'(x) = \cos x$, that $f^{(6)}(x) = f^{(2)}(x) = -\sin x$, and so forth, repeating every four terms. If we let $x = 0$, we get

$$f(0) = \sin (0) = 0,$$
$$f'(0) = \cos (0) = 1,$$
$$f''(0) = f^{(2)}(0) = -\sin (0) = 0,$$
$$f^{(3)}(0) = -\cos (0) = -1,$$
$$f^{(4)}(0) = \sin (0) = 0,$$
$$f^{(5)}(0) = f^{(1)}(0) = 1,$$

and so on. In other words, the derivatives, when evaluated at $x = 0$, produce the sequence $0, 1, 0, -1, 0, 1, 0, -1, 0, 1, \ldots$, in which the four numbers $0, 1, 0, -1$ (in that order) are continually repeated.

It is helpful to be able to describe this sequence concisely. The even-numbered terms are all equal to 0. This is equivalent to $f^{(j)}(0) = 0$ for $j = 2k$, or $f^{(2k)}(0) = 0$ for all k. If we let $j = 2k + 1$, then, as k takes on the values $0, 1, 2, 3, \ldots, j$ takes on the values $1, 3, 5, 7, \ldots$. Using this, we can write $f^{(j)}(0) = (-1)^k$ for $j = 2k + 1$, or $f^{(2k+1)}(0) = (-1)^k$. Note that as k takes on the values $0, 1, 2, \ldots$, the quantity $(-1)^k$ is equal to $1, -1, 1, -1, \ldots$.

Finally, since $c_j = f^{(j)}(0)/j!$, we see that for j even (that is, $j = 2k$), we have $c_{2k} = 0$, whereas for j odd ($j = 2k + 1$), we have $c_{2k+1} = (-1)^k/(2k + 1)!$. Thus the Taylor series for $\sin x$ at $x = 0$ is

$$\sum_{k=0}^{\infty} \frac{(-1)^k x^{2k+1}}{(2k + 1)!},$$

or

$$x - \frac{x^3}{3!} + \frac{x^5}{5!} - \frac{x^7}{7!} + \frac{x^9}{9!} - \cdots.$$

2. Compute the Taylor series of

$$f(x) = \cos x \quad \text{at } x = 0.$$

3. Compute the Taylor series of

$$f(x) = 7x^3 - 5x^2 + 2x + 4$$

at $x = 0$.

DO EXERCISE 2.

Example 3 Compute the Taylor series of $f(x) = 2x^2 - 3x + 5$ at $x = 0$.

Solution We have

$$f(0) = 2(0)^2 - 3(0) + 5 = 5.$$

Since $f'(x) = 4x - 3$, we have

$$f'(0) = 4(0) - 3 = -3.$$

Next, $f^{(2)}(x) = 4$, so that $f^{(2)}(0) = 4$. Observe that $f^{(3)}(x)$ and all higher-order derivatives of $f(x)$ are equal to 0. So

$$c_0 = \frac{f(0)}{0!} = \frac{5}{1} = 5,$$

$$c_1 = \frac{f^{(1)}(0)}{1!} = \frac{-3}{1} = -3,$$

and

$$c_2 = \frac{f^{(2)}(0)}{2!} = \frac{4}{2} = 2,$$

with all other coefficients equal to 0. Thus the Taylor series for $f(x) = 2x^2 - 3x + 5$ at $x = 0$ is $5 - 3x + 2x^2 + 0 \cdot x^3 + 0 \cdot x^4 + \cdots$, or $5 - 3x + 2x^2$, which is what we started with.

DO EXERCISE 3.

Example 4 Compute the Taylor series of $f(x) = \ln x$ at $x = 1$.

Solution We have $f'(x) = 1/x = x^{-1}$. So, differentiating again, we get

$$f^{(2)}(x) = f''(x) = (-1)x^{-2} = -1/x^2.$$

Continuing, we have

$$f^{(3)}(x) = (-1)(-2x^{-3}) = 2x^{-3} = \frac{2}{x^3},$$

$$f^{(4)}(x) = 2(-3x^{-4}) = -(3!)x^{-4},$$

and

$$f^{(5)}(x) = -(3!)(-4x^{-5}) = (4!)x^{-5} = \frac{4!}{x^5}.$$

To see the pattern, consider the following:

$$f^{(1)}(x) = (0!)x^{-1},$$
$$f^{(2)}(x) = -(1!)x^{-2},$$
$$f^{(3)}(x) = (2!)x^{-3},$$
$$f^{(4)}(x) = -(3!)x^{-4},$$
$$f^{(5)}(x) = (4!)x^{-5}.$$

We see that $f^{(j)}(x) = (-1)^{j+1}(j-1)!x^{-j}$ [here, the $(-1)^{j+1}$ term takes care of the alternating signs of the derivatives]. We then have

$$f^{(j)}(1) = (-1)^{j+1}(j-1)!(1)^{-j} = (-1)^{j+1}(j-1)! \quad \text{for } j > 0,$$

so that

$$c_j = \frac{f^{(j)}(1)}{j!} = \frac{(-1)^{j+1}(j-1)!}{j!} = \frac{(-1)^{j+1}(j-1)!}{j(j-1)!}$$
$$= \frac{(-1)^{j+1}}{j} \quad \text{for } j > 0.$$

Recalling that

$$c_0 = \frac{f^{(0)}(1)}{0!} = \frac{\ln(1)}{1} = 0,$$

we see that the Taylor series of $\ln x$ at $x = 1$ is

$$(x-1) - \tfrac{1}{2}(x-1)^2 + \tfrac{1}{3}(x-1)^3 - \tfrac{1}{4}(x-1)^4 + \cdots$$
$$= \sum_{j=1}^{\infty} \frac{(-1)^{j+1}(x-1)^j}{j}.$$

Example 5 Compute the Taylor series of $f(x) = 1/(1-x)$ at $x = 0$.

Solution Writing $f(x) = (1-x)^{-1}$, we see that

$$f'(x) = -1(1-x)^{-2}(-1) = (1-x)^{-2}.$$

So

$$f^{(2)}(x) = f''(x) = -2(1-x)^{-3}(-1) = 2(1-x)^{-3}$$

and

$$f^{(3)}(x) = 2(-3)(1-x)^{-4}(-1) = 3!(1-x)^{-4}.$$

We see that

$$f^{(j)}(x) = (j!)(1-x)^{-(j+1)} \quad \text{for all } j > 0.$$

Thus,

$$f^{(j)}(0) = (j!)(1-0)^{-(j+1)} = (j!)(1)^{-(j+1)} = j!.$$

4. Compute the Taylor series of

$$f(x) = 1/(1 + x^2) \quad \text{at } x = 0.$$

We get

$$c_j = \frac{f^{(j)}(0)}{j!} = \frac{j!}{j!} = 1 \quad \text{for all } j \geqslant 0,$$

and the Taylor series for $1/(1 - x)$ at $x = 0$ is

$$1 + x + x^2 + \cdots + x^n + \cdots = \sum_{j=0}^{\infty} x^j.$$

Note that this is a geometric series.

DO EXERCISE 4.

In Example 5, we showed that the Taylor series for $1/(1 - x)$ at $x = 0$ is given by $1 + x + x^2 + \cdots + x^n + \cdots$. In Section 10.3, we showed that the geometric series $1 + r + r^2 + \cdots + r^n + \cdots$ converges to $1/(1 - r)$ if $-1 < r < 1$, and diverges otherwise. This shows that the Taylor series for $f(x) = 1/(1 - x)$ is *equal to* $f(x)$ for values of x between -1 and 1, that is,

$$\frac{1}{1 - x} = 1 + x + x^2 + \cdots + x^n + \cdots, \qquad -1 < x < 1,$$

and that this formula makes no sense for other values of x.

It turns out that this example illustrates the general case. The series $\sum_{j=0}^{\infty} c_j(x - a)^j$ always converges at $x = a$, since all the terms except the first are 0 in this case. Also, there is a nonnegative number R so that the series converges on the interval $(a - R, a + R)$ and diverges if $|x - a| > R$ (where R may be 0 or $+\infty$). If R is positive and finite, then the series converges on the following interval:

and diverges for $x > a + R$ and $x < a - R$ (convergence at the endpoints $a + R$ and $a - R$ must be decided on a case-by-case basis). If $R = 0$, then the series converges only at $x = a$ (where it always converges) and diverges for all other values of x. If $R = +\infty$, then the series converges for all real numbers x. Fortunately, for a large class of functions, it turns out that R (the *radius of convergence*) is positive and $f(x)$ is equal to its Taylor series on $(a - R, a + R)$, the *interval of convergence* of the series.

If $f(x) = \sum_{j=0}^{\infty} c_j(x - a)^j$ on $(a - R, a + R)$ for R positive, then this formula can be differentiated or integrated termwise within the given interval.

5. Compute the Taylor series of

$$f(x) = 1/(1 + x)^2 \quad \text{at } x = 0.$$

We can also make appropriate substitutions. These techniques may make the computation of Taylor series much less tedious. We illustrate with several examples.

Example 6 Compute the Taylor series of $f(x) = 1/(1 - x)^2$ at $x = 0$.

Solution We use

$$(1 - x)^{-1} = \frac{1}{1 - x} = 1 + x + x^2 + \cdots, \qquad \text{valid for } -1 < x < 1,$$

from Example 5. Differentiating $(1 - x)^{-1}$ gives

$$(-1)(1 - x)^{-2}(-1) = (1 - x)^{-2}.$$

Differentiating the series $1 + x + x^2 + x^3 + \cdots$ term by term gives $0 + 1 + 2x + 3x^2 + \cdots$. Since the formula with which we started was correct for all x between -1 and 1, so too is the new formula. That is,

6. Compute the Taylor series of

$$f(x) = 1/(1 - x)^3 \quad \text{at } x = 0.$$

$$(1 - x)^{-2} = \frac{1}{(1 - x)^2} = 1 + 2x + 3x^2 + \cdots,$$

for all x with $-1 < x < 1$.

DO EXERCISES 5 AND 6.

Example 7 Compute the Taylor series of $f(x) = 1/(1 + x^2)$ at $x = 0$.

Solution We could compute the Taylor series directly, but those calculations are lengthy (see Margin Exercise 4). For a shortcut, we use

7. Compute the Taylor series of

$$f(x) = 1/(1 - x^3) \quad \text{at } x = 0.$$

$$\frac{1}{1 - t} = 1 + t + t^2 + t^3 + \cdots, \qquad -1 < t < 1.$$

Letting $t = -x^2$, we get

$$\frac{1}{1 - (-x^2)} = 1 + (-x^2) + (-x^2)^2 + (-x^2)^3 + \cdots$$

or

$$\frac{1}{1 + x^2} = 1 - x^2 + x^4 - x^6 + \cdots.$$

This formula is valid for $-1 < x^2 < 1$ or, in other words, for $-1 < x < 1$.

DO EXERCISE 7.

8. Compute the Taylor series of

$$f(x) = \sin (x^2) \quad \text{at } x = 0.$$

Example 8   Compute the Taylor series of $f(x) = e^{-x^2}$ at $x = 0$.

Solution   We use the fact that e^x is equal to its Taylor series at $x = 0$ for *all* values of x. That is,

$$e^t = 1 + t + \frac{t^2}{2!} + \frac{t^3}{3!} + \cdots \quad \text{for all } t.$$

Letting $t = -x^2$, we get

$$e^{-x^2} = 1 - x^2 + \frac{x^4}{2!} - \frac{x^6}{3!} + \frac{x^8}{4!} - \frac{x^{10}}{5!} + \cdots \quad \text{for all } x.$$

DO EXERCISES 8 AND 9.

9. Compute the Taylor series of

$$f(x) = e^{-5x} \quad \text{at } x = 0.$$

Example 9   Compute the Taylor series of $f(x) = \ln (1 + x)$ at $x = 0$.

Solution   Letting $x = -t$ in the series for $1/(1 - x)$ yields

$$\frac{1}{1 + t} = 1 - t + t^2 - t^3 + \cdots, \qquad -1 < t < 1.$$

Integrating both sides from 0 to x, we obtain

$$\int_0^x \frac{1}{1 + t} \, dt = \int_0^x (1 - t + t^2 - t^3 + \cdots) \, dt$$

$$= \int_0^x dt - \int_0^x t \, dt + \int_0^x t^2 \, dt - \int_0^x t^3 \, dt + \cdots,$$

or

10. Compute the Taylor series of

$$f(x) = \ln (1 - x) \quad \text{at } x = 0.$$

$$[\ln (1 + t)]_0^x = [t]_0^x - \left[\frac{t^2}{2}\right]_0^x + \left[\frac{t^3}{3}\right]_0^x - \left[\frac{t^4}{4}\right]_0^x + \cdots,$$

or

$$\ln (1 + x) - \ln (1 + 0) = (x - 0) - \left(\frac{x^2}{2} - 0\right)$$

$$+ \left(\frac{x^3}{3} - 0\right) - \left(\frac{x^4}{4} - 0\right) + \cdots.$$

Thus,

$$\ln (1 + x) = x - \frac{x^2}{2} + \frac{x^3}{3} - \frac{x^4}{4} + \cdots, \qquad -1 < x < 1.$$

DO EXERCISE 10.

Example 10 Compute the Taylor series of $f(x) = \tan^{-1} x$ at $x = 0$.

Solution Recall that the derivative of $\tan^{-1} x$ is $1/(1 + x^2)$ (see Section 8.4). In Example 7, we found that the Taylor series at $x = 0$ of $1/(1 + x^2)$ is

$$\frac{1}{1 + x^2} = 1 - x^2 + x^4 - x^6 + \cdots, \qquad -1 < x < 1.$$

Replacing x by t and integrating from $t = 0$ to $t = x$ yields

$$\int_0^x \frac{1}{1 + t^2}\, dt = \int_0^x \{1 - t^2 + t^4 - t^6 + \cdots\}\, dt, \qquad -1 < x < 1,$$

or

$$[\tan^{-1} t]_0^x = \left[t - \frac{t^3}{3} + \frac{t^5}{5} - \frac{t^7}{7} + \cdots \right]_0^x, \qquad -1 < x < 1$$

and

$$\tan^{-1} x - \tan^{-1} 0 = \left(x - \frac{x^3}{3} + \frac{x^5}{5} - \frac{x^7}{7} + \cdots \right) - 0.$$

Since $\tan^{-1} 0 = 0$, we have

$$\tan^{-1} x = x - \frac{x^3}{3} + \frac{x^5}{5} - \frac{x^7}{7} + \cdots, \qquad -1 < x < 1.$$

Example 11 Use Taylor series to estimate

$$\int_0^{1/2} \frac{1}{1 + x^3}\, dx$$

with an error less than 0.001.

Solution We use a combination of several techniques from this chapter. First, we compute the Taylor series for $1/(1 + x^3)$ at $x = 0$ by letting $t = x^3$ in the expression for $1/(1 + t)$ to get

$$\frac{1}{1 + x^3} = 1 - x^3 + x^6 - x^9 + \cdots, \qquad -1 < x < 1.$$

Integrating both sides from 0 to $\frac{1}{2}$ yields

$$\int_0^{1/2} \frac{1}{1 + x^3}\, dx = \left[x - \frac{x^4}{4} + \frac{x^7}{7} - \frac{x^{10}}{10} + \cdots \right]_0^{1/2},$$

11. Use Taylor series to compute

$$\int_0^{1/2} \frac{1}{1+x^4}\, dx$$

with an error less than 0.001.

or

$$\int_0^{1/2} \frac{1}{1+x^3}\, dx = \frac{1}{2} - \frac{1}{4}\left(\frac{1}{2}\right)^4 + \frac{1}{7}\left(\frac{1}{2}\right)^7 - \frac{1}{10}\left(\frac{1}{2}\right)^{10} + \cdots.$$

Before proceeding, note that the series on the right is an alternating series. Thus if we were to approximate the integral by using, say, $\frac{1}{2} - \frac{1}{4}(\frac{1}{2})^4$, the resulting approximation would be smaller than the actual value of the integral, and the absolute value of error would be less than the next term,

$$\frac{1}{7}\left(\frac{1}{2}\right)^7 = \frac{1}{7\cdot 2^7} = \frac{1}{7\cdot 128} = \frac{1}{896}.$$

If we use the sum of the first three terms to approximate the integral, then we will have an overestimate of the integral, and the error will be less than

$$\frac{1}{10}\cdot\left(\frac{1}{2}\right)^{10} = \frac{1}{10\cdot 2^{10}} = \frac{1}{10,240}.$$

Since we want an error of less than $0.001 = \frac{1}{1000}$, and since $\frac{1}{896} > \frac{1}{1000}$, we must use three terms. We get

$$\int_0^{1/2} \frac{1}{1+x^3}\, dx \approx \frac{1}{2} - \frac{1}{4}\left(\frac{1}{2}\right)^4 + \frac{1}{7}\left(\frac{1}{2}\right)^7$$

$$= \frac{1}{2} - \frac{1}{64} + \frac{1}{896},$$

or

$$\int_0^{1/2} \frac{1}{1+x^3}\, dx \approx 0.485491$$

with an error of less than $1/10,000 = 0.0001$.

DO EXERCISE 11.

Example 12   Compute the Taylor series expansion of $f(x) = x^2 e^x$ at $x = 0$.

Solution   We could calculate the coefficients of the Taylor series of $x^2 e^x$ by directly computing the derivatives of $f(x)$ (see Exercise Set 10.4). There is, however, an easier way. From Example 1, we have

$$e^x = 1 + x + \frac{1}{2}x + \frac{1}{3!}x^3 + \cdots,$$

12. Compute the Taylor series of

$$f(x) = x^2 \cos x \quad \text{at } x = 0.$$

valid for all values of x. Simply multiplying both sides of this equation by x^2 yields

$$x^2 e^x = x^2 \left(1 + x + \frac{1}{2} x^2 + \frac{1}{3!} x^3 + \cdots \right)$$

$$= x^2 + x^3 + \frac{1}{2} x^4 + \frac{1}{3!} x^5 + \cdots,$$

valid for all real numbers x.

DO EXERCISES 12 AND 13.

13. Compute the Taylor series of

$$f(x) = x \tan^{-1} x \quad \text{at } x = 0.$$

Example 13 Calculate the Taylor series expansion of $f(x)$ at $x = 0$, where $f(x)$ is given by

$$f(x) = \begin{cases} \dfrac{\sin x}{x}, & x \neq 0, \\ 1, & x = 0. \end{cases}$$

Solution The formula $(\sin x)/x$ makes sense only for $x \neq 0$. From Example 2, we get

$$\sin x = x - \frac{1}{3!} x^3 + \frac{1}{5!} x^5 - \frac{1}{7!} x^7 + \cdots,$$

valid for all real numbers x. If x is not equal to 0, then we can divide both sides of this equation by x, getting

$$\frac{\sin x}{x} = 1 - \frac{1}{3!} x^2 + \frac{1}{5!} x^4 - \frac{1}{7!} x^6 + \cdots,$$

valid for all x (except $x = 0$, of course). Note that as x approaches 0 in the formula above, the right-hand side will approach 1. That is,

$$\lim_{x \to 0} \frac{\sin x}{x} = 1,$$

a fact that we will use in Section 10.5. For the present, note that we have shown that

$$f(x) = 1 - \frac{x^2}{3!} + \frac{x^4}{5!} - \frac{x^6}{7!} + \cdots$$

for *all* values of x.

14. Compute the Taylor series of $f(x)$ at $x = 0$ if

$$f(x) = \begin{cases} \dfrac{e^x - 1}{x}, & x \neq 0, \\ 1, & x = 0 \end{cases}.$$

DO EXERCISE 14.

EXERCISE SET 10.4

Compute the Taylor series expansion of $f(x)$ at the prescribed value of x by directly calculating the coefficients.

1. $f(x) = e^{4x}; x = 0$

2. $f(x) = \sin(5x); x = 0$

3. $f(x) = e^{1-x}; x = 1$

4. $f(x) = \cos(x - 2); x = 2$

5. $f(x) = x^3 - 1; x = -1$

6. $f(x) = x^2 - 5x + 1; x = 3$

7. $f(x) = \dfrac{1}{1 - 2x}; x = 0$

8. $f(x) = \dfrac{1}{2 - x}; x = 0$

Using any method you wish, compute the Taylor series expansion of $f(x)$ for the prescribed value of x.

9. $f(x) = e^{6x}; x = 0$

10. $f(x) = \cos(x^2); x = 0$

11. $f(x) = \dfrac{1}{1 - x^5}; x = 0$

12. $f(x) = \dfrac{1}{1 - bx}\ (b \neq 0); x = 0$

13. $f(x) = \ln(1 + 4x); x = 0$

14. $f(x) = \dfrac{1}{(1 - bx)^2}\ (b \neq 0); x = 0$

15. $f(x) = \dfrac{1}{2 + x}; x = -1\left(\text{Hint: } \dfrac{1}{2 + x} = \dfrac{1}{1 + (1 + x)}\right)$

16. $f(x) = \dfrac{1}{(2 + x)^2}; x = -1$

17. $f(x) = \ln(2 + x); x = -1$

18. $f(x) = \dfrac{1}{(1 + x)^3}; x = 0$

19. $f(x) = \dfrac{5x^4}{1 - x^5}; x = 0$ (See Exercise 11 above.)

20. $f(x) = \ln(1 - x^5); x = 0$

21. $f(x) = \begin{cases} \dfrac{\cos x - 1}{x^2}, & x \neq 0, \\ -\dfrac{1}{2}, & x = 0 \end{cases}$

22. $f(x) = x^3 e^{-x}; x = 0$

23. $f(x) = x^5 \sin(x^3); x = 0$

24. $f(x) = \begin{cases} \dfrac{\sin(x^3)}{x}, & x \neq 0, \\ 0, & x = 0 \end{cases}$

Use the Taylor series and the alternating series test to find an approximate value for each of the following integrals (use three terms of the series and give an error estimate).

25. $\int_0^1 \cos(x^2)\, dx$ (See Exercise 10 above.)

26. $\int_0^1 x^5 \sin(x^3)\, dx$ (See Exercise 23 above.)

27. $\int_0^1 e^{-x^3}\, dx$

28. $\int_0^{0.1} \ln(1 + x)\, dx$

29. $\int_0^{0.5} x^5 \sin(x^3)\, dx$

30. $\int_0^{0.25} \cos(x^2)\, dx$

EXTENSION EXERCISE

31. *Probability theory.* Let

$$F(t) = \int_0^t e^{(-x^2)/2} \, dx.$$

a) Find the Taylor series expansion of $F(t)$ at $x = 0$.

b) Use your answer to part (a) to calculate

$$P(0 \leqslant t \leqslant 0.1) = \frac{1}{\sqrt{2\pi}} F(0.1)$$

for the normal distribution.

c) Check your answer to (b) in Table 2.

What we have given here is a method for constructing that table.

OBJECTIVE

a) Determine whether or not a given limit expression is an indeterminate form. If it is, use a version of l'Hôpital's Rule to evaluate it.

1. Evaluate

$$\lim_{x \to 3} \frac{x + 1}{x^2 + 2}.$$

10.5

INDETERMINATE FORMS AND l'HÔPITAL'S RULE

We will lay the groundwork for this section with three preliminary examples. You may wish to review the definition of continuous function given in Section 2.1.

Example 1 Evaluate

$$\lim_{x \to 0} \frac{x^2 + 1}{x + 3}.$$

Solution Let $f(x) = x^2 + 1$ and $g(x) = x + 3$. Then $f(x)$ and $g(x)$ are continuous functions for all x. It follows that

$$\frac{f(x)}{g(x)} = \frac{x^2 + 1}{x + 3}$$

is continuous except when $g(x) = 0$ (that is, when $x = -3$). So we can evaluate the given limit as follows:

$$\lim_{x \to 0} \frac{x^2 + 1}{x + 3} = \frac{\lim\limits_{x \to 0} (x^2 + 1)}{\lim\limits_{x \to 0} (x + 3)} = \frac{(0)^2 + 1}{0 + 3} = \frac{1}{3}.$$

DO EXERCISE 1.

Evaluate.

2. $\lim\limits_{x \to 0} \dfrac{1 - \cos x}{x^2 + 2}$

3. $\lim\limits_{x \to 1} \dfrac{\ln x}{x}$

Example 2 Evaluate

$$\lim_{x \to 0} \frac{\sin x}{x + 3}.$$

Solution Let $f(x) = \sin x$ and $g(x) = x + 3$. Then $f(x)$ and $g(x)$ are continuous functions. It follows that $f(x)/g(x)$ is continuous except at $x = -3$. So, just as in Example 1, we have

$$\lim_{x \to 0} \frac{\sin x}{x + 3} = \frac{\lim\limits_{x \to 0} \sin x}{\lim\limits_{x \to 0} x + 3}$$

$$= \frac{\sin (0)}{0 + 3} = \frac{0}{3} = 0.$$

Note that the fact that the *numerator* of the resulting fraction is 0 does not affect the validity of the steps that were used.

DO EXERCISES 2 AND 3.

Example 3 Evaluate

$$\lim_{x \to 1} \frac{x^2}{x - 1}.$$

Solution Let $f(x) = x^2$ and $g(x) = x - 1$. Then, just as before, $f(x)$ and $g(x)$ are continuous functions. Also, just as before,

$$\frac{f(x)}{g(x)} = \frac{x^2}{x - 1}$$

is continuous for all values of x such that $g(x) \neq 0$ (in this case, except for $x = 1$). But this is exactly the point at which we wish to evaluate the limit. We cannot use the techniques of Examples 1 and 2 since the expression $1/0$ is meaningless.

To try to understand what is happening, we compute a few values of $x^2/(x - 1)$ for choices of x near 1 (but different from 1). For $x = 1.1$, we obtain

$$\frac{(1.1)^2}{1.1 - 1} = \frac{1.21}{0.1} = 12.1.$$

For $x = 1.01$, we get

$$\frac{(1.01)^2}{1.01 - 1} = \frac{1.0201}{0.01} = 102.01.$$

4. Compute the ratio $x^2/(x-1)$ for $x = 0.93$, $x = 0.991$, and $x = 0.999$.

CSS

For $x = 1.001$, we get

$$\frac{(1.001)^2}{1.001 - 1} = \frac{1.002001}{0.001} = 1002.001.$$

Note that as we take values of x that are closer and closer to 1, the numerator of $x^2/(x-1)$ gets closer and closer to 1 while the denominator gets closer and closer to 0, so that the resulting quotients are larger and larger positive numbers.

The values of x used above were all larger than 1. If we had used values of x smaller than 1 (and getting closer and closer to 1), the quotients would be negative numbers with large absolute values.

In any event, the quotients do not get closer to a real number. In this case, we say that the limit *does not exist*.

DO EXERCISES 4 AND 5.

In investigating

$$\lim_{x \to a} \frac{f(x)}{g(x)},$$

5. Evaluate

$$\lim_{x \to -1} \frac{x}{x+1}.$$

where $f(x)$ and $g(x)$ are continuous functions, we have seen that we can compute the limit by substitution so long as $g(a) \neq 0$. If the denominator approaches 0 as x approaches a, while the numerator approaches a nonzero number, then the limit does not exist. This leaves unresolved only the case in which both the numerator and the denominator are both 0 at $x = a$. This is the most interesting case.

The Indeterminate Form 0/0

Example 4   Evaluate

$$\lim_{x \to 0} \frac{x^2 + 3x}{5x}.$$

Solution   This limit is of the form

$$\lim_{x \to a} \frac{f(x)}{g(x)},$$

where $a = 0$, $f(x) = x^2 + 3x$, and $g(x) = 5x$. Substitution in the numerator yields $f(0) = (0)^2 + 3(0) = 0$. Substitution in the denominator yields $g(0) = 5(0) = 0$. Since substitution of $x = a$ in both numerator

6. Evaluate

$$\lim_{x \to 0} \frac{x^2 - 7x}{x}.$$

and denominator yields 0, we say that

$$\lim_{x \to 0} \frac{x^2 + 3x}{5x}$$

is an indeterminate form of the type $\frac{0}{0}$.

In this case, we cannot evaluate the limit by direct substitution. Note, however, that we can factor an x out of $x^2 + 3x$, obtaining

$$\lim_{x \to 0} \frac{x^2 + 3x}{5x} = \lim_{x \to 0} \frac{x(x + 3)}{5x}.$$

If x is not zero,

$$\frac{x(x + 3)}{5x} = \frac{x + 3}{5}.$$

Thus we have

$$\lim_{x \to 0} \frac{x^2 + 3x}{5x} = \lim_{x \to 0} \frac{x + 3}{5}.$$

Note that the limit on the right-hand side of the last equation is *not* an indeterminate form. In fact, direct substitution yields

$$\lim_{x \to 0} \frac{x + 3}{5} = \frac{(0) + 3}{5} = \frac{3}{5}.$$

So

$$\lim_{x \to 0} \frac{x^2 + 3x}{5x} = \frac{3}{5}.$$

DO EXERCISE 6.

Example 5   Evaluate

$$\lim_{x \to -1} \frac{x^2 + 4x + 3}{x + 1}.$$

Solution   Substitution in the numerator yields

$$(-1)^2 + 4(-1) + 3 = 1 - 4 + 3 = 0.$$

Since substitution of $x = -1$ in the denominator also yields 0, this is an indeterminate form. We can write the numerator in the form $x^2 + 4x + 3 = (x + 1)(x + 3)$. For values of x that are not equal to -1, we have

$$\frac{x + 4x + 3}{x + 1} = \frac{(x + 1)(x + 3)}{x + 1} = x + 3.$$

7. Evaluate

$$\lim_{x \to -3} \frac{x^2 + 4x + 3}{x + 3}.$$

So

$$\lim_{x \to -1} \frac{x^2 + 4x + 3}{x + 1} = \lim_{x \to -1} (x + 3) = (-1) + 3 = 2.$$

DO EXERCISE 7.

Example 6   Evaluate

$$\lim_{x \to 0} \frac{\sin x}{x}.$$

Solution   Both the numerator and the denominator are continuous functions of x, but substitution of $x = 0$ yields $\sin (0) = 0$ in the numerator and 0 in the denominator. This limit is an indeterminate form. In this case, we recall from Section 10.4 that $f(x) = \sin x$ can be represented as

$$\sin x = x - \frac{x^3}{3!} + \frac{x^5}{5!} - \frac{x^7}{7!} + \cdots.$$

This formula is valid for all values of x. Dividing both sides by x, we see that if x is not equal to 0, then

$$\frac{\sin x}{x} = 1 - \frac{x^2}{3!} + \frac{x^4}{5!} - \frac{x^6}{7!} + \cdots.$$

8. Evaluate

$$\lim_{x \to 0} \frac{1 - \cos x}{x^2}.$$

So

$$\lim_{x \to 0} \frac{\sin x}{x} = \lim_{x \to 0} \left(1 - \frac{x^2}{3!} + \frac{x^4}{5!} - \frac{x^6}{7!} + \cdots \right)$$

$$= 1 - \frac{(0)^2}{3!} + \frac{(0)^4}{5!} - \frac{(0)^6}{7!} + \cdots = 1.$$

DO EXERCISE 8.

So far, we have evaluated several indeterminate forms using algebraic techniques that are often tedious and time-consuming. Fortunately, there is an easier way to do many of these problems.
Consider

$$\lim_{x \to a} \frac{f(x)}{g(x)},$$

where we assume that $f(a) = 0$ and $g(a) = 0$, so that this limit is an indeterminate form. Suppose that $f(x)$ and $g(x)$ are both represented by their Taylor series expansion (at $x = a$) on an interval containing

$x = a$. Then

$$f(x) = f(a) + f'(a)(x - a) + \frac{f''(a)}{2!}(x - a)^2 + \cdots.$$

Since $f(a) = 0$, we have

$$f(x) = f'(a)(x - a) + \frac{f''(a)}{2!}(x - a)^2 + \frac{f'''(a)}{3!}(x - a)^3 + \cdots.$$

Factoring the common term $x - a$ from the right-hand side yields

$$f(x) = (x - a)\left[f'(a) + \frac{f''(a)}{2!}(x - a) + \frac{f'''(a)}{3!}(x - a)^2 + \cdots\right].$$

Likewise,

$$g(x) = (x - a)\left[g'(a) + \frac{g''(a)}{2!}(x - a) + \frac{g'''(a)}{3!}(x - a)^2 + \cdots\right],$$

since $g(a) = 0$. So, for values of x that are not equal to a, we have

$$\frac{f(x)}{g(x)} = \frac{(x - a)\left[f'(a) + \dfrac{f''(a)}{2!}(x - a) + \cdots\right]}{(x - a)\left[g'(a) + \dfrac{g''(a)}{2!}(x - a) + \cdots\right]}$$

$$= \frac{f'(a) + \dfrac{f''(a)}{2!}(x - a) + \cdots}{g'(a) + \dfrac{g''(a)}{2!}(x - a) + \cdots}.$$

Note that substituting $x = a$ in the right-hand side yields $f'(a)/g'(a)$. This suggests the following rule.

THEOREM 6

l'Hôpital's Rule

If

$$\lim_{x \to a} \frac{f(x)}{g(x)}$$

is an indeterminate form of the type $\frac{0}{0}$ and if

$$\lim_{x \to a} \frac{f'(x)}{g'(x)}$$

exists, then

$$\lim_{x \to a} \frac{f(x)}{g(x)} = \lim_{x \to a} \frac{f'(x)}{g'(x)}.$$

9. Use l'Hôpital's Rule to evaluate

$$\lim_{x \to -1} \frac{x^2 + 4x + 3}{x + 1}.$$

Example 7   Use l'Hôpital's Rule to evaluate

$$\lim_{x \to 0} \frac{x^2 + 3x}{5x}.$$

Solution   This is an indeterminate form of the type $\frac{0}{0}$, with $f(x) = x^2 + 3x$ and $g(x) = 5x$. Here, $f'(x) = 2x + 3$ and $g'(x) = 5$. So

$$\lim_{x \to 0} \frac{x^2 + 3x}{5x} = \lim_{x \to 0} \frac{2x + 3}{5}$$

$$= \frac{2(0) + 3}{5} = \frac{3}{5}.$$ CSS

This is the correct answer, as we saw in Example 4.

DO EXERCISES 9 AND 10.

10. Use l'Hôpital's Rule to evaluate

$$\lim_{x \to -3} \frac{x^2 + 4x + 3}{x + 3}.$$

Example 8   Evaluate

$$\lim_{x \to 0} \frac{e^{2x} - 1}{x}.$$

Solution   Substitution of $x = 0$ in $f(x) = e^{2x} - 1$ yields

$$f(0) = e^{2(0)} - 1 = e^0 - 1 = 1 - 1 = 0.$$

Since the same substitution into $g(x) = x$ also yields 0, this is an indeterminate form. Since $f'(x) = 2e^{2x}$ and $g'(x) = 1$, we have

$$\lim_{x \to 0} \frac{e^{2x} - 1}{x} = \lim_{x \to 0} \frac{2e^{2x}}{1} = \frac{2e^{2(0)}}{1} = 2.$$

11. Use l'Hôpital's Rule to evaluate

$$\lim_{x \to 0} \frac{\sin x}{x}.$$

DO EXERCISE 11.

Example 9   Evaluate

$$\lim_{x \to 0} \frac{1 - \cos x}{x^2}.$$

Solution   This is an indeterminate form of the type $\frac{0}{0}$. Since $f(x) = 1 - \cos x$ and $g(x) = x^2$, we have $f'(x) = \sin x$ and $g'(x) = 2x$. Now, both $\sin x$ and $2x$ tend to 0 as x approaches 0. In other words,

$$\lim_{x \to 0} \frac{f'(x)}{g'(x)} = \lim_{x \to 0} \frac{\sin x}{2x}$$

is *another* indeterminate form. But we may apply l'Hôpital's Rule

12. Use l'Hôpital's Rule to evaluate

$$\lim_{x \to 1} \frac{1 + \ln x - x}{(x - 1)^2}.$$

again and again, stopping as soon as we obtain a limit that is not an indeterminate form. In this case, we obtain

$$\lim_{x \to 0} \frac{f''(x)}{g''(x)} = \lim_{x \to 0} \frac{\cos x}{2}$$

$$= \frac{\cos(0)}{2} = \frac{1}{2}.$$

Thus,

$$\lim_{x \to 0} \frac{f(x)}{g(x)} = \lim_{x \to 0} \frac{f'(x)}{g'(x)} = \lim_{x \to 0} \frac{f''(x)}{g''(x)} = \frac{1}{2}.$$

DO EXERCISE 12.

Example 10   Use l'Hôpital's Rule to evaluate

$$\lim_{x \to 0} \frac{e^x}{\sin x}.$$

13. Use l'Hôpital's Rule, if possible, to evaluate

$$\lim_{x \to 2} \frac{e^x}{x - 1}.$$

Solution   As x approaches 0, substitution shows that the numerator approaches $e^0 = 1$, while the denominator approaches $\sin(0) = 0$. This is *not* an indeterminate form of the type $\frac{0}{0}$ and l'Hôpital's Rule does *not* apply. In fact, this limit is similar to the limit in Example 3. So

$$\lim_{x \to 0} \frac{e^x}{\sin x} \quad \text{does not exist.}$$

DO EXERCISE 13.

Example 11   Evaluate

$$\lim_{x \to 0} \frac{\sin x}{x^2}.$$

Solution   This is an indeterminate form of the type $\frac{0}{0}$. Here, $f(x) = \sin x$ and $g(x) = x^2$, so that $f'(x) = \cos x$ and $g'(x) = 2x$. But

$$\lim_{x \to 0} \frac{\cos x}{2x}$$

does not exist since as x approaches 0, the numerator approaches $\cos(0) = 1$ while the denominator approaches 0. L'Hôpital's Rule, as stated earlier, does not apply. It turns out, though, that if the

14. Evaluate

$$\lim_{x \to 0} \frac{e^x - 1}{x^3}.$$

numerator $f'(x)$ has a nonzero limit and the denominator $g'(x)$ approaches 0 as x approaches a, then the original limit,

$$\lim_{x \to a} \frac{f(x)}{g(x)},$$

does not exist either. Thus,

$$\lim_{x \to 0} \frac{\sin x}{x^2}$$

does not exist.

DO EXERCISE 14.

Indeterminate Forms of the Form ∞/∞

In applications, one encounters limits of the form

$$\lim_{x \to a} \frac{f(x)}{g(x)},$$

where both $f(x)$ and $g(x)$ tend to infinity as x approaches a. In this case, the following version of l'Hôpital's Rule applies.

THEOREM 7

If $f(x)$ and $g(x)$ are continuous and if $f(x)$ and $g(x)$ approach infinity as x approaches a, then

$$\lim_{x \to a} \frac{f(x)}{g(x)} = \lim_{x \to a} \frac{f'(x)}{g'(x)}$$

if this latter limit exists. If $f'(x)/g'(x)$ tends to infinity as x approaches a, then so does $f(x)/g(x)$.

Example 12 Evaluate

$$\lim_{x \to +\infty} \frac{2x - 3}{x + 1}.$$

Solution Both numerator and denominator tend to infinity as x does. Applying l'Hôpital's Rule, we have

$$\lim_{x \to +\infty} \frac{2x - 3}{x + 1} = \lim_{x \to +\infty} \frac{2}{1} = 2.$$

15. Evaluate

$$\lim_{x \to +\infty} \frac{5x - 1}{4 - 3x}.$$

16. Evaluate

$$\lim_{x \to +\infty} \frac{7e^{3x}}{5x}.$$

17. Evaluate

$$\lim_{x \to +\infty} \frac{3x^2 - 5x + 1}{x^2 + 1}. \quad \boxed{\text{CSS}}$$

Thus,

$$\lim_{x \to +\infty} \frac{2x - 3}{x + 1} = 2.$$

DO EXERCISE 15.

Example 13   Evaluate

$$\lim_{x \to +\infty} \frac{e^x}{x}.$$

Solution   Again, both numerator and denominator tend to infinity as x does. By l'Hôpital's Rule,

$$\lim_{x \to +\infty} \frac{e^x}{x} = \lim_{x \to +\infty} \frac{e^x}{1} = +\infty.$$

That is, the limit does not exist.

DO EXERCISE 16.

Example 14   Evaluate

$$\lim_{x \to +\infty} \frac{3x^2 - 5x + 1}{x^2 - 3x + 7}.$$

Solution   One application of l'Hôpital's Rule yields

$$\lim_{x \to +\infty} \frac{3x^2 - 5x + 1}{x^2 - 3x + 7} = \lim_{x \to +\infty} \frac{6x - 5}{2x - 3},$$

which is again an indeterminate form of the type ∞/∞. A second application, however, gives

$$\lim_{x \to +\infty} \frac{6x - 5}{2x - 3} = \lim_{x \to +\infty} \frac{6}{2} = 3.$$

Thus,

$$\lim_{x \to +\infty} \frac{3x^2 - 5x + 1}{x^2 - 3x + 7} = 3.$$

DO EXERCISE 17.

EXERCISE SET 10.5

Find the limit, if possible, *without* using l'Hôpital's Rule.

1. $\lim\limits_{x \to 1} \dfrac{1}{x}$

2. $\lim\limits_{x \to 2} \dfrac{x-2}{x^2}$

3. $\lim\limits_{x \to -1} \dfrac{3}{x+1}$

4. $\lim\limits_{x \to -5} \dfrac{7}{x+5}$

5. $\lim\limits_{x \to -1} \dfrac{x+1}{3}$

6. $\lim\limits_{x \to -5} \dfrac{x+3}{7}$

7. $\lim\limits_{x \to 0} \dfrac{7x+3}{x^2}$

8. $\lim\limits_{x \to 0} \dfrac{x^2}{7x+3}$

9. $\lim\limits_{x \to 0} \dfrac{7x+3x^2}{x}$

10. $\lim\limits_{x \to -1} \dfrac{x+1}{x^2+6x+5}$

11. $\lim\limits_{x \to 2} \dfrac{x-2}{x^2-4}$

12. $\lim\limits_{x \to 1} \dfrac{x-1}{x^3-1}$

13. $\lim\limits_{x \to +\infty} \dfrac{5}{x}$

14. $\lim\limits_{x \to 1} \dfrac{x^3}{x-1}$

15. $\lim\limits_{x \to +\infty} \dfrac{2x+1}{x-6}$

16. $\lim\limits_{x \to 0} \dfrac{7+3x}{x}$

Evaluate the limit, if it exists, using l'Hôpital's Rule, if it applies.

17. $\lim\limits_{x \to 1} \dfrac{x-1}{x^3-1}$

18. $\lim\limits_{x \to 2} \dfrac{x-2}{x^2-4}$

19. $\lim\limits_{x \to 0} \dfrac{1-e^x}{x}$

20. $\lim\limits_{x \to 0} \dfrac{\sin(ax)}{x}$

21. $\lim\limits_{x \to -1} \dfrac{x+1}{x^3+3x^2-x-3}$

22. $\lim\limits_{x \to 1} \dfrac{\ln x}{x-1}$

23. $\lim\limits_{x \to \pi} \dfrac{\sin x}{x-\pi}$

24. $\lim\limits_{x \to 0} \dfrac{x \cos x}{\sin(2x)}$

25. $\lim\limits_{x \to 0} \dfrac{e^x}{\cos x}$

26. $\lim\limits_{x \to +\infty} \dfrac{2x+1}{x-6}$

27. $\lim\limits_{x \to 1} \dfrac{\ln x - x + 1}{x-1}$

28. $\lim\limits_{x \to 0} \dfrac{e^x - x - 1}{x^2}$

29. $\lim\limits_{x \to +\infty} \dfrac{x^2}{7}$

30. $\lim\limits_{x \to +\infty} \dfrac{x^2}{e^x}$

31. $\lim\limits_{x \to 0} \dfrac{x^2-2x+2}{3x+4}$

32. $\lim\limits_{x \to 1} \dfrac{\ln x}{(x-1)^3}$

33. $\lim\limits_{x \to +\infty} \dfrac{x^2-5x+1}{3x^2-7x+2}$

34. $\lim\limits_{x \to +\infty} \dfrac{x^3}{x^3-1}$

35. $\lim\limits_{x \to +\infty} \dfrac{e^{5x}}{x^3}$

36. $\lim\limits_{x \to +\infty} \dfrac{\tan^{-1} x}{x}$

37. $\lim\limits_{x \to 0} \dfrac{e^{5x}}{x^3}$

38. $\lim\limits_{x \to 0} \dfrac{e^{3x}-1}{x}$

EXTENSION EXERCISES

39. In a certain chemical reaction, the amount of the product at time t is given by

$$x(t) = ab\left\{\frac{r-pe^{kt}}{pa-be^{kt}}\right\},$$

where a, b, k, p, and r are positive constants. Use l'Hôpital's Rule to find the limiting amount of the product present, that is, to evaluate

$$\lim\limits_{t \to +\infty} x(t).$$

40. The velocity v of a parachutist t seconds after ejection from a plane is given by

$$v(t) = a \left[\frac{e^{bt} - e^{-bt}}{e^{bt} + e^{-bt}} \right],$$

where a and b are positive constants. Determine the limiting velocity of the parachutist,

$$\lim_{t \to +\infty} v(t).$$

41. If f is continuous on $(-\infty, +\infty)$, find

$$\lim_{x \to 0} \frac{1}{x} \int_0^x f(t) \, dt.$$

SUMMARY AND REVIEW: CHAPTER 10

The following contains a summary of what you should be able to do after completing this chapter. The review exercises are for practice. Answers are at the back of the book. If you miss an exercise, restudy the section indicated alongside the answer.

You should be able to:

Expand a polynomial function in powers of $x - a$ (for any real number a).

1. Expand $x^2 - 5x + 3$ in powers of $x - 1$.

Compute the Taylor polynomial of degree n for $f(x)$ at $x = a$.

Find $p_2(x)$ and $p_3(x)$ for the given function and the specified value of a.

2. $y = x^3$; $x = -1$ **3.** $y = e^{x^2}$; $x = 0$ **4.** $y = \sin(x - 2)$; $x = 2$

Use a Taylor polynomial to find an approximate value for $f(a)$.

5. $\sin(0.05)$ **6.** $e^{0.05}$

Tell whether a geometric series converges or diverges by calculating its ratio. If it converges, calculate its sum.

7. Find the sum of $\frac{8}{5} - \frac{6}{5} + \frac{9}{10} - \frac{27}{40} + \frac{81}{160} - \frac{243}{640} + \cdots$.   **8.** Find the sum of $1 - \frac{3}{2} + \frac{9}{4} - \frac{27}{8} + \cdots$.

9. Evaluate the sum of $\frac{1}{3} + \frac{1}{5} + \frac{3}{25} + \cdots + \left(\frac{1}{3}\right)\left(\frac{3}{5}\right)^{10}$.   **10.** Write $1.0641\overline{41}$ as a rational fraction.

Tell whether an alternating series converges or diverges, and if it converges, approximate its sum by one of its partial sums and estimate the error involved in this approximation.

Determine whether the given series converges or diverges. If it converges, estimate its sum with an error of less than 0.01.

11. $1 - \dfrac{1}{3 \cdot 5} + \dfrac{1}{5 \cdot 5^2} - \dfrac{1}{7 \cdot 5^3} + \cdots = \displaystyle\sum_{i=0}^{\infty} \dfrac{(-1)^i}{[2i + 1] \cdot 5^i}$

12. $1 - 3 + 5 - 7 + 9 - 11 + \cdots$

13. $\dfrac{2}{5} + \dfrac{3}{6} + \dfrac{4}{7} + \dfrac{5}{8} + \dfrac{6}{9} + \cdots$

14. $\dfrac{2}{5} - \dfrac{3}{6} + \dfrac{4}{7} - \dfrac{5}{8} + \dfrac{6}{9} - \cdots$

15. $1 - \dfrac{1}{12} + \dfrac{1}{144} - \dfrac{1}{1728} + \dfrac{1}{12^4} - \dfrac{1}{12^5} + \cdots$

Compute the Taylor series expansion for a given function (at a given point).

Expand each of the following functions in a Taylor series using the prescribed value of x.

16. $y = x^3 \sin (x); x = 0$

17. $y = e^{(x + 1)^2}; x = -1$

18. $y = (1 - 3x)^{-1}; x = 0$

19. Use a Taylor series to evaluate $\displaystyle\lim_{x \to 0} \dfrac{\sin (x^3)}{x^3}$.

Determine whether or not a given limit expression is an indeterminate form and, if it is, use l'Hôpital's Rule to evaluate it.

20. $\displaystyle\lim_{x \to 0} \dfrac{1 - \cos x}{3x \sin x}$

21. $\displaystyle\lim_{x \to 0} \dfrac{1 - \cos x}{2e^x}$

22. $\displaystyle\lim_{x \to \pi} \dfrac{\sin x}{x - \pi}$

23. $\displaystyle\lim_{x \to +\infty} \dfrac{x^{2n}}{e^{5x}}$,

where n is a positive integer

24. $\displaystyle\lim_{x \to +\infty} \dfrac{x^2 - 3x + 5}{3x^2 - 7x + 1}$

EXTENSION EXERCISES

25. A certain stock pays a dividend of $6 per year. Assuming that the dividend remains constant forever and that interest rates are constant at 10%, so that the present value of the dividend to be paid n years from today is $6(1.10)^{-n}$, find the sum of the present value of all future dividend payments, assuming that the first payment is one year from today.

26. Find a function $f(x)$ if there is one such that $f(0) = 0$ and $f^{(n)}(0) = 1/n^2$ for all positive integers n.

TEST: CHAPTER 10

Find $p_2(x)$ and $p_3(x)$ for the given function and the specified value of x.

1. $y = x^3; x = -2$

2. $y = e^{x^3}; x = 0$

3. $y = \sin (x - 1); x = 1$

Use a Taylor polynomial of degree 3 to find an approximate value for each of the following.

4. $\sin(0.04)$

5. $e^{0.07}$

6. Find the sum of the infinite series

$$\frac{3}{5} - \frac{2}{5} + \frac{4}{15} - \frac{8}{45} + \frac{16}{135} - \frac{32}{405} + \cdots.$$

7. Evaluate

$$\frac{1}{2} + \frac{1}{7} + \frac{2}{49} + \cdots + \frac{1}{2}\left(\frac{2}{7}\right)^{12}.$$

8. Write $2.431313\overline{31}$ as a rational fraction.

Determine whether the given infinite series converges or diverges. If it converges, estimate its sum with an error of less than 0.01.

9. $1 - \dfrac{1}{3 \cdot 4} + \dfrac{1}{5 \cdot 4^2} - \dfrac{1}{7 \cdot 4^3} + \dfrac{1}{9 \cdot 4^4} - \cdots$

10. $1 - 2 + 4 - 8 + 16 - 32 + \cdots$

11. $\dfrac{2}{3} - \dfrac{3}{4} + \dfrac{4}{5} - \dfrac{5}{6} + \dfrac{6}{7} - \dfrac{7}{8} + \cdots$

12. $1 - \dfrac{1}{16} + \dfrac{1}{256} - \dfrac{1}{4^6} + \dfrac{1}{4^8} - \cdots$

Expand each of the following functions in a Taylor series using the prescribed value of x.

13. $y = x^2 \sin x$; $x = 0$

14. $y = e^{(x-1)^2}$; $x = 1$

15. $y = \dfrac{1}{1 - 5x}$; $x = 0$

16. Use a Taylor series to evaluate

$$\lim_{x \to 0} \frac{1 - \cos(x^2)}{x^4}.$$

Evaluate each of the following limits.

17. $\lim\limits_{x \to 0} \dfrac{x \sin x}{\cos x - 1}$

18. $\lim\limits_{x \to 1} \dfrac{3x \ln x}{e^x}$

19. $\lim\limits_{x \to \pi/2} \dfrac{\cos x}{x - \pi/2}$

20. $\lim\limits_{x \to +\infty} \dfrac{x^n}{e^x}$ (n is a positive integer)

21. $\lim\limits_{x \to +\infty} \dfrac{x^3 - 5x + 6}{2x^3 - 7x^2 + 1}$

EXTENSION EXERCISES

22. A certain stock pays a dividend of $2 per year. Assuming that the dividend remains constant forever and that interest rates are constant at 10% so that the present value of the dividend payment n years from today is $2(1.10)^{-n}$, find the sum of the present value of all future dividend payments, assuming that the first payment is one year from today.

23. In Problem 22, if the dividend payment is D and the interest rate is i, find the present value of all future dividend payments.

24. Find a function $f(x)$ (if there is one) such that $f(0) = 0$ and $f^{(n)}(0) = 1/n$ for all positive integers n.

11

NUMERICAL
TECHNIQUES

Many interesting problems lead to the
solution of an algebraic equation, to the
evaluation of a definite integral, or to
the solution of a differential equation.
Unfortunately, it may be difficult (or
impossible) to find the exact solution in
one of these situations. This chapter deals
with finding approximate solutions to the
problems described above. The techniques
are used to solve several interesting applied
problems.

AN APPLICATION

An investor buys an annuity (a
sequence of future payments) for
$2700. The annuity will return $1000
per year for three years, beginning
one year from today. Find the yield
rate on this investment.

THE MATHEMATICS

If i is the yield rate, then it is
more convenient to first find
$v = (1 + i)^{-1}$. It turns out that
elementary algebra will lead us to
the equation

$$v^3 + v^2 + v - 2.7 = 0,$$

an equation that is difficult to solve
exactly. Using Newton's method, a
technique for finding approximate
solutions to such equations, we will
obtain the solution quickly and
easily.

11.1

SOLVING EQUATIONS NUMERICALLY: ITERATION

The solutions of many equations that arise in practical applications cannot be found exactly using standard algebraic techniques. An example of an equation of this sort is $\cos x = x$, which we consider below. We use the following general procedure.

To solve a given equation, we start by making an *initial approximation*, or *initial guess* (denoted x_1), of the solution to the equation. Then we use our initial guess x_1 to construct a new value x_2 (which we expect to be closer to the solution than x_1 is). Continuing, we use x_2 to construct a new value x_3, and so forth. In theory, we construct an infinite sequence $x_1, x_2, \ldots, x_n, \ldots$, which converges to a solution, say $\bar{x}$, of the given equation. In practice, we must stop the process after a finite number of steps N, using the approximate solution x_N instead of the actual solution $\bar{x}$.

Example 1 Solve $\cos x = x$.

Solution The graphs of $y = \cos x$ and $y = x$ are plotted as shown below. They appear to intersect at one point only. The x-coordinate of that point, $\bar{x}$, is the solution to our equation. Judging from the graph, $x_1 = \frac{1}{2} = 0.5000000$ is not an unreasonable initial guess.

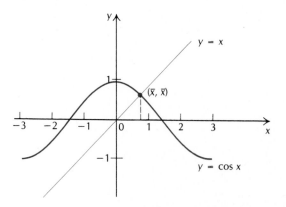

We'll construct x_2 using what is called the *method of iteration*. In this case, it means that $x_2 = \cos(x_1) = \cos(0.5)$. Computing (using a calculator in radian mode), we get $x_2 \approx 0.8775826$. We continue by computing $x_3 = \cos(x_2) = \cos(0.8775826)$. We get $x_3 \approx 0.6390125$.

1. Using $x_1 = 1.8$ and $x_{n+1} = \sqrt{x_n}$ for $n = 1, 2, \ldots$, solve $\sqrt{x} = x$.

Then, with $x_4 = \cos(x_3)$, we find that $x_4 \approx 0.8026851$. We continue with $x_{n+1} = \cos(x_n)$, for all n. We list a few values (check them on your calculator):

$$x_5 \approx 0.6947780,$$
$$x_6 \approx 0.7681958,$$
$$x_7 \approx 0.7191654,$$
$$x_8 \approx 0.7523558,$$
$$x_9 \approx 0.7300811,$$
$$x_{10} \approx 0.7451203,$$
$$\vdots$$
$$x_{17} \approx 0.7387045,$$
$$x_{18} \approx 0.7393415,$$
$$x_{19} \approx 0.7389124,$$
$$\vdots$$
$$x_{43} \approx 0.7390851,$$
$$x_{44} \approx 0.7390851,$$

with the readings unchanged after that. We conclude that $\bar{x} \approx 0.7390851$. As a check, we substitute this value into the original equation to get

$$\cos(\bar{x}) \approx 0.7390851 \approx \bar{x}.$$

DO EXERCISE 1.

We will apply the method of iteration only to equations of the form $g(x) = x$, where $g(x)$ is a continuous function. Remember that $x_{n+1} = g(x_n)$ for all n. If the sequence $x_1, x_2, \ldots x_n, \ldots$, has a limit, say $\bar{x}$, then

$$\bar{x} = \lim_{x \to \infty} x_n.$$

For continuous functions g,

$$g(\bar{x}) = \lim_{n \to \infty} g(x_n) = \lim_{n \to \infty} x_{n+1} = \bar{x},$$

or

$$g(\bar{x}) = \bar{x}.$$

Thus if the sequence of iterates $x_1, x_2, \ldots x_n, \ldots$, has a limit $\bar{x}$ and if g is continuous, then $\bar{x}$ is a solution to $g(x) = x$. The catch, of course, is that the sequence of iterates may not converge.

2. Try to solve $x^2 = x$ using $x_1 = 1.1$.

Example 2   Solve $x^2 = x$ using $x_1 = 1.5$.

Solution   Here, $g(x) = x^2$. So $x_{n+1} = x_n^2$ for $n = 1, 2, \ldots$. Using $x_1 = 1.5$, we obtain $x_2 = (1.5)^2 = 2.25$, $x_3 = 5.0625$, $x_4 \approx 25.62891$, and so on. The sequence of iterates diverges. Note that $x = 0$ and $x = 1$ are the only solutions to $x^2 = x$.

DO EXERCISES 2 AND 3.

3. Try to solve $x^2 = x$ using each of the following.

     a) $x_1 = \frac{1}{3}$

     b) $x_1 = -\frac{1}{2}$

     c) $x_1 = \frac{2}{3}$

Example 3   Solve $x^2 - 5x + 6 = 0$.

Solution   We write $x^2 + 6 = 5x$, or $\frac{1}{5}x^2 + \frac{6}{5} = x$. Letting $g(x) = \frac{1}{5}x^2 + \frac{6}{5}$, we have $g(x) = x$.

     The original equation can be factored as $(x - 2)(x - 3) = 0$, so that $\bar{x} = 2$ and $\bar{x} = 3$ are the solutions. Alternatively, we could have used the quadratic formula to obtain the solutions.

     Ignoring the fact that we can easily find the solutions in this case, we start with $x_1 = 1$. Then

$$x_2 = g(x_1) = \tfrac{1}{5}(1)^2 + \tfrac{6}{5} = 1.4.$$

Continuing, we have

$$x_3 = g(x_2) = \tfrac{1}{5}(1.4)^2 + \tfrac{6}{5} = 1.592$$

4. What happens in Example 3 if we use $x_1 = 2.9$?

and

$$x_4 = g(x_3) = \tfrac{1}{5}(1.592)^2 + \tfrac{6}{5} \approx 1.7068928.$$

Similarly, $x_5 \approx 1.782697$, $x_6 \approx 1.835601$, $x_7 \approx 1.873887$,   and $x_8 \approx 1.902290$. We find that $x_{14} \approx 1.976536$ and, after a few more iterations, $x_{20} \approx 1.99398$. It seems clear that these iterates have the limit $\bar{x} = 2$.

     Now, what happens if we start with the initial guess $x_1 = 3.2$? Then $x_2 = \frac{1}{5}(3.2)^2 + \frac{6}{5} = 3.248$. Continuing, we have $x_3 = g(x_2) = \frac{1}{5}(3.248)^2 + \frac{6}{5} \approx 3.30990$, a bad sign, since $x_3 - x_2$ is larger than $x_2 - x_1$. Continuing, we find that $x_4 \approx 3.391089$, $x_5 \approx 3.499896$, $x_6 \approx 3.649855$, and $x_7 \approx 3.864288$, with the sequence of iterates diverging.

5. Solve $x^2 + 1 = 4x$ using $x_1 = 2$.

DO EXERCISES 4 AND 5.

     Examples 2 and 3 and Margin Exercises 2, 3, and 4 show that the sequence of iterates may diverge for one choice of x_1 and converge for another choice. The following theorem clarifies the situation.

6. Solve $x^2 + 1 = 4x$ using $x_1 = 2$ and $g(x) = \sqrt{4x - 1}$.

THEOREM 1

If $g(x)$ and $g'(x)$ are continuous and $|g'(x)| < 1$ on an interval containing *both* x_1 and an $\bar{x}$ satisfying $g(\bar{x}) = \bar{x}$, and if we put

$$x_{n+1} = g(x_n) \quad \text{for } n = 1, 2, \ldots,$$

then the sequence $x_1, x_2, \ldots, x_n, \ldots$, converges to $\bar{x}$.

We reconsider Examples 2 and 3. In Example 2, we had $g(x) = x^2$, so that $g'(x) = 2x$. If x is between $-\frac{1}{2}$ and $\frac{1}{2}$, then $g'(x) = 2x$ will be between -1 and 1. Since the interval contains $\bar{x} = 0$, the theorem guarantees us that if we pick x_1 between $-\frac{1}{2}$ and $\frac{1}{2}$ and define $x_{n+1} = x_n^2$ for $n = 1, 2, \ldots$, then the iterates will converge to $\bar{x} = 0$.

In Example 3, we had $g(x) = \frac{1}{5}x^2 + \frac{6}{5}$. Thus $g'(x) = \frac{2}{5}x$. In order to guarantee that $-1 < \frac{2}{5}x < 1$, we must choose x so that $-\frac{5}{2} < x < \frac{5}{2}$ (multiply the original inequality by $\frac{5}{2}$). This interval contains $\bar{x} = 2$. It follows that if we select x_1 between $-\frac{5}{2}$ and $\frac{5}{2}$, then the iterates will converge to 2. For this function, we can show that the iterates will converge to 2 if we select any x_1 between -3 and 3.

Example 4 Use iteration to solve $x^2 - 5x + 6 = 0$.

Solution Yes! This is the same equation that appeared in Example 3 above. But there may be many ways to rewrite $x^2 - 5x + 6 = 0$ in the form $g(x) = x$. For instance, let's write $x^2 = 5x - 6$. Then, taking the principal square roots of both sides (assuming that x and $5x - 6$ are nonnegative), we get $x = \sqrt{5x - 6}$. Let $g(x) = \sqrt{5x - 6}$. If we define

$$x_{n+1} = g(x_n) = \sqrt{5x_n - 6},$$

then we obtain a new and different iteration formula for this equation. If we pick $x_1 = 4$, for example, then

$$x_2 = \sqrt{5x_1 - 6} = \sqrt{5(4) - 6} = \sqrt{14} \approx 3.741657.$$

Continuing, we get $x_3 = \sqrt{5x_2 - 6} \approx \sqrt{12.708287} \approx 3.564868$, $x_4 \approx 3.438654$, $x_5 \approx 3.345635$, $x_6 \approx 3.275389, \ldots, x_{24} \approx 3.008502$, $x_{25} \approx 3.007077$, $x_{26} \approx 3.005892$, and so forth. Theorem 1 can be applied to show that the iterates converge to $\bar{x} = 3$, although the convergence is very slow. Examples 3 and 4 show that in solving an equation by iteration, some experimentation may be required to determine a workable iteration formula.

DO EXERCISE 6.

EXERCISE SET 11.1

Solve the given equation by iteration, using the indicated initial guess. Check your answers.

1. $x = \frac{1}{3}x^2 + \frac{1}{3}; x_1 = 0$

2. $x = \frac{1}{3}x^2 + \frac{1}{3}; x_1 = 3$

3. $x = \sin x + \frac{1}{2}; x_1 = 1$
   (x in radians)

4. $x = \frac{1}{2}\cos x; x_1 = 1$

5. $x = 2 \cos x; x_1 = 1$

6. $x = \sqrt{3x - 1}; x_1 = 2$

7. $x = \frac{1}{4}x^3 + 1; x_1 = 0$

8. $x = \sqrt[3]{3x - 3}; x_1 = 0$

Use iteration to find the solution to each of the following. Check your answers.

9. $x - e^{-x} = 0$

10. $x^5 - 5x + 5 = 0$

11. $3 \cos x - 7x = 0$

12. $\cos x - 2x^2 = 0$

EXTENSION EXERCISE

13. Define $x_{n+1} = \frac{1}{3}x_n^2 + \frac{1}{3}$ for all n, as in Exercise 2 above.
    a) Show that if $x_n > 3$, then $x_{n+1} > \frac{10}{3} > 3$. Conclude that if $x_1 > 3$, then all the succeeding x_n's are greater than $\frac{10}{3}$.
    b) By writing

    $$\frac{x_{n+1} + 1}{x_n} = \frac{1}{3}x_n + \frac{1}{3x_n},$$

    show that if $x_n > \frac{10}{3}$, then

    $$\frac{x_{n+1}}{x_n} > \frac{10}{9}.$$

    c) By combining parts (a) and (b), show that if $x_1 > 3$, then $x_2 > \frac{10}{9}$ and $x_{k+2} > (\frac{10}{9})^k \cdot x_2$, which in turn shows that the iterates in Exercise 2 diverge.

11.2

OBJECTIVE

a) Solve equations using Newton's method.

NEWTON'S METHOD

We introduce Newton's method, another technique for solving equations numerically. This method is fairly easy to apply and usually yields more rapid convergence than the method of iteration. To apply this method, we write the given equation in the form $f(x) = 0$. Let x_1 be our initial approximation, as shown in the figure on the following page.

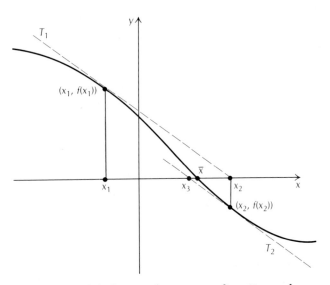

Assuming that $f'(x_1) \neq 0$, the tangent line T_1 to the graph of $y = f(x)$ at the point $(x_1, f(x_1))$ will intersect the x-axis in exactly one point. We call this point $(x_2, 0)$. The x-coordinate of this point, x_2, will be our second approximation. Next we locate $(x_2, f(x_2))$ on the graph of $y = f(x)$ and construct the tangent line T_2 to the graph at this new point. Then the tangent line T_2 will intersect the x-axis at $(x_3, 0)$ (assuming that $f'(x_2) \neq 0$). The x-coordinate of this point, x_3, is our third approximation to the solution $\bar{x}$. Continuing, we can generate as many approximations as we need.

We now have a geometric method for constructing x_{n+1} from x_n. To obtain a formula relating x_n and x_{n+1}, we return to the tangent line T_1. The slope of T_1 is $f'(x_1)$, the value of the derivative of $f(x)$ at $x = x_1$. Now T_1 passes through the point $(x_1, f(x_1))$. Applying the point–slope formula from Section 1.4, we see that the equation of T_1 is

$$y - f(x_1) = f'(x_1)(x - x_1).$$

The point $(x_2, 0)$ is on T_1 and so its coordinates must satisfy the equation of T_1. Thus, setting $x = x_2$ and $y = 0$, we have

$$0 - f(x_1) = f'(x_1)(x_2 - x_1),$$

or

$$-f(x_1) = f'(x_1)(x_2 - x_1).$$

Dividing both sides by $f'(x_1)$ (recall that $f'(x_1) \neq 0$), we get

$$\frac{-f(x_1)}{f'(x_1)} = x_2 - x_1.$$

Solving for x_2, we have

$$x_2 = x_1 - \frac{f(x_1)}{f'(x_1)}.$$

CSS

In general, we have the following:

$$x_{n+1} = x_n - \frac{f(x_n)}{f'(x_n)}. \tag{1}$$

Example 1   Use Newton's method with $x_1 = 1.5$ to solve $x^2 = x$.

Solution   To apply Newton's method, we must write the equation in the form $f(x) = 0$. So we write $x^2 - x = 0$. If we put $f(x) = x^2 - x$, then the equation takes the form $f(x) = 0$ and we can apply formula (1). Since $f(x) = x^2 - x$, we get $f'(x) = 2x - 1$. So formula (1) becomes

$$x_{n+1} = x_n - \frac{f(x_n)}{f'(x_n)}$$

$$= x_n - \frac{x_n^2 - x_n}{2x_n - 1}.$$

If $n = 1$, we get

$$x_2 = x_1 - \frac{x_1^2 - x_1}{2x_1 - 1}$$

and, with $x_1 = 1.5$,

$$x_2 = 1.5 - \frac{(1.5)^2 - 1.5}{2(1.5) - 1} = 1.5 - \frac{2.25 - 1.5}{3 - 1}$$

$$= 1.5 - \frac{0.75}{2} = 1.5 - 0.375$$

$$= 1.125.$$

Then

$$x_3 = x_2 - \frac{x_2^2 - x_2}{2x_2 - 1} = 1.125 - \frac{(1.125)^2 - (1.125)}{2(1.125) - 1}$$

$$= 1.125 - \frac{1.265625 - 1.125}{2.25 - 1} = 1.125 - \frac{0.140625}{1.25}$$

or

$$x_3 = 1.125 - 0.1125 = 1.0125.$$

1. Using Newton's method with $x_1 = \frac{1}{3}$ to solve $x^2 = x$.

Then

$$x_4 = 1.0125 - \frac{(1.0125)^2 - 1.0125}{2(1.0125) - 1}$$

$$\approx 1.0125 - 0.1234756$$

$$\approx 1.000152,$$

and

$$x_5 \approx 1.000152 - \frac{(1.000152)^2 - (1.000152)}{2(1.000152) - 1}$$

$$\approx 1.0000000.$$

With $x_5 = 1$ (correct to seven decimal places), we get $x_6 = 1, \ldots$, and we see that 1 is a solution to $f(x) = 0$.

Note the graph of $f(x) = x^2 - x$ below. It's a parabola and the initial guess x_1, the tangent line T_1, and x_2 are all shown. If we pick x_1 greater than 0.5, then the iterates given by Newton's method will converge to 1. If x_1 is less than 0.5, then the iterates will converge to 0. Note that we may not choose $x_1 = 0.5$ since $f'(x_1) = 2x_1 - 1 = 2(0.5) - 1 = 0$ and the tangent line at $x = 0.5$ is horizontal.

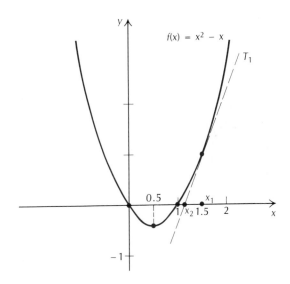

DO EXERCISE 1.

Example 2 Solve $\cos x = x$.

2. Use Newton's method to solve $\cos x = 2x$. Check your answer.

Solution   To put the equation in the correct form, we write $x - \cos x = 0$ and let $f(x) = x - \cos x$. Then $f'(x) = 1 + \sin x$ and formula (1) becomes

$$x_{n+1} = x_n - \frac{x_n - \cos(x_n)}{1 + \sin(x_n)}.$$

If $x_1 = 0.5$, then

$$x_2 = x_1 - \frac{x_1 - \cos(x_1)}{1 + \sin(x_1)}$$

$$= 0.5 - \frac{(0.5) - \cos(0.5)}{1 + \sin(0.5)}$$

$$= 0.5 - \frac{0.5 - (0.8775826)}{1 + 0.4794255}$$

$$= 0.5 - \frac{-0.3775826}{1.4794255}$$

$$\approx 0.5 + 0.2552224$$

$$\approx 0.7552224.$$

So,

$$x_3 \approx 0.7552224 - \frac{(0.7552224) - \cos(0.7552224)}{1 + \sin(0.7552224)}$$

$$\approx 0.7552224 - \frac{0.0271033}{1.685451}$$

$$\approx 0.7391417.$$

Continuing, we have

$$x_4 \approx 0.7391417 - \frac{(0.7391417) - \cos(0.7391417)}{1 + \sin(0.7391417)}$$

$$\approx 0.7391417 - \frac{0.000094672}{1.6736538}$$

$$\approx 0.7391417 - 0.0000565661$$

$$\approx 0.7390851,$$

which is correct to seven decimal places, as we saw in Example 1 of Section 11.1.

Be sure to note the difference in the speed of convergence of the two methods. Newton's method generally gives much faster convergence than simple iteration does.

DO EXERCISE 2.

Newton's method doesn't always work. One equation on which it must fail is $x^2 + 1 = 0$, because the equation has no real solutions.

Even if the equation has a solution, though, the iterates generated by Newton's method may not converge to it. Consider the graph below. The function has no particular formula—call it $f(x)$.

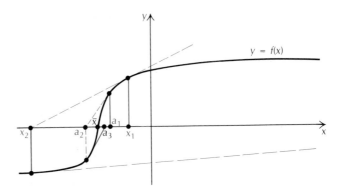

The equation $f(x) = 0$ has exactly one solution, $\bar{x}$. But if x_1 is chosen as shown, then x_2 will be farther from $\bar{x}$ than x_1 is. In fact, it appears from the graph that successive estimates will get worse and worse. Note that if the initial guess had been a_1 instead of x_1, the successive approximations would seem to converge rapidly to $\bar{x}$.

The following theorem gives us a way to make a correct choice.

THEOREM 2

Newton's Method

If there exists an interval containing a solution $\bar{x}$ to $f(x) = 0$ on which $f(x), f'(x)$, and $f''(x)$ are continuous, $f'(x) \neq 0$, and

$$\left| \frac{f(x)f''(x)}{(f'(x))^2} \right| < 1,$$

and if x_1 is chosen in this interval, then the sequence given by

$$x_{n+1} = x_n - \frac{f(x_n)}{f'(x_n)}$$

converges to $\bar{x}$.

We omit the proof. To see how to apply it, consider the following example.

Example 3　Use Newton's method to solve $x^2 - 3 = 0$.

Solution　We set $f(x) = x^2 - 3$. Then

$$f(1) = (1)^2 - 3 = -2 \quad \text{and} \quad f(2) = (2)^2 - 3 = 1,$$

so that the equation $f(x) = 0$ has a solution $\bar{x}$ in the interval from 1 to 2.

In this case, $f'(x) = 2x$ and $f''(x) = 2$ so that

$$\frac{f(x)f''(x)}{(f'(x))^2} = \frac{(x^2 - 3)(2)}{(2x)^2}$$

$$= \frac{2x^2 - 6}{4x^2} = \frac{1}{2} - \frac{3}{2x^2}.$$

This expression is always less than 1 for x between 1 and 2. To keep this ratio between -1 and 1, as the theorem requires, we need only choose x so that

$$-1 < \frac{2x^2 - 6}{4x^2}$$

or

$$-4x^2 < 2x^2 - 6.$$

Rearranging terms, we get

$$6 < 2x^2 + 4x^2,$$

or

$$6 < 6x^2,$$

which is satisfied if x is greater than 1 (or less than -1). Now $f'(x) = 2x$ is 0 only when $x = 0$. So, if we choose x_1 to be greater than 1, then all the requirements of the theorem are met and the iterates will converge to the solution $\bar{x}$, which lies between 1 and 2.

If $x_1 = 2$, for example, then we have

$$x_{n+1} = x_n - \frac{x_n^2 - 3}{2x_n}$$

$$= \frac{2x_n^2}{2x_n} - \frac{x_n^2 - 3}{2x_n}$$

$$= \frac{2x_n^2 - x_n^2 + 3}{2x_n}$$

$$= \frac{x_n^2 + 3}{2x_n}.$$

3. Using Newton's method with $x_1 = -2$, solve $x^3 - 3x + 3 = 0$. Check your answer.

So

$$x_2 = \frac{(2)^2 + 3}{2 \cdot 2} = \frac{4 + 3}{4} = 1.75.$$

Then

$$x_3 = \frac{(1.75)^2 + 3}{2 \cdot (1.75)} = \frac{3.0625 + 3}{3.5},$$

or

$$x_3 \approx 1.732143.$$

Continuing, we have

$$x_4 \approx \frac{(1.732143)^2 + 3}{2 \cdot (1.732143)} \approx 1.732051.$$

Another iteration yields $x_5 \approx x_4$. If we put $\bar{x} = 1.732051$, then $(\bar{x})^2 \approx 3$.

For this particular function, the iterates given by Newton's method will converge for any initial choice x_1 except $x_1 = 0$ (the *one* value of x for which $f'(x) = 0$). If $x_1 > 0$, the iterates will converge to $\sqrt{3}$, whereas if $x_1 < 0$, they will converge to $-\sqrt{3}$ (see the following figure).

4. Use Newton's method to solve $x^3 = 10$. Check your answer.

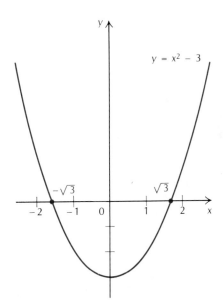

DO EXERCISES 3 AND 4.

EXERCISE SET 11.2

Use the indicated choice of x_1 and Newton's method to solve the given equation. Check your answers.

1. $x = \frac{1}{3}x^2 + \frac{1}{3};\ x_1 = 0$ 2. $x = \frac{1}{3}x^2 + \frac{1}{3};\ x_1 = 3$ CSS

3. $x = \sin x + \frac{1}{2}$ (x in radians); $x_1 = 1$ 4. $x = 2\cos x;\ x_1 = 1$

5. $x^3 - 3x + 3 = 0;\ x_1 = -3$ 6. $x^4 - 4x + 1 = 0;\ x_1 = 0$

Use Newton's method to find all real solutions to the given equation.

7. $x^3 = e^{-x}$ 8. $x^4 - x - 2 = 0$

9. $2x^3 + 4x^2 - 6x - 1 = 0$ 10. $x^4 + 4x^2 + 1 = 0$

11. $x^7 - 7x + 8 = 0$ 12. $3\cos x - 7x = 0$

13. $\cos x - 2x^2 = 0$ 14. $x + \ln x = 0$

EXTENSION EXERCISES

15. **a)** Show that applying Newton's method to

$$x^2 - a = 0$$

yields the iteration scheme

$$x_{n+1} = \frac{x_n}{2} + \frac{a}{2x_n},$$

provided $x_n \neq 0$.

b) Use the formula above to approximate $\sqrt{19}$.

16. **a)** Sketch the graph of $y = f(x) = \tan^{-1} x$ to show that the equation $\tan^{-1} x = 0$ has only the solution $\bar{x} = 0$.

b) Calculate

$$\left| \frac{f(x)f''(x)}{(f'(x))^2} \right|.$$

c) Use Newton's method and $x_1 = 4$ to try to find the solution.

d) Repeat part (c) using $x_1 = 0.5$.

e) Explain the different results in parts (c) and (d) by referring to Theorem 2 and part (b) above.

17. **a)** Use Newton's method to show that the iterates

$$x_{n+1} = \frac{k-1}{k} x_n + \frac{a}{kx_n^{k-1}} \qquad (x_1, a > 0)$$

converge to $a^{1/k}$.

b) Use the algorithm above to calculate $\sqrt[10]{20}$.

11.3

AN APPLICATION: YIELD RATE (INTERNAL RATE OF RETURN)

OBJECTIVES

a) Set up the equation for the yield rate for a (finite) sequence of investments and returns on those investments.

b) Solve the equations resulting from (a).

1. If $8000 is deposited in a savings account at $8\frac{1}{4}\%$ interest compounded annually, find its accumulated value after three years.

Which is the better deal: a payment of $1000 immediately or a payment of $1100 one year from today? (It's understood here that inflation and taxes are *not* to be taken into account in our analysis.) Consider the following. If we deposit $1000 in a savings account (or money-market fund) at 12% interest compounded annually, then, at the end of one year, we will have $1000 plus 12% of $1000, that is, $1120. In this situation (12% interest), we would be better off taking the $1000 now and depositing it at 12% interest, since, at the end of one year, we would have $1120 instead of $1100. On the other hand, if the interest rate is 8%, then our $1000 would earn $(0.08) \times (\$1000) = \$80$ interest. At the end of one year, we would have only $1080. In this case (8% interest), we would be better off taking $1100 at the end of one year.

A little thought shows that if the interest rate i is greater than 10%, then we should take the $1000 immediately. If i is less than 10%, we should take the $1100 at the end of the year. If i equals 10%, then the choices are equivalent.

If the interest rate is 8%, that is, if $i = 0.08$, then $1000 will accumulate to $1080. A deposit of $P will accumulate to $P + \$(0.08)P = \$(1.08)P$. In general, $P deposited at a rate of interest i (compounded annually) will accumulate to $P(1 + i)$ at the end of one year and $P(1 + i)^n$ at the end of n years.

Example 1   If $1400 is deposited in a money-market fund at $11\frac{1}{2}\%$ interest (compounded annually), find its accumulated value after two years.

Solution   Here, $P = 1400$, $i = 0.115$, and $n = 2$. If A denotes the accumulated value, then

$$A = (1400)(1 + 0.115)^2$$
$$= (1400)(1.115)^2$$
$$= \$1740.52.$$

DO EXERCISE 1.

2. Find the present value of $1100, payable one year from today, at a rate of interest of 12%.

To find the *present value* of an amount $A payable n years from today at a rate of interest i, we solve the equation $P(1 + i)^n = A$ for P. In other words, we find the amount P, which, if deposited at a rate of interest i, would accumulate to $A in n years. Dividing both sides by $(1 + i)^n$, we have

$$P = \frac{A}{(1 + i)^n},$$

or

$$P = A(1 + i)^{-n}.$$

DEFINITION

The *present value* $P of a payment of $A to be made n years from today at a rate of interest i (compounded annually) is

$$P = A(1 + i)^{-n}.$$

Unless otherwise indicated, assume in this section that interest is compounded annually.

Example 2   Find the present value of $1100, payable one year from today, at a rate of interest of 8%.

Solution   Here, $A = 1100$, $i = 0.08$, and $n = 1$. Thus,

$$P = (1100)(1 + 0.08)^{-1} = (1100)(1.08)^{-1}$$
$$= (1100)(0.9259259) \approx \$1018.52.$$

This means that $1018.52, invested at 8%, will yield $1100 after one year.

DO EXERCISE 2.

If we revisit the problem of the present payment of $1000 versus the payment of $1100 one year from today, we see, in light of Example 2 and Margin Exercise 2, that the present value of the future payment of $1100 one year from today will be less than $1000 if the rate of interest is greater than 10% and more than $1000 if the rate of interest is less than 10%. Although the actual numbers are different, the results of the analysis are identical.

We define the *present value* of a series of future payments as the *sum of the present values of the payments.*

3. An annuity will pay $250 at the end of each year for the next four years. Find the present value of the annuity if the interest rate is 12%.

Example 3 An annuity will pay $1000 at the end of each year for the next three years. Find the present value of the annuity if the interest rate is 8%.

Solution The present value of the first payment (one year from today) is $(1000)(1.08)^{-1}$. The present value of the second payment is $(1000)(1.08)^{-2}$, and the present value of the third payment is $(1000)(1.08)^{-3}$. The present value P of the annuity is the sum

$$P = (1000)(1.08)^{-1} + (1000)(1.08)^{-2} + (1000)(1.08)^{-3}$$
$$= 925.93 + 857.34 + 793.83 = \$2577.10.$$

DO EXERCISE 3.

In the preceding problems, we were given the dollar amount of a series of future payments and the interest rate and were asked to find the *present value* of the future payments. In practice, we may be given the dollar amount of the future payments and the present value of the future payments and asked to find an interest rate i. In this case, we can call i the *yield rate,* or the *internal rate of return,* of the initial investment.

Example 4 An investor makes two investments of $1000 each. The first returns $1050 at the end of one year, while the second returns $1210 at the end of two years. Find the yield rate i.

Solution Note what happens if we consider the investments separately. The first investment of $1000 returns $1050 at the end of one year. Solving the equation

$$1000 = 1050(1 + i_1)^{-1}$$

for i_1 gives us $i_1 = 5\%$. The second investment of $1000 returns $1210 at the end of two years. Solving

$$1000 = 1210(1 + i_2)^{-2}$$

for i_2 gives us $i_2 = 10\%$. We have two interest rates, each coming as a part of the investment. What we want is one interest rate that represents the entire investment. This is the *yield rate.*

We define the yield rate i as the value of i at which the sum of the present value of the returns equals the sum of the present value of the investments. In this case, the present value of the investments is $2000. The present value of the returns is $(1050)(1 + i)^{-1}$ and $(1210)(1 + i)^{-2}$, respectively. Therefore, the value of i that we are looking for is the solution to

$$2000 = (1050)(1 + i)^{-1} + (1210)(1 + i)^{-2}.$$

To simplify the equation, let

$$v = \frac{1}{1 + i} = (1 + i)^{-1}.$$

Then we get

$$2000 = 1050v + 1210v^2,$$

a quadratic equation in v. Writing it in the form

$$1210v^2 + 1050v - 2000 = 0$$

and applying the quadratic formula gives

$$v = \frac{-1050 \pm \sqrt{(1050)^2 - 4(1210)(-2000)}}{2(1210)}$$

$$= \frac{-1050 \pm \sqrt{1,102,500 + 9,680,000}}{2420}$$

$$= \frac{-1050 \pm 3283.67}{2420}.$$

We get $v_1 \approx 0.923005$ and $v_2 \approx -1.790773$. Note that even if i is the worst case possible (-100%, or $i = -1$, at which point we lose all our money), $1 + i$ is a nonnegative number. In this situation, then, negative values of v are irrelevant and v_2 can be discarded. To complete the problem, observe that since

$$v = \frac{1}{1 + i},$$

we can take reciprocals to get

$$\frac{1}{v} = 1 + i.$$

Then,

$$i = \frac{1}{v} - 1 = \frac{1}{v} - \frac{v}{v} = \frac{1 - v}{v}.$$

Here,

$$i \approx \frac{1 - 0.923005}{0.923005} \approx 0.083418.$$

The yield rate of this investment "package" is approximately 8.34%. We now have one interest rate that represents the entire investment.

4. An investor invests a total of $4000 in two projects. The first returns $2000 at the end of one year, while the second returns $2500 at the end of two years. Find the investor's yield rate on this package.

DO EXERCISE 4.

Example 5 An investor buys an annuity for $2700. The annuity will return $1000 per year for three years, beginning one year from today. Find the yield rate on this investment.

Solution The present value of the three payments, $2700, is given by

$$2700 = 1000(1 + i)^{-1} + 1000(1 + i)^{-2} + 1000(1 + i)^{-3}.$$

Letting

$$v = \frac{1}{1 + i} = (1 + i)^{-1}$$

(v is called *the discount factor*), we see that the equation to be solved is

$$2700 = 1000v + 1000v^2 + 1000v^3,$$

or, if we rearrange and divide by 1000,

$$v^3 + v^2 + v - 2.7 = 0.$$

To solve this, we'll use Newton's method. First, though, observe that since i is generally a small positive number, $v = 1/(1 + i)$ will be close to 1 (and usually less than 1). So, $v_1 = 1$ is a good first approximation.

Newton's method yields

$$v_{j+1} = v_j - \frac{f(v_j)}{f'(v_j)}, \qquad j = 1, 2, \ldots .$$

Since $f(v) = v^3 + v^2 + v - 2.7$, differentiation yields $f'(v) = 3v^2 + 2v + 1$. Thus,

$$v_{j+1} = v_j - \frac{v_j^3 + v_j^2 + v_j - 2.7}{3v_j^2 + 2v_j + 1}.$$

If we let $v_1 = 1$, we have

$$v_2 = 1 - \frac{(1)^3 + (1)^2 + 1 - 2.7}{3(1)^2 + 2(1) + 1}$$

$$= 1 - \frac{1 + 1 + 1 - 2.7}{3 + 2 + 1}$$

$$= 1 - \frac{0.3}{6} = 1 - 0.05 = 0.95.$$

CURRENT DIVIDEND

5½%

INTEREST DAILY
PAID MONTHLY

MT. VERNON CO-OPERATIVE BANK

Does this yield rate conform to the notion of yield rate introduced in this section?

5. An investor pays $5000 for three annual payments of $2000 per year beginning one year from today. Find the yield rate on this annuity.

Using $v_2 = 0.95$ gives us

$$v_3 = 0.95 - \frac{(0.95)^3 + (0.95)^2 + (0.95) - 2.7}{3(0.95)^2 + 2(0.95) + 1}$$

or

$$v_3 = 0.95 - \frac{0.857375 + 0.9025 + 0.95 - 2.7}{2.7075 + 1.90 + 1}$$

$$= 0.95 - \frac{0.009875}{5.6075}$$

$$= 0.95 - 0.00176,$$

or

$$v_3 \approx 0.948239.$$

Using this value for v yields

$$i = \frac{1 - v}{v} \approx \frac{1 - 0.948239}{0.948239} \approx 0.0545865.$$

Therefore, $i \approx 5.46\%$ on this annuity.

DO EXERCISE 5.

It is interesting to compare the results of Examples 3 and 5. The series of payments is identical. Note that as the price (present value) of the annuity increased from $2577.10 to $2700, the yield rate *dropped* from 8% to roughly 5.46%. This illustrates the general principle that as interest rates rise, the prices of annuities (and bonds) fall, while as interest rates fall, the prices of annuities (and bonds) rise.

Note too that in Example 4, an initial investment with return payments after one and two years led to a quadratic equation in v, whereas in Example 5 an investment with return payments in one, two, and three years led to a cubic equation. The general situation of an initial investment followed by return payments at the end of the year for a period of n years will lead to a polynomial equation of degree n. If payments are made at times other than the end of the year, fractional exponents may be required to solve the resulting equation (see Exercise Set 11.3). In any case, a technique such as Newton's method (interpolation in tables is another way) is generally needed.

The Method of Bisection

We revisit the equation $f(x) = x^3 - 3x + 3 = 0$, which we solved using Newton's method (in Margin Exercise 3 of Section 11.2). We note that the solution $\bar{x}$ to $f(x) = 0$ lies between -2 and -3. In fact,

$$f(-3) = (-3)^3 - 3(-3) + 3 = -15$$ `CSS`

and

$$f(-2) = (-2)^3 - 3(-2) + 3 = 1,$$

so that, by continuity, a solution to $f(x) = 0$ must lie between -2 and -3. Now, we bisect the interval from -3 to -2. The midpoint is -2.5. Then

$$f(-2.5) = (-2.5)^3 - 3(-2.5) + 3 = -5.125.$$

Since $f(-2.5)$ is negative, while $f(-2)$ is positive, there must be a root between -2.5 and -2. (There *may* also be a root, or even several roots, between -3 and -2.5, but the argument *guarantees* one between -2.5 and -2.) We repeat the procedure. Bisecting the interval yields -2.25. We compute

$$f(-2.25) = (-2.25)^3 - 3(-2.25) + 3 = -1.640625,$$

which is negative. Arguing as before, we see that a solution to $f(x) = 0$ must lie between -2.25 and -2. Bisecting again, we arrive at -2.125. Calculating, we now have

$$f(-2.125) = (-2.125)^3 - 3(-2.125) + 3 \approx -0.22070313.$$

A solution lies between -2.125 and -2. Note that had we wanted to stop at this step, we could have used the midpoint of the interval, -2.0625, for $\bar{x}$ with the assurance that the error involved here is less than one-half the length of the interval (in this case, 0.0625).

The method of bisection is simple, but usually much too laborious for hand calculations. It is, however, ideal for high-speed computing machinery since it is easy to program and gives an error estimate at each stage of the calculation.

EXERCISE SET 11.3

Find the accumulated value of an initial amount P for n years at a rate of interest i (compounded annually).

1. $P = 500, n = 7, i = 10\%$
2. $P = 500, n = 8, i = 10\%$
3. $P = 1000, n = 30, i = 12\%$
4. $P = 1000, n = 30, i = 10\%$
5. $P = 1000, n = 30, i = 8\%$
6. $P = 1000, n = 30, i = 6\%$

Find the present value of a payment of \$$A$ to be made n years from today at a rate of interest i (compounded annually).

7. $A = \$1000, n = 3, i = 10\%$
8. $A = \$1000, n = 5, i = 10\%$
9. $A = \$1000, n = 20, i = 10\%$
10. $A = \$1000, n = 10, i = 12\%$

11. $A = \$1000, n = 10, i = 9\%$ 12. $A = \$1000, n = 10, i = 6\%$

13. $A = \$10{,}062.66, n = 30, i = 8\%$ 14. $A = \$974.36, n = 7, i = 10\%$

An annuity pays $\$R$ per year for n years. Find the present value of the annuity if the rate of interest is i (compounded annually).

15. $R = \$1000, n = 2, i = 12\%$ 16. $R = \$1000, n = 2, i = 9\%$ 17. $R = \$1000, n = 2, i = 6\%$

18. $R = \$1000, n = 2, i = 19\%$ 19. $R = \$200, n = 5, i = 10\%$ 20. $R = \$5000, n = 6, i = 11\frac{1}{2}\%$

21. An investor makes two different investments of $5000, and receives a payment of $6000 after one year on the first and $5800 after two years on the second. Find the equation that determines the yield rate i (in terms of the discount factor $v = (1 + i)^{-1}$) for this pair of investments. Then calculate i.

22. An investor makes two investments, one of $3000, the second of $5000. The first returns $3500 after one year, the second $6500 after four years. Find the equation that determines the yield rate and use Newton's method to solve it.   `CSS`

23. An investor invests a total of $9000 in two projects, receiving return payments of $4000 after two years and $7000 after five years. Find the internal rate of return on the investment.

24. An investor pays $3000 for four annual payments of $1000, beginning one year from today. Find the yield rate of this annuity. Check your answer.

25. An investor must choose between two annuities. The first pays $1000 a year for three years and sells for $2400. The second pays $800 a year for five years and sells for $2900. Which should the investor buy?

EXTENSION EXERCISES

26. An investor makes an initial investment of $2000 in a company, receives a payment of $4300 one year later, and makes another investment of $2310 in the company two years after the initial investment. Set up the equation for the investor's yield rate and show that the equation has two solutions, that is, that the yield rate is not well defined.

27. Repeat Exercise 26 given a second investment of $2500 (instead of $2310).

28. An investor makes an initial investment of $\$P$ and receives payment of $R_1, R_2, \ldots, R_k$ dollars after $1, 2, \ldots, k$ years. Find the equation that determines the investor's yield rate and use calculus to show that there is *exactly one* positive number v that solves the equation.

11.4

OBJECTIVE

a) Find a numerical estimate for $\int_a^b f(x)\,dx$ using each of the following.

 i) Riemann sums

 ii) The Trapezoidal Rule

 iii) Simpson's Rule

NUMERICAL INTEGRATION

In Chapters 5 and 6, we saw that many interesting problems in business, medicine, probability, and the physical sciences lead to the evaluation of definite integrals, that is, expressions of the form

$$\int_a^b f(x)\,dx.$$

In many cases, $f(x)$ has an antiderivative that we can find and evaluate.

Thus, for example,

$$\int_0^5 3x^2\, dx = [x^3]_0^5 = 5^3 - 0^3 = 125.$$

In many other cases, though, this won't be so. As we saw in Chapter 6, this happens with the function $f(x) = e^{-(1/2)x^2}$, a function that occurs naturally in many probability applications, but has no elementary antiderivative.

In cases where $f(x)$ has no elementary antiderivative, we may have to evaluate $\int_a^b f(x)\, dx$ approximately. In this section, we will consider several approximation schemes for estimating integrals.

Let $f(x)$ be continuous on $[a, b]$ and consider $\int_a^b f(x)\, dx$. If we subdivide $[a, b]$ into n equal pieces (see the figure below) of length Δx, then $\Delta x = (b - a)/n$. Let $a = x_1$ and choose $x_2, x_3, \ldots, x_n, x_{n+1}$ so that $x_{i+1} - x_i = \Delta x$ for $i = 1, 2, \ldots, n$, that is, so that the points are equally spaced. Just as in Section 5.8, we see that the area of the first rectangle is $f(x_1) \cdot \Delta x$, and so forth, so that the sum of the areas of the n inscribed rectangles indicated in the figure below is

$$\sum_{i=1}^n f(x_i)\, \Delta x.$$

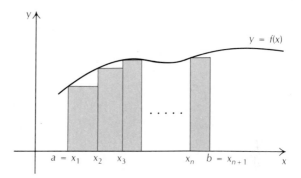

In Section 5.8, we saw that

$$\int_a^b f(x)\, dx = \lim_{n \to \infty} \sum_{i=1}^n f(x_i)\, \Delta x.$$

For large values of n, then, we expect that $\sum_{i=1}^n f(x_i)\, \Delta x$ can yield a good approximation to $\int_a^b f(x)\, dx$.

Example 1 Approximate $\int_0^1 (1 + x^2)\, dx$ using $\sum_{i=1}^n f(x_i)\, \Delta x$ with $n = 5$ and compare your approximation with the exact answer.

Solution The integral represents the area shown below.

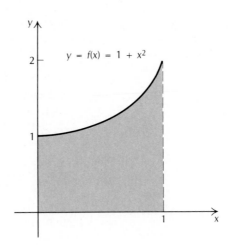

We can find its exact value as follows:

$$\int_0^1 (1 + x^2)\, dx = \left[x + \frac{1}{3} x^3 \right]_0^1$$

$$= \left(1 + \frac{1}{3}\, (1)^3 \right) - \left(0 + \frac{1}{3}\, (0)^3 \right)$$

$$= 1 + \frac{1}{3} = \frac{4}{3} \approx 1.333333.$$

If $n = 5$, then $\Delta x = (1 - 0)/5 = \frac{1}{5} = 0.2$. The points of subdivision are $x_1 = 0$, $x_2 = 0.2$, $x_3 = 0.4$, $x_4 = 0.6$, $x_5 = 0.8$, and $x_6 = 1$. Since $f(x) = 1 + x^2$, we have

$$f(x_1) = f(0) = 1 + (0)^2 = 1,$$
$$f(x_2) = f(0.2) = 1 + (0.2)^2 = 1.04,$$
$$f(x_3) = f(0.4) = 1 + (0.4)^2 = 1.16,$$
$$f(x_4) = f(0.6) = 1 + (0.6)^2 = 1.36,$$
$$f(x_5) = f(0.8) = 1 + (0.8)^2 = 1.64.$$

Therefore,

$$\sum_{i=1}^{5} f(x_i)\, \Delta x = f(x_1)\, \Delta x + f(x_2)\, \Delta x + \cdots + f(x_5)\, \Delta x$$

$$= [f(x_1) + f(x_2) + \cdots + f(x_5)]\, \Delta x$$
$$= [1 + (1.04) + (1.16) + (1.36) + (1.64)] \cdot \frac{1}{5}$$
$$= [6.2](0.2)$$
$$= 1.24.$$

1. Approximate $\int_0^1 (1 + x)\,dx$ using $\sum_{i=1}^n f(x_i)\,\Delta x$ with $n = 4$, and compare your approximation with the exact answer.

This sum equals the sum of the areas of the rectangles shown in the following figure.

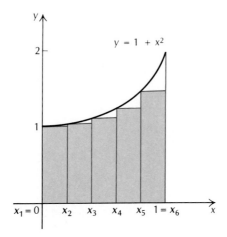

The error involved in our approximation is about $(1.333333) - (1.24) \approx 0.09$.

DO EXERCISE 1.

In Example 1, we used the values of $f(x)$ at the left-hand endpoints of the subintervals, $f(x_1), f(x_2), \ldots, f(x_5)$, for the altitudes of the approximating rectangles. If, instead, we had used the values of $f(x)$ at the right-hand endpoints of the subintervals, $f(x_2), f(x_3), \ldots, f(x_6)$, we would have obtained the rectangles shown below.

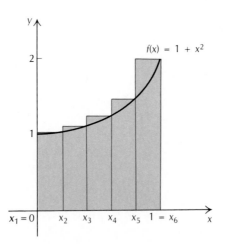

2. Approximate $\int_0^1 (1 + x) \, dx$ using $\Sigma_{i=1}^n f(x_{i+1}) \, \Delta x$ with $n = 4$, and compare your approximation with the exact answer.

Example 2 Approximate $\int_0^1 (1 + x^2) \, dx$ by using $\Sigma_{i=1}^n f(x_{i+1}) \, \Delta x$ with $n = 5$, as shown above. Compare your result with the exact answer.

Solution We have

$$\sum_{i=1}^{5} f(x_{i+1}) \, \Delta x = [f(x_2) + f(x_3) + \cdots + f(x_6)] \, \Delta x$$

$$= [(1.04) + (1.16) + (1.36) + (1.64) + 2] \cdot \tfrac{1}{5}$$

$$= [7.2](0.2) = 1.44.$$

The error involved is about $(1.44) - (1.333333) \approx 0.106667$.

DO EXERCISE 2.

In each of Examples 1 and 2, we approximated $\int_0^1 (1 + x^2) \, dx$ by using the sum of areas of rectangles. In Example 1, we used the values of $f(x)$ at the left-hand endpoints for the altitudes of the rectangles, whereas in Example 2 we used the right-hand endpoints. Both cases are examples of *approximation by Riemann sums*. In the general case, we pick any number c_1 in $[x_1, x_2]$ and use $f(c_1)$ for the altitude of the first rectangle (see the following figure).

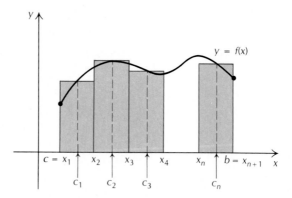

We pick c_2 between x_2 and x_3 and use $f(c_2) \, \Delta x$ for the area of the second rectangle. Continuing, we pick $c_3, c_4, \ldots, c_n$ so that c_i is between x_i and x_{i+1} for $i = 1, 2, \ldots, n$. Then we can approximate $\int_a^b f(x) \, dx$ by the sum $\Sigma_{i=1}^n f(c_i) \, \Delta x$. This sum is the sum of the areas of the rectangles shown in the figure above. Any sum of the form

$$\sum_{i=1}^{n} f(c_i) \, \Delta x$$

is called a *Riemann sum for $\int_a^b f(x) \, dx$*.

The sums $\Sigma_{i=1}^{n} f(x_i) \, \Delta x$ (using the left-hand endpoints of the sub-intervals for c_i) and $\Sigma_{i=1}^{n} f(x_{i+1}) \, \Delta x$ (using the right-hand endpoints) are both special cases of Riemann sums. We might use the midpoints of the subintervals or any other convenient choice of $c_1, c_2, \ldots, c_n$.

In general, though, Riemann sums are not an efficient way to approximate integrals. For example, suppose that we were to approximate $\int_0^1 (1 + x^2) \, dx$ using left-hand endpoints and $n = 100$ (think of the computations required!). It turns out that we would obtain

$$\sum_{i=1}^{100} f(x_i) \, \Delta x = 1.32835.$$

Note that even with 100 subintervals, the error (which is about $1.333333 - 1.32835 \approx 0.005$) is rather large.

The Trapezoidal Rule

There is an easy and obvious way of improving on approximation by Riemann sums. As before, we partition $[a, b]$ into n equal subintervals. Instead of using rectangles, however, we use trapezoids, as shown below.

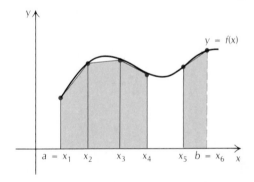

Note that the sum of the areas of the trapezoids in the figure, although not equal to $\int_a^b f(x) \, dx$, certainly appears to yield a better approximation than the inscribed rectangles would.

Note that the area A of a trapezoid whose base is b and whose altitudes are h_1 and h_2 (see the figure) is

$$\left(\begin{array}{c} \text{Area of} \\ \text{the rectangle} \end{array} \right) + \left(\begin{array}{c} \text{Area of} \\ \text{the triangle} \end{array} \right) = bh_1 + \frac{1}{2} b(h_2 - h_1).$$

Then

$$A = bh_1 + \tfrac{1}{2}b(h_2 - h_1) = b[h_1 + \tfrac{1}{2}(h_2 - h_1)]$$
$$= b[h_1 + \tfrac{1}{2}h_2 - \tfrac{1}{2}h_1] = b[\tfrac{1}{2}h_1 + \tfrac{1}{2}h_2] = \tfrac{1}{2}b[h_1 + h_2].$$

Applying these calculations to the situation shown in the figure, we see that each of the trapezoids will have base equal to Δx. The first trapezoid has altitudes $f(x_1)$ and $f(x_2)$, the values of $f(x)$ at the right- and left-hand endpoints of its base. Thus the area of the first trapezoid will be $\frac{1}{2}(f(x_1) + f(x_2)) \cdot \Delta x$. The area of the second trapezoid will be $\frac{1}{2}(f(x_2) + f(x_3)) \cdot \Delta x$. The area of the last ($n$th) trapezoid will be $\frac{1}{2}(f(x_n) + f(x_{n+1})) \cdot \Delta x$. Let T_n denote the sum of the areas of all n trapezoids. Then,

$$
\begin{aligned}
T_n &= \tfrac{1}{2}(f(x_1) + f(x_2)) \cdot \Delta x + \tfrac{1}{2}(f(x_2) + f(x_3)) \cdot \Delta x \\
&\quad + \cdots + \tfrac{1}{2}(f(x_n) + f(x_{n+1})) \cdot \Delta x \\
&= \tfrac{1}{2}[(f(x_1) + f(x_2)) + (f(x_2) + f(x_3)) \\
&\quad + \cdots + (f(x_n) + f(x_{n+1}))] \cdot \Delta x \\
&= \tfrac{1}{2}[f(x_1) + 2f(x_2) + 2f(x_3) + \cdots + 2f(x_n) + f(x_{n+1})] \cdot \Delta x \\
&= [\tfrac{1}{2}f(a) + f(x_2) + f(x_3) + \cdots + f(x_n) + \tfrac{1}{2}f(b)] \cdot \Delta x.
\end{aligned}
$$

Note that the endpoints a and b of the interval receive "special treatment" in the formula since they occur only once as endpoints of subintervals, whereas the other points occur twice.

Example 3 Approximate $\int_0^1 (1 + x^2)\, dx$ using the Trapezoidal Rule with $n = 5$, that is, T_5. Compare your answer with the exact value of the integral.

Solution As in the solution to Example 1, we have $x_1 = 0$, $x_2 = 0.2$, $x_3 = 0.4$, $x_4 = 0.6$, $x_5 = 0.8$, and $x_6 = 1$. Also,

$$
\begin{aligned}
f(x_1) &= 1 + (x_1)^2 = 1 + (0)^2 \ \ = 1.00, \\
f(x_2) &= 1 + (x_2)^2 = 1 + (0.2)^2 = 1.04, \\
f(x_3) &= 1 + (x_3)^2 = 1 + (0.4)^2 = 1.16, \\
f(x_4) &= 1 + (x_4)^2 = 1 + (0.6)^2 = 1.36, \\
f(x_5) &= 1 + (x_5)^2 = 1 + (0.8)^2 = 1.64,
\end{aligned}
$$

and

$$
f(x_6) = 1 + (x_6)^2 = 1 + (1)^2 = 2.00.
$$

Thus,

$$
\begin{aligned}
T_5 &= [\tfrac{1}{2}f(x_1) + f(x_2) + \cdots + f(x_5) + \tfrac{1}{2}f(x_6)] \cdot \Delta x \\
&= [\tfrac{1}{2}(1) + (1.04) + (1.16) + (1.36) + (1.64) + \tfrac{1}{2}(2)] \cdot \tfrac{1}{5} \\
&= [6.70](0.2) \\
&= 1.34.
\end{aligned}
$$

The error involved is about $(1.34) - (1.333333) = 0.006667$. Note

3. Approximate $\int_0^1 (1 + x)\, dx$ using the Trapezoidal Rule with $n = 4$, that is, by T_4. Compare your result with the exact answer.

that this compares with the error we get with $n = 100$ using the rectangle method (Riemann sums) with left-hand endpoints.

DO EXERCISE 3.

In Margin Exercise 3, the Trapezoidal Rule is exact. This is always the case if $f(x) = c_0 + c_1 x$ is a polynomial of degree 1. In fact, we have the following theorem.

THEOREM 3

The Trapezoidal Rule

If $f(x)$ and its first and second derivatives are continuous on $[a, b]$, and if

$$T_n = [\tfrac{1}{2}f(x_1) + f(x_2) + \cdots + f(x_n) + \tfrac{1}{2}f(x_{n+1})] \cdot \Delta x,$$

then

$$\int_a^b f(x)\, dx \approx T_n$$

and the error involved in this approximation is

$$E = \frac{b - a}{12} f''(c)(\Delta x)^2,$$

where $a < c < b$.

---

4. Use the error estimate in Theorem 3 to calculate the error in approximating $\int_0^1 (1 + x)\, dx$ by T_4 in Margin Exercise 3.

In particular, if $f(x) = c_0 + c_1 x$, then $f''(x) = 0$ and E must be 0. If we let $K = [(b - a)/12]f''(c)$ and observe that K is a constant, we see that the error is $K(\Delta x)^2$. The error is proportional to $1/n^2$.

Example 4 Use the error estimate in Theorem 3 to calculate the error in approximating $\int_0^1 (1 + x^2)\, dx$ by T_5 in Example 3.

Solution We have $a = 0$, $b = 1$, $\Delta x = \tfrac{1}{5}$, and $f(x) = 1 + x^2$. So $f'(x) = 2x$ and $f''(x) = 2$. Thus $f''(c) = 2$, independent of c. The error E is then given by

$$\frac{b - a}{12} f''(c)(\Delta x)^2 = \frac{1 - 0}{12} \cdot 2 \cdot \left(\frac{1}{5}\right)^2$$

$$= \frac{1}{12} \cdot 2 \cdot \frac{1}{25} = \frac{1}{150} \approx 0.006667.$$

From Example 3, we know that this is correct.

DO EXERCISE 4.

5. Use T_{10} to approximate $\int_0^1 x^3 \, dx$. Estimate the error involved.

Margin Exercise 3 and Example 3 suggest that the Trapezoidal Rule is simply the average of the Riemann sums corresponding to the right- and left-hand endpoints, and that is correct.

Example 5 Use T_{10} to approximate $\int_1^2 (1/x) \, dx$. Estimate the error involved.

Solution The points of subdivision are $x_1 = 1, x_2 = 1.1, x_3 = 1.2, \ldots,$ $x_{10} = 1.9,$ and $x_{11} = 2.0.$ Also, $\Delta x = 0.1.$ Applying the Trapezoidal Rule gives us

$$T_{10} = \left[\frac{1}{2} \cdot \frac{1}{1} + \frac{1}{1.1} + \frac{1}{1.2} + \cdots + \frac{1}{1.9} + \frac{1}{2} \cdot \frac{1}{2.0} \right] (0.1)$$

$$= [6.9377140](0.1)$$

$$= 0.69377140.$$

According to Theorem 3, the error, call it E, is given by

$$E = \frac{b - a}{12} \cdot f''(c) \cdot (\Delta x)^2,$$

where $a < c < b.$ Here, $a = 1, b = 2,$ and $\Delta x = 0.1.$ Since $f(x) = 1/x = x^{-1},$ we have $f'(x) = -1x^{-2}$ and $f''(x) = (-1)(-2)x^{-3} = 2/x^3.$ So we get

$$E = \frac{2 - 1}{12} \cdot \frac{2}{c^3} \cdot (0.1)^2 = \frac{1}{12} \cdot \frac{2}{c^3} \cdot (0.01) = \frac{1}{600} \cdot \frac{1}{c^3},$$

where $1 < c < 2.$ Since $1 < c < 2,$ we have $1 < c^3 < 8$ and $\frac{1}{8} < 1/c^3 < 1.$ Thus $E < \frac{1}{600} \approx 0.0016667.$ In fact,

$$\int_1^2 \frac{1}{x} \, dx = [\ln x]_1^2$$

$$= \ln 2 - \ln 1$$

$$= \ln 2.$$

So, to seven decimal places,

$$\int_1^2 \frac{1}{x} \, dx = \ln 2$$

$$\approx 0.6931472.$$

The actual error is about 0.0006242, which, of course, is less than 0.00166667, the error estimate obtained above.

DO EXERCISE 5.

Simpson's Rule

In approximating $\int_a^b f(x)\,dx$ by Riemann sums, we partitioned $[a, b]$ into n equal subintervals and used $f(c_i)$ to approximate $f(x)$ on the ith subinterval. In approximating the same integral using the Trapezoidal Rule, we approximated $f(x)$ on the ith subinterval by means of the line segment joining $(x_i, f(x_i))$ to $(x_{i+1}, f(x_{i+1}))$. This line segment is the top of the ith trapezoid in the figure on page 691.

To obtain Simpson's Rule, we partition $[a, b]$ into n equal subintervals and fit a parabola to $f(x)$ on each subinterval. Since it takes three points to uniquely determine a parabola, we use $(x_i, f(x_i))$, $(x_{i+1}, f(x_{i+1}))$ and the point on $y = f(x)$ corresponding to the midpoint of the subinterval,

$$\left(\frac{x_i + x_{i+1}}{2}, f\left(\frac{x_i + x_{i+1}}{2}\right)\right),$$

as shown in the following figure. (The parabolas here are P_1, P_2, and P_3).

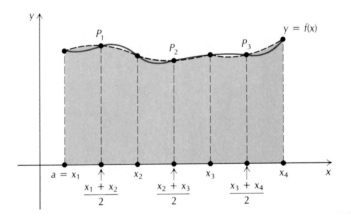

Calculating the area under each parabolic arch, summing the areas, and simplifying yields

$$S_n = \frac{1}{6}\left[f(x_1) + 4f\left(\frac{x_1 + x_2}{2}\right) + 2f(x_2) + 4f\left(\frac{x_2 + x_3}{2}\right) + 2f(x_3)\right.$$

$$\left. + \cdots + 2f(x_n) + 4f\left(\frac{x_n + x_{n+1}}{2}\right) + f(x_{n+1})\right] \cdot \Delta x.$$

We reiterate this result and the error estimate in the following theorem.

THEOREM 4   CSS

Simpson's Rule

If $f(x)$ and its first four derivatives are continuous on $[a, b]$, and if

$$S_n = \frac{1}{6}\left[f(x_1) + 4f\left(\frac{x_1 + x_2}{2}\right) + 2f(x_2)\right.$$
$$\left. + \cdots + 4f\left(\frac{x_n + x_{n+1}}{2}\right) + f(x_{n+1}) \right] \cdot \Delta x,$$

then

$$\int_a^b f(x)\, dx \approx S_n,$$

and the error involved is

$$\frac{(b - a)}{180} \cdot f^{(4)}(c) \cdot (\Delta x)^4,$$

where $a < c < b$.

Example 6 Approximate $\int_1^2 (1/x)\, dx$ using Simpson's Rule with $n = 5$. Estimate the error involved.

Solution The points of subdivision are $x_1 = 1.0$, $x_2 = 1.2$, $x_3 = 1.4$, $x_4 = 1.6$, $x_5 = 1.8$, and $x_6 = 2.0$. Since $f(x) = 1/x$, we can arrange the calculations in the form of the following table.

x	$f(x) = \dfrac{1}{x}$	Multiplying factor	Product
1.0	1	1	1.000000
1.1	0.9090909	4	3.636364
1.2	0.8333333	2	1.666667
1.3	0.7692308	4	3.076923
1.4	0.7142857	2	1.428571
1.5	0.6666667	4	2.666667
1.6	0.6250000	2	1.250000
1.7	0.5882353	4	2.352941
1.8	0.5555556	2	1.111111
1.9	0.5263158	4	2.105263
2.0	0.5000000	1	0.500000

Summing the products on the right, we get

$$S_5 \approx \tfrac{1}{6}(20.794507) \cdot (\tfrac{1}{5}) \approx 0.6931502.$$

6. Use S_5 to approximate $\int_0^1 x^3\,dx$. Estimate the error involved.

So

$$\int_1^2 \frac{1}{x}\,dx \approx 0.6931502.$$

According to Theorem 4, the error E is given by

$$\frac{(b-a)}{180}\cdot f^{(4)}(c)\cdot(\Delta x)^4,$$

where $a < c < b$. In this case, $a = 1$, $b = 2$, and $\Delta x = \frac{1}{5}$. Since $f(x) = x^{-1}$, $f'(x) = x^{-2}$, and $f''(x) = 2x^{-3}$, we get

$$f^{(3)}(x) = 2(-3)x^{-4}$$
$$= -6x^{-4}$$

and

$$f^{(4)}(x) = -6(-4)x^{-5} = \frac{24}{x^5}.$$

Thus,

$$E = \frac{(2-1)}{180}\cdot\frac{24}{c^5}\cdot\left(\frac{1}{5}\right)^4$$
$$= \frac{1}{180}\cdot\frac{24}{c^5}\cdot\frac{1}{625}$$
$$= \frac{2}{9375c^5}.$$

Here, since $1 < c < 2$, we have $\frac{1}{32} < 1/c^5 < 1$. Therefore,

$$E < \tfrac{2}{9375} \approx 0.0002133.$$

In fact, the actual error is about 0.000003, which is less than the error estimate computed above.

DO EXERCISE 6.

EXERCISE SET 11.4

Approximate each of the given integrals by the Riemann sum for the given Δx and using (a) left-hand endpoints and (b) right-hand endpoints.

1. $\int_0^1 (2x+1)\,dx;\ \Delta x = 0.1$

2. $\int_{-1}^2 x^2\,dx;\ \Delta x = 0.5$

3. $\int_0^1 \frac{1}{1+x^2}\,dx;\ \Delta x = 0.2$

4. $\int_0^1 x^4\,dx;\ \Delta x = 0.25$

Use the Trapezoidal Rule with the indicated value of n to approximate the integral.

5. $\int_0^1 (2x + 1)\, dx; n = 10$

6. $\int_{-1}^2 x^2\, dx; n = 6$

7. $\int_0^1 \frac{1}{1 + x^2}\, dx; n = 5$

8. $\int_0^1 x^4\, dx; n = 4$

Use the Trapezoidal Rule with the indicated value of n to approximate the integral. Then use the result in Theorem 3 to estimate the error involved in your approximation.

9. $\int_0^1 x^5\, dx; n = 10$

10. $\int_1^2 \frac{1}{\sqrt{x}}\, dx; n = 10$

11. $\int_1^2 \ln x\, dx; n = 10$

12. $\int_0^1 (x - x^3)\, dx; n = 5$

Use Simpson's Rule with the indicated value of n to approximate the integral.

13. $\int_0^1 (2x + 1)\, dx; n = 5$

14. $\int_{-1}^2 x^2\, dx; n = 6$

15. $\int_0^1 \frac{1}{1 + x^2}\, dx; n = 5$

16. $\int_0^1 x^4\, dx; n = 4$

Use Simpson's Rule with the indicated value of n to approximate the integral. Then use the result in Theorem 4 to estimate the error involved in your approximation.

17. $\int_0^1 x^3\, dx; n = 4$ **18.** $\int_0^1 x^5\, dx; n = 5$ **19.** $\int_1^2 \ln x\, dx; n = 5$ **20.** $\int_0^1 (x - x^3)\, dx; n = 5$

11.5

OBJECTIVES

a) Use simple iteration and Aitken acceleration to solve equations.

b) Find a numerical estimate for $\int_a^b f(x)\, dx$ using the Trapezoidal Rule, Simpson's Rule, and extrapolation (that is, Romberg integration).

EXTRAPOLATION: SPEEDING UP CONVERGENCE*

In Section 11.1 we saw that the method of iteration, although a convenient way to solve equations of the form $g(x) = x$, is often ineffective because the iterates converge so slowly. Similar difficulties arose in Section 11.4, where our extensive computations sometimes yielded approximations to integrals that were valid to only one or two decimal places. In this section, we will investigate two methods that will "speed up" convergence. Both methods are examples of *extrapolation* techniques.

We reconsider the equation $g(x) = x$. Let $\bar{x}$ be a solution to the equation and suppose that $x_1, x_2, \ldots, x_n, \ldots$, is the sequence of iterates generated by an initial estimate x_1. Suppose also that our sequence $x_1, \ldots, x_n, \ldots$, converges to $\bar{x}$. Then for all n, $x_{n+1} = g(x_n)$. For the moment, let $e_n = x_n - \bar{x}$ denote the error involved in approximating

---

* This section can be omitted without loss of continuity.

$\bar{x}$ by x_n (for all n). It turns out that if both x_1 and $\bar{x}$ lie in an interval on which the derivative of $g(x)$ satisfies $|g'(x)| < K$ (where $K < 1$), then we can show that

$$|e_2| < K|e_1|,$$
$$|e_3| < K|e_2| < K(K|e_1|) = K^2|e_1|,$$

and so forth, yielding

$$|e_{n+1}| < K|e_n| < \cdots < K^n|e_1|.$$

In particular, e_n tends to 0 as n tends to infinity. Assume, for the moment, that equality holds above, so that $e_n = K^{n-1}e_1$. Then

$$x_n = \bar{x} + K^{n-1}e_1. \tag{1}$$

If we also assume that $e_{n+1} = K^n e_1$ and $e_{n+2} = K^{n+1}e_1$, then we get

$$x_{n+1} = \bar{x} + K^n e_1 \tag{2}$$

and

$$x_{n+2} = \bar{x} + K^{n+1}e_1. \tag{3}$$

We could solve Eqs. (1), (2), and (3) simultaneously to eliminate K and e_1. After some simplification (see Exercise Set 11.5), we would have

$$\bar{x} = \frac{x_n x_{n+2} - (x_{n+1})^2}{x_{n+2} - 2x_{n+1} + x_n}.$$

If we only assume that Eqs. (1), (2), and (3) are approximately true, then we obtain

$$\bar{x} \approx \frac{x_n x_{n+2} - (x_{n+1})^2}{x_{n+2} - 2x_{n+1} + x_n}.$$

So we can use three values of the iterates to generate a new approximation. This technique is sometimes called *Aitken acceleration*.

Example 1 Use $x_1 = 0.5$, iteration, and Aitken acceleration to solve $\cos x = x$.

Solution As in Example 1 of Section 11.1, we obtain

$$x_1 = 0.5,$$
$$x_2 = \cos(x_1) = \cos(0.5) \approx 0.8775826,$$

and

$$x_3 = \cos(x_2) \approx 0.63901249.$$

Applying Aitken acceleration, we obtain a new estimate, call it x_4 (this

1. Using $x_1 = 4$ and $x_{n+1} = \sqrt{5x_n - 6}$, compute x_2 and x_3 by iteration and x_4 by Aitken acceleration. Then compute x_5 and x_6 (from x_4) by iteration and use them to compute x_7 by Aitken's method. Compare your results with those of Example 4 in Section 11.1.

is *not* the same x_4 as in Example 1 of Section 11.1), as follows:

$$x_4 = \frac{x_1 x_3 - (x_2)^2}{x_3 - 2x_2 + x_1}$$

$$\approx \frac{(0.5)(0.63901249) - (0.8775826)^2}{(0.63901249) - 2(0.8775826) + 0.5}$$

$$\approx 0.7313852.$$

Of course, we don't have to stop here. Using x_4 and iteration, we get

$$x_5 = \cos(x_4) \approx 0.7442499$$

and

$$x_6 = \cos(x_5) \approx 0.7355962.$$

Using Aitken acceleration to generate x_7, we get

$$x_7 = \frac{x_4 x_6 - (x_5)^2}{x_6 - 2x_5 + x_4}$$

$$\approx \frac{(0.7313852)(0.7355962) - (0.7442499)^2}{(0.7355962) - 2(0.7442499) + (0.7313852)}$$

$$\approx 0.7390763.$$

Compare this with the results of Example 1 of Section 11.1.

DO EXERCISE 1.

We obtained Aitken's acceleration formula above because we were able to estimate the error in successive applications of the method of iteration. The technique used above, called *extrapolation*, can be used in numerical integration, too.

Suppose that we would like to evaluate the integral $I = \int_a^b f(x)\,dx$ numerically. We might fix n, partition $[a, b]$ into n subintervals of equal length, and use the Trapezoidal Rule to obtain an estimate T_n for I. According to Theorem 3, the error here, $|I - T_n|$, is less than $K(\Delta x)^2$, where K is a constant and $\Delta x = (b - a)/n$ is the length of each subinterval. The key here is going to be this error estimate. To see what happens, suppose, for a moment, that we repeat the process above using twice as many points. We obtain a new estimate, T_{2n}, for I. How much better is our new estimate? Since we doubled the number of subintervals, it follows that the length of each of the new subintervals is exactly one-half the length of those used the first time. In other words, we simply replaced Δx by $\Delta x/2$. Our new error estimate will look like

$$K\left(\frac{\Delta x}{2}\right)^2 = K\frac{(\Delta x)^2}{4} = \frac{1}{4}K(\Delta x)^2.$$

This suggests that doubling the number of points used should improve the accuracy by a factor of 4. But it gives us a lot more than that.

If we pretend for a moment that our error estimates are exact, then we get

$$\tfrac{1}{4}(I - T_n) = I - T_{2n},$$

or

$$I - T_n = 4(I - T_{2n}).$$

Thus,

$$I - T_n = 4I - 4T_{2n}.$$

Solving for I, we get

$$3I = 4T_{2n} - T_n,$$

or

$$I = \frac{4T_{2n} - T_n}{3}.$$

Of course, the error estimates are only approximations, so that we write

$$I \approx \frac{4T_{2n} - T_n}{3}.$$

Example 2 Find an approximate value for $\int_0^1 x^4 \, dx$.

Solution We know that the exact value of the integral is $\tfrac{1}{5}$. Using the Trapezoidal Rule with $n = 4$ and $f(x) = x^4$, we have

$$T_4 = \left[\frac{1}{2} f(0) + f\left(\frac{1}{4}\right) + f\left(\frac{1}{2}\right) + f\left(\frac{3}{4}\right) + \frac{1}{2} f(1) \right] \cdot \frac{1}{4}$$

$$= \left[\frac{1}{2} (0)^4 + \left(\frac{1}{4}\right)^4 + \left(\frac{1}{2}\right)^4 + \left(\frac{3}{4}\right)^4 + \frac{1}{2} (1)^4 \right] \cdot \frac{1}{4}$$

$$= \left[0 + \frac{1}{256} + \frac{1}{16} + \frac{81}{256} + \frac{1}{2} \right] \cdot \frac{1}{4}$$

$$= [0.8828125] \cdot \frac{1}{4} = 0.220703125.$$

Similarly,

$$T_8 = \left[\frac{1}{2} (0)^4 + \left(\frac{1}{8}\right)^4 + \left(\frac{2}{8}\right)^4 + \cdots + \left(\frac{7}{8}\right)^4 + \frac{1}{2} (1)^4 \right] \cdot \frac{1}{8}$$

$$\approx [1.6416016] \cdot \frac{1}{8} \approx 0.2052002.$$

2. Approximate $\int_1^2 (1/x)\,dx$ using T_5. Use T_5 and the value of T_{10} from Example 5 of Section 11.4 to compute a new approximation to the integral.

Using the formula above, we get

$$I \approx \frac{4T_8 - T_4}{3}$$

$$= \frac{4(0.2052002) - (0.220703125)}{3}$$

$$= 0.2000325521.$$

We see that this is an excellent approximation.

DO EXERCISE 2.

Compare the answer to Margin Exercise 2 to that in Example 6 of Section 11.4. The answers are the same. This is no coincidence. The answer that we have obtained from T_n and T_{2n} by extrapolation (or, as it is usually called, *Romberg integration*) is S_n. In other words,

$$S_n = \frac{4T_{2n} - T_n}{3}.$$

This will be proved in Exercise Set 11.5. In the meantime, note that we can use our earlier argument, along with the error estimate from Theorem 4, to obtain even better numerical estimates.

If $I = \int_a^b f(x)\,dx$ and S_n is the Simpson's Rule approximation to I, then the absolute error $|I - S_n|$ is less than $M(\Delta x)^4$, where M is a constant and Δx is the length of the subintervals used. Here, if we double the number of points used, then we halve the length of each subinterval and, replacing Δx by $\Delta x/2$ in our error estimate, we get

$$M\left(\frac{\Delta x}{2}\right)^4 = M \cdot \frac{(\Delta x)^4}{16}.$$

Doubling the number of points should improve the accuracy by a factor of 16. Thus,

$$I - S_{2n} \approx \tfrac{1}{16}(I - S_n)$$

or

$$16I - 16S_{2n} \approx I - S_n$$

or

$$15I \approx 16S_{2n} - S_n.$$

Solving, we get

$$I \approx \frac{16S_{2n} - S_n}{15}.$$

Define the Romberg sum R_n by

$$R_n = \frac{16S_{2n} - S_n}{15}$$

for all positive integers n.

Example 3   Estimate

$$\int_0^1 \frac{x}{1 + x^2}\, dx$$

by S_2, S_4, and the Romberg sum

$$R_2 = \frac{16S_4 - S_2}{15}.$$

Solution   Here,

$$f(x) = \frac{x}{1 + x^2}.$$

Thus,

$$S_2 = \tfrac{1}{6} \cdot [f(0) + 4f(\tfrac{1}{4}) + 2f(\tfrac{1}{2}) + 4f(\tfrac{3}{4}) + f(1)] \cdot \Delta x$$

$$= \tfrac{1}{6} \cdot [0 + 4(\tfrac{4}{17}) + 2(\tfrac{2}{5}) + 4(\tfrac{12}{25}) + \tfrac{1}{2}] \cdot \tfrac{1}{2}$$

$$\approx \tfrac{1}{6} \cdot [4.161176] \cdot \tfrac{1}{2}$$

$$\approx 0.3467647.$$

Similarly,

$$S_4 = \tfrac{1}{6}[f(0) + 4f(\tfrac{1}{8}) + 2f(\tfrac{1}{4}) + \cdots + 4f(\tfrac{7}{8}) + f(1)] \cdot \Delta x$$

$$= \tfrac{1}{6} \cdot [0 + 4 \cdot \tfrac{8}{65} + 2 \cdot \tfrac{4}{17} + \cdots + 4 \cdot \tfrac{56}{113} + \tfrac{1}{2}] \cdot \tfrac{1}{4}$$

$$\approx \tfrac{1}{24} \cdot [8.3180181]$$

$$\approx 0.3465841.$$

Now,

$$R_2 = \frac{16S_4 - S_2}{15}$$

$$\approx \frac{16(0.3465841) - (0.3467647)}{15}$$

$$\approx \frac{5.1985809}{15}$$

$$\approx 0.34657206.$$

3. Estimate

$$\int_0^1 \frac{1}{1+x}\, dx$$

using S_2, S_4, and the Romberg sum

$$R_2 = \frac{16S_4 - S_2}{15}.$$

To get an idea of the accuracy, we write

$$I = \int_0^1 \frac{x}{1 + x_2}\, dx$$

$$= \left[\frac{1}{2} \ln (1 + x^2) \right]_0^1 = \frac{1}{2} \ln 2.$$

So, using a calculator, we get $I = \frac{1}{2} \ln 2 \approx 0.34657359$.

DO EXERCISE 3.

EXERCISE SET 11.5

Use the initial guess x_1 and iteration to compute x_2 and x_3; then calculate x_4 using Aitken's method.

1. $x = \frac{1}{3}x^2 + \frac{1}{3};\ x_1 = 0$

2. $x = \frac{1}{3}x^2 + \frac{1}{3};\ x_1 = \frac{1}{2}$

3. $x = \sin x + \frac{1}{2}$ (x in radians); $x_1 = 1$

4. $x = \frac{1}{2} \cos x;\ x_1 = 1$

5. $x = \sqrt[3]{3x - 3};\ x_1 = -1$

6. $x = \sqrt{3x - 1};\ x_1 = 2$

Calculate T_2, T_4, and T_8 for each of the following integrals. Then use Romberg integration to calculate S_2 and S_4. Finally, use Romberg integration to calculate R_2 from S_2 and S_4.

7. $\int_0^1 x^2\, dx$

8. $\int_0^1 x^4\, dx$

9. $\int_0^1 \frac{1}{1+x^2}\, dx$

10. $\int_0^1 \frac{1}{1+x^6}\, dx$

EXTENSION EXERCISES

11. Let

$$G = \frac{x_n x_{n+2} - (x_{n+1})^2}{x_{n+2} - 2x_{n+1} + x_n}$$

for a fixed positive integer n. Also, let

$$x_n = \bar{x} + K^{n-1}e_1, \tag{1}$$

$$x_{n+1} = \bar{x} + K^n e_1, \tag{2}$$

and

$$x_{n+2} = \bar{x} + K^{n+1}e_1, \tag{3}$$

where K, $\bar{x}$, and e_1 are constants. Substitute Eqs. (1), (2), and (3) into the expression for G, expand, and simplify to obtain $G = \bar{x}$.

12. Let $f(x)$ be a continuous function on $[a, b]$, and let T_n and S_n be as in Sections 11.4 and 11.5. Show that

$$\frac{4T_{2n} - T_n}{3} = S_n \quad \text{for all } n.$$

11.6

EULER'S METHOD

a) Find an approximate solution to a first-order differential equation with initial condition using either Euler's method or the three-term method.

b) Find an approximate solution to a system of two first-order differential equations with initial conditions using Euler's method.

In Chapter 9, we saw that many differential equations can be solved directly using various methods that depend on the form of the equations. Unfortunately, many applications give rise to differential equations that cannot be solved directly. In these cases, it is necessary to have methods for generating accurate, but approximate, solutions to the given equation.

We first need to introduce some notation. In Example 3 of Section 9.2, we solved the differential equation

$$y' = x - xy.$$

Note that this equation involves variables x and y, with y being a function of x. It can then be thought of as a function of two variables and can be written

$$y' = x - xy = f(x, y).$$

The solution of the equation is

$$y = 1 + C_1 e^{-x^2/2}.$$

Now we know that y is a function of x. Though we have not done so in the past, we can express this as $y = y(x)$. With the initial condition $y(0) = 2$, the solution can be written as

$$y = 1 + e^{-x^2/2}.$$

In this section we will be considering first-order differential equations (no derivative will occur higher than the first)

$$y' = f(x, y), \tag{1}$$

with initial condition

$$y(x_0) = y_0. \tag{2}$$

If we assume (and we always will) that f and $\partial f/\partial y$ are continuous functions on some rectangle in the plane containing (x_0, y_0), then we can show that there is an interval containing x_0 on which there is a unique solution $y = g(x)$ to Eq. (1) that also satisfies Eq. (2). In the specific example above, these conditions are certainly satisfied.

The first numerical technique to be considered is Euler's method. Let's suppose that we must solve

$$y' = f(x, y), \tag{1}$$
$$y(x_0) = y_0, \tag{2}$$

where f and its partials are so smooth (have continuous partial derivatives) as to guarantee a unique solution to the given system on some interval $[x_0, b]$. We will construct an approximate solution as follows. First, we select n. We then partition $[x_0, b]$ into n equal subintervals (of length $(b - x_0)/n$, called the *step size*). Let $x_1, x_2, \ldots, x_{n-1}, x_n = b$ denote the points in the partition. Now, condition (2) tells us that the point (x_0, y_0) is on the graph of the solution (see the following figure). If we substitute $x = x_0$ and $y = y_0$ into Eq. (1), we can find the slope of the exact solution $y = g(x)$ at the point (x_0, y_0). The slope will be $f(x_0, y_0) = y'|_{(x_0, y_0)}$. With this information, we can compute the equation of the tangent line to $y = g(x)$ at (x_0, y_0). The equation will be

$$y - y_0 = f(x_0, y_0)(x - x_0).$$

We use this equation to find the y-value of our approximation when $x = x_1$. Graphically, we find the point of intersection of the tangent line T_0 with the vertical line $x = x_1$ (see the figure). The y-coordinate of the point, call it y_1, will be the value that we use. If we let $x = x_1$ in the equation of the tangent line and solve for y, we get

$$y = y_0 + f(x_0, y_0)(x_1 - x_0).$$

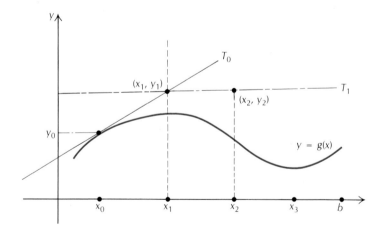

We now define y_1 by this formula, so that

$$y_1 = y_0 + f(x_0, y_0)(x_1 - x_0).$$

But now we can repeat the whole process. We substitute the coordinates (x_1, y_1) into $y' = f(x, y)$ to obtain $f(x_1, y_1)$. The equation of the line T_1 through (x_1, y_1) with slope $f(x_1, y_1)$ is

$$y - y_1 = f(x_1, y_1)(x - x_1).$$

Letting $x = x_2$ and solving for $y = y_2$ yields

$$y_2 = y_1 + f(x_1, y_1)(x_2 - x_1).$$

Note that if we let $\Delta x = x_2 - x_1$ denote the step size, then we get

$$y_2 = y_1 + f(x_1, y_1) \cdot \Delta x.$$

Continuing, we obtain the general formula

$$y_{n+1} = y_n + f(x_n, y_n) \cdot \Delta x$$

for all nonnegative integers n.

Example 1   Apply Euler's method to $y' = y, y(0) = 2$ on the interval $[0, 1]$ with $n = 5$.

Solution   We have deliberately chosen a differential equation whose solution we know, so that we can compare our approximate solution with the exact solution, which is $y = g(x) = 2e^x$.

First, since $n = 5$ and our interval has length 1, note that the step size is $\frac{1}{5} = 0.2 = \Delta x$. The points to be used as x-coordinates are $x_1 = 0.2, x_2 = 0.4, x_3 = 0.6, x_4 = 0.8$, and, of course, $x_5 = 1.0$.

The initial condition is $y(0) = 2$. In other words, $x_0 = 0$ and $y_0 = 2$. So, using $\Delta x = 0.2$ and $f(x, y) = y$, we get

$$\begin{aligned}
y_1 &= y_0 + f(x_0, y_0) \cdot \Delta x \\
&= 2 + (y_0) \cdot (0.2) \\
&= 2 + 2 \cdot (0.2) \\
&= 2 + 0.4 = 2.4.
\end{aligned}$$

Next,

$$\begin{aligned}
y_2 &= y_1 + f(x_1, y_1) \, \Delta x \\
&= 2.4 + y_1 \cdot (0.2) \\
&= 2.4 + (2.4) \cdot (0.2) \\
&= 2.88.
\end{aligned}$$

Continuing, we have

$$\begin{aligned}
y_3 &= 2.88 + (2.88)(0.2) \\
&= 3.456,\\
y_4 &= 3.456 + (3.456)(0.2) \\
&= 4.1472,
\end{aligned}$$

and

$$\begin{aligned}
y_5 &= 4.1472 + (4.1472)(0.2) \\
&= 4.97664.
\end{aligned}$$

1. a) By separating the variables, solve $y' = -2xy$, $y(0) = 1$ explicitly.

b) Using Euler's method with $n = 5$ on $[0, 1]$, find an approximate solution to $y' = -2xy$, $y(0) = 1$.

c) Repeat part (b) with a step size of 0.1.

If these points were plotted against the graph of the exact solution $y = 2e^x = g(x)$, then we would find that on the left-hand side of the interval, the approximations aren't so bad, but when we reach $x_5 = 1$, the approximate value, $y_5 = 4.97664$, bears little resemblance to the actual value,

$$g(1) = 2e^1 = 2e \approx 5.4365637.$$

The great defect in Euler's method (sometimes called the *tangent-line method*) is that it isn't very accurate.

DO EXERCISE 1.

There are many ways to alter Euler's method so as to improve its accuracy. To obtain one such improvement, return to the figure on page 706. There we obtained our initial approximations by use of the tangent line T_1 to the graph of the exact solution $y = g(x)$. To obtain better accuracy, we could use the "tangent quadratic" to the graph of $y = g(x)$. Suppose for the moment that $y = g(x)$ is represented by its Taylor series at $x = x_0$. Then,

$$g(x) = g(x_0) + g'(x_0)(x - x_0) + \frac{g''(x_0)}{2!}(x - x_0)^2 + \cdots.$$

We want to use this formula to compute $y_1 = g(x_1)$. Setting $x = x_1$ and ignoring all terms after the third one, we obtain

$$y_1 = g(x_1) \approx g(x_0) + g'(x_0)(x_1 - x_0) + \frac{g''(x_0)}{2!}(x_1 - x_0)^2.$$

Now $x_1 - x_0 = \Delta x$, the step size. So,

$$y_1 \approx g(x_0) + g'(x_0) \cdot \Delta x + \frac{g''(x_0)}{2!}(\Delta x)^2.$$

But $g(x_0) = y_0$ and $g'(x_0)$ is just $y'(x_0) = f(x_0, y_0)$, which is given. Substituting, we get

$$y_1 \approx y_0 + f(x_0, y_0)\, \Delta x + \tfrac{1}{2}g''(x_0)(\Delta x)^2.$$

Every quantity on the right-hand side above is given except Δx, the step size, which we are free to choose, and $g''(x_0)$. The last quantity may or may not be easy to compute. We illustrate with two examples.

Example 2 Compute $g''(x_0) = y''(x_0)$ for the initial-value problem,

$$y' = xy^2,$$
$$y(1) = 2.$$

2. Compute y'' for the initial-value problem $y' = y \sin(xy)$, $y(1) = 1$.

Solution In this problem, $f(x, y) = xy^2$, $x_0 = 1$, and $y_0 = 2$. Thus,

$$y'(x_0) = f(x_0, y_0) = f(1, 2) = 1 \cdot (2)^2 = 4.$$

To get y'', we must differentiate $y' = g(x, y) = xy^2$ with respect to x. Using the Product Rule, we obtain

$$y'' = x(y^2)' + y^2(x)'$$
$$= x(y^2)' + y^2,$$

and, applying the Chain Rule, we get

$$y'' = x(2yy') + y^2.$$

So $y'' = 2xyy' + y^2$. Replacing y' by xy^2 gives

$$y'' = 2xy(xy^2) + y^2,$$

or

$$y'' = 2x^2y^3 + y^2.$$

Even a very simple choice of $f(x, y)$ results in a fairly complicated expression for y''.

DO EXERCISE 2.

When y'' is relatively simple, however, this method (sometimes called the *three-term method*) yields superior results. The general iteration formula will be as follows:

$$y_{n+1} = y_n + f(x_n, y_n) \cdot \Delta x + \tfrac{1}{2} y''(x_n, y_n)(\Delta x)^2.$$

Example 3 Apply the three-term method to $y' = y$, $y(0) = 2$, on the interval $[0, 1]$ with $n = 5$.

Solution This is the problem from Example 1 given earlier. To apply the method, we must compute y''. Here, since $y' = y$, we get $y'' = y'$ by differentiation. Combining the two equations, we have $y'' = y$. Since $f(x, y) = y$ and $y'' = g'' = y$, the iteration formula becomes

$$y_{n+1} = y_n + y_n \cdot (\Delta x) + \tfrac{1}{2} y_n \cdot (\Delta x)^2,$$

or

$$y_{n+1} = y_n(1 + \Delta x + \tfrac{1}{2}(\Delta x)^2).$$

Here, $\Delta x = 0.2$. So,

$$y_{n+1} = y_n(1 + (0.2) + \tfrac{1}{2}(0.2)^2)$$
$$= y_n(1 + 0.2 + 0.02) = 1.22 y_n.$$

3. Use the three-term method to find an approximate solution to $y' = -2xy$, $y(0) = 1$, on the interval $[0, 1]$ using a step size of 0.2.

Using $y(x_0) = y(0) = 2$, we find that

$$y_1 = y(x_1) = 1.22y_0 = 1.22 \cdot (2)$$
$$= 2.44.$$

Next, we have

$$y_2 = 1.22y_1 = 1.22(2.44),$$

or

$$y_2 = 2.9768.$$

Similarly,

$$y_3 = 1.22(2.9768) = 3.631696,$$
$$y_4 = 1.22(3.631696) \approx 4.4306691,$$

and

$$y_5 \approx 1.22(4.4306691) \approx 5.4054163.$$

DO EXERCISE 3.

We compare the result of using Euler's method with the three-term method on the initial-value problem $y' = y$, $y(0) = 2$, in the table below. The exact values in the right-hand column were calculated using $y = 2e^x$.

4. Compare the results of Margin Exercises 1 and 3 with the corresponding values of the exact solution $y = e^{-x^2}$ as in the table at right.

	Euler	Three-term method	Exact
$y_0 = y(0)$	2	2.	2
$y_1 = y(0.2)$	2.4	2.44	2.4428055
$y_2 = y(0.4)$	2.88	2.9768	2.9836494
$y_3 = y(0.6)$	3.456	3.631696	3.6442376
$y_4 = y(0.8)$	4.1472	4.4306691	4.4510819
$y_5 = y(1)$	4.97664	5.4054163	5.4365637

DO EXERCISE 4.

As expected, the three-term method yields greater accuracy (although more computation is involved) than Euler's method.

We can also apply these techniques to systems of first-order differential equations. They could be used if, for instance, we have

$$x' = f(x, y, t) \quad \text{and} \quad y' = g(x, y, t)$$

with the variables x and y being thought of as functions of the third variable t. If the initial conditions are

$$x(t_0) = x_0,$$
$$y(t_0) = y_0,$$

then Euler's method becomes

$$x_{n+1} = x_n + f(x_n, y_n, t_n) \cdot \Delta t$$

and

$$y_{n+1} = y_n + g(x_n, y_n, t_n) \cdot \Delta t$$

for all n.

Example 4   Use Euler's method to find approximate values for $x(t)$ and $y(t)$ on $[0, 1]$, where x and y satisfy

$$x' = x - 5y,$$
$$y' = 2x - 6y$$

with

$$x(0) = 6,$$
$$y(0) = 3$$

using $\Delta t = 0.2$.

Solution   Here, $f(x, y, t) = x - 5y$ and $g(x, y, t) = 2x - 6y$. Our formulas become

$$x_{n+1} = x_n + (x_n - 5y_n) \cdot \Delta t$$

and

$$y_{n+1} = y_n + (2x_n - 6y_n) \cdot \Delta t,$$

or, if we use $\Delta t = 0.2$,

$$x_{n+1} = x_n + (x_n - 5y_n) \cdot (0.2)$$

and

$$y_{n+1} = y_n + (2x_n - 6y_n) \cdot (0.2).$$

Simplification yields

$$x_{n+1} = 1.2x_n - y_n$$

and

$$y_{n+1} = 0.4x_n - 0.2y_n.$$

5. a) Show that $x(t) = 5e^{-t} + e^{-4t}$ and $y(t) = 2e^{-t} + e^{-4t}$ are the exact solutions to the initial-value problem in Example 4.

   b) Use part (a) to compute the exact values of $y(0.4), x(0.6)$, and $y(1)$.

6. Use Euler's method to find two approximate values for $t = 0.2$ and $t = 0.4$ for the system

$$x' = x - 2y,$$
$$y' = -2x + y,$$

with initial conditions $x(0) = 2$, $y(0) = 0$. Use $\Delta t = 0.2$.

Using $x_0 = x(0) = 6$ and $y_0 = y(0) = 3$, we find that

$$x_1 = 1.2x_0 - y_0$$
$$= (1.2)(6) - 3$$
$$= 4.2,$$

whereas

$$y_1 = 0.4x_0 - 0.2y_0$$
$$= (0.4)(6) - (0.2)(3)$$
$$= 1.8.$$

Continuing, we get

$$x_2 = (1.2)x_1 - y_1 = (1.2)(4.2) - (1.8) = 3.24,$$
$$y_2 = (0.4)x_1 - (0.2)y_1 = (0.4)(4.2) - (0.2)(1.8) = 1.32,$$
$$x_3 = (1.2)(3.24) - (1.32) = 2.568,$$
$$y_3 = (0.4)(3.24) - (0.2)(1.32) = 1.032,$$
$$x_4 = (1.2)(2.568) - (1.032) = 2.0496,$$
$$y_4 = (0.4)(2.568) - (0.2)(1.032) = 0.8208,$$

and

$$x_5 = (1.2)(2.0496) - (0.8208) = 1.63872,$$
$$y_5 = (0.4)(2.0496) - (0.2)(0.8208) = 0.65568.$$

DO EXERCISES 5 AND 6.

EXERCISE SET 11.6

Find the solution to the initial-value problem using Euler's method on $[0, 1]$ with the indicated step size Δx.

1. $y' = -1, y(0) = 1, \Delta x = 0.1$

2. $y' = 2x, y(0) = -1, \Delta x = 0.1$

3. $y' = 2x, y(0) = 0, \Delta x = 0.25$

4. $y' = 3x^2 - 1, y(0) = 1, \Delta x = 0.2$

5. $y' = 2xy, y(0) = 1, \Delta x = 0.2$

6. $y' = -y, y(0) = 1, \Delta x = 0.25$

7. $y' = x\sqrt{y}, y(0) = 1, \Delta x = 0.2$

Find the solutions to the initial-value problem using the three-term method on $[0, 1]$ with the indicated step size Δx.

8. $y' = -1, y(0) = 1, \Delta x = 0.1$

9. $y' = 2x, y(0) = -1, \Delta x = 0.1$

10. $y' = 2x, y(0) = 0, \Delta x = 0.25$

11. $y' = 3x^2 - 1, y(0) = 1, \Delta x = 0.2$

12. $y' = 2xy, y(0) = 1, \Delta x = 0.2$

13. $y' = -y, y(0) = 1, \Delta x = 0.25$

14. $y' = x\sqrt{y}, y(0) = 1, \Delta x = 0.2$

For each of the following initial-value problems, find the exact solution $y = g(x)$. Then make a table comparing the exact values with the approximate values obtained using Euler's method and the three-term method.

15. $y' = -1, y(0) = 1$

16. $y' = 2x, y(0) = -1$

17. $y' = 2xy, y(0) = 1$

18. $y' = -y, y(0) = 1$

Use Euler's method with the indicated value of Δt to approximate the solution to the given system of differential equations on $[0, 1]$. Compare these with the exact solutions, where possible.

19. $x'(t) = x + y,$
$\quad y'(t) = 3x - y,$
$\quad x(0) = 3, y(0) = -1, \Delta t = 0.2$

20. $x'(t) = x + y,$
$\quad y'(t) = 4x - 2y,$
$\quad x(0) = 3, y(0) = -2, \Delta t = 0.25$

21. $x'(t) = y,$
$\quad y'(t) = -2x + 3y,$
$\quad x(0) = 0, y(0) = 1, \Delta t = 0.2$

22. $x'(t) = -3x + 4y,$
$\quad y'(t) = -2x + 3y,$
$\quad x(0) = 4, y(0) = 3, \Delta t = 0.1$

23. $x' = x^3,$
$\quad y' = -y,$
$\quad x(0) = 0, y(0) = 1, \Delta t = 0.1$

24. $x' = x + ty,$
$\quad y' = xy,$
$\quad x(0) = 1, y(0) = 1, \Delta t = 0.2$

APPLICATIONS

25. *Accumulated present value of a continuous cash flow.* If money flows continuously into an investment at the rate $R(t)$ (in thousands of dollars), then the *accumulated present value* $V(t)$ satisfies the initial-value problem (see Exercise Set 6.3)

$$V' = R(t)e^{-kt}, \qquad V(0) = 0,$$

where k is the current interest rate. Using step size $\Delta t = 0.2$, calculate $V(1)$ using Euler's method if $R(t) = 1/(1 + t^2)$ and $k = 8\%$.

26. In an experiment (see Exercise 3 of Exercise Set 9.3) it was found that the rate of growth of yeast cells in a laboratory satisfied the initial-value problem

$$\frac{dP}{dt} = 0.0008P(700 - P), \qquad P(0) = 10,$$

where $P(t)$ is the number of cells present after time t (in hours). Using step size 1, calculate $P(5)$ and $P(8)$ using (a) Euler's method; (b) the three-term method; and (c) the exact solution.

27. *Fick's law in biology* is given by

$$\frac{dC}{dt} = \frac{kA}{V} (C_T - C),$$

where $C(t)$ is the amount of a substance that passes through a tube with cross-sectional area A and volume V at time t (see Exercise 11 of Exercise Set 9.4), k is a constant, and C_T is the total amount of substance. Assuming that $k = 1$, $A = 3$, $V = 20$, $C_T = 1000$, and $C(0) = C_0 = 2$, use step size $\Delta t = 1$ to estimate $C(5)$ using (a) Euler's method; (b) the three-term method; and (c) the exact solution.

Yeast cells growing in a Petri dish.

28. *Accumulated present value of dividends.* Suppose that $d(t)$ represents the instantaneous dividend payment of a stock at time t. Then $d(t)e^{-mt}$ is the present value of that payment, where m is the current interest rate. Then the accumulated value $D_p(t)$ of the dividends at time t satisfies the initial-value problem

$$D_p'(t) = d(t)e^{-mt}, \qquad D(0) = 0.$$

Use step size $\Delta t = 1$ to estimate $D_p(10)$ if $m = 8\%$ and $d(t) = \$10$ using (a) Euler's method and (b) the three-term method.

30. *Commodities prices.* The price (in dollars)· of two commodities at time t (in weeks) are $P(t)$ and $Q(t)$, respectively. Given that the prices satisfy the system

$$P' = -0.5P + 0.5Q + e^{-t^2},$$
$$Q' = 0.25P - 0.25Q - 3e^{-t^2},$$
$$P(0) = 225,$$
$$Q(0) = 100,$$

use Euler's method with step size $\Delta t = 1$ to estimate $P(10)$ and $Q(10)$.

31. Let $F(x)$ be defined by

$$F(x) = \int_0^x f(t)\, dt$$

where $f(t)$ is continuous on $[0, 10]$.

a) Show that $F(x)$ satisfies the initial-value problem

$$F'(x) = f(x), \qquad F(0) = 0$$

for all x in $[0, 10]$.

b) Use Euler's method with step size $\Delta x = 1$ to estimate $F(10)$.

29. *Population.* The population of two species satisfies the system

$$\frac{dx}{dt} = 0.10x - 1.5y,$$
$$\frac{dy}{dt} = 0.015x + 0.3y,$$
$$x(0) = 1600,$$
$$y(0) = 75,$$

where $x(t)$ and $y(t)$ are the populations of the two species at time t (in years), respectively. Use Euler's method with step size $\Delta t = 1$ to estimate $x(5)$ and $y(5)$.

Commodities listing.

Predator–prey. Suppose that $x(t)$ and $y(t)$ denote the number of hares and the number of foxes, respectively, present at time t in a certain enclosed wilderness area. If we assume that plenty of food is provided for the hares, then it seems reasonable that their birth rate should be a (large) constant that is independent of time. Their death rate, on the other hand, ought to be proportional to the number of encounters between foxes and hares. Since the rate of change of the hare population will be the difference between the birth and death rates, we assume that

$$\frac{dx}{dt} = ax - bxy,$$

where a and b are positive constants.

For the foxes, the birth rate tends to increase with the supply of food (hares). Here, we might take the birth rate to be proportional to xy. As the fox population increases, there will be shortages of food and the death rate of the foxes will be proportional to the number of foxes present. We get

$$\frac{dy}{dt} = cxy - dy,$$

where c and d are positive constants. Thus we have the (nonlinear) system of equations

$$\frac{dx}{dt} = ax - bxy,$$

$$\frac{dy}{dx} = cxy - dy.$$

We will omit the derivation but we can show that these equations lead to the equation

$$\frac{dy}{dx} = \frac{cxy - dy}{ax - bxy} = \frac{y(cx - d)}{x(a - by)}.$$

32. a) Use separation of variables to solve the equation in the predator–prey discussion above.

b) Evaluate the constant of integration in part (a) in terms of a, b, c, d, and the initial populations $x_0 = x(0)$ and $y_0 = y(0)$.

Although the equation relating $x(t)$ and $y(t)$ that we obtained in the predator–prey discussion above cannot generally be solved for y or x as a function of the other variable, the graph of the relation can be plotted, as shown at right.

Note that if we start at (x_0, y_0) and proceed counter-clockwise around the figure, then the number of hares increases at first, as does the number of foxes. This results in a dramatic increase in the number of foxes (starting at P). Eventually, the foxes begin to devour the hares and we have a long period in the cycle during which the hare population drops steadily, while the fox population remains constant (arc BQ). Finally, there are so few hares that the foxes begin to starve (at Q) and their population dwindles. As the number of foxes declines, the hare population begins to grow (R), and the cycle begins anew.

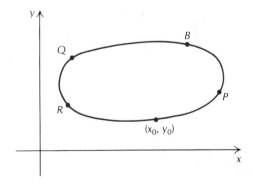

33. In a certain forest, the population of hares and foxes (in thousands) is determined by the system

$$\frac{dx}{dt} = 0.75x - 0.04xy, \qquad x(0) = 60,$$

$$\frac{dy}{dt} = 0.004xy - 0.25y, \qquad y(0) = 20,$$

Use Euler's method with step size $\Delta t = 1$ to find an approximation for **(a)** $x(5)$ and $y(5)$ and **(b)** $x(10)$ and $y(10)$.

SUMMARY AND REVIEW: CHAPTER 11

The following contains a summary of what you should be able to do after completing this chapter. The review exercises are for practice. Answers are at the back of the book. If you miss an exercise, restudy the section indicated alongside the answer.

You should be able to:

Solve equations using the method of iteration.

Use iteration and the suggested initial guesses to solve each of the given equations.

1. $x = \sqrt{5x - 6}; x_1 = 2.5$ **2.** $x = (3x - 2)^{1/3}; x_1 = 2$ **3.** $x = ([2x + 5]/9)^{1/4}; x_1 = 1$

4. $2x = \cos(x); x_1 = 0.8$ **5.** $x = 2\cos(x); x_1 = 0.8$

Solve equations using Newton's method.

Use Newton's method to find all solutions in $[0, 2]$ to each of the following equations.

6. $x^3 - 4x + 1 = 0$ **7.** $5x^4 - x - 3 = 0$ **8.** $x^8 - 4x + 2 = 0$

9. Find the equilibrium points for the supply and demand functions $S(x) = x^3 + 6x + 2$ and $D(x) = 10x^{-2}, x > 0$.

Set up and solve the equations for the yield rate of a (finite) sequence of investments and returns on these investments.

10. If interest is compounded at 8.75% annually, find the present value of a payment of $100,000 to be made twenty years from today.

11. An investor makes an initial investment of $4000 followed by an additional $1000 one year later. The investment returns $7000 four years after the original investment was made.

     a) Set up the equation for the yield rate of this investment.

     b) Find the investor's yield rate.

12. An investor must choose between two investment packages. The first requires an initial investment of $1000 and returns $1750 after three years. The second requires an initial investment of $1000, with another $1000 after one year, and returns $9500 twelve years after the initial investment was made. Find the respective yield rates.

Find a numerical estimate for $\int_a^b f(x)\,dx$ using Riemann sums, the Trapezoidal Rule, or Simpson's Rule.

13. Let

$$f(x) = \begin{bmatrix} \dfrac{1 - \cos(x)}{x}, & \text{if } x > 0, \\ 0, & \text{if } x = 0 \end{bmatrix}.$$

Use the Trapezoidal Rule with $n = 2$, 4, and 8 to approximate $\int_0^1 f(x)\,dx$.

14. Use Simpson's Rule to calculate S_2 and S_4 for Problem 13.

Use simple iteration and Aitken acceleration to solve equations.

15. Use the initial guess $x_1 = 0.8$ and iteration to obtain x_2 and x_3 for the equation $x = 2\cos(x)$. Then use Aitken acceleration to calculate x_4, etc. Compare your answer here with your answer to Problem 5 above.

16. Repeat Problem 15 with $2x = \cos(x)$.

Find a numerical estimate for $\int_a^b f(x)\,dx$ using the Trapezoidal Rule, Simpson's Rule, and Romberg integration.

17. Use Romberg integration and the results of Problem 14 to find an approximate value for $\int_0^1 f(x)\,dx$, where $f(x)$ is as in Problems 13 and 14.

Find an approximate solution to a first-order differential equation with initial condition using either Euler's method or the three-term method.

18. Use Euler's method with step size $\Delta x = 0.2$ to find an approximate solution to $y' = -y$ with $y(0) = 3$ on the interval $[0, 1]$.

19. Use the three-term method with $\Delta x = 0.2$ to find an approximate solution to Problem 18.

20. a) Separate the variables to find the exact solution to the equation in Problem 18.

b) Compare the exact values obtained above with the approximate values obtained in your solution to Problems 18 and 19.

Find an approximate solution to a system of two first-order differential equations with initial conditions using Euler's method.

21. Use Euler's method with step size $\Delta t = 0.2$ on the interval $[0, 1]$ to approximate a solution to

$$\frac{dx}{dt} = -4x + y, \qquad x(0) = 3,$$

$$\frac{dy}{dt} = -3x, \qquad y(0) = 1.$$

EXTENSION EXERCISES

22. Using the techniques of Section 9.8, find the exact solution to the system of differential equations in Problem 21.

23. Use the exact solution obtained in Problem 22 to compute x- and y-values at $t = 0, 0.2, 0.4, 0.6, 0.8$, and 1. Then compare these results with the approximations obtained in Problem 21.

TEST: CHAPTER 11

Use iteration and the suggested initial guesses to solve each of the given equations.

1. $x = \sqrt{3x - 2}; x_1 = 1.5$
2. $x = \sqrt[3]{3x - 2}; x_1 = 0$
3. $x = \sqrt[4]{\dfrac{x + 3}{6}}; x_1 = 1$

4. $x = \frac{2}{5} \cos x; x_1 = 0.75$
5. $x = \frac{5}{2} \cos x; x_1 = 0.75$

Use Newton's method to find all solutions in $[0, 2]$ of each of the given equations.

6. $2x^3 - 5x + 1 = 0$
7. $6x^4 - x - 3 = 0$
8. $x^9 - 4x + 2 = 0$

9. Find the equilibrium point for the supply and demand functions

$$S(x) = x^3 + 6x + 1 \quad \text{and} \quad D(x) = \frac{12}{x^2}.$$

10. If interest is compounded annually at 9.25%, find the present value of a payment of $100,000 to be made twenty years from today.

11. An investor makes an initial investment of $3000 followed by an additional $1000 one year later. The investment returns $6000 four years after the original investment was made.

    a) Set up the equation for the yield rate for this investment.

    b) Find the investor's yield rate.

12. An investor must choose between two investment packages. The first requires an initial investment of $1000 and returns $1500 after three years. The second requires an initial investment of $1000, another $1000 after one year, and returns $10,000 12 years after the initial investment was made. Find the respective yield rates.

13. Let

$$f(x) = \begin{cases} \dfrac{\sin x}{x}, & \text{if } x > 0, \\ 1, & \text{if } x = 0. \end{cases}$$

Use the Trapezoidal Rule with $n = 2$, 4, and 8 to approximate

$$\int_0^1 f(x) \, dx.$$

14. Use Simpson's Rule to calculate S_2 and S_4 for Problem 13.

15. Use the initial guess $x_1 = 0.75$ and iteration to obtain x_2 and x_3 for $x = \frac{5}{2} \cos x$. Then use Aitken acceleration to calculate x_4, etc. Compare your answer here with your answer to Problem 5 above.

16. Repeat Problem 15 with $x = \frac{2}{5} \cos x$.

17. Use Romberg integration and the results of Problem 14 to find an approximate value for

$$\int_0^1 f(x) \, dx.$$

18. Use Euler's method with step size $\Delta x = 0.2$ to find an approximate solution to $y' = -y^2$ with $y(0) = 1$ on the interval $[0, 1]$.

19. Use the three-term method with $\Delta x = 0.2$ to find an approximate solution to Problem 18.

20. a) Separate the variables to find the exact solution to Problem 18.

    b) Compare the exact values with the approximate values obtained in Problems 18 and 19.

21. Use Euler's method with step size $\Delta t = 0.2$ on the interval $[0, 1]$ to approximate a solution to

$$\frac{dx}{dt} = -3x + 2y, \qquad x(0) = 1,$$

$$\frac{dy}{dt} = -x, \qquad y(0) = 2.$$

EXTENSION EXERCISES

22. Show that the present value V of a series of (equal) payments of R dollars per year for n years (starting one year from today) at a rate of interest i is

$$V = Rv \cdot \frac{1 - v^n}{1 - v},$$

where $v = (1 + i)^{-1}$.

23. Show that Aitken's formula can be written in the form

$$x_n - \frac{(x_{n+1} - x_n)^2}{x_{n+2} - 2x_{n+1} + x_n}.$$

CUMULATIVE REVIEW

1. Write an equation of the line with slope -4 and containing the point $(-7, 1)$.

2. For $f(x) = x^2 - 5$, find $f(x + h)$.

3. a) Graph:
$$f(x) = \begin{cases} 5 - x, & \text{for } x \neq 2, \\ -3, & \text{for } x = 2. \end{cases}$$

   b) Find $\lim_{x \to 2} f(x)$.

   c) Find $f(2)$.

   d) Is f continuous at 2?

Find each limit, if it exists.

4. $\lim\limits_{x \to -4} \dfrac{x^2 - 16}{x + 4}$

5. $\lim\limits_{x \to 1} \sqrt{x^3 + 8}$

6. $\lim\limits_{x \to 3} \dfrac{4}{x - 3}$

7. $\lim\limits_{x \to \infty} \dfrac{12x - 7}{3x + 2}$

8. $\lim\limits_{h \to 0} \dfrac{3x^2 + h}{x - 2h}$

Differentiate.

9. $y = -9x + 3$

10. $y = x^2 - 7x + 3$

11. $y = x^{1/4}$

12. $f(x) = x^{-6}$

13. $f(x) = (x - 3)(x + 1)^5$

14. $f(x) = \dfrac{x^3 - 1}{x^5}$

15. $y = \dfrac{e^x + x}{e^x}$

16. $y = \ln(x^2 + 5)$

17. $y = e^{\ln x}$

18. $y = e^{3x} + x^2$

19. $y = e^{-x} \sin(x^2 + 3)$

20. $y = \ln|\cos x|$

21. For $y = \tan x$, find d^2y/dx^2.

22. Differentiate implicitly to find dy/dx if $x^3 + x/y = 7$.

Find maximum and minimum values, if they exist, over the indicated interval. If no interval is indicated, use the real line.

23. $f(x) = 3x^2 - 6x - 4$

24. $f(x) = -5x + 1$

25. $f(x) = \frac{1}{3}x^3 - x^2 - 3x + 5; \, [-2, 0]$

26. For a certain product the total-revenue and total-cost functions are given by

$$R(x) = 4x^2 + 11x + 110,$$
$$C(x) = 4.2x^2 + 5x + 10.$$

Find the number of units that must be produced and sold in order to maximize profit.

28. A certain population of furbearing animals has the reproduction curve

$$f(P) = P(400 - P),$$

where P is measured in thousands. Find the population at which the maximum sustainable harvest occurs, and find the maximum sustainable harvest.

30. Sketch the graph of $y = x^3 - 3x + 1$.

32. Find the elasticity: $p = D(x) = 80 - 30 \ln (x + 1)$.

Integrate.

33. $\int 3x^5 \, dx$

35. $\int \dfrac{x}{(7 - 3x)^2} \, dx$ (Use Table 1.)

37. $\int (x + 3) \ln x \, dx$

39. $\int e^x \cos (e^x) \, dx$

40. Find the area under the graph of $y = x^2 + 3x$ on the interval $[1, 5]$.

42. Find the accumulated present value of an investment over a 50-year period where there is a continuous money flow of $5400 and the current interest rate is 9%, compounded continuously.

27. An appliance store sells 450 pocket radios each year. It costs $4 to store one radio for one year. To order radios there is a fixed cost of $1 plus $0.75 for each radio. How many times per year should the store reorder radios, and in what lot size, in order to minimize inventory costs?

29. For $f(x) = 3x^2 - 7$, $x = 5$, and $\Delta x = 0.1$, find Δy and $f'(x) \, \Delta x$.

31. The demand for oil in the world is increasing at the rate of 10% per year; that is,

$$\frac{dA}{dt} = 0.1A,$$

where $A =$ the amount of oil used and $t =$ the time in years from 1980.

a) Given that 66,164 million barrels were used in 1980, find the solution to the equation, assuming $A_0 = 66{,}164$ and $k = 0.1$.

b) How much oil will be used in 1990?

34. $\int_{-1}^{0} (2e^x + 1) \, dx$

36. $\int x^3 e^x \, dx$ (Use substitution. Do not use Table 1.)

38. $\int 2x \sin (x^2 + 3) \, dx$

41. What annual payment should be made so that the amount of an annuity after 30 years at 9%, compounded continuously, will be $81,000?

43. Determine whether this improper integral is convergent or divergent, and calculate its value if convergent:

$$\int_3^{\infty} \frac{1}{x^7} \, dx.$$

44. Given the probability density function

$$f(x) = \frac{3}{2x^2} \quad \text{over } [1, 3],$$

find $E(x)$.

46. Given the demand and supply functions

$$D(x) = (x - 20)^2 \quad \text{and} \quad S(x) = x^2 + 10x + 50,$$

find the equilibrium point and the consumer's surplus.

48. Solve the differential equation $dy/dx = xy$.

49. In an advertising experiment it was determined that

$$\frac{dP}{dt} = k(L - P),$$

where $P = $ the percentage of the people in the city who bought the product after the ad was run t times and $L = 100\%$, or 1.

a) Express $P(t)$ in terms of $L = 1$.

b) It was determined that $P(10) = 70\%$. Use this to determine k. Round your answer to the nearest hundredth.

c) Rewrite $P(t)$ in terms of k.

d) Use the equation in part (c) to find $P(20)$.

Given $f(x, y) = e^y + 4x^2y^3 + 3x$, find each of the following.

50. f_x

52. Find the relative maximum and minimum values of $f(x, y) = 8x^2 - y^2$.

54. Evaluate

$$\int_0^3 \int_{-1}^2 e^x \, dy \, dx.$$

Find each of the following.

55. $\sin \dfrac{\pi}{4}$

56. $\cos \dfrac{\pi}{3}$

57. $\tan \pi$

Integrate.

58. $\displaystyle\int \sec 8x \, dx$

59. $\displaystyle\int \dfrac{dx}{1 + 9x^2}$

Differentiate.

60. $y = \cot (\sin x^3)$

61. $f(x) = \sin^{-1} x$

Find.

62. dw when $w = 3x^5y^2$

63. dw when $w = 7e^x \sin (yz)$

45. Let x be a continuous random variable that is normally distributed with mean $\mu = 3$ and standard deviation $\sigma = 5$. Using Table 2, find $P(-2 \leqslant x \leqslant 8)$.

47. Find the volume of the solid of revolution generated by rotating the region under the graph of

$$y = e^{-x} \quad \text{from} \quad x = 0 \text{ to } x = 5$$

about the x-axis.

51. f_{yy}

53. Maximize $f(x, y) = 4x + 2y - x^2 - y^2 + 4$, subject to the constraint $x + 2y = 9$.

64. A rectangular box has dimensions 7.01 feet by 2.03 feet by 0.97 feet. Use differentials to approximate the volume of the box.

65. Evaluate
$$\int_0^1 \left\{ \int_0^1 x^3 y^2 \, dx \right\} dy.$$

66. Find the volume of the solid whose base is the triangular region
$$G = \{(x, y) \,|\, x \leqslant y \leqslant 2 - x, \quad 0 \leqslant x \leqslant 1\}$$
and whose height at each point is $2xy$.

67. Evaluate $\iint_G x^4 y$ where G is the unit disk,
$$G = \{(x, y) \,|\, x^2 + y^2 \leqslant 1\}.$$

Solve.

68. $(y - 3) \, dx + (x + 3y^2) \, dy = 0$

69. $y' + (\cos x)y = 0$

70. $y''' + 4y' = 0$

71. $y'' - 14y' + 49y = 0$

72. $y'' + 4y' + 3y = 15$

73. $x' = -4x + y$
    $y' = -15x + 4y$

74. Expand $5x^3 - 13x^2 + 8x - 5$ in powers of $x - 1$.

75. Use the Taylor polynomial of degree 3 for $y = \sin x$ at $a = 0$ to find an approximate value for $\sin (0.07)$.

76. Find the sum of $\dfrac{1}{5} - \dfrac{1}{35} + \dfrac{1}{245} - \dfrac{1}{1715} + \cdots$.

77. Write $0.679\overline{679}$ as a rational fraction.

78. Does $\dfrac{1}{9} - \dfrac{1}{16} + \dfrac{1}{25} - \dfrac{1}{36} + \cdots$ converge or diverge?

79. Compute the Taylor series expansion of $f(x) = x \sin (x^4)$ at $x = 0$.

80. Find an approximate value for $\int_0^1 x \sin (x^4) \, dx$.

Evaluate.

81. $\lim\limits_{x \to 1} \dfrac{2x}{x - 1}$

82. $\lim\limits_{x \to 1} \dfrac{x^3 - 5x^2 + 3x + 1}{x^2 - 1}$

83. $\lim\limits_{x \to 0} \dfrac{e^{4x} - 1}{x}$

84. $\lim\limits_{x \to 1} \dfrac{2x}{x + 1}$

85. Use iteration to solve $5x - 2 \cos x = 0$. Then check your answer.

86. Use Newton's Method to find the solution of $x^3 - 6x + 4 = 0$ that lies between 0 and 1.

87. Find the present value of a payment of $1000 to be made 8 years from today at a rate of interest of 10%.

88. An investor pays $7000 for three annual payments of $3000 each, beginning one year from today. Find the yield rate of this investment and check your answer.

89. Use the Trapezoidal Rule with $n = 5$ to estimate
$$\int_0^1 \frac{dx}{x + 2}.$$
Compare your value with the exact answer.

90. Use Simpson's Rule with $n = 5$ to estimate
$$\int_0^1 \frac{dx}{x + 2}.$$

91. a) Use Simpson's Rule with $n = 2$ and $n = 4$ to estimate
$$I = \int_0^1 x^8 \, dx.$$

b) Then use S_2 and S_4 to compute R_2.

TABLES

TABLE 1 INTEGRATION FORMULAS

(Whenever $\ln X$ is used, it is assumed that $X > 0$.)

1. $\int x^n \, dx = \dfrac{x^{n+1}}{n+1} + C, n \neq -1$

2. $\int \dfrac{dx}{x} = \ln x + C$

3. $\int u \, dv = uv - \int v \, du$

4. $\int e^x \, dx = e^x + C$

5. $\int e^{ax} \, dx = \dfrac{1}{a} \cdot e^{ax} + C$

6. $\int xe^{ax} \, dx = \dfrac{1}{a^2} \cdot e^{ax}(ax - 1) + C$

7. $\int x^n e^{ax} \, dx = \dfrac{x^n e^{ax}}{a} - \dfrac{n}{a} \int x^{n-1} e^{ax} \, dx$

8. $\int \ln x \, dx = x \ln x - x + C$

9. $\int (\ln x)^n \, dx = x(\ln x)^n - n \int (\ln x)^{n-1} \, dx, n \neq -1$

10. $\int x^n \ln x \, dx = x^{n+1}\left[\dfrac{\ln x}{n+1} - \dfrac{1}{(n+1)^2}\right] + C, n \neq -1$

11. $\int a^x \, dx = \dfrac{a^x}{\ln a} + C, a > 0, a | 1$

12. $\int \dfrac{1}{\sqrt{x^2 + a^2}} \, dx = \ln(x + \sqrt{x^2 + a^2}) + C$

13. $\int \dfrac{1}{\sqrt{x^2 - a^2}} \, dx = \ln(x + \sqrt{x^2 - a^2}) + C$

14. $\int \dfrac{1}{x^2 - a^2} \, dx = \dfrac{1}{2a} \ln\left(\dfrac{x - a}{x + a}\right) + C$

15. $\int \dfrac{1}{a^2 - x^2} \, dx = \dfrac{1}{2a} \ln\left(\dfrac{a + x}{a - x}\right) + C$

16. $\int \dfrac{1}{x\sqrt{a^2 + x^2}} \, dx = -\dfrac{1}{a} \ln\left(\dfrac{a + \sqrt{a^2 + x^2}}{x}\right) + C$

17. $\int \dfrac{1}{x\sqrt{a^2 - x^2}} \, dx = -\dfrac{1}{a} \ln\left(\dfrac{a + \sqrt{a^2 - x^2}}{x}\right) + C, 0 < x < a$

18. $\int \dfrac{x}{ax + b} \, dx = \dfrac{b}{a^2} + \dfrac{x}{a} - \dfrac{b}{a^2} \ln(ax + b) + C$

19. $\displaystyle\int \frac{x}{(ax+b)^2}\,dx = \frac{b}{a^2(ax+b)} + \frac{1}{a^2}\ln(ax+b) + C$

20. $\displaystyle\int \frac{1}{x(ax+b)}\,dx = \frac{1}{b}\ln\left(\frac{x}{ax+b}\right) + C$

21. $\displaystyle\int \frac{1}{x(ax+b)}\,dx = \frac{1}{b(ax+b)} + \frac{1}{b^2}\ln\left(\frac{x}{ax+b}\right) + C$

22. $\displaystyle\int \sqrt{x^2 \pm a^2}\,dx = \tfrac{1}{2}[x\sqrt{x^2 \pm a^2} \pm a^2\ln(x+\sqrt{x^2 \pm a^2})] + C$

23. $\displaystyle\int \sin x\,dx = -\cos x + C, \int \cos x\,dx = \sin x + C$

24. $\displaystyle\int \sec^2 x\,dx = \tan x + C, \int \csc^2 x\,dx = -\cot x + C$

25. $\displaystyle\int \tan x \sec x\,dx = \sec x + C, \int \cot x \csc x = -\csc x + C$

26. $\displaystyle\int \sec x\,dx = \ln|\sec x + \tan x| + C, \int \csc x\,dx = -\ln|\csc x + \cot x| + C$

27. $\displaystyle\int \frac{1}{1+x^2}\,dx = \tan^{-1} + C$

28. $\displaystyle\int \frac{1}{\sqrt{1-x^2}}\,dx = \sin^{-1} x + C, \int \frac{-1}{\sqrt{1-x^2}}\,dx = \cos^{-1} x + C$

29. $\displaystyle\int x\sqrt{a+bx}\,dx = \frac{2}{15b^3}(3bx - 2a)(a+bx)^{3/2} + C$

30. $\displaystyle\int x^2\sqrt{a+bx}\,dx = \frac{2}{105b^3}(15b^2x^2 - 12abx + 8a^2)(a+bx)^{3/2} + C$

31. $\displaystyle\int \frac{x\,dx}{\sqrt{a+bx}} = \frac{2}{3b^2}(bx - 2a)\sqrt{a+bx} + C$

32. $\displaystyle\int \frac{x^2\,dx}{\sqrt{a+bx}} = \frac{2}{15b^3}(3b^2x^2 - 4abx + 8a^2)\sqrt{a+bx} + C$

33. $\displaystyle\int \sin^n x\,dx = -\frac{1}{n}\sin^{n-1} x \cos x + \frac{n-1}{n}\int \sin^{n-2} x\,dx + C$

34. $\displaystyle\int \cos^n x\,dx = \frac{1}{n}\cos^{n-1} x \sin x + \frac{n-1}{n}\int \cos^{n-2} x\,dx + C$

35. $\displaystyle\int e^{ax} \sin x\,dx = \frac{e^{ax}}{a^2+b^2}(a\sin bx - b\cos bx) + C$

36. $\displaystyle\int e^{ax} \cos x\,dx = \frac{e^{ax}}{a^2+b^2}(a\cos bx + b\sin bx) + C$

TABLE 2
AREAS FOR A STANDARD NORMAL DISTRIBUTION

Entries in the table represent area under the curve between $t = 0$ and a positive value of t. Because of the symmetry of the curve, area under the curve between $t = 0$ and a negative value of t would be found in a similar manner.

Area =
Probability = $P(0 \leqslant x \leqslant t)$

$$= \int_0^t \frac{1}{\sqrt{2\pi}} e^{-x^2/2} \, dx$$

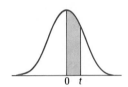

0   t

t	0.00	0.01	0.02	0.03	0.04	0.05	0.06	0.07	0.08	0.09
0.0	.0000	.0040	.0080	.0120	.0160	.0199	.0239	.0279	.0319	.0359
0.1	.0398	.0438	.0478	.0517	.0557	.0596	.0636	.0675	.0714	.0753
0.2	.0793	.0832	.0871	.0910	.0948	.0987	.1026	.1064	.1103	.1141
0.3	.1179	.1217	.1255	.1293	.1331	.1368	.1406	.1443	.1480	.1517
0.4	.1554	.1591	.1628	.1664	.1700	.1736	.1772	.1808	.1844	.1879
0.5	.1915	.1950	.1985	.2019	.2054	.2088	.2123	.2157	.2190	.2224
0.6	.2257	.2291	.2324	.2357	.2389	.2422	.2454	.2486	.2517	.2549
0.7	.2580	.2611	.2642	.2673	.2704	.2734	.2764	.2794	.2823	.2852
0.8	.2881	.2910	.2939	.2967	.2995	.3023	.3051	.3078	.3106	.3133
0.9	.3159	.3186	.3212	.3238	.3264	.3289	.3315	.3340	.3365	.3389
1.0	.3413	.3438	.3461	.3485	.3508	.3531	.3554	.3577	.3599	.3621
1.1	.3643	.3665	.3686	.3708	.3729	.3749	.3770	.3790	.3810	.3830
1.2	.3849	.3869	.3888	.3907	.3925	.3944	.3962	.3980	.3997	.4015
1.3	.4032	.4049	.4066	.4082	.4099	.4115	.4131	.4147	.4162	.4177
1.4	.4192	.4207	.4222	.4236	.4251	.4265	.4279	.4292	.4306	.4319
1.5	.4332	.4345	.4357	.4370	.4382	.4394	.4406	.4418	.4429	.4441
1.6	.4452	.4463	.4474	.4484	.4495	.4505	.4515	.4525	.4535	.4545
1.7	.4554	.4564	.4573	.4582	.4591	.4599	.4608	.4616	.4625	.4633
1.8	.4641	.4649	.4656	.4664	.4671	.4678	.4686	.4693	.4699	.4706
1.9	.4713	.4719	.4726	.4732	.4738	.4744	.4750	.4756	.4761	.4767
2.0	.4772	.4778	.4783	.4788	.4793	.4798	.4803	.4808	.4812	.4817
2.1	.4821	.4826	.4830	.4834	.4838	.4842	.4846	.4850	.4854	.4857
2.2	.4861	.4864	.4868	.4871	.4875	.4878	.4881	.4884	.4887	.4890
2.3	.4893	.4896	.4898	.4901	.4904	.4906	.4909	.4911	.4913	.4916
2.4	.4918	.4920	.4922	.4925	.4927	.4929	.4931	.4932	.4934	.4936
2.5	.4938	.4940	.4941	.4943	.4945	.4946	.4948	.4949	.4951	.4952
2.6	.4953	.4955	.4956	.4957	.4959	.4960	.4961	.4962	.4963	.4964
2.7	.4965	.4966	.4967	.4968	.4969	.4970	.4971	.4972	.4973	.4974
2.8	.4974	.4975	.4976	.4977	.4977	.4978	.4979	.4979	.4980	.4981
2.9	.4981	.4982	.4982	.4983	.4984	.4984	.4985	.4985	.4986	.4986
3.0	.4987	.4987	.4987	.4988	.4988	.4989	.4989	.4989	.4990	.4990

PHOTO CREDITS

ANSWERS

Margin Exercises, Section 1.1, pp. 2–9

1. $3 \cdot 3 \cdot 3 \cdot 3$, or 81   2. $(-3)(-3)$, or 9
3. $1.02 \times 1.02 \times 1.02$, or 1.061208 4. $\frac{1}{4} \cdot \frac{1}{4}$, or $\frac{1}{16}$
5. 1   6. $5t$ 7. 1   8. m 9. $\frac{1}{4}$ 10. 1
11. $\dfrac{1}{2 \cdot 2 \cdot 2 \cdot 2}$, or $\dfrac{1}{16}$ 12. $\dfrac{1}{10 \cdot 10}$, or $\dfrac{1}{100}$, or 0.01

13. 64   14. $\dfrac{1}{t^7}$ 15. $\dfrac{1}{e^t}$ 16. $\dfrac{1}{M}$ 17. $\dfrac{1}{(x+1)^2}$
18. t^9 19. t^{-3} 20. $50e^{-13}$ 21. t^{-6} 22. $24b^3$
23. x^4 24. x^{-4} 25. 1   26. e^{2-k} 27. e^{12} 28. e^2
29. x^{-12} 30. e^4 31. e^{3x} 32. $25x^6y^{10}$
33. $\frac{1}{256}x^{20}y^{24}z^{-8}$, or $\dfrac{x^{20}y^{24}}{256z^8}$ 34. $2x + 14$ 35. $P - Pi$

36. $x^2 + 3x - 28$ 37. $a^2 - 2ab + b^2$ 38. $a^2 - b^2$
39. $x^2 - 2xh + h^2$ 40. $9x^2 + 6xt + t^2$ 41. $25t^2 - m^2$
42. $P(1 - i)$ 43. $(x + 5y)^2$ 44. $4(x + 5)(x + 2)$
45. $(5c - d)(5c + d)$ 46. $(4y + 7)(3y - 2)$ 47. 1.01
48. $\$1188.10$ 49. $\$1378.84$

Exercise Set 1.1, pp. 9–11

1. $5 \cdot 5 \cdot 5$, or 125   3. $(-7)(-7)$, or 49   5. 1.0201

7. $\frac{1}{16}$ 9. 1   11. t 13. 1   15. $\dfrac{1}{3^2}$, or $\dfrac{1}{9}$ 17. 8

19. 0.1 21. $\dfrac{1}{e^b}$ 23. $\dfrac{1}{b}$ 25. x^5 27. x^{-6}, or $\dfrac{1}{x^6}$

29. $35x^5$ 31. x^4 33. 1   35. x^3 37. x^{-3}, or $\dfrac{1}{x^3}$

39. 1   41. e^{t-4} 43. t^{14} 45. t^2 47. t^{-6}, or $\dfrac{1}{t^6}$

49. e^{4x} 51. $8x^6y^{12}$ 53. $\frac{1}{81}x^8y^{20}z^{-16}$, or $\dfrac{x^8y^{20}}{81z^{16}}$

55. $9x^{-16}y^{14}z^4$, or $\dfrac{9y^{14}z^4}{x^{16}}$ 57. $5x - 35$ 59. $x - xt$

61. $x^2 - 7x + 10$ 63. $a^3 - b^3$ 65. $2x^2 + 3x - 5$
67. $a^2 - 4$ 69. $25x^2 - 4$ 71. $a^2 - 2ah + h^2$
73. $25x^2 + 10xt + t^2$ 75. $5x^5 + 30x^3 + 45x$
77. $a^3 + 3a^2b + 3ab^2 + b^3$ 79. $x^3 - 15x^2 + 75x - 125$
81. $x(1 - t)$ 83. $(x + 3y)^2$ 85. $(x - 5)(x + 3)$
87. $(x - 5)(x + 4)$ 89. $(7x - t)(7x + t)$
91. $4(3t - 2m)(3t + 2m)$ 93. $ab(a + 4b)(a - 4b)$
95. $(a^4 + b^4)(a^2 + b^2)(a + b)(a - b)$
97. $10x(a + 2b)(a - 2b)$
99. $2(1 + 4x^2)(1 + 2x)(1 - 2x)$ 101. $(9x - 1)(x + 2)$
103. $(x + 2)(x^2 - 2x + 4)$
105. $(y - 4t)(y^2 + 4yt + 16t^2)$
107. (a) 0.81; (b) 0.0801; (c) 0.008001
109. (a) $\$1160$; (b) $\$1166.40$; (c) $\$1169.86$;
(d) $\$1173.47$, assuming 365 days in a year; (e) $\$1173.51$
111. $\$353.52$

Margin Exercises, Section 1.2, pp. 13–17

1. $\frac{56}{9}$ 2. $725   3. $0, -2, \frac{3}{2}$ 4. $-4, 3$ 5. $0, -1, 1$
6. $x < \frac{11}{5}$ 7. $\frac{20}{17} \leqslant x$ 8. More than 19,975 suits
9. (a) $(-1, 3)$; (b) $(1, 4)$ 10. (a) $(-1, 4)$;
(b) $(-\frac{1}{4}, \frac{1}{4})$ 11. (a) $[-1, 4]$; (b) $(-1, 4]$; (c) $[-1, 4)$;
(d) $(-1, 4)$ 12. (a) $(-\sqrt{2}, \sqrt{2})$; (b) $[0, 1)$;
(c) $(-6.7, -4.2]$; (d) $[3, 7\frac{1}{2}]$ 13. (a) $(-\infty, 5]$;
(b) $(4, \infty)$; (c) $(-\infty, 4.8)$; (d) $[3, \infty)$ 14. (a) $[8, \infty)$;
(b) $(-\infty, -7)$; (c) $(10, \infty)$; (d) $(-\infty, -0.78]$

Exercise Set 1.2, pp. 18–19

1. $\frac{7}{4}$ 3. -8 5. 120 7. 200 9. 480 lb   11. $650
13. $810,000$ 15. $0, -3, \frac{4}{5}$ 17. $0, 2$ 19. $0, 3$
21. $0, 7$ 23. $0, \frac{1}{3}, -\frac{1}{3}$ 25. 1 27. $-\frac{4}{5} \leqslant x$
29. $x > -\frac{1}{12}$ 31. $x > -\frac{4}{7}$ 33. $x \leqslant -3$ 35. $x > \frac{2}{3}$
37. $x < -\frac{2}{5}$ 39. $2 < x < 4$ 41. $\frac{3}{2} \leqslant x \leqslant \frac{11}{2}$
43. $-1 \leqslant x \leqslant \frac{14}{5}$ 45. More than 7000 units
47. $60\% \leqslant x < 100\%$ 49. $(0, 5)$ 51. $[-9, -4)$
53. $[x, x + h]$ 55. (p, ∞) 57. $[-3, 3]$
59. $[-14, -11)$ 61. $(-\infty, -4]$

Margin Exercises, Section 1.3, pp. 21–31

1.

2. (a) Yes; (b) no

3.

4.

5.

6.

Inputs	Outputs
5	$\frac{1}{5}$
$-\frac{2}{3}$	$-\frac{3}{2}$
$\frac{1}{4}$	4
$\frac{1}{a}$	a
k	$\frac{1}{k}$
$1 + t$	$\frac{1}{1+t}$

7. $f(5) = \frac{1}{5}, f(-2) = -\frac{1}{2}$,
$f\left(\frac{1}{4}\right) = 4, f\left(\frac{1}{a}\right) = a$,
$f(k) = \frac{1}{k}, \quad f(1 + t) = \frac{1}{1+t}$,
$f(x + h) = \frac{1}{x+h}$

8. $t(5) = 30, t(-5) = 20$,
$t(x + h) = x + h + x^2 + 2xh + h^2$
9. (a) All real numbers except 3, since an input of 3
would result in division by 0; (b) $f(5) = \frac{1}{2}, f(4) = 1$,
$f(2.5) = -2, f(x + h) = \dfrac{1}{x + h - 3}$
10. Same as Margin Exercise 3, only labeled
$f(x) = -2x + 1$
11. Same as Margin Exercise 4, only labeled
$g(x) = x^2 - 3$
12. (c), (d)
13.
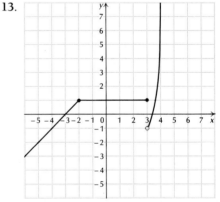

Exercise Set 1.3, pp. 32–34

1. (a)

Inputs	Outputs
4.1	11.2
4.01	11.02
4.001	11.002
4	11

(b) $f(5) = 13$,
$f(-1) = 1$,
$f(k) = 2k + 3$,
$f(1 + t) = 2t + 5$,
$f(x + h) = 2x + 2h + 3$

3. $g(-1) = -2, g(0) = -3, g(1) = -2, g(5) = 22,$
$g(u) = u^2 - 3, g(a + h) = a^2 + 2ah + h^2 - 3,$
$g(1 - h) = h^2 - 2h - 2$
5. **(a)** $f(4) = 1, f(-2) = 25, f(0) = 9, f(a) = a^2 - 6a + 9,$
$f(t + 1) = t^2 - 4t + 4, f(t + 3) = t^2,$
$f(x + h) = x^2 + 2xh + h^2 - 6x - 6h + 9;$
(b) Take an input, square it, subtract 6 times the input, add 9.

7.

9.

11.

13.

15. Yes   **17.** Yes
19. No   **21.** No

23. **(a)**

(b) No

25. $f(x + h) = x^2 + 2xh + h^2 - 3x - 3h$
27. $R(10) = \$70; R(100) = \$250$
29.

31.

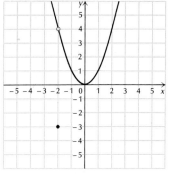

33. $y = 5$; a function   **35.** $y = \pm\sqrt{x}$; not a function

Margin Exercises, Section 1.4, pp. 35–46

1. **(a)** Horizontal line through $(0, 3)$; **(b)** yes
2. **(a)** Vertical line through $(1, 0)$; **(b)** no
3. **(a)**

(b) yes; **(c)** -2

4. **(a)** A, B; **(b)** C, D, E; **(c)** A; **(d)** E
5. **(a)** $T = \frac{1}{36}h$; **(b)** 4.5
6. **(a)**

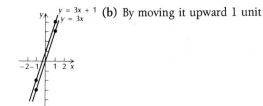

(b) By moving it upward 1 unit

7. $m = -\frac{2}{3}$; y-intercept: $(0, 2)$ **8.** $y = -4x + 1$
9. $y + 7 = -4(x - 2)$, or $y = -4x + 1$ **10.** 2
11. $\frac{1}{8}$ **12.** $-\frac{17}{2}$ **13.** 0   **14.** 0   **15.** No slope
16. **(a)**

(b) $C(100) = \$18,000, C(400) = \$27,000$;
(c) $C(400) - C(100) = \$9000$

17. (a), (b)

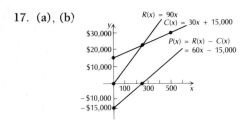

(c) Break-even is 250

Exercise Set 1.4, pp. 46–49

1. Horizontal line through $(0, -4)$
3. Vertical line through $(4.5, 0)$

5.

7.

9.

11.

13. $m = -2$; y-intercept: $(0, 2)$
15. $m = -1$; y-intercept: $(0, -\frac{5}{2})$
17. $y + 5 = -5(x - 1)$, or $y = -5x$
19. $y - 3 = -2(x - 2)$, or $y = -2x + 7$
21. $y = \frac{1}{2}x - 6$ **23.** $y = 3$ **25.** $\frac{3}{2}$ **27.** $\frac{1}{2}$
29. No slope **31.** 0 **33.** 3 **35.** 2
37. $y - 1 = \frac{3}{2}(x + 2)$, or $y + 2 = \frac{3}{2}(x + 4)$, $y = \frac{3}{2}x + 4$
39. $y + 4 = \frac{1}{2}(x - 2)$, or $y + 3 = \frac{1}{2}(x - 4)$, or $y = \frac{1}{2}x - 5$
41. $x = 3$ **43.** $y = 3$ **45.** $y = 3x$ **47.** $y = 2x + 3$
49. (a) $R = 4.17T$; **(b)** $R = 25.02$ **51. (a)** $B = 0.025W$;
(b) $B = 2.5\%W$. The weight of the brain is 2.5% of the body weight. **(c)** 3 lb
53. (a) $A = P + 14\%P = P + 0.14P = 1.14P$; **(b)** \$114;
(c) \$240 **55. (a)** $D(0°) = 115$ ft, $D(-20°) = 75$ ft, $D(10°) = 135$ ft, $D(32°) = 179$ ft;

(b)

(c) Temperature below $-57.5°$ would yield a negative stopping distance, which has no meaning here. For temperatures above $32°$ there would be no ice.
57. (a) $A(0) = 2$, $A(1) = 3.1$, $A(4) = 6.4$, $A(10) = 13$;
(b) straight line through $(0, 2)$ and $(10, 13)$;
(c) The area is measured only from the time the organism is released. Thus only nonnegative values of t would be used as inputs. **59. (a)** $C(x) = 20x + 100,000$;
(b) $R(x) = 45x$; **(c)** $P(x) = R(x) - C(x) = 25x - 100,000$;
(d) \$3,650,000, a profit; **(e)** 4000 **61. (a)** 200.69 cm;
(b) 195.23 cm

Margin Exercises, Section 1.5, pp. 51–63

1.

2. (a)

(b) By moving it to the right 3 units

3. $\dfrac{-1 \pm \sqrt{22}}{3}$

4.

5. (a) All real numbers except -4; **(b)** all real numbers except -5, 1; **(c)** all real numbers except 5

6.

7. 160,000

8.

9.

10. $[-\frac{3}{2}, \infty)$ **11.** $t^{3/4}$ **12.** $y^{1/5}$ **13.** $x^{-2/5}$
14. $t^{-1/3}$ **15.** x^3 **16.** $x^{7/2}$ **17.** $\sqrt[7]{y}$ **18.** $\sqrt{x^3}$
19. $\dfrac{1}{\sqrt{t^3}}$ **20.** $\dfrac{1}{b}$ **21.** 32 **22.** 9 **23.** $(2, \$9)$

Exercise Set 1.5, pp. 63–65

1.

3. See Margin Exercise 1 for the graph of $y = x^2$. Move it to the right 1 unit to get the graph of $y = (x - 1)^2$.

5. See Margin Exercise 1 for the graph of $y = x^2$. Move it to the left 1 unit to get the graph of $y = (x + 1)^2$.

7. $y = |x + 3|$ $\quad$ $y = |x|$

9. See Exercise Set 1.3, Exercise 13 for the graph of $y = x^3$. Move it up 1 unit for the graph of $y = x^3 + 1$.

11. See Margin Exercise 9 for the graph of $y = \sqrt{x}$. Move it to the left 1 unit for the graph of $y = \sqrt{x + 1}$.

13. $y = x^2 - 4x + 3$

15. $y = -x^2 + 2x - 1$

17.

19. $y = \frac{-2}{x}$

21.

x	-2	-1	$-\frac{1}{2}$	$\frac{1}{2}$	1	2
y	$\frac{1}{4}$	1	4	4	1	$\frac{1}{4}$

23.

25. $1 \pm \sqrt{3}$ $\quad$ **27.** $-3 \pm \sqrt{10}$ $\quad$ **29.** $\dfrac{1 \pm \sqrt{2}}{2}$

31. $\dfrac{-4 \pm \sqrt{10}}{3}$ $\quad$ **33.** $x^{3/2}$ $\quad$ **35.** $a^{3/5}$ $\quad$ **37.** $t^{1/7}$

39. $t^{-4/3}$ $\quad$ **41.** $t^{-1/2}$ $\quad$ **43.** $(x^2 + 7)^{-1/2}$ $\quad$ **45.** $\sqrt[5]{x}$

47. $\sqrt[3]{y^2}$ $\quad$ **49.** $\dfrac{1}{\sqrt[5]{t^2}}$ $\quad$ **51.** $\dfrac{1}{\sqrt[3]{b}}$ $\quad$ **53.** $\dfrac{1}{\sqrt[6]{e^{17}}}$ $\quad$ **55.** $\dfrac{1}{\sqrt{x^2 - 3}}$

57. 27 $\quad$ **59.** 16 $\quad$ **61.** 8 $\quad$ **63.** All real numbers except 5
65. All real numbers except 2, 3 $\quad$ **67.** $[-\frac{4}{5}, \infty)$
69. (2, \$4) $\quad$ **71.** (1, \$4) $\quad$ **73.** (2, \$4) $\quad$ **75.** \$2.27
77. 84, 220, 364

Margin Exercises, Section 1.6, pp. 65–71

1. $3.77744 \approx 3{:}46.6$ $\quad$ **2.** 1993 $\quad$ **3.** (a) $T = 0.5x - 946$;
(b) 48.5¢; 1.5¢ more than in part (b) of Example 1
4. (a) $A = \frac{131}{300}x^2 - 39\frac{7}{10}x + 1039\frac{1}{3}$, or
$A = 0.437x^2 - 39.7x + 1039.333$; (b) 516

Exercise Set 1.6, pp. 71–73

1. (b) Linear; (c) $y = 32x + 9998$; (d) \$10,126; \$10,318;
(e) $C = 29.5x + 10,000$; (f) \$10,118; \$10,295
3. (b) Quadratic; (c) $N = -50t^2 + 350t$; (d) 163, 413, 563
5. (b) Constant, linear; (c) $S = -20x + 100,330$;
(d) Let $S = b$, where b is the average of sales totals:
\$100,296.25.

Summary and Review: Chapter 1, pp. 73–75

1. [1.1] $(2.01)(2.01)$ $\quad$ **2.** [1.1] y^3 $\quad$ **3.** [1.1] y^{-7}
4. [1.1] $243t^{-10}m^{20}$ $\quad$ **5.** [1.1] $9x^2 - 6xt + t^2$
6. [1.1] $9x^2 - t^2$ $\quad$ **7.** [1.1] $12x^2 + 11xt - 5t^2$
8. [1.1] $(x - 8)(x + 6)$ $\quad$ **9.** [1.1] $(5x - 4t)(5x + 4t)$
10. [1.1] $(a - 4b)^2$ $\quad$ **11.** [1.1] $(3x - 1)(7x - 4)$
12. [1.1] \$1212.75 $\quad$ **13.** [1.1] \$5017.60
14. [1.2] $x \geqslant 1$ $\quad$ **15.** [1.2] 1 $\quad$ **16.** [1.2] $\frac{5}{4}, -\frac{5}{4}$
17. [1.2] $0, 3, -\frac{5}{2}$ $\quad$ **18.** [1.2] $(-6, 1]$ $\quad$ **19.** [1.2] $(0, \infty)$
20. [1.3] 13 $\quad$ **21.** [1.3] $2h^2 + 3h + 4$ $\quad$ **22.** [1.3] 3

23. [1.3] 36   **24.** [1.3] $h^2 - 2h + 1$ **25.** [1.3] 9
26. [1.3]

27. [1.3]

28. [1.3] No   **29.** [1.3] Yes   **30.** [1.3] Yes
31. [1.4]

32. [1.4]

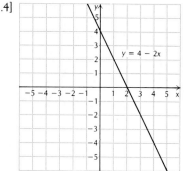

33. [1.4] $y = -\frac{7}{11}x + \frac{6}{11}$, or $7x + 11y = 6$
34. [1.4] $y = 8x + 7$ **35.** [1.4] $m = -\frac{1}{6}$; y-intercept: $(0, 3)$
36. [1.4] 72 lb
37. [1.4] **(a)** $R(x) = 28x$; **(b)** $C(x) = 16x + 80,000$;
(c) $P(x) = 12x - 80,000$; **(d)** \$16,000 profit; **(e)** 6667

38. [1.5] $\sqrt[6]{y}$ **39.** [1.5] $x^{3/20}$ **40.** [1.5] 125
41. [1.5] $(-\infty, 3]$
42. [1.5] All real numbers except 1 and -1
43. [1.5] (3, \$27)   **44.** [1.6] $y = -\frac{11}{6}x + 10$
45. [1.6] $y = x^2 - 2x + 5$
46. [1.6]

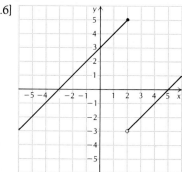

47. [1.3] $\frac{1}{32}$

Test: Chapter 1, pp. 75–76

1. $\frac{1}{e^k}$ **2.** e^{-13}, or $\frac{1}{e^{13}}$ **3.** $x^2 + 2xh + h^2$
4. $(5x - t)(5x + t)$ **5.** \$920   **6.** $x > -4$
7. **(a)** $f(-3) = 5$; **(b)** $x^2 + 2xh + h^2 - 4$
8. $m = -3$; y-intercept: $(0, 2)$
9. $y + 5 = \frac{1}{4}(x - 8)$, or $y = \frac{1}{4}x - 7$ **10.** $m = 6$
11. $F = \frac{2}{3}W$ **12.** **(a)** $C(x) = 0.5x + 10,000$;
(b) $R(x) = 1.3x$; **(c)** $P(x) = R(x) - C(x) = 0.8x - 10,000$;
(d) 12,500   **13.** (3, \$16)
14.

15. $t^{-1/2}$ **16.** $\frac{1}{\sqrt[5]{t^3}}$ **17.** All real numbers except 2, -7
18. $[-2, \infty)$ **19.** $y = 4x - 1$
20. $y = -4.5x^2 + 17.5x - 8$ **21.** [c, d)

22.

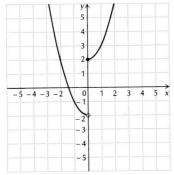

23. (a) $9.25; **(b)** 525

CHAPTER 2

Margin Exercises, Section 2.1, pp. 79–88

1. (a) 14, 16.4, 16.7, 16.97, 16.997; 17.003, 17.03, 17.3, 18.2, 20; **(b)** 17;

(c)

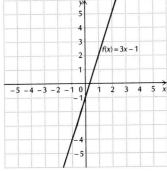

(d) −4; **(e)** 5; **(f)** −1

2. (a) −0.5, −1, −10, −100, −1000; 1000, 100, 10, 1.25, 1; **(b)** does not exist;

(c)

(d) does not exist; **(e)** −0.5; **(f)** 1   **3. (a)** 4; **(b)** −3; **(c)** does not exist; **(d)** 0; **(e)** 0; **(f)** 0; **(g)** 3; **(h)** 3; **(i)** 3

4. (a)

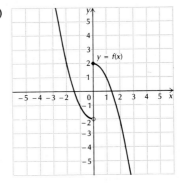

(b) does not exist; **(c)** 2

5. (a)

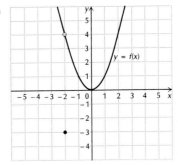

(b) 4; **(c)** 4; **(d)** 4; **(e)** 4; **(f)** 4; **(g)** no; **(h)** yes
6. (a) 2; **(b)** 3; **(c)** yes; **(d)** no   **7. (a), (b)**
8. (a) No, yes; **(b)** yes, no   **9. (a)** yes; 17; **(b)** yes; 17; **(c)** yes; **(d)** yes   **10. (a)** No; **(b)** yes   **11.** No   **12.** No
13. $\sqrt[3]{x}$ is continuous by (ii); 7 is continuous by (i) and x^2 is continuous by (ii), so $7x^2$ is continuous by (iii). Then $\sqrt[3]{x} - 7x^2$ is continuous by (iii). Now x is continuous by (ii) and 2 is continuous by (i), so $x - 2$ is continuous by (iii) and $\dfrac{1}{x-2}$ is continuous by (iv). Thus $\dfrac{\sqrt[3]{x} - 7x^2}{x - 2}$ is continuous by (iii).

Exercise Set 2.1, pp. 88–90

1. No   **3.** Yes   **5. (a)** −1, 2, does not exist; **(b)** −1; **(c)** no; **(d)** 3; **(e)** 3; **(f)** yes   **7. (a)** 2; **(b)** 2; **(c)** yes; **(d)** 0; **(e)** 0; **(f)** yes   **9.** No, yes, no, yes
11. 22, 39, does not exist   **13.** 56¢

Margin Exercises, Section 2.2, pp. 90–94

1. 53　2. $\sqrt{8}$　3. -8　4. $\frac{1}{6}$
5. (a) $2x + 1, 2x + 0.7, 2x + 0.4, 2x + 0.1, 2x + 0.01,$
$2x + 0.001$; (b) $2x$
6. (a) 4.00, 4.50, 5.14, 6.00, 7.20, 9.00, 12.00, 18.00, 36.00,
54.00, 108.00; (b) ∞; (c) ∞
7. (a) 3.25, 2.25, 2.0625, 2.025, 2.005, 2.0005; (b) 2　8. $\frac{2}{3}$

Exercise Set 2.2, pp. 94–96

1. -2　3. Does not exist　5. 11　7. -10　9. $-\frac{5}{2}$
11. 5　13. 2　15. 3　17. Does not exist　19. $\frac{2}{7}$
21. $\frac{5}{4}$　23. $\frac{1}{2}$　25. $6x^2$　27. $\dfrac{-2}{x^3}$　29. $\frac{2}{5}$　31. 5
33. $\frac{1}{2}$　35. $\frac{2}{3}$　37. (a) 1; (b) 1; (c) 1; (d) 1; (e) yes; no
39. (a) $800, $736, $677.12, $622.95, $573.11;
(b) $5656.12; (c) $10,000　41. 0, no　43. -0.25
45. 0

Margin Exercises, Section 2.3, pp. 97–103

1. (a) 20 suits/hr, 35 suits/hr, 9 suits/hr, 36 suits/hr;
(b) 11 A.M. to 12 P.M.; (c) workers were anticipating the
lunch break; answers may vary; (d) 10 A.M. to 11 A.M.;
(e) workers were fatigued or took longer breaks than they
should have; answers may vary; (f) 25 suits/hr
2. (a) 21; (b) 7; (c) 28; (d) 21　3. (a) $\frac{1}{2}$; (b) $\frac{1}{2}$; (c) $\frac{1}{2}$

5. (a) $f(x + h) = 4x^2 + 8xh + 4h^2$;
(b) $f(x + h) - f(x) = 8xh + 4h^2$;
(c) $\dfrac{f(x+h) - f(x)}{h} = 4(2x + h)$;
(d) 36, 40, 44, 47.6, 47.96, 47.996
6. (a) $4(3x^2 + 3xh + h^2)$; (b) 28, 45.64, 47.7604,
47.976004　7. (a) $\dfrac{-1}{x(x+h)}$;
(b) -0.1; -0.1667, -0.2381, -0.2488, -0.2499

Exercise Set 2.3, pp. 103–106

1. (a) 70, 39, 29, 23 pleasure units/unit of product;
(b) the more you get, the less pleasure you get from each
additional unit
3. (a) 1.25 words/min, 1.25 words/min, 0.625 words/min,
0 words/min, 0 words/min; (b) you have reached
a saturation point; you cannot memorize any more
5. (a) 125 million people/yr for each; (b) no;
(c) A: 290 million people/yr, -40 million people/yr,
-50 million people/yr, 300 million people/yr;
B: 125 million people/yr in all intervals; (d) A
7. (a) $300,093.99; (b) $299,100; (c) $993.99;
(d) $993.99　9. (a) 144 ft; (b) 400 ft; (c) 128 ft/sec
11. approximately 44,182 marriages/yr
13. (a) $13,151,000 per yr; (b) $54,566,500 per yr
15. (a) $7(2x + h)$; (b) 70, 63, 56.7, 56.07
17. (a) $-7(2x + h)$; (b) $-70, -63, -56.7, -56.07$
19. (a) $7(3x^2 + 3xh + h^2)$; (b) 532, 427, 344.47,
336.8407

4.

x	h	$x + h$	$f(x)$	$f(x + h)$	$f(x + h) - f(x)$	$\dfrac{f(x+h) - f(x)}{h}$
3	2	5	36	100	64	32
3	1	4	36	64	28	28
3	0.1	3.1	36	38.44	2.44	24.4
3	0.01	3.01	36	36.2404	0.2404	24.04
3	0.001	3.001	36	36.024004	0.024004	24.004

21. (a) $\dfrac{-5}{x(x+h)}$; (b) $-0.2083, -0.25, -0.3049,$ -0.3117 23. (a) -2; (b) all -2 25. (a) $2x+h-1$; (b) $9, 8, 7.1, 7.01$ 27. $2ax+b+ah$ 29. $\dfrac{1}{\sqrt{x+h}+\sqrt{x}}$

31. $\dfrac{-2x-h}{x^2(x+h)^2}$ 33. $\dfrac{1}{(x+1)(x+1+h)}$

Margin Exercises, Section 2.4, pp. 108–116

1. (a) L_2, L_3, L_4, L_6; (b) $-2, -1, 0, 1, 2$; (c) $m(x) = 2x$
2. $f'(5) = 10$ 3. $f'(x) = 8x, f'(5) = 40 =$ the slope of the tangent line at $(5, f(5))$, or $(5, 100)$
4. $f'(x) = 12x^2, f'(-5) = 300, f'(0) = 0$
5. $f'(x) = \dfrac{-1}{x^2}, f'(-10) = -\dfrac{1}{100} = -0.01,$
$f'(-2) = \frac{1}{4} = -0.25$ 6. x_2, x_4, x_5, x_6

Exercise Set 2.4, p. 117

1. $f'(x) = 10x, f'(-2) = -20, f'(-1) = -10, f'(0) = 0,$
$f'(1) = 10, f'(2) = 20$ 3. $f'(x) = -10x, f'(-2) = 20,$
$f'(-1) = 10, f'(0) = 0, f'(1) = -10, f'(2) = -20$
5. $f'(x) = 15x^2, f'(-2) = 60, f'(-1) = 15, f'(0) = 0,$
$f'(1) = 15, f'(2) = 60$ 7. $f'(x) = 2$, all 2
9. $f'(x) = -4$, all -4 11. $f'(x) = 2x+1, f'(-2) = -3,$
$f'(-1) = -1, f'(0) = 1, f'(1) = 3, f'(2) = 5$
13. $f'(x) = \dfrac{-4}{x^2}, f'(-2) = -1, f'(-1) = -4,$
$f'(0)$ does not exist, $f'(1) = -4, f'(2) = -1$
15. $f'(x) = m$, all m 17. $x_0, x_3, x_4, x_6, x_{12}$
19. $x = -3$ 21. $\dfrac{-2}{x^3}$ 23. $\dfrac{1}{(1+x)^2}$

Margin Exercises, Section 2.5, pp. 119–126

1. (a) $3x^2$; (b) $3x^2$; (c) 48 2. $\dfrac{dy}{dx} = 6x^5$

3. $\dfrac{dy}{dx} = -7x^{-8}$, or $-\dfrac{7}{x^8}$ 4. $\dfrac{dy}{dx} = \dfrac{1}{3}x^{-2/3}$, or $\dfrac{1}{3\sqrt[3]{x^2}}$

5. $\dfrac{dy}{dx} = -\dfrac{1}{4}x^{-5/4}$, or $\dfrac{-1}{4\sqrt[4]{x^5}}$ 6. $g'(x) = 0$

7. $\dfrac{dy}{dx} = 100x^{19}$ 8. $\dfrac{dy}{dx} = \dfrac{3}{x^2}$

9. $\dfrac{dy}{dx} = -4x^{-1/2}$, or $\dfrac{-4}{\sqrt{x}}$ 10. $\dfrac{dy}{dx} = x^{5.25}$

11. $\dfrac{dy}{dx} = -\dfrac{1}{4}$ 12. $\dfrac{dy}{dx} = 28x^3 + 12x$

13. $\dfrac{dy}{dx} = 30x - \dfrac{4}{x^2} + \dfrac{1}{2\sqrt{x}}$ 14. $(2, \frac{8}{3})$

15. $(2+\sqrt{3}, \frac{8}{3} + \sqrt{3}), (2-\sqrt{3}, \frac{8}{3} - \sqrt{3})$

Exercise Set 2.5, pp. 126–127

1. $7x^6$ 3. 0 5. $600x^{149}$ 7. $3x^2 + 6x$ 9. $\dfrac{4}{\sqrt{x}}$

11. $0.07x^{-0.93}$ 13. $\dfrac{2}{5 \cdot \sqrt[5]{x}}$ 15. $\dfrac{-3}{x^4}$ 17. $6x - 8$

19. $\dfrac{1}{4\sqrt[4]{x^3}} + \dfrac{1}{x^2}$ 21. $1.6x^{1.5}$ 23. $\dfrac{-5}{x^2} - 1$

25. 4 27. 4 29. x^3 31. $-0.02x - 0.5$
33. $-2x^{-5/3} + \frac{3}{4}x^{-1/4} + \frac{6}{5}x^{1/5} - 24x^{-4}$ 35. $(0, 0)$
37. $(0, 0)$ 39. $(\frac{5}{6}, \frac{23}{12})$ 41. $(-25, 76.25)$
43. There are none.
45. The tangent is horizontal at all points on the graph.
47. $(\frac{5}{3}, \frac{148}{27}), (-1, -4)$
49. $(\sqrt{3}, 2 - 2\sqrt{3}), (-\sqrt{3}, 2 + 2\sqrt{3})$
51. $(\frac{19}{2}, \frac{399}{4})$ 53. $(60, 150)$
55. $(-2+\sqrt{3}, \frac{4}{3} - \sqrt{3}), (-2 - \sqrt{3}, \frac{4}{3} + \sqrt{3})$
57. $(0, -4), (\sqrt{\frac{2}{3}}, -\frac{40}{9}), (-\sqrt{\frac{2}{3}}, -\frac{40}{9})$

59. $2x - 1$ 61. $2x + 1$ 63. $3x^2 - \dfrac{1}{x^2}$

65. $-192x^2$ 67. $\dfrac{2}{3 \cdot \sqrt[3]{x^2}}$ 69. $3x^2 + 6x + 3$

71. $\dfrac{F(x+h) - F(x)}{h} =$
$\dfrac{[f(x+h) - g(x+h)] - [f(x) - g(x)]}{h} =$
$\dfrac{f(x+h) - f(x)}{h} - \dfrac{g(x+h) - g(x)}{h}$. As $h \to 0$, the two terms on the right approach $f'(x)$ and $g'(x)$, respectively, so their difference approaches $f'(x) - g'(x)$. Thus $F'(x) = f'(x) - g'(x$

Margin Exercises, Section 2.6, pp. 129–133

1. (a) 70 mi/hr; (b) 100 mi/hr   2. (a) $v(t) = 32t$;
(b) $v(2) = 64$ ft/sec; (c) $v(10) = 320$ ft/sec
3. $a(t) = 32$ ft/sec^2 4. (a) $V'(s) = 3s^2$;
(b) $V'(10) = 300$ ft^2 5. (a) $P'(t) = 9700 + 20{,}000t$;
(b) 308,500; 109,700 bacteria/hr;
(c) 428,200; 129,700 bacteria/hr
6. (a) $P(x) = 40x - 0.5x^2 - 3$; (b) $R(40) = \$1200$,
$C(40) = \$403$, $P(40) = \$797$; (c) $R'(x) = 50 - x$,
$C'(x) = 10$, $P'(x) = 40 - x$; (d) $R'(40) = \$10$ per unit,
$C'(40) = \$10$ per unit, $P'(40) = \$0$ per unit; (e) no

Exercise Set 2.6, pp. 134–135

1. (a) $v(t) = 3t^2 + 1$; (b) $a(t) = 6t$;
(c) $v(4) = 49$ ft/sec, $a(4) = 24$ ft/sec^2

3. (a) $\dfrac{dV}{dh} = 0.61/\sqrt{h}$; (b) 244 miles;

(c) 0.00305 miles per foot   5. $A'(t) = 0.08$

7. $\dfrac{dC}{dr} = 2\pi$ 9. (a) $T'(t) = -0.2t + 1.2$; (b) $100.175°$;

(c) 0.9 degrees/day   11. $\dfrac{dB}{dx} = 0.1x - 0.9x^2$

13. $\dfrac{dT}{dW} = 1.31W^{0.31}$

15. (a) $P(x) = -0.001x^2 + 3.8x - 60$;
(b) $R(100) = \$500$, $C(100) = \$190$, $P(100) = \$310$;
(c) $R'(x) = 5$, $C'(x) = 0.002x + 1.2$,
$P'(x) = -0.002x + 3.8$; (d) $R'(100) = \$5$ per unit,
$C'(100) = \$1.4$ per unit, $P'(100) = \$3.6$ per unit

Margin Exercises, Section 2.7, pp. 137–139

1. $f'(x) = 54x^{17}$
2. $f'(x) = (9x^3 + 4x^2 + 10)(-14x + 4x^3) +$
$(27x^2 + 8x)(-7x^2 + x^4)$ 3. $f'(x) = 4x^3$

4. $\dfrac{3x^2 - 5}{x^6}$

5. $\dfrac{-x^4 + 3x^2 + 2x}{(x^3 + 1)^2}$

6. (a) $R(x) = x(200 - x) = 200x - x^2$;
(b) $R'(x) = 200 - 2x$

Exercise Set 2.7, pp. 140–141

1. $11x^{10}$ 3. $\dfrac{1}{x^2}$ 5. $3x^2$

7. $(8x^5 - 3x^2 + 20)\left(32x^3 - \dfrac{3}{2\sqrt{x}}\right) +$
$(40x^4 - 6x)(8x^4 - 3\sqrt{x})$ 9. $300 - 2x$

11. $\dfrac{300}{(300 - x)^2}$ 13. $\dfrac{17}{(2x + 5)^2}$

15. $\dfrac{-x^4 - 3x^2 - 2x}{(x^3 - 1)^2}$ 17. $\dfrac{1}{(1 - x)^2}$

19. $\dfrac{2}{(x + 1)^2}$ 21. $\dfrac{-1}{(x - 3)^2}$

23. $\dfrac{-2x^2 + 6x + 2}{(x^2 + 1)^2}$ 25. $\dfrac{-18x + 35}{x^8}$

27. (a) $R(x) = x(400 - x) = 400x - x^2$;
(b) $R'(x) = 400 - 2x$ 29. (a) $R(x) = 4000 + 3x$;

(b) $R'(x) = 3$ 31. $A'(x) = \dfrac{xC'(x) - C(x)}{x^2}$

33. $\dfrac{5x^3 - 30x^2\sqrt{x}}{2\sqrt{x}(\sqrt{x} - 5)^2}$.   35. $\dfrac{-3(1 + 2v)}{(1 + v + v^2)^2}$

37. $\dfrac{2t^3 - t^2 + 1}{(1 - t + t^2 - t^3)^2}$ 39. $\dfrac{5x^3 + 15x^2 + 2}{2x\sqrt{x}}$

41. $[x(9x^2 + 6) + (3x^3 + 6x - 2)](3x^4 + 7) +$
$12x^4(3x^3 + 6x - 2)$ 43. $\dfrac{6t^2(t^5 + 3)}{(t^3 + 1)^2} + \dfrac{5t^4(t^3 - 1)}{t^3 + 1}$

45. $\dfrac{(x^7 - 2x^6 + 9)[(2x^2 + 3)(12x^2 - 7) + 4x(4x^3 - 7x + 2)]}{- (7x^6 - 12x^5)(2x^2 + 3)(4x^3 - 7x + 2)}{(x^7 - 2x^6 + 9)^2}$

Margin Exercises, Section 2.8, pp. 142–147

1. $20x(1 + x^2)^9$ 2. $\dfrac{-x}{\sqrt{1 - x^2}}$

3. $-2x(1 + x^2)(1 + 3x^2)$ 4. $2(x - 4)^4(6 - x)^2(21 - 4x)$

5. $\left(\dfrac{x + 5}{x - 4}\right)^{-2/3} \cdot \dfrac{-3}{(x - 4)^2}$

6. $3(x^2 - 1)$, $9x^2 - 1$ 7. $4 \cdot \sqrt[3]{x} + 5$, $\sqrt[3]{4x + 5}$
8. $2u$, $3x^2$, $6x^2(x^3 + 2)$ 9. (a) $f(x) = \sqrt[3]{x}$, $g(x) = x^2 + 1$;

(b) $f(x) = \dfrac{1}{x}$, $g(x) = (x + 5)^4$

Exercise Set 2.8, pp. 148–149

1. $-55(1-x)^{54}$ 3. $\dfrac{4}{\sqrt{1+8x}}$ 5. $\dfrac{3x}{\sqrt{3x^2-4}}$

7. $-240x(3x^2-6)^{-41}$

9. $\sqrt{2x+3}+\dfrac{x}{\sqrt{2x+3}}$, or $\dfrac{3(x+1)}{\sqrt{2x+3}}$

11. $2x\sqrt{x-1}+\dfrac{x^2}{2\sqrt{x-1}}$, or $\dfrac{5x^2-4x}{2\sqrt{x-1}}$ 13. $\dfrac{-6}{(3x+8)^3}$

15. $(1+x^3)^2(-3x^2-12x^5)$, or $-3x^2(1+x^3)^2(1+4x^3)$

17. $4x-400$ 19. $2(x+6)^9(x-5)^3(7x-13)$

21. $4(x-4)^7(3-x)^3(10-3x)$

23. $4(2x-3)^2(3-8x)$ 25. $\left(\dfrac{1-x}{1+x}\right)^{-1/2}\cdot\dfrac{-1}{(x+1)^2}$

27. $\dfrac{1}{2}u^{-1/2}, 2x, \dfrac{x}{\sqrt{x^2-1}}$

29. (a) $\dfrac{2x-3x^2}{(1+x)^6}$; (b) $\dfrac{2x-3x^2}{(1+x)^6}$; (c) same

31. $C'(x)=\dfrac{1500x^2}{\sqrt{x^3+2}}$ 33. $\$3000(1+i)^2$

35. $12x^2-12x+5, 6x^2+3$ 37. $\dfrac{16}{x^2}-1, \dfrac{2}{4x^2-1}$

39. x^4-2x^2+2, x^4+2x^2 41. $f(x)=x^5, g(x)=3x^2-7$

43. $f(x)=\dfrac{x+1}{x-1}, g(x)=x^3$ 45. $\dfrac{x^2-2}{\sqrt[3]{(x^3-6x+1)^2}}$

47. $\dfrac{x-2}{2(x-1)^{3/2}}$ 49. $\dfrac{-4(1+2v)^3}{v^5}$

51. $\dfrac{1}{\sqrt{1-x^2}(1-x)}$, or $\dfrac{\sqrt{1-x^2}}{(1-x^2)(1-x)}$

53. $3\left(\dfrac{x^2-x-1}{x^2+1}\right)^2\cdot\dfrac{x^2+4x-1}{(x^2+1)^2}$, or

$\dfrac{3(x^2-x-1)^2(x^2+4x-1)}{(x^2+1)^4}$ 55. $\dfrac{1}{\sqrt{t}(1+\sqrt{t})^2}$

Margin Exercises, Section 2.9, pp. 150–151

1. $f'(x)=12x^5-5x^4, f''(x)=60x^4-20x^3$,
$f'''(x)=240x^3-60x^2, f^{(4)}(x)=720x^2-120x$,
$f^{(5)}(x)=1440x-120, f^{(6)}(x)=1440$

2. (a) $\dfrac{dy}{dx}=7x^6-3x^2$; (b) $\dfrac{d^2y}{dx^2}=42x^5-6x$;

(c) $\dfrac{d^3y}{dx^3}=210x^4-6$; (d) $\dfrac{d^4y}{dx^4}=840x^3$

3. $\dfrac{d^2y}{dx^2}=\dfrac{4}{x^3}$ 4. $y'=60(x^2-12x)^{29}(x-6)$;

$y''=60(x^2-12x)^{28}[59x^2-708x+2088]$

5. $a(t)=12t^2$

Exercise Set 2.9, p. 152

1. 0 3. $-\dfrac{2}{x^3}$ 5. $-\dfrac{3}{16}x^{-7/4}$ 7. $12x^2+\dfrac{8}{x^3}$ 9. $\dfrac{12}{x^5}$

11. $n(n-1)x^{n-2}$ 13. $12x^2-2$ 15. $-\dfrac{1}{4}(x-1)^{-3/2}$,

or $\dfrac{-1}{4\sqrt{(x-1)^3}}$

17. $2a$ 19. 24 21. $720x$ 23. $20(x^2-5)^8[19x^2-5]$

25. $a(t)=6t+2$ 27. $P''(t)=200{,}000$

29. $y'=x^{-2}-2x^{-3}$,
$y''=2x^{-3}+6x^{-4}, y'''=-6x^{-4}-24x^{-5}$

31. $y'=\dfrac{1+2x^2}{\sqrt{1+x^2}}, y''=\dfrac{2x^3+3x}{(1+x^2)^{3/2}}$,

$y'''=\dfrac{3}{(1+x^2)^{5/2}}$ 33. $y'=\dfrac{11}{(2x+3)^2}$,

$y''=\dfrac{-44}{(2x+3)^3}, y'''=\dfrac{264}{(2x+3)^4}$

35. $y'=\dfrac{x-2}{2(x-1)^{3/2}}, y''=\dfrac{4-x}{4(x-1)^{5/2}}, y'''=\dfrac{3(x-6)}{8(x-1)^{7/2}}$

37. $\dfrac{2}{(x-1)^3}$

Summary and Review: Chapter 2, pp. 153–155

1. [2.1, 2.2] -4 2. [2.1, 2.2] 10 3. [2.1, 2.2] -10

4. [2.2] 5 5. [2.1] No   6. [2.1] Yes   7. [2.1] -4

8. [2.1] -4 9. [2.1] Yes   10. [2.1] Does not exist

11. [2.1] -2 12. [2.1] No   13. [2.3] 1 14. [2.3] -3

15. [2.3] $4x+2h$ 16. [2.5] $(4,5)$ 17. [2.5] $(5,-108)$

18. [2.5] $20x^4$ 19. [2.5] $x^{-2/3}$ 20. [2.5] $64x^{-9}$

21. [2.5] $6x^{-3/5}$ 22. [2.5] $0.7x^6-12x^3-3x^2$

23. [2.5] x^5-32x^3-5 24. [2.7] $2x$

25. [2.7] $\dfrac{-x^2+16x+8}{(8-x)^2}$

26. [2.8] $2(5-x)(2x-1)^4(26-7x)$

27. $[2.8]$ $35x^4(x^5 - 2)^6$
28. $[2.8]$ $3x^2(4x + 3)^{-1/4} + 2x(4x + 3)^{3/4}$
29. $[2.9]$ $240x^{-6}$ **30.** $[2.9]$ $840x^3$ **31.** $[2.6]$ **(a)** $1 + 4t^3$;
(b) $12t^2$; **(c)** $33, 48$ **32.** $[2.6]$ **(a)** $-8x^2 + 47x + 10$;
(b) $800, 3050, -2250$; **(c)** $40, 16x - 7, -16x + 47$;
(d) $40, 313, -273$ **33.** $[2.6]$ **(a)** $100t$; **(b)** $30,000$;
(c) 2000 per yr   **34.** $[2.8]$ $4x^2 - 4x + 6, -2x^2 - 9$
35. $[2.2]$ 0 **36.** $[2.8]$ $\dfrac{-\frac{9}{2}x^4 - 2x^3 + \frac{9}{2}x + 1}{(1 + x^3)^2\sqrt{1 + 3x}}$

Test: Chapter 2, pp. 155–156

1. Yes   **2.** No   **3.** Does not exist   **4.** 1   **5.** No
6. 3   **7.** 3   **8.** Yes   **9.** 6   **10.** $\frac{1}{2}$ **11.** Does not exist
12. 4   **13.** $3(2x + h)$ **14.** $(0, 0), (2, -4)$ **15.** $84x^{83}$
16. $\dfrac{5}{\sqrt{x}}$ **17.** $\dfrac{10}{x^2}$ **18.** $\dfrac{5}{4}x^{1/4}$, or $\dfrac{5}{4} \cdot \sqrt[4]{x}$
19. $-x + 0.61$ **20.** $x^2 - 2x + 2$ **21.** $\dfrac{-6x + 20}{x^5}$
22. $\dfrac{5}{(5 - x)^2}$ **23.** $(x + 3)^3(7 - x)^4(-9x + 13)$
24. $-5(x^5 - 4x^3 + x)^{-6}(5x^4 - 12x^2 + 1)$
25. $\sqrt{x^2 + 5} + \dfrac{x^2}{\sqrt{x^2 + 5}}$, or $\dfrac{2x^2 + 5}{\sqrt{x^2 + 5}}$ **26.** $24x$
27. **(a)** $P(x) = -0.001x^2 + 48.8x - 60$;
(b) $R(10) = \$500, C(10) = \$72.10, P(10) = \$427.90$;
(c) $R'(x) = 50, C'(x) = 0.002x + 1.2$,
$P'(x) = -0.002x + 48.8$; **(d)** $R'(10) = \$50$ per unit,
$C'(10) = \$1.22$ per unit, $P'(10) = \$48.78$
28. **(a)** $\dfrac{dM}{dt} = -0.003t^2 + 0.2t$; **(b)** 9;
(c) 1.7 words/min   **29.** $x^3 + x^6, x^3 + 3x^5 + 3x^4 + x^6$
30. 12   **31.** $-2\left(\dfrac{1 + 3x}{1 - 3x}\right)^{1/3} + \left(\dfrac{1 - 3x}{1 + 3x}\right)^{2/3}$

CHAPTER 3

Margin Exercises, Section 3.1, pp. 159–165

1. **(a)** $[a, b], [c, d]$; **(b)** $[a, b), (c, d)$; **(c)** $[b, c], [d, e]$;
(d) $(b, c), (d, e)$ **2.** **(a)** $x_1, x_3, x_5, x_6, x_8, x_{10}$;
(b) x_4, x_7, x_9; **(c)** $x_1, x_3, x_4, x_5, x_6, x_7, x_8, x_9, x_{10}$

3. Answers will vary as to the graph, but it must have at least one critical point.

4. It is not possible.
5. **(a)** Maximum at c_2, minimum at b;
(b) maximum at c_1, minimum at c_2
6. Maximum $= 4$ at $x = 2$; minimum $= 1$ at $x = 1$ and $x = -1$
7. Maximum $= 176$ at $x = 6$; minimum $= 97$ at $x = 5$

Exercise Set 3.1, p. 166

1. **(a)** 41 mph; **(b)** 80 mph; **(c)** 13.5 mpg; **(d)** 16.5 mpg;
(e) about 22%
3. Maximum $= 5\frac{1}{4}$ at $x = \frac{1}{2}$; minimum $= 3$ at $x = 2$
5. Maximum $= 4$ at $x = 2$; minimum $= 1$ at $x = 1$
7. Maximum $= \frac{59}{27}$ at $x = -\frac{1}{3}$; minimum $= 1$ at $x = -1$
9. Maximum $= 1$ at $x = 1$; minimum $= -5$ at $x = -1$
11. Maximum $= 15$ at $x = -2$; minimum $= -13$ at $x = 5$
13. Maximum $=$ minimum $= -5$ for all x in $[-1, 1]$

Margin Exercises, Section 3.2, pp. 167–175

1. **(a)** $[b, c], [d, e]$; **(b)** $[b, c], [d, e]$; **(c)** $[a, b], [c, d]$;
(d) $[a, b], [c, d]$ **2.** **(a)** $(0, b]$; **(b)** $[a, 0)$ **3.** R, T, V
4. Minimum $= -4$ at $x = 2$; there is no maximum
5. Minimum $= -4$ at $x = 2$; maximum $= 0$ at $x = 0$ and $x = 4$ **6.** There are none.
7. **(a)** Maximum $= 8$ at $x = 2$; minimum $= -8$ at $x = -2$
8. Minimum $= \dfrac{10}{\sqrt{10}} + \sqrt{10}$, or $2\sqrt{10}$ at $x = \dfrac{1}{\sqrt{10}}$

Exercise Set 3.2, pp. 176–177

1. Maximum $= 1225$ at $x = 35$
3. Minimum $= 200$ at $x = 10$ **5.** Maximum $= \frac{1}{3}$ at $x = \frac{1}{2}$
7. Maximum $= \frac{289}{4}$ at $x = \frac{17}{2}$
9. Maximum $= 2\sqrt{3}$ at $x = -\sqrt{3}$; minimum $= -2\sqrt{3}$ at $x = \sqrt{3}$ **11.** Maximum $= 5700$ at $x = 2400$
13. Minimum $= -55\frac{1}{3}$ at $x = 1$

15. Maximum = 2000 at $x = 20$; minimum = 0 at $x = 0$ and $x = 30$ **17.** Minimum = 24 at $x = 6$
19. Minimum = 108 at $x = 6$
21. Maximum = 3 at $x = -1$; minimum = $-\frac{3}{8}$ at $x = \frac{1}{2}$
23. Maximum = 2 at $x = 8$; minimum = 0 at $x = 0$
25. None
27. Maximum = -1 at $x = 1$; minimum = -5 at $x = -1$
29. None   **31.** 22,506; $150,000
33. (a) 179 ft at $32°$; (b) 0 ft at $-57.5°$ **35.** 61.64 mph
37. Maximum = $3\sqrt{6}$ at $x = 3$; minimum = -2 at $x = -2$
39. Maximum = 1 at $x = -1$ and $x = 1$;
minimum = 0 at $x = 0$ **41.** None
43. Maximum = $-\frac{10}{3} + 2\sqrt{3}$ at $x = 2 - \sqrt{3}$;
minimum = $-\frac{10}{3} - 2\sqrt{3}$ at $x = 2 + \sqrt{3}$
45. Minimum = -1 at $x = -1$ and $x = 1$ **47.** 7

Margin Exercises, Section 3.3, pp. 182–188

1. Relative maximum at $(-1, \frac{13}{6})$; relative minimum at $(2, -\frac{7}{3})$

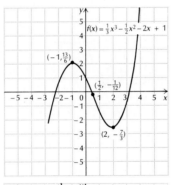

2. Relative maximum at $(0, 2)$; relative minimum at $(-1, 0)$ and $(1, 0)$

3. Relative minimum at $(0, 0)$

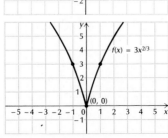

4. Relative maximum at $(-1, -2)$; relative minimum at $(1, 2)$

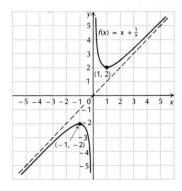

Exercise Set 3.3, p. 189

1. Relative maximum at $(0, 2)$

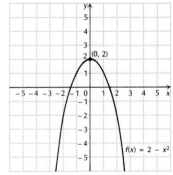

3. Relative minimum at $(-\frac{1}{2}, -\frac{5}{4})$

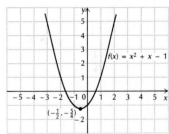

5. Relative maximum at $(-\frac{1}{2}, 1)$; relative minimum at $(\frac{1}{2}, -\frac{1}{3})$

7.

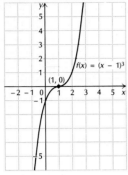

9. Relative minimum at $(-1, 0)$

11. Relative maximum at $(0, 0)$; relative minimum at $(-\sqrt{3}, -9)$ and $(\sqrt{3}, -9)$

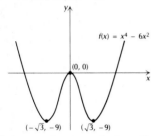

13. Relative maximum at $(-3, -6)$; relative minimum at $(3, 6)$

15. Relative maximum at $\left(-\frac{2}{3}, \frac{121}{27}\right)$; relative minimum at $(2, -5)$

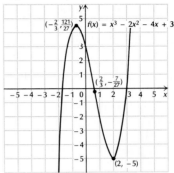

17. Relative minimum at $(-1, -1)$

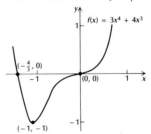

19. Relative maximum at $(0, 4)$; relative minimum at $(2, 0)$

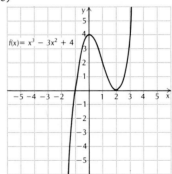

21. Relative maximum at $(-5, 400)$; relative minimum at $(9, -972)$

23.

25.

27.

29.

31.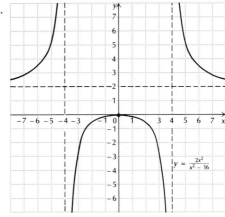

Margin Exercises, Section 3.4, pp. 190–199

1. (a) y: 20, 16, 13.5, 12, 10, 8, 6.8, 0; A: 0, 64, 87.75, 96, 100, 96, 89.76, 0;

(b)   **(c)** yes;

(d) maximum = 100 ft^2 at $x = 10$ ft

2. 25 ft by 25 ft; 625 ft^2

3. (a) $R(x) = x(200 - x) = 200x - x^2$;

(b) $P(x) = -x^2 + 192x - 5000$; **(c)** 96; **(d)** \$4216;

(e) $200 - 96$, or \$104 per unit

4. (a) h: 4, 3.5, 3, 2.5, 2, 1.7, 1.5, 1, 0.6, 0.5, 0; V: 0, 3.5, 12, 22.5, 32, 35.972, 37.5, 36, 27.744, 24.5, 0;

(b)   **(c)** about 38; between 5 and 6, maybe 5.2

5. Maximum volume $= 74\frac{2}{27}$ in^3; dimensions $6\frac{2}{3}$ in. by $6\frac{2}{3}$ in. by $1\frac{2}{3}$ in.

6. Minimum area $= 300$ cm^2; dimensions 10 cm by 10 cm by 5 cm   **7.** 40¢

Exercise Set 3.4, pp. 199–203

1. 25 and 25; maximum product $= 625$
3. No; $Q = x(50 - x)$ has no minimum
5. 2 and -2; minimum product $= -4$
7. $x = \frac{1}{2}, y = \sqrt{\frac{1}{2}}$; maximum $= \frac{1}{4}$
9. $x = 10, y = 10$; minimum $= 200$
11. $x = 2, y = \frac{32}{3}$; maximum $= \frac{64}{3}$
13. 30 yd by 60 yd; maximum area $= 1800$ yd^2
15. 13.5 ft by 13.5 ft; 182.25 ft^2
17. 46 units; maximum profit $= \$1048$
19. 70 units; maximum profit $= \$19$
21. Approximately 1667 units; maximum profit $\approx \$5500$
23. (a) $R(x) = 150x - 0.5x^2$;
(b) $P(x) = -0.75x^2 + 150x - 4000$; (c) 100; (d) $3500;
(e) \$100
25. 20 in. by 20 in. by 5 in.; maximum $= 2000$ in^3
27. 5 in. by 5 in. by 2.5 in.; minimum $= 75$ in^2
29. \$5.75, 72,500 (Will the stadium hold that many?)
31. 25   33. 14 in. by 14 in. by 28 in.   35. $\sqrt[3]{\frac{1}{10}}$

37. 4 ft by 4 ft by 20 ft   39. $x = y = \dfrac{24}{4 + \pi}$ 41. 9%

43. Minimum at $x = \dfrac{24\pi}{\pi + 4} \approx 10.55$,

$24 - x = \dfrac{96}{\pi + 4} \approx 13.45$; there is no maximum if the

string is to be cut. One would interpret the maximum to be at the endpoint, with the string uncut and used to form a circle.
45. S should be about 4.25 mi downshore from A
47. (a) $C'(x) = 8 + \dfrac{3x^2}{100}$; (b) $A(x) = 8 + \dfrac{20}{x} + \dfrac{x^2}{100}$;

(c) $A'(x) = \dfrac{x}{50} - \dfrac{20}{x^2}$; (d) minimum $= 11$ at $x_0 = 10$,

$C'(10) = 11$; (e) $A(10) = 11$, $C'(10) = 11$
49. Minimum $= 6 - 4\sqrt{2}$ at $x = 2 - \sqrt{2}$ and
$y = -1 + \sqrt{2}$

Margin Exercises, Section 3.5, pp. 206–208

1.

x	$\dfrac{2500}{x}$	$\dfrac{x}{2}$	$10 \cdot \dfrac{x}{2}$	$20 + 9x$	$(20 + 9x)\dfrac{2500}{x}$	$10 \cdot \dfrac{x}{2} + (20 + 9x)\dfrac{2500}{x}$
2500	1	1250	$12,500	$22,520	$22,520	$35,020
1250	2	625	$6,250	$11,270	$22,540	$28,790
500	5	250	$2,500	$4,520	$22,600	$25,100
250	10	125	$1,250	$2,270	$22,700	$23,950
167	15	84	$840	$1,523	$22,845	$23,685
125	20	63	$630	$1,145	$22,900	$23,530
100	25	50	$500	$920	$23,000	$23,500
90	28	45	$450	$830	$23,240	$23,690
50	50	25	$250	$470	$23,500	$23,750

Lot size ≈ 100 to minimize cost; 25 orders.

2. 15 times per year at lot size 40
3. 19 times per year at lot size 31

Exercise Set 3.5, pp. 208–209

1. Reorder 5 times per year; lot size = 20
3. Reorder 12 times per year; lot size = 30
5. Reorder about 13 times per year; lot size = 28
7. $x = \sqrt{\dfrac{2bQ}{a}}$

Margin Exercises, Section 3.6, pp. 210–214

1. (a)

(b)

2. (a)

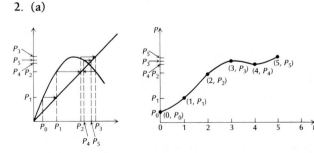

(b) oscillating

3. 4.0, 4.08, 4.16, 4.24, 4.32, 4.40, 4.48, 4.57, 4.66, 4.75,
4.84; these were rounded each time to the nearest hundredth
of a billion   **4.** $f(P) = 1.0825P$ **5.** $f(P) = 1.01P$
6. 3500; maximum sustainable harvest = 12,250

Exercise Set 3.6, p. 215

1. 9500; maximum sustainable harvest = 90,250
3. 60,000; maximum sustainable harvest = 90,000
5. 400,000; maximum sustainable harvest = 400,000
7. There is none.

Margin Exercises, Section 3.7, pp. 217–221

1. $\Delta y = -0.59$ **2.** $\Delta y = 19$ **3.** 8.1875; 8.185 in Table 1
4. (a) $\Delta C = \$4.11$, $C'(5) = \$4.10$;
(b) $\Delta C = \$6.01$, $C'(100) = \$6$
5. (a) $dy = (3x^2 - 10x - 4)\, dx$; **(b)** $dy = -0.7$

Exercise Set 3.7, pp. 221–222

1. $\Delta y = 0.0401$; $f'(x)\, \Delta x = 0.04$ **3.** 0.2816, 0.28
5. $-0.556, -1$
7. 6, 6   **9.** $\Delta C = \$2.01$, $C'(70) = \$2$
11. $\Delta R = \$2$, $R'(70) = \$2$
13. (a) $P(x) = -0.01x^2 + 1.4x - 30$;
(b) $\Delta P = -\$0.01$, $P'(70) = 0$ **15.** 4.375   **17.** 10.1
19. 2.167   **21.** 2.512 cm³   **23.** $dy = 9x^2(2x^3 + 1)^{1/2}\, dx$
25. $dy = \frac{1}{5}(x + 27)^{-4/5}\, dx$
27. $dy = (4x^3 - 6x^2 + 10x + 3)\, dx$ **29.** $dy = 3.1$
31. $5/\pi \approx 1.6$ ft

Margin Exercises, Section 3.8, pp. 223–226

1. $\dfrac{dy}{dx} = \dfrac{2}{5y^4}$ **2. (a)** $\dfrac{dy}{dx} = \dfrac{-3x^2 - 2xy^4}{3y^2 + 4x^2y^3}$; **(b)** 0
3. $\dfrac{dp}{dx} = \dfrac{-\sqrt{p}}{50}$ **4.** $24\pi \dfrac{\text{ft}^2}{\text{sec}} \approx 75 \dfrac{\text{ft}^2}{\text{sec}}$
5. (a) $\dfrac{dR}{dt} = \$0$ per day; **(b)** $\dfrac{dC}{dt} = \$64$ per day;
(c) $\dfrac{dP}{dt} = -\$64$ per day

Exercise Set 3.8, pp. 226–229

1. $\dfrac{1-y}{x+2}$; $-\dfrac{1}{9}$ 3. $-\dfrac{x}{y}$; $-\dfrac{1}{\sqrt{3}}$

5. $\dfrac{6x^2 - 2xy}{x^2 - 3y^2}$; $-\dfrac{36}{23}$ 7. $-\dfrac{y}{x}$ 9. $\dfrac{x}{y}$

11. $\dfrac{3x^2}{5y^4}$ 13. $\dfrac{-3x^2y^4 - 2xy^3}{3x^2y^2 + 4x^3y^3}$

15. $\dfrac{dp}{dx} = \dfrac{-2}{2p + 1}$ 17. $\dfrac{dp}{dx} = -\dfrac{p+4}{x+3}$

19. 0.1728π cm^3/day ≈ 0.54 cm^3/day
21. \$400 per day, \$80 per day, \$320 per day
23. \$16 per day, \$8 per day, \$8 per day

25. 65 mph   27. (a) $1000R \cdot \dfrac{dR}{dt}$;

(b) -0.01125 29. $-\dfrac{3}{4}$ 31. $-\dfrac{\sqrt{y}}{\sqrt{x}}$

33. $\dfrac{2}{3y^2(x+1)^2}$ 35. $-\dfrac{9}{4}\sqrt[3]{y}\sqrt{x}$

37. $\dfrac{dy}{dx} = \dfrac{1+y}{2-x}$, $\dfrac{d^2y}{dx^2} = \dfrac{2+2y}{(2-x)^2}$

39. $\dfrac{dy}{dx} = \dfrac{x}{y}$, $\dfrac{d^2y}{dx^2} = \dfrac{y^2 - x^2}{y^3}$

Summary and Review: Chapter 3, pp. 229–231

1. [3.2] None   2. [3.1] Maximum $= 32$ at $x = -2$;
minimum $= -17$ at $x = 5$ 3. [3.1] Maximum $= 19$ at
$x = -1$; minimum $= -35$ at $x = 2$
4. [3.1] Maximum $= -2$ at $x = 1$; minimum $= -12$
at $x = -1$ 5. [3.2] Minimum $= 3$ at $x = -1$
6. [3.2] None   7. [3.1] Maximum $= 163$ at $x = -3$;
minimum $= -1$ at $x = -1$
8. [3.2] Minimum $= 10$ at $x = 1$
9. [3.1] Maximum $= 256$ at $x = 4$; minimum $= 0$
at $x = 0$ and $x = 5$ 10. [3.2] Maximum $= \frac{53}{4}$ at $x = \frac{5}{2}$
11. [3.4] 30, 30   12. [3.4] $Q = -1$ at $x = -1, y = -1$
13. [3.4] 30 units; \$451 profit
14. [3.4] 10 ft $\times$ 10 ft $\times$ 25 ft; \$1500
15. [3.5] 12 times; 30 bicycles   16. [3.6] 35,000;
1,225,000   17. [3.7] $\Delta y = -10.875$; $f'(x)\,\Delta x = -13$
18. [3.7] (a) $dy = (3x^2 - 1)\,dx$; (b) $dy = 0.11$
19. [3.7] 8.3125

20. [3.3] Relative
maximum at
$(-\sqrt{2}, -\sqrt{2})$; relative
minimum at $(\sqrt{2}, \sqrt{2})$

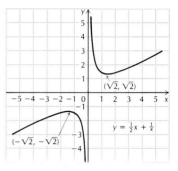

21. [3.3] Relative
maximum at $(-1, 4)$;
relative minimum at
$(\frac{1}{3}, \frac{76}{27})$

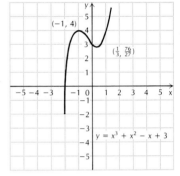

22. [3.8] $\dfrac{2x^2 + 3y}{-2y^2 - 3x}$; $\dfrac{4}{5}$ 23. [3.8] $-1\frac{3}{4}$ ft/sec
24. [3.8] \$600 per day, \$450 per day, \$150 per day
25. [3.4] Minimum $= 0$ at $x = 3$
26. [3.8] $\dfrac{6x^5 - 8x^3 - 24xy^2}{24x^2y + 8y^3 - 6y^5}$
27. [3.3] Relative minimum at $(-9, -9477)$; relative
maximum at $(0, 0)$; relative minimum at $(15, -37,125)$

Test: Chapter 3, pp. 231–232

1. Maximum $= 9$ at $x = 3$
2. Maximum $= 2$ at $x = -1$; minimum $= -1$ at $x = -2$
3. Maximum $= 28.49$ at $x = 4.3$
4. Maximum $= 7$ at $x = -1$; minimum $= 3$ at $x = 1$
5. None   6. Minimum $= -\frac{13}{12}$ at $x = \frac{1}{6}$
7. Minimum $= 48$ at $x = 4$ 8. 4 and -4
9. $x = 5, y = -5$; minimum $= 50$
10. Maximum profit $= \$24,980$; 500 units
11. 40 in. by 40 in. by 10 in.;
maximum volume $= 16,000$ in^3
12. 35 times at lot size 35
13. 49,500; maximum sustainable harvest $= 2,450,250$

14. $\Delta y = 1.01, f'(x)\,\Delta x = 1$ **15.** 10.2

16. (a) $\dfrac{x\,dx}{\sqrt{x^2+3}}$; (b) 0.0075593

17.

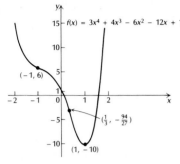

18. $-\dfrac{x^2}{y^2}, -\dfrac{1}{4}$ **19.** -0.96 ft/sec

20. Maximum $= \frac{1}{3}\cdot 2^{2/3}$ at $x = \sqrt[3]{2}$; minimum $= 0$ at $x = 0$

21. (a) $A(x) = \dfrac{C(x)}{x} = 100 + \dfrac{100}{\sqrt{x}} + \dfrac{\sqrt{x}}{100}$;

(b) minimum $= 102$ at $x = 10{,}000$

CHAPTER 4

Margin Exercises, Section 4.1, pp. 235–243

1. 8, 8.574188, 8.815241, 8.821353, 8.824411, 8.824962; $2^{\pi} \approx 8.82$

2. (a)

x	0	$\frac{1}{2}$	1	2	-1	-2
3^x	1	1.7	3	9	$\frac{1}{3}$	$\frac{1}{9}$

(b)

3. (a)

x	0	$\frac{1}{2}$	1	2	-1	-2
$(\frac{1}{3})x$	1	0.6	$\frac{1}{3}$	$\frac{1}{9}$	3	9

(b)

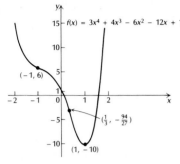

4. (a) 0.8284, 0.7568, 0.7369, 0.7083, 0.7007, 0.6934; (b) 0.7

5. (a) 1.4641, 1.2642, 1.2113, 1.1372, 1.1177, 1.09928; (b) 1.1

6. (a)

x	-3	-2	-1	0	1	2	3
2^x	0.125	0.25	0.5	1	2	4	8
$(0.7)2^x$	0.09	0.18	0.35	0.7	1.4	2.8	5.6

(b) See Figure 1 in the text.

7. (a)

x	-2	-1	0	1	2
3^x	0.11	0.33	1	3	9
$(1.1)3^x$	0.12	0.36	1.1	3.3	9.9

(b) See Figure 2 in the text.

8. $2, \$2.25, \$2.370370, \$2.441406, \$2.488320, \$2.704814, \$2.714567, \$2.718121, \$2.718869 \quad$ **9.** $6e^x$

10. $e^x(x^3 + 3x^2)$, or $x^2e^x(x+3)$ **11.** $\dfrac{e^x(x-2)}{x^3}$

12. $-4e^{-4x}$ **13.** $(3x^2 + 8)e^{x^3 + 8x}$ **14.** $\dfrac{xe^{\sqrt{x^2+5}}}{\sqrt{x^2+5}}$

15.

$f(x) = 2e^{-x}$

16. (a)

x	0	$\frac{1}{2}$	1	2	3	4
$f(x)$	0	0.39	0.63	0.86	0.95	0.98

(b)

$f(x) = 1 - e^{-x}$

Exercise Set 4.1, pp. 244–245

1.

$y = 4^x$

3.

$y = (0.4)^x$

5.

$x = 4^y$

7. $3e^{3x}$ **9.** $-10e^{-2x}$
11. e^{-x} **13.** $-7e^x$
15. e^{2x} **17.** $x^3e^x(x+4)$
19. $\dfrac{e^x(x-4)}{x^5}$

21. $(-2x + 7)e^{-x^2 + 7x}$ **23.** $-xe^{-x^2/2}$

25. $\dfrac{e^{\sqrt{x-7}}}{2\sqrt{x-7}}$ **27.** $\dfrac{e^x}{2\sqrt{e^x - 1}}$

29. $(1 - 2x)e^{-2x} - e^{-x} + 3x^2$ **31.** e^{-x} **33.** ke^{-kx}

35.

$f(x) = e^{2x}$

37.

$f(x) = e^{-2x}$

39.

$f(x) = 1 - e^{-x}$

41. (a) $C'(t) = 50e^{-t}$; **(b)** $50 million; **(c)** $.916 million

43. $15(e^{3x} + 1)^4 e^{3x}$ **45.** $-e^{-t} - 3e^{3t}$ **47.** $\dfrac{e^x(x-1)^2}{(x^2+1)^2}$

49. $\dfrac{e^{\sqrt{x}}}{2\sqrt{x}} + \dfrac{1}{2}\sqrt{e^x}$ **51.** $\dfrac{1}{2}e^{x/2}\left[\dfrac{x}{\sqrt{x-1}}\right]$ **53.** $\dfrac{4}{(e^x + e^{-x})^2}$

55. 2, 2.25, 2.48832, 2.59374, 2.71692
57. Maximum $= 4e^{-2} \approx 0.54$ at $x = 2$

59.

$y = x^2 e^{-x}$

Margin Exercises, Section 4.2, pp. 246–260

1. (a) $b^T = P$; **(b)** $9^{1/2} = 3$; **(c)** $10^3 = 1000$;
(d) $10^{-1} = 0.1$
2. (a) $\log_e T = k$; **(b)** $\log_{16} 2 = \frac{1}{4}$; **(c)** $\log_{10} 10{,}000 = 4$;
(d) $\log_{10} 0.001 = -3$ **3.**

$y = \log_3 x$

4. (a) 1000; **(b)** 3; **(c)** 0.699; **(d)** ≈ 5.000 **5.** 0.602

6. 1   7. -0.398 8. 0.398   9. -0.699 10. $\frac{3}{2}$
11. 1.699   12. 1.204   13. (a) 4.4977;
(b) -0.0419; (c) 1.8954; (d) 0.8954; (e) -0.1046;
(f) -1.1046; (g) -2.1046 14. 2.3025   15. -0.9163
16. 0.9163   17. 2.7724   18. 2.6094   19. $\frac{1}{2}$
20. -1.6094 21. 0.693147   22. 2.995732
23. 4.605170   24. -2.599375 25. 0.076961
26. -0.000100 27. $t \approx 4$ 28. $t \approx 44$

29. (a)

x	0.5	1	2	3	4
$\ln x$	-0.7	0	0.7	1.1	1.4

(b) same as text figure   30. $\dfrac{5}{x}$ 31. $x^2(1 + 3 \ln x) + 4$

32. $\dfrac{1 - 2 \ln x}{x^3}$ 33. $\dfrac{1}{x}$ 34. $\dfrac{6x}{3x^2 + 4}$ 35. $\dfrac{1}{x \ln 5x}$

36. $\dfrac{4x^5 + 2}{x(x^5 - 2)}$ 37. (a) 500; (b) $N'(a) = \dfrac{200}{a}$;

(c) minimum $= 500$ at $a = 1$ 38. 87 days

Exercise Set 4.2, pp. 261–264

1. $2^3 = 8$ 3. $8^{1/3} = 2$ 5. $a^J = K$ 7. $b^v = T$
9. $\log_e b = M$ 11. $\log_{10} 100 = 2$ 13. $\log_{10} 0.1 = -1$
15. $\log_M V = p$ 17. 2.708   19. 0.51   21. -1.609
23. $\frac{3}{2}$ 25. 2.609   27. 3.218   29. 2.9957
31. 0.2231   33. -1.3863 35. 2.3863   37. 4
39. 2.7726   41. 8.681690   43. -4.006334
45. 0.631272   47. -3.973898 49. 6.809039
51. -4.509860 53. $t \approx 5$ 55. $t \approx 4$ 57. $t \approx 2$

59. $t \approx 141$ 61. $-\dfrac{6}{x}$ 63. $x^3(1 + 4 \ln x) - x$

65. $\dfrac{1 - 4 \ln x}{x^5}$ 67. $\dfrac{1}{x}$ 69. $\dfrac{10x}{5x^2 - 7}$ 71. $\dfrac{1}{x \ln 4x}$

73. $\dfrac{x^2 + 7}{x(x^2 - 7)}$ 75. $e^x\left(\dfrac{1}{x} + \ln x\right)$ 77. $\dfrac{e^x}{e^x + 1}$

79. $\dfrac{2 \ln x}{x}$ 81. (a) 68%; (b) 36%; (c) 3.6%; (d) 5%;

(e) $-\dfrac{20}{t + 1}$; (f) maximum $= 68\%$ at $t = 0$

83. (a) 1000; (b) $N'(a) = \dfrac{200}{a}$, $N'(10) = 20$;

(c) minimum $= 1000$ at $a = 1$
85. (a) 2.37 ft/sec; (b) 3.37 ft/sec;

(c) $v'(p) = \dfrac{0.37}{p}$, $v'(p) =$ the acceleration of the walker

87. 58 days   89. (a) 18.1%, 69.9%; (b) $P'(t) = 0.2e^{-0.2t}$;
(c) 11.5   91. (a) \$58.69, \$78.00; (b) \$63.80$e^{-1.1t}$; (c) 2.7

93. $\dfrac{-4(\ln x)^{-5}}{x}$ 95. $\dfrac{15t^2}{t^3 + 1}$ 97. $\dfrac{4[\ln(x + 5)]^3}{x + 5}$

99. $\dfrac{5t^4 - 3t^2 + 6t}{(t^3 + 3)(t^2 - 1)}$ 101. $\dfrac{24x + 25}{8x^2 + 5x}$

103. $\dfrac{2(1 - \ln t^2)}{t^3}$ 105. $x^n \ln x$ 107. $\dfrac{1}{\sqrt{1 + t^2}}$

109. $\dfrac{3}{x \ln x}$ 111. 1

113. $\sqrt[e]{e} \approx 1.444667861$; $\sqrt[e]{e} > \sqrt[x]{x}$ for any $x > 0$ such that

$x \neq e$ 115. $-\dfrac{1}{2e}$ 117. $t = \dfrac{\ln P - \ln P_0}{k}$ 119. ∞

121. Let $a = \ln x$. Then $e^a = x$, so $\log x = \log e^a = a \log e$.

Then $a = \ln x = \dfrac{\log x}{\log e} \approx \dfrac{\log x}{0.4343} \approx 2.3026 \log x$.

Margin Exercises, Section 4.3, pp. 265–272

1. (a) $\dfrac{dy}{dx} = 20e^{4x}$; (b) $\dfrac{dy}{dx} = 4y$

2. (a) $N(t) = ce^{kt}$; (b) $f(t) = ce^{kt}$
3. (a) Should be about $\frac{1}{8}$ in., or 0.125 in.;
(b) 0.032, 0.064, 0.128; (c) about 2 mi
4. (a) $P(t) = P_0e^{0.13t}$; (b) \$1138.83; (c) 5.3 yr
5. 2%, 34.7; 6.9%, 10; 14%, 5.0; 4.6%, 15; 1%, 69.3
6. (a) $P(t) = 241e^{0.009t}$; (b) 248 million; (c) 77 yr (2063)
7. (a) $C(t) = 80e^{kt}$; (b) $k = 0.145613$;
(c) $C(t) = 80e^{0.145613t}$; (d) \$4718 thousand

Exercise Set 4.3, pp. 273–276

1. $Q(t) = Q_0e^{kt}$ 3. (a) $P(t) = P_0e^{0.09t}$;
(b) \$1094.17, \$1197.22; (c) 7.7 yr   5. 19.8 yr   7. 6.9%
9. (a) $P(t) = 209e^{0.01t}$; (b) 312 million; (c) 69.3 yr

11. 0.20%   **13.** (a) $N(t) = 50e^{0.1t}$; (b) 369; (c) 6.9 yr
15. 6.9 yr (1986)
17. (a) $k = 0.024464$, $P(t) = 107e^{0.024464t}$;
(b) 124 thousand
19. (a) $k = 0.060295$; $C(t) = \$0.58e^{0.060295t}$; (b) \$6.47;
(c) 11.5 yr (1996)
21. (a) $k = 0.061248$, $P(t) = \$100e^{0.061248t}$; (b) \$340.40;
(c) 11.3 yr (1978)   **23.** $\approx \$90,650,000,000,000$
25. $\approx 2.2\%$ **27.** $k = 0.16778$ or 16.778%; $\approx \$710,200$;

$\approx \$2,720,000$ **29.** 15.03%   **31.** 9%   **33.** $T_3 = \dfrac{\ln 3}{k}$

35. Answers depend on particular data.

37. $\dfrac{\ln 2}{24} \approx 2.9\%$ per hr

Margin Exercises, Section 4.4, pp. 276–283

1. See Exercises 35 and 37 of Exercise Set 4.1.
2. (a) $N(t) = N_0 e^{-0.14t}$; (b) 246.6 g; (c) 5 days
3. 30 yr   **4.** 5.3%   **5.** 13,356 yr
6. (a) $a = 130°$; (b) $k = 0.01$; (c) 188°; (d) 256 min
7. Answers will vary. "Theoretically" it is never possible for
the temperature of the water to be the same as the room
temperature.   **8.** 10:00 A.M.

Exercise Set 4.4, pp. 284–286

1. 7.2 days   **3.** 23% per min   **5.** 10.1 g   **7.** 19,109 yr
9. 7604 yr   **11.** (a) $A = A_0 e^{-kt}$; (b) 9 hr
13. (a) 0.8%; (b) 79%W_0
15. (a) 11 watts; (b) 173 days; (c) 402 days; (d) 50 watts
17. (a) \$40,000; (b) \$5413   **19.** (a) 25%I_0, 6%I_0, 1.5%I_0;
(b) 0.00008%I_0 **21.** (a) $a = 25°$; (b) $k = 0.05$;
(c) 84.2°; (d) 32 min   **23.** 7 P.M.
25. (a) $k = 0.0162649$, $P(t) = 453,000e^{-0.0162649t}$, where
t = number of years since 1970; (b) 327,208;
(c) 801 yr (2050)   **27.** \$2018.97
29. (a) $k = 0.03$, $P(t) = \$100e^{-0.03t}$, where t = number of
years that precede 1967; (b) \$13.40

Margin Exercises, Section 4.5, pp. 286–290

1. $e^{6.9315}$ **2.** $e^{0.6931x}$ **3.** $(\ln 5)5^x$ **4.** $(\ln 4)4^x$

5. $(\ln 4.3)(4.3)^x$ **6.** $\dfrac{1}{\ln 2} \cdot \dfrac{1}{x}$ **7.** $-\dfrac{7}{x \ln 10}$

8. $x^5 \left(\dfrac{1}{\ln 10} + 6 \log x \right)$

Exercise Set 4.5, pp. 290–292

1. $e^{6.4378}$ **3.** $e^{12.238}$ **5.** $e^{k \cdot \ln 4}$ **7.** $e^{kT \cdot \ln 8}$
9. $(\ln 6)6^x$ **11.** $(\ln 10)10^x$ **13.** $(6.2)^x[x \ln 6.2 + 1]$
15. $10^x x^2[x \ln 10 + 3]$ **17.** 5
19. (a) $I = 10^7 \cdot I_0$; (b) $I = 10^8 \cdot I_0$;
(c) the intensity in part (b) is 10 times that in part (a);
(d) $\dfrac{dI}{dR} = (I_0 \cdot \ln 10) \cdot 10^R$ **21.** $\dfrac{1}{\ln 4} \cdot \dfrac{1}{x}$ **23.** $\dfrac{2}{\ln 10} \cdot \dfrac{1}{x}$

25. $\dfrac{1}{\ln 10} \cdot \dfrac{1}{x}$ **27.** $x^2 \left[\dfrac{1}{\ln 8} + 3 \log_8 x \right]$ **29.** $\dfrac{1}{\ln 10} \cdot \dfrac{1}{I}$

31. $\dfrac{m}{\ln 10} \cdot \dfrac{1}{x}$ **33.** $-250,000 (\ln 4) \left(\dfrac{1}{4} \right)^t$

35. $2(\ln 3)3^{2x}$ **37.** $(\ln x + 1)x^x$ **39.** $x^{e^x} e^x \left(\ln x + \dfrac{1}{x} \right)$

41. $\dfrac{1}{\ln a} \cdot \dfrac{f'(x)}{f(x)}$

Margin Exercises, Section 4.6, p. 295

1. (a) $E(x) = \dfrac{300 - x}{x}$; (b) $E(100) = 2$, $E(200) = \dfrac{1}{2}$;
(c) $x = 150$; (d) $R(x) = 300x - x^2$; (e) $x = 150$

Exercise Set 4.6, pp. 296–297

1. (a) $E(x) = \dfrac{400 - x}{x}$; (b) $x = 200$

3. (a) $E(x) = \dfrac{50 - x}{x}$; (b) $x = 25$

5. (a) $E(x) = 1$, for all $x > 0$;
(b) total revenue $= R(x) = 400$, for all $x > 0$. It has 400
as a maximum for all $x > 0$.

7. (a) $E(x) = \dfrac{1000 - 2x}{x}$; (b) $x = \dfrac{1000}{3}$

9. (a) $E(x) = \dfrac{4}{x}$; (b) $x = 4$

11. (a) $E(x) = \dfrac{x+3}{2x}$; (b) $x = 3$

13. (a) $E(x) = \dfrac{1}{n}$; (b) no, E is a constant $\dfrac{1}{n}$;

(c) only when $n = 1$ 15. $E(x) = -\dfrac{1}{x} \cdot \dfrac{1}{L'(x)}$

Summary and Review: Chapter 4, pp. 297–299

1. [4.2] $\dfrac{1}{x}$ 2. [4.1] e^x 3. [4.2] $\dfrac{4x^3}{x^4+5}$ 4. [4.1] $\dfrac{e^{2\sqrt{x}}}{\sqrt{x}}$

5. [4.2] $\dfrac{6}{x}$ 6. [4.1] $4e^{4x} + 4x^3$ 7. [4.2] $\dfrac{1 - 3\ln x}{x^4}$

8. [4.1, 4.2] $e^{x^2}\left(\dfrac{1}{x} + 2x \ln 4x\right)$ 9. [4.1, 4.2] $4e^{4x} - \dfrac{1}{x}$

10. [4.2] $8x^7 - \dfrac{8}{x}$ 11. [4.1] $\dfrac{1-x}{e^x}$ 12. [4.2] 6.9300

13. [4.2] -3.2698 14. [4.2] 8.7601   15. [4.2] 3.2698
16. [4.2] 2.54995   17. [4.2] -3.6602
18. [4.3] $Q(t) = Q_0 e^{kt}$ 19. [4.3] 4.3%   20. [4.3] 8.7 yr
21. [4.3] (a) $k = 0.050991$, $C(t) = \$4.65e^{0.050991t}$;
(b) \$16.63, \$27.70
22. [4.3] (a) $N(t) = 60e^{0.12t}$; (b) 97; (c) 5.8 yr
23. [4.4] 5.3 yr   24. [4.4] 18.2% per day
25. [4.4] (a) $A(t) = 800e^{-0.07t}$; (b) 197 grams; (c) 9.9 days

26. [4.5] $(\ln 3)3^x$ 27. [4.5] $\dfrac{1}{\ln 15} \cdot \dfrac{1}{x}$

28. [4.6] (a) $\dfrac{x+4}{2x}$; (b) $x = 4$ 29. [4.1] $\dfrac{-8}{(e^{2x} - e^{-2x})^2}$

30. [4.2] $-\dfrac{1}{1024e}$

Test: Chapter 4, pp. 299–300

1. e^x 2. $\dfrac{1}{x}$ 3. $-2xe^{-x^2}$ 4. $\dfrac{1}{x}$ 5. $e^x - 15x^2$

6. $3e^x\left(\dfrac{1}{x} + \ln x\right)$ 7. $\dfrac{e^x - 3x^2}{e^x - x^3}$

8. $\dfrac{\dfrac{1}{x} - \ln x}{e^x}$, or $\dfrac{1 - x \ln x}{xe^x}$ 9. 1.0674   10. 0.5554

11. 0.4057   12. $M(t) = M_0 e^{kt}$ 13. 17.3% per hr
14. 10 yr   15. (a) 0.0330326; $C(t) = \$0.54e^{0.0330326t}$;
(b) \$2.72, \$3.79
16. (a) $A(t) = 3e^{-0.1t}$; (b) 1.1 cc; (c) 6.9 hr   17. 63 days

18. 0.000069% per yr   19. $(\ln 20)20^x$ 20. $\dfrac{1}{\ln 20} \cdot \dfrac{1}{x}$

21. (a) $E(x) = \dfrac{5}{x}$; (b) $x = 5$ 22. $(\ln x)^2$

23. Maximum $= 256e^{-4} \approx 4.69$ at $x = 4$; minimum $= 0$ at $x = 0$

CHAPTER 5

Margin Exercises, Section 5.1, pp. 303–308

1. $y = 7x$, $y = 7x - \tfrac{1}{2}$, $y = 7x + C$ (answers can vary)
2. $y = -2x$, $y = -2x + 27$, $y = -2x + C$

(answers can vary)   3. $\dfrac{x^2}{2} + C$ 4. $\dfrac{x^4}{4} + C$ 5. $e^x + C$

6. $\ln x + C$ 7. $\dfrac{x^4}{4} + C$ 8. $\dfrac{x^2}{2} + C$ 9. $\ln x + C$

10. $\tfrac{7}{5}x^5 + x^2 + C$ 11. $e^x - \tfrac{5}{7}x^{7/5} + C$
12. $5 \ln x - 7x - \tfrac{1}{5}x^{-5} + C$

13.

14. $f(x) = \dfrac{x^3}{3} + \dfrac{23}{3}$

15. $g(x) = x^2 - 4x + 13$

16. $C(x) = \tfrac{1}{3}x^3 + \tfrac{5}{2}x^2 + 35$

17. $s(t) = t^4 + 13$ 18. $s(t) = 2t^4 - 6t^2 + 7t + 8$

Exercise Set 5.1, pp. 309–311

1. $\dfrac{x^7}{7} + C$ 3. $2x + C$ 5. $\dfrac{4}{5}x^{5/4} + C$

7. $\dfrac{x^3}{3} + \dfrac{x^2}{2} - x + C$ 9. $\dfrac{t^3}{3} - t^2 + 3t + C$ 11. $5e^x + C$

13. $\dfrac{x^4}{4} - \dfrac{7}{15}x^{15/7} + C$ 15. $1000 \ln x + C$

17. $-x^{-1} + C$ 19. $f(x) = \dfrac{x^2}{2} - 3x + 13$

21. $f(x) = \dfrac{x^3}{3} - 4x + 7$ 23. $C(x) = \dfrac{x^4}{4} - x^2 + 100$

25. (a) $R(x) = \dfrac{x^3}{3} - 3x$;

(b) If you sell no products, you make no money.
27. $s(t) = t^3 + 4$ 29. $v(t) = 2t^2 + 20$
31. $s(t) = -\frac{1}{3}t^3 + 3t^2 + 6t + 10$
33. $s(t) = -16t^2 + v_0 t + s_0$ 35. $\frac{1}{4}$mi
37. (a) $E(t) = 30t - 5t^2 + 32$;
(b) $E(3) = 77\%$, $E(5) = 57\%$
39. (a) $A(t) = 43.4t^{-1} - 3.7$; (b) 2.5 cm²

41. $f(t) = \dfrac{t^{\sqrt{3}+1}}{\sqrt{3}+1} + 8$ 43. $\dfrac{x^6}{6} - \dfrac{2}{5}x^5 + \dfrac{1}{4}x^4 + C$

45. $\dfrac{2}{5}t^{5/2} + 4t^{3/2} + 18\sqrt{t} + C$ 47. $\dfrac{t^4}{4} + t^3 + \dfrac{3}{2}t^2 + t + C$

49. $\dfrac{b}{a}e^{ax} + C$ 51. $\dfrac{12}{7}x^{7/3} + C$ 53. $\dfrac{t^3}{3} - t^2 + 4t + C$

Margin Exercises, Section 5.2, pp. 312–320

1. (a) $A(x) = 3x$; (b) $A(1) = 3, A(2) = 6, A(5) = 15$;
(c)   (d) $A(x)$ is an antiderivative of $f(x)$
2. (a) $C(x) = 50x$; (b) $A(x) = 50x$;
(c) As x increases, the area over $[0, x]$ increases.
3. (a) $5000 + 4000 + 3000 + 2000$, or $\$14,000$;
(b) 14,000, same   4. (a) $A(x) = \frac{3}{2}x^2$; (b) $A(1) = \frac{3}{2}$;
(c) $A(2) = 6$; (d) $A(3.5) = 18.375$; (e)
(f) $A(x)$ is an antiderivative of $f(x)$

5. (a) $C(x) = -0.05x^2 + 50x$; (b) $A(x) = -0.05x^2 + 50x$;
(c) $\$12,000$; less than in Margin Exercise 3   6. $5\frac{1}{3}$
7. $13\frac{1}{2}$ 8. (a) Yes; (b) $\$4100$

Exercise Set 5.2, pp. 320–321

1. 8   3. 8   5. $41\frac{2}{3}$ 7. $\frac{1}{4}$ 9. $10\frac{2}{3}$
11. $e^3 - 1 \approx 19.086$ 13. $\ln 3 \approx 1.0986$ 15. 51
17. An antiderivative, velocity
19. An antiderivative, energy used in time t
21. An antiderivative, total revenue
23. An antiderivative, amount of drug in blood
25. (a) $s(t) = t^3 + t^2$; (b) 150
27. (a) $C(x) = 100x - 0.1x^2$; (b) $R(x) = 100x + 0.1x^2$;
(c) $P(x) = 0.2x^2$; (d) $P(1000) = \$200,000$ 29. $3\frac{1}{2}$
31. $359\frac{7}{15}$ 33. $6\frac{3}{4}$

Margin Exercises, Section 5.3, pp. 323–329

1. $\displaystyle\int_a^b 2x \, dx = b^2 - a^2$ 2. $\displaystyle\int_a^b e^x \, dx = e^b - e^a$

3. 8   4. $1 - \dfrac{1}{e^2}$ 5. $\dfrac{2}{3}$ 6. $e + e^3 - 3$ 7. $5\dfrac{1}{3}$ 8. $13\dfrac{1}{2}$

9. $\ln 7$ 10. $-\frac{1}{3}b^{-3} + \frac{1}{3}$ 11. 75   12. $17\frac{1}{3}$
13. (a) $\$595,991.60$; (b) $k = 7.3$, so it is the 8th day;
(c) $\$4,185,966.53$

Exercise Set 5.3, pp. 329–331

1. $\dfrac{1}{6}$ 3. $\dfrac{4}{15}$ 5. $e^b - e^a$ 7. $b^3 - a^3$ 9. $\dfrac{e^2}{2} + \dfrac{1}{2}$

11. $\dfrac{2}{3}$ 13. $\dfrac{5}{34}$ 15. 4   17. $9\dfrac{5}{6}$ 19. 12   21. $e^5 - \dfrac{1}{e}$

23. $7\frac{1}{3}$ 25. $17\frac{1}{3}$ 27. (a) $\$2948.26$; (b) $\$2913.90$;
(c) $k = 6.9$, so it will be on the 7th day   29. 90
31. $\$3600$ 33. 9   35. $\frac{307}{6}$ 37. $\frac{15}{4}$ 39. 8   41. 12

Margin Exercises, Section 5.4, pp. 333–338

1. 4   2. 4   3. (a) $-\frac{1}{6}$; (b) $\frac{5}{6}$; (c) $\frac{4}{6}$
4. (a) -4; (b) 4; (c) 0   5. (a) 0; (b) $A - A$, or 0
6. (a) Negative; (b) $A - 2A$, or $-A$

7. **(a)** Positive; **(b)** $3A - A$, or $2A$
8. **(a)** Negative; **(b)** $-3A$ 9. $\frac{1}{6}$
10. **(a)** $P(k) = 5k^2 + 10k$; **(b)** $P(5) = 175$

Exercise Set 5.4, pp. 338–339

1. $\frac{1}{4}$ 3. $\frac{9}{2}$ 5. $\frac{125}{6}$ 7. $\frac{9}{2}$ 9. $\frac{1}{6}$ 11. $\frac{125}{3}$ 13. $\frac{32}{3}$

15. 3   17. **(a)** B; **(b)** 2   19. $\dfrac{(e-1)^2}{e}$ 21. $\frac{2}{3}$ 23. 16

Margin Exercises, Section 5.5, pp. 340–344

1. $dy = (12x + 1)\,dx$ 2. $du = dx$ 3. $e^{x^3} + C$

4. $\ln(5 + x^2) + C$ 5. $-\dfrac{1}{3 + x^2} + C$ 6. $\dfrac{(\ln x)^2}{2} + C$

7. $\frac{1}{3}e^{x^3} + C$ 8. $\frac{1}{5}e^{5x} + C$

9. $\frac{1}{0.02}e^{0.02x} + C$, or $50e^{0.02x} + C$ 10. $-e^{-x} + C$

11. $\frac{1}{80}(x^4 + 5)^{20} + C$

12. $\displaystyle\int \frac{\ln x\,dx}{x} = \frac{(\ln x)^2}{2} + C$, so $\displaystyle\int_1^e \frac{\ln x\,dx}{x} = \left[\frac{(\ln x)^2}{2}\right]_1^e =$

$\dfrac{(\ln e)^2}{2} - \dfrac{(\ln 1)^2}{2} = \dfrac{1}{2} - 0 = \dfrac{1}{2}$

Exercise Set 5.5, pp. 344–346

1. $\ln(7 + x^3) + C$ 3. $\frac{1}{4}e^{4x} + C$ 5. $2e^{x/2} + C$

7. $\frac{1}{4}e^{x^4} + C$ 9. $-\frac{1}{3}e^{-t^3} + C$ 11. $\dfrac{(\ln 4x)^2}{2} + C$

13. $\ln(1 + x) + C$ 15. $-\ln(4 - x) + C$

17. $\frac{1}{24}(t^3 - 1)^8 + C$ 19. $\frac{1}{8}(x^4 + x^3 + x^2)^8 + C$

21. $\ln(4 + e^x) + C$ 23. $\frac{1}{4}(\ln x^2)^2 + C$, or $(\ln x)^2 + C$

25. $\ln(\ln x) + C$ 27. $\dfrac{2}{3a}(ax + b)^{3/2} + C$

29. $\dfrac{b}{a}e^{ax} + C$ 31. $e - 1$ 33. $\dfrac{21}{4}$

35. $\ln 4 - \ln 2 = \ln\dfrac{4}{2} = \ln 2$ 37. $\ln 19$ 39. $1 - \dfrac{1}{e^b}$

41. $1 - \dfrac{1}{e^{mb}}$ 43. $\dfrac{208}{3}$

45. **(a)** $\dfrac{100,000}{0.025}(e^{2.2} - 1) \approx 32,100,054$;

(b) $\dfrac{100,000}{0.025}(e^{2.2} - e^2) \approx 6,543,830$

47. $-\frac{5}{12}(1 - 4x^2)^{3/2} + C$ 49. $-\frac{1}{3}e^{-x^3} + C$

51. $-e^{1/t} + C$ 53. $-\frac{1}{3}(\ln x)^{-3} + C$

55. $\frac{2}{9}(x^3 + 1)^{3/2} + C$ 57. $\frac{3}{4}(x^2 - 6x)^{2/3} + C$

59. $\frac{1}{8}[\ln(t^4 + 8)]^2 + C$ 61. $x + \dfrac{9}{x + 3} + C$

63. $t - \ln(t - 4) + C$ 65. $-\ln(1 + e^{-x}) + C$

67. $\dfrac{1}{n + 1}(\ln x)^{n+1} + C$ 69. $\ln[\ln(\ln x)] + C$

71. $\dfrac{9(7x^2 + 9)^{n+1}}{14(n + 1)} + C$

Margin Exercises, Section 5.6, pp. 348–349

1. $xe^{3x} - \frac{1}{3}e^{3x} + C$

2. $\dfrac{x^2}{2}\ln x - \dfrac{x^2}{4} + C$. Let $u = \ln x$, $dv = x\,dx$.

3. $\frac{2}{3}x(x + 3)^{3/2} - \frac{4}{15}(x + 3)^{5/2} + C$ 4. $2\ln 2 - \frac{3}{4}$

Exercise Set 5.6, pp. 349–350

1. $xe^{5x} - \dfrac{1}{5}e^{5x} + C$ 3. $\dfrac{1}{2}x^6 + C$ 5. $\dfrac{x}{2}e^{2x} - \dfrac{1}{4}e^{2x} + C$

7. $-\dfrac{x}{2}e^{-2x} - \dfrac{1}{4}e^{-2x} + C$ 9. $\dfrac{x^3}{3}\ln x - \dfrac{x^3}{9} + C$

11. $\dfrac{x^2}{2}\ln x^2 - \dfrac{x^2}{2} + C$

13. $(x + 3)\ln(x + 3) - x + C$. Let $u = \ln(x + 3)$, $dv = dx$, and choose $v = x + 3$ for an antiderivative of v.

15. $\left(\dfrac{x^2}{2} + 2x\right)\ln x - \dfrac{x^2}{4} - 2x + C$

17. $\left(\dfrac{x^2}{2} - x\right)\ln x - \dfrac{x^2}{4} + x + C$

19. $\frac{2}{3}x(x + 2)^{3/2} - \frac{4}{15}(x + 2)^{5/2} + C$

21. $\dfrac{x^4}{4}\ln 2x - \dfrac{x^4}{16} + C$ 23. $x^2e^x - 2xe^x + 2e^x + C$

25. $\frac{1}{2}x^2e^{2x} + \frac{1}{4}e^{2x} - \frac{1}{2}xe^{2x} + C$ 27. $\frac{8}{3}\ln 2 - \frac{7}{9}$

29. $9\ln 9 - 5\ln 5 - 4$ 31. 1

33. **(a)** $10[e^{-T}(-T - 1) + 1]$; **(b)** ≈ 9.085

35. $\dfrac{2}{9}x^{3/2}[3\ln x - 2] + C$ 37. $\dfrac{e^t}{t + 1} + C$

39. $2\sqrt{x}\ln x - 4\sqrt{x} + C$

41. Let $u = x^n$ and $dv = e^x\,dx$. Then $du = nx^{n-1}\,dx$ and $v = e^x$. Then use integration by parts.

Margin Exercises, Section 5.7, pp. 351–353

1. $\dfrac{1}{7}\ln\left(\dfrac{x}{2x+7}\right) + C$ **2.** $\dfrac{15}{4} - \dfrac{3x}{2} - \dfrac{15}{4}\ln(5-2x) + C$

3. $\dfrac{5}{4(5-2x)} + \dfrac{1}{4}\ln(5-2x) + C$

4. $\frac{5}{2}y\sqrt{y^2 - \frac{4}{25}} - \frac{2}{5}\ln\left(y + \sqrt{y^2 - \frac{4}{25}}\right) + C$

5. $\dfrac{1}{7}\ln\left(\dfrac{7+x}{7-x}\right) + C$ **6.** $x(\ln x)^2 - 2(x\ln x - x) + C$

Exercise Set 5.7, pp. 353–354

1. $\dfrac{1}{9}e^{-3x}(-3x-1) + C$ **3.** $\dfrac{5^x}{\ln 5} + C$

5. $\dfrac{1}{8}\ln\left(\dfrac{4+x}{4-x}\right) + C$ **7.** $5 - x - 5\ln(5-x) + C$

9. $\dfrac{1}{5(5-x)} + \dfrac{1}{25}\ln\left(\dfrac{x}{5-x}\right) + C$

11. $(\ln 3)x + x\ln x - x + C$

13. $\dfrac{x^4 e^{5x}}{5} - \dfrac{4}{25}x^3 e^{5x} + \dfrac{12}{125}x^2 e^{5x} - \dfrac{24}{3125}e^{5x}(5x-1) + C$

15. $x^4\left(\frac{1}{4}\ln x - \frac{1}{16}\right) + C$ **·17.** $\ln(x + \sqrt{x^2+7}) + C$

19. $\dfrac{2}{5-7x} + \dfrac{2}{5}\ln\left(\dfrac{x}{5-7x}\right) + C$

21. $-\dfrac{5}{4}\ln\left(\dfrac{x-\frac{1}{2}}{x+\frac{1}{2}}\right) + C$

23. $m\sqrt{m^2+4} + 4\ln(m + \sqrt{m^2+4}) + C$

25. $\dfrac{5\ln x}{2x^2} + \dfrac{5}{4x^2} + C$ **27.** $x^3 e^x - 3x^2 e^x + 6xe^x - 6e^x + C$

29. $-4\ln\left(\dfrac{x}{3x-2}\right) + C$

31. $\dfrac{1}{-2(x-2)} + \dfrac{1}{4}\ln\left(\dfrac{x}{x-2}\right) + C$

33. $\dfrac{-3}{(e^{-x}-3)} + \ln(e^{-x}-3) + C$

Margin Exercises, Section 5.8, pp. 355–361

1. $\displaystyle\sum_{i=1}^{6} i^2$ **2.** $\displaystyle\sum_{i=1}^{4} e^i$ **3.** $\displaystyle\sum_{i=1}^{38} P(x_i)\,\Delta x$

4. $4^1 + 4^2 + 4^3$, or 84   **5.** $e^1 + 2e^2 + 3e^3 + 4e^4 + 5e^5$

6. $t(x_1)\,\Delta x + t(x_2)\,\Delta x + \cdots + t(x_{20})\,\Delta x$ **7.** \$11,250

8. (a) 3.0667; (b) 2.4501; (c) $\displaystyle\int_1^7 \dfrac{1}{x}\,dx = \ln 7 \approx 1.9459$

9. 2   **10.** (a) $\frac{95}{3} \approx 31.7°$; (b) $-10°$; (c) 46.25°

11. \$257

Exercise Set 5.8, pp. 361–363

1. (a) 1.4914; (b) 0.8571   **3.** 0   **5.** $e - 1$ **7.** $\frac{4}{3}$

9. 13   **11.** $\dfrac{1}{n+1}$ **13.** (a) 100 after 10 hr; (b) $33\frac{1}{3}$

15. 252 million   **17.** (a) 3 cc; (b) ≈ 2.7 cc   **19.** 47.5

Summary and Review: Chapter 5, pp. 364–365

1. [5.1] $\frac{8}{5}x^5 + C$ **2.** [5.1] $3e^x + 2x + C$

3. [5.1] $t^3 + \frac{7}{2}t^2 + \ln t + C$ **4.** [5.4] 6   **5.** [5.2, 5.3] $\frac{62}{3}$

6. [5.2] Total number of words written in t minutes; an antiderivative

7. [5.2] Total amount of sales at the end of t days; an antiderivative

8. [5.3] $\frac{1}{6}(b^6 - a^6)$ **9.** [5.4] $-\frac{2}{5}$ **10.** [5.3] $e - \frac{1}{2}$

11. [5.3] $3\ln 3$ **12.** [5.4] 0   **13.** [5.4] Negative

14. [5.4] Positive   **15.** [5.4] $\frac{27}{2}$ **16.** [5.5] $\frac{1}{4}e^{x^4} + C$

17. [5.5] $\ln(4t^6 + 3) + C$ **18.** [5.5] $\frac{1}{4}(\ln 4x)^2 + C$

19. [5.5] $-\frac{2}{3}e^{-3x} + C$ **20.** [5.6] $e^{3x}(x - \frac{1}{3}) + C$

21. [5.6] $x\ln x^7 - 7x + C$ **22.** [5.6] $x^3\ln x - \dfrac{x^3}{3} + C$

23. [5.7] $\dfrac{1}{14}\ln\left(\dfrac{7+x}{7-x}\right) + C$

24. [5.7] $\frac{1}{5}x^2 e^{5x} - \frac{2}{25}xe^{5x} + \frac{2}{125}e^{5x} + C$

25. [5.7] $\dfrac{1}{49} + \dfrac{x}{7} - \dfrac{1}{49}\ln(7x+1) + C$

26. [5.7] $\ln(x + \sqrt{x^2 - 36}) + C$

27. [5.7] $x^7\left(\dfrac{\ln x}{7} - \dfrac{1}{49}\right) + C$

28. [5.7] $\frac{1}{64}e^{8x}(8x-1) + C$ **29.** [5.8] 3.667

30. [5.8] $\frac{1}{2}(11 - e^{-2})$ **31.** [5.3] 80
32. [5.5] $e^{12} - 1$, or about \$162,754
33. [5.5, 5.7] $\frac{1}{10}[\ln (t^5 + 3)]^2 + C$
34. [5.5] $-\frac{1}{2} \ln (1 + 2e^{-x}) + C$
35. [5.6, 5.7] $(\ln \sqrt{x})^2 + C$

36. [5.6, 5.7] $\dfrac{x^{92}}{92} \left(\ln x - \dfrac{1}{92} \right) + C$

37. [5.6] $(x - 3) \ln (x - 3) - (x - 4) \ln (x - 4) + C$

38. [5.5] $-\dfrac{1}{3 (\ln x)^3} + C$

Test: Chapter 5, pp. 365–366

1. $x + C$ **2.** $200x^5 + C$ **3.** $e^x + \ln x + \frac{8}{11}x^{11/8} + C$
4. $\frac{1}{6}$ **5.** $4 \ln 3$ **6.** An antiderivative, total number of

words typed in t minutes   **7.** 12   **8.** $-\dfrac{1}{2} \left(\dfrac{1}{e^2} - 1 \right)$

9. $\ln b - \ln a$ **10.** Positive   **11.** $\ln (x + 8) + C$
12. $-2e^{-0.5x} + C$ **13.** $\frac{1}{40}(t^4 + 1)^{10} + C$
14. $\dfrac{x}{5} e^{5x} - \dfrac{e^{5x}}{25} + C$ **15.** $\dfrac{x^4}{4} \ln x^4 - \dfrac{x^4}{4} + C$

16. $\dfrac{2^x}{\ln 2} + C$ **17.** $\dfrac{1}{7} \ln \left(\dfrac{x}{7 - x} \right) + C$ **18.** 6   **19.** $\dfrac{1}{3}$

20. 95   **21.** \$49,000   **22.** 94 words
23. $\dfrac{(\ln x)^4}{4} - \dfrac{4}{3} (\ln x)^3 + 5 \ln x + C$

24. $(x + 3) \ln (x + 3) - (x + 5) \ln (x + 5) + C$

CHAPTER 6

Margin Exercises, Section 6.1, pp. 370–373

1. \$53.33   **2.** \$1.17
3. (a) $(2, \$16)$; (b) \$18.67; (c) \$7.33

Exercise Set 6.1, p. 374

1. (a) $(6, \$5)$; (b) \$15; (c) \$9   **3.** (a) $(1, \$9)$;
(b) \$3.33; (c) \$1.67   **5.** (a) $(3, \$9)$; (b) \$36; (c) \$18
7. (a) $(5, \$0.61)$; (b) \$86.36; (c) \$2.45

Margin Exercises, Section 6.2, pp. 375–379

1. \$1161.83   **2.** \$29,786.67   **3.** \$215.41
4. 1,136,884 million barrels   **5.** After 7 yr (1987)

Exercise Set 6.2, pp. 379–380

1. \$131   **3.** \$5610.72   **5.** \$340,754.12   **7.** \$949.94
9. \$259.37   **11.** 1,250,427,016 tons
13. After 31 yr (2011)   **15.** \$33,535.73
17. \$5,136,016.50

Margin Exercises, Section 6.3, pp. 381–382

1. \$2231.30   **2.** \$24,261.23   **3.** \$15,563.96

Exercise Set 6.3, p. 383

1. \$826.50   **3.** \$22,973.57   **5.** \$17,802.91
7. \$511,471.15   **9.** 19.994 lb   **11.** \$74.23   **13.** \$68.83

Margin Exercises, Section 6.4, pp. 384–387

1. $\frac{2}{3}, \frac{9}{10}, \frac{99}{100}, \frac{199}{200}$ **2.** $\frac{1}{2}$ **3.** ∞ **4.** Convergent, 1
5. Divergent   **6.** \$36,000

Exercise Set 6.4, pp. 388–389

1. $\frac{1}{3}$ **3.** Divergent   **5.** 1   **7.** $\frac{1}{2}$ **9.** Divergent   **11.** 5
13. Divergent   **15.** Divergent   **17.** Divergent   **19.** 1
21. \$45,000   **23.** 33,333 lb   **25.** Divergent   **27.** 2

29. $\frac{1}{2}$ **31.** \$0.93   **33.** $\dfrac{1}{k^2}$; the total dose of the drug

Margin Exercises, Section 6.5, pp. 390–399

1. (a) $\frac{7}{20}$; (b) $\frac{6}{20}$, or $\frac{3}{10}$; (c) $\frac{4}{20}$, or $\frac{1}{5}$; (d) 0
2. (a) $\frac{1}{2}$; (b) $\frac{1}{8}$; (c) $\frac{1}{4}$; (d) $\frac{1}{8}$ **3.** [15, 25]   **4.** $[0, \infty)$

5. $\dfrac{2}{3}$ **6.** $\displaystyle\int_1^2 \dfrac{2}{3} x \, dx = \left[\dfrac{2}{3} \cdot \dfrac{1}{2} x^2 \right]_1^2 = \dfrac{1}{3} (2^2 - 1) = 1$

7. (a) $\int_3^6 \frac{24}{t^3}\, dt = \left[24\left(-\frac{t^{-2}}{2}\right)\right]_3^6 = -12\left(\frac{1}{6^2} - \frac{1}{3^2}\right) = 1;$

(b) $\frac{64}{75}$; (c) $\frac{5}{12}$ 8. $\frac{3}{26}$ 9. 4   10. $\frac{1}{4}$ 11. $\frac{1}{4}$

12. $\int_0^\infty ke^{-kx}\, dx = \lim_{b \to \infty} \int_0^b ke^{-kx}\, dx = \lim_{b \to \infty} \left[\frac{k}{-k}e^{-kx}\right]_0^b =$

$\lim_{b \to \infty} [-e^{-kx}]_0^b = \lim_{b \to \infty} [-e^{-kb} - (-e^{-k(0)})] =$

$\lim_{b \to \infty}\left[-\frac{1}{e^{kb}} + 1\right] = 0 + 1 = 1$ 13. 0.3297

Exercise Set 6.5, pp. 399–400

1. $\int_0^1 2x\, dx = [x^2]_0^1 = 1^2 - 0^2 = 1$

3. $\int_4^7 \frac{1}{3}\, dx = \left[\frac{1}{3}x\right]_4^7 = \frac{1}{3}(7 - 4) = 1$

5. $\int_1^3 \frac{3}{26}x^2\, dx = \left[\frac{3}{26}\cdot\frac{x^3}{3}\right]_1^3 = \frac{1}{26}(3^3 - 1^3) = 1$

7. $\int_1^e \frac{1}{x}\, dx = [\ln x]_1^e = \ln e - \ln 1 = 1 - 0 = 1$

9. $\int_{-1}^1 \frac{3}{2}x^2\, dx = \left[\frac{3}{2}\cdot\frac{1}{3}x^3\right]_{-1}^1 = \frac{1}{2}(1^3 - (-1)^3) =$

$\frac{1}{2}(1 + 1) = 1$ 11. $\int_0^\infty 3e^{-3x}\, dx = \lim_{b \to \infty} \int_0^b 3e^{-3x}\, dx =$

$\lim_{b \to \infty}\left[\frac{3}{-3}e^{-3x}\right]_0^b = \lim_{b \to \infty} [-e^{-3x}]_0^b =$

$\lim_{b \to \infty} [-e^{-3b} - (-e^{-3\cdot 0})] = \lim_{b \to \infty}\left(1 - \frac{1}{e^{3b}}\right) = 1$

13. $k = \frac{1}{4}$ 15. $k = \frac{3}{2}$ 17. $k = \frac{1}{5}$ 19. $k = \frac{1}{2}$

21. $k = \frac{1}{\ln 3}$ 23. $k = \frac{1}{e^3 - 1}$ 25. $\frac{8}{25}$ 27. $\frac{1}{2}$

29. 0.3297   31. 0.99995   33. 0.9502   35. 0.3935

37. $b = \sqrt[4]{4}$, or $\sqrt{2}$

Margin Exercises, Section 6.6, pp. 403–410

1. $E(x) = \frac{2}{3}$, $E(x^2) = \frac{1}{2}$ 2. $\mu = \frac{2}{3}$, $\sigma^2 = \frac{1}{18}$, $\sigma = \sqrt{\frac{1}{18}} = \frac{1}{3}\sqrt{\frac{1}{2}}$
3. (a) 0.4850; (b) 0.4608; (c) 0.9559; (d) 0.2720;
(e) 0.1102; (f) 0.0307   4. 0.1210

Exercise Set 6.6, pp. 410–412

1. $\mu = E(x) = \frac{7}{2}$, $E(x^2) = 13$, $\sigma^2 = \frac{3}{4}$, $\sigma = \frac{1}{2}\sqrt{3}$
3. $\mu = E(x) = 2$, $E(x^2) = \frac{9}{2}$, $\sigma^2 = \frac{1}{2}$, $\sigma = \sqrt{\frac{1}{2}}$
5. $\mu = E(x) = \frac{14}{9}$, $E(x^2) = \frac{5}{2}$, $\sigma^2 = \frac{13}{162}$, $\sigma = \sqrt{\frac{13}{162}} = \frac{1}{9}\sqrt{\frac{13}{2}}$
7. $\mu = E(x) = -\frac{5}{4}$, $E(x^2) = \frac{11}{5}$, $\sigma^2 = \frac{51}{80}$, $\sigma = \sqrt{\frac{51}{80}} = \frac{1}{4}\sqrt{\frac{51}{5}}$
9. $\mu = E(x) = \frac{2}{\ln 3}$, $E(x^2) = \frac{4}{\ln 3}$, $\sigma^2 = \frac{4\ln 3 - 4}{(\ln 3)^2}$,

$\sigma = \frac{2}{\ln 3}\sqrt{\ln 3 - 1}$ 11. 0.4964

13. 0.3665   15. 0.6442   17. 0.0078   19. 0.1716
21. 0.0013   23. (a) 0.6826; (b) 68.26%   25. 0.2898
27. 0.4514   29. (a) 0.3413; (b) 0.1610; (c) 0.3336

31. 0.62%   33. $\mu = E(x) = \frac{a+b}{2}$, $E(x^2) =$

$\frac{b^3 - a^3}{3(b-a)}$, or $\frac{b^2 + ba + a^2}{3}$, $\sigma^2 = \frac{(b-a)^2}{12}$, $\sigma = \frac{b-a}{2\sqrt{3}}$

35. $\sqrt{2}$ 37. $\frac{\ln 2}{k}$

Margin Exercises, Section 6.7, pp. 413–414

1. $\frac{\pi}{3}$ 2. $\frac{\pi}{2}(e^2 - e^{-4})$

Exercise Set 6.7, pp. 414–415

1. $\frac{9\pi}{2}$ 3. $\frac{7\pi}{3}$ 5. $\frac{\pi}{2}(e^{10} - e^{-4})$ 7. $\frac{2\pi}{3}$ 9. $\pi \ln 3$

11. 32π 13. $\frac{32\pi}{5}$ 15. 56π 17. $\frac{32\pi}{3}$ 19. $2\pi e^3$

21. (a) Divergent; (b) π

Summary and Review: Chapter 6, pp. 415–417

1. [6.1] (2, $16)   2. [6.1] $18.67   3. [6.1] $5.33
4. [6.2] $83,560.54   5. [6.2] $251.50
6. [6.2] 835,821 thousand tons   7. [6.2] 165 yr (2143)
8. [6.3] $247.88   9. [6.3] $44,517   10. [6.4] $53,333
11. [6.4] Convergent, 1   12. [6.4] Divergent

13. [6.4] Convergent, $\frac{1}{2}$ 14. [6.5] $k = \frac{8}{3}$
15. [6.5] 0.6   16. [6.6] $\frac{3}{5}$ 17. [6.6] $\frac{3}{4}$ 18. [6.6] $\frac{3}{4}$

19. [6.6] $\frac{3}{80}$ 20. [6.6] $\frac{\sqrt{15}}{20}$ 21. [6.6] 0.4678

22. [6.6] 0.8827   23. [6.6] 0.1002   24. [6.6] 0.5000

25. [6.6] 0.0918   26. [6.7] $\frac{127\pi}{7}$ 27. [6.7] $\frac{\pi}{6}$

28. [6.5] $c = \sqrt[9]{4.5}$ 29. [6.4] Convergent, $-\frac{1}{5}$
30. [6.4] Convergent, 3

Test: Chapter 6, pp. 417–418

1. (3, $16)   2. $45   3. $22.50   4. $29,192
5. $344.66   6. 13,941,765 thousand tons
7. After 33 years (2013)   8. $2465.97   9. $30,717.71
10. $34,545.45   11. Convergent, $\frac{1}{4}$ 12. Divergent
13. $k = \frac{1}{4}$ 14. 0.8647   15. $E(x) = \frac{13}{6}$ 16. $E(x^2) = 5$
17. $\mu = \frac{13}{6}$ 18. $\sigma^2 = \frac{11}{36}$ 19. $\sigma = \frac{\sqrt{11}}{6}$ 20. 0.4332
21. 0.4420   22. 0.9071   23. 0.4207   24. $\pi \ln 5$
25. $\frac{5\pi}{2}$ 26. $b = \sqrt[6]{6}$ 27. Convergent, $-\frac{1}{4}$

CHAPTER 7

Margin Exercises, Section 7.1, pp. 420–426

1. (a) 128; the profit from selling 14 items of the first
product and 12 of the second is $128; (b) 48; the profit
from selling none of the first product and 8 items of the
second is $48   2. 7.2, 94.5, 0.60   3. 2000
4. (a) 352; (b) 15.4   5. $125,656.83   6. (a) 4; (b) 4
7.

Exercise Set 7.1, pp. 427–428

1. $f(0, -2) = 0, f(2, 3) = -8, f(10, -5) = 200$
3. $f(0, -2) = 1, f(-2, 1) = -13\frac{8}{9}, f(2, 1) = 23$
5. $f(e, 2) = \ln e + 2^3 = 1 + 8 = 9, f(e^2, 4) = 66,$
$f(e^3, 5) = 128$ 7. $f(-1, 2, 3) = 6, f(2, -1, 3) = 12$
9. 105, 95   11. 7.7%   13. $151,571.66   15. $0°$
17. $-22°$

Margin Exercises, Section 7.2, pp. 428–433

1. (a) $f(x, 4) = -x^2 - 15$; (b) $-2x$ 2. $\frac{\partial f}{\partial x} = -2x$

3. (a) $\frac{\partial z}{\partial x} = 6xy + 15x^2$; (b) $\frac{\partial z}{\partial y} = 3x^2$

4. (a) $\frac{\partial t}{\partial x} = y + z + 2x$; (b) $\frac{\partial t}{\partial y} = x + 3y^2$; (c) $\frac{\partial t}{\partial z} = x$

5. (a) $f_x = 9x^2y + 2y$; (b) $f_x(-4, 1) = 146$;
(c) $f_y = 3x^3 + 2x$; (d) $f_y(2, 6) = 28$

6. $f_x = \frac{1}{x} + ye^x, f_y = \frac{1}{y} + e^x$

7. (a) 108,000 units; (b) $\frac{\partial p}{\partial x} = 600 \left(\frac{y}{x}\right)^{1/4}$,

$\frac{\partial p}{\partial y} = 200 \left(\frac{x}{y}\right)^{3/4}$; (c) 1000, 43.2

Exercise Set 7.2, pp. 433–435

1. $\frac{\partial z}{\partial x} = 2 - 3y, \frac{\partial z}{\partial y} = -3x, \left.\frac{\partial z}{\partial y}\right|_{(-2, -3)} = 11, \left.\frac{\partial z}{\partial y}\right|_{(0, -5)} = 0$

3. $\frac{\partial z}{\partial x} = 6x - 2y, \frac{\partial z}{\partial y} = -2x + 1, \left.\frac{\partial z}{\partial x}\right|_{(-2, -3)} = -6,$

$\left.\frac{\partial z}{\partial y}\right|_{(0, -5)} = 1$ 5. $f_x = 2, f_y = -3, f_x(-2, 4) = 2,$

$f_y(4, -3) = -3$ 7. $f_x = \frac{x}{\sqrt{x^2 + y^2}}, f_y = \frac{y}{\sqrt{x^2 + y^2}},$

$f_x(-2, 1) = \frac{-2}{\sqrt{5}}, f_y(-3, -2) = \frac{-2}{\sqrt{13}}$

9. $f_x = 2e^{2x+3y}, f_y = 3e^{2x+3y}$ 11. $f_x = ye^{xy}, f_y = xe^{xy}$

13. $f_x = \frac{y}{x + y}, f_y = \frac{y}{x + y} + \ln(x + y)$

15. $f_x = 1 + \ln xy$, $f_y = \dfrac{x}{y}$ **17.** $f_x = \dfrac{1}{y} + \dfrac{y}{x^2}$, $f_y = -\dfrac{x}{y^2} - \dfrac{1}{x}$

19. $f_x = 12(2x + y - 5)$, $f_y = 6(2x + y - 5)$

21. $\dfrac{\partial f}{\partial b} = 12m + 6b - 30$, $\dfrac{\partial f}{\partial m} = 28m + 12b - 64$

23. $f_x = 3y - 2\lambda$, $f_y = 3x - \lambda$, $f_\lambda = -(2x + y - 8)$
25. $f_x = 2x - 10\lambda$, $f_y = 2y - 2\lambda$, $f_\lambda = -(10x + 2y - 4)$

27. (a) 3,888,064 units; **(b)** $\dfrac{\partial p}{\partial x} = 1117.8 \left(\dfrac{y}{x}\right)^{0.379}$

$\dfrac{\partial p}{\partial y} = 682.2 \left(\dfrac{x}{y}\right)^{0.621}$; **(c)** 965.8, 866.8

29. $f_x = \dfrac{-4xt^2}{(x^2 - t^2)^2}$, $f_t = \dfrac{4x^2 t}{(x^2 - t^2)^2}$

31. $f_x = \dfrac{1}{\sqrt{x}\,(1 + 2\sqrt{t})}$, $f_t = \dfrac{-1 - 2\sqrt{x}}{\sqrt{t}\,(1 + 2\sqrt{t})^2}$

33. $f_x = 4x^{-1/3} - 2x^{-3/4} t^{1/2} + 6x^{-3/2} t^{3/2}$,
$f_t = -4x^{1/4} t^{-1/2} - 18x^{-1/2} t^{1/2}$

Margin Exercises, Section 7.3, pp. 435–436

1. (a) $\dfrac{\partial z}{\partial y} = 6xy + 2x$; **(b)** $\dfrac{\partial}{\partial x}\left(\dfrac{\partial z}{\partial y}\right) = 6y + 2$;

(c) $\dfrac{\partial}{\partial y}\left(\dfrac{\partial z}{\partial y}\right) = 6x$ **2. (a)** $f_y = 6xy + 2x$;

(b) $f_{yx} = 6y + 2$; **(c)** $f_{yy} = 6x$ **3.** $\dfrac{\partial^2 f}{\partial x^2} = 2$,

$\dfrac{\partial^2 f}{\partial y\, \partial x} = 6y + 2 + \dfrac{1}{y}$, $\dfrac{\partial^2 f}{\partial x\, \partial y} = 6y + 2 + \dfrac{1}{y}$, $\dfrac{\partial^2 f}{\partial y^2} = 6x - \dfrac{x}{y^2}$

Exercise Set 7.3, p. 437

1. $\dfrac{\partial^2 f}{\partial x^2} = 6$, $\dfrac{\partial^2 f}{\partial y\, \partial x} = \dfrac{\partial^2 f}{\partial x\, \partial y} = -1$, $\dfrac{\partial^2 f}{\partial y^2} = 0$

3. $\dfrac{\partial^2 f}{\partial x^2} = 0$, $\dfrac{\partial^2 f}{\partial y\, \partial x} = \dfrac{\partial^2 f}{\partial x\, \partial y} = 3$, $\dfrac{\partial^2 f}{\partial y^2} = 0$

5. $\dfrac{\partial^2 f}{\partial x^2} = 20x^3 y^4 + 6xy^2$, $\dfrac{\partial^2 f}{\partial y\, \partial x} = \dfrac{\partial^2 f}{\partial x\, \partial y} = 20x^4 y^3 + 6x^2 y$,

$\dfrac{\partial^2 f}{\partial y^2} = 12x^5 y^2 + 2x^3$ **7.** $f_{xx} = 0$, $f_{yx} = 0$, $f_{xy} = 0$, $f_{yy} = 0$

9. $f_{xx} = 4y^2 e^{2xy}$, $f_{yx} = f_{xy} = 4xye^{2xy} + 2e^{2xy}$,
$f_{yy} = 4x^2 e^{2xy}$ **11.** $f_{xx} = 0$, $f_{yx} = f_{xy} = 0$, $f_{yy} = e^y$

13. $f_{xx} = -\dfrac{y}{x^2}$, $f_{yx} = f_{xy} = \dfrac{1}{x}$, $f_{yy} = 0$

15. $f_{xx} = \dfrac{-6y}{x^4}$, $f_{yx} = f_{xy} = \dfrac{2(y^3 - x^3)}{x^3 y^3}$, $f_{yy} = \dfrac{6x}{y^4}$

17. $\dfrac{\partial^2 f}{\partial x^2} = \dfrac{2y^2 - 2x^2}{(x^2 + y^2)^2}$, $\dfrac{\partial^2 f}{\partial y^2} = \dfrac{2x^2 - 2y^2}{(x^2 + y^2)^2}$, so the sum is 0

19. (a) $-y$; **(b)** x;
(c) $f_{yx}(0, 0) = 1$, $f_{xy}(0, 0) = -1$; so $f_{yx}(0, 0) \neq f_{xy}(0, 0)$

Margin Exercises, Section 7.4, pp. 442–445

1. Minimum $= -9$ at $(-3, 3)$
2. Maximum $= \frac{1}{108}$ at $(\frac{1}{6}, \frac{1}{6})$
3. 5 thousand of the \$15 calculator;
7 thousand of the \$20 calculator

Exercise Set 7.4, pp. 445–447

1. Minimum $= -\frac{1}{3}$ at $(-\frac{1}{3}, \frac{2}{3})$
3. Maximum $= \frac{4}{27}$ at $(\frac{2}{3}, \frac{2}{3})$ **5.** Minimum $= -1$ at $(1, 1)$
7. Minimum $= -7$ at $(1, -2)$
9. Minimum $= -5$ at $(-1, 2)$ **11.** None
13. 6 (thousand) of the \$17 radio and 5 (thousand) of
the \$21 radio
15. Maximum of $P = 35$ (million dollars) when $a = 10$
(million dollars) and $p = \$3$
17. (a) $R(p_1, p_2) = 78p_1 - 6p_1^2 - 6p_1 p_2 + 66p_2 - 6p_2^2$;
(b) $p_1 = 5$ (\$50), $p_2 = 3$ (\$30);
(c) $q_1 = 78 - 6 \cdot 5 - 3 \cdot 3 = 39$ (hundreds),
$q_2 = 33$ (hundreds);
(d) $R = 50 \cdot 3900 + 30 \cdot 3300 = \$294,000$
19. None; saddle point $(0, 0)$
21. Minimum $= 0$ at $(0, 0)$; saddle points at $(2, 1)$ and
$(-2, 1)$

Margin Exercises, Section 7.5, pp. 447–453

1. $dz = e^y \, dx + xe^y \, dy$ **2.** $\dfrac{1}{x} \, dx + \dfrac{1}{y} \, dy$ **3.** $9 \, dx + 12 \, dy$

4. -0.66 5. $dw = \left(-\dfrac{1}{x^2} - y\right) dx - x\, dy + 2\, dz$

6. $\dfrac{x\, dx + y\, dy + z\, dz}{(x^2 + y^2 + z^2)^{1/2}}$

7. $2(x_1\, dx_1 + x_2\, dx_2 + x_3\, dx_3 + x_4\, dx_4)$
8. $dw = -2\, dx + 3\, dy - 2\, dz$ 9. 0.3 10. $\Delta z = 0.7947$
11. 0.78 12. $\Delta w = 0.2226,\ dw = 0.22$
13. $\ln([1.01][0.97]) \approx -0.02$ 14. 0.04
15. 28.875 psi

Exercise Set 7.5, pp. 453–455

1. $dz = 2x\, dx + 8y\, dy$ 3. $e^{4y}\, dx + 4xe^{4y}\, dy$

5. $\left(4x^3 + \dfrac{1}{y}\right) dx + \left(3y^2 - \dfrac{x}{y^2}\right) dy$

7. $2x\, dx - 2y\, dy + 6z\, dz$
9. $yz^3[yz\, dx + 2xz\, dy + 4xy\, dz]$
11. $dw = 2x_1\, dx_1 + x_3\, dx_2 + x_2\, dx_3 - 2x_4\, dx_4$
13. $\Delta z = 0.5805,\ dz = 0.58$
15. $\Delta z = 0.010553,\ dz = 0.0105$
17. $\Delta w = 0.0801,\ dw = 0.08$
19. $\Delta w = -0.062608,\ dw = -0.06$ 21. 167.66
23. 4.09 25. 4.53025 sq ft   27. 39.06 sq ft
29. (a) $P(500, 100) = -50,000 + 300(100) +$
$1000(500) - 3(100)^2 - (500)^2 + (100)(500) =$
$\$-50,000 + 30,000 + 500,000 - 30,000 - 250,000 +$
$50,000 = \$250,000$; (b) $\$300$ 31. 0.00125

Margin Exercises, Section 7.6, pp. 456–462

1. (a) $\$6.2$ billion; (b) $\$6.8$ billion;
(c) they differ by $\$0.6$ billion
2. (a) $y = 53,600x + 124,500$; (b) $\$660,500; \$1,035,700$
3. Same answers   4. (a) $y = 148,241.87e^{0.211321x}$;
(b) $\$1,226,666; \$5,384,621$

Exercise Set 7.6, pp. 462–464

1. (a) $y = 2.29x + 33.67$; (b) $\$54.28, \$70.31$
3. (a) $y = 1.068421x - 1.236842$; (b) 85.3
5. (a) $y = -0.0059x + 15.4997$, (b) $3{:}47.3; 3{:}42.0$;
(c) According to the regression line, the record should have
been $3{:}47.3$, so Cram beat the regression line prediction.
7. (a) $y = 33.2781e^{0.0613x}$; (b) $\$57.78, \$88.74$

Margin Exercises, Section 7.7, pp. 467–470

1. (a) $A(x, y) = xy$, subject to $x + y = 50$;
(b) maximum $= 625$ at $(25, 25)$
2. Maximum $= 125$ at $(2.5, 10)$
3. $r \approx 1.8$ in., $h \approx 3.6$ in.; surface area is about 61.04 in^2

Exercise Set 7.7, pp. 470–472

1. Maximum $= 8$ at $(2, 4)$
3. Maximum $= -16$ at $(2, 4)$
5. Minimum $= 20$ at $(4, 2)$
7. Minimum $= -96$ at $(8, -12)$
9. Minimum $= \frac{3}{2}$ at $(1, \frac{1}{2}, -\frac{1}{2})$ 11. 35 and 35
13. 3 and -3 15. $9\frac{3}{4}$ in., $9\frac{3}{4}$ in.; $95\frac{1}{16}$ in^2; no

17. $r = \sqrt[3]{\dfrac{27}{2\pi}} \approx 1.6$ ft; $h = 2 \cdot r \approx 3.2$ ft;

minimum surface area ≈ 48.3 ft^2
19. Maximum of $S = 800$ at $L = 20$, $M = 60$
21. (a) $C(x, y, z) = 7xy + 6yz + 6xz$; (b) $x = 60$ ft,
$y = 60$ ft, $z = 70$ ft; $\$75,600$ 23. $10,000$ on A, 100 on B
25. Minimum $= -\frac{155}{128}$ at $\left(-\frac{7}{16}, -\frac{3}{4}\right)$

27. Maximum $= \dfrac{1}{27}$ at eight triples $\left(\dfrac{\pm 1}{\sqrt{3}}, \dfrac{\pm 1}{\sqrt{3}}, \dfrac{\pm 1}{\sqrt{3}}\right)$

29. Maximum $= 2$ at $(\frac{1}{2}, \frac{1}{2}, \frac{1}{2}, \frac{1}{2})$ 31. $\lambda = \dfrac{p_x}{c_1} = \dfrac{p_y}{c_2}$

Margin Exercises, Section 7.8, p. 477

1. $h \approx 184$ ft, $k \approx 74$ ft; dimensions are 74 ft by
74 ft by 184 ft

Exercise Set 7.8, p. 477

1. $h = \sqrt[3]{10,240,000} \approx 217$ ft, $k \approx 54$ ft; dimensions are
54 ft by 54 ft by 225 ft

3. $h = \sqrt[3]{\dfrac{Aca^2}{b^2}}$, $k = \sqrt[3]{\dfrac{Abc}{a}}$; dimensions are k by k by $h + c$

Margin Exercises, Section 7.9, pp. 482–487

1. $4y$ **2.** $\frac{1}{2}y^2$ **3.** $\frac{1}{3}x$ **4.** $\frac{3}{2}x + 3x^2$ **5.** $2rx^2 + 2y$
6. $\frac{1}{6}$ **7.** $\frac{25}{6}$ **8.** $\frac{1}{6}$; they are equal **9.** $\frac{25}{6}$; equal
10. (a) $4x^2$; (b) $\frac{224}{3}$ **11.** 72 **12.** $\frac{3}{2}(e-1)$ **13.** $\frac{7}{6}$
14. $\frac{5}{6}$

Exercise Set 7.9, pp. 487–488

1. $\frac{1}{3}y$ **3.** $\frac{8}{3}$ **5.** $\frac{4}{3}$ **7.** 1 **9.** 12 **11.** $\frac{1}{3}\ln 2$
13. 2 **15.** $8\ln 2$ **17.** $f_{av} = 1$

Margin Exercises, Section 7.10, pp. 490–499

1. $1 - x^3$ **2.** $8x$ **3.** $\frac{3}{4}$ **4.** 4
5. (a) 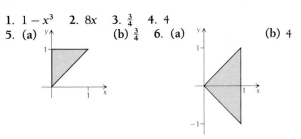 (b) $\frac{3}{4}$ **6.** (a) (b) 4

7. (a) 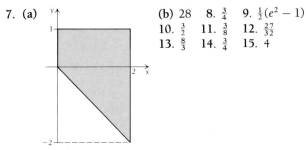 (b) 28 **8.** $\frac{3}{4}$ **9.** $\frac{1}{2}(e^2 - 1)$
10. $\frac{3}{2}$ **11.** $\frac{3}{8}$ **12.** $\frac{27}{32}$
13. $\frac{8}{3}$ **14.** $\frac{3}{4}$ **15.** 4

Exercise Set 7.10, pp. 499–501

1. 1 **3.** 0 **5.** 6 **7.** $\frac{3}{20}$ **9.** 4
11. 2; **13.** $\frac{1}{2}$;

15. $\frac{4}{15}$ **17.** $\frac{375}{4}$ **19.** $\frac{14}{3}$

21. (a) $\int_1^{e^2}\left[\int_{\ln y}^2 dx\right]dy$; (b) $e^2 - 3$. **23.** 1 **25.** $\frac{5}{12}$
27. 39 **29.** $\frac{13}{240}$ **31.** $\frac{1}{8}$

Summary and Review: Chapter 7, pp. 501–503

1. [7.2] $3y^3$ **2.** [7.2] $e^y + 9xy^2 + 2$ **3.** [7.3] $9y^2$
4. [7.3] $9y^2$ **5.** [7.3] 0 **6.** [7.3] $e^y + 18xy$
7. [7.2] $2x\ln y$ **8.** [7.2] $\frac{x^2}{y} + 4y^3$ **9.** [7.3] $\frac{2x}{y}$
10. [7.3] $\frac{2x}{y}$ **11.** [7.3] $2\ln y$ **12.** [7.3] $-\frac{x^2}{y^2} + 12y^2$
13. [7.4] Minimum $= -\frac{549}{20}$ at $(5, \frac{27}{2})$
14. [7.4] Minimum $= -4$ at $(0, -2)$
15. [7.4] Maximum $= \frac{45}{4}$ at $(\frac{3}{2}, -3)$
16. [7.4] Minimum $= 27$ at $(-1, 2)$
17. [7.5] (a) $y = 2.5x + 21.67$; (b) $46.67, 74.17
18. [7.5] (a) $y = 0.6x + 6.7$; (b) 9.1 million
19. [7.6] Minimum $= \frac{125}{4}$ at $(-\frac{3}{2}, -3)$
20. [7.6] Maximum $= 300$ at $(5, 10)$ **21.** [7.10] $\frac{97}{30}$
22. [7.10] $\frac{1}{60}$ **23.** [7.5] $dz = 10x\,dx - 2e^{2y}\,dy$
24. [7.5] $dz = 4xy\,dx + 2x^2\,dy$ **25.** [7.5] 1.1 ft^2
26. [7.5] 0.72 ft^3 **27.** [7.10] $\frac{1}{60}$ **28.** [7.9] 4
29. [7.10] 16 **30.** [7.10] $\frac{8}{3}$ **31.** [7.10] 0
32. [7.10] The cylindrical container

Test: Chapter 7, pp. 503–504

1. $\dfrac{\partial f}{\partial x} = e^x + 6x^2y$ **2.** $\dfrac{\partial f}{\partial y} = 2x^3 + 1$

3. $\dfrac{\partial^2 f}{\partial x^2} = e^x + 12xy$ **4.** $\dfrac{\partial^2 f}{\partial x\,\partial y} = 6x^2$ **5.** $\dfrac{\partial^2 f}{\partial y\,\partial x} = 6x^2$

6. $\dfrac{\partial^2 f}{\partial y^2} = 0$ **7.** Minimum $= -\frac{7}{16}$ at $(\frac{3}{4}, \frac{1}{2})$

8. None **9.** (a) $y = \frac{9}{2}x + \frac{17}{3}$; (b) $23.67
10. Maximum $= -19$ at $(4, 5)$ **11.** 1
12. $dz = 3x^2\,dx + e^y\,dy$ **13.** $dz = 3y^2\,dx + 6xy\,dy$
14. $dV = 0.59$ ft^3 **15.** $\frac{4}{3}$ **16.** $\frac{3}{20}$
17. $400,000 for labor, $200,000 for capital
18. $f_x = \dfrac{-x^4 + 4xt + 6x^2t}{(x^3 + 2t)^2}, f_t = \dfrac{-2x^2(x + 1)}{(x^3 + 2t)^2}$

CHAPTER 8

Margin Exercises, Section 8.1, pp. 507–516

1. (a) I; (b) III; (c) IV; (d) III; (e) I; (f) IV; (g) I

2. (a) $\dfrac{3}{4}\pi$; (b) $\dfrac{7}{4}\pi$; (c) $-\dfrac{\pi}{2}$; (d) 4π; (e) $-\dfrac{5\pi}{4}$;

(f) $-\dfrac{7\pi}{4}$; (g) $\dfrac{9\pi}{4}$; (h) $\dfrac{8\pi}{3}$

3. (a) $60°$; (b) $135°$; (c) $450°$; (d) $1800°$; (e) $-210°$; (f) $54{,}000°$; (g) $-48{,}600°$; (h) $1125°$

4. (a) $-\dfrac{\sqrt{3}}{2}$; (b) $\dfrac{1}{2}$; (c) $-\dfrac{\sqrt{3}}{2}$; (d) $\dfrac{1}{2}$

5. (a) 1; (b) $\sqrt{3}$; (c) 0; (d) $\sqrt{2}$; (e) 2; (f) 1

6. (a) $\sqrt{3}$; (b) undefined; (c) $\dfrac{\sqrt{3}}{3}$; (d) 2; (e) undefined;

(f) $\sqrt{2}$ 7. $1 + \cot^2 t = \csc^2 t$ 8. $\dfrac{\sqrt{6} - \sqrt{2}}{4}$

9. $\dfrac{\sqrt{2} + \sqrt{6}}{4}$ 10. $2 \sin u \cos u$

Exercise Set 8.1, pp. 516–517

1. I   3. III   5. $\dfrac{\pi}{6}$ 7. $\dfrac{\pi}{3}$ 9. $\dfrac{5\pi}{12}$ 11. $270°$

13. $-45°$ 15. $1440°$ 17. $\left(\dfrac{180}{\pi}\right)° \approx 57.3°$ 19. $\dfrac{\sqrt{3}}{2}$

21. 0   23. $\dfrac{1}{2}$ 25. Undefined   27. $\dfrac{\sqrt{3}}{3}$ 29. -1

31. $\tan(u+v) = \dfrac{\sin(u+v)}{\cos(u+v)}$. Then use identities (5) and (3)   33. 0.5210 35. -0.8632 37. -0.3327 39. Undefined   41. $101.6°, 102.7°, 103.7°, 104.6°, 98.6°$

Margin Exercises, Section 8.2, pp. 520–526

1. $\cot x = \dfrac{\cos x}{\sin x}$,

---

so $\dfrac{d}{dx} \cot x = [\sin x(-\sin x) - \cos x(\cos x)] \div \sin^2 x =$

$\dfrac{-(\sin^2 x + \cos^2 x)}{\sin^2 x} = -\dfrac{1}{\sin^2 x} = -\csc^2 x$ 2. $\dfrac{3 \sin^2 x}{\cos^4 x}$

3. $(4x^3 + 6x)\cos(x^4 + 3x^2)$

4. $-2xe^{x^2}\sin(e^{x^2})\sin x^3 + 3x^2\cos(e^{x^2})\cos x^3$

5. (a) Amplitude $= 4$, period $= \pi$, phase shift $= \dfrac{\pi}{4}$;

(b) 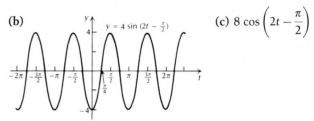   (c) $8\cos\left(2t - \dfrac{\pi}{2}\right)$

6. $\dfrac{dy}{dx} = 0.03594\pi \cos 1.198\pi x$

7. $\dfrac{dL}{dt} = 2\pi Af \cos\left(2\pi fT - \dfrac{d}{\omega}\right)$

8. (a) $L = (a^{2/3} + b^{2/3})^{3/2}$; (b) 27

Exercise Set 8.2, pp. 526–528

1. $\sec x = \dfrac{1}{\cos x}$, so $\dfrac{d}{dx}\sec x = \dfrac{\cos x \cdot 0 - (-\sin x)\cdot 1}{\cos^2 x} =$

$\dfrac{\sin x}{\cos^2 x} = \dfrac{\sin x}{\cos x}\cdot\dfrac{1}{\cos x} = \tan x \sec x$ 3. $x\cos x + \sin x$

5. $e^x(\cos x + \sin x)$ 7. $\dfrac{x\cos x - \sin x}{x^2}$ 9. $2\sin x \cos x$

11. $\cos^2 x - \sin^2 x$ 13. $\dfrac{1}{1 + \cos x}$ 15. $\dfrac{2 \sin x}{\cos^3 x}$

17. $\dfrac{-\sin x}{2\sqrt{1 + \cos x}}$ 19. $-x^2 \sin x$ 21. $\cos x \cdot e^{\sin x}$

23. $-\sin x$ 25. $(2x + 3x^2)\cos(x^2 + x^3)$

27. $(4x^3 - 5x^4)\sin(x^5 - x^4)$ 29. $-\tfrac{1}{2}x^{-1/2}\cdot \sin\sqrt{x}$

31. $-\sin x \cdot \cos(\cos x)$ 33. $\dfrac{-2\sin 4x}{\sqrt{\cos 4x}}$

35. $\tfrac{2}{3}[\csc^2(5 - 2x)^{1/3}](5 - 2x)^{-2/3}$

37. $-12(\tan 3x)(\sec 3x)^2$ **39.** $\cot x$ **41.** $\dfrac{3\pi}{8}\cos\dfrac{\pi}{8}t$

43. $-\dfrac{1000\pi}{9}\left[\sin\dfrac{\pi}{45}(t-10)\right]$

45. (a) Amplitude $= 5$, period $= \dfrac{\pi}{2}$, phase shift $= -\dfrac{\pi}{4}$;

(b) $20\cos(4t+\pi)$ **47.** $40{,}000(\cos t - \sin t)$
49. $8\sqrt{8}$ **51.** $90°$
53. $f_x = -ye^x$, $f_y = 2\cos 2y - e^x$, $f_{xx} = -ye^x$,
$f_{yx} = f_{xy} = -e^x$, $f_{yy} = -4\sin 2y$
55. $f_x = -2\sin(2x+3y)$, $f_y = -3\cdot\sin(2x+3y)$,
$f_{xx} = -4\cos(2x+3y)$, $f_{yx} = f_{xy} = -6\cos(2x+3y)$,
$f_{yy} = -9\cos(2x+3y)$
57. $f_x = 5x^3y\cdot\sec^2(5xy) + 3x^2\cdot\tan(5xy)$,
$f_y = 5x^4\cdot\sec^2(5xy)$, $f_{xx} = 50x^3y^2\cdot\sec^2(5xy)\times$
$\tan(5xy) + 30x^2y\sec^2(5xy) + 6x\tan(5xy)$,
$f_{yx} = f_{xy} = 50x^4y\cdot\sec^2(5xy)\cdot\tan(5xy) + 20x^3\times$
$\sec^2(5xy)$, $f_{yy} = 50x^5\cdot\sec^2(5xy)\cdot\tan(5xy)$

59. $\dfrac{\cos y}{1 + x\sin y}$

Margin Exercises, Section 8.3, pp. 529–531

1. 2 **2.** $\dfrac{(\sin x)^4}{4} + C$ **3.** $\dfrac{1}{4}\sin 4x + C$

4. $\ln|x^2 + \cos x| + C$ **5.** $\sin x - x\cos x + C$
6. $-\ln|\csc x + \cot x| + C$

Exercise Set 8.3, pp. 531–533

1. $\dfrac{1}{2}$ **3.** $\dfrac{(\sin x)^5}{5} + C$ **5.** $\dfrac{(\cos x)^3}{3} + C$

7. $\sin(x+3) + C$ **9.** $-\dfrac{1}{2}\cos 2x + C$
11. $\dfrac{1}{2}\sin x^2 + C$ **13.** $-\cos(e^x) + C$
15. $-\ln|\cos x| + C$, or $\ln|\sec x| + C$
17. $\dfrac{1}{4}x\sin 4x + \dfrac{1}{16}\cos 4x + C$ **19.** $3x\sin x + 3\cos x + C$
21. $2x\sin x - (x^2 - 2)\cos x + C$ **23.** $\tan x - x + C$
25. $\$84$ thousand **27.** $\dfrac{1}{2}x[\sin(\ln x) - \cos(\ln x)] + C$

29. $\dfrac{e^x}{2}(\cos x + \sin x) + C$ **31.** $\dfrac{1}{7}\tan 7x + C$

33. $\tan x + \sec x + C$ **35.** $-\csc u + C$
37. $\cos x - \ln|\csc x + \cot x| + C$
39. $x + 2\ln|\sec x + \tan x| + \tan x + C$

Margin Exercises, Section 8.4, pp. 533–535

1. $\dfrac{\pi}{6}$ **2.** 0 **3.** $-\dfrac{\pi}{3}$ **4.** $\dfrac{\pi}{3}$ **5.** $\dfrac{\pi}{6}$ **6.** $\dfrac{2\pi}{3}$

7. $\dfrac{1}{5}\tan^{-1}(5x) + C$ **8.** $\dfrac{1}{2}\cos^{-1}(2x) + C$

Exercise Set 8.4, p. 536

1. $\dfrac{\pi}{4}$ **3.** $\dfrac{\pi}{2}$ **5.** $\dfrac{\pi}{4}$ **7.** $-\dfrac{\pi}{6}$ **9.** $\tan^{-1}(e^t) + C$

11. Use substitution. Let $u = \sin^{-1} x$; then $x = \sin u$ and

$dx = \cos u\,du$. Then the integral $\displaystyle\int \dfrac{1}{\sqrt{1-x^2}}\,dx =$

$$\int \dfrac{1}{\sqrt{1-\sin^2 u}}\cos u\,du = \int \dfrac{1}{\cos u}\cdot\cos u\,du =$$

$\displaystyle\int du = u + C = \sin^{-1} x + C.$ **13.** 1.412 radians

15. 0.049 radian **17.** $x\tan^{-1}x - \dfrac{1}{2}\ln(1+x^2) + C$

Summary and Review: Chapter 8, pp. 536–537

1. [8.1] $\dfrac{4\pi}{3}$ **2.** [8.1] $\dfrac{7\pi}{4}$ **3.** [8.1] $150°$ **4.** [8.1] $420°$

5. [8.1] $\dfrac{\sqrt{3}}{2}$ **6.** [8.1] 0 **7.** [8.1] 1 **8.** [8.1] Undefined

9. [8.2] $(6x-1)\sec^2(3x^2-x)$ **10.** [8.2] $6\sin^5 x\cos x$

11. [8.2] $\dfrac{1}{\cos x\sin x}$ **12.** [8.2] $3\cos 3x\,e^{\sin 3x}$

13. [8.2] $\dfrac{-2(1+\cos 2x)}{\sin^2 2x}$ **14.** [8.2] $\dfrac{\sin\sqrt{t}}{2\sqrt{t}} - \dfrac{\sin t}{2\sqrt{\cos t}}$

15. [8.2] $\dfrac{\sec^2 x}{2\sqrt{\tan x}}$ **16.** [8.2] $e^x(\cos x + \sin x)$

17. [8.2] $3(\cos^2 3x - \sin^2 3x)$ **18.** [8.2] $x\sin x$

19. [8.2] $-9\sin(3t - 2\pi)$ **20.** [8.2] $\dfrac{5\pi}{3}\cos\dfrac{\pi}{3}t$

21. [8.3] $\dfrac{1}{2}$ **22.** [8.3] $\sin e^x + C$
23. [8.3] $-\cos(x-1) + C$ **24.** [8.3] $\dfrac{1}{8}\sin 8x + C$
25. [8.3] $\dfrac{3}{2}x\sin 2x + \dfrac{3}{4}\cos 2x + C$
26. [8.3] $-\dfrac{1}{3}x\cos 3x + \dfrac{1}{9}\sin 3x + C$
27. [8.4] $\dfrac{1}{8}\sin^{-1}8t + C$ **28.** [8.4] $\ln|5x - \sin x| + C$
29. [8.2] $e^{\sin x}\cos x(\cos^2 x - 3\sin x - 1)$
30. [8.2] $f_x = -\sin y\sin x$, $f_{xy} = -\sin x\cos y$

Test: Chapter 8, p. 538

1. $\dfrac{2\pi}{3}$ 2. $150°$ 3. $\dfrac{1}{2}$ 4. -1 5. $-\sin t$

6. $(6x - 5)\cos(3x^2 - 5x)$ 7. $\dfrac{\sin x - x \cdot \cos x}{\sin^2 x}$

8. $2\tan t \cdot \sec^2 t$ 9. $\dfrac{\cos x - \sin x}{2\sqrt{\sin x + \cos x}}$

10. $\dfrac{-2}{1 - 2\sin x \cos x}$ 11. $S'(t) = \dfrac{5\pi}{2}\cos\dfrac{\pi}{8}t$

12. $1 - \dfrac{\sqrt{3}}{2}$ 13. $\dfrac{1}{10}(\sin x)^{10} + C$ 14. $\dfrac{1}{5}\sin 5t + C$

15. $\frac{1}{5}\sin 5x - x\cos 5x + C$ 16. $\frac{1}{4}\tan^{-1} 4t + C$

17. $\ln|2x - \sin x| + C$

18. $\dfrac{dy}{dx} = -2(\sin x)e^{\cos x}$, $\dfrac{d^2 y}{dx^2} = 2e^{\cos x}(\sin^2 x - \cos x)$

19. $f_x = -\dfrac{\sin x}{\cos y}$, $f_{xy} = \dfrac{-\sin x \sin y}{\cos^2 y}$

CHAPTER 9

Margin Exercises, Section 9.1, pp. 541–542

1. $y = x^3 + C$ 2. (a) $y = x^3 + C$; (b) $y = x^3 - 7$,
$y = x^3 + \frac{1}{2}$, $y = x^3$. Answers may vary.   3. $y = x^2 + 6$
4. $f(x) = \ln x - x^2 + \frac{2}{3}x^{3/2} + \frac{13}{3}$
5. $y' = 2e^x - 21e^{3x}$, $y'' = 2e^x - 63e^{3x}$. Then

$$y'' - 4y' + 3y = 0$$

$$\begin{array}{c|c} 2e^x - 63e^{3x} - 4(2e^x - 21e^{3x}) + 3(2e^x - 7e^{3x}) & 0 \\ 2e^x - 63e^{3x} - 8e^x + 84e^{3x} + 6e^x - 21e^{3x} & \\ & 0 \end{array}$$

6. $\dfrac{dy}{dx} = 2xe^{2x} + e^{2x}$. Then

$$\dfrac{dy}{dx} - 2y = e^{2x}$$

$$\begin{array}{c|c} (2xe^{2x} + e^{2x}) - 2(xe^{2x}) & e^{2x} \\ & e^{2x} \end{array}$$

Exercise Set 9.1, pp. 542–543

1. $y = x^4 - C$; $y = x^4 + 3$, $y = x^4$, $y = x^4 - 796$;
answers may vary.
3. $y = \frac{1}{2}e^{2x} + \frac{1}{2}x^2 + C$; $y = \frac{1}{2}e^{2x} + \frac{1}{2}x^2 - 5$,
$y = \frac{1}{2}e^{2x} + \frac{1}{2}x^2 + 7$; $y = \frac{1}{2}e^{2x} + \frac{1}{2}x^2$; answers may vary.
5. $y = 3\ln x - \frac{1}{3}x^3 + \frac{1}{6}x^6 + C$;
$y = 3\ln x - \frac{1}{3}x^3 + \frac{1}{6}x^6 - 15$, $y = 3\ln x - \frac{1}{3}x^3 + \frac{1}{6}x^6 - 7$,
$y = 3\ln x - \frac{1}{3}x^3 + \frac{1}{6}x^6$; answers may vary.
7. $y = \frac{1}{3}x^3 + x^2 - 3x + 4$ 9. $y = \frac{3}{5}x^{5/3} - \frac{1}{2}x^2 - \frac{61}{10}$

11. $y'' = \dfrac{1}{x}$. Then

$$y'' - \dfrac{1}{x} = 0$$

$$\begin{array}{c|c} \dfrac{1}{x} - \dfrac{1}{x} & 0 \\ 0 & \end{array}$$

13. $y' = 4e^x + 3xe^x$, $y'' = 7e^x + 3xe^x$. Then

$$y'' - 2y' + y = 0$$

$$\begin{array}{c|c} (7e^x + 3xe^x) - 2(4e^x + 3xe^x) + (e^x + 3xe^x) & 0 \\ 7e^x + 3xe^x - 8e^x - 6xe^x + e^x + 3xe^x & \\ & 0 \end{array}$$

15. $C(x) = 2.6x - 0.01x^2 + 120$, $A(x) = 2.6 - 0.01x + \dfrac{120}{x}$

17. (a) $P(C) = \dfrac{400}{(C + 3)^{1/2}} - 40$; (b) $\$97$

Margin Exercises, Section 9.2, pp. 544–549

1. $y = C_1 e^{x^3}$, where $C_1 = e^C$ 2. $y = \sqrt[3]{x^2 + 339}$
3. $y = \sqrt{10x + C_1}$, $y = -\sqrt{10x + C_1}$, where $C_1 = 2C$
4. $y = -2 + C_1 e^{x^2/2}$, where $C_1 = e^C$ 5. $p = Cx^{-1/3}$
6. $R = C_1 \cdot S^k$, where $C_1 = e^C$

Exercise Set 9.2, pp. 549–550

1. $y = C_1 e^{x^4}$, where $C_1 = e^C$ 3. $y = \sqrt[3]{\dfrac{5}{2}x^2 + C}$

5. $y = \sqrt{2x^2 + C_1}$, $y = -\sqrt{2x^2 + C_1}$, where $C_1 = 2C$
7. $y = \sqrt{6x + C_1}$, $y = -\sqrt{6x + C_1}$, where $C_1 = 2C$

9. $y = -3 + 8e^{x^2/2}$ 11. $y = \sqrt[3]{15x - 3}$
13. $y = C_1 e^{3x}$, where $C_1 = e^C$
15. $P = C_1 e^{2t}$, where $C_1 = e^C$
17. (a) $P = C_1 e^{kt}$ where $C_1 = e^C$; (b) $P = P_0 e^{kt}$
19. (a) $R = k \cdot \ln (S + 1) + C$; (b) $R = k \cdot \ln (S + 1)$;
(c) no units, no pleasure from them   21. $p(x) = 200 - x$
23. $p(x) = C_1 x^{-1/2}$

Margin Exercises, Section 9.3, pp. 553–558

1. (a)

t	-3	-2	-1	0	1	2	3
$P(t)$	0.05	0.12	0.27	0.5	0.73	0.88	0.95

(b)

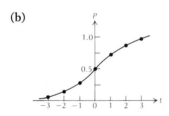

2. (a) $P = \dfrac{\dfrac{P_0}{L - P_0} \cdot Le^{Lkt}}{1 + \dfrac{P_0}{L - P_0} e^{Lkt}} = \dfrac{P_0 Le^{Lkt}}{(L - P_0) + P_0 e^{Lkt}}$;

(b) $\dfrac{P_0 Le^{Lkt}}{(L - P_0) + P_0 e^{Lkt}} \cdot \dfrac{e^{-Lkt}}{e^{-Lkt}} = \dfrac{P_0 L}{P_0 + e^{-Lkt}(L - P_0)}$
(add exponents)

3. 3312   4. (a) $P(t) = \dfrac{80,000}{20 + 3980e^{-2t}}$;

(b)

(c) $t \approx 1.95$ mo   5. $Q(y) = \dfrac{Q_0 M}{Q_0 + e^{-Mry}(M - Q_0)}$

Exercise Set 9.3, pp. 559–562

1. 900   3. (a) $P(t) = \dfrac{7000}{10 + 690e^{-0.56t}}$;

(b)

(c) $t \approx 9.197$ hr   5. (a) $N(t) = \dfrac{4800}{6 + 794e^{-800kt}}$;

(b) $k = 0.0005$; (c) $t \approx 17.08$ min

7. (a) $P(w) = \dfrac{0.04}{0.04 + 0.96e^{-kw}}$; (b) $k = 1.06$;

(c) $P(w) = \dfrac{0.04}{0.04 + 0.96e^{-1.06w}}$; (d) $w \approx 4.3$ (about 106 g)

Margin Exercises, Section 9.4, pp. 563–565

1. (a) Undefined, 5%, 2.5%, 1.2%, 0.7%, 0.3%;

(b)

(c)

(d) no; (e) no; (f) 100%; (g) decreasing

2. (a)

t	0	1	2	3	4	5
$P(t)$	0	1.9	2.6	2.85	2.95	2.98

(b)

3. (a) $k = 0.05$; (b) $P(t) = 1 - e^{-0.05t}$;
(c) $P(60) = 0.9502$, or 95.02%; (d) $t \approx 92$

Exercise Set 9.4, pp. 565–568

1.   **3.**

5. (a) $P(t) = 0.5(1 - e^{-kt})$; **(b)** $k = 0.07$;
(c) $P(t) = 0.5(1 - e^{-0.07t})$; **(d)** $P(30) = 0.439$, or 43.9%;
(e) $t \approx 56$ **7. (a)** $P(t) = 1 - e^{-kt}$; **(b)** $k = 0.18$;
(c) $P(t) = 1 - e^{-0.18t}$; **(d)** $P(10) = 0.8347$; **(e)** 13

9. $M(t) = \dfrac{p}{k}(1 - e^{-kt})$

11. $C(t) = C_T - (C_T - C_0)e^{-(kA/v)t}$

Margin Exercises, Section 9.5, pp. 569–571

1. It checks.   **2.** $x\,dx + y\,dy = 0$ **3.** $P_y = 0 = Q_x$
4. $P_y = 6x - 2 = Q_x$ **5. (a)** $2y\,dx + x\,dy = 0$;
(b) $2 = P_y \neq Q_x = 1$ **6.** $x^2 y = C$ **7.** $xy = C$
8. $(3x^2 - 2x + 1)y = C$ **9.** $y = \dfrac{x \pm \sqrt{x^2 - 4x^2(5x - 7)}}{2x^2}$

10. $x^2 y^3 = 8$ **11.** $y = \dfrac{-(x+1)}{2x}$

Exercise Set 9.5, p. 572

1. Exact   **3.** Exact   **5.** Exact   **7.** Exact
9. $x^2 + 3x + 3y^2 = C$ **11.** $x \sin y = C$
13. $x^3 + 3xy + y^3 = 5$ **15.** $x \cos y + y \cos x = 1$
17. $x^2 - 2xy + 2y^2 = 1$ **19.** Exact, $P^2 + 3Pu = 100$,
$P = 2$ when $u = 16$

Margin Exercises, Section 9.6, pp. 574–579

1. $y = C_1 x$ **2.** It checks.   **3.** $y = C_1 e^{-x^2}$
4. $y = 2 + Cx^{-4}, x > 0$ **5.** $y = \frac{1}{5}e^{3x} + Ce^{-2x}$
6. $A(t) = 1200 - Ce^{-t/240}$
7. $A(t) = 720 - 320e^{-t/240}$; $A(300) \approx 628.32$ lb

Exercise Set 9.6, pp. 579–580

1. $y = Ce^{3x}$ **3.** $y = Cx^{-3}$ **5.** $m(x) = e^{-3x}$

7. $m(x) = e^{5/x}$ **9.** $y = Ce^{3x}$ **11.** $y = Ce^{x^2}$
13. $y = 7e^{-5x}$ **15.** $y = Ce^{x^2} - 1$

17. $y = \dfrac{2x^2 + x^4 + 4C}{4(1 + x^2)}$ **19.** $y = \dfrac{x^3}{4} + \dfrac{6}{x}$

21. (a) 201.16 lb; **(b)** 249.99 lb
23. (a) $T(t) = M + Ce^{-kt}$; **(b)** 88.21 min

Margin Exercises, Section 9.7, pp. 582–591

1. They all check.   **2.** $y = C_1 e^{4x} + C_2 e^{2x}$
3. $y = C_1 e^{(1/2)x} + C_2 e^{3x}$ **4.** $y = C_1 e^x + C_2 e^{-2x} + C_3$
5. $r^2 - 7r + 12 = 0$ **6.** $r^3 - 6r^2 + 11r - 6 = 0$
7. $y = C_1 e^{3x} + C_2 e^{4x}$ **8.** $y = C_1 e^{2x} + C_2 xe^{2x}$
9. $y = C_1 e^{3x} + C_2 xe^{3x}$ **10.** $y = C_1 e^{-(1/2)x} + C_2 xe^{-(1/2)x}$
11. $y = C_1 e^{-x} + C_2 xe^{-x} + C_3 x^2 e^{-x}$
12. $y = C_1 \sin 3x + C_2 \cos 3x$
13. $y = C_1 e^{2x} \sin 5x + C_2 e^{2x} \cos 5x$
14. $y = C_1 e^{-x} \sin 2x + C_2 e^{-x} \cos 2x$
15. $y = C_1 + C_2 e^x \sin(\sqrt{7}x) + C_3 e^x \cos(\sqrt{7}x)$
16. $y = C_1 \sin 2x + C_2 \cos 2x + C_3 x \sin 2x + C_4 x \cos 2x$
17. It checks.   **18.** $y = C_1 e^{-(1/2)x} + C_2 xe^{-(1/2)x} + 3$
19. $y = C_1 e^{3x} + C_2 e^{-3x} - \frac{1}{9}x^2 - \frac{2}{9}x - \frac{2}{81}$
20. $y = C_1 + C_2 e^{-x} + 7x$
21. $y = C_1 + C_2 x + C_3 e^{-2x} + \frac{1}{2}x^3 - \frac{7}{4}x^2$
22. $y = \frac{9}{2}e^x - \frac{17}{2}e^{-x} + 5$

Exercise Set 9.7, pp. 591–592

1. $y = C_1 e^x + C_2 e^{5x}$ **3.** $y = C_1 e^{-x} + C_2 e^{2x}$
5. $y = C_1 e^{-x} + C_2 e^{-2x}$ **7.** $y = C_1 e^{(1/2)x} + C_2 e^{2x}$
9. $y = C_1 e^{3x} + C_2 e^{-3x}$ **11.** $y = C_1 e^{-5x} + C_2 xe^{-5x}$
13. $y = C_1 e^{-(3/2)x} + C_2 xe^{-(3/2)x}$
15. $y = C_1 e^{2x} + C_2 xe^{2x} + C_3 e^{-3x}$
17. $y = C_1 e^{-x} + C_2 xe^{-x} + C_3 x^2 e^{-x} + C_4$
19. $y = C_1 e^x + C_2 e^{4x} + C_3 + C_4 x$
21. $y = C_1 \sin 6x + C_2 \cos 6x$
23. $y = C_1 e^{-4x} \sin 5x + C_2 e^{-4x} \cos 5x$
25. $y = C_1 e^{-x} \sin 2x + C_2 e^{-x} \cos 2x + C_3$
27. $y = C_1 e^x + C_2 e^{-x} \sin 4x + C_3 e^{-x} \cos 4x$
29. $y = C_1 \sin x + C_2 \cos x + 7$
31. $y = C_1 e^x + C_2 xe^x + 3$
33. $y = C_1 e^{-2x} + C_2 xe^{-2x} - 3x + 5$
35. $y = C_1 e^{-x} + C_2 e^{-3x} + 2x^2 - \frac{16}{3}x + \frac{40}{9}$
37. $y = C_1 + C_2 e^{3x} - \frac{4}{3}x$
39. $y = C_1 + C_2 x + C_3 e^{-x} - x^2$

41. $y = C_1 + C_2 e^{-2x} \sin 4x + C_3 e^{-2x} \cos 4x + x^2 - x$
43. $y = 1 - e^x$ **45.** $y = 2e^x - e^{-x}$
47. $y = 6e^x - 2e^{-x} - 3$
49. $y = \sin 3x + 2 \cos 3x - x + 1$
57. $y = C_1 e^{7x} + C_2 e^{9x} + C_3 e^{2x} \sin 4x + C_4 e^{2x} \cos 4x + C_5 x e^{2x} \sin 4x + C_6 x e^{2x} \cos 4x$

Margin Exercises, Section 9.8, pp. 593–602

1. $x = C_1 e^{4t}, y = C_2 e^t$
2. $x = C_1 e^{4t} + C_2 e^{-4t}, y = 2C_1 e^{4t} - 2C_2 e^{-4t}$
3. $x = C_1 \sin 2t + C_2 \cos 2t, y = -2C_2 \sin 2t + 2C_1 \cos 2t$
4. They check.
5. $x = C_1 e^{2t} + C_2 e^{-2t}, y = -\frac{2}{3}C_1 e^{2t} - 2C_2 e^{-2t}$
6. $x = e^{2t} + 4te^{2t}, y = 3e^{2t} - 4te^{2t}$
7. $x = 2e^t + e^{2t}, y = 4e^t + e^{2t}$
8. $x = C_1 e^{2t} + C_2 e^{5t} + 5, y = -2C_1 e^{2t} + C_2 e^{5t} - 20$
9. $A(t) = 1200 + 400e^{-t}, B(t) = 1200 - 400e^{-t}$
10. $A(t) = 1100, B(t) = 1100$
11. $P(t) = \sin 8t + 4 \cos 8t + 20$,
$Q(t) = 40 - 2 \cos 8t + 8 \sin 8t$

Exercise Set 9.8, pp. 603–604

1. $x = C_1 e^{2t}, y = C_2 e^{3t}$
3. $x = C_1 e^t + C_2 e^{2t}, y = C_1 e^t + 2C_2 e^{2t}$
5. $x = C_1 e^{5t} + C_2 e^t, y = 3C_1 e^{5t} - C_2 e^t$
7. $x = C_1 e^{(1/2)t} + C_2 e^{-t}, y = C_1 e^{(1/2)t} - \frac{1}{2}C_2 e^{-t}$
9. $x = C_1 e^{3t} + C_2 te^{3t}, y = (C_2 - C_1)e^{3t} - C_2 te^{3t}$
11. $x = C_1 \sin 6t + C_2 \cos 6t, y = -3C_2 \sin 6t + 3C_1 \cos 6t$
13. $x = C_1 e^t \sin t + C_2 e^t \cos t$,
$y = (\frac{2}{5}C_1 + \frac{1}{5}C_2)e^t \sin t + (-\frac{1}{5}C_1 + \frac{2}{5}C_2)e^t \cos t$
15. $x = e^{3t} + 2e^{4t}, y = -e^{3t} - 4e^{4t}$ **17.** $x = e^{5t}, y = e^{5t}$
19. $x = \sin 2t + \cos 2t, y = -2 \sin 2t + 2 \cos 2t$
21. $x = C_1 e^{3t} - C_2 e^{4t} - 1, y = 2C_1 e^{3t} + C_2 e^{4t}$
23. $x = C_1 \sin t + C_2 \cos t + 4t$,
$y = -C_2 \sin t + C_1 \cos t - 2t + 1$
25. $x = e^{3t} + e^{-3t} - 2, y = e^{3t} - 5e^{-3t} + 1$
27. $A' = -\frac{1}{5}A + \frac{1}{5}B, B' = \frac{1}{5}A - \frac{1}{5}B, A(0) = 200, B(0) = 100$;
$A(t) = 150 + 50e^{-(2/5)t}, B(t) = 150 - 50e^{-(2/5)t}$;
Limiting concentration in both is 0.3 lb/gal
29. $P(t) = 13 \sin 5t + 13 \cos 5t + 36$,
$Q(t) = -6 \sin 5t + 4 \cos 5t + 28$
31. $x = C_1 e^{2t} + C_2 e^{-3t}, y = 2C_1 e^{2t} - \frac{1}{2}C_2 e^{-3t} - e^t$

Summary and Review: Chapter 9, pp. 604–606

1. [9.1] $y = 2x^3 + C$ **2.** [9.2] $y = C_1 e^{2x^3}$, where $C_1 = e^C$
3. [9.2] $y = 5 - C_1 e^{-x^3/3}$, where $C_1 = e^{-C}$
4. [9.1] $y = \frac{2}{3}x^3 - 2x^2 + 3x - 1$
5. [9.2] $y = \pm(8t + 4C)^{1/4}$ **6.** [9.1] $y = x^4 - x^2 + 1$
7. [9.1] $x(2x) - 2(x^2) = 0$
8. [9.3]

9. [9.3] **(a)** $N(t) = \dfrac{9600}{8 + 1192e^{-0.24t}}$;
(b)

(c) $t \approx 23.7$ minutes
10. [9.4] **(a)** $P(t) = 1 - e^{-kt}$; **(b)** $k \approx 0.16$;
(c) $P(t) = 1 - e^{-0.16t}$; **(d)** $P(12) \approx 0.85$; **(e)** $t \approx 18.7$
11. [9.4] 750   **12.** [9.5] $x^2 + y^3 = 8$
13. [9.5] $x^4 + 3x^2y - y^4 = -1$ **14.** [9.6] $y = 3x^{-4}, x > 0$
15. [9.6] $y = -e^{3x} + Ce^{2x}$ **16.** [9.7] $y = C_1 e^{2x} + C_2 e^{5x}$
17. [9.7] $y = C_1 e^{3x} + C_2 x e^{3x}$
18. [9.7] $y = C_1 \sin (5x) + C_2 \cos (5x) + 4$
19. [9.8] $x = C_1 e^{6t} + C_2 e^{-4t}, y = \frac{2}{3}C_1 e^{6t} - C_2 e^{-4t}$
20. [9.8] $x = C_1 \sin (6t) + C_2 \cos (6t)$,
$y = -\frac{1}{3} \sin (6t) + \frac{1}{3} \cos (6t)$
21. [9.8] $x = C_1 e^{-2t} + C_2 te^{-2t}$,
$y = (2C_1 + C_2)e^{-2t} + 2C_2 te^{-2t}$
22. [9.8] $x = e^{4t} + e^{2t}, y = e^{4t} - e^{2t}$
23. [9.8] $x = 3e^{2t}, y = 6e^{2t}$

Test: Chapter 9, pp. 606–607

1. $y = \frac{1}{3}x^3 + \frac{3}{2}x^2 - 5x + 7$ **2.** $y = C_1 e^{x^8}$, where $C_1 = e^C$
3. $y = \sqrt{18x + C_1}, y = -\sqrt{18x + C_1}$, where $C_1 = 2C$
4. $y = C_1 e^{6t}$, where $C_1 = e^C$.   **5.** 1360
6. **(a)** $P(t) = \dfrac{16{,}000}{20 + 780e^{-0.72t}}$;
(b)

7.

(c) $t = 5.798$
8. **(a)** $S(t) = 20(1 - e^{-kt})$; **(b)** $k = 0.11$;
(c) $S(t) = 20(1 - e^{-0.11t})$; **(d)** $13.343 million;
(e) $t = 12.6$ yr, or the 13th yr   **9.** $x^3 + y^2 = 1$

10. $x^3 + 2x^2y - y^5 = -1$ 11. $y = \dfrac{4}{x^3}, x > 0$

12. $y = e^{4x} + Ce^{3x}$ 13. $y = C_1 e^{2x} + C_2 e^{3x}$
14. $y = C_1 e^{6x} + C_2 x e^{6x}$
15. $y = C_1 \sin 6x + C_2 \cos 6x + 2$
16. $x = C_1 e^{5t} + C_2 e^{-3t}, y = \frac{3}{5}C_1 e^{5t} - C_2 e^{-3t}$
17. $x = C_1 \sin 12t + C_2 \cos 12t,$
$y = -2C_2 \sin 12t + 2C_1 \cos 12t$
18. $x = C_1 e^{-3t} + C_2 t e^{-3t}, y = (C_1 + \frac{1}{3}C_2)e^{-3t} + C_2 t e^{-3t}$
19. $x = e^{-5t} - e^{-4t}, y = e^{-5t} - 2e^{-4t}$

CHAPTER 10

Margin Exercises, Section 10.1, pp. 611–619

1. $2(x-3) + 4 + \dfrac{1}{x-3}$ 2. $x - 2 + \dfrac{2}{x-2} + \dfrac{4}{(x-2)^2}$

3. $(x-3)^2 + 2(x-3) - 1$
4. (a) 24; (b) 8; (c) -5; (d) 1
5. $2(x-1)^2 + 3(x-1) + 4$
6. $(x-3)^3 + 9(x-3)^2 + 26(x-3) + 24$
7. $(x-4)^2 + 5(x-4) + 5$ 8. $(x-2)^2 + (x-2) - 1$
9. $(x-1)^2 - (x-1) + 2$
10. $3(x+1)^3 + 2(x+1)^2 - \frac{1}{2}(x+1) + 1$
11. $-\frac{1}{2}x^5 + \frac{1}{2}x^4 + 7x^3 - \frac{1}{2}x^2 + 2x + 1$

Exercise Set 10.1, pp. 619–620

1. $x - 4 + \dfrac{3}{x}$ 3. $2x - 12 + \dfrac{7}{x-3}$

5. $(x-2)^2 + 7 + \dfrac{4}{x-2}$ 7. $(x-2)^2 + 2(x-2) + 3$

9. $(x-2)^2 - (x-2) + 1$
11. $(x+1)^2 - 7(x+1) + 13$
13. $(x+2)^2 - 7(x+2) + 14$
15. $(x-1)^3 + (x-1)^2 + 3(x-1) + 4$
17. $(x+1)^3 - 5(x+1)^2 + 11(x+1) - 6$
19. $2(x-3)^2 + 6(x-3) + 3$
21. $3(x-1)^4 + 12(x-1)^3 + 11(x-1)^2 - 2(x-1) - 3$
23. $(x-1)^5 - 10(x-1)^3 - 20(x-1)^2 - 10(x-1)$
27. $x + 1 - \dfrac{4}{x-1}$ 29. $x - 1 + \dfrac{2}{x+1}$

31. $(x-1)^4 + 3$

Margin Exercises, Section 10.2, pp. 621–625

1. $2 - 6x + 9x^2$
2. (a) $(x-4)^2 + 5(x-4) + 5$; (b) they are equal.
3. $p_2(x) = x, p_3(x) = x - \frac{1}{6}x^3$ 4. $1 + x^2$
5. (a) $1 + x + \frac{1}{2}x^2 + \frac{1}{6}x^3$; (b) 1.105167   6. 0.100333

Exercise Set 10.2, pp. 626–627

1. $1 - x + \frac{1}{2}x^2$ 3. x^2 5. $-4x^2 + x - 1$ 7. $1 + \frac{1}{2}x^2$
9. x^2 11. $1 + \frac{1}{2}x + \frac{1}{8}x^2 + \frac{1}{48}x^3$
13. $1 + 3(x-1) + 3(x-1)^2 + (x-1)^3$
15. $-x - \frac{1}{2}x^2 - \frac{1}{3}x^3$
17. $p_4(x) = x - \frac{1}{6}x^3; p_5(x) = x - \frac{1}{6}x^3 + \frac{1}{120}x^5$
19. $p_4(x) = 1 - \frac{1}{2}x^2 + \frac{1}{24}x^4 = p_5(x)$

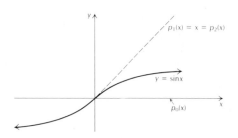

21. $p_0(x) = 0; p_1(x) = x = p_2(x)$
23. $p_0(x) = 1, p_1(x) = 1 - x, p_2(x) = 1 - x + \frac{1}{2}x^2$

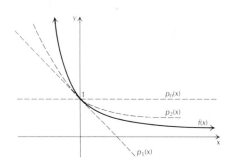

25. 0.1296338   27. 0.9512292   29. -0.0512917
31. (a) $P(1) \approx 0.16473, P(2) \approx 0.30232, P(3) \approx 0.41725$;
(b) $p_3(t) = 0.18t - 0.0162t^2 + 0.000972t^3$;
(c) $p_3(1) = 0.164772, p_3(2) = 0.302976, p_3(3) = 0.420444$
33. (a) $p_2(t) = 5 + 0.1(t-3) - 0.001(t-3)^2$;
(b) $P(3.5) \approx 5.04975, P(4) \approx 5.099$

Margin Exercises, Section 10.3, pp. 629–637

1. (a) $a_n = \dfrac{2}{3^n}$;

(b) $s_1 = \frac{2}{3} \approx 0.667$, $s_2 = \frac{8}{9} \approx 0.889$, $s_3 = \frac{26}{27} \approx 0.963$,
$s_4 = \frac{80}{81} \approx 0.988$, $s_5 = \frac{242}{243} \approx 0.996$, $s_6 = \frac{728}{729} \approx 0.999$
2. (a) $a_n = 2n - 1$;
(b) $s_1 = 1$, $s_2 = 4$, $s_3 = 9$, $s_4 = 16$, $s_5 = 25$;
(c) $s_n = n^2$; (d) no, it diverges.

3. (a) $a_n = \dfrac{2}{5^{n-1}} = 2\left(\dfrac{1}{5}\right)^{n-1}$; (b) $a_5 = \frac{2}{625}$, $a_6 = \frac{2}{3125}$

4. $a = \frac{1}{5}$, $r = \frac{1}{3}$ 5. $s_n = 3[1 - (\frac{1}{3})^n]$ 6. $\frac{5}{2}$ 7. $\frac{3}{10}$
8. Diverges $(r = -2)$ 9. $\frac{5}{7}$ 10. \$20,000
11. \$6,666.67   12. $\frac{4}{9}$ 13. $\frac{17}{37}(= \frac{459}{999})$
14. 2, 1.3333, 1.5556, 1.4815, 1.5062, and 1.4979
15. Converges
16. $0.9498457 \approx s_3 \geqslant s \geqslant s_4 \approx 0.9459394$
17. (a) $0.8998629 \leqslant s \leqslant 0.9012346$; (b) $\frac{9}{10}$

Exercise Set 10.3, pp. 637–639

1. $a_n = 5^{-n} = \left(\dfrac{1}{5}\right)^n$ 3. $\dfrac{n}{n+1}$ 5. $\left(\dfrac{2}{3}\right)^n$
7. $n(-1)^{n+1} = n(-1)^{n-1}$

9. $a = 1$, $r = \dfrac{1}{3}$, $s_{15} = \dfrac{3}{2}\left[1 - \left(\dfrac{1}{3}\right)^{15}\right]$

11. $a = 1$, $r = -\dfrac{1}{3}$, $s_{15} = \dfrac{3}{4}\left[1 + \left(\dfrac{1}{3}\right)^{15}\right]$

13. $a = \dfrac{1}{8}$, $r = -\dfrac{2}{3}$, $s_{15} = \dfrac{3}{40}\left[1 + \left(\dfrac{2}{3}\right)^{15}\right]$

15. $a = \dfrac{5}{3}$, $r = -\dfrac{2}{5}$, $s_{15} = \dfrac{25}{21}\left[1 + \left(\dfrac{2}{5}\right)^{15}\right]$

17. Converges; $s = \frac{8}{7}$ 19. Converges; $s = \frac{9}{8}$
21. Converges; $s = \frac{3}{5}$ 23. Diverges   25. $\frac{7}{9}$
27. $2 + \frac{38}{99} = \frac{236}{99}$ 29. $\frac{1}{7}$ 31. Diverges   33. Converges
35. Converges   37. Converges   39. Converges
41. $0.7407407 \leqslant s \leqslant 0.7530864$
43. $0.9459394 \leqslant s \leqslant 0.9475394$
45. $0.0791736 \leqslant s \leqslant 0.0874381$ 47. 32 billion dollars
49. (a) 5 mg, $5 + 5(0.4) = 7$ mg,
$5 + 5(0.4) + 5(0.4)^2 = 7.8$ mg; (b) 8.3311488 mg;

(c) $\dfrac{25}{3} \approx 8.333333$ mg   51. $V \approx \$37.04$ 55. $\dfrac{d}{p}$ mg

57. $\dfrac{r}{i}$

Margin Exercises, Section 10.4, pp. 641–649

1. $1 - x + \dfrac{x^2}{2!} - \dfrac{x^3}{3!} + \cdots = \displaystyle\sum_{j=0}^{\infty} \dfrac{(-1)^j x^j}{j!}$ 2. $\displaystyle\sum_{j=0}^{\infty} \dfrac{(-1)^j x^{2j}}{(2j)!}$

3. $4 + 2x - 5x^2 + 7x^3$ 4. $\displaystyle\sum_{j=0}^{\infty} (-1)^j x^{2j}$

5. $\displaystyle\sum_{j=0}^{\infty} (-1)^j(j+1)x^j$, $|x| < 1$

6. $1 + 3x + 6x^2 + 10x^3 + \cdots$, $|x| < 1$

7. $1 + x^3 + x^6 + \cdots = \displaystyle\sum_{j=0}^{\infty} x^{3j}$, $|x| < 1$

8. $x^2 - \dfrac{x^6}{3!} + \dfrac{x^{10}}{5!} - \dfrac{x^{14}}{7!} + \cdots = \displaystyle\sum_{j=0}^{\infty} (-1)^j \dfrac{x^{4j+2}}{(2j+1)!}$, all x

9. $e^{-5x} = 1 - 5x + \dfrac{25}{2}x^2 - \dfrac{125}{3!}x^3 + \cdots$, all x

10. $\ln(1-x) = -x - \dfrac{x^2}{2} - \dfrac{x^3}{3} - \dfrac{x^4}{4} \cdots = \displaystyle\sum_{j=0}^{\infty} -\dfrac{x^{j+1}}{j+1}$

11. 0.494   12. $x^2 \cos x = x^2 - \dfrac{1}{2!}x^4 + \dfrac{1}{4!}x^6 - \dfrac{1}{6!}x^8 + \cdots$

13. $x \tan^{-1} x = x^2 - \dfrac{x^4}{3} + \dfrac{x^6}{5} - \dfrac{x^8}{7} + \cdots$, $|x| \leqslant 1$

14. $f(x) = 1 + \dfrac{1}{2!}x + \dfrac{1}{3!}x^2 + \dfrac{1}{4!}x^3 + \cdots$

Exercise Set 10.4, pp. 650–651

1. $1 + 4x + 8x^2 + \dfrac{32}{3}x^3 + \dfrac{4^4}{4!}x^4 + \cdots$

3. $1 - (x-1) + \dfrac{1}{2!}(x-1)^2 - \dfrac{1}{3!}(x-1)^3 + \cdots$

5. $-2 + 3(x+1) - 3(x+1)^2 + (x+1)^3$
7. $1 + 2x + 4x^2 + 8x^3 + \cdots$

9. $e^{6x} = 1 + (6x) + \dfrac{1}{2!}(6x)^2 + \dfrac{1}{3!}(6x)^3 + \cdots$

11. $\dfrac{1}{1-x^5} = 1 + x^5 + x^{10} + x^{15} + \cdots$, $|x| < 1$

13. $\ln(1 + 4x) = 4x - 8x^2 + \frac{64}{3}x^3 - 64x^4 + \cdots$, $|x| < \frac{1}{4}$

15. $\frac{1}{2 + x} = 1 - (x + 1) + (x + 1)^2 - (x + 1)^3 + \cdots$, $-2 < x < 0$

17. $\ln(2 + x) = (x + 1) - \frac{1}{2}(x + 1)^2 + \frac{1}{3}(x + 1)^3 - \frac{1}{4}(x + 1)^4 + \cdots$, $-2 < x < 0$

19. $\frac{5x^4}{1 - x^5} = 5x^4 + 5x^9 + 5x^{14} + \cdots$, $|x| < 1$

21. $f(x) = -\frac{1}{2} + \frac{1}{4!}x^2 - \frac{1}{6!}x^4 + \cdots$, all x

23. $x^5 \sin(x^3) = x^8 - \frac{1}{3!}x^{14} + \frac{1}{5!}x^{20} - \frac{1}{7!}x^{26} + \cdots$, all x

25. $\int_0^1 \cos(x^2)\,dx = 1 - \frac{1}{10} + \frac{1}{216} - \frac{1}{9360} + \cdots \approx 0.90463$ (error $< \frac{1}{9360} \leqslant 0.0002$)   27. 0.82143 (error $< \frac{1}{60} < 0.02$)
29. 0.0002167 (error $< 10^{-13}$)
31. (a) $F(t) = t - \frac{1}{6}t^3 + \frac{1}{40}t^5 - \frac{1}{336}t^7 + \cdots$;
(b) $P(0 \leqslant t \leqslant 0.1) \approx 0.0398278$; (c) it checks.

Margin Exercises, Section 10.5, pp. 651–660

1. $\frac{4}{11}$ 2. 0   3. 0   4. -12.356, -109.12, and -998.00
5. Does not exist   6. -7 7. -2 8. $\frac{1}{2}$ 9. 2
10. -2 11. 1   12. $-\frac{1}{2}$
13. e^2 (l'Hôpital's Rule doesn't apply.)   14. Does not exist
15. $-\frac{5}{3}$ 16. Does not exist   17. 3

Exercise Set 10.5, pp. 661–662

1. 1   3. Does not exist   5. 0   7. Does not exist
9. 7   11. $\frac{1}{4}$ 13. 0   15. 2   17. $\frac{1}{3}$ 19. -1
21. $-\frac{1}{4}$ 23. -1 25. 1   27. 0   29. Does not exist
31. $\frac{1}{2}$ 33. $\frac{1}{3}$ 35. Does not exist   37. Does not exist
39. ap 41. $f(0)$

Summary and Review: Chapter 10, pp. 662–663

1. [10.1] $x^2 - 5x + 3 = (x - 1)^2 - 3(x - 1) - 1$
2. [10.2] $p_2(x) = -3(x + 1)^2 + 3(x + 1) - 1$;
$p_3(x) = (x + 1)^3 - 3(x + 1)^2 + 3(x + 1) - 1$
3. [10.2] $p_2(x) = p_3(x) = 1 + x^2$
4. [10.2] $p_2(x) = x - 2$; $p_3(x) = (x - 2) - \frac{1}{6}(x - 2)^3$
5. [10.2] 0.049979 6. [10.2] 1.051271 7. [10.3] $\frac{32}{35}$

8. [10.3] Diverges   9. [10.3] $\frac{5}{6}[1 - (\frac{3}{5})^{11}]$ 10. [10.3] $\frac{2107}{1980}$
11. [10.3] Converges; $s \approx s_3 \approx 0.941333$
12. [10.3] Diverges   13. [10.3] Diverges
14. [10.3] Diverges
15. [10.3] Converges; $s = \frac{12}{13} \approx 0.9230769$

16. [10.4] $y = x^4 - \frac{x^6}{3!} + \frac{x^8}{5!} - \frac{x^{10}}{7!} + \cdots$, for all x

17. [10.4] $y = 1 + (x + 1)^2 + \frac{(x + 1)^4}{2!} + \frac{(x + 1)^6}{3!} + \cdots$, for all x
18. [10.4] $y = 1 + 3x + 9x^2 + 27x^3 + \cdots$, for $|x| < \frac{1}{3}$
19. [10.4] 1   20. [10.5] $\frac{1}{6}$
21. [10.5] 0 (not an indeterminate form)   22. [10.5] -1
23. [10.5] 0   24. [10.5] $\frac{1}{3}$ 25. [10.3] $50

26. [10.4] $\sum\limits_{n=1}^{\infty} \frac{x^n}{n^2 \cdot n!}$

Test: Chapter 10, pp. 663–664

1. $p_2(x) = -6(x + 2)^2 + 12(x + 2) - 8$;
$p_3(x) = (x + 2)^3 - 6(x + 2)^2 + 12(x + 2) - 8$
2. $p_2(x) = 1$; $p_3(x) = 1 + x^3$
3. $p_2(x) = x - 1$; $p_3(x) = (x - 1) - \frac{1}{6}(x - 1)^3$
4. 0.039989 5. 1.072507 6. $\frac{9}{25}$ 7. $\frac{7}{10}[1 - (\frac{2}{7})^{13}]$
8. $\frac{2407}{990}$ 9. Converges; $s \approx s_3 \approx 0.929167$ 10. Diverges
11. Diverges   12. Converges; $s = \frac{16}{17} \approx 0.941176$

13. $y = x^3 - \frac{x^5}{3!} + \frac{x^7}{5!} - \frac{x^9}{7!} + \cdots$, all x

14. $y = 1 + (x - 1)^2 + \frac{(x - 1)^4}{2!} + \frac{(x - 1)^6}{3!} + \cdots$, all x

15. $y = 1 + 5x + 25x^2 + 125x^3 + \cdots$, $|x| < \frac{1}{5}$ 16. $\frac{1}{2}$
17. -2 18. 0 (not indeterminate)   19. -1 20. 0

21. $\frac{1}{2}$ 22. $20   23. $\frac{D}{i}$ 24. $\sum\limits_{n=1}^{\infty} \frac{x^n}{n \cdot n!}$

CHAPTER 11

Margin Exercises, Section 11.1, pp. 667–669

1. 1   2. The iterates diverge.   3. $\bar{x} = 0$ (all 3 cases)
4. 2   5. $2 - \sqrt{3} \approx 0.267949$ 6. $2 + \sqrt{3} \approx 3.732051$

Exercise Set 11.1, p. 670

1. $\bar{x} = \frac{3}{2} - \frac{1}{2}\sqrt{5} \approx 0.381966$ 3. 1.497300
5. They diverge.   7. They diverge.   9. 0.567143
11. 0.395489

Margin Exercises, Section 11.2, pp. 673–677

1. $x_2 = -\dfrac{1}{3}, x_3 = -\dfrac{1}{15}, x_4 = \dfrac{-1}{255}, x_5 = \dfrac{-1}{65,535},$
$x_6 \approx x_7 \approx 0 = \bar{x}$ 2. 0.450184 3. -2.103803
4. 2.154435

Exercise Set 11.2, p. 678

1. $\bar{x} = \frac{3}{2} - \frac{1}{2}\sqrt{5} \approx 0.381966$ 3. 1.497300
5. -2.103803 7. 0.772914
9. $\bar{x}_1 \approx -0.152368; \bar{x}_2 \approx 1.109643; \bar{x}_3 \approx -2.957275$
11. -1.518662 13. 0.6345599
15. (b) $\sqrt{19} \approx 4.358899$ 17. (b) $\sqrt[10]{20} \approx 1.349283$

Margin Exercises, Section 11.3, pp. 679–684

1. $\$10,147.84$ 2. $\$982.14$ 3. $\$759.34$
4. $i \approx 7.9156\%$ 5. $i \approx 9.7011\%$

Exercise Set 11.3, pp. 685–686

1. $\$974.36$ 3. $\$29,959.92$ 5. $\$10,062.66$ 7. $\$751.31$
9. $\$148.64$ 11. $\$422.41$ 13. $\$1000$ 15. $\$1690.05$
17. $\$1833.39$ 19. $\$758.16$
21. (a) $5800v^2 + 6000v - 10000 = 0$; (b) $i \approx 11.853\%$
23. $i \approx 5.3445\%$
25. The first one (12.044% against 11.775%)
27. (a) $2500v^2 - 4300v + 2000 = 0$; (b) no real roots

Margin Exercises, Section 11.4, pp. 689–697

1. Approx. $= 1.375$; exact $= 1.5$
2. Approx. $= 1.625$; exact $= 1.5$

3. $T_4 = 1.5$; exact $= 1.5$ 4. Error $= 0$ since $f'' = 0$
5. $T_{10} = 0.2525; E \leqslant \frac{1}{200} = 0.005$ 6. $S_5 = 0.25; E = 0$

Exercise Set 11.4, pp. 697–698

1. (a) 1.9; (b) 2.1 3. (a) 0.833732; (b) 0.733732
5. $T_{10} = 2$ 7. $T_5 = 0.7837315$
9. $T_{10} = 0.170825; E \leqslant 0.01667$
11. $T_{10} \approx 0.385878; |E| < 0.000833$ 13. $S_5 = 2$
15. 0.7853982 17. $0.25; E = 0$
19. $0.3862934; E \leqslant 0.0000533$

Margin Exercises, Section 11.5, pp. 700–704

1. $x_2 \approx 3.741657; x_3 \approx 3.564868, x_4 \approx 3.181632;$
$x_5 \approx 3.147723; x_6 \approx 3.120675; x_7 \approx 3.014054$
2. $T_5 \approx 0.695635; T_{10} \approx 0.6937714; I \approx 0.6931502$
3. $S_2 \approx 0.6932539; S_4 \approx 0.6931545; R_2 \approx 0.6931479$

Exercise Set 11.5, p. 704

1. $x_2 = \frac{1}{3}, x_3 = \frac{10}{27}, x_4 = \frac{3}{8}$
3. $x_2 \approx 1.341471, x_3 \approx 1.47382, x_4 \approx 1.55758$
5. $x_2 \approx -1.817121, x_3 \approx -2.036927, x_4 \approx -2.117815$
7. $T_2 = \frac{3}{8} = 0.375; T_4 = \frac{11}{32} = 0.34375;$
$T_8 = \frac{43}{128} \approx 0.3359375; S_2 = \frac{1}{3}; S_4 = \frac{1}{3}; R_2 = \frac{1}{3}$
9. $T_2 = 0.775; T_4 \approx 0.7827941; T_8 \approx 0.7847471;$
$S_2 \approx 0.78539216; S_4 \approx 0.7853981; R_2 \approx 0.7853985$

Margin Exercises, Section 11.6, pp. 705–712

1. (a) $y = g(x) = e^{-x^2}$; (b) $y_0 = 1, y_1 = 1, y_2 = 0.92,$
$y_3 = 0.7728, y_4 = 0.587328, y_5 \approx 0.399383$; (c) $y_0 = 1,$
$y_1 = 1, y_2 = 0.98, y_3 = 0.9408, y_4 = 0.884352,$
$y_5 \approx 0.813604, y_6 \approx 0.7322435, y_7 \approx 0.644374,$
$y_8 \approx 0.554162, y_9 \approx 0.465496, y_{10} \approx 0.381707$
2. $y'' = y^2 \cos(xy) + xy^2 \sin(xy) \cos(xy) + y \sin^2(xy)$
3. (a) $y'' = y(4x^2 - 2)$;
(b) $y_{n+1} = y_n[0.96 - (0.4)x_n + (0.08)x_n^2]$; (c) $y_0 = 1,$
$y_1 = 0.96, y_2 = 0.847872, y_3 \approx 0.689150, y_4 \approx 0.5160358,$
$y_5 \approx 0.356684$

4.

	Euler	Three-term	Exact
$y_0 = y(0)$	1	1	1
$y_1 = y(0.2)$	1	0.96	0.960789
$y_2 = y(0.4)$	0.92	0.847872	0.852144
$y_3 = y(0.6)$	0.7728	0.689150	0.697676
$y_4 = y(0.8)$	0.587328	0.516036	0.527292
$y_5 = y(1)$	0.399383	0.356684	0.367879

5. (b) $y(0.4) \approx 1.5425366$; $x(0.6) \approx 2.8347761$; $y(1) \approx 0.7540745$

6. $x_1 = 2.4$, $y_1 = -0.8$, $x_2 = 3.2$, $y_2 = -1.92$

Exercise Set 11.6, pp. 712–716

1. 1, 0.9, 0.8, 0.7, 0.6, 0.5, 0.4, 0.3, 0.2, 0.1, and 0
3. 0, 0, 0.125, 0.375, and 0.75
5. 1, 1, 1.08, 1.2528, 1.553472, and 2.050583
7. 1, 1, 1.04, 1.121584, 1.248670, and 1.427460
9. -1, -0.99, -0.96, -0.91, -0.84, -0.75, -0.64, -0.51, -0.36, -0.19, and 0
11. 1, 0.8, 0.648, 0.592, 0.68, and 0.96
13. 1, 0.78125, 0.6103516, 0.476837, and 0.372529
15. $y = g(x) = 1 - x$. Both methods yield the exact values.
17. $y = g(x) = e^{x^2}$. For Euler, see number 5 above. Three-term gives 1, 1.04, 1.168218, 1.4167056, 1.854184, and 2.616625.
19. $x = 2e^{2t} + e^{-2t}$, $y = 2e^{2t} - 3e^{-2t}$ are the exact solutions. Euler gives $x_0 = 3$, $y_0 = -1$, $x_1 = 3.4$, $y_1 = 1$, $x_2 = 4.28$, $y_2 = 2.84$, $x_3 = 5.704$, $y_3 = 4.84$, $x_4 = 7.8128$, $y_4 = 7.2944$, $x_5 = 10.83424$, and $y_5 = 10.5232$.
21. $x = e^{2t} - e^t$, $y = 2e^{2t} - e^t$ are the exact solutions. Euler gives $x_0 = 0$, $y_0 = 1$, $x_1 = 0.2$, $y_1 = 1.6$, $x_2 = 0.52$, $y_2 = 2.48$, $x_3 = 1.016$, $y_3 = 3.76$, $x_4 = 1.768$, $y_4 = 5.6096$, $x_5 = 2.88992$, and $y_5 = 8.26816$.
23. $x_0 = x_1 = \cdots = x_{10} = 0$. $y_0 = 1$, $y_1 = 0.9$, $y_2 = 0.81$, $y_3 = 0.729$, $y_4 = 0.6561$, $y_5 = 0.59049$, $y_6 = 0.531441$, $y_7 = 0.478297$, $y_8 = 0.430467$, $y_9 = 0.387420$, and

$y_{10} = 0.348678$ **25.** \$810.80
27. (a) 557.18; **(b)** 527.09; **(c)** 528.58
29. $x(5) \approx 860.94$; $y(5) \approx 491.42$
31. (b) $F(10) = \sum_{i=0}^{9} f(i)$
33. (a) $x(5) \approx 54.7273$, $y(5) \approx 17.3924$;
(b) $x(10) \approx 72.204$, $y(10) \approx 18.782$

Summary and Review: Chapter 11, pp. 716–717

1. [11.1] 3 **2.** [11.1] 1 **3.** [11.1] 0.9346927
4. [11.1] 0.4501836 **5.** [11.1] Diverges
6. [11.2] $\bar{x}_1 \approx 0.2541017$; $\bar{x}_2 \approx 1.8608058$
7. [11.2] 0.9423132
8. [11.2] $\bar{x}_1 \approx 0.5009922$; $\bar{x}_2 \approx 1.1202946$
9. [11.2] $\bar{x}_E \approx 1.0356017$ **10.** [11.3] \$18,681.63
11. [11.3] **(a)** $7000v^4 = 4000 + 1000v$;
(b) $i \approx 9.2407275\%$
12. [11.3] $i_1 \approx 20.51\%$, $i_2 \approx 14.49\%$.
13. [11.4] $T_2 \approx 0.2373419$, $T_4 \approx 0.2391956$, $T_8 \approx 0.2396578$
14. [11.4] $S_2 \approx 0.2398134$, $S_4 \approx 0.2398118$
15. [11.5] $x_2 \approx 1.3934134$, $x_3 \approx 0.3529083$, $x_4 \approx 1.0155184$, $x_5 \approx 1.0543590$, $x_6 \approx 0.9875705$, $x_7 \approx 1.0298004$, $x_8 \approx 1.0299800$, $x_9 \approx 1.0296720$, $x_{10} \approx 1.0298665$
16. [11.5] $x_2 \approx 0.3483533$, $x_3 \approx 0.4699680$, $x_4 \approx 0.4441681$, $x_5 \approx 0.45148424$, $x_6 \approx 0.4499003$, $x_7 \approx 0.4501571$, $x_8 \approx 0.45018938$, $x_9 \approx 0.45018236$, $x_{10} \approx 0.450180624$ **17.** [11.5] $R_2 \approx 0.2398117$
18. [11.6] $y(0) = 3$, $y(0.2) = 2.4$, $y(0.4) = 1.92$, $y(0.6) = 1.536$, $y(0.8) = 1.2288$, and $y(1) = 0.99304$
19. [11.6] $y(0) = 3$, $y(0.2) = 2.46$, $y(0.4) = 2.0172$, $y(0.6) = 1.654104$, $y(0.8) \approx 1.3563653$, $y(1) \approx 1.1122195$
20. [11.6] **(a)** $y = 3e^{-x}$; **(b)** $y(0) = 3$, $y(0.2) = 2.4561923$, $y(0.4) = 2.0109601$, $y(0.6) = 1.6464349$, $y(0.8) = 1.3479869$, $y(1) = 1.1036383$
21. [11.6] $x_0 = 3$, $y_0 = 1$, $x_1 = 0.8$, $y_1 = -0.8$, $x_2 = 0$, $y_2 = -1.28$, $x_3 = -0.256$, $y_3 = -1.28$, $x_4 = -0.3072$, $y_4 = -1.1264$, $x_5 = -0.28672$, and $y_5 = -0.94208$
22. [9.8] $x = 4e^{-3t} - e^{-t}$, $y = 4e^{-3t} - 3e^{-t}$
23. [11.6] $x_0 = x(0) = 3$, $y_0 = y(0) = 1$, $x(0.2) \approx 1.376516$, $y(0.2) \approx -0.260946$, $x(0.4) \approx 0.534457$, $y(0.4) \approx -0.806183$, $x(0.6) \approx 0.112384$, $y(0.6) \approx -0.985239$, $x(0.8) \approx -0.086457$, $y(0.8) \approx -0.985115$, $x(1) \approx -0.168731$, $y(1) \approx -0.90449$

Test: Chapter 11, pp. 718–719

1. 2   2. -2 3. 0.8977725   4. 0.372559
5. The iterates diverge.
6. $\bar{x}_1 \approx 0.2033642$; $\bar{x}_2 \approx 1.469617$ 7. 0.8977725
8. $\bar{x}_1 \approx 0.50049263$; $\bar{x}_2 \approx 1.1027159$
9. $x_E \approx 1.1366384$ 10. \$17,043.98
11. (a) $6000v^4 \overset{\&}{=} 3000 + 1000v$; (b) $i \approx 11.386791\%$
12. $i_1 \approx 14.47\%$; $i_2 \approx 14.99\%$
13. $T_2 \approx 0.9397933$; $T_4 \approx 0.94451352$; $T_8 \approx 0.94569087$
14. $S_2 \approx 0.94608694$; $S_4 \approx 0.94608331$
15. $x_{10} \approx 1.1105271$ 16. $x_{10} \approx 0.3725596$
17. $R_2 \approx 0.9460837$ 18. $y(0) = 1$, $y(0.2) = 0.8$,
$y(0.4) = 0.672$, $y(0.6) = 0.581683$, $y(0.8) = 0.514012$,
and $y(1) = 0.461170$
19. $y(0) = 1$, $y(0.2) = 0.84$, $y(0.4) = 0.722588$,
$y(0.6) = 0.633253$, $y(0.8) = 0.563209$, and
$y(1) = 0.506914$ 20. $y = (1 + x)^{-1}$
21. $x_0 = 1$, $y_0 = 2$, $x_1 = 1.2$, $y_1 = 1.8$, $x_2 = 1.2$,
$y_2 = 1.56$, $x_3 = 1.104$, $y_3 = 1.32$, $x_4 = 0.9696$,
$y_4 = 1.0992$, $x_5 = 0.82752$, and $y_5 = 0.90528$.

CUMULATIVE REVIEW, pp. 721–724

1. $y = -4x - 27$ 2. $x^2 + 2xh + h^2 - 5$
3. (a)

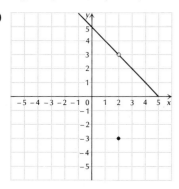

(b) 3; (c) -3;
(d) no

4. -8 5. 3   6. Does not exist   7. 4   8. $3x$
9. -9 10. $2x - 7$ 11. $\frac{1}{4}x^{-3/4}$ 12. $-6x^{-7}$
13. $(x - 3)5(x + 1)^4 + (x + 1)^5$, or
$2(x + 1)^4(3x - 7)$ 14. $\dfrac{5 - 2x^3}{x^6}$ 15. $\dfrac{1 - x}{e^x}$
16. $\dfrac{2x}{x^2 + 5}$ 17. 1   18. $3e^{3x} + 2x$
19. $e^{-x}[2x \cos (x^2 + 3) - \sin (x^2 + 3)]$ 20. $-\tan x$

21. $2(\sec^2 x)(\tan x)$ 22. $3xy^2 + \dfrac{y}{x}$
23. Minimum $= -7$ at $x = 1$ 24. None
25. Maximum $= 6\frac{2}{3}$ at $x = -1$; minimum $= 4\frac{1}{3}$ at $x = -2$
26. 15   27. 30 times; lot size 15
28. $P = 199,500$; maximum sustainable
harvest $= 39,800,250$ 29. $\Delta y = 3.03$, $f'(x)\,\Delta x = 3$
30.

31. (a) $A(t) = 66,164e^{0.1t}$; (b) 179,852.4 million barrels
32. $E(x) = \dfrac{(x + 1)[80 - 30 \ln (x + 1)]}{30x}$
33. $\dfrac{1}{2}x^6 + C$ 34. $3 - \dfrac{2}{e}$
35. $\dfrac{7}{9(7 - 3x)} + \dfrac{1}{9} \ln (7 - 3x) + C$ 36. $\dfrac{1}{4}e^{x^4} + C$
37. $\left(\dfrac{x^2}{2} + 3x\right) \ln x - \dfrac{x^2}{4} - 3x + C$
38. $-\cos (x^2 + 3) + C$ 39. $\sin (e^x) + C$ 40. $\dfrac{232}{3}$
41. \$525.23   42. \$49,444.55   43. Convergent, $\frac{1}{4374}$
44. $\frac{3}{2} \ln 3$ 45. 0.6826   46. $(7, \$169)$; \$751.33
47. $-\dfrac{\pi}{2}\left(\dfrac{1}{e^{10}} - 1\right)$ 48. $y = C_1 e^{x^2/2}$, where $C_1 = e^C$
49. (a) $P(t) = 1 - e^{-kt}$; (b) $k = 0.12$;
(c) $P(t) = 1 - e^{-0.12t}$; (d) $P(20) = 0.9093$, or 90.93%
50. $8xy^3 + 3$ 51. $e^y + 24x^2y$ 52. None
53. Maximum $= 4$ at $(3, 3)$ 54. $3(e^3 - 1)$
55. $\dfrac{\sqrt{2}}{2}$ 56. $\dfrac{1}{2}$ 57. 0   58. $\dfrac{1}{8} \ln |\sec 8x + \tan 8x| + C$
59. $\frac{1}{3} \tan^{-1} (3x) + C$
60. $[-\csc^2 (\sin (x^3))][\cos (x^3)][3x^2]$ 61. $\dfrac{1}{\sqrt{1 - x^2}}$

62. $dw = 15x^4 y^2 \, dx + 6x^5 y \, dy$

63. $dw = 7e^x[\sin{(yz)} \, dx + z \cos{(yz)} \, dy + y \cos{(yz)} \, dz]$

64. 13.81 cubic feet   **65.** $\dfrac{1}{12}$ **66.** $\dfrac{2}{3}$ **67.** 0

68. $xy - 3x + y^3 = C$ **69.** $y = Ce^{-\sin x}$

70. $y = C_1 + C_2 \sin 2x + C_3 \cos 2x$

71. $y = C_1 e^{7x} + C_2 x e^{7x}$ **72.** $y = 5 + C_1 e^{-x} + C_2 e^{-3x}$

73. $x = C_1 e^t + C_2 e^{-t}; y = 5C_1 e^t + 3C_2 e^{-t}$

74. $5(x - 1)^3 + 2(x - 1)^2 - 3(x - 1) - 5$

75. $\sin{(0.07)} \approx (0.07) - \dfrac{(0.07)^3}{6} = 0.069942833$

76. $\dfrac{7}{40} = 0.175$ **77.** $\dfrac{679}{999}$ **78.** Converges

79. $x^5 - \dfrac{x^{13}}{3!} + \dfrac{x^{21}}{5!} - \dfrac{x^{29}}{7!} + \cdots = \displaystyle\sum_{k=0}^{\infty} \dfrac{(-1)^k x^{8k+5}}{(2k+1)!}$

80. 0.32180075   **81.** Does not exist.   **82.** -2 **83.** 4

84. 1   **85.** 0.3725595   **86.** 0.7320508

87. \$466.51   **88.** 13.7%   **89.** $T_5 \approx 0.4059274$;

exact $= \ln{(1.5)} \approx 0.4054651$ **90.** $s_5 \approx 0.4054653$

91. (a) $s_2 \approx 0.1173604; s_4 \approx 0.1115501$;

$I = \dfrac{1}{9} \approx 0.1111111$; **(b)** $R_2 \approx 0.1111627$

INDEX

SUMMARY OF IMPORTANT FORMULAS

18. $\int k \, dx = kx + C$

19. $\int x^r \, dx = \dfrac{x^{r+1}}{r+1} + C, \quad r \neq -1$

   $\int (r+1)x^r \, dx = x^{r+1} + C, \quad r \neq -1$

20. $\int \dfrac{dx}{x} = \ln x + C, \quad x > 0;$

   $\int \dfrac{dx}{x} = \ln |x| + C, \quad x < 0;$

21. $\int e^x \, dx = e^x + C$

22. $\int k f(x) \, dx = k \int f(x) \, dx$

23. $\int [f(x) \pm g(x)] \, dx = \int f(x) \, dx \pm \int g(x) \, dx$

24. $\int u \, dv = uv - \int v \, du$

Further integration formulas occur in Table 1 at the back of the book.